T0178370

Lecture Notes in Artificial Intelligence   11439

Subseries of Lecture Notes in Computer Science

More information about this series at http://www.springer.com/series/1244

Qiang Yang · Zhi-Hua Zhou ·
Zhiguo Gong · Min-Ling Zhang ·
Sheng-Jun Huang (Eds.)

Advances in Knowledge Discovery and Data Mining

23rd Pacific-Asia Conference, PAKDD 2019
Macau, China, April 14–17, 2019
Proceedings, Part I

 Springer

Editors
Qiang Yang
Hong Kong University of Science
and Technology
Hong Kong, China

Zhiguo Gong
University of Macau
Taipa, Macau, China

Sheng-Jun Huang
Nanjing University of Aeronautics
and Astronautics
Nanjing, China

Zhi-Hua Zhou
Nanjing University
Nanjing, China

Min-Ling Zhang
Southeast University
Nanjing, China

ISSN 0302-9743   ISSN 1611-3349   (electronic)
Lecture Notes in Artificial Intelligence
ISBN 978-3-030-16147-7   ISBN 978-3-030-16148-4   (eBook)
https://doi.org/10.1007/978-3-030-16148-4

Library of Congress Control Number: 2019934768

LNCS Sublibrary: SL7 – Artificial Intelligence

This Springer imprint is published by the registered company Springer Nature Switzerland AG
The registered company address is: Gewerbestrasse 11, 6330 Cham, Switzerland

PC Chairs' Preface

It is our great pleasure to introduce the proceedings of the 23rd Pacific-Asia Conference on Knowledge Discovery and Data Mining (PAKDD 2019). The conference provides an international forum for researchers and industry practitioners to share their new ideas, original research results, and practical development experiences from all KDD-related areas, including data mining, data warehousing, machine learning, artificial intelligence, databases, statistics, knowledge engineering, visualization, decision-making systems, and the emerging applications.

We received 567 submissions to PAKDD 2019 from 46 countries and regions all over the world, noticeably with submissions from North America, South America, Europe, and Africa. The large number of submissions and high diversity of submission demographics witness the significant influence and reputation of PAKDD. A rigorous double-blind reviewing procedure was ensured via the joint efforts of the entire Program Committee consisting of 55 Senior Program Committee (SPC) members and 379 Program Committee (PC) members.

The PC Co-Chairs performed an initial screening of all the submissions, among which 25 submissions were desk rejected due to the violation of submission guidelines. For submissions entering the double-blind review process, each one received at least three quality reviews from PC members or in a few cases from external reviewers (with 78.5% of them receiving four or more reviews). Furthermore, each valid submission received one meta-review from the assigned SPC member who also led the discussion with the PC members. The PC Co-Chairs then considered the recommendations and meta-reviews from SPC members, and looked into each submission as well as its reviews and PC discussions to make the final decision. For borderline papers, additional reviews were further requested and thorough discussions were conducted before final decisions.

As a result, 137 out of 567 submissions were accepted, yielding an acceptance rate of 24.1%. We aim to be strict with the acceptance rate, and all the accepted papers are presented in a total of 20 technical sessions. Each paper was allocated 15 minutes for oral presentation and 2 minutes for Q/A. The conference program also featured three keynote speeches from distinguished data mining researchers, five cutting-edge workshops, six comprehensive tutorials, and one dedicated data mining contest session.

We wish to sincerely thank all SPC members, PC members and external reviewers for their invaluable efforts in ensuring a timely, fair, and highly effective paper review and selection procedure. We hope that readers of the proceedings will find that the PAKDD 2019 technical program was both interesting and rewarding.

February 2019

Zhiguo Gong
Min-Ling Zhang

General Chairs' Preface

On behalf of the Organizing Committee, it is our great pleasure to welcome you to Macau, China for the 23rd Pacific-Asia Conference on Knowledge Discovery and Data Mining (PAKDD 2019). Since its first edition in 1997, PAKDD has well established as one of the leading international conferences in the areas of data mining and knowledge discovery. This year, after its four previous editions in Beijing (1999), Hong Kong (2001), Nanjing (2007), and Shenzhen (2011), PAKDD was held in China for the fifth time in the fascinating city of Macau, during April 14–17, 2019.

First of all, we are very grateful to the many authors who submitted their work to the PAKDD 2019 main conference, satellite workshops, and data mining contest. We were delighted to feature three outstanding keynote speakers: Dr. Jennifer Neville from Purdue University, Professor Hui Xiong from Baidu Inc., and Professor Josep Domingo-Ferrer from Universitat Rovira i Virgili. The conference program was further enriched with six high-quality tutorials, five workshops on cutting-edge topics, and one data mining contest on AutoML for lifelong machine learning.

We would like to express our gratitude to the contributions of the SPC members, PC members, and external reviewers, led by the PC Co-Chairs, Zhiguo Gong and Min-Ling Zhang. We are also very thankful to the other Organizing Committee members: Workshop Co-Chairs, Hady W. Lauw and Leong Hou U, Tutorial Co-Chairs, Bob Durrant and Yang Yu, Contest Co-Chairs, Hugo Jair Escalante and Wei-Wei Tu, Publicity Co-Chairs, Yi Cai, Xiangnan Kong, Gang Li, and Yasuo Tabei, Proceedings Chair, Sheng-Jun Huang, and Local Arrangements Chair, Andrew Jiang. We wish to extend our special thanks to Honorary Co-Chairs, Hiroshi Motoda and Lionel M. Ni, for their enlightening support and advice throughout the conference organization.

We appreciate the hosting organization University of Macau, and our sponsors Macao Convention & Exhibition Association, Intel, Baidu, for their institutional and financial support of PAKDD 2019. We also appreciate the Fourth Paradigm Inc., ChaLearn, Microsoft, and Amazon for sponsoring the PAKDD 2019 data mining contest. We feel indebted to the PAKDD Steering Committee for its continuing guidance and sponsorship of the paper award and student travel awards.

Last but not least, our sincere thanks go to all the participants and volunteers of PAKDD 2019—there would be no conference without you. We hope you enjoy PAKDD 2019 and your time in Macau, China.

February 2019

Qiang Yang
Zhi-Hua Zhou

Organization

Organizing Committee

Honorary Co-chairs

Hiroshi Motoda — Osaka University, Japan
Lionel M. Ni — University of Macau, SAR China

General Co-chairs

Qiang Yang — Hong Kong University of Science and Technology, SAR China
Zhi-Hua Zhou — Nanjing University, China

Program Committee Co-chairs

Zhiguo Gong — University of Macau, China
Min-Ling Zhang — Southeast University, China

Workshop Co-chairs

Hady W. Lauw — Singapore Management University, Singapore
Leong Hou U — University of Macau, China

Tutorial Co-chairs

Bob Durrant — University of Waikato, New Zealand
Yang Yu — Nanjing University, China

Contest Co-chairs

Hugo Jair Escalante — INAOE, Mexico
Wei Wci Tu — The Fourth Paradigm Inc., China

Publicity Co-chairs

Yi Cai — South China University of Technology, China
Xiangnan Kong — Worcester Polytechnic Institute, USA
Gang Li — Deakin University, Australia
Yasuo Tabei — RIKEN, Japan

Proceedings Chair

Sheng-Jun Huang — Nanjing University of Aeronautics and Astronautics, China

Local Arrangements Chair

Andrew Jiang   Macao Convention & Exhibition Association, China

Steering Committee

Co-chairs

Ee-Peng Lim   Singapore Management University, Singapore
Takashi Washio   Institute of Scientific and Industrial Research,
                               Osaka University, Japan

Treasurer

Longbing Cao   Advanced Analytics Institute, University
                               of Technology, Sydney, Australia

Members

Dinh Phung   Monash University, Australia (Member since 2018)
Geoff Webb   Monash University, Australia (Member since 2018)
Jae-Gil Lee   Korea Advanced Institute of Science & Technology,
                               Korea (Member since 2018)
Longbing Cao   Advanced Analytics Institute,
                               University of Technology, Sydney, Australia
                               (Member since 2013, Treasurer since 2018)
Jian Pei   School of Computing Science,
                               Simon Fraser University (Member since 2013)
Vincent S. Tseng   National Cheng Kung University,
                               Taiwan (Member since 2014)
Gill Dobbie   University of Auckland,
                               New Zealand (Member since 2016)
Kyuseok Shim   Seoul National University, Korea (Member since 2017)

Life Members

P. Krishna Reddy   International Institute of Information Technology,
                               Hyderabad (IIIT-H), India (Member since 2010,
                               Life Member since 2018)
Joshua Z. Huang   Shenzhen University, China (Member since 2011,
                               Life Member since 2018)
Ee-Peng Lim   Singapore Management University, Singapore
                               (Member since 2006, Life Member since 2014,
                               Co-chair 2015–2017, Chair 2018–2020)
Hiroshi Motoda   AFOSR/AOARD and Osaka University, Japan
                               (Member since 1997, Co-chair 2001–2003,
                               Chair 2004–2006, Life Member since 2006)

Rao Kotagiri | University of Melbourne, Australia (Member since
1997, Co-chair 2006–2008, Chair 2009–2011,
Life Member since 2007, Co-sign since 2006)

Huan Liu | Arizona State University, USA (Member since 1998,
Treasurer 1998–2000, Life Member since 2012)

Ning Zhong | Maebashi Institute of Technology,
Japan (Member since 1999, Life Member
since 2008)

Masaru Kitsuregawa | Tokyo University, Japan (Member since 2000,
Life Member since 2008)

David Cheung | University of Hong Kong, SAR China (Member since
2001, Treasurer 2005–2006, Chair 2006–2008,
Life Member since 2009)

Graham Williams | Australian National University, Australia
(Member since 2001, Treasurer 2006–2017, Co-sign
since 2006, Co-chair 2009–2011, Chair 2012–2014,
Life Member since 2009)

Ming-Syan Chen | National Taiwan University, Taiwan (Member since
2002, Life Member since 2010)

Kyu-Young Whang | Korea Advanced Institute of Science & Technology,
Korea (Member since 2003, Life Member
since 2011)

Chengqi Zhang | University of Technology Sydney, Australia
(Member since 2004, Life Member since 2012)

Tu Bao Ho | Japan Advanced Institute of Science and Technology,
Japan (Member since 2005, Co-chair 2012–2014,
Chair 2015–2017, Life Member since 2013)

Zhi-Hua Zhou | Nanjing University, China (Member since 2007,
Life Member since 2015)

Jaideep Srivastava | University of Minnesota, USA (Member since 2006,
Life Member since 2015)

Takashi Washio | Institute of Scientific and Industrial Research, Osaka
University (Member since 2008, Life Member since
2016, Co-chair 2018–2020)

Thanaruk Theeramunkong | Thammasat University, Thailand (Member since 2009,
Life Member since 2017)

Past Members

Hongjun Lu | Hong Kong University of Science and Technology,
SAR China (Member 1997–2005)

Arbee L. P. Chen | National Chengchi University,
Taiwan (Member 2002–2009)

Takao Terano | Tokyo Institute of Technology,
Japan (Member 2000–2009)

Tru Hoang Cao   Ho Chi Minh City University of Technology,
                                  Vietnam (Member 2015–2017)
Myra Spiliopoulou   Information Systems, Otto-von-Guericke-University
                                  Magdeburg (Member 2013–2019)

Senior Program Committee

James Bailey   University of Melbourne, Australia
Albert Bifet   Telecom ParisTech, France
Longbin Cao   University of Technology Sydney, Australia
Tru Cao   Ho Chi Minh City University of Technology, Vietnam
Peter Christen   Australian National University, Australia
Peng Cui   Tsinghua University, China
Guozhu Dong   Wright State University, USA
Benjamin C. M. Fung   McGill University, Canada
Bart Goethals   University of Antwerp, Belgium
Geoff Holmes   University of Waikato, New Zealand
Qinghua Hu   Tianjin University, China
Xia Hu   Texas A&M University, USA
Sheng-Jun Huang   Nanjing University of Aeronautics and Astronautics,
                                  China
Shuiwang Ji   Texas A&M University, USA
Kamalakar Karlapalem   IIIT Hyderabad, India
George Karypis   University of Minnesota, USA
Latifur Khan   University of Texas at Dallas, USA
Byung S. Lee   University of Vermont, USA
Jae-Gil Lee   KAIST, Korea
Gang Li   Deakin University, Australia
Jiuyong Li   University of South Australia, Australia
Ming Li   Nanjing University, China
Yu-Feng Li   Nanjing University, China
Shou-De Lin   National Taiwan University, Taiwan
Qi Liu   University of Science and Technology of China, China
Weiwei Liu   University of New South Wales, Australia
Nikos Mamoulis   University of Ioannina, Greece
Wee Keong Ng   Nanyang Technological University, Singapore
Sinno Pan   Nanyang Technological University, Singapore
Jian Pei   Simon Fraser University, Canada
Wen-Chih Peng   National Chiao Tung University, Taiwan
Rajeev Raman   University of Leicester, UK
Chandan K. Reddy   Virginia Tech, USA
Krishna P. Reddy   IIIT Hyderabad, India
Kyuseok Shim   Seoul National University, Korea
Myra Spiliopoulou   Otto-von-Guericke-University Magdeburg, Germany
Masashi Sugiyama   RIKEN/The University of Tokyo, Japan
Jiliang Tang   Michigan State University, USA

Kai Ming Ting	Federation University, Australia
Hanghang Tong	Arizona State University, USA
Vincent S. Tseng	National Chiao Tung University, Taiwan
Fei Wang	Cornell University, USA
Jianyong Wang	Tsinghua University, China
Jie Wang	University of Science and Technology of China, China
Wei Wang	University of California at Los Angeles, USA
Takashi Washio	Osaka University, Japan
Jia Wu	Macquarie University, Australia
Xindong Wu	Mininglamp Software Systems, China
Xintao Wu	University of Arkansas, USA
Xing Xie	Microsoft Research Asia, China
Jeffrey Xu Yu	Chinese University of Hong Kong, SAR China
Osmar R. Zaiane	University of Alberta, Canada
Zhao Zhang	Soochow University, China
Feida Zhu	Singapore Management University, Singapore
Fuzhen Zhuang	Institute of Computing Technology, CAS, China

Program Committee

Saurav Acharya	University of Vermont, USA
Swati Agarwal	BITS Pilani Goa, India
David Albrecht	Monash University, Australia
David Anastasiu	San Jose State University, USA
Luiza Antonie	University of Guelph, Canada
Xiang Ao	Institute of Computing Technology, CAS, China
Sunil Aryal	Deakin University, Australia
Elena Baralis	Politecnico di Torino, Italy
Jean Paul Barddal	Pontifícia Universidade Católica do Paraná, Brazil
Arnab Basu	Indian Institute of Management Bangalore, India
Gustavo Batista	Universidade de São Paulo, Brazil
Bettina Berendt	KU Leuven, Belgium
Raj K. Bhatnagar	University of Cincinnati, USA
Arnab Bhattacharya	Indian Institute of Technology, Kanpur, India
Kevin Bouchard	Université du Quebec a Chicoutimi, Canada
Krisztian Buza	Eotvos Lorand University, Hungary
Lei Cai	Washington State University, USA
Rui Camacho	Universidade do Porto, Portugal
K. Selcuk Candan	Arizona State University, USA
Tanmoy Chakraborty	Indraprastha Institute of Information Technology Delhi, India
Shama Chakravarthy	University of Texas at Arlington, USA
Keith Chan	Hong Kong Polytechnic University, SAR China
Chia Hui Chang	National Central University, Taiwan
Bo Chen	Monash University, Australia
Chun-Hao Chen	Tamkang University, Taiwan

Lei Chen	Nanjing University of Posts and Telecommunications, China
Meng Chang Chen	Academia Sinica, Taiwan
Rui Chen	Samsung Research America, USA
Shu-Ching Chen	Florida International University, USA
Songcan Chen	Nanjing University of Aeronautics and Astronautics, China
Yi-Ping Phoebe Chen	La Trobe University, Australia
Yi-Shin Chen	National Tsing Hua University, Taiwan
Zhiyuan Chen	University of Maryland Baltimore County, USA
Jiefeng Cheng	Tencent Cloud Security Lab, China
Yiu-ming Cheung	Hong Kong Baptist University, SAR China
Silvia Chiusano	Politecnico di Torino, Italy
Jaegul Choo	Korea University, Korea
Kun-Ta Chuang	National Cheng Kung University, Taiwan
Bruno Cremilleux	Université de Caen Normandie, France
Chaoran Cui	Shandong University of Finance and Economics, China
Lin Cui	Nanjing University of Aeronautics and Astronautics, China
Boris Cule	University of Antwerp, Belgium
Bing Tian Dai	Singapore Management University, Singapore
Dao-Qing Dai	Sun Yat-Sen University, China
Wang-Zhou Dai	Nanjing University, China
Xuan-Hong Dang	IBM T.J. Watson Research Center, USA
Jeremiah Deng	University of Otago, New Zealand
Zhaohong Deng	Jiangnan University, China
Lipika Dey	Tata Consultancy Services, India
Bolin Ding	Data Analytics and Intelligence Lab, Alibaba Group, China
Steven H. H. Ding	McGill University, Canada
Trong Dinh Thac Do	University of Technology Sydney, Australia
Gillian Dobbie	University of Auckland, New Zealand
Xiangjun Dong	Qilu University of Technology, China
Dejing Dou	University of Oregon, USA
Bo Du	Wuhan University, China
Boxin Du	Arizona State University, USA
Lei Duan	Sichuan University, China
Sarah Erfani	University of Melbourne, Australia
Vladimir Estivill-Castro	Griffith University, Australia
Xuhui Fan	University of Technology Sydney, Australia
Rizal Fathony	University of Illinois at Chicago, USA
Philippe Fournier-Viger	Harbin Institute of Technology (Shenzhen), China
Yanjie Fu	Missouri University of Science and Technology, USA
Dragan Gamberger	Rudjer Boskovic Institute, Croatia
Niloy Ganguly	Indian Institute of Technology Kharagpur, India
Junbin Gao	University of Sydney, Australia

Wei Gao	Nanjing University, China
Xiaoying Gao	Victoria University of Wellington, New Zealand
Angelo Genovese	Università degli Studi di Milano, Italy
Arnaud Giacometti	University Francois Rabelais of Tours, France
Heitor M. Gomes	Telecom ParisTech, France
Chen Gong	Nanjing University of Science and Technology, China
Maciej Grzenda	Warsaw University of Technology, Poland
Lei Gu	Nanjing University of Posts and Telecommunications, China
Yong Guan	Iowa State University, USA
Himanshu Gupta	IBM Research, India
Sunil Gupta	Deakin University, Australia
Michael Hahsler	Southern Methodist University, USA
Yahong Han	Tianjin University, China
Satoshi Hara	Osaka University, Japan
Choochart Haruechaiyasak	National Electronics and Computer Technology Center, Thailand
Jingrui He	Arizona State University, USA
Shoji Hirano	Shimane University, Japan
Tuan-Anh Hoang	Leibniz University of Hanover, Germany
Jaakko Hollmén	Aalto University, Finland
Tzung-Pei Hong	National University of Kaohsiung, Taiwan
Chenping Hou	National University of Defense Technology, China
Michael E. Houle	National Institute of Informatics, Japan
Hsun-Ping Hsieh	National Cheng Kung University, Taiwan
En-Liang Hu	Yunnan Normal University, China
Juhua Hu	University of Washington Tacoma, USA
Liang Hu	University of Technology Sydney, Australia
Wenbin Hu	Wuhan University, China
Chao Huang	University of Notre Dame, USA
David Tse Jung Huang	University of Auckland, New Zealand
Jen-Wei Huang	National Cheng Kung University, Taiwan
Nam Huynh	Japan Advanced Institute of Science and Technology, Japan
Akihiro Inokuchi	Kwansei Gakuin University, Japan
Divyesh Jadav	IBM Research, USA
Sanjay Jain	National University of Singapore, Singapore
Szymon Jaroszewicz	Polish Academy of Sciences, Poland
Songlei Jian	University of Technology Sydney, Australia
Meng Jiang	University of Notre Dame, USA
Bo Jin	Dalian University of Technology, China
Toshihiro Kamishima	National Institute of Advanced Industrial Science and Technology, Japan
Wei Kang	University of South Australia, Australia
Murat Kantarcioglu	University of Texas at Dallas, USA
Hung-Yu Kao	National Cheng Kung University, Taiwan

Shanika Karunasekera	University of Melbourne, Australia
Makoto P. Kato	Kyoto University, Japan
Chulyun Kim	Sookmyung Women University, Korea
Jungeun Kim	Korea Advanced Institute of Science and Technology, Korea
Kyoung-Sook Kim	Artificial Intelligence Research Center, Japan
Yun Sing Koh	University of Auckland, New Zealand
Xiangnan Kong	Worcester Polytechnic Institute, USA
Irena Koprinska	University of Sydney, Australia
Ravi Kothari	Ashoka University, India
P. Radha Krishna	National Institute of Technology, Warangal, India
Raghu Krishnapuram	Indian Institute of Science Bangalore, India
Marzena Kryszkiewicz	Warsaw University of Technology, Poland
Chao Lan	University of Wyoming, USA
Hady Lauw	Singapore Management University, Singapore
Thuc Duy Le	University of South Australia, Australia
Ickjai J. Lee	James Cook University, Australia
Jongwuk Lee	Sungkyunkwan University, Korea
Ki Yong Lee	Sookmyung Women's University, Korea
Ki-Hoon Lee	Kwangwoon University, Korea
Sael Lee	Seoul National University, Korea
Sangkeun Lee	Korea University, Korea
Sunhwan Lee	IBM Research, USA
Vincent C. S. Lee	Monash University, Australia
Wang-Chien Lee	Pennsylvania State University, USA
Yue-Shi Lee	Ming Chuan University, Taiwan
Zhang Lei	Anhui University, China
Carson K. Leung	University of Manitoba, Canada
Bohan Li	Nanjing University of Aeronautics and Astronautics, China
Jianmin Li	Tsinghua University, China
Jianxin Li	Deakin University, Australia
Jundong Li	Arizona State University, USA
Nan Li	Alibaba, China
Peipei Li	Hefei University of Technology, China
Qian Li	University of Technology Sydney, Australia
Rong-Hua Li	Beijing Institute of Technology, China
Shao-Yuan Li	Nanjing University, China
Sheng Li	University of Georgia, USA
Wenyuan Li	University of California, Los Angeles, USA
Wu-Jun Li	Nanjing University, China
Xiaoli Li	Institute for Infocomm Research, A*STAR, Singapore
Xue Li	University of Queensland, Australia
Yidong Li	Beijing Jiaotong University, China
Zhixu Li	Soochow University, China

Defu Lian	University of Electronic Science and Technology of China, China
Sungsu Lim	Chungnam National University, Korea
Chunbin Lin	Amazon AWS, USA
Hsuan-Tien Lin	National Taiwan University, Taiwan
Jerry Chun-Wei Lin	Western Norway University of Applied Sciences, Norway
Anqi Liu	California Institute of Technology, USA
Bin Liu	IBM Research, USA
Jiajun Liu	Renmin University of China, China
Jiamou Liu	University of Auckland, New Zealand
Jie Liu	Nankai University, China
Lin Liu	University of South Australia, Australia
Liping Liu	Tufts University, USA
Shaowu Liu	University of Technology Sydney, Australia
Zheng Liu	Nanjing University of Posts and Telecommunications, China
Wenpeng Lu	Qilu University of Technology, China
Jun Luo	Machine Intelligence Lab, Lenovo Group Limited, China
Wei Luo	Deakin University, Australia
Huifang Ma	Northwest Normal University, China
Marco Maggini	University of Siena, Italy
Giuseppe Manco	ICAR-CNR, Italy
Silviu Maniu	Universite Paris-Sud, France
Naresh Manwani	International Institute of Information Technology, Hyderabad, India
Florent Masseglia	Inria, France
Tomoko Matsui	Institute of Statistical Mathematics, Japan
Michael Mayo	The University of Waikato, New Zealand
Stephen McCloskey	The University of Sydney, Australia
Ernestina Menasalvas	Universidad Politécnica de Madrid, Spain
Xiangfu Meng	Liaoning Technical University, China
Xiaofeng Meng	Renmin University of China, China
Jun-Ki Min	Korea University of Technology and Education, Korea
Nguyen Le Minh	Japan Advanced Institute of Science and Technology, Japan
Leandro Minku	The University of Birmingham, UK
Pabitra Mitra	Indian Institute of Technology Kharagpur, India
Anirban Mondal	Ashoka University, India
Taesup Moon	Sungkyunkwan University, Korea
Yang-Sae Moon	Kangwon National University, Korea
Yasuhiko Morimoto	Hiroshima University, Japan
Animesh Mukherjee	Indian Institute of Technology Kharagpur, India
Miyuki Nakano	Advanced Institute of Industrial Technology, Japan
Mirco Nanni	ISTI-CNR, Italy

Richi Nayak	Queensland University of Technology, Australia
Raymond Ng	University of British Columbia, Canada
Wilfred Ng	Hong Kong University of Science and Technology, SAR China
Cam-Tu Nguyen	Nanjing University, China
Hao Canh Nguyen	Kyoto University, Japan
Ngoc-Thanh Nguyen	Wroclaw University of Science and Technology, Poland
Quoc Viet Hung Nguyen	Griffith University, Australia
Arun Reddy Nelakurthi	Arizona State University, USA
Thanh Nguyen	Deakin University, Australia
Thin Nguyen	Deakin University, Australia
Athanasios Nikolakopoulos	University of Minnesota, USA
Tadashi Nomoto	National Institute of Japanese Literature, Japan
Eirini Ntoutsi	Leibniz University of Hanover, Germany
Kouzou Ohara	Aoyama Gakuin University, Japan
Kok-Leong Ong	La Trobe University, Australia
Shirui Pan	University of Technology Sydney, Australia
Yuangang Pan	University of Technology Sydney, Australia
Guansong Pang	University of Adelaide, Australia
Dhaval Patel	IBM T.J. Watson Research Center, USA
Francois Petitjean	Monash University, Australia
Hai Nhat Phan	New Jersey Institute of Technology, USA
Xuan-Hieu Phan	University of Engineering and Technology, VNUHN, Vietnam
Vincenzo Piuri	Università degli Studi di Milano, Italy
Vikram Pudi	International Institute of Information Technology, Hyderabad, India
Chao Qian	University of Science and Technology of China, China
Qi Qian	Alibaba Group, China
Tang Qiang	Luxembourg Institute of Science and Technology, Luxembourg
Biao Qin	Renmin University of China, China
Jie Qin	Eidgenössische Technische Hochschule Zürich, Switzerland
Tho Quan	Ho Chi Minh City University of Technology, Vietnam
Uday Kiran Rage	University of Tokyo, Japan
Chedy Raissi	Inria, France
Vaibhav Rajan	National University of Singapore, Singapore
Santu Rana	Deakin University, Australia
Thilina N. Ranbaduge	Australian National University, Australia
Patricia Riddle	University of Auckland, New Zealand
Hiroshi Sakamoto	Kyushu Institute of Technology, Japan
Yücel Saygin	Sabanci University, Turkey
Mohit Sharma	Walmart Labs, USA
Hong Shen	Adelaide University, Australia

Wei Shen	Nankai University, China
Xiaobo Shen	Nanjing University of Science and Technology, China
Victor S. Sheng	University of Central Arkansas, USA
Chuan Shi	Beijing University of Posts and Telecommunications, China
Motoki Shiga	Gifu University, Japan
Hiroaki Shiokawa	University of Tsukuba, Japan
Moumita Sinha	Adobe, USA
Andrzej Skowron	University of Warsaw, Poland
Yang Song	University of New South Wales, Australia
Arnaud Soulet	University of Tours, France
Srinath Srinivasa	International Institute of Information Technology, Bangalore, India
Fabio Stella	University of Milan-Bicocca, Italy
Paul Suganthan	University of Wisconsin-Madison, USA
Mahito Sugiyama	National Institute of Informatics, Japan
Guangzhong Sun	University of Science and Technology of China, China
Yuqing Sun	Shandong University, China
Ichigaku Takigawa	Hokkaido University, Japan
Mingkui Tan	South China University of Technology, China
Ming Tang	Institute of Automation, CAS, China
Qiang Tang	Luxembourg Institute of Science and Technology, Luxembourg
David Taniar	Monash University, Australia
Xiaohui (Daniel) Tao	University of Southern Queensland, Australia
Vahid Taslimitehrani	PhysioSigns Inc., USA
Maguelonne Teisseire	Irstea, France
Khoat Than	Hanoi University of Science and Technology, Vietnam
Lini Thomas	International Institute of Information Technology, Hyderabad, India
Hiroyuki Toda	NTT Corporation, Japan
Son Tran	New Mexico State University, USA
Allan Tucker	Brunel University London, UK
Jeffrey Ullman	Stanford University, USA
Dinusha Vatsalan	Data61, CSIRO, Australia
Ranga Vatsavai	North Carolina State University, USA
Joao Vinagre	LIAAD—INESC TEC, Portugal
Bay Vo	Ho Chi Minh City University of Technology, Vietnam
Kitsana Waiyamai	Kasetsart University, Thailand
Can Wang	Griffith University, Australia
Chih-Yu Wang	Academia Sinica, Taiwan
Hongtao Wang	North China Electric Power University, China
Jason T. L. Wang	New Jersey Institute of Technology, USA
Lizhen Wang	Yunnan University, China
Peng Wang	Southeast University, China
Qing Wang	Australian National University, Australia

Shoujin Wang	Macquarie University, Australia
Sibo Wang	Chinese University of Hong Kong, SAR China
Suhang Wang	Pennsylvania State University, USA
Wei Wang	University of New South Wales, Australia
Wei Wang	Nanjing University, China
Weiqing Wang	Monash University, Australia
Wendy Hui Wang	Stevens Institute of Technology, USA
Wenya Wang	Nanyang Technological University, Singapore
Xiao Wang	Beijing University of Posts and Telecommunications, China
Xiaoyang Wang	Zhejiang Gongshang University, China
Xin Wang	University of Calgary, Canada
Xiting Wang	Microsoft Research Asia, China
Yang Wang	Dalian University of Technology, China
Yue Wang	AcuSys, USA
Zhengyang Wang	Texas A&M University, USA
Zhichao Wang	University of Technology Sydney, Australia
Lijie Wen	Tsinghua University, China
Jorg Wicker	University of Auckland, New Zealand
Kishan Wimalawarne	Kyoto University, Japan
Raymond Chi-Wing Wong	Hong Kong University of Science and Technology, SAR China
Brendon J. Woodford	University of Otago, New Zealand
Fangzhao Wu	Microsoft Research Asia, China
Huifeng Wu	Hangzhou Dianzi University, China
Le Wu	Hefei University of Technology, China
Liang Wu	Arizona State University, USA
Lin Wu	University of Queensland, Australia
Ou Wu	Tianjin University, China
Qingyao Wu	South China University of Technology, China
Shu Wu	Institute of Automation, CAS, China
Yongkai Wu	University of Arkansas, USA
Yuni Xia	Indiana University—Purdue University Indianapolis (IUPUI), USA
Congfu Xu	Zhejiang University, China
Guandong Xu	University of Technology Sydney, Australia
Jingwei Xu	Nanjing University, China
Linli Xu	University of Science and Technology China, China
Miao Xu	RIKEN, Japan
Tong Xu	University of Science and Technology of China, China
Bing Xue	Victoria University of Wellington, New Zealand
Hui Xue	Southeast University, China
Shan Xue	University of Technology Sydney, Australia
Pranjul Yadav	Criteo, France
Takehisa Yairi	University of Tokyo, Japan
Takehiro Yamamoto	Kyoto University, Japan

Chun-Pai Yang	National Taiwan University, Taiwan
De-Nian Yang	Academia Sinica, Taiwan
Guolei Yang	Facebook, USA
Jingyuan Yang	George Mason University, USA
Liu Yang	Tianjin University, China
Ming Yang	Nanjing Normal University, China
Shiyu Yang	East China Normal University, China
Yiyang Yang	Guangdong University of Technology, China
Lina Yao	University of New South Wales, Australia
Yuan Yao	Nanjing University, China
Zijun Yao	IBM Research, USA
Mi-Yen Yeh	Academia Sinica, Taiwan
feng Yi	Institute of Information Engineering, CAS, China
Hongzhi Yin	University of Queensland, Australia
Jianhua Yin	Shandong University, China
Minghao Yin	Northeast Normal University, China
Tetsuya Yoshida	Nara Women's University, Japan
Guoxian Yu	Southwest University, China
Kui Yu	Hefei University of Technology, China
Yang Yu	Nanjing University, China
Long Yuan	University of New South Wales, Australia
Shuhan Yuan	University of Arkansas, USA
Xiaodong Yue	Shanghai University, China
Reza Zafarani	Syracuse University, USA
Nayyar Zaidi	Monash University, Australia
Yifeng Zeng	Teesside University, UK
De-Chuan Zhan	Nanjing University, China
Daoqiang Zhang	Nanjing University of Aeronautics and Astronautics, China
Du Zhang	California State University, Sacramento, USA
Haijun Zhang	Harbin Institute of Technology (Shenzhen), China
Jing Zhang	Nanjing University of Science and Technology, China
Lu Zhang	University of Arkansas, USA
Mengjie Zhang	Victoria University of Wellington, New Zealand
Quangui Zhang	Liaoning Technical University, China
Si Zhang	Arizona State University, USA
Wei Emma Zhang	Macquarie University, Australia
Wei Zhang	East China Normal University, China
Wenjie Zhang	University of New South Wales, Australia
Xiangliang Zhang	King Abdullah University of Science and Technology, Saudi Arabia
Xiuzhen Zhang	RMIT University, Australia
Yudong Zhang	University of Leicester, UK
Zheng Zhang	University of Queensland, Australia
Zili Zhang	Southwest University, China
Mingbo Zhao	Donghua University, China

Peixiang Zhao	Florida State University, USA
Pengpeng Zhao	Soochow University, China
Yanchang Zhao	CSIRO, Australia
Zhongying Zhao	Shandong University of Science and Technology, China
Zhou Zhao	Zhejiang University, China
Huiyu Zhou	University of Leicester, UK
Shuigeng Zhou	Fudan University, China
Xiangmin Zhou	RMIT University, Australia
Yao Zhou	Arizona State University, USA
Chengzhang Zhu	University of Technology Sydney, Australia
Huafei Zhu	Nanyang Technological University, Singapore
Pengfei Zhu	Tianjin University, China
Tianqing Zhu	University of Technology Sydney, Australia
Xingquan Zhu	Florida Atlantic University, USA
Ye Zhu	Deakin University, Australia
Yuanyuan Zhu	Wuhan University, China
Arthur Zimek	University of Southern Denmark, Denmark
Albrecht Zimmermann	Université de Caen Normandie, France

External Reviewers

Ji Feng	Zheng-Fan Wu
Xuan Huo	Yafu Xiao
Bin-Bin Jia	Yang Yang
Zhi-Yu Shen	Meimei Yang
Yanping Sun	Han-Jia Ye
Xuan Wu	Peng Zhao

Sponsoring Organizations

 University of Macau

 Macao Convention & Exhibition Association

 Intel

 Baidu Inc.

Contents – Part I

Spatio-Temporal and Stream Data Mining

Factor and Tensor Analysis

Healthcare, Bioinformatics and Related Topics

Clustering and Anomaly Detection

Contents – Part II

Weakly Supervised Learning

Recommender System

Social Network and Graph Mining

Data Pre-processing and Feature Selection

Contents – Part III

Mining Unstructured and Semi-structured Data

Behavioral Data Mining

Visual Data Mining

Knowledge Graph and Interpretable Data Mining

Classification and Supervised Learning

Classification and Supervised Learning

Multitask Learning for Sparse Failure Prediction

Simon Luo[1,2]([✉]), Victor W. Chu[3], Zhidong Li[2,4], Yang Wang[2,4],
Jianlong Zhou[2,4], Fang Chen[2,4], and Raymond K. Wong[5]

[1] The University of Sydney, Sydney, Australia
sluo4225@uni.sydney.edu.au
[2] Data61, CSIRO, Sydney, Australia
[3] Nanyang Technological University, Singapore, Singapore
wchu@ntu.edu.sg
[4] University of Technology Sydney, Ultimo, Australia
{zhidong.li,yang.wang,jianlong.zhou,fang.chen}@uts.edu.au
[5] The University of New South Wales, Kensington, Australia
wong@cse.unsw.edu.au

Abstract. Sparsity is a problem which occurs inherently in many real-world datasets. Sparsity induces an imbalance in data, which has an adverse effect on machine learning and hence reducing the predictability. Previously, strong assumptions were made by domain experts on the model parameters by using their experience to overcome sparsity, albeit assumptions are subjective. Differently, we propose a multi-task learning solution which is able to automatically learn model parameters from a common latent structure of the data from related domains. Despite related, datasets commonly have overlapped but dissimilar feature spaces and therefore cannot simply be combined into a single dataset. Our proposed model, namely hierarchical Dirichlet process mixture of hierarchical beta process (HDP-HBP), learns *tasks* with a common model parameter for the failure prediction model using hierarchical Dirichlet process. Our model uses recorded failure history to make failure predictions on a water supply network. Multi-task learning is used to gain additional information from the failure records of water supply networks managed by other utility companies to improve prediction in one network. We achieve superior accuracy for sparse predictions compared to previous state-of-the-art models and have demonstrated the capability to be used in risk management to proactively repair critical infrastructure.

Keywords: Multi-task learning · Sparse predictions ·
Dirichlet process · Beta process · Failure predictions

1 Introduction

Sparsity is an undesirable property which occurs in many real-world datasets and is known to degrade the prediction performance of machine learning models.

© Springer Nature Switzerland AG 2019
Q. Yang et al. (Eds.): PAKDD 2019, LNAI 11439, pp. 3–14, 2019.
https://doi.org/10.1007/978-3-030-16148-4_1

Sparsity may arise intrinsically and/or extrinsically, e.g., high dimensionality of feature space or label space, and a small number of observations respectively. It is easy to say that we can improve a model by providing more training data, but there are many situations in which it is infeasible, e.g., infrastructure failure history. In the past, clustering is used to gain additional information from data points with similar characteristics. However, sparsity also degrades the quality of clustering which in turn also degrades the quality of predictions [10]. Indeed, sparse prediction problem is difficult to overcome using a single dataset because there is simply not enough data to build a reliable model. Previous solutions have been relying on the assistance from domain experts and using strong assumptions in model construction [9]. Differently, we propose a multi-task learning solution to automatically learn the model parameters for a group of tasks from multiple domains to be applied to a failure prediction model such as the hierarchical beta process (HBP) [14].

HBP is a Bayesian non-parametric model (BNP) which is well suited to make sparse predictions by using information obtained from clustering. However, HBP model itself in the plain form is unable to assign data points to a cluster. Previously, there have been proposals to apply a Dirichlet prior to create a generalised hierarchical beta process (G-HBP) [4,10] to automatically learn the model parameters. However, this does not overcome the sparsity issue because it is difficult to produce low variance clusters in a sparse dataset, and it is also difficult to select the model parameters for each *task*. Differently, we propose a solution which uses a BNP approach to automatically select the model parameters previously selected by domain experts by learning the *task* for a subset of data. Specifically, we adapt hierarchical Dirichlet process (HDP) [13] to form a multi-task learning framework to simultaneously learn the model parameters and prediction *tasks*. The proposed model can be used as a stand-alone transfer learning model for clustering sparse datasets or can be connected to HBP to form our proposed hierarchical Dirichlet process mixture of hierarchical beta process (HDP-HBP) to learn the model parameters for a group of *tasks* to make sparse predictions.

Our proposed model is applied to make water pipe failure predictions to demonstrate its capability performing multi-task learning to improve sparse predictions. American Society for Civil Engineers estimated that the United States needs to invest \$3.6 trillion by 2020 to maintain its infrastructure [7]. Failures could lead to significant social and economic cost in a city a single failure could cost around 10 times the repair cost. Prioritising the correct water pipes to repair with a limited budget could have huge financial savings by preventing financial loss from water pipe failure. The datasets used to make water pipe failure predictions are extremely sparse because the feature space is large. Moreover, there is only a very short observation period available and critical water main (CWM) have a low failure rate of about 0.5%. However, despite the low failure rate, the failure of a single CWM could lead to major disruptions and heavy financial loses in a society. Predicting failure of CWM is very difficult because CWM components only make up only 10% of the network. Due to the low event rate and the small number of components in each network, it is very difficult to train a model with a single dataset without taking any strong assumptions. Currently, most of these critical infrastructures are repaired reactively. Our multi-task learning provides a solution by using additional datasets supplied by other utility companies (i.e. different feature space) to automatically learn model parameters and *tasks* to make predictions for critical components. A task in our case study is defined to be a group of pipes with the same subset of features (i.e. A cluster). This subset of features is created by Dirichlet Process (DP) to cluster on features with similar failure rates. It is defined to be a task because they will have the same model-parameters and hyper-parameters, but different observations when making the failure prediction.

2 Related Work

2.1 Failure Predictions for Sparse Data

A number of statistical models have previously been used to make failure prediction. Some example used in the past include, survival analysis such as Cox proportional hazard model [3,10] and Weibull models [5]. However, survival analysis may not be well suited for this task because in practice, only a small section of the component is repair instead of replacing the entire asset. More recent formulations of infrastructure failure prediction attempt to estimate the latent failure rate given the historical failure records for each asset. Unlike the approached used in survival analysis, this approach does not assume a 'life' for each asset, but instead group assets using features to estimate the failure rate. An example is hierarchical beta process [8,14], where the failure rates are estimated by using a combination of individual observations for each pipe and the average group observations. When the model was originally proposed, a heuristic provided by domain experts was used to group water pipes with similar characteristics together based on the intrinsic pipe failures [8]. Later on, techniques such as stochastic block models [11] and DPs [10] have been proposed to automatically create clusters with similar failure rate. However, none of these methods are able to accurate predictions for extremely sparse components, such as a critical water main, where there are only very few components for any water pipe network. Our proposed approach aims to solve this issue by automatically selecting the model-parameters by clustering across *domains* to learn a common *task* for the failure prediction model.

2.2 Multi-task Learning in Hierarchical BNP Models

Hierarchical model are a natural way to formulate a multi-task learning frameworks. Hierarchical models are able to apply a constraint to ensure that the distribution between the domains learn a common marginal probability distribution by learning a common parameter or prior [12]. For example, multi-task learning in Gaussian process (GP) a common kernel is learned to share information between domains [1]. Dirichlet Process (DP) and Beta Process (BP) have also been explored in multi-task learning by using it as a prior in hierarchical Bayesian models [4, 15]. Multi-task learning in DP and BP attempts to learn the parameters for a group of common tasks. The parameters learnt by the DP and BP can then be applied to the prediction model to perform each task. We use the multi-task learning techniques previously used in BNP and apply it to HBP to formulate a solution which is able to simultaneously learn a group of tasks along with the model parameters (i.e. q_k) for failure prediction.

3   Preliminary Model

3.1   Stick-Breaking Process

Beta Process. The BP, $B \sim \mathrm{BP}(c, B_0)$ is defined as a Lévy process over a general space Ω with a random measure B_0. The BP can also be derived from the stick-breaking process. BP uses the part of the stick which has been broken off to continue the process. This process can be represented as,

$$H(\omega) = \sum_i \pi_i \delta_{\omega_i}(\omega), \qquad \pi_i \sim \mathrm{Beta}(cq_i, c(1 - q_i)), \qquad \delta_{\omega_i} \sim B_0, \qquad (1)$$

where c is the concentration parameter, H_0 is the mean parameter and B_0 is the discrete form of the base measure defined by $B_0 = \sum_i q_i \delta_{\omega_i}$. The beta process can be used as a natural prior to the Bernoulli process to make prediction on binary data.

3.2   Chinese Restaurant Process

The Chinese restaurant process (CRP) is a perspective on DP which uses the Pólya urn modelling scheme representation. The name CRP comes from the analogy used to explain the model. Imagine a restaurant which has tables that can support an infinite number of customers. When a customer (i.e. a new data point) enters the restaurant (i.e. the *domain*), the probability for the new customer to be assigned to a table (i.e. the cluster) is proportional to the number of customers sitting on that particular table. Since each table attracts a new customer proportional to the number of customers on the table. The equation for CRP is given by,

$$P(z_i = a | z_1, \ldots, z_{i-1}) = \mathrm{CRP}(\gamma) = \begin{cases} \frac{n_a}{N - 1 + \gamma} & \text{If } k \leq K \\ \frac{\gamma}{N - 1 + \gamma} & \text{Otherwise} \end{cases} \qquad (2)$$

Where n_a is the number of data points in each table, k is the cluster index in K, N is the total number of customers in the restaurant and γ is the hyperparameter controlling the probability of the new customer sitting on a new table.

The CRP can be constructed in hierarchy of each other to create hierarchical Dirichlet process (HDP) from a CRP perspective, this is a Chinese restaurant franchise (CRF) [13]. Explaining HDP with the same analogy as CRP, each customer enters restaurant franchise with a shared menu (i.e. a list of dishes). At each table, one dish (i.e. model parameter) is ordered from the menu and is shared among all customers who sit at that table. Each restaurant can serve the same dish at different table, the same dish can be served at different restaurants.

Both CRP and CRF are often used as a prior for clustering and mixture models because the exchangeability property provides a computational efficient method to solve DP.

4   Our Proposed Model for Sparse Predictions

This section presents our proposed model for failure prediction Hierarchical Beta Process and our parameter selection via HDP.

4.1   Problem Definition Based on HBP Model

For a given water pipe dataset, $X = \{x_{\ell j} \in \{0,1\}\}$ represents a sparse dataset with recorded failure history for all water pipe ℓ in year j. When a failure occurs, only a small section where the failure has occurred of the water pipe is repaired. The entire water pipe is not replaced with a new asset (i.e. newly repaired pipe does not mean lower chance of failure). For CWM, the failure rate of these water pipes is extremely low (on average less than 0.5% per year). Therefore an assumption has been made that the failures are independent from each other, where the event of a failure is drawn from a Bernoulli process $x_{\ell j} \sim \text{BerP}\,(\pi_{\ell j})$. $\pi_{\ell j}$ represents the predicted failure rate on a beta process which is a natural prior to the Bernoulli process. The plate notation of the Hierarchical Beta Process (HBP) is shown in Fig. 1a and the algebraic formulation is given as follows,

$$q_{kj} \sim \text{BetaP}\,(c_0 q_0, c_0\,(1 - q_0))\,, \pi_{\ell j} \sim \text{BetaP}\,(c_{kj} q_{kj}, c_{kj}\,(1 - q_{kj}))\,, x_{\ell j} \sim \text{BerP}\,(\pi_{\ell j})\,. \quad (3)$$

where c_0 (concentration) and q_0 (mean) are the hyper-parameters of the upper-layer of the beta process. c_{kj} and q_{kj} are the latent parameters for the middle layer beta process. Each k represents a cluster of feature, where each feature combination represents a group of pipes with the same features. Making a prediction directly from the beta-Bernoulli process $P\,(\pi_{\ell j}|x_{\ell j})$ will have a high bias for sparse observations. HBP overcomes this issue by making a prediction for $\pi_{\ell j}$ using the clustering information and the individual observation to reduce the bias in the prediction (i.e. $P(\pi_{\ell j}|q_{kj} X, c_0, q_0)$). However, HBP does not provide a method to learn q_{kj}. Previous approaches such as self-taught clustering [2], flexible grouping approach [11] and stochastic block models [6] are all prone to learning a bias in the model parameter q_{kj} in sparse datasets. Our proposed approach attempts to reduce the bias using multi-task learning by learning q_{kj} from $P\,(q_{kj}|X_{\mathcal{T}}, \{X_{\mathcal{S}}\}, c_0, q_0)$, where $\mathcal{T}$ represents the target (sparse) domain and $\mathcal{S}$ represents the source domain. An assumption used in multi-task learning is applied to the marginal probability distribution to enforce that the source and target domain are equal. Each tasks learnt in the model share a common parameter q_{kj}. The next section will discuss how we learn q_{kj} using a multi-task learning framework to reduce the bias in the model by learning the model parameters for a group of task.

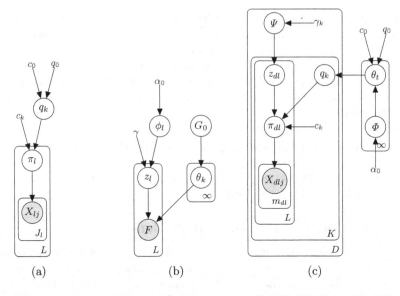

Fig. 1. Plate diagrams: (a) Hierarchical beta process (HBP), (b) Hierarchical Dirichlet process mixture model constructed, and (c) Our proposed hierarchical Dirichlet process mixture of hierarchical beta process (HDP-HBP).

4.2 Sharing Parameter Estimation with HDP-HBP Model

We propose to use hierarchical Dirichlet process mixture model (HDPMM) to perform the clustering to group across *domain* and use the HBP as the mixture component to make the failure prediction. The aim of HDPMM framework is to perform cross *domain* clustering to find a better representation of the structure of the failure patterns in a given network.

$$
\begin{aligned}
\Theta &\sim \text{Beta}\left(c_0 q_0, c_0\left(1-q_0\right)\right), &\Theta = \{\theta_{1\ldots\infty}\}\,; \\
\Phi &\sim \text{DP}\left(\alpha_0\right), \Phi = \{\phi_{1\ldots\infty}\} \\
t &\sim \text{Mul}\left(\Phi\right)\,; \\
\Psi_d &\sim \text{DP}\left(\gamma_d\right), &\Psi_d \sim \{\psi_{d,1\ldots\infty}\}\,; \\
z_{d\ell} &\sim \text{Mul}\left(\psi_d\right)\,; \\
q_{z_{d\ell}} &= \theta_t, \\
\pi_{d\ell} &\sim \text{Beta}\left(c_{z_{d\ell}} q_{z_{d\ell}}, c_{z_{d\ell}}\left(1-q_{z_{d\ell}}\right)\right), \\
X_{d\ell j} &\sim \text{BerP}\left(\pi_{d\ell}\right)\,;
\end{aligned}
\tag{4}
$$

Formulating the problem in this way allows HDP-HBP to have the ability to learn different *tasks* from *domains* with different feature space $\mathcal{X}$. In this formulation, each table in the CRP represents a *task*. Each *task* to learn its own model parameter q_k for its prediction function. The dishes from the global menu provides a constraint for sparse data to prevent assignment to new tables.

Each *domain* has a parameter c_k which controls contribution between the beta-Bernoulli process and the group level from q_k. This allows parameter c_k to be used to control the variance of the clustering. Later on we introduce Theorem 1 to shows how c_k changes the behaviour of the prediction model.

We will use the analogy used to explain the CRP's representation of HDP (discussed in Sect. 3.2) to explain the information is transferred:

Our approach treats the denser *domain* as a restaurant which can assign new tables and add new dishes to the global menu shared amongst the restaurants. Both the dense and sparse *domains* are able to assign new dishes into the global menu. This gives the opportunity for both sparse and dense *domains* to create new clusters. The amount of clusters created by each *domain* is controlled by the concentration parameter. For this approach it is not required to know in advance which *domain* is sparse and which *domain* is dense.

4.3   Inference Algorithm

The objective of the inferencing algorithm is a combination of both clustering and estimating the latent feature rate in the target dataset. This can be seen as a maximum a posteriori (MAP) estimation problem (i.e. $P(\mathbf{z}, \mathbf{q}|\mathbf{X}_\mathcal{T}, \{\mathbf{X}_\mathcal{S}\}, c_0, q_0, \alpha_0, c_k, f(\cdot)))$. Gibbs sampling is used to update the upper layer of the Dirichlet process (dish level).

$$
\begin{aligned}
P\left(q_k = \theta_t | \mathbf{X}_\mathcal{T}, \{\mathbf{X}_\mathcal{S}\}, \alpha_0, -q_k, \Theta\right) &= \mathrm{CRP}\left(\alpha_0\right) \frac{P\left(x_{d1}, x_{d,-1}|q_t, c_t\right)}{P\left(x_{-1}|q_t, c_t\right)} \\
&= \mathrm{CRP}\left(\alpha_0\right) \prod_{\ell \in 1} P\left(x_{d1}|q_t, c_t\right)
\end{aligned}
\tag{5}
$$

The mean of each cluster q_k is then determined by,

$$
P\left(z_{d1} = k|q_{z_k}, \mathbf{X}_\mathcal{T}, \{\mathbf{X}_\mathcal{S}\}, \gamma_k, -\mathbf{z}_{d1}\right) = \mathrm{CRP}\left(\gamma_k\right) \prod_{1:z_{d1}=k} P\left(x_{d1}|q_k, c_k\right)
\tag{6}
$$

The likelihood functions $P\left(x_1|q_k, c_k\right)$ and $\Gamma\left(x_1|\theta, c_k\right)$ are easily solved because beta-Bernoulli are conjugate priors.

The value of q_k can be approximated by Lemma 1, where $s_{d\ell} = \sum_{d,j} X_{d,\ell,j}$ and m is the number of observations.

Lemma 1 *(Mixture component of the Hierarchical Beta Process). The mixture component of the Hierarchical Beta Process q_k can be approximated as a beta distribution.*

$$
P\left(q_k|c_k, \delta_{z_{d\ell}=k}, X_{d\ell,1...m}\right) \sim Beta\left(c_0 q_0 + \sum_\ell s_{d\ell}, c_0\left(1 - q_0\right) + \sum_\ell \sum_t^{m-s_{d\ell}-1} \frac{c_k}{c_k + t}\right).
\tag{7}
$$

Finally π can be sampled directly from its conditional distribution by using the group index generated by $\theta_{d\ell}$ as follows

$$
P\left(\pi_{d\ell}|q_{z_{d\ell}}, z_{d\ell}, c_{z_{d\ell}}, X_{d\ell,1...m}\right) \sim Beta\left(c_{z_{d\ell}} q_{z_{d\ell}} + s_{d\ell}, c_{z_{d\ell}}\left(1 - q_{z_{d\ell}}\right) + m - s_{d\ell}\right).
\tag{8}
$$

All of the updating steps outlined above are repeated iteratively until convergence has been reached. The inferencing algorithm above does not outline a method to select the parameters for c_k. Theorem 1 provides some guidelines on how to select c_k.

Theorem 1 (*Variance of the Prediction by the Hierarchical Beta Process*). *The variance of the prediction in the Hierarchical Beta Process is monotonically decreasing with respect to number of observation m, if $c_{z_{d\ell}} \geq s$.*

Theorem 1 applies for Eqs. 7 and 8. From this result a reasonable choice for the concentration parameter should be $c_{z_{d\ell}} \geq m$. The value to select for $c_{z_{d\ell}}$ is non-trivial and is subjected to further research in future work. For our experiment, the value of $c_{z_{d\ell}}$ is fixed to m as $s \leq m$ to ensure that there is a decreasing variance with an increasing number of observation.

5 Experiments

5.1 Synthetic Data

For our synthetic data experiment, we attempt to create a typical dataset from a utility company to make infrastructure failure prediction. We create a synthetic infrastructure dataset with categorical features that has 512 unique features combinations and assign a true failure rate $P_{\text{true\_fr}}$ to each of these features using a beta distribution. Another beta distribution to assign the true distribution on unique features $P_{\text{true\_f}}$. Samples are then from $P_{\text{true\_f}}$ and then draw the samples for the observations for each feature sample from $P_{\text{true\_fr}}$. A dense domain with sample size $N = 10^6$ is also created using the same marginal probability distribution.

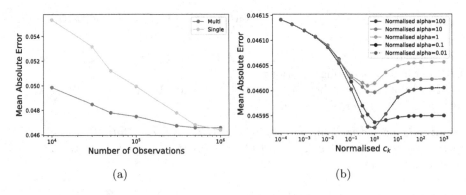

(a)   (b)

Fig. 2. (a) Synthetic data experiment generated using a beta distribution with parameters $\alpha = 0.5, \beta = 10$, comparing HDP-HBP learning from a single domain and multiple domains for varying sample size. (b) Sensitive of the hyper-parameters for a sample size of $N = 10^4$.

We create a scenarios using a beta distribution with parameters $\alpha = 0.5$, $\beta = 10$ to simulate the failure observations in a water pipe network. For each sample size, we create 20 independent datasets and take the mean value of the mean averaged error (MAE). The result in Fig. 2a has shown that multi-task learning is able to produce a lower error prediction compare to DPMHBP for the single task. As the dataset increases, the error of just using a single *domain* reduces to the same order of magnitude as using multi-task learning. This is because for large sample size, the dataset is no longer considered sparse, therefore there is less of a improvement when using multi-task learning.

A sample size of $N = 10^4$ to study the sensitively of the concentration parameter α_0, γ_k and c_k. For this experiment, for simplicity, we set $\alpha_0 = \gamma_k$ and all values of c_k are equal. The parameters are then normalised by dividing c_k by the number of observations and α_0 and γ_k are normalised by dividing with the number of features. Figure 2b shows that decreasing the concentration parameter in DP allow for a lower MAE, however, the model becomes more sensitive to c_k. This is because a lower concentration from DP allows more clusters to be formed closer to each data point, however, the increase in model complexity increases the difficulty to tune the parameters.

5.2   Case Study: Water Pipe Failure Prediction

The data collected for each metropolitan area includes the water pipe network data and water pipe failure data. Three sets of data are used and referred as Region A (a central business district, population:210,000, area:25 km^2), Region B (a prominent suburb, population:230,000, area:80 km^2) and Region C (a large rural city, population:205,000, area:685 km^2). The water pipe network data contains all pipe information in the network. This consists of a unique pipe ID to identify the pipe, pipe attributes and location. The pipe attributes represent intrinsic features belonging to each pipe. These include the year the pipe is laid, protective coating, diameter, Materials and Methods to join pipe segments together. These particular features are factors which may be correlated to the failure rate. Additionally, there are also external factors such as, tree canopy converge, soil corrosion, water supply demand and temperature which are also identified as external factors which may be correlated with the failure rate. The failure data is a time series data which contains the pipe ID, failure dates and the failure location. A new record is added to the dataset each time a failure has been observed in the water supply network.

For the dataset collected for the three metropolitan areas, the pipes are laid between years 1884–2011. These pipes are often split into two main categories, reticulation water main (RWM) and critical water main (CWM). The categories of these pipes are defined by domain experts according to the impact of each water main, such as the financial cost of compensation and the social effect during an event of a failure. The ratio of these pipes for each region is summarized in Table 1. As the role of RWM and CWM is different, it is expected that the failure behaviour of RWM and CWM is also different. Therefore, RWM and CWM are considered as separate *domains*. Quite typically, the water supply network is

Table 1. Summary of pipe network and pipe failure data

Region	Type	No. pipes	No. failures	Avg. failure per. year	Total length[m]
Region A	RWM	23 926	3 782	0.99%	879 050
	CWM	2 945	182	0.39%	229 915
Region B	RWM	62 089	10 717	1.08%	3 878 255
	CWM	7 119	435	0.38%	671 884
Region C	RWM	45 030	10 719	1.49%	2 593 412
	CWM	5 001	606	0.76%	477 094

Table 2. Area Under Curve (AUC)

		Dense-to-sparse					
		HDP-HBP	DPMHBP	HBP	Cox	SVM	Boost
AUC	Region A	**67.00%**	62.45%	62.38%	60.94%	61.07%	66.19%
	Region B	**89.46%**	84.10%	86.62%	63.33%	77.91%	63.95%
	Region C	87.81%	86.93%	87.24%	75.96%	82.98%	**89.92%**

comprised of approximately 10% CWM and 90%. The average failure rate each year for the CWM across the entire network is approximately 0.5%. The failure history recorded has an observation period of 16 years spanning from 1997 to 2012 which is very short given the failure rate. Due to a combination these factors the dataset used to make failure predictions on water pipe is extremely sparse.

Our experiments focus on a dense-to-sparse transfer scenario. The dense-to-sparse scenario attempts to improve the learning performance by transferring information from RWM to CWM. RWM is considerably denser compared to CWM as there are fewer pipes and failures observed, which makes it much more difficult for a model to use a single CWM dataset to learn the failure patterns as there are fewer training examples. We demonstrate the ability for HDP-HBP to use the information from RWM and CWM to improve the failure prediction performance. Figure 3 shows the results of the dense-to-sparse transfer.

In Fig. 3 the behavior of HBP based models are similar on the left-hand side due to the contribution from the beta-Bernoulli process. If there is a high number of observations of failure for the water pipe, these water pipes are generally ranked with the highest probability of failure in each cluster. As a result, on the left-hand side all variants of HBP model, HBP, DPMHBP and HDP-HBP have very similar failure prediction performance. In Fig. 3, Region A has shown to have detected very little failure across all HBP models.

Manually inspecting these water pipes in the dataset, these water pipes have historically had a high failure rate. Towards the right-hand side, these pipes generally have either no observations or very few observations of failure during its observation period. In this region, our proposed model is shown to perform much better by providing better clustering and parameter selection. In Region A

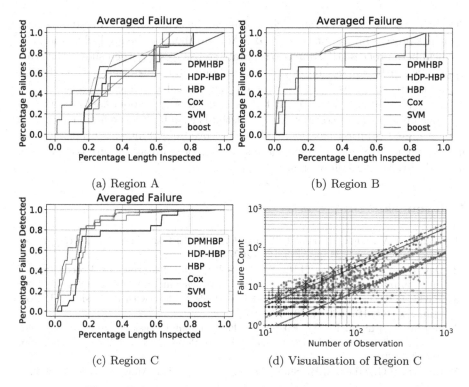

Fig. 3. Failure prediction results for each region in 2012.

and Region B, HDP-HBP has shown that it is superior in learning the model parameters for each group of *task*. The AUC can be used to quantify the average model performance. The AUC for each model is shown in Table 2. A visualising of the clustering in Region C. The top three clusters are shown to avoid clutter in the visualisation. The solid line represents the model mean, the dotted line represents the empirical mean. 'X' denotes data from the sparse domain, while '.' denotes data from a dense domain. Grey points are other feature combinations which are not in the top 3 clusters. The improvement in performance by the hierarchical clustering can be seen clearly in the magenta line, where there is a difference in the failure rate (the exponential of the gradient).

6   Conclusion

This paper has proposed a multi-task learning solution using a Bayesian non-parametric approach to construct a model which can automatically learn the model parameters for a group of *tasks* to improve the failure prediction model for sparse datasets by using data obtained from other domains with different feature space. The cross *domain* clustering learns a common latent structure to be used as the model parameters for failure prediction for each *task*. We have demonstrated how cross *domain* clustering can be achieved by using hierarchical Dirichlet process mixture model (HDPMM) and then can be applied to the hierarchical beta process (HBP) in two separate steps. We have then shown that this approach can be further improved by combining the clustering and prediction model to create a new framework called hierarchical Dirichlet process mixture of hierarchical beta process (HDP-HBP). Our proposed model has been able to improve upon previous state-of-the-art methods with higher prediction accuracy. Our proposed solution has provided an improved capability for utility companies to proactively inspect extremely sparse critical infrastructure components instead of repairing failures reactively.

References

1. Bonilla, E.V., Chai, K.M.A., Williams, C.K.: Multi-task Gaussian process prediction. In: NIPs, vol. 20, pp. 153–160 (2007)
2. Dai, W., Yang, Q., Xue, G.R., Yu, Y.: Self-taught clustering. In: Proceedings of the 25th International Conference on Machine Learning, pp. 200–207. ACM (2008)
3. David, C.R., et al.: Regression models and life tables (with discussion). J. Roy. Stat. Soc. **34**, 187–220 (1972)
4. Gupta, S., Phung, D., Venkatesh, S.: Factorial multi-task learning: a Bayesian nonparametric approach. In: International conference on machine learning, pp. 657–665 (2013)
5. Ibrahim, J.G., Chen, M.H., Sinha, D.: Bayesian Survival Analysis. Wiley, Hoboken (2005)
6. Kemp, C., Tenenbaum, J.B., Griffiths, T.L., Yamada, T., Ueda, N.: Learning systems of concepts with an infinite relational model. In: AAAI, vol. 3, p. 5 (2006)
7. Kumar, A., et al.: Using machine learning to assess the risk of and prevent water main breaks. arXiv preprint arXiv:1805.03597 (2018)
8. Li, B., Zhang, B., Li, Z., Wang, Y., Chen, F., Vitanage, D.: Prioritising water pipes for condition assessment with data analytics (2015)
9. Li, Z., et al.: Water pipe condition assessment: a hierarchical beta process approach for sparse incident data. Mach. Learn. **95**(1), 11–26 (2014)
10. Lin, P., et al.: Data driven water pipe failure prediction: a Bayesian nonparametric approach. In: Proceedings of the 24th ACM International on Conference on Information and Knowledge Management, pp. 193–202. ACM (2015)
11. Luo, S., Chu, V.W., Zhou, J., Chen, F., Wong, R.K., Huang, W.: A multivariate clustering approach for infrastructure failure predictions. In: 2017 IEEE International Congress on Big Data (BigData Congress), pp. 274–281. IEEE (2017)
12. Schwaighofer, A., Tresp, V., Yu, K.: Learning Gaussian process kernels via hierarchical bayes. In: Advances in Neural Information Processing Systems, pp. 1209–1216 (2005)
13. Teh, Y.W., Jordan, M.I., Beal, M.J., Blei, D.M.: Hierarchical dirichlet processes. J. Am. Stat. Assoc. **101**(476), 1566–1581 (2006)
14. Thibaux, R., Jordan, M.I.: Hierarchical beta processes and the Indian buffet process. In: AISTATS, vol. 2, pp. 564–571 (2007)
15. Xue, Y., Liao, X., Carin, L., Krishnapuram, B.: Multi-task learning for classification with dirichlet process priors. J. Mach. Learn. Res. **8**(Jan), 35–63 (2007)

Cost Sensitive Learning in the Presence of Symmetric Label Noise

Sandhya Tripathi$^{(\boxtimes)}$ and Nandyala Hemachandra

IEOR, Indian Institute of Technology Bombay, Mumbai 400076, India
{sandhya.tripathi,nh}@iitb.ac.in

Abstract. In binary classification framework, we are interested in making cost sensitive label predictions in the presence of uniform/symmetric label noise. We first observe that 0–1 Bayes classifiers are not (uniform) noise robust in cost sensitive setting. To circumvent this impossibility result, we present two schemes; unlike the existing methods, our schemes do not require noise rate. The first one uses α-weighted γ-uneven margin squared loss function, $l_{\alpha,usq}$, which can handle cost sensitivity arising due to domain requirement (using user given α) or class imbalance (by tuning γ) or both. However, we observe that $l_{\alpha,usq}$ Bayes classifiers are also not cost sensitive and noise robust. We show that regularized ERM of this loss function over the class of linear classifiers yields a cost sensitive uniform noise robust classifier as a solution of a system of linear equations. We also provide a performance bound for this classifier. The second scheme that we propose is a re-sampling based scheme that exploits the special structure of the uniform noise models and uses in-class probability estimates. Our computational experiments on some UCI datasets with class imbalance show that classifiers of our two schemes are on par with the existing methods and in fact better in some cases w.r.t. Accuracy and Arithmetic Mean, without using/tuning noise rate. We also consider other cost sensitive performance measures viz., F measure and Weighted Cost for evaluation.

1 Introduction

We are interested in cost sensitive label predictions when only noise corrupted labels are available. The labels might be corrupted when the data has been collected by crowd scouring with not so high labeling expertise. We consider the basic case when the induced label noise is independent of the class or example, i.e., symmetric/uniform noise model. In real world, there are various applications requiring differential misclassification costs due to class imbalance or domain requirement or both; we list these below and elaborate in the long version of the paper. The long version also has details of some examples, proofs and some computations.

A long version of this paper along with Supplementary material is available at https://arxiv.org/abs/1901.02271.

© Springer Nature Switzerland AG 2019
Q. Yang et al. (Eds.): PAKDD 2019, LNAI 11439, pp. 15–28, 2019.
https://doi.org/10.1007/978-3-030-16148-4_2

In *case 1*, there is no explicit need for different penalization of classes but the data has imbalance. Example: Predicting whether the age of an applicant for a vocational training course is above or below 15 years. Here, asymmetric cost should be learnt from data.

In *case 2*, there is no imbalance in data but one class's misclassification cost is higher than that of other. Example: Product recommendation by paid web advertisements. Here, the misclassification cost has to come from the domain.

In *case 3*, there is both imbalance and need for differential costing. Example: Rare (imbalance) disease diagnosis where the cost of missing a patient with disease is higher than cost of wrongly diagnosing a person with disease. The model should incorporate both the cost from domain and the cost due to imbalance.

In Sect. 1.1, we provide a summary of how these 3 cases are handled. For cost and uniform noise, we have considered real datasets belonging to cases 2 and 3.

Contributions

- Show that, unlike 0–1 loss, weighted 0–1 loss is not cost sensitive uniform noise robust.
- Show α-weighted γ-uneven margin squared loss $l_{\alpha,usq}$ with linear classifiers is both uniform noise robust and handles cost sensitivity. Present a performance bound of a classifier obtained from $l_{\alpha,usq}$ based regularized ERM.
- Propose a re-sampling based scheme for cost sensitive label prediction in the presence of uniform noise using in-class probability estimates.
- Unlike existing work, both the proposed schemes do not need true noise rate.
- Using a balanced dataset (Bupa) which requires domain cost too, we demonstrate that tuning γ on such corrupted datasets can be beneficial.

Related Work. For classification problems with label noise, particularly in Empirical Risk Minimization framework, the most recent work [3,4,7,8,13] aims to make the loss function noise robust and then develop algorithms. Cost sensitive learning has been widely studied by [2,5,10] and many more. An extensive empirical study on the effect of label noise on cost sensitive learning is presented in [14]. The problem of cost sensitive uniform noise robustness is considered in [6] where asymmetric misclassification cost α is tuned and class dependent noise rates are cross validated over corrupted data. However, our work incorporates cost due to both imbalance (γ) and domain requirement (α) with the added benefit that there is no need to know the true noise rate.

Organization. Section 1.1 has some details about weighted uneven margin loss functions and in-class probability estimates. In Sect. 2, weighted 0–1 loss is shown to be non cost sensitive uniform noise robust. Sections 3 and 4 present two different schemes that make cost sensitive predictions in the presence of uniform label noise. Section 5 has empirical evidence of the performance of proposed methods. Some discussion and future directions are presented in Sect. 6.

Notations. Let D be the joint distribution over $\mathbf{X} \times Y$ with $\mathbf{X} \in \mathcal{X} \subseteq \mathbb{R}^n$ and $Y \in \mathcal{Y} = \{-1, 1\}$. Let the in-class probability and class marginal on D be denoted by $\eta(\mathbf{x}) := P(Y = 1|\mathbf{x})$ and $\pi := P(Y = 1)$. Let the decision function be $f : \mathbf{X} \mapsto \mathbb{R}$, hypothesis class of all measurable functions be $\mathcal{H}$ and class of linear hypothesis be $\mathcal{H}_{lin} := \{(\mathbf{w}, b), \mathbf{w} \in \mathbb{R}^n, b \in \mathbb{R} : \|\mathbf{w}\|_2 \leq W\}$ for a given W. Let $\tilde{D}$ denote the distribution on $\mathbf{X} \times \tilde{Y}$ obtained by inducing noise to D with $\tilde{Y} \in \{-1, 1\}$. The corrupted sample is $\tilde{S} = \{(\mathbf{x}_1, \tilde{y}_1), \ldots, (\mathbf{x}_m, \tilde{y}_m)\} \sim \tilde{D}^m$. The noise rate $\rho := P(\tilde{Y} = -y|Y = y, \mathbf{X} = \mathbf{x})$ is constant across classes and the model is referred to as Symmetric Label Noise (SLN) model. In such cases, the corrupted in-class probability is $\tilde{\eta}(\mathbf{x}) := P(\tilde{Y} = 1|\mathbf{x}) = (1 - 2\rho)\eta(\mathbf{x}) + \rho$ and the corrupted class marginal is $\tilde{\pi} := P(\tilde{Y} = 1) = (1 - 2\rho)\pi + \rho$. Symmetric and uniform noise are synonymous in this work.

1.1 Some Relevant Background

The first choice of loss function for cost sensitive learning is that of α-weighted 0–1 loss defined as follows:

$$l_{0-1,\alpha}(f(\mathbf{x}), y) = (1 - \alpha)\mathbf{1}_{\{Y=1, f(\mathbf{x}) \leq 0\}} + \alpha\mathbf{1}_{\{Y=-1, f(\mathbf{x}) > 0\}}, \quad \forall \alpha \in (0, 1) \quad (1)$$

Let the α-weighted 0-1 risk be $R_{D,\alpha} := E_D[l_{0-1,\alpha}(f(\mathbf{x}), y)]$. The minimizer of this risk is $f_\alpha^*(\mathbf{x}) = sign(\eta(\mathbf{x}) - \alpha)$ and referred to as cost-sensitive Bayes classifier. The corresponding surrogate l_α based risk and the minimizer is defined as $R_{D,l_\alpha} := E_D[l_\alpha(f(\mathbf{x}), y)]$ and $f_{l_\alpha}^* \in \mathcal{H}$ respectively. Consider the following notion of α-classification calibration.

Definition 1 (α-Classification Calibration [10]). *For $\alpha \in (0, 1)$ and a loss function l, define the α-weighted loss:*

$$l_\alpha(f(\mathbf{x}), y) = (1 - \alpha)l_1(f(\mathbf{x})) + \alpha l_{-1}(f(\mathbf{x})), \quad (2)$$

where $l_1(\cdot) := l(\cdot, 1)$ and $l_{-1}(\cdot) := l(\cdot, -1)$. l_α is α-classification calibrated (α-CC) iff there exists a convex, non-decreasing and invertible transformation ψ_{l_α}, with $\psi_{l_\alpha}^{-1}(0) = 0$, such that

$$\psi_{l_\alpha}(R_{D,\alpha}(f) - R_{D,\alpha}(f_\alpha^*)) \leq R_{D,l_\alpha}(f) - R_{D,l_\alpha}(f_{l_\alpha}^*). \quad (3)$$

If the classifiers obtained from α-CC losses are consistent w.r.t l_α-risk then they are also consistent w.r.t α-weighted 0–1 risk. We consider the α-weighted uneven margin squared loss [10] which is by construction α-CC and defined as follows:

$$l_{\alpha,usq}(f(\mathbf{x}), y) = (1 - \alpha)\mathbf{1}_{\{y=1\}}(1 - f(\mathbf{x}))^2 + \alpha\mathbf{1}_{\{y=-1\}}\frac{1}{\gamma}(1 + \gamma f(\mathbf{x}))^2, \quad \gamma > 0 \quad (4)$$

Interpretation of α and γ. The role of α and γ can be related to the three cases of differential costing described at the start of this paper. In *case 1*, there are 3 options: fix $\alpha = 0.5$ and let tuned γ pick up the imbalance; fix $\gamma = 1$ and tune α; tune both α and γ. Our experimental results suggest that latter two perform equally good. For *case 2*, α is given and γ can be fixed at 1. However, we observe that even in this case tuning γ can be more informative. For *case 3*, γ is tuned and α is given a priori. There would be a trade-off between α and γ, i.e., for a given α, there would be an optimal γ in a suitable sense. Above observations are based on the experiments described in Supplementary material Section D.

Choice of η Estimation Method. As η estimates are required in re-sampling scheme, we investigated the performance of 4 methods: *Lk-fun* [9], uses classifier to get the estimate $\hat{\eta}$; interpreting η as a conditional expectation and obtaining it by a suitable squared deviation minimization; LSPC [11], density ratio estimation by l_2 norm minimization; KLIEP [12], density ratio estimation by KL divergence minimization. We chose to use *Lk-fun* with logistic loss l_{log} and squared loss l_{sq}, LSPC, and $\hat{\eta}_{norm}$, a normalized version of KLIEP because in re-sampling algorithm we are concerned with label prediction and these estimators performed equally well on Accuracy measure. A detailed study is available in Supplementary material Section F.

2   Cost Sensitive Bayes Classifiers Using $l_{0-1,\alpha}$ and $l_{\alpha,usq}$ Need Not Be Uniform Noise Robust

The robustness notion for risk minimization in cost insensitive scenarios was introduced by [4]. They also proved that cost insensitive 0–1 loss based risk minimization is SLN robust. We extend this definition to cost-sensitive learning.

Definition 2 (Cost sensitive noise robust). *Let $f^*_{\alpha,A}$ and $\tilde{f}^*_{\alpha,A}$ be obtained from clean and corrupted distribution D and $\tilde{D}$ using any arbitrary scheme A, then the scheme A is said to be cost sensitive noise robust if*

$$R_{D,\alpha}(\tilde{f}^*_{\alpha,A}) = R_{D,\alpha}(f^*_{\alpha,A}).$$

If $f^*_{\alpha,A}$ and $\tilde{f}^*_{\alpha,A}$ are obtained from a cost sensitive loss function l and noise induced is symmetric, then l is said to be cost sensitive uniform noise robust.

Let the $l_{0-1,\alpha}$ risk on $\tilde{D}$ be denoted by $R_{\tilde{D},\alpha}(f)$. If one is interested in cost sensitive learning with noisy labels, then the sufficient condition of [3] becomes $(1-\alpha)\mathbf{1}_{[f(\mathbf{x})\leq 0]} + \alpha\mathbf{1}_{[f(\mathbf{x})>0]} = K$. This condition is satisfied if and only if $(1-\alpha) = K = \alpha$ implying that it cannot be a sufficient condition for SLN robustness if there is a differential costing of $(1-\alpha,\alpha)$, $\alpha \neq 0.5$.

Let f^*_α and $\tilde{f}^*_\alpha$ be the minimizers of $R_{D,\alpha}(f)$ and $R_{\tilde{D},\alpha}(f)$. Then, it is known that they have the following form:

$$f^*_\alpha(\mathbf{x}) = sign\left(\eta(\mathbf{x}) - \alpha\right) \tag{5}$$

$$\tilde{f}^*_\alpha(\mathbf{x}) = sign(\tilde{\eta}(\mathbf{x}) - \alpha) = sign\left(\eta(\mathbf{x}) - \frac{\alpha - \rho}{(1 - 2\rho)}\right) \tag{6}$$

The last equality in (6) follows from the fact that $\tilde{\eta} = (1 - 2\rho)\eta + \rho$. In Example 1 below, we show that $l_{0-1,\alpha}$ is not cost sensitive uniform noise robust with $\mathcal{H}$.

Example 1. Let Y has a Bernoulli distribution with parameter $p = 0.2$. Let $\mathbf{X} \subset \mathbb{R}$ be such that $\mathbf{X}|Y = 1 \sim Uniform(0, p)$ and $\mathbf{X}|Y = -1 \sim Uniform(1 - p, 1)$. Then, the in-class probability $\eta(\mathbf{x})$ is given as follows:

$$\eta(\mathbf{x}) = P(Y = 1|\mathbf{X} = \mathbf{x}) = p = 0.2$$

Suppose $\rho = 0.3$. Then, $\tilde{\eta}(\mathbf{x}) = (1 - 2\rho)\eta(\mathbf{x}) + \rho = 0.38$. If $\alpha = 0.25$, $f_\alpha^*(\mathbf{x}) = -1$ and $\tilde{f}_\alpha^*(\mathbf{x}) = 1, \forall \mathbf{x} \in \mathbf{X}$. Consider the α-weighted 0–1 risk of $f_\alpha^*(\mathbf{x})$ and $\tilde{f}_\alpha^*(\mathbf{x})$:

$$R_{D,\alpha}(f_\alpha^*) = E_D[l_{0-1,\alpha}(f_\alpha^*(\mathbf{x}), y)] = (1 - \alpha)p, \qquad \text{since } f_\alpha^*(\mathbf{x}) \leq 0 \ \forall \mathbf{x}$$
$$R_{D,\alpha}(\tilde{f}_\alpha^*) = E_D[l_{0-1,\alpha}(\tilde{f}_\alpha^*(\mathbf{x}), y)] = \alpha(1 - p), \qquad \text{since } \tilde{f}_\alpha^*(\mathbf{x}) > 0 \ \forall \mathbf{x}$$

Therefore, $R_{D,\alpha}(f_\alpha^*) \neq R_{D,\alpha}(\tilde{f}_\alpha^*)$ implying that the α-weighted 0–1 loss function $l_{0-1,\alpha}$ is not uniform noise robust with $\mathcal{H}$. Details are in Supplementary material Section B.1. Note that due to $p < 0.5$, D is linearly separable; a linearly inseparable variant can be obtained by $p > 0.5$. Another linearly inseparable distribution based counter example is available in Supplementary material Section B.4.

In view of the above example, one can try to use the principle of inductive bias, i.e., consider a strict subset of the above set of classifiers; however, Example 2 below says that the set of linear class of classifiers need not be cost sensitive uniform noise robust.

Example 2. Consider the training set $\{(3, -1), (8, -1), (12, 1)\}$ with uniform probability distribution. Let the linear classifier be of the form $fl = sign(\mathbf{x} + b)$. Let $\alpha = 0.3$ and the uniform noise be $\rho = 0.42$. Then,

$$fl_\alpha^* = \arg\min_{fl} E_D[l_{0-1,\alpha}(y, fl)] = b^* \in (-8, -12) \quad \text{with} \quad R_{\alpha,D}(fl_\alpha^*) = 0.$$
$$\tilde{fl}_\alpha^* = \arg\min_{fl} E_{\tilde{D}}[l_{0-1,\alpha}(\tilde{y}, \tilde{fl})] = \tilde{b}^* \in (-3, \infty) \quad \text{with} \quad R_{\alpha,D}(\tilde{fl}_\alpha^*) = 0.2.$$

Details of Example 2 are available in Supplementary material Section B.2. To avoid above counter examples, we resort to convex surrogate loss function and a type of regularization which restricts the hypothesis class. Consider an α-weighted uneven margin loss functions $l_{\alpha,un}$ [10] with its optimal classifiers on D and $\tilde{D}$ denoted by $f_{l_{\alpha,un}}^*$ and $\tilde{f}_{l_{\alpha,un}}^*$ respectively. Regularized risk minimization defined below is known to avoid over-fitting.

$$R_{D,l_{\alpha,un}}^r(f) = E_D[l_{\alpha,un}(f(\mathbf{x}), y)] + \lambda\|f\|_2^2, \qquad \text{where} \quad \lambda > 0 \tag{7}$$

Let the regularized risk of $l_{\alpha,un}$ on $\tilde{D}$ be $R_{\tilde{D},l_{\alpha,un}}^r(f)$. Also, let the minimizers of clean and corrupted $l_{\alpha,un}$-regularized risks be $f_{r,l_{\alpha,un}}^*$ and $\tilde{f}_{r,l_{\alpha,un}}^*$. Now, Definition 2 can be specialized to $l_{\alpha,un}$ to assure cost sensitivity, classification calibration and uniform noise robustness as follows:

Definition 3 $((\alpha, \gamma, \rho)$-robustness of risk minimization). *For a loss function $l_{\alpha,un}$ and classifiers $\tilde{f}^*_{l_{\alpha,un}}$ and $f^*_{l_{\alpha,un}}$, risk minimization is said to be (α, γ, ρ)-robust if*

$$R_{D,\alpha}(\tilde{f}^*_{l_{\alpha,un}}) = R_{D,\alpha}(f^*_{l_{\alpha,un}}). \tag{8}$$

Further, if the classifiers in Eq. (8) are $f^*_{r,l_{\alpha,un}}$ and $\tilde{f}^*_{r,l_{\alpha,un}}$ then, we say that regularized risk minimization under $l_{\alpha,un}$ is (α, γ, ρ)-robust.

Due to squared loss's SLN robustness property [4], we check whether $l_{\alpha,usq}$ is (α, γ, ρ) robust or not. It is not with $\mathcal{H}$ as shown in Example 3; details of Example 3 are available in Supplementary material Section B.3.

Example 3. Consider the settings as in Example 1. Let $\alpha = 0.25$ and $\gamma = 0.4$. Then, for all $\mathbf{x}$,

$$f^*_{l_{\alpha,usq}}(\mathbf{x}) = \frac{\eta(\mathbf{x}) - \alpha}{\eta(\mathbf{x})(1 - \alpha) + \gamma\alpha(1 - \eta(\mathbf{x}))} = -0.21, \quad \tilde{f}^*_{l_{\alpha,usq}}(\mathbf{x}) = \frac{\tilde{\eta}(\mathbf{x}) - \alpha}{\tilde{\eta}(\mathbf{x})(1 - \alpha) + \gamma\alpha(1 - \tilde{\eta}(\mathbf{x}))} = 0.37.$$

And, $\quad R_{D,\alpha}(f^*_{l_{\alpha,usq}}) = (1 - \alpha)p = 0.15, \quad R_{D,\alpha}(\tilde{f}^*_{l_{\alpha,usq}}) = \alpha(1 - p) = 0.2.$

Hence, $R_{D,\alpha}(f^*_{l_{\alpha,usq}}) \neq R_{D,\alpha}(\tilde{f}^*_{l_{\alpha,usq}})$, implying that $l_{\alpha,usq}$ based ERM may not be cost sensitive uniform noise robust.

We again have a negative result with $l_{\alpha,usq}$ when we consider hypothesis class $\mathcal{H}$. In next section, we present a positive result and show that regularized risk minimization under loss function $l_{\alpha,usq}$ is (α, γ, ρ)-robust if the hypothesis class is restricted to $\mathcal{H}_{lin}$.

3 $(l_{\alpha,usq}, \mathcal{H}_{lin})$ Is (α, γ, ρ) Robust

In this section, we consider the weighted uneven margin squared loss function $l_{\alpha,usq}$ from Eq. (4) with restricted hypothesis class $\mathcal{H}_{lin}$ and show a positive result that regularized cost sensitive risk minimization under loss function $l_{\alpha,usq}$ is (α, γ, ρ)-robust. A proof is available in Supplementary material Section A.1.

Theorem 1. $(l_{\alpha,usq}, \mathcal{H}_{lin})$ *is (α, γ, ρ)-robust, i.e., linear classifiers obtained from α-weighted γ-uneven margin squared loss $l_{\alpha,usq}$ based regularized risk minimization are SLN robust.*

Remark 1. The above results relating to counter examples and Theorem 1 about cost sensitive uniform noise robustness can be summarized as follows: There are two loss functions, $l_{0-1,\alpha}$ and $l_{\alpha,usq}$ and two hypothesis classes, $\mathcal{H}_{lin}$ and $\mathcal{H}$. Out of the four combinations of loss functions and hypothesis classes only $l_{\alpha,usq}$ with $\mathcal{H}_{lin}$ is cost sensitive uniform noise robust, others are not.

Next, we provide a closed form expression for the classifier learnt on corrupted data by minimizing empirical $l_{\alpha,usq}$-regularized risk. We also provide a performance bound on the clean risk of this classifier.

3.1 $l_{\alpha,usq}$ Based Classifier from Corrupted Data and Its Performance

In this subsection, we present a descriptive scheme to learn a cost-sensitive linear classifier in the presence of noisy labels, by minimizing $l_{\alpha,usq}$ based regularized empirical risk, i.e.,

$$\hat{f}_r := \arg\min_{f \in \mathcal{H}_{lin}} \hat{R}^r_{\tilde{D},l_{\alpha,usq}}(f), \tag{9}$$

where $\hat{R}^r_{\tilde{D},l_{\alpha,usq}}(f) := \frac{1}{m} \sum_{i=1}^{m} l_{\alpha,usq}(f(\mathbf{x}_i), \tilde{y}_i) + \lambda \|f\|_2^2$, α is user given, γ and regularization parameter $\lambda > 0$ are to be tuned by cross validation. A proof is available in Supplementary material Section A.2.

Proposition 1. *Consider corrupted regularized empirical risk* $\hat{R}^r_{\tilde{D},l_{\alpha,usq}}(f)$ *of* $l_{\alpha,usq}$*. Then, the optimal* (α,γ,ρ)*-robust linear classifier* $\hat{f}_r = (\mathbf{w}, b) \in \mathcal{H}_{lin}$ *with* $\mathbf{w} \in \mathbb{R}^n$ *has the following form:*

$$\hat{f}_r = \bar{\mathbf{w}}^* = (A + \lambda\mathbf{I})^{-1}\mathbf{c}, \quad \lambda > 0 \tag{10}$$

where $\bar{\mathbf{w}} = [w_1, w_2, \ldots, w_n, b]^T$, *a* $n+1$ *dimensional vector of variables;* A, *a* $(n+1 \times n+1)$ *dimensional known symmetric matrix and* $\mathbf{c}$, *a* $n+1$ *dimensional known vector are as follows:*

$$A + \lambda I = \begin{bmatrix} \sum\limits_{i=1}^{m} x_{i1}^2 a_i + \lambda & \sum\limits_{i=1}^{m} x_{i1}x_{i2}a_i & \cdots & \sum\limits_{i=1}^{m} x_{i1}x_{in}a_i & \sum\limits_{i=1}^{m} x_{i1}a_i \\ \sum\limits_{i=1}^{m} x_{i2}x_{i1}a_i & \sum\limits_{i=1}^{m} x_{i2}^2 a_i + \lambda & \cdots & \sum\limits_{i=1}^{m} x_{i2}x_{in}a_i & \sum\limits_{i=1}^{m} x_{i2}a_i \\ \vdots & \vdots & \ddots & \vdots & \vdots \\ \sum\limits_{i=1}^{m} x_{in}x_{i1}a_i & \sum\limits_{i=1}^{m} x_{in}x_{i2}a_i & \cdots & \sum\limits_{i=1}^{m} x_{in}^2 a_i + \lambda & \sum\limits_{i=1}^{m} x_{in}a_i \\ \sum\limits_{i=1}^{m} x_{i1}a_i & \sum\limits_{i=1}^{m} x_{i2}a_i & \cdots & \sum\limits_{i=1}^{m} x_{in}a_i & \sum\limits_{i=1}^{m} a_i + \lambda \end{bmatrix}, \quad \mathbf{c} = \begin{bmatrix} \sum\limits_{i=1}^{m} x_{i1}c_i \\ \sum\limits_{i=1}^{m} x_{i2}c_i \\ \vdots \\ \sum\limits_{i=1}^{m} x_{in}c_i \\ \sum\limits_{i=1}^{m} c_i \end{bmatrix}$$

with $a_i = (\mathbf{1}_{[\tilde{y}_i=1]}(1-\alpha) + \gamma\alpha\mathbf{1}_{[\tilde{y}_i=-1]})$ *and* $c_i = (\mathbf{1}_{[\tilde{y}_i=1]}(1-\alpha) - \alpha\mathbf{1}_{[\tilde{y}_i=-1]})$.

Next, we provide a result on the performance of $\hat{f}_r$ in terms of the Rademacher complexity of the function class $\mathcal{H}_{lin}$. For this, we need Lemmas 1 and 2 whose proofs are available in Supplementary material Section A.3 and A.4 respectively.

Lemma 1. *Consider the* α*-weighted uneven margin squared loss* $l_{\alpha,usq}(f(\mathbf{x}), y)$ *which is locally L-Lipschitz with* $L = 2a + 2$ *where* $|f(\mathbf{x})| \leq a$, *for some* $a \geq 0$. *Then, with probability at least* $1 - \delta$,

$$\max_{f \in \mathcal{H}_{lin}} |\hat{R}_{\tilde{D},l_{\alpha,usq}}(f) - R_{\tilde{D},l_{\alpha,usq}}(f)| \leq 2L\mathfrak{R}(\mathcal{H}_{lin}) + \sqrt{\frac{log(1/\delta)}{2m}},$$

where $\mathfrak{R}(\mathcal{H}_{lin}) := E_{\mathbf{X},\sigma}\left[\sup_{f \in \mathcal{H}_{lin}} \frac{1}{m} \sum_{i=1}^{m} \sigma_i f(\mathbf{x}_i)\right]$ *is the Rademacher complexity of the function class* $\mathcal{H}_{lin}$ *with* σ_i*'s as independent uniform random variables taking values in* $\{-1, 1\}$.

Lemma 2. *For a classifier $f \in \mathcal{H}$ and user given $\alpha \in (0,1)$, the $l_{\alpha,usq}$ risk on clean and corrupted distribution satisfy the following equation:*

$$R_{\tilde{D},l_{\alpha,usq}}(f) = R_{D,l_{\alpha,usq}}(f) + 4\rho E_D[yf(\mathbf{x})[(1-\alpha)\mathbf{1}_{[y=1]} + \alpha\mathbf{1}_{[y=-1]}]]. \quad (11)$$

Theorem 2. *Under the settings of Lemma 1, with probability at least $1 - \delta$,*

$$R_{D,l_{\alpha,usq}}(\hat{f}_r) \leq \min_{f \in \mathcal{H}_{lin}} R_{D,l_{\alpha,usq}}(f) + 4L\mathfrak{R}(\mathcal{H}_{lin}) + 2\sqrt{\frac{log(1/\delta)}{2m}} + 2\lambda W^2$$

$$+ \frac{4\rho}{(1-2\rho)} E_{\mathbf{x}}[(\tilde{fl}^*_{l_{\alpha,usq}}(\mathbf{x}) - (1-2\rho)\hat{f}_r(\mathbf{x}))(\eta(\mathbf{x}) - \alpha)]$$

where $\tilde{fl}^*_{l_{\alpha,usq}}$ is the linear minimizer of $R_{\tilde{D},l_{\alpha,usq}}$ and $\eta(\mathbf{x})$ is the in-class probability for $\mathbf{x}$. Furthermore, as $l_{\alpha,usq}$ is α-CC, there exists a non-decreasing and invertible function $\psi_{l_{\alpha,usq}}$ with $\psi^{-1}_{l_{\alpha,usq}}(0) = 0$ such that,

$$R_{D,\alpha}(\hat{f}_r) - R_{D,\alpha}(f^*_\alpha) \leq \psi^{-1}_{l_{\alpha,usq}} \left(\min_{f \in \mathcal{H}_{lin}} R_{D,l_{\alpha,usq}}(f) - \min_{f \in \mathcal{H}} R_{D,l_{\alpha,usq}}(f) + 4L\mathfrak{R}(\mathcal{H}_{lin}) \right.$$

$$+ 2\sqrt{\frac{log(1/\delta)}{2m}} + \frac{4\rho}{(1-2\rho)} E_{\mathbf{x}}[(\tilde{fl}^*_{l_{\alpha,usq}}(\mathbf{x}) - \hat{f}_r(\mathbf{x}))(\eta(\mathbf{x}) - \alpha)]$$

$$\left. + \frac{8\rho^2}{(1-2\rho)} E_{\mathbf{x}}[\hat{f}_r(\mathbf{x})(\eta(\mathbf{x}) - \alpha)] + 2\lambda W^2 \right). \quad (12)$$

A proof of Theorem 2 is available in Supplementary material Section A.5. The first two terms (involving the difference) in the right hand side of Eq. (12) denotes the approximation error which is small if $\mathcal{H}_{lin}$ is large and the third term involving the Rademacher complexity denotes the estimation error which is small if $\mathcal{H}_{lin}$ is small. The fourth term denotes the sample complexity which vanishes as the sample size increases. The bound in (12) can be used to show consistency of $l_{\alpha,usq}$ based regularized ERM if the argument of $\psi^{-1}_{l_{\alpha,usq}}$ tends to zero as sample size increases. However, in this case, it is not obvious because the last two terms involving noise rates may not vanish with increasing sample size. In spite of this, our empirical experience with this algorithm is very good.

4   A Re-sampling Based Algorithm $(\tilde{\eta}, \alpha)$

In this section, we present a cost sensitive label prediction algorithm based on re-balancing (which is guided by the costs) the noisy training set given to the learning algorithm. Let us consider uneven margin version of α-weighted 0–1 loss from Eq. (1) defined as follows:

$$l_{0-1,\alpha,\gamma}(f(x),y) = (1-\alpha)\mathbf{1}_{\{Y=1,f(x)\leq 0\}} + \frac{\alpha}{\gamma}\mathbf{1}_{\{Y=-1,\gamma f(x)>0\}}, \quad \forall \alpha \in (0,1)$$

where α is user given cost and γ, tunable cost handles the class imbalance. This definition is along the lines of the uneven margin losses defined in [10]. Let $l_{0-1,\alpha,\gamma}$-risk on D be $R_{D,\alpha,\gamma}(f)$ and corresponding optimal classifier be $f^*_{0-1,\alpha,\gamma}$:

$$f^*_{0-1,\alpha,\gamma} = \arg\min_{f \in \mathcal{H}} R_{D,\alpha,\gamma}(f) = sign\left(\eta(\mathbf{x}) - \frac{\alpha}{\gamma + (1-\gamma)\alpha}\right). \quad (13)$$

Also, let $l_{0-1,\alpha,\gamma}$-risk on $\tilde{D}$ be $R_{\tilde{D},\alpha,\gamma}(f)$ and the corresponding optimal classifier be $\tilde{f}^*_{0-1,\alpha,\gamma}$ as given below:

$$\tilde{f}^*_{0-1,\alpha,\gamma} = sign\left(\tilde{\eta}(\mathbf{x}) - \frac{\alpha}{\gamma + (1-\gamma)\alpha}\right). \tag{14}$$

We propose Algorithm $(\tilde{\eta}, \alpha)$ which is mainly based on two ideas: (i) predictions based on a certain threshold (p^*) can correspond to predictions based on threshold (p_0) if the number of negative examples in the training set is multiplied by $r^* = \frac{p^*}{1-p^*}\frac{1-p_0}{p_0}$ (Theorem 1 of [2]) (ii) for a given $\mathbf{x}$, $\tilde{\eta}(\mathbf{x})$ and $\eta(\mathbf{x})$ lie on the same side of threshold 0.5 when noise rate is ρ. We first formalize the latter idea in terms of a general result. A proof is available in Supplementary material Section A.6.

Lemma 3. *In SLN models, for a given noise rate $\rho < 0.5$, the clean and corrupted class marginals π and $\tilde{\pi}$ satisfy the following condition:*

$$\pi \lessgtr 0.5 \Rightarrow \tilde{\pi} \lessgtr 0.5.$$

Further, the above monotonicity holds for $\eta(\mathbf{x})$ and $\tilde{\eta}(\mathbf{x})$ too.

In our case, the cost sensitive label prediction requires the desired threshold to be $\frac{\alpha}{\gamma+(1-\gamma)\alpha}$ $(= p^*)$ but the threshold which we can use is 0.5 $(= p_0)$ implying that for us $r^* = \frac{\alpha}{\gamma(1-\alpha)}$. If m_+ and m_- are number of positive and negative examples in m_{tr}, then we should re-sample such that the size of balanced dataset is $m_{tr,b} = m_+ + \left\lfloor \frac{\alpha m_-}{\gamma(1-\alpha)} \right\rfloor$. As we have access to only corrupted data, the learning scheme is: re-balance the corrupted data using r^* and then threshold $\tilde{\eta}$ at 0.5. Since, for SLN model, predictions made by thresholding $\tilde{\eta}$ at 0.5 are same as the predictions made by thresholding η at 0.5, for a test point $\mathbf{x}_0$ from D, predicted label is $sign(\tilde{\eta}(\mathbf{x}_0) - 0.5)$. The main advantage of this algorithm is that it doesn't require the knowledge of true noise rates. Also, unlike Sect. 3's scheme involving $l_{\alpha,usq}$ based regularized ERM, this algorithm uses $\tilde{\eta}$ estimates and hence is a generative learning scheme.

Since, we do not want to lose any minority (rare) class examples, we reassign positive labels to the minority class WLOG, if needed, implying that negative class examples are always under-sampled. The performance of Algorithm $(\tilde{\eta}, \alpha)$ is majorly dependent on sampling procedure and $\tilde{\eta}$ estimation methods used.

Remark 2. Algorithm $(\tilde{\eta}, \alpha)$ exploits the fact that $sign(\tilde{\eta} - \frac{\alpha}{\gamma+(1-\gamma)\alpha}) = sign(\tilde{\eta}_b - 0.5)$ where $\tilde{\eta}_b$ is learnt on re-sampled data. This implies $R_{D,\alpha}(sign(\tilde{\eta}_b - 0.5)) = R_{D,\alpha}(sign(\tilde{\eta} - \frac{\alpha}{\gamma+(1-\gamma)\alpha}))$ but due to counter examples in Sect. 2, these risks may not be equal to $R_{D,\alpha}(sign(\eta - \frac{\alpha}{\gamma+(1-\gamma)\alpha}))$. Hence, this scheme is not in contradiction to Sect. 2. However, as $\tilde{\eta}$ estimation methods use a subset of $\mathcal{H}$ (e.g., LSPC and KLIEP use linear combinations of finite Gaussian kernels as basis functions), these risks may be equal to $R_{D,\alpha}(sign(\tilde{\eta}_e - \frac{\alpha}{\gamma+(1-\gamma)\alpha}))$ where $\tilde{\eta}_e$ is estimate of $\tilde{\eta}$ obtained from strict subset of hypothesis class. Also,

---

Algorithm $(\tilde{\eta}, \alpha)$. $\tilde{\eta}$ based scheme to make cost sensitive label predictions from uniform noise corrupted data

---

Input: Training data $\tilde{S}_{tr} = \{(x_1, \tilde{y}_i)\}_{i=1}^{m_{tr}}$, test data $S_{te} = \{(x_i, y_i)\}_{i=1}^{m_{te}}$, cost α, performance measure $PM \in \{Acc, AM, F, WC\}$
Output: Predicted labels and $\tilde{\eta}$ estimate on test data S_{te}
1: $\gamma_0 = \frac{\alpha}{1-\alpha} + 0.001$,　　since for under-sampling $r^* = \frac{\alpha}{\gamma(1-\alpha)} < 1 \Rightarrow \gamma_0 > \frac{\alpha}{1-\alpha}$
2: $\Gamma = \{\frac{\alpha}{1-\alpha} + i \times 0.05 \quad i = 1, \ldots, 12\}$
3: **for** $\gamma \in \Gamma$ **do**
4:　　Under-sample the $-ve$ class to get $\tilde{S}_{tr,b}$ such that $|\tilde{S}_{tr,b}| = m_+ + \left\lfloor \frac{\alpha m_-}{\gamma(1-\alpha)} \right\rfloor$.
5:　　Use 5-fold CV to estimate $\tilde{\eta}$ from $\tilde{S}_{tr,b}$ via *Lk-fun* or LSPC or KLIEP.
6:　　Compute 5-fold cross validated PM from the partitioned data.
7: **end for**
8: $\gamma^* = \arg\max\limits_{\gamma \in \Gamma} PM$ if $PM = \{Acc, AM, F\}$ otherwise $\gamma^* = \arg\min\limits_{\gamma \in \Gamma} WC$.
9: Under-sample the $-ve$ class to get $\tilde{S}_{tr,b}$ such that $|\tilde{S}_{tr,b}| = m_+ + \left\lfloor \frac{\alpha m_-}{\gamma^*(1-\alpha)} \right\rfloor$.
10: Estimate $\tilde{\eta}$ from $\tilde{S}_{tr,b}$ using *Lk-fun* method or LSPC or KLIEP.
11: **for** $i = 1, 2, \ldots, m_{te}$ **do**
12:　　Compute $\tilde{\eta}(\mathbf{x}_i)$
13:　　$\hat{y}_i = sign(\tilde{\eta}(\mathbf{x}_i) - 0.5)$
14: **end for**

---

based on very good empirical performance of the scheme, we believe that $R_{D,\alpha}(sign(\tilde{\eta}_e - \frac{\alpha}{\gamma + (1-\gamma)\alpha})) = R_{D,\alpha}(sign(\eta_e - \frac{\alpha}{\gamma + (1-\gamma)\alpha}))$ where η_e is an estimate of η.

5　Comparison of $l_{\alpha,usq}$ Based Regularized ERM and Algorithm $(\tilde{\eta}, \alpha)$ to Existing Methods on UCI Datasets

In this section, we consider some UCI datasets [1] and demonstrate that $l_{\alpha,usq}$ is (α, γ, ρ)-robust. Also, we demonstrate the performance of Algorithm $(\tilde{\eta}, \alpha)$ with $\tilde{\eta}$ estimated using *Lk-fun*, LSPC and KLIEP. In addition to Accuracy (Acc), Arithmetic mean (AM) of True positive rate (TPR) and True negative rate (TNR), we also consider two measures suited for evaluating classifiers learnt on imbalanced data, viz., F measure and Weighted cost (WC) defined as below:

$$F = \frac{2TP}{2TP + FP + FN} \quad \text{and} \quad WC = (1-\alpha)FN + \frac{\alpha}{\gamma}FP$$

where TP, TN, FP, FN are number of true positives, true negatives, false positives and false negatives for a classifier.

To account for randomness in the flips to simulate a given noise rate, we repeat each experiment 10 times, with independent corruptions of the data set for same noise (ρ) setting. In every trial, the data is partitioned into train and test with 80-20 split. Uniform noise induced 80% data is used for training and validation (if there are any parameters to be tuned like γ). Finally, 20% clean test data is used for evaluation. Regularization parameter, λ is tuned over the set

$\Lambda = \{0.01, 0.1, 1, 10\}$. On a synthetic dataset, we observed that the performance of our methods and cost sensitive Bayes classifier on clean data w.r.t. Accuracy, AM, F and Weighted Cost measure is comparable for moderate noise rates; details in Supplementary material Section E.3. In all the tables, values within 1% of the best across a row are in bold.

Class Imbalance and Domain Requirement of Cost. We report the accuracy and AM values of logistic loss based unbiased estimator (MUB) approach and approach of surrogates for weighted 0–1 loss (S-W0-1), as it is from the work of [6] and compare them to Accuracy and AM for our cost sensitive learning schemes. It is to be noted that [6] assumes that the true noise rate ρ is known and cost α is tuned. We are more flexible and user friendly as we don't need the noise rate ρ and allow for user given misclassification cost α and tune γ.

It can be observed in Table 1 that as far as Accuracy is concerned Algorithm $(\tilde{\eta}, \alpha)$ and $l_{\alpha, usq}$ based regularized ERM have comparable values to that from MUB and S-W0-1 on all datasets. As depicted in Table 2, the proposed algorithms have marginally better values of AM measure than that of MUB and S-W0-1 method. Due to lack of a benchmark w.r.t. F and WC, on these measures, we compared our schemes to the SVMs trained on clean data and observed that our schemes fare well w.r.t to these measures too. However, due to space constraint the details are presented in Supplementary material Section E.1.

Table 1. Averaged Acc ($\pm$ s.d.) of the cost sensitive predictions made by Algorithm $(\tilde{\eta}, \alpha)$ and $l_{\alpha, usq}$ based regularized ERM on UCI datasets corrupted by uniform noise.

Dataset	Cost	ρ	Algorithm $(\tilde{\eta}, \alpha)$: $\tilde{\eta}$ estimate from				$l_{\alpha, usq}$ RegERM	MUB	S-W0-1
(n, m_+, m_-)	α		Lk-fun with l_{sq}	Lk-fun with l_{log}	LSPC	KLIEP		l_{log}	l_{log}
Breast cancer (9,77,186)	0.2	0.2	0.71 ± 0.06	$\mathbf{0.72 \pm 0.04}$	$\mathbf{0.72 \pm 0.04}$	0.61 ± 0.09	$\mathbf{0.72 \pm 0.04}$	0.70	0.66
		0.4	0.45 ± 0.15	0.46 ± 0.16	0.52 ± 0.11	0.54 ± 0.13	$\mathbf{0.66 \pm 0.06}$	$\mathbf{0.67}$	0.56
Pima diabetes (8,268,500)	0.16	0.2	0.74 ± 0.02	$0.74 + 0.03$	0.71 ± 0.04	0.63 ± 0.03	$\mathbf{0.77 \pm 0.03}$	$\mathbf{0.76}$	0.73
		0.4	0.70 ± 0.06	$\mathbf{0.71 \pm 0.04}$	0.51 ± 0.09	0.57 ± 0.07	$\mathbf{0.71 \pm 0.04}$	0.65	0.66
German (20,300,700)	0.3	0.2	$\mathbf{0.74 \pm 0.02}$	$\mathbf{0.74 \pm 0.02}$	0.72 ± 0.01	0.70 ± 0.03	$\mathbf{0.74 \pm 0.03}$	0.66	0.71
		0.4	0.64 ± 0.05	0.64 ± 0.05	0.58 ± 0.04	$\mathbf{0.66 \pm 0.02}$	0.64 ± 0.03	0.55	$\mathbf{0.67}$
Thyroid (5,65,150)	0.25	0.2	0.84 ± 0.04	0.85 ± 0.05	$\mathbf{0.89 \pm 0.03}$	0.76 ± 0.13	0.84 ± 0.03	0.87	0.82
		0.4	0.70 ± 0.15	0.67 ± 0.15	0.61 ± 0.07	0.70 ± 0.12	$\mathbf{0.82 \pm 0.06}$	$\mathbf{0.83}$	0.76

Class Balanced and Domain Requirement of Cost. We consider the Bupa dataset [1] with 6 features where the label is +1 if the value of feature 6 is greater than 3 otherwise -1 ($m_+ = 176, m_- = 169$). We learn a cost sensitive classifier by implementing Algorithm $(\tilde{\eta}, \alpha)$ with $\alpha = 0.25$ and $\tilde{\eta}$ estimated

Table 2. Averaged AM ($\pm$ *s.d.*) of the cost sensitive predictions made by Algorithm $(\tilde{\eta}, \alpha)$ and $l_{\alpha,usq}$ based regularized ERM on UCI datasets corrupted by uniform noise.

Dataset	Cost	ρ	Algorithm $(\tilde{\eta}, \alpha)$: $\tilde{\eta}$ estimate from				$l_{\alpha,usq}$ RegERM	MUB	S-W0-1
(n, m_+, m_-)	α		*Lk-fun* with l_{sq}	*Lk-fun* with l_{log}	LSPC	KLIEP		l_{log}	l_{log}
Breast cancer (9,77,186)	0.2	0.2	0.62 ± 0.06	$\mathbf{0.64 \pm 0.05}$	0.60 ± 0.07	0.56 ± 0.09	0.59 ± 0.04	0.59	**0.63**
		0.4	0.49 ± 0.04	0.50 ± 0.03	0.54 ± 0.05	0.52 ± 0.05	$\mathbf{0.56 \pm 0.05}$	0.51	**0.56**
Pima diabetes (8,268,500)	0.16	0.2	0.71 ± 0.02	0.72 ± 0.02	0.71 ± 0.02	0.65 ± 0.01	$\mathbf{0.73 \pm 0.03}$	0.63	**0.74**
		0.4	$\mathbf{0.67 \pm 0.06}$	$\mathbf{0.68 \pm 0.07}$	0.54 ± 0.03	0.57 ± 0.06	$\mathbf{0.68 \pm 0.05}$	0.56	**0.67**
German (20,300,700)	0.3	0.2	0.67 ± 0.02	0.66 ± 0.01	0.63 ± 0.03	0.59 ± 0.04	0.65 ± 0.04	0.67	**0.69**
		0.4	0.57 ± 0.06	0.56 ± 0.06	0.57 ± 0.03	0.53 ± 0.04	$\mathbf{0.60 \pm 0.03}$	0.51	0.56
Thyroid (5,65,150)	0.25	0.2	0.78 ± 0.06	0.80 ± 0.05	$\mathbf{0.85 \pm 0.06}$	0.72 ± 0.18	0.75 ± 0.05	0.82	0.78
		0.4	0.63 ± 0.14	0.62 ± 0.11	0.64 ± 0.06	0.63 ± 0.19	$\mathbf{0.74 \pm 0.08}$	0.53	0.72

using *Lk-fun* with logistic loss. When comparing tuned γ case (from the set $\Gamma = \{0.5, 0.8, 1, 1.2, 1.5\}$) and $\gamma = 1$ case, we observed in Table 3 that tuning γ is favorable for all measures. Increase in noise rates from $(0.1/0.2/0.3)$ to 0.4 leads to change in γ values from 0.5 to 1 or 1.5 due to imbalance induced by noise. Results for $\tilde{\eta}$ estimated using *Lk-fun* with l_{sq} are available in Supplementary material E.2.

Table 3. Dataset: BUPA Liver Disorder $(6, 176, 169)$. The above table depicts the performance of Algorithm $(\tilde{\eta}, \alpha)$ when $\tilde{\eta}$ estimate is from *Lk-fun* with l_{log} based regularized ERM (λ tuned via CV). Here, $\alpha = 0.25$ and γ is tuned.

	Algorithm $(\tilde{\eta}, \alpha)$: $\tilde{\eta}$ estimate from *Lk-fun* with l_{log}							
	γ tuned				$\gamma = 1$			
ρ	Acc	AM	F	WC	Acc	AM	F	WC
0.0	$\mathbf{1.0 \pm 0.0}$	$\mathbf{1.0 \pm 0.0}$	$\mathbf{1.0 \pm 0.0}$	$\mathbf{0.0 \pm 0.0}$	0.99 ± 0.00	0.99 ± 0.005	0.99 ± 0.005	0.05 ± 0.1
0.1	$\mathbf{0.96 \pm 0.02}$	$\mathbf{0.96 \pm 0.03}$	$\mathbf{0.96 \pm 0.04}$	2.36 ± 1.01	0.88 ± 0.88	0.88 ± 0.07	0.89 ± 0.05	$\mathbf{2.22 \pm 1.29}$
0.2	$\mathbf{0.91 \pm 0.03}$	$\mathbf{0.90 \pm 0.03}$	$\mathbf{0.91 \pm 0.02}$	4.09 ± 1.33	0.68 ± 0.11	0.68 ± 0.11	0.77 ± 0.06	5.95 ± 1.88
0.3	$\mathbf{0.88 \pm 0.03}$	$\mathbf{0.88 \pm 0.03}$	$\mathbf{0.82 \pm 0.1}$	5.47 ± 0.32	0.55 ± 0.05	0.55 ± 0.05	0.69 ± 0.02	8.02 ± 0.96
0.4	$\mathbf{0.61 \pm 0.11}$	$\mathbf{0.58 \pm 0.09}$	$\mathbf{0.67 \pm 0.00}$	5.64 ± 0.29	0.51 ± 0.01	0.51 ± 0.01	$\mathbf{0.67 \pm 0.01}$	8.3 ± 0.475

6 Discussion

We considered the binary classification problem of cost sensitive learning in the presence of uniform label noise. We are interested in the scenarios where there can be two separate costs: α, the one fixed due to domain requirement and γ that can be tuned to capture class imbalance. We first show that weighted 0–1 loss is neither uniform noise robust with measurable class of classifiers nor with linear classifiers. In spite of this, we propose two schemes to address the problem in consideration, without requiring the knowledge of true noise rates.

For the first scheme, we show that linear classifiers obtained using weighted uneven margin squared loss $l_{\alpha,usq}$ is uniform noise robust and incorporates cost sensitivity. This classifier is obtained by solving $l_{\alpha,usq}$ based regularized ERM, that only requires a matrix inversion. Also, a performance bound with respect to clean distribution for such a classifier is provided. One possible issue here could be weighted uneven margin squared loss function's susceptibility to outliers. However, in our experiments, this scheme performed well. The second scheme is a re-sampling based algorithm using the corrupted in-class probability estimates. It handles class imbalance in the presence of uniform label noise but it can be heavily influenced by the quality of $\tilde{\eta}$ estimates. However, we chose the $\tilde{\eta}$ estimation method based on their label prediction ability (as discussed in Supplementary material Section F.6) and obtained good empirical results.

We provide empirical evidence for the performance of our schemes. We observed that the Accuracy and AM values for our schemes are comparable to those given by [6]. Our schemes have other measures like F and Weighted Cost comparable to that of SVMs trained on clean data; details are in Supplementary material Section E.1. The two proposed schemes have comparable performance measures among themselves.

An interesting direction to explore would be cost sensitive learning in Class Conditional Noise (CCN) models without using/tuning noise rates ρ_+ and ρ_-.

References

1. Dheeru, D., Karra Taniskidou, E.: UCI Machine Learning Repository (2017)
2. Elkan, C.: The foundations of cost-sensitive learning. In: International Joint Conference on Artificial Intelligence, vol. 17, pp. 973–978 (2001)
3. Ghosh, A., Manwani, N., Sastry, P.S.: Making risk minimization tolerant to label noise. Neurocomputing **160**, 93–107 (2015)
4. Manwani, N., Sastry, P.S.: Noise tolerance under risk minimization. IEEE Trans. Cybern. **43**(3), 1146–1151 (2013)
5. Masnadi-Shirazi, H., Vasconcelos, N.: Risk minimization, probability elicitation, and cost-sensitive SVMs. In: International Conference on Machine Learning, pp. 759–766 (2010)
6. Natarajan, N., Dhillon, I.S., Ravikumar, P., Tewari, A.: Cost-sensitive learning with noisy labels. J. Mach. Learn. Res. **18**(155), 1–33 (2018)
7. Natarajan, N., Dhillon, I.S., Ravikumar, P.K., Tewari, A.: Learning with noisy labels. In: Advances in Neural Information Processing Systems, pp. 1196–1204 (2013)
8. Patrini, G., Nielsen, F., Nock, R., Carioni, M.: Loss factorization, weakly supervised learning and label noise robustness. In: International Conference on Machine Learning, pp. 708–717 (2016)
9. Reid, M.D., Williamson, R.C.: Composite binary losses. J. Mach. Learn. Res. **11**(Sep), 2387–2422 (2010)
10. Scott, C.: Calibrated asymmetric surrogate losses. Electron. J. Stat. **6**, 958–992 (2012)

11. Sugiyama, M.: Superfast-trainable multi-class probabilistic classifier by least-squares posterior fitting. IEICE Trans. Inf. Syst. **93**(10), 2690–2701 (2010)
12. Sugiyama, M., Nakajima, S., Kashima, H., Buenau, P.V., Kawanabe, M.: Direct importance estimation with model selection and its application to covariate shift adaptation. In: Advances in Neural Information Processing Systems, pp. 1433–1440 (2008)
13. Van Rooyen, B., Menon, A., Williamson, R.C.: Learning with symmetric label noise: the importance of being unhinged. In: Advances in Neural Information Processing Systems, pp. 10–18 (2015)
14. Zhu, X., Wu, X., Khoshgoftaar, T.M., Shi, Y.: An empirical study of the noise impact on cost-sensitive learning. In: International Joint Conference on Artificial Intelligence, pp. 1168–1174 (2007)

Semantic Explanations in Ensemble Learning

Md. Zahidul Islam$^{(\boxtimes)}$, Jixue Liu, Lin Liu, Jiuyong Li, and Wei Kang

School of Information Technology and Mathematical Sciences (ITMS),
University of South Australia, Adelaide, SA 5095, Australia
md_zahidul.islam@mymail.unisa.edu.au,
{Jixue.Liu,Lin.Liu,Jiuyong.Li,Wei.Kang}@unisa.edu.au

Abstract. A combination method is an integral part of an ensemble classifier. Existing combination methods determine the combined prediction of a new instance by relying on the predictions made by the majority of base classifiers. This can result in incorrect combined predictions when the majority predict the incorrect class. It has been noted that in group decision-making, the decision by the majority, if lacking consistency in the reasons for the decision provided by its members, could be less reliable than the minority's decision with higher consistency in the reasons of its members. Based on this observation, in this paper, we propose a new combination method, EBCM, which considers the consistency of the features, i.e. explanations of individual predictions for generating ensemble classifiers. EBCM firstly identifies the features accountable for each base classifier's prediction, and then uses the features to measure the consistency among the predictions. Finally, EBCM combines the predictions based on both the majority and the consistency of features. We evaluated the performance of EBCM with 16 real-world datasets and observed substantial improvement over existing techniques.

1 Introduction

With the rapid development of artificial intelligence, we are increasingly relying on machine-generated decisions such as approval of bank loans or credit cards. Due to their higher prediction accuracy, ensemble classifiers are often used for facilitating such automated decision making [5,6]. Ensemble classifiers use group decision-making process to make predictions using a set of base classifiers.

Traditional ensemble classifiers combine the predictions using majority voting (MV) or arithmetic operations such as sum, average, product, and max [11,17]. The main idea in these methods is that the class predicted by most base classifiers is most likely to be the true class [12]. However, in many cases, reliance on the majority voting leads to incorrect combined predictions, because if the majority of classifiers make incorrect predictions, their combination will be incorrect as well. To illustrate, Fig. 1 shows an ensemble classifier using MV for sentiment analysis of patients' reviews on their doctors. Each review is lablled as positive or negative based on the expressed sentiment [21]. For the given example, though

© Springer Nature Switzerland AG 2019
Q. Yang et al. (Eds.): PAKDD 2019, LNAI 11439, pp. 29–41, 2019.
https://doi.org/10.1007/978-3-030-16148-4_3

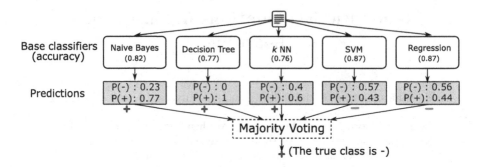

Fig. 1. Majority voting predicts the incorrect class.

the true sentiment is negative, the majority (three out of five) predict positive sentiment. Consequently, the ensemble fails to predict the true sentiment.

Moreover, favouring the majority may not be reliable when the probabilities of the predicted classes are very close. For example, if $P(-) = 0.51$ and $P(+) = 0.49$ for all three majority classifiers, and $P(-) = 0.48$ and $P(+) = 0.52$ for both minority classifiers then the ensemble output is "$-$", the true class which is not reliable as the probabilities of different classes are not significantly different.

To improve combined predictions, researchers have proposed various methods to assign weights to the base classifiers, based on their performance, and thus more accurate base classifiers contribute more to the combined prediction [3, 14]. However, the weights cannot resolve the issues depicted in Fig. 1, as the majority of the base classifiers have predicted "$+$", and the weighted aggregated probability of "$+$" is still higher than that of "$-$". Moreover, the weights are computed once and the same weights are applied to all instances, but in reality a classifier can make better predictions for a certain type of data even though its overall accuracy is poor.

To solve the above problems, we revisit the human group decision-making process. Research shows that conflicting decisions (e.g. some people in a group is supporting "$+$" and some "$-$") often lead to errors [8]. Hence, when a decision is made by a group of people, we ask not only for each persons's decision, but also the reason *why* such a decision is made. For example, when three people vote for "$+$", but giving inconsistent reasons for their decisions, while two people vote for "$-$" with consistent reasons, it is reasonable to have the final decision as "$-$" even fewer people support it. The reason is that inconsistent explanations among the majority group has lowered their confidence on the decision.

Inspired by these observations, we propose a novel Explanation-Based Combination Method (EBCM) to make use of the semantic explanations to base classifiers' predictions to facilitate the generation of the final prediction. To the best our knowledge, this is the first work which combines multiple predictions based on their semantic explanations.

The main contributions of our work can be summarized as follows:

- By analysing the human group decision-making process we propose the idea of including semantic explanations in ensemble learning.
- We propose EBCM for incorporating semantic explanations in ensemble learning and two measures for determining the most consistent predictions.

2   Problem Definition

Let $C = \{Cl_1, Cl_2, \ldots, Cl_T\}$ be a set of classifiers and $W = \{\omega_1, \omega_2, \ldots, \omega_K\}$ be a set of target classes. Given an instance $\mathbf{x}$, Cl_t derives a probability $p_{k,t}$ for ω_k and makes a prediction P_t. The goal of a combination method is to combine $\mathcal{P} = \{P_t : (1 \leq t \leq T)\}$ to produce a final prediction (the ensemble prediction).

Let $SC_k = \{Cl_t \mid P_t = \omega_k, 1 \leq t \leq T\}, 1 \leq k \leq K$ be the set of classifiers whose predictions are ω_k, or supporting ω_k. The prediction $P_t = \omega_k$ if $\forall i \in \{1, \ldots, K\}$, $p_{k,t} \geq p_{i,t}$. We use $\mathcal{Q}$ to denote the set of maximum class probabilities by all the classifiers respectively over the K classes, i.e. $\mathcal{Q} = \{p_{k*,t} \mid p_{k*,t} = \max\{p_{i,t} : 1 \leq i \leq K\}, 1 \leq t \leq T\}$.

Our problem is to define a combination measure $\Psi_k, 1 \leq k \leq K$, which is a function from SC_k to a real number i.e., $\Psi_k : SC_k \to \mathbb{R}$. The combined predication then can be represented as:

$$\omega^* = \arg \max_{k \in \{1,2,\ldots,K\}} \{\Psi_k\} \tag{1}$$

Traditional combination methods define Ψ_k based on the base classifiers' predictions supporting class ω_k. In this paper, we follow the idea of human group decision-making, and we expect Ψ_k to reflect the bases (or explanations) on which the predictions are made. Though the classifiers in SC_k agree on the class ω_k, we cannot differentiate whether their agreement is based on the same ground or not from $p_{k*,t}$ only.

Therefore Ψ_k should consider two factors in the decision making. The first factor should reflect the probabilities of the class ω_k, denoted as μ_k. The second factor should reflect the consistency of the explanations to the prediction made by the base classifiers in SC_k, denoted by c_k. Following [20], Ψ_k can be obtained by multiplying the two factors, that is,

$$\Psi_k = \mu_k \times c_k \tag{2}$$

Now the question is how to obtain c_k based on the set of classifiers SC_k. This question can be divided into the following two sub-questions:

1. Given a set of classifiers SC_k and their prediction ω_k, how to extract the explanations to the prediction by all classifiers in SC_k, denoted as E_k?
2. Given E_k, how to measure the consistency of the explanations of all classifiers in SC_k, i.e. how to define c_k?

3   Explanation-Based Combination Method (EBCM)

The flow of EBCM consists of three steps. Firstly, for each supporting classifiers set SC_k, we compute the aggregated probabilities μ_k from $\mathcal{Q}$ following [11,17]. Secondly, we extract the explanations for the predictions (Sub-Problem 1). Finally, we compute the consistency score c_k among the explanations supporting ω_k (Sub-Problem 2).

An overview of EBCM, expanding the example in Fig. 1, is given in Fig. 2. We use NB, DT and LR to refer to the Naïve Bayes, Decision Tree, and Regression classifiers respectively. Hence, $\mathcal{W} = \{-, +\}$, $\mathcal{Q} = \{0.77, 1, 0.6, 0.57, 0.56\}$ the supporting classifiers sets are $SC_- = \{\text{SVM}, \text{LR}\}$ and $SC_+ = \{\text{NB}, \text{DT}, k\text{NN}\}$. Accordingly, the averaged probabilities of the corresponding predicted classes are $\mu_- = 0.57$ and $\mu_+ = 0.79$.

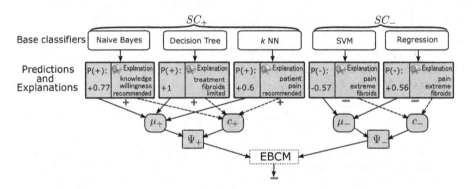

Fig. 2. An overview of EBCM.

3.1 Explanation Extraction

In this step, we extract explanations for the predictions and then calculate the consistency scores $c_k, 1 \leq k \leq K$ from the explanations. Each classifier Cl_t learns different weights for the features during the training phase. For a new instance, Cl_t maps the features in the instance to the target classes using the learned weights. We extract such features as the explanation $e_{k,t}$ for each prediction corresponding to the probabilities in $\mathcal{Q}$ such that features in $e_{k,t}$ are most weighted to generate $p_{k^*,t}$. Most of the classifiers do not provide explanations [1] i.e., the features used by Cl_t to derive the prediction are often not available directly.

To extract $e_{k,t}$, we use LIME [18] which gives the most valued features for a prediction. Given any classifier Cl_t and instance $\mathbf{x}$, LIME creates a sampled training set by randomly sampling with a probability inversely proportional to the distance to $\mathbf{x}$ and obtaining the labels by applying Cl_t to the samples. We note that LIME is independent of the type of Cl_t and works on an explainable feature space (transformed from the original features) to return easy to understand explanations. Hence, the interpretability of Cl_t is not important. For text classification, bag-of-words features (used in our experiments) are self explanatory and hence, we do not use any transformation. LIME selects a set of features using Lasso (linear model fitted to the sampled and relabelled dataset) explaining the prediction made by Cl_t for an instance. We empirically elect the number of features in the returned explanation word set. We take the explanation word set for the predictions of the classifiers in SC_k to constitute E_k.

For the example in Fig. 2, we extract the following explanations using LIME.

- $e_{+,\mathrm{NB}} = \{\mathrm{recommended, willingness, knowledge}\}$
- $e_{+,\mathrm{DT}} = \{\mathrm{treatment, fibroids, limited}\}$
- $e_{+,k\mathrm{NN}} = \{\mathrm{patient, recommended, pain}\}$
- $e_{-,\mathrm{SVM}} = \{\mathrm{pain, extreme, fibroids}\}$
- $e_{-,\mathrm{LR}} = \{\mathrm{pain, extreme, fibroids}\}$

The explanations provide insights regarding the predictions supporting a certain class. For example, $e_{+,\mathrm{NB}} = \{\mathrm{recommended, willingness, knowledge}\}$ discloses that NB predicts the *negative* class based on the features *recommended, willingness* and *knowledge*. Once we have the explanations, we evaluate them to determine the most reliable supporting classifiers SC_k by computing c_- from $E_- = \{e_{-,\mathrm{SVM}}, e_{-,\mathrm{LR}}\}$, and c_+ from $E_+ = \{e_{+,\mathrm{NB}}, e_{+,\mathrm{DT}}, e_{+,k\mathrm{NN}}\}$ as follows.

3.2   Consistency Measurement

Our intuition is that for the same input, the classifiers predicting the same class with similar explanations should be weighted more towards the combined prediction (larger value for c_k). Similarly, if multiple predictions support the same class based on different features, then we should downweight their contributions towards the combined prediction (smaller value for c_k). To illustrate our intuition, let us consider the explanations showed in Fig. 2 where the explanations in E_- have the same set of features whereas the explanations in E_+ do not have a single common feature. In this example, the E_- is more consistent than the E_+, therefore the c_- should be greater than the c_+.

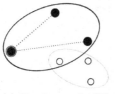

(a) Consistency in SC_k.

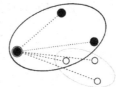

(b) Consistency in SC_k w.r.t the classifiers supporting the other class.

Fig. 3. Intuitions for cohesion-based and cohesion and separation-based consistency measures. Explanations corresponding to two target classes are presented by filled and unfilled circles. (Color figure online)

Based on the above intuition, we propose the following two consistency measures to compute the consistency score c_k for each E_k.

Cohesion-Based Measure (C): This consistency measure is based on the closeness of the explanations within the same cluster. The closer, the higher the consistency. The consistency is measured by the similarities among the explanations (cohesion). We calculate the pairwise similarities for all pairs of explanations in E_k and take their average. For example, Fig. 3(a) shows the pairwise similarities from the explanation marked in the red circle to the other explanations supporting the same class. Considering $sim(e_{k,i}, e_{k,j})$ be the similarity between the explanation $e_{k,i}$ and $e_{k,j}$, we define the consistency score c_k in Eq. 3. We use any similarity measure (e.g., cosine, Jaccard etc.) to calculate $sim(e_{k,i}, e_{k,j})$.

$$c_k = \frac{|E_k|}{|C|} \sum_{e_{k,x}, e_{k,y} \in E_k \land x \neq y} sim(e_{k,x}, e_{k,y}) \tag{3}$$

Note that if all base classifiers predict the same class, then the consistency scores corresponding to the other classes are 0.

Cohesion and Separation-Based Measure (CS): Our second consistency measure is based on the distance among the explanations in E_k (cohesion) compared to the explanations corresponding to the classifiers supporting other classes (separation). If the explanations of classifier set SC_k are far away from the explanations belonging to other classifier sets, the consistency is higher.

To define this measure, we use concepts from clustering where the measure silhouette coefficient [19] assigns a score to each point indicating how well it suits in the corresponding cluster. A large score (closer to 1) indicates that the point is well clustered, 0 means that it is on the border of two clusters and a negative score means that it is better suited in a different cluster.

In our context, the explanations E_k form a cluster corresponding to the classifiers predicting the same class ω_k. We derive a fitness score for each $e_{k,t}$ which describes how an explanation in E_k is similar to other explanations. As shown in Eq. 4, it is derived from $e_{k,t}$'s average distance $a(e_{k,t})$ within the cluster of explanations and its average distance $b(e_{k,t})$ to the explanations outside the cluster. Here, $d(e_{k,t}, e_{l,j})$ is the distance between the two explanations.

$$f(e_{k,t}) = \frac{b(e_{k,t}) - a(e_{k,t})}{\max\left(a(e_{k,t}), b(e_{k,t})\right)} \tag{4}$$

where,

$$a(e_{k,t}) = \frac{1}{|E_k| - 1} \sum_{e_{k,i} \in E_k} d(e_{k,t}, e_{k,i}) \text{ and}$$

$$b(e_{k,t}) = \min_{e_{k,t} \notin E_l} \left\{ \frac{1}{|E_l|} \sum_{e_{l,j} \in E_l} d(e_{k,t}, e_{l,j}) \right\}$$

In this measure, the consistency score c_k is then derived from $f(e_{k,t})$ as shown in Eq. 5. It is the average of the fitness scores of the explanations in the cluster. When a cluster is more compact, it is far away from other clusters, the measure is closer to 1. Otherwise, it is closer to 0.

$$c_k = \frac{1}{|E_k|} \sum_{e_{k,t} \in E_k} f(e_{k,t}) \tag{5}$$

Note that if all base classifiers predict the same class, then $c_k = -1$ as $b(e_{k,t}) = 0$ in Eq. 4 and consistency scores corresponding to the other classes are set to 0.

To illustrate, the fitness score of the explanation marked in the red circle in Fig. 3(b) is computed from its average distance from the explanations supporting the same class (shown in blue dotted lines) and the average distance from the explanations supporting the other class (shown in orange dashed lines). The first average distance is the $a(e_{k,t})$ and the minimum of the second average distance is the $b(e_{k,t})$. The fitness scores of the explanations are computed using Eq. 4. The fitness scores of the other explanations supporting the same class are computed in a similar fashion and their average is the desired c_k.

Our method is summarized in Algorithm 1. The EBCM divides the classifiers into K supporting sets where each classifier $Cl_i \in SC_k$ supports a target class $\omega_k \in W$. For each SC_k, the EBCM accumulates the probability of each target class μ_k from Q, extracts explanations E_k corresponding to the predictions and measures a consistency score c_k using one of the consistency measures described above. Finally, the combined prediction ω^* is made by selecting the class with the maximum value of $\Psi_k = c_k \times \mu_k$.

Applying Algorithm 1 to Fig. 2, we get $c_- = 1.0$ and $c_+ = 0.11$ using the C measure. Similarly, using the CS measure, we get $c_- = 1.0$ and $c_+ = -0.06$. Combining the consistency scores with the supports $\mu_- = 0.57$ and $\mu_+ = 0.79$, EBCM correctly predicts the "$-$" class as the combined prediction using both consistency measures, although three out of five classifiers predict the "$+$" class.

Algorithm 1. Explanation-Based Combination Method (EBCM)

Input: A set of classifiers $\mathcal{C}$, predictions $\mathcal{P}$, target classes $\mathcal{W}$ and a test instance $\mathbf{x}$.
Output: Combined prediction $\omega*$.
1: **for** $t \in \{1, 2, \ldots, |\mathcal{C}|\}$ **do**
2:    $\mathcal{Q} \leftarrow \mathcal{Q} \cup (p_{k^*,t} = \max\{p_{k,t} : (1 \le k \le |\mathcal{W}|)\})$
3: **end for**
4: Find the sets of supporting classifiers $\{SC_k : (1 \le k \le |\mathcal{W}|)\}$
5: **for** $k \in \{1, 2, \ldots, |\mathcal{W}|\}$ **do**
6:    $\mu_k \leftarrow \frac{1}{|SC_k|} \sum_{t=1}^{|SC_k|} p_{k^*,t}$
7:    $E_k \leftarrow$ extracted explanations for the predictions of SC_k
8:    $c_k \leftarrow$ using Eq. 3 or Eq. 5 from E_k
9:    $\Psi_k \leftarrow \mu_k \times c_k$
10: **end for**
11: **return** $\omega* = \arg\max_{1 \le k \le |\mathcal{W}|}\{\Psi_k\}$

---

4   Experiments

We apply EBCM to create an ensemble of five base classifiers for sentiment analysis, and evaluate the method with 16 real-world datasets, in comparison with five baseline methods for ensemble creation.

4.1   Experiment Setup

Datasets. Table 1 provides a summary of the 16 real-world social media datasets used in the experiments (after some preprocessing as described in the next paragraph). These datasets can be divided into the following three categories:

1. User reviews on products or services, including *TVShow*, *Radio*, *Camera*, *Camp*, *Music*, *Drug*, *Doctor* and *Movies*.
2. Tweets on various events or entities, including *Sanders*, *HCR*, *OMD*, *SSTwitter*, *TwiSent*, and *SE2017*.
3. User comments on YouTube videos, including *SSYouTube* and *SenTube*.

For the original datasets with more than two classes in our experiments, we only take the instances with positive and negative labels. For the datasets where numerical ratings are provided, we categorize the numerical values to positive and negative labels. For a dataset with imbalanced numbers of positive and negative instances (i.e., unequal numbers of instances from both class), after taking all the samples in the minor class, we randomly select the equal number of samples from the major class to form a balanced dataset. We use bag-of-words with binary weights [15] as features.

Table 1. Summary of the datasets.

Dataset	#Instance	#Features	#AW	Source
TVShow	190	1704	22	http://cs.coloradocollege.edu/~mwhitehead
Radio	460	3406	25	http://cs.coloradocollege.edu/~mwhitehead
Camera	468	4556	63	http://cs.coloradocollege.edu/~mwhitehead
Camp	534	3839	34	http://cs.coloradocollege.edu/~mwhitehead
Music	550	7311	70	http://cs.coloradocollege.edu/~mwhitehead
Drug	662	3877	36	http://cs.coloradocollege.edu/~mwhitehead
Doctor	876	4727	27	http://cs.coloradocollege.edu/~mwhitehead
Movies	2000	38952	313	http://www.cs.cornell.edu/people/pabo/movie-review-data/
Sanders	852	2868	16	http://www.sananalytics.com/
HCR	1050	3596	19	https://bit.ly/2yfmZiL
OMD	1080	3002	16	https://bit.ly/2QAWyKT
SSTwitter	1638	5910	17	http://sentistrength.wlv.ac.uk/
TwiSent	2160	7250	18	http://www.mpi-inf.mpg.de/~smukherjee/data/
SE2017	6034	12655	20	http://alt.qcri.org/semeval2017/task4/
SSYouTube	628	4879	18	http://sentistrength.wlv.ac.uk/
SenTube	4684	11500	19	http://ikernels-portal.disi.unitn.it/projects/sentube/

#AW: average words per instance.

Base Classifiers and Baseline Combination Methods. To evaluate EBCM, we apply it to build an ensemble of five base classifiers - Naïve Bayes (NB), Decision Tree (DT), k-Nearest Neighbour (kNN), Support Vector Machine (SVM), and Logistic Regression (LR). We compare EBCM with the other five ensemble classifiers built using the same base classifiers and the five commonly used combination methods - Majority Vote (MV), Averaging (AVG), Weighted Averaging (WAvg), Maximum (Max), and Sum, respectively [11, 17].

We use the implementation of the base classifiers in the Python Scikit-learn machine learning library (http://scikit-learn.org/) with their default parameters. Then we implement the combination methods using Python.

Evaluation Approach. To measure the performance of EBCM, we firstly compare EBCM with the baseline methods in terms of their performance for the *close instances* where two among the five base classifiers predict a class and three predict the other class. For non-close instances, EBCM works in the same way as MV, thus we evaluate EBCM for close instances to see how it improves the prediction in such cases. Then we compare the prediction accuracy of EBCM with the baseline combination methods in order to evaluate the general performance of the ensemble classifier built using EBCM as the combination method. We also include the prediction accuracy of the base classifiers in the comparison.

4.2   Results

EBCM Significantly Reduces Errors in Close Instances. Table 2 shows the experimental results in the close instances. In each cell of Table 2, the content $R_1/R_2/R_3$ represents the number of incorrect predictions (or misclassified instances) by a baseline method (R_1), by EBCM using the cohesion-based (C) consistency measure (R_2), and by EBCM using cohesion and separation-based (CS) consistency measure (R_3). For example, $10/6/2$ in the cell for dataset TVShow and baseline MV means that among all the close instances for this dataset, MV misclassified 10 instances, whereas EBCM misclassified only 6 and 2 close instances using the C and CS consistency measures respectively. In other words, EBCM recovers 8 of 10 errors using CS consistency measure over the MV method, achieving 80% improvement.

From Table 2, we see that in all cases, except for the Camp dataset, the CS measure reduces more errors than the C measure. EBCM significantly reduces errors by making correct predictions for the close instances.

A deeper analysis of the results confirms that EBCM achieves such a significant reduction in errors due to the contribution of the consistency scores. For illustration, two instances are presented in Fig. 4. Three out of the five base classifiers predict the incorrect class in both instances. For Fig. 4(a) the true class is "+" and for Fig. 4(b) the true class is "−". MV fails due to the incorrect predictions of the majority classifiers. Moreover, the aggregated probability of the incorrect class is greater than that of the correct class i.e., for Fig. 4(a) $\mu_- > \mu_+$ and for Fig. 4(b) $\mu_+ > \mu_-$. Therefore, AVG, WAvg, Max, and Sum make errors. A close look at the explanations in Fig. 4 shows in Fig. 4(a) the explanations or

Table 2. # misclassified close instances per dataset by the baselines and EBCM.

Dataset	MV/C/CS	AVG/C/CS	WAvg/C/CS	Max/C/CS	Sum/C/CS
TVShow	10/6/2	12/5/1	12/5/1	24/6/2	12/5/1
Radio	46/38/31	49/23/21	43/22/20	54/16/16	49/23/21
Camera	40/25/23	33/18/17	33/19/17	60/12/17	33/18/17
Camp	26/19/20	21/11/11	22/14/15	37/5/6	21/11/11
Music	58/37/33	70/20/18	68/22/20	87/13/16	70/20/18
Drug	69/52/41	70/29/20	71/34/23	93/20/17	70/29/20
Doctor	45/29/25	46/17/13	46/24/20	72/12/11	46/17/13
Movies	154/74/72	158/46/43	148/50/47	333/47/47	158/46/53
Sanders	56/44/32	64/31/17	60/34/19	99/27/15	64/31/17
HCR	77/68/49	73/40/28	71/40/28	72/27/19	73/40/28
OMD	84/70/55	92/45/34	85/48/32	124/38/25	92/45/34
SSTwitter	136/120/92	152/76/59	140/80/62	213/58/47	152/76/59
TwiSent	200/168/130	209/92/73	208/96/77	252/60/53	209/92/73
SE2017	360/245/184	417/158/117	391/164/117	590/112/83	417/158/117
SSYouTube	45/40/33	46/28/19	45/27/18	49/15/4	46/28/19
SenTube	427/363/273	434/265/174	424/271/178	540/189/119	434/265/174

extracted features (words) of the two classifiers which predicted "+" are more consistent than those for the "−" class. In Fig. 4(b), the explanations for the two classifiers which predicted "−" class are more consistent. This is also reflected in the consistency scores (using CS measure), as for Fig. 4(a) $c_+ > c_-$ and for Fig. 4(b) $c_- > c_+$. Hence, EBCM was able to correct the errors of the baseline methods by combining the consistency scores c with μ.

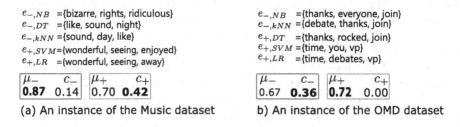

$e_{-,NB}$ ={bizarre, rights, ridiculous}
$e_{-,DT}$ ={like, sound, night}
$e_{-,kNN}$ ={sound, day, like}
$e_{+,SVM}$={wonderful, seeing, enjoyed}
$e_{+,LR}$ ={wonderful, seeing, away}

μ_-	c_-	μ_+	c_+
0.87	0.14	0.70	**0.42**

(a) An instance of the Music dataset

$e_{-,NB}$ ={thanks, everyone, join}
$e_{-,kNN}$ ={debate, thanks, join}
$e_{+,DT}$ ={thanks, rocked, join}
$e_{+,SVM}$ ={time, you, vp}
$e_{+,LR}$ ={time, debates, vp}

μ_-	c_-	μ_+	c_+
0.67	**0.36**	**0.72**	0.00

b) An instance of the OMD dataset

Fig. 4. Examples of correct predictions of EBCM.

EBCM Achieves Better Prediction Accuracy. Figure 5 shows the prediction accuracy of EBCM and the baselines, as well as the average prediction accuracy of the five base classifiers (i.e. the left-most bar labelled with Indv). The prediction accuracy of each ensemble classifier and each base classifier is the average accuracy over the 10-fold cross-validations of the classifier with the

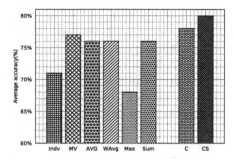 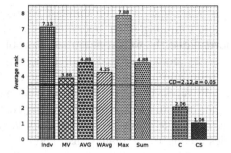

Fig. 5. The prediction accuracies of the methods over all 16 datasets.

Fig. 6. Average ranks from the Friedman test. The lower the better.

16 datasets. From Fig. 5 we see that all baselines (except Max) perform better than the base classifiers, and among the combination methods, EBCM achieves the highest accuracy using both C and CS consistency measures. Overall, our approach outperforms all the compared methods. Among the two consistency measures, CS achieves better accuracy than C.

The statistical analysis involving the accuracies of the compared methods confirms that they are significantly different. We apply the Friedman test followed by a post-hoc Bonferroni-Dunn test [4,7]. From the Friedman test, we obtain the p-value $= 4.58e-11$, i.e., the differences in the prediction accuracies of the methods (EBCM, baselines and average of the base classifiers) are statistically significant. The average ranks of the methods, obtained during the Friedman test and a cut line from the Bonferroni-Dunn test [7] at $\alpha = 0.05$ (critical difference **CD** $= 2.12$) are illustrated in Fig. 6. The methods whose ranks are above the cut line are worse than the methods below the line. Thus, Fig. 6 confirms that the compared methods are significantly worse than EBCM.

5   Related Work

The combination methods can be divided into *trainable* and *non-trainable* methods [10,17]. The trainable methods either create a set of reference points or a new classifier form the predictions of the base classifiers on the training dataset. Given a new instance, the combined prediction is obtained by comparing the predictions of the base classifiers with the reference points [9,13] or from the output of the new classifier [2,22]. In contrast, the non-trainable methods aggregate the predictions by counting the frequency of the target classes or applying an arithmetic operator. Our work belongs to the non-trainable methods.

Among the non-trainable methods, Majority Voting (MV) selects the most frequent class as the output ignoring the probabilities of the target classes provided by the base classifiers. To combine the probabilities, there are several methods based on arithmetic operators such as summation (Sum), average (Avg), product (Prod), maximum (Max), and minimum (Min) [11,23].

Some non-trainable combination methods use weights to emphasize the predictions from more accurate classifiers. Generally, the weights are calculated from the training set and the simplest strategy is to consider performance as weights [11,17]. Among other approaches to determine weights, some researchers have formulated the weight assignment problem as an optimization problem [3,14].

The aforementioned combination methods use information available in the outputs and but do not benefit from the valuable insights hidden in the underlying features considered for generating particular predictions. Our work looks at a new way to incorporate the consistencies among the classifiers' predictions by extracting features explaining individual predictions and measuring consistencies among the explanations. There are some studies incorporating knowledge from additional sources to improve ensemble classifiers [16,24]. However, their approaches provide additional information to the predictions, but cannot distinguish the predictions based on the reasoning.

6   Conclusion

In this article, we have presented a new combination method for ensemble classifiers. Unlike previous combination methods that usually combine the predictions based on the predictions of the majority of the classifiers, our method extracts feature level explanations for each individual prediction and incorporate consistency among explanations in combination. We have evaluated our method on sentiment classification task on 16 real-world social media datasets. Experimental results have shown that our proposed method outperforms the alternative methods. Although our experiments were performed on sentiment classification, the proposed method can be used in any multi-classifier system with an appropriate consistency measure. In future, we plan to investigate the semantic explanations for evaluating base classifiers in order to improve ensemble systems by combining only the reliable predictions.

Acknowledgements. We acknowledge the University of South Australia and Data to Decisions CRC (D2DCRC) for partially funding this research.

References

1. Baehrens, D., Schroeter, T., Harmeling, S., Kawanabe, M., Hansen, K., Müller, K.R.: How to explain individual classification decisions. JMLR **11**, 1803–1831 (2010)
2. Breiman, L.: Stacked regressions. Mach. Learn. **24**(1), 49–64 (1996)
3. Cevikalp, H., Polikar, R.: Local classifier weighting by quadratic programming. IEEE Trans. Neural Netw. **19**(10), 1832–1838 (2008)
4. Demšar, J.: Statistical comparisons of classifiers over multiple data sets. JMLR **7**, 1–30 (2006)
5. Dietterich, T.G.: Ensemble methods in machine learning. In: Kittler, J., Roli, F. (eds.) MCS 2000. LNCS, vol. 1857, pp. 1–15. Springer, Heidelberg (2000). https://doi.org/10.1007/3-540-45014-9_1
6. Friedman, J., Hastie, T., Tibshirani, R.: Additive logistic regression: a statistical view of boosting. Ann. Stat. **28**(2), 337–407 (2000)

7. García, S., Fernández, A., Luengo, J., Herrera, F.: A study of statistical techniques and performance measures for genetics-based machine learning: accuracy and interpretability. Soft Comput. **13**(10), 959 (2009)
8. Herrera, F., Herrera-Viedma, E., Verdegay, J.L.: A rational consensus model in group decision making using linguistic assessments. Fuzzy Sets Syst. **88**(1), 31–49 (1997)
9. Huang, Y.S., Suen, C.Y.: The behavior-knowledge space method for combination of multiple classifiers. In: Proceedings of CVPR, pp. 347–352. IEEE (1993)
10. Jain, A.K., Duin, R.P.W., Mao, J.: Statistical pattern recognition: a review. IEEE Trans. Pattern Anal. Mach. Intell. **22**(1), 4–37 (2000)
11. Kuncheva, L.I.: Combining Pattern Classifiers: Methods and Algorithms, 2nd edn. Wiley, New York (2014)
12. Kuncheva, L.I.: A theoretical study on six classifier fusion strategies. IEEE Trans. Pattern Anal. Mach. Intell. **24**(2), 281–286 (2002)
13. Kuncheva, L.I., Bezdek, J.C., Duin, R.P.: Decision templates for multiple classifier fusion: an experimental comparison. Pattern Recognit. **34**(2), 299–314 (2001)
14. Onan, A., Korukoğlu, S., Bulut, H.: A multiobjective weighted voting ensemble classifier based on differential evolution algorithm for text sentiment classification. Expert. Syst. Appl. **62**, 1–16 (2016)
15. Pang, B., Lee, L., Vaithyanathan, S.: Thumbs up?: sentiment classification using machine learning techniques. In: Proceedings of EMNLP, pp. 79–86. ACL (2002)
16. Perikos, I., Hatzilygeroudis, I.: Recognizing emotions in text using ensemble of classifiers. Eng. Appl. Artif. Intell. **51**, 191–201 (2016)
17. Polikar, R.: Ensemble based systems in decision making. IEEE Circuits Syst. Mag. **6**(3), 21–45 (2006)
18. Ribeiro, M.T., Singh, S., Guestrin, C.: "Why should I trust you?": explaining the predictions of any classifier. In: The 22nd ACM SIGKDD, pp. 1135–1144 (2016)
19. Rousseeuw, P.J.: Silhouettes: a graphical aid to the interpretation and validation of cluster analysis. Comput. Appl. Math. **20**, 53–65 (1987)
20. Tofallis, C.: Add or multiply? a tutorial on ranking and choosing with multiple criteria. INFORMS Trans. Educ. **14**(3), 109–119 (2014)
21. Whitehead, M., Yaeger, L.: Sentiment mining using ensemble classification models. In: Sobh, T. (ed.) Innovations and Advances in Computer Sciences and Engineering, pp. 509–514. Springer, Dordrecht (2010). https://doi.org/10.1007/978-90-481-3658-2 89
22. Wolpert, D.H.: Stacked generalization. Neural Netw. **5**(2), 241–259 (1992)
23. Woźniak, M., Graña, M., Corchado, E.: A survey of multiple classifier systems as hybrid systems. Inf. Fusion **16**, 3–17 (2014)
24. Yan, Y., Yang, H., Wang, H.: Two simple and effective ensemble classifiers for twitter sentiment analysis. In: 2017 Computing Conference, pp. 1386–1393 (2017)

Latent Gaussian-Multinomial Generative Model for Annotated Data

Shuoran Jiang, Yarui Chen[✉], Zhifei Qin, Jucheng Yang, Tingting Zhao, and Chuanlei Zhang

Tianjin University of Science and Technology, Tianjin 300457, China
shaunbysn@gmail.com, {yrchen,zfqin,jcyang,tingting,97313114}@tust.edu.cn

Abstract. Traditional generative models annotate images by multiple instances independently segmented, but these models have been becoming prohibitively expensive and time-consuming along with the growth of Internet data. Focusing on the annotated data, we propose a latent Gaussian-Multinomial generative model (LGMG), which generates the image-annotations using a multimodal probabilistic models. Specifically, we use a continuous latent variable with prior of Normal distribution as the latent representation summarizing the high-level semantics of images, and a discrete latent variable with prior of Multinomial distribution as the topics indicator for annotation. We compute the variational posteriors from a mapping structure among latent representation, topics indicator and image-annotation. The stochastic gradient variational Bayes estimator on variational objective is realized by combining the reparameterization trick and Monte Carlo estimator. Finally, we demonstrate the performance of LGMG on LabelMe in terms of held-out likelihood, automatic image annotation with the state-of-the-art models.

Keywords: Annotated data · Gaussian-Multinomial ·
Multimodal generative models · Latent representation ·
Topics indicator

1 Introduction

Today, the multimedia data produced by the Internet devices is not mere collections of single type data, but is assemble of related text, images, audio, video and so on. Annotating image is an essential task in the information retrieval, but artificial annotating has been becoming prohibitively expensive and time-consuming with the increasing growth of the Internet data. We take interest in finding automatic methods which find probabilistic model over image-annotations instead annotating all images explicitly.

In the past few years, a number of methods had been proposed to model annotated data. Some earlier works view image annotation as a classification problem [1,2] and make a yes/no decision on every word in vocabulary. However, the discriminative function cannot estimate the probabilities of annotations. Capturing this probability would be helpful for further annotating new

© Springer Nature Switzerland AG 2019
Q. Yang et al. (Eds.): PAKDD 2019, LNAI 11439, pp. 42–54, 2019.
https://doi.org/10.1007/978-3-030-16148-4_4

images [1]. The probabilistic generative models take the image annotation as a multiple instances learning problem [3,5], in which each image is represented by a bag of regions (i.e. instances) artificially segmented from image, and corresponding annotation is represented by a bag of words (i.e. annotations). In this research field, Blei had explored a range of models for annotated data, among which Gaussian-Multinomial mixture model (GM-Mixture) and correspondence latent Dirichlet allocation (cLDA) are very popular [3]. GM-Mixture assumes each image-annotation is generated with a common topic, so the resulting Multinomial over annotations and Gaussian over images are correlative. But the independent generative processes for segmented regions and annotated words lose abundant mutual information between specific regions and words. cLDA is a Bayesian probabilistic model over a mixture of topics. Given a corpus, cLDA firstly sample a topic indicator and then draws the word in annotation from a Multinomial model. cLDA learns a range of probabilistic distributions on the bag-of-regions for each word in annotation, but it ignores the correlative relationship among instances in common image to some independent instances, which is one important cause led to incorrect image annotation.

In recent years, deep learning models, including CNN-based regression model [6], deep multiple instance learning [7] and Recurrent Neural Network Fisher Vector [8] make use of features extracted from neural network and embedding vectors to represent the associated tags between image and annotation, then learn the relationships between the two-types features. These promising researches have enlightening reference value for further studies. However, traditional deep learning models cannot calculate the precise probability which is an insufficiency. Most noteworthy, the neural variational inference [9] has been becoming popular to optimize latent variational models. In this frame, variational autoencoders (VAEs) [10,11] are typical models, and they suppose each observation has an unique latent representation used to summarize the high-level semantics of the observation. If the observation is image, the latent representation can summarize the structure, color, light and so on. VAEs use a inference network mapping from observation to compute the variational posterior of latent representation, and calculate the parameters of generative model on observations from a neural network with sampling latent representations. The inference network and generative model are jointly optimized by stochastic gradient variational Bayes estimator on variational objective.

In this paper, we propose a latent Gaussian-Multinomial generative model (LGMG) for the annotated data, which is inspired by the neural variational inference and a higher expressive architecture. Unlike the multiple instances learning methods, LGMG doesn't segment image into multiple regions, but builds a multimodal generative models for annotated data. In the generative models, a mixture of Gaussian or Bernoulli is used to generate images and a mixture of Multinomial is used to generate annotation. Both generative models share a same topic and are computed from neural networks with same latent representation, so the images and annotations are tighter correlative than GM-Mixture and cLDA.

This paper is organized as follow. We briefly describe the latent Gaussian-Multinomial generative model and variational autoencoders in Sect. 2. The details of latent Gaussian-Multinomial generative model and its variational inference will be described in Sect. 3. Experiments and analysis are presented in Sect. 4. Finally, we conclude with a brief summary in Sect. 5.

2 Related Works

2.1 Gaussian-Multinomial Mixture Model

Gaussian-Multinomial mixture model (GM-Mixture) is a sample finite mixture model over annotated data [12–14], and its generative model over annotation is multiple instances learning. In GM-Mixture model, a discrete latent variable c with prior of multinomial $\mathbf{Mult}(\boldsymbol{\lambda})$ is used to represent a joint topic for image-annotation pair $\{r, w\}$. Each $\{r, w\}$ in dataset is assumed to be generated by first choosing a value $c \sim \mathbf{Mult}(\boldsymbol{\lambda})$, and then repeatedly sampling N region descriptions $(r_1, r_2, ..., r_N)$ from a Gaussian $\mathcal{N}(\boldsymbol{\mu}, \boldsymbol{\sigma})$ and M annotation words $(w_1, w_2, ..., w_M)$ from a multinomial $\mathbf{Mult}(\boldsymbol{\beta})$. These generative models over regions and annotations are both conditional on the chosen value c_n. The value of latent topic variable is sampled once on per image-annotation, and is held fixed during the generative process. Given the parameters $\boldsymbol{\mu}$, $\boldsymbol{\sigma}$, $\boldsymbol{\beta}$, the joint distribution of GM-Mixture is given by,

$$p(c, r, w) = p(c|\boldsymbol{\lambda}) \prod_{n=1}^{N} p(r_n|c, \boldsymbol{\mu}, \boldsymbol{\sigma}) \prod_{m=1}^{M} p(w_m|c, \boldsymbol{\beta}). \tag{1}$$

Given a predefined dimension on latent topic c and a dataset of image-annotations, the parameters of GM-Mixture model are estimated by the EM algorithm. Since each image and its annotation are assumed to have been generated with the condition of same latent topic, the resulting Multinomial and Gaussian will be corresponding. However, regions and annotations are generated independently, that results in the lack of mutual information between specific regions and annotations.

2.2 Variational Autoencoders

Given a dataset $\mathbf{X} = \{\mathbf{x}_n\}_{n=1}^{N}$ consisting of N i.i.d. observations, variational autoencoders (VAEs) suppose each sample $\mathbf{x}_n$ is generated by following process: (1) a latent representation $\mathbf{z}_n$ is generated from distribution $\mathcal{N}(\mathbf{0}, \boldsymbol{I})$; (2) a sample $\mathbf{x}_n$ is generated from a conditional distribution $p_\theta(\mathbf{x}|\mathbf{z}_n)$. VAEs use a disentangled inference model $q_\phi(z|x)$ as a proxy to intractable true posterior $p_\theta(z|x)$, and its inference and generative models are computed by neural networks. The model is optimized with maximizing the evidence lower bound (ELBO) [10,11],

$$\log p(x) \geq \mathcal{L}_{ELBO} = \mathbb{E}_{z \sim q}[\log p_\theta(x|z)] - \mathcal{D}_{KL}(q_\phi(z|x)\|p(z)), \tag{2}$$

where sample z_n is yield from the inference model $q_\phi(z|x) = \mathcal{N}(z; \mu_\phi(x_n), \Sigma_\phi(x_n)$ by reparameterization trick [15, 16].

$$z_n = \frac{1}{L} \sum_{l=1}^{L} \mu_\phi(x_n) + \Sigma_\phi(x_n) * \epsilon_n \quad \text{with} \quad \epsilon_n \sim \mathcal{N}(0, I). \qquad (3)$$

The parameters ϕ, θ can be jointly optimized through maximizing the ELBO $\mathcal{L}(\theta, \phi; x)$ by stochastic gradient descent (SGD) method [17, 18]. And using values sampled from the inference network, we will be able to compute gradient estimates for a large class of models and inference networks with higher expressive architectures.

3   Latent Gaussian-Multinomial Generative Model

In consideration of the limitations of artificially segmenting image used in traditional generative models over annotated data, we propose a latent Gaussian-Multinomial generative model (LGMG) for annotated data and to realize automatic image annotation. We use a continuous latent variable in LGMG to represent the abstract semantics of images, and approximate it using a inference network computed from a neural network with images. Additionally, we assume the latent topic shared by image and annotation has a prior of Multinomial, and use a Dirichlet parameterized by a neural network with latent representation as its inference network. Furthermore, we also build up a multimodal generative models on the image-annotation pair $\{x, w\}$ respectively computed from neural networks. The variational objective is calculated by the combination of Monte Carlo estimator and reparameterization trick, and it is able to be backpropagated through neural network.

3.1   Generative Models

Given a dataset $\mathcal{D}$ containing N image-annotation pairs $\{x_n, w_n\}_{n=1}^{N}$, we consider the annotation w_n is generated from a mixture of Multinomial, and the image x_n is generated from a mixture of Gaussian or Bernoulli. The generative process is,

1. For each image-annotation pair $\{x_n, w_n\}$ in dataset $\mathcal{D}$:
   Choose a topic c_n from $\mathbf{Mult}(\pi)$
2. For image x_n:
   Sample $z_n \sim \mathcal{N}(0, I)$
   Generate image x_n from $p_\Theta(x|z_n, c_n) = \prod_{k=1}^{K} p_{\theta_k}(x|z_n)^{c_k}$
3. For annotation w_n:
   Generate annotation w_n from $p_\eta(w|c_n) = \prod_{k=1}^{K} p_{\eta_k}(w)^{c_k}$

where $\Theta = \{\theta_1, \theta_2, ..., \theta_K\}$ and $\eta = \{\eta_1, \eta_2, ..., \eta_K\}$. The k-dimensional variable c represents the latent topic with a prior of Multinomial distribution $\mathbf{Mult}(\pi)$,

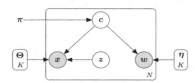

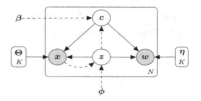

Fig. 1. The generative process of image-annotation $\{x, w\}$. Each image x in dataset is generated from a mixture of Gaussian or Bernoulli $p_\Theta(x|z) = \prod_{k=1}^{K}[p_{\theta_k}(x|z)]^{c_k}$, and its annotation w is generated from a mixture of Multinomial $p_\eta(w|z, c) = \prod_{k=1}^{K}[p_{\eta_k}(w)]^{c_k}$.

Fig. 2. The graphical models on the generative process and inference networks in LGMG. In which the solid lines denote the generative model, dashed lines denote the inference models on latent representation z and π respectively.

the image and annotation in a pair share a common topic. The generative model $p_\Theta(x|z)$ over image is a mixture of Gaussian for continue data or a mixture of Bernoulli for discrete data. And each image has an unique latent representation z generated from Normal distribution $\mathcal{N}(0, I)$. Each annotation w_n can be generated from a mixture of Multinomial $p_\eta(w|c) = \prod_{k=1}^{K}[p_{\eta_k}(w)]^{c_k}$. The generative process of image-annotation based in LGMG is shown in Fig. 1.

Given the parameters Θ, η, π, the joint distribution of LGMG is given by,

$$p(x, w, z, c|\Theta, \eta, \pi) = \sum_{k=1}^{K} p(c_k = 1|\pi)p_{\theta_k}(x|z)p(z)p_{\eta_k}(w) \qquad (4)$$

To estimate the LGMG model, we study a structure among the image x, latent representation z and latent topic c by neural networks, which can explore the complex relationship between image and annotation and realize the automatic image annotation.

3.2   Neural Variational Inference

• Inference Networks over Latent Variables

For the latent representation $z \sim \mathcal{N}(0, I)$, we choose the inference network $q_\phi(z|x) = \mathcal{N}(z; \mu_\phi(x), \Sigma_\phi(x))$ as a proxy for its intractable posterior $p_\Theta(z|x)$, where parameters $\mu_\phi(x)$ and $\Sigma_\phi(x)$ are neural networks parameterized by ϕ.

The latent representation z can summarize high-level semantics of images, so it is useful for the annotations. We choose a inference network $q_\beta(c|z) = \mathbf{Dir}(f(z, \beta))$ as a proxy of the true posterior $p_{\Theta,\eta}(c|x, w)$. In which, $f(z, \beta)$ is a neural network with K-dimensional output and sigmoid activation. Stick-Breaking sampler is one popular construction for Dirichlet distribution $c \sim Dir(\alpha)$, which bases on a key point that each marginal distribution $p(c_k)$ is a $\mathbf{Beta}(\alpha_k, \sum_{j<k} \alpha_j)$. So we consider the probability of latent topic as,

$$q_\beta(c_k = 1|f(z, \beta)) = \frac{\exp(f_k(z, \beta))}{\sum_{j<k}[1 + \exp(f_j(z, \beta))]}. \qquad (5)$$

• **Generative Networks over Image-Annotations**

In each topic, the generative model over images $p_{\theta_k}(x|z)$, $k = 1, 2, ..., K$, is computed from a fully-connected neural network parameterized by θ_k. In the case of binary data, we use

$$\log p_{\theta_k}(x|z) = \sum_{i=1}^{D}[x_i \log y_{k,i} + (1 - x_i) \log(1 - y_{k,i})] \quad \text{with} \quad y_k = f(z, \theta_k).$$

(6)

In the case of real-value data, we use

$$\log p_{\theta_k}(x|z) = \log \mathcal{N}(x; \mu_k, \sigma_k^2 I)$$
$$\text{with} \quad \mu_k = f_\mu(z, \theta_k) \quad \text{and} \quad \log \sigma_k^2 = f_\sigma(z, \theta_k).$$

(7)

where f, f_μ and f_σ are the elementwise sigmoid activation function, and θ is the set of weights and biases of these neural networks.

In generative model over annotations, each topic is generated by a mapping from latent representation with sufficiently complicated function. We use a fully-connected neural network parameterized by η_k as the complicated function,

$$\log p_{\eta_k}(w|z) = \sum_{v=1}^{V} w_v \log \varpi_v \quad \text{with} \quad \varpi = f_\varpi(z, \eta_k), \qquad (8)$$

where f_ϖ is a neural network with elementwise sigmoid activation, and it outputs V-dimensional vector as the generated annotation. The parameters set η_k represents weights and biases of this neural network.

Figure 2 shows the complete directed graphical models for all inference networks on latent variables z, c and generative networks on image-annotation $\{x, w\}$. The solid lines represent the generative process, and dashed lines represent the inference networks.

3.3 The Variational Bound

In the LGMG, the marginal likelihood over each image annotation $\{x, w\}$ can be written as a combination of KL divergence and ELBO,

$$\log p(x, w) = \mathcal{D}_{KL}(q_\phi(z|x)\|p_\Theta(z|x, c)) + \mathcal{D}_{KL}(q_\beta(c|z)\|p_{\Theta,\eta}(c|x, w)) + \mathcal{L}(\phi, \beta, \Theta, \eta),$$

where both $\mathcal{D}_{KL}$ and $\log p(x, w)$ are onstant, and the lower bound $\mathcal{L}(\phi, \beta, \Theta, \eta)$ on the marginal likelihoods of x and w can be rewritten as,

$$\mathcal{L}(\phi, \beta, \theta, \eta) = \mathbb{E}_{q_\phi(z|x)q_\beta(c|z)}\left[\sum_{k=1}^{K} c_k \log p_{\theta_k}(x|z)\right] + \mathbb{E}_{q_\beta(c|z)}\left[\sum_{k=1}^{K} c_k \log p_{\eta_k}(w|z)\right]$$
$$- \mathcal{D}_{KL}(q_\phi(z|x)\|p(z)) - \mathcal{D}_{KL}(q_\beta(c|z)\|p_\pi(c))$$

The parameters $\phi, \beta, \Theta, \eta$ can be jointly optimized by maximizing $\mathcal{L}(\phi, \beta, \Theta, \eta)$ with the stochastic optimization method, that is,

$$\{\phi, \beta, \Theta, \eta\} \leftarrow \arg\max \mathcal{L}(\phi, \beta, \Theta, \eta) \qquad (9)$$

3.4 SGVB Estimator

The stochastic gradient variational Bayes (SGVB) estimator [22,23] which evaluates the intractable integration by averaged sampling, and we use it to optimize the lower bound. For $\mathcal{L}(\phi, \beta, \theta, \eta)$, we use two practical sampling methods to yield samples from $q_\phi(z|x)$ and $q_\beta(c, \pi|z)$ respectively.

Firstly, we use reparameterization trick to sample value $z^{(i)}$ from the inference model $q_\phi(z|x) = \mathcal{N}(z; \mu_\phi(x), \Sigma(x))$. Given sample $x^{(i)}$, we sample latent representation $z^{(i)}$ by noise samples $\epsilon^{(l)} \sim \mathcal{N}(0, I)$, that is,

$$z^{(i)} = \frac{1}{L} \sum_{l=1}^{L} z^{(i,l)} \quad \text{with} \quad z^{(i,l)} = \mu_\phi(x^{(i)}) + \Sigma_\phi^{\frac{1}{2}}(x^{(i)} * \epsilon^{(l)}). \tag{10}$$

The Monte Carlo estimates of some function $f(z)$ can be calculated by,

$$f(z^{(i)}) = \mathbb{E}_{p(\epsilon)}[f(z^{(i)})] \approx \frac{1}{L} \sum_{l=1}^{L} f(z^{(i,l)}). \tag{11}$$

Secondly, we sample π and the probability of c from inference model $q_\beta(c|z) = Dir(f(z, \beta))$. Its Monte Carlo estimates [20] can be calculated by sampling parameters $f(z, \beta)$, that is,

$$\pi_k^{(i)} = q_\beta(c_k^{(i)} = 1|z^{(i)}) = \mathbb{E}_{q_\phi(z|x^{(i)})}[q_\beta(c_k^{(i)} = 1|z^{(i)})]$$

$$\approx \frac{1}{L} \sum_{l=1}^{L} \frac{\exp(f_k(z^{(i,l),\beta}))}{\sum_{j<k}[1 + \exp(f_j(z^{(i,j)}, \beta))]} \tag{12}$$

$$c_k^{(i)} = \begin{cases} 1 & \text{where} \quad q_\beta(c_k^{(i)} = 1|z^{(i)}) > q_\beta(c_j^{(i)} = 1|z^{(i)}) \quad k \neq j \\ 0 & \text{otherwise} \end{cases} \tag{13}$$

We use minibatch technique to train our model, and Kingma [10] had verified that as long as the minibatch size M is large enough the number L of samples in reparameterization trick can be set as 1. In this paper we set the parameters $M = 500$ and $L = 1$. For the minibatch $X^M = \{x^{(i)}\}_{i=1}^M$ randomly drawn from dataset, our SGVB estimator on ELBO is shown as follows,

$$\mathcal{L}(\phi, \beta, \theta, \eta) \approx -\frac{1}{M} \sum_{i=1}^{M} [\log q_\beta(c^{(i)}|z^{(i)}) - \log p(c^{(i)}) + \log q_\phi(z^{(i)}|x^{(i)}) - \log p(z^{(i)})]$$

$$+ \frac{1}{M} \sum_{i=1}^{M} \sum_{k=1}^{K} c_k^{(i)} [\log p_\eta(w^{(i)}|z^{(i)}) + \log p_{\theta_k}(x^{(i)}|z^{(i)})].$$

We can derivate the gradient $\nabla_{\phi, \beta, \theta, \eta} \mathcal{L}(\phi, \beta, \theta, \eta)$ used in conjunction with stochastic gradient descend (SGD) method. The generative performance can be qualified by the negative log-likelihood (NLL) of input $\{x^{(i)}, w^{(i)}\}$:

$$NLL = -\frac{1}{M} \sum_{i=1}^{M} \sum_{k=1}^{K} c_k^{(i)} [\log p_\eta(w^{(i)}|z^{(i)}) + \log p_{\theta_k}(x^{(i)}|z^{(i)})]. \tag{14}$$

4   Experimental Analysis

We evaluated the proposed LGMG by several experiments comparing with tr-mmLDA [4], cLDA [2], VAEs [10,11] and cVAE [21], among which LGMG, VAEs and cVAE were all trained end-to-end with minibatch size $M = 500$, epochs 200 and learning rate 0.001. The dataset used in our experiments is LabelMe [12] consisting of image-annotation pairs. The inference models on latent representation z in LGMG, VAEs and cVAE were all fixed with same architecture. Specifically, the preceding two hidden layers were CNN (Convolutional neural network) [23], and the next two hidden layers were fully connected network with two hidden layers (each layer with 500 units and activation function Tanh). The generative networks over image x in LGMG, VAEs and cVAE were an invertible architecture of the inference network. In LGMG, the generative network $p_{\eta_k}(w|z)$ in each topic was a fully connected neural network with two hidden layes, each layer had 500 units with activation function Tanh, and the output layer in which had 198 (the amount of words in vocabulary of LabelMe) units with activation function sigmoid. Inference network $q_\beta(c|z)$ of latent topic indictor was a fully connected neural network with two hidden layers, each hidden layer has 500 units with activation function Tanh, and the output layer had three (the amount of topics) units with activation function sigmoid. In addition, in cLDA and LDA the topic indicator was set as 3-dimensions, and the annotation corpus was set as 198-dimensions.

In Sect. 4.1, we compared and analyzed the generative likelihood performances in all comparative models on LabelMe. In Sect. 4.2, we compared the latent representative spaces between LGMG and VAEs. Finally, we analyzed the results of automatic image annotation in models of LGMG, tr-mmLDA and cLDA in Sect. 4.3.

4.1   Automatic Image Annotation

After training LGMG on dataset, we can sample the value $z^{(i)}$ from inference network $q_\phi(z|x)$ to generate new image $x^{(i)}$, then sample the topic indictor $c^{(i)}$ from the posterior $q_\beta(c|z)$ and generate annotation $w^{(i)} \sim \prod_{k=1}^{K}[p_{\eta_k}(w|z)]^{c_k}$. In order to evaluate the performance of automatic image annotation, we compared the generated annotations from cLDA, tr-mmLDA and LGMG on the selected six images (Fig. 3). We presented the words annotated by three models in Fig. 3, and these annotations all had more superior likelihoods. In addition, we presented the likelihoods of annotations generated from LGMG in Table 2.

From Fig. 3 we can see that our model annotated the optimal results among three models. In comparison, the annotations generated by tr-mmLDA and cLDA both occurred redundant words and missed some truth words. The reason for this results is the probabilistic distributions over annotations in tr-mmLDA and cLDA are both based on independent instances segmented from image, which ignored the important information of correlations between these instances. If two instances (for example, the sea and sky) have similar feature, the results of annotations in these two models perhaps be confused. LGMG was not depending on

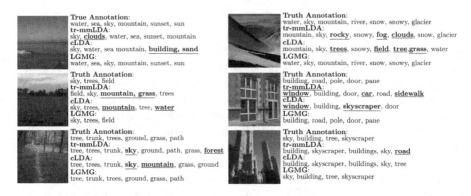

Fig. 3. The results of automatic annotated annotations in models of LGMG, tr-mmLDA and cLDA on selected images from testing set of LabelMe. The error annotations are labeled by blue bold font. (Color figure online)

segmented instances, but used a latent representation to summarize the abstract semantics (for example the color, light, lines, shape and so on) for image. More importantly, LGMG generated two types both from neural networks with same latent representation, which connected the correspondence between image and annotation. For example, an image is likely the sky if it occurred an airplane together with some clouds, while it would be the sea if it occurred some chairs together with motorboats. These results showed the LDAbased models ignored the important relevant information between instances. While the recognition model of LGMG could learn an abstract representation on whole image, and the network-based recognition and generative models are more flexible than tr-mmLDA and cLDA.

4.2   Generative Likelihood Performance

In this experiment, we compared LGMG with other comparative models in term of the generative likelihoods on images and annotations. For images, we compared the negative log-likelikoods among LGMG, VAEs and cVAE with different dimensions of latent representation z. These results were presented in Table 1. For annotations, we randomly selected six images (Fig. 3) from LabelMe and recorded the annotations learned by LGMG, tr-mmLDA and cLDA. The probabilities of generated annotations were shown in Table 2.

Table 1. Negative log-likelihoods in models of LGMG, VAEs and cVAE for different dimensions of z on LabelMe.

Latent representation	VAEs	cVAEs	LGMG (ours)
$D_z = 2$	−2667.909	−2567.42	−2634.722
$D_z = 5$	−2441.334	−2398.442	−2435.358
$D_z = 10$	−2383.964	−2321.336	−2382.739

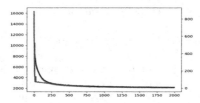

Fig. 4. The 2000-epochs training of LGMG with 20-dimensional latent representation. The blue line represents negative KL divergences over topics indicator scaled by the right-axis, the red line represents NLLs over images scaled by the left y-axis. (Color figure online)

Table 2. The probabilities of annotations generated from LGMG for selected images from LabelMe.

Images	Latent representation	Generated likelihoods for annotations							
Image 1		water	sea	sky	mountain	sunset	sun	others	
	$D_z = 2$	0.663	0.337	0.466	0.473	0.529	0.40	≤ 0.14	
	$D_z = 5$	0.987	0.949	0.970	0.942	0.870	0.908	≤ 0.027	
	$D_z = 10$	0.972	0.966	0.969	0.929	0.876	0.904	≤ 0.04	
	$D_z = 20$	0.995	0.962	0.994	0.980	0.965	0.972	≤ 0.02	
	$D_z = 40$	0.985	0.994	0.937	0.938	0.960	0.941	≤ 0.015	
Image 2		sky	trees	field	others				
	$D_z = 2$	0.52	0.07	0.143	≤ 0.15				
	$D_z = 5$	0.957	0.913	0.920	≤ 0.015				
	$D_z = 10$	0.984	0.970	0.971	≤ 0.04				
	$D_z = 20$	0.999	0.972	0.965	≤ 0.03				
	$D_z = 40$	0.994	0.785	0.84	≤ 0.01				
Image 3		tree	trunk	trees	ground	grass	path	others	
	$D_z = 2$	0.742	0.821	0.917	0.934	0.964	0.963	≤ 0.55	
	$D_z = 5$	0.995	0.971	0.913	0.952	0.947	0.923	≤ 0.15	
	$D_z = 10$	0.962	0.972	0.971	0.950	0.961	0.968	≤ 0.04	
	$D_z = 20$	0.991	0.964	0.976	0.951	0.954	0.950	≤ 0.015	
	$D_z = 40$	0.992	0.901	0.965	0.952	0.958	0.963	≤ 0.025	
Image 4		water	sky	mountain	river	snow	snowy	glacier	others
	$D_z = 2$	0.735	0.823	0.888	0.872	0.723	0.861	0.894	≤ 0.20
	$D_z = 5$	0.979	0.965	0.977	0.962	0.960	0.965	0.964	≤ 0.20
	$D_z = 10$	0.984	0.968	0.959	0.971	0.943	0.931	0.954	≤ 0.03
	$D_z = 20$	0.974	0.970	0.984	0.977	0.978	0.913	0.969	≤ 0.02
	$D_z = 40$	0.972	0.983	0.984	0.953	0.946	0.957	0.958	≤ 0.016
Image 5		building	road	pole	door	pane	others		
	$D_z = 2$	0.909	0.718	0.644	0.498	0.878	≤ 0.20		
	$D_z = 5$	0.970	0.978	0.942	0.967	0.978	≤ 0.16		
	$D_z = 10$	0.95	0.938	0.951	0.922	0.917	≤ 0.02		
	$D_z = 20$	0.987	0.996	0.963	0.969	0.954	≤ 0.15		
	$D_z = 40$	0.940	0.914	0.977	0.991	0.902	≤ 0.17		

(continued)

Table 2. (*continued*)

Images	Latent representation	Generated likelihoods for annotations						
Image 6		**sky**	**building**	**tree**	**skycraper**	**others**		
	$D_z = 2$	0.629	0.70	0.774	0.781	$\leq$0.20		
	$D_z = 5$	0.965	0.99	0.968	0.97	$\leq$0.04		
	$D_z = 10$	0.95	0.973	0.939	0.936	$\leq$0.04		
	$D_z = 20$	0.959	0.999	0.961	0.965	$\leq$0.015		
	$D_z = 40$	0.926	0.99	0.983	0.98	$\leq$0.02		

From Table 1 we can see that the generative likelihoods among LGMG, VAEs and cVAE were very closed. From this results we observe that in different dimensions of latent representation z, the negative log-likelihoods generated by LGMG were all slightly better than VAEs. This results verified that the LGMG can keep the generative likelihood performance on image with VAEs.

Figure 4 showed the training process of LGMG on LabelMe, in which the blue line represents the negative log-density on images and its scales on the left, and the red line represents the negative log-density on annotations and its scales on the right. As we can see in this result, the NLL of annotations generated from LGMG attained to 0 when the epoch beyonds 1500. That is to say, the annotations generated from LGMG closed to truth values. We recorded the likelihoods for annotation generated by a series LGMG with different dimensions D_z of latent representation. We can see from these results that along with increase of D_z the generated likelihoods for annotations became larger, and all big than 0.9 when $D_z \geq 20$. In addition, when $D_z \geq 20$ the generated likelihoods for untruth annotations are all less that 0.03.

5   Conclusion

In this paper, we have proposed the LGMG model for annotated data, which used a latent representation with prior of Normal distribution to summarize the abstract semantics of images. We assumed the image was generated from a mixture of Gaussian or Bernoulli, and its annotation was generated from a mixture of Multinomial. Two generative models shared a common topic and latent representation and computed from neural networks. Besides, we organized a mapping structure to calculate the variational posteriors of latent representation and topics indicator. Furthermore, we optimized the variational objective through sampling the latent representation and discrete latent indicator from reparameterization trick and Monte Carlo estimator respectively. The annotation for a new image was inferred by the probabilities computed from multinomial generative models. The experimental results have shown that LGMG improved the accuracy of automatic image annotation, meanwhile it kept likelihood performance on images with state-of-the-art generative models. We believe that the latent Gaussian-Multinomial generative model is a promising direction for the application of image-annotation.

Ackonwledgement. This work has been partly supported by National Natural Science Foundation of China (61402332, 61502339, 61502338, 61402331); Tianjin Municipal Science and Technology Commission (17JCQNJC00400, 18JCZDJC32100); the Foundation of Tianjin University of Science and Technology (2017LG10); the Key Laboratory of food safety intelligent monitoring technology, China Light Industry; research Plan Project of Tianjin Municipal Education Commission (2017KJ034, 2017KJ035, 2018KJ106).

References

1. Jeon, J., Lavrenko, V., Manmatha, R.: Automatic image annotation and retrieval using cross-media relevance models. In: Proceedings of the 26th Annual International ACM SIGIR Conference on Research and Development in Information Retrieval, pp. 119–126. ACM (2003). https://doi.org/10.1145/860435.860459
2. Barnard, K., Duygulu, P., Forsyth, D., et al.: Matching words and pictures. J. Mach. Learn. Res. **3**(Feb), 1107–1135 (2003)
3. Blei, D.M., Jordan, M.I.: Modeling annotated data. In: Proceedings of the 26th Annual International ACM SIGIR Conference on Research and Development in Informaion Retrieval, pp. 127–134. ACM (2003). https://doi.org/10.1145/860435.860460
4. Putthividhy, D., Attias, H.T., Nagarajan, S.S.: Topic regression multi-modal latent dirichlet allocation for image annotation (2010). https://doi.org/10.1109/CVPR.2010.5540000
5. Huang, S.J., Gao, W., Zhou, Z.H.: Fast multi-instance multi-label learning. IEEE Trans. Pattern Anal. Mach. Intell.(2018)
6. Murthy, V.N., Maji, S., Manmatha, R.: Automatic image annotation using deep learning representations. In: Proceedings of the 5th ACM on International Conference on Multimedia Retrieval, pp. 603–606. ACM (2015). https://doi.org/10.1145/2671188.2749391
7. Wu, J., Yu, Y., Huang, C., et al.: Deep multiple instance learning for image classification and auto-annotation. In: Proceedings of the IEEE Conference on Computer Vision and Pattern Recognition, pp. 3460–3469 (2015). https://doi.org/10.1109/CVPR.2015.7298968
8. Lev, G., Sadeh, G., Klein, B., Wolf, L.: RNN fisher vectors for action recognition and image annotation. In: Leibe, B., Matas, J., Sebe, N., Welling, M. (eds.) ECCV 2016. LNCS, vol. 9910, pp. 833–850. Springer, Cham (2016). https://doi.org/10.1007/978-3-319-46466-4_50
9. Mnih, A., Gregor, K.: Neural variational inference and learning in belief networks. In: International Conference on Machine Learning, pp. 1791–1799 (2014)
10. Kingma, D.P., Welling, M.: Auto-Encoding Variational Bayes. Stat 1050:1 (2014)
11. Doersch, C.: Tutorial on variational autoencoders. Stat 1050:13 (2016)
12. Russell, B.C., Torralba, A., Murphy, K.P., et al.: LabelMe: a database and web-based tool for image annotation. Int. J. Comput. Vis. **77**(1–3), 157–173 (2008). https://doi.org/10.1007/s11263-007-0090-8
13. Uricchio, T., Ballan, L., Seidenari, L., et al.: Automatic image annotation via label transfer in the semantic space. Pattern Recogn. **71**, 144–157 (2017). https://doi.org/10.1016/j.patcog.2017.05.019

14. Kumar, R.: Natural language processing. In: Machine Learning and Cognition in Enterprises, pp. 65–73. Apress, Berkeley (2017). https://doi.org/10.1007/978-1-4842-3069-5_5

15. Murphy, K.P.: Machine Learning: A Probabilistic Perspective. MIT Press, Cambridge (2012)

16. Blei, D.M., Kucukelbir, A., McAuliffe, J.D.: Variational inference: a review for statisticians. J. Am. Stat. Assoc. **112**(518), 859–877 (2017). https://doi.org/10.1080/01621459.2017.1285773

17. Kingma, D.P.: Variational inference & deep learning: a new synthesis (2017)

18. Bottou, L.: Stochastic gradient descent tricks. In: Montavon, G., Orr, G.B., Müller, K.-R. (eds.) Neural Networks: Tricks of the Trade. LNCS, vol. 7700, pp. 421–436. Springer, Heidelberg (2012). https://doi.org/10.1007/978-3-642-35289-8_25

19. Blei, D.M.: Probabilistic topic models. Commun. ACM **55**(4), 77–84 (2012). https://doi.org/10.1145/2133806.2133826

20. van Ravenzwaaij, D., Cassey, P., Brown, S.D.: A simple introduction to Markov Chain Monte-Carlo sampling. Psychon. Bull. Rev. **25**(1), 143–154 (2018)

21. Pu, Y., Gan, Z., Henao, R., et al.: Variational autoencoder for deep learning of images, labels and captions. In: Advances in Neural Information Processing Systems, pp. 2352–2360 (2016). https://doi.org/10.3758/s13423-016-1015-8

22. Kinga, D., Adam, J.B.: A method for stochastic optimization. In: International Conference on Learning Representations (ICLR), p. 5 (2015)

23. Krizhevsky, A., Sutskever, I., Hinton, G.E.: Imagenet classification with deep convolutional neural networks. In: Advances in Neural Information Processing Systems, pp. 1097–1105 (2012). https://doi.org/10.1145/3065386

Investigating Neighborhood Generation Methods for Explanations of Obscure Image Classifiers

Riccardo Guidotti[1,2(✉)], Anna Monreale[2], and Leonardo Cariaggi[2]

[1] ISTI-CNR, Pisa, Italy
guidotti@isti.cnr.it
[2] University of Pisa, Pisa, Italy
anna.monreale@di.unipi.it, leonardocariaggi@gmail.com

Abstract. Given the wide use of machine learning approaches based on opaque prediction models, understanding the reasons behind decisions of black box decision systems is nowadays a crucial topic. We address the problem of providing meaningful explanations in the widely-applied image classification tasks. In particular, we explore the impact of changing the neighborhood generation function for a local interpretable model-agnostic explanator by proposing four different variants. All the proposed methods are based on a grid-based segmentation of the images, but each of them proposes a different strategy for generating the neighborhood of the image for which an explanation is required. A deep experimentation shows both improvements and weakness of each proposed approach.

1 Introduction

In the last years, automated decision systems are widely used in all those situations in which classification and prediction tasks are the main concern [20]. All these systems exploit machine learning techniques to extract the relationships between input and output. Input variables can be of any type, as long as it is possible to find a convenient representation for them. For instance, we can represent images by matrices of pixels or by a set of features that correspond to specific areas or patterns of the image [5,16].

We talk about "black box" classifiers when dealing with classifiers having an opaque, hidden internal structure whose comprehension is not our main concern [12]. The typical example of black box is a neural network, one of the most used machine learning approaches due to its excellent performance. Therefore, the recent interest in explanations derives from the fact that we constantly use decision systems that we cannot understand. How can we prove that an image classifier built to recognize poisonous mushrooms actually focuses on the mushrooms themselves and not on the background?

Another reason for the recent interest in black box explanations is the *General Data Protection Regulation* approved by the European Parliament in May 2018. Besides giving people control over their personal data, it also provides restrictions

© Springer Nature Switzerland AG 2019
Q. Yang et al. (Eds.): PAKDD 2019, LNAI 11439, pp. 55–68, 2019.
https://doi.org/10.1007/978-3-030-16148-4_5

and guidelines for automated decision-making processes which, for the first time, introduce a right of explanation. This means that an individual has the right to obtain meaningful explanations about the logic involved when automated decision making takes place [13,18,27].

In this work we study the reasons that lead classifiers to make certain predictions. A common approach for "opening" black boxes is to focus on the predictions themselves by understanding the predictions *a-posteriori* and comprehend on *what* the black box focused for returning the prediction [8]. A recently established approach consists in generating a neighborhood composed of both similar and different instances from the one to be explained [11,21]. Then, observing the behavior of the black box on the neighborhood is possible to understand which are the features used for the prediction.

In this work, we propose four variants of the LIME method [21] that enable the explanation of image classifications based on sparse linear approaches. In particular, we design alternative ways for generating the neighborhood of the classified image for which an explanation is required. All the proposed methods use a grid-based approach for the image segmentation instead of the segmentation based on the *quickshift* [26] algorithm typical of LIME. However, each method differs in the strategy adopted for generating a neighborhood of images as perturbation of the image for which an explanation is required. The idea behind our proposals is to obtain neighbors by replacing some parts of the image to be explained with parts of other images and not by simply obscuring the original pixels. We experiment the impact of these approaches to understand which are the informative image regions and the overall quality of the explanations by introducing a systematic approach for the evaluation of explanations. Our evaluation highlights for each proposed approach both improvements and deficiencies.

The rest of this work is organized as follows. In Sect. 2 we provide an overview of state-of-art methods for explaining predictions in image classification. Section 3 summarizes LIME and provides evidence for some inconsistencies. In Sect. 4 we present the details of the proposed explanation methods. Section 5 contains a deep experimentation of the proposed approaches. Finally, Sect. 6 concludes the paper by discussing strengths and weaknesses of the proposed solutions and future research directions.

2   Related Work

In this section we provide an overview of the state-of-the-art for explaining the predictions of a black box image classifier. According to [12], the problem faced in this work is the *outcome explanation problem* that aims at returning a local explanation for an individual instance. The common strategy for the methods solving this problem is to provide a locally interpretable model (i.e. that can be clearly understood by a human). In case of image classification, such interpretable model can be an heatmap, defining the most important regions of the image that contribute to the prediction [21], or a mask, defining the minimal amount of information that, when deleted, causes the prediction to change drastically [8]. Other approaches, instead, aim at providing as explanation a prototypical image that clearly illustrates the treats for the given outcome [15].

Approaches based on saliency masks highlights which are the parts of an image that contribute the most to the prediction. A *saliency mask* is a subset of the record to be explained, i.e. a specific part of an image in our case, that causes the black box to make that specific prediction. The strategy adopted by [29] consists in the generation of attention maps for a CNN (Convolutional Neural Network), which highlight salient regions of an image and localize various categories of objects. Such maps are generated by using the backpropagation scheme. In this setting, each neuron of a convolutional layer can be matched with a specific area of an image. Also in the approaches presented in [23,32] neuron activations are incorporated in their visual explanations. For this reason, the only family of models that can be explained by these approaches are CNN.

In [8] is presented a model-agnostic framework that defines saliency masks as *meaningful perturbations*. The goal is to study the effect of deleting specific regions from and image and find the smallest *deletion mask* or *artifact* that, when applied, causes the accuracy to drop significantly. Such deletion mask are presented as explanation. A problem with artifacts is that they can look unintuitive and unnatural and this impacts heavily on the quality of the visual explanation. The authors of [8] suggest to not focus on the specific details of the mask and apply a random jitter.

In [6] is developed a detection method performed by single forward pass (rather than iteratively, like in [31]) that produces high quality and sharp masks. The problem of artifacts is solved by cropping the input image rather than masking it: the goal is to find the tightest rectangular crop that contains the entire salient region of the image.

All the aforementioned methods define, with different strategies, an explanation at *pixel level* which identifies "unstructured" relevant areas of an image. The union of these areas may highlight a mixture of features that does not express a well defined concept but a blend of many aspects. In contrast, the aim of [30] is to produce explanation at *object level* that specify clear, distinct and highly interpretable parts of an image. The approach is based on the definition of *interpretable CNNs* by modifying the way features are represented inside filters of convolutional layers.

Our approach shares some properties with [23,29,32] with respect to the definition of masks highlighting the salient regions of an image for a certain prediction. However, in these approaches the masks are created by digging into the architecture of specific CNN and they can not be used for other types of black boxes (see [12] for more details). On the other hand, since we propose a set of methods extending LIME, they are black box agnostic by design. Finally, the proposed methods differ both from [21] and [8] because, as detailed in the following, the neighborhood is not generated by just blurring or obfuscating part of the image for which an explanation is required, but rather by replacing some feature of this image with features of other images.

As last remark we point out that, with respect to the way images are represented, our methods differs from traditional matrix of pixels, bags of pixels [14], and bag-of-words [5,17,28]. Since we intent to highlight, cut and replace contiguous patches of pixels, our natural choice is to use grids of RGB values obtained by aggregating the pixels in the same cell. Further details are provided in Sect. 4.

3   Background

In this section we summarize *LIME* [21], its main logic, and we provide evidence for some inconsistencies. LIME is a local *model-agnostic* explainer. It can provide an explanation for individual predictions of any classifier without making any assumptions on its internal structure. LIME's explanations consists in providing feedback that can help to understand which are the relationships between the input and the outcome. In the specific case of image classification, LIME's explanation is a saliency mask highlighting the areas that the black box looks at when taking its decision.

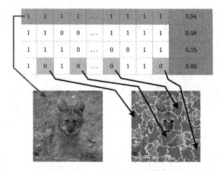

Fig. 1. LIME neighborhood generation process. Left: original image x. Right neighbor image $z \in N_x$. Top neighborhood of IDRs weights, first row in blue original image weights, last column in red black box probability of labeling the image as *red fox* (best displayed in colors). (Color figure online)

For every image to explain x, LIME builds an interpretable model that mimics the black box in the vicinity of x. To ensure interpretability, LIME builds ad-hoc *interpretable data representations* (IDR) $\hat{x}$ for x by "shattering" x into a set of areas made of contiguous pixels (regardless of their shape). The IDR $\hat{x}$ is therefore defined as an array of bits telling whether or not the single areas are present[1]. In order to build the local interpretable model for x, LIME generates a neighborhood N_x by randomly changing $\hat{x}$. Then, N_x is provided to the black box for approximating its local behaviour and seeing how it reacts to variations

---

[1] IDRs *do not* necessarily correspond to the features used by the black box in the prediction process. Indeed, such features may be not suitable for being shown as explanation.

of the input image x. To do that, a certain number m of samples is drawn uniformely at random from the domain of the IDRs in the form of binary vectors $w \in \{0,1\}^k$ where k is the number of features (contiguous patches of pixels[2]) in the image x and each term of w indicates the fact that the corresponding i-th feature is included, if $w_i = 1$, or not[3]. More formally, the neighborhood is defined as $N_x = \{\hat{x} \times w_i | w \in \{0,1\}^k\}$ where w is drawn uniformally at random from $\{0,1\}^k$.

The visual version of an image $z \in N_x$ has a plain color area (e.g. grey, white or black) for the features equals to 0 and the same areas of x for the features equals to 1. Figure 1 shows how a neighborhood image looks like. In the neighborhood image we highlight in yellow all the areas the image has been split into. They are contiguous patches of pixels with an irregular shape. LIME adopts *quickshift* [26] an algorithm based on an approximation of mean-shift [4] for obtaining the contiguous areas.

Keeping the label of the real image as a reference, LIME applies the black box to the images z in N_x obtaining an array $p \in [0,1]^m$ of prediction probabilities (highlighted in red in Fig. 1), where m is the number of samples in N_x. Each value is the probability that the black box assigns the original label to a neighborhood image. Then, LIME trains a comprehensible regression model c where the input are the IDRs in N_x and the targets are the corresponding values of p. If a weight of c is *positive*, it means that the associated area of the image contributes positively to the prediction of the black box, if the weight is *negative*, then the corresponding feature contributes negatively. Hence, for LIME as well as for the methods we propose, an explanation e is a vector where e_i is the importance of the i-th feature of the image. Finally, LIME returns as explanation the top n features in e (ordering them by their weights, in descending order) and highlights the corresponding areas in the image. Intuitively, those are the areas where the black box looks when making such a prediction. An example of neighborhood images and explanation can be found in Fig. 2-*(top left)*.

4   Explanation Methods

In this paper we try to overcome a conceptual problem with the neighborhood generation of LIME. Similarly to [8], LIME does not really generate images with different information, but it only randomly removes some features, i.e., it suppresses the presence of an information rather than modifying it. On the

---

[2] In the rest of this work, *feature, patch, area, piece* are used to denote the same concept.

[3] For the neighborhood N_x LIME generates m vectors w uniformly at random, assigning to them a weight that is proportional to their distance from the original image. The distance is used for assigning less importance to noisy images (that are too far away to be considered neighbors) and for focusing on the samples that are close to the original picture.

other hand, with respect to tabular data, LIME behaves differently. It generates the neighborhood not by information suppression, but by changing the feature values with other values of the domain: e.g., if the value of the *age* feature for a customer is 24, then LIME might replace it with 18 or 36. Our objective is to understand if an actual modification of the image to generate the neighborhood can lead to any improvement in the explanation itself.

Therefore, the goal of this work is to explore the impact of changing the neighborhood generation of LIME when it tries to provide an explanation for an image classification obtained by a black box. To this end we propose four different variants of the original LIME approach which are described in details in the following.

LIME# adopts a grid-based tessellation function that leads to patches with a regular shape. More formally, given an image x this function returns the vector $\hat{x}$ representing a grid with size $k = k_w \times k_h$. Any element $\hat{x}_i$ is a cell corresponding to the i-th feature of the original image. Given the IDR $\hat{x}$ and the vector w (introduced in the Sect. 3), the feature replacement function of LIME# works as in LIME, i.e., it replaces all the pixels in the feature x_i with $w_i = 0$ with a plain color (e.g. gray, black or white). The *grid size* k is a parameter of LIME# and has to be specified by the user, for example in Fig. 2-*(top right)* is fixed to 8×8. This figure shows the result of the application of this approach. The explanation illustrated by the green cells is simple and clear.

R-LIME# applies the same grid-based tessellation of LIME#. However, it completely changes the feature replacement function. In particular, it uses an image-based replacement, that substitutes any feature $\hat{x}_i$ having $w_i = 0$ with the corresponding feature $\hat{z}_i$ of an image z randomly selected among a set of given images I, arbitrary provided by the user. As a consequence R-LIME# takes as input the image to be explained and a set of images I which will be used

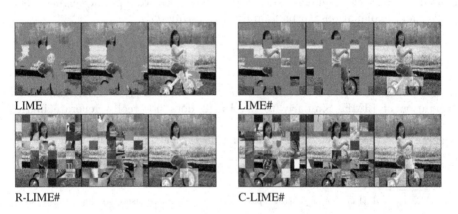

Fig. 2. Examples of neighborhood images generated by LIME, LIME#, R-LIME#, C-LIME#, respectively from left to right, from top to bottom. The first two images are an example of neighbors while the third one highlights the explanation for the label *tricycle* for the four methods. (Color figure online)

for the image-based replacement. Such kind of random generation would have been not possible in a straightforward way without the grid-based segmentation providing regular (and hence easily replaceable) features. For each image the algorithm computes the grid-based segmentation and generates the neighbors as described above. Figure 2-*(bottom left)* shows the application of R-LIME#, i.e., some images in the neighborhood and the explanation for a specific image.

C-LIME# differs from R-LIME# in the definition of the image-based replacement function. The basic idea of C-LIME# is to use images similar to that one to be explained for its neighborhood generation. Given an image x, and an initial set of images I, C-LIME# opportunely constructs the image pool, to be used for the neighborhood generation of x by selecting only images similar to x itself. To this aim, C-LIME# applies on I a clustering algorithm that returns groups of similar images $C = \{C_1, \ldots C_t\}$. Let f be the function that given an image x assigns it to the closer cluster. Then, for explaining a specific image x, C-LIME# builds the image pool using only images of the cluster $f(x)$. Once the image pool is built, the image-based replacement function works as R-LIME#. Figure 2-*(bottom right)* shows the impact of applying this method for the neighborhood generation. The neighborhood images still contain pieces of other images, but we can notice some interesting details such as wheels of other vehicles.

The clustering of images is computed by applying a process of undersampling of each image that depends on the size of the grid. Given the grid size $k_w \times k_h$ the clustering is performed on images undersampled to $k_w \times k_h$ pixels. The process first applies the grid-based segmentation to each image. Then, before the set of images is passed to the clustering algorithm they are flattened into a monodimensional vector. Clearly, C-LIME# is parametric with respect to the clustering algorithm and its own parameters.

Lastly, we propose **H-LIME#**, an *hybrid* approach that as C-LIME# applies a clustering algorithm on the set of images I (input of the algorithm) for finding groups of visually similar images $C = \{C_1, \ldots C_t\}$. However, the image pool is not composed only by images belonging to the same cluster of the image x to be explained. Instead, it includes the whole set of images, where each image x has associated the information about the cluster label $f(x)$. Thus, the image-based replacement function, given the image x to be explained, substitutes any feature $\hat{x}_i$ having $w_i = 0$ applying the following steps: (i) following a uniform probability distribution, it randomly decides if using an image belonging to the closest cluster $f(x)$ or one of the other images[4]; (ii) then, it randomly draws an image $\hat{z}$ from the selected group and replaces $\hat{x}_i$ with $\hat{z}_i$.

---

[4] For the sake of space we do not report experiments varying the probability of selection.

Table 1. Selected classes (from ILSVRC2012) for the dataset used in the experiments. For each of the 20 classes, we picked all the 50 corresponding images.

Food	Animals	Clocks	Shops	Vehicles
Cheeseburger	Timber wolf	Analog clock	Tobacco shop	Motor scooter
Hotdog	White wolf	Digital clock	Barbershop	Mountain bike
	Red wolf	Digital watch	Bookshop	Unicycle
	Coyote	Wall clock	Toyshop	Tricycle
	Dingo			Bicycle-for-two

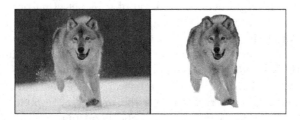

Fig. 3. Original image (left), Reference area (right).

5   Experiments

In this section we evaluate the methods described in Sect. 4. First, we present the dataset used for the evaluation of the explanations, then we define the measures used for the evaluation and finally, we present the results. The main points we address are the following: (i) what is the impact of the neighborhood generation? (ii) is it useful to generate neighborhood using visually similar images when replacing features? (iii) how well do the explanations approximate the area of the image containing the label? (iv) what is the ideal number of features?

The experiments[5] were performed on Ubuntu 16.04.5 64 bit, 504 GB RAM, 3.6 GHz Intel Xeon CPU E5-2698 v4.

5.1   Dataset and Black Box

Answering the previous questions requires a dataset of images with the areas responsible for the image classification, so that they can be compared with those of the explanations. In our experiments we use a subset of the images of the *ILSVRC2012* dataset [22]. We selected a subset of 1000 images belonging to 20 classes from the test set, keeping them equally distributed (50 images per class). This set of images represent the images requiring an explanation. Table 1 shows the single classes, grouped by category. We grouped together similar objects to test the model's ability to distinguish between small particulars. In these images we isolated by hand their *reference areas* by selecting the parts that best

---

[5] Source code and dataset can be found at: https://github.com/leqo-c/Tesi.

represent the object specified in the label provided by the black box and filling all the remaining parts with a plain white color. The result of this operation is an enriched dataset that can be used as a reference when evaluating explanations. A sample image from such dataset is shown in Fig. 3.

As black box we adopt the *Inception v3* [24]. Inception v3 is a CNN that operates on salient parts of the image. If we consider a picture of a bicycle, we may think of a feature as one of its wheels. Hence, the areas of interest identified by the black box fit the type of explanations provided by LIME and by the variants we propose.

5.2   Evaluation Measures

In order to compare the different methods, we define some *quality* measures among explanations. Intuitively, we want to measure the ability of the explanations to cover the reference area: the more the explanation covers the reference area and do not cover not-reference areas, the more it can be trusted. Given the n most important features of an explanation e, namely $e^{(n)}$, and the reference area r for a certain image x, we redefine well-known *precision, recall* and *F-measure* as follows:

$$pre(e^{(n)}, r) = \frac{|\{e_i^{(n)} \triangleright r | e_i^{(n)} \in e^{(n)}\}|}{n} \qquad rec(e^{(n)}, r) = \frac{pixels(\{e_i^{(n)} \triangleright r | e_i^{(n)} \in e^{(n)}\})}{pixels(r)}$$

$$F\text{-}measure(e^{(n)}r) = (1 + \beta^2) \cdot \frac{pre(e_n, r) \cdot rec(e_n, r)}{(\beta^2 \cdot pre(e_n, r)) + rec(e_n, r)}$$

where $e_i^{(n)} \triangleright r$ means that a feature among the top n in the saliency mask of the explanation is completely contained in the reference area, and $pixels(\cdot)$ returns the number of pixels in a certain area. The precision measures the ability to highlight only the relevant parts in the image, the recall measures the ability to highlight the overall number of pixels of the reference area, while the F-measure with $\beta = 0.5$ puts more emphasis on precision than recall [3] as we want to focus on quantifying correct explanations [10].

5.3   Assessing Explanation Quality

In this section we analyze the results of the different methods in terms of precision and F-measure. The results are presented by varying the following parameters[6]:

- the grid size k that takes values in the power of 2, i.e., 8×8, 16×16, 32×32, etc.[7]

---

[6] For the sake of simplicity of exposure and due to length constraints, we analyze both parameters in the same plots and we remand interested readers to the repository for further details.

[7] We do not report results using grid size lower than 8×8 (i.e., 2×2, 4×4, 6×6) or higher than 32×32 (i.e., 64×64, 128×128) has they have poor performance compared to those reported.

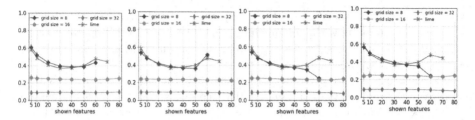

Fig. 4. Precision of LIME#, R-LIME#, C-LIME#, H-LIME# (from left to right) w.r.t. LIME varying the number of shown features, with different lines for different values of the grid size.

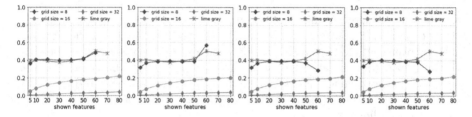

Fig. 5. F-measure of LIME#, R-LIME#, C-LIME#, H-LIME# (from left to right) w.r.t. LIME varying the number of shown features, with different lines for different values of the grid size.

– the number of top n features shown as explanation that takes values in $[1, 100]$.

As clustering algorithm for C-LIME# and H-LIME# we tested an array of approaches, i.e., K-Means [25], DBSCAN [7], Spectral [19]. We evaluated their performance both using internal validation measures (e.g., SSE [25]), and also by checking the cluster purity with respect to the label of the images clustered together. As result, we selected K-Means with 20 clusters as best performer to be used for C-LIME# and H-LIME#.

In Figs. 4 and 5 we report the average precision and F-measure, respectively. In these plots we vary the number of shown features n on the x-axis, and we report different lines for different values of the grid size k. Each plot reports the results with one of the proposed methods, from left to right LIME#, R-LIME#, C-LIME#, H-LIME#, and compares them with the original version of LIME.

The precision reaches a maximum of about 0.6 using LIME# (Fig. 4 leftmost plot) with $k = 8 \times 8$ and considering 5 features in both. For *shown features* ≤ 50 LIME# outperforms LIME. Higher values of *grid size*, have a negative impact on the precision, which stabilizes around low values (about 0.25 and 0.1 respectively). R-LIME#, 2nd plot in Fig. 4, has a precision slightly worse than those of LIME#, except when *shown features* = 60. C-LIME#, 3rd plot in Fig. 4, is outperformed by R-LIME# in terms of precision. This is signaling that generating a neighborhood where the neighbors are created by replacing a part of an image with a part of a similar image (belonging to the same cluster) drops the ability of the explainer in finding which are the important features. Indeed, as we replace a patch with a visually similar patches, the probability of disrupting the prediction is lower, causing the features not to be considered relevant. H-LIME# (rightmost plot in Fig. 4) tries to mitigate the weaknesses of C-LIME# by partly using a random replacement similar to R-LIME#. As result H-LIME# outperforms the other methods when *shown features* is in the range [10, 40]. Concerning the F-measure shown in Fig. 5, we can notice the following aspects. First, with a number of *shown features* n which does not cover nearly the whole image (i.e., $n < 50$), the F-measure reaches a maximum of about 0.42 with LIME# with $n = 15$ and $k = 8 \times 8$ overcoming LIME. Second, *grid size* $= 8 \times 8$ always outperforms the methods with a different setting. Third, for *gridsize* $\in \{16 \times 16, 32 \times 32\}$ the F-measure rapidly increases for *shown features* lower than 30, while remains almost constant for LIME.

An evident result is that high values of *grid size* inevitably lead to low-quality explainers. This is because the more we reduce the size of the features, the less we are likely to capture meaningful patterns inside an image. On the other hand, lower values noticeably increase the chance of highlighting informative regions. We should however keep in mind that overlying extended areas (e.g. *grid size* $= 4 \times 4$) would lead to dispersive and coarse explanations. Moreover, it is worth to analyze some information about the sizes of the superpixels generated by LIME and those of the cells of the grid adopted by the proposed methods. The total number of pixels in an image of the dataset is about $90,000$ (300×300). The average number of pixels in a superpixel generated by LIME is $1,405 \pm 221$. On the other hand, the number of pixels in a cell of the grid depends on the size of the grid. Using grid size $k = 8 \times 8$ we obtain $300/8 \times 300/8 = 1,406.25$. Therefore, a *grid size* $k = 8 \times 8$ on average generates cells that resemble the dimension of the superpixels. This high correspondence of pixels in a feature is a plausible justification for the similar performance got by the proposed methods with respect to LIME.

As a consequence, the *LIME# family of methods is generally more precise than LIME itself. This statement is validated by the deeper analysis we present in the following. Considering that the object of an image, i.e., its reference area, on average covers more than one third of the image itself, we can suppose that for covering the relevant area we need a number of features between 15 and 30. Using *grid size* $= 8 \times 8$, in Tables 2 and 3 we report the mean and standard deviation of precision and F-measure for the analyzed methods, respectively. We report in bold the best method for each row, an up-arrow ↑ highlights if a method outperforms the original version of LIME. From Table 2 we can notice that the members of the *LIME# family of methods are (nearly) always more precise than LIME, and H-LIME# is the most precise and stable method. On the other hand, Table 3 underlines that (*i*) only LIME# and H-LIME# overtake LIME in terms of F-measure, (*ii*) LIME# as the highest F-measure for the *shown features* observed, (*iii*) H-LIME# is still the most stable approach.

Table 2. Precision (mean ± stdev) of the methods evaluated for grid size $k = 8 \times 8$). For each row, the best method is reported in bold, an up-arrow ↑ highlights if a method outperforms LIME.

n	LIME	LIME#	R-LIME#	C-LIME#	H-LIME#
10	.485 ± .260	**.521 ± .256** ↑	.477 ± .260	.473 ± .257	.495 ± .252 ↑
15	.437 ± .243	.469 ± .243 ↑	.443 ± .246 ↑	.440 ± .244 ↑	**.478 ± .239** ↑
20	.406 ± .237	.436 ± .237 ↑	.416 ± .236 ↑	.417 ± .238 ↑	**.438 ± .231** ↑
25	.384 ± .234	.411 ± .231 ↑	.396 ± .232 ↑	.401 ± .234 ↑	**.415 ± .227** ↑
30	.367 ± .232	.393 ± .227 ↑	.384 ± .230 ↑	.384 ± .231 ↑	**.396 ± .224** ↑

Table 3. F-measure (mean ± stdev) of the methods evaluated for grid size $= 8 \times 8$). For each row, the best method is reported in bold, an up-arrow ↑ highlights if a method outperforms LIME.

n	LIME	LIME#	R-LIME#	C-LIME#	H-LIME#
10	.403 ± .181	**.405 ± .161** ↑	.368 ± .163	.365 ± .161	.382 ± .159
15	.398 ± .183	**.411 ± .172** ↑	.386 ± .173	.383 ± .172	.399 ± .170 ↑
20	.391 ± .191	**.409 ± .183** ↑	.389 ± .182	.390 ± .184	.403 ± .180 ↑
25	.384 ± .200	**.404 ± .191** ↑	.388 ± .192 ↑	.392 ± .195 ↑	.401 ± .189 ↑
30	.376 ± .207	**.399 ± .199** ↑	.389 ± .201 ↑	.389 ± .203 ↑	.397 ± .197 ↑

6   Conclusion

We have presented an array of methods based on the re-design of the neighborhood generation process of LIME. We have focused on the concept of interpretability of an explanation returned in terms of saliency mask, and we have tested the impact of using different techniques in the definition of the images' features. In particular, we argued that replacing patches with a solid color is not a natural way for producing a significant perturbed image. As additional contribution we have defined a systematic measure for assessing the quality of an explanation (i.e. what is its level of comprehensibility) by creating an annotated dataset, that contains for each image an ideal reference area, i.e., the part of the image that would be highlighted by an explanation provided by a human.

Our results show that replacing features with similar patterns (rather than random ones) like for C-LIME# prevents the explanation system from individuating relevant areas in the original image. On the other hand, H-LIME# which is based on both random patches and patches similar to the image to be explained, can help in improving the precision of an explanation. Nevertheless, LIME# which simply suppresses the information of a feature, has an overall better accuracy.

The findings of this paper are a solid starting point for a work that extends the methods presented by removing the constraint of the grid tessellation that, while simplifying the replacement process, it anchors all the approaches to behave too similarly to the original LIME method. Indeed, a fascinating research direction would be to adopt other techniques for embedding images into vectors (e.g., by means of image histograms [2], bag of visual words from SIFT key-points [16], or the embedding provided by another neural network [9]), to produce the random perturbations on these vectors and then, to reconstruct the corresponding neighbor images. Another interesting evolution of this work consists in extending the image neighborhood generation process using more sophisticated generation techniques, such as genetic programming [11], or an approach based on the minimum descriptive set [1], rather than drawing instances completely at random. Finally, even though an experimentation on different datasets and neural networks would not probably provide further insights relatively to the goodness of the neighborhood generation processes due to the nature of CNN generally used for classification of images, it would be interesting to check empirically these assumptions.

Acknowledgements. This work is partially supported by the European Community H2020 program under the funding scheme "INFRAIA-1-2014-2015: Research Infrastructures" G.A. 654024 *"SoBigData"*, http://www.sobigdata.eu, by the European Unions H2020 program under G.A. 78835, *"Pro-Res"*, http://prores-project.eu/, and by the European Unions H2020 program under G.A. 780754, *"Track & Know"*.

References

1. Angelino, E., Larus-Stone, N., Alabi, D., Seltzer, M., Rudin, C.: Learning certifiably optimal rule lists. In: KDD, pp. 35–44. ACM (2017)
2. Chapelle, O., Haffner, P., Vapnik, V.N.: Support vector machines for histogram-based image classification. IEEE Trans. Neural Netw. **10**(5), 1055–1064 (1999)
3. Christopher, D.M., Prabhakar, R., Hinrich, S.: Introduction to information retrieval **151**(177), 5 (2008)
4. Comaniciu, D., Meer, P.: Mean shift: a robust approach toward feature space analysis. IEEE Trans. Pattern Anal. Mach. Intell. **24**, 603–619 (2002)
5. Csurka, G., et al.: Visual categorization with bags of keypoints. In: Workshop on Statistical Learning in Computer Vision, ECCV, Prague, vol. 1, pp. 1–2 (2004)
6. Dabkowski, P., Gal, Y.: Real time image saliency for black box classifiers. In: Advances in NIPS 30, pp. 6967–6976. Curran Associates Inc. (2017)
7. Ester, M., Kriegel, H.-P., Sander, J., Xu, X., et al.: A density-based algorithm for discovering clusters in large spatial databases with noise. In: KDD, vol. 96, pp. 226–231 (1996)
8. Fong, R., Vedaldi, A.: Interpretable explanations of black boxes by meaningful perturbation. CoRR, abs/1704.03296 (2017)
9. Gal, Y., et al.: A theoretically grounded application of dropout in recurrent neural networks. In: Advances in Neural Information Processing Systems, pp. 1019–1027 (2016)
10. Guidotti, R., Monreale, A., Nanni, M., Giannotti, F., Pedreschi, D.: Clustering individual transactional data for masses of users. In: KDD, pp. 195–204. ACM (2017)

11. Guidotti, R., Monreale, A., Ruggieri, S., Pedreschi, D., Turini, F., Giannotti, F.: Local rule-based explanations of black box decision systems. arXiv preprint arXiv:1805.10820 (2018)
12. Guidotti, R., Monreale, A., Turini, F., Pedreschi, D., Giannotti, F.: A survey of methods for explaining black box models. CoRR, abs/1802.01933 (2018)
13. Guidotti, R., Soldani, J., Neri, D., Brogi, A., Pedreschi, D.: Helping your docker images to spread based on explainable models. In: Brefeld, U., et al. (eds.) ECML PKDD 2018. LNCS (LNAI), vol. 11053, pp. 205–221. Springer, Cham (2019). https://doi.org/10.1007/978-3-030-10997-4_13
14. Jebara, T.: Images as bags of pixels. In: ICCV, Washington, D.C., USA, pp. 265–272. IEEE (2003)
15. Kim, B., et al. :The Bayesian case model: a generative approach for case-based reasoning and prototype classification. In: Advances in NIPS, pp. 1952–1960 (2014)
16. Lowe, D.G.: Object recognition from local scale-invariant features. In: Computer Vision, vol. 2, pp. 1150–1157. IEEE (1999)
17. Lu, Z., Wang, L., Wen, J.: Image classification by visual bag-of-words refinement and reduction. CoRR, abs/1501.04292:197–206 (2015)
18. Malgieri, G., Comandé, G.: Why a right to legibility of automated decision-making exists in the general data protection regulation. Int. Data Priv. Law 7(4), 243–265 (2017)
19. Ng, A.Y., Jordan, M.I., Weiss, Y.: On spectral clustering: analysis and an algorithm. In: Advances in Neural Information Processing Systems, pp. 849–856 (2002)
20. Pedreschi, D., Giannotti, F., Guidotti, R., et al.: Open the black box data-driven explanation of black box decision systems. arXiv preprint arXiv:1806.09936 (2018)
21. Ribeiro, M.T., Singh, S., Guestrin, C.: Why should i trust you?: explaining the predictions of any classifier. In: KDD, pp. 1135–1144. ACM (2016)
22. Russakovsky, O., et al.: ImageNet large scale visual recognition challenge. Int. J. Comput. Vis. (IJCV) 115(3), 211–252 (2015)
23. Selvaraju, R.R., et al.: Grad-CAM: why did you say that? Visual explanations from deep networks via gradient-based localization. CoRR, abs/1610.02391 (2016)
24. Szegedy, C., et al.: Rethinking the inception architecture for computer vision. In: Proceedings of the IEEE Conference on Computer Vision and Pattern Recognition, pp. 2818–2826 (2016)
25. Tan, P.-N., et al.: Introduction to Data Mining. Pearson Education, New Delhi (2007)
26. Vedaldi, A., Soatto, S.: Quick shift and kernel methods for mode seeking. In: Forsyth, D., Torr, P., Zisserman, A. (eds.) ECCV 2008. LNCS, vol. 5305, pp. 705–718. Springer, Heidelberg (2008). https://doi.org/10.1007/978-3-540-88693-8_52
27. Wachter, S., et al.: Why a right to explanation of automated decision-making does not exist in the general data protection regulation. Int. Data Priv. Law 7(2), 76–99 (2017)
28. Yang, J., et al.: Evaluating bag-of-visual-words representations in scene classification. In: International Workshop on Multimedia Information Retrieval, pp. 197–206. ACM (2007)
29. Zhang, J., et al.: Top-down neural attention by excitation backprop. CoRR, 1608.00507 (2016)
30. Zhang, Q., Wu, Y.N., Zhu, S.-C.: Interpretable convolutional neural networks. arXiv preprint arXiv:1710.00935 (2017). 2(3), 5
31. Zhou, B., Khosla, A., Lapedriza, A., Oliva, A., Torralba, A.: Object detectors emerge in deep scene CNNs. arXiv preprint arXiv:1412.6856 (2014)
32. Zhou, B., Khosla, A., Lapedriza, À., Oliva, A., Torralba, A.: Learning deep features for discriminative localization. CoRR, abs/1512.04150:2921–2929 (2015)

On Calibration of Nested Dichotomies

Tim Leathart$^{(\boxtimes)}$, Eibe Frank, Bernhard Pfahringer, and Geoffrey Holmes

Department of Computer Science, University of Waikato, Hamilton, New Zealand
tml15@students.waikato.ac.nz, {eibe,bernhard,geoff}@waikato.ac.nz

Abstract. Nested dichotomies (NDs) are used as a method of transforming a multiclass classification problem into a series of binary problems. A tree structure is induced that recursively splits the set of classes into subsets, and a binary classification model learns to discriminate between the two subsets of classes at each node. In this paper, we demonstrate that these NDs typically exhibit poor probability calibration, even when the binary base models are well-calibrated. We also show that this problem is exacerbated when the binary models are poorly calibrated. We discuss the effectiveness of different calibration strategies and show that accuracy and log-loss can be significantly improved by calibrating both the internal base models and the full ND structure, especially when the number of classes is high.

1 Introduction

As the amount of data collected online continues to grow, modern datasets utilised in machine learning are increasing in size. Not only do these datasets exhibit a large number of examples and features, but many also have a very high number of classes. It is not uncommon in some application areas to see datasets containing tens of thousands or even millions of classes [2,10].

An attractive option to handle datasets with such large label spaces is to induce a binary tree structure over the label space. At each split node k, the set of classes present, $\mathcal{C}_k$, is split into two disjoint subsets $\mathcal{C}_{k1}$ and $\mathcal{C}_{k2}$. Then, a binary classification model is trained to distinguish between these two subsets of classes. Many algorithms have been proposed that fit this general description, for example [3,5,8,9]. Often, a greedy inference approach is taken in these tree structures, i.e., test examples only take a single path from the root node to leaf nodes. This has the inherent drawback that a single mistake along the path to a leaf node results in an incorrect prediction [5,11].

In this paper, we consider methods with probabilistic classifiers at the internal nodes, called *nested dichotomies* (NDs) in the literature [15]. Utilising probabilistic binary classifiers to make routing decisions for test examples has several advantages over simply taking a hard 0/1 classification. For example, multiclass class probability estimates can be computed in a natural way by taking the product of binary probability estimates on the path from the root to the leaf node [14]. However, although hard classification decisions are avoided, small errors in the binary probability estimates can accumulate over this product,

Q. Yang et al. (Eds.): PAKDD 2019, LNAI 11439, pp. 69–80, 2019.
https://doi.org/10.1007/978-3-030-16148-4_6

resulting in inaccurate predictions. Datasets with more classes result in deeper trees, exacerbating this issue.

In this paper, we investigate approaches to reduce the impact of the accumulation of errors by utilising *probability calibration* techniques. Probability calibration is the task of transforming the probabilities output by a model to reflect their true empirical distribution; for the group of test examples that are predicted to belong to some class with probability 0.8, we expect about 80% of them to actually belong to that class if our model is well-calibrated. Our main hypothesis is that the overall predictive performance of NDs can be improved by calibrating the individual binary models at internal nodes (which we refer to as *internal calibration*). However, we also observe that significant performance gains can be achieved by calibrating the predictions made from the entire ND (referred to as *external calibration*), even if the internal models are well-calibrated.

This paper is structured as follows. First, we briefly review NDs and probability calibration. We then discuss internal and external calibration, providing theoretical motivation and showing experimental results for each method. Finally, we conclude and discuss future research directions.

2 Nested Dichotomies

NDs are used as a binary decomposition method for multiclass problems [15]. In this paper, we only consider the case where a single ND structure is built, although generally superior performance can be achieved by training an ensemble of NDs with different structures. Ensembles of NDs have been shown to outperform binary decomposition methods like one-vs-all [33], one-vs-one [16] and error-correcting output codes [12], on some classification problems [15].

The structure of an ND can have a large impact on the predictive performance, training time and prediction time. To this end, several methods have been proposed for deciding the structure of NDs [13,22,23,26,35]. In this paper, we focus on a simple method that randomly splits the class set into two at each internal node.

As previously stated, a useful feature of NDs is the ability to produce multiclass probability estimates $\hat{\mathbf{p}}_i$ for a test instance $(\mathbf{x}_i, y_i)$ from the product of binary estimates on the path $\mathcal{P}_c$ to the leaf node corresponding to class c:

$$\hat{\mathbf{p}}_i^{(c)} = p(y_i = c|\mathbf{x}_i)$$
$$= \prod_{k \in \mathcal{P}_c} \left(\mathbb{I}(c \in \mathcal{C}_{k1})p(c \in \mathcal{C}_{k1}|\mathbf{x}_i, y_i \in \mathcal{C}_k) + \mathbb{I}(c \in \mathcal{C}_{k2})p(c \in \mathcal{C}_{k2}|\mathbf{x}_i, y_i \in \mathcal{C}_k) \right)$$

where $\mathbb{I}(\cdot)$ is the indicator function, $\mathcal{C}_k$ is the set of classes present at node k and $\mathcal{C}_{k1}, \mathcal{C}_{k2} \subset \mathcal{C}_k$ are the sets of classes present at the left and right child of node k, respectively. If desired, one can still perform greedy inference by taking the most promising branch at each split point, but this is not guaranteed to find the best solution [5]. Having binary class probability estimates facilitates efficient tree search techniques [11,20,27] for better inference, as well as top-k prediction.

3   Probability Calibration

Probability calibration is the task of transforming the outputs of a classifier to accurate probabilities. It is useful in a range of settings, such as cost-sensitive classification and scenarios where the outputs of a model are used as inputs for another. It is also important in real world decision making systems to know when a prediction from a model is likely to be incorrect.

Some models, like logistic regression, tend to be well-calibrated out-of-the-box, while other models like naïve Bayes and boosted decision trees usually exhibit poor calibration, despite high classification accuracy [30]. Some other models such as support vector machines cannot output probabilities at all, but calibration can be applied to produce a probability estimate. Calibration is typically applied as a post-processing step—a calibration model is trained to transform the output score from a model into a well-calibrated probability.

3.1   Calibration Methods

The most commonly used calibration methods in practice are Platt scaling (PS) [32] and isotonic regression (IR) [36]. Both of these methods are only directly applicable to binary problems, but standard multiclass transformation techniques can be used to apply them to multiclass problems [37].

Platt Scaling is a technique that fits a sigmoid curve

$$\sigma(z_i) = \frac{1}{1 + e^{\alpha z_i + \beta}}$$

from the output of a binary classifier z_i to the true labels. The parameters α and β are fitted using logistic regression. PS was originally proposed for scaling the output of SVMs, but has been shown to be an effective calibration technique for a range of models [30]. Usually, PS is applied to the log-odds (sometimes called logits) of the positive class, rather than the probability.

Matrix and Vector Scaling are simple extensions of PS for multiclass problems [17]. In matrix scaling (MS), instead of single parameters α and β, a matrix $\mathbf{W}$ and bias vector $\mathbf{b}$ are learned:

$$\sigma(\mathbf{z}_i) = \frac{1}{1 + e^{\mathbf{W}\mathbf{z}_i + \mathbf{b}}}$$

where $\mathbf{z}_i$ is the vector of the log-odds of each class for instance i. MS is equivalent to a standard multiple logistic regression model applied to the log-odds. It is expensive for datasets with many classes, as the weight matrix $\mathbf{W}$ grows quadratically with the number of classes. Vector scaling (VS) is designed to overcome this. It is a variant where $\mathbf{W}$ is restricted to be a diagonal matrix to achieve scaling that is linear in the number of classes.

Isotonic Regression is a non-parametric technique for probability calibration [36]. It fits a piecewise constant function to minimise the mean squared error between the estimated class probabilities and the true labels. IR is a more general method than PS because no assumptions are made about the function used to map classifier outputs, other than that the function is non-decreasing (isotonicity). IR has been found to work well as a calibration model, but the flexibility of the fitted function means it can overfit on small samples.

Other Related Work. In this paper, efficiency is a concern as there are many models to be calibrated. For this reason, we opt for the simple calibration methods mentioned above in our experiments. However, there are several more expressive (and expensive) calibration methods in the literature. Zhong and Kwok [38] and Jiang et al. [19] propose methods for creating a smooth spline from the piecewise constant function produced in IR. Naeini et al. [29] propose a method for performing Bayesian averaging over all possible binning schemes—schemes that split the probability space into several bins and establish a calibrated probability value per bin. Leathart et al. [21] split the feature space into regions using a decision tree and build a localised calibration model in each region.

3.2   Measuring Miscalibration

The level of probability calibration that a model exhibits is frequently measured by the negative log-likelihood (NLL):

$$\text{NLL} = -\frac{1}{n} \sum_{i=1}^{n} \mathbf{y}_i \log \hat{\mathbf{p}}_i$$

where n is the number of examples, $\mathbf{y}_i$ is the one-hot true label for an instance i and $\hat{\mathbf{p}}_i$ is the estimated probability distribution. NLL heavily penalises probability estimates that are far from the true label. For this reason, models which optimise NLL in training tend to be well-calibrated, although interestingly it has been shown recently that the kinds of architectures used in modern neural networks can also produce poorly calibrated models [17]. NLL is also commonly used as a general measure of model accuracy.

Probability calibration for classification tasks can be visualised through reliability diagrams [28]. In reliability diagrams, the probability range $[0, 1]$ is discretised into K bins $B_1, \ldots, B_k$. These bins are chosen such that they have equal width, or equal numbers of examples. The *confidence* of each bin is given as the average estimated probability of examples that fall inside the bin, while the *accuracy* of each bin is the empirical accuracy:

$$\text{conf}(B_k) = \frac{1}{|B_k|} \sum_{i \in B_k} \hat{p}_i, \qquad \text{acc}(B_k) = \frac{1}{|B_k|} \sum_{i \in B_k} \mathbb{I}(\hat{y}_i = y_i)$$

where y_i is the true binary label, $\hat{y}_i$ is the predicted label, and $\hat{p}_i$ is the estimated probability for an instance i [17]. Intuitively, a well-calibrated classifier should have comparable confidence and accuracy for each bin. The confidence and accuracy are plotted against each other for each bin, producing a straight diagonal line for a well-calibrated classifier.

Naeini et al. applied the same idea to give a direct quantitative measure of calibration [29], called the *expected calibration error* (ECE). This is simply a weighted average of the residuals in a reliability diagram, weighted by the number of instances that fall inside each bin.

4   Internal Calibration

In this section, we investigate the effect of calibrating the internal models of NDs. Our hypothesis is that by improving the quality of the binary probability estimates, the final multiclass predictive performance will be improved. This is due to the fact that multiclass probability estimates are produced by computing the product of a series of binary probability estimates. If the binary probability estimates are not well-calibrated, then these errors will accumulate throughout the calculation.

4.1   Theoretical Motivation

It seems reasonable that improving the calibration of internal models will result in superior probability estimates for the ND, but can we theoretically quantify this improvement? It turns out that reducing the binary NLL of any internal model by some amount δ strictly reduces the multiclass NLL of the ND, and depending on the depth of the internal model being calibrated, the reduction in multiclass NLL can be as high as δ.

Proposition 1. *The NLL of an instance under an ND is equal to the sum of NLLs of the instance under the binary models on the path from the root node to the leaf node.*

Proof. The NLL of an instance i is given by

$$\text{NLL} = -\mathbf{y}_i \log \hat{\mathbf{p}}_i = -\log \hat{\mathbf{p}}_i^{(c)} \tag{1}$$

where $\hat{\mathbf{p}}_i^{(c)}$ is the probability estimate for the true class c. Let $\mathcal{P}_c$ be the set of internal nodes on the path from the root to the leaf corresponding to class c. Then, $\hat{\mathbf{p}}_i^{(c)}$ can be expressed as

$$\hat{\mathbf{p}}_i^{(c)} = \prod_{k \in \mathcal{P}_c} \tilde{y}_{ik} \hat{p}_{ik} + (1 - \tilde{y}_{ik})(1 - \hat{p}_{ik}) \tag{2}$$

where $\tilde{y}_{ik} \in \{0, 1\}$ is the binary meta-label and $\hat{p}_{ik} \in [0, 1]$ is the estimated probability of the positive meta-label for instance i for the binary model at node k respectively. Because $\tilde{y}_{ik} \in \{0, 1\}$, it is equivalent to write

$$\hat{\mathbf{p}}_i^{(c)} = \prod_{k \in \mathcal{P}_c} \hat{p}_{ik}^{\tilde{y}_{ik}} (1 - \hat{p}_{ik})^{(1-\tilde{y}_{ik})}. \tag{3}$$

Plugging this into (1) yields

$$\text{NLL} = -\log \prod_{k \in \mathcal{P}_c} \hat{p}_{ik}^{\tilde{y}_{ik}} (1 - \hat{p}_{ik})^{(1-\tilde{y}_{ik})} \tag{4}$$

$$= -\sum_{k \in \mathcal{P}_c} \log \left(\hat{p}_{ik}^{\tilde{y}_{ik}} (1 - \hat{p}_{ik})^{(1-\tilde{y}_{ik})} \right) \tag{5}$$

$$= -\sum_{k \in \mathcal{P}_c} \tilde{y}_{ik} \log \hat{p}_{ik} + (1 - \tilde{y}_{ik}) \log(1 - \hat{p}_{ik}), \tag{6}$$

the sum of NLLs from the models $k \in \mathcal{P}_c$.   □

It directly follows that reducing the binary NLL for the model at internal node k by some amount δ results in a reduction of the multiclass NLL by δ for each class corresponding to the leaf nodes that are descendants of node k. This means that a calibration resulting in a binary NLL reduction of δ for some internal node k reduces the multiclass NLL by $\delta(n_k/n)$, where n_k is the number of examples in classes whose corresponding leaf nodes are descendants of k.

5 External Calibration

As well as calibrating each internal model, we also consider external calibration of the entire ND. Even models like logistic regression are usually not perfectly calibrated in practice. We hypothesise that these minor miscalibrations accumulate as the ND gets deeper, which can be rectified by an external calibration model. More specifically, the accumulated miscalibration is likely to be realised as *under-confident* predictions. This is because the final multiclass probability estimates are established by computing a product of probability estimates along the path from the root to the leaf. For example, consider an ND of depth six for some multiclass problem. If each binary model on the path is highly confident with a probability estimate of 0.9, the correct class will be assigned a relatively low probability estimate of $0.9^6 = 0.531$. Naturally, this effect is greater for problems with more classes, as the paths to leaf nodes will be longer.

As an illustrative investigation into the effect of ND depth on their calibration, we built an ND with logistic regression base learners for the ALOI dataset (see Table 1). Figure 1 shows reliability plots for versions of this ND that have been "cut-off" at incrementally increasing depths. A test example is considered to be classified correctly at depth d if its actual class is in the subset of classes $\mathcal{C}_k$ of the node k with highest probability and maximum depth d. Limited to a depth of one, the ND is simply a single binary logistic regression model, which exhibits good calibration. However, as the depth cut-off limit increases, it is clear that the ND becomes increasingly under-confident, i.e., bins that have high accuracy often have low confidence (Fig. 1, top row). This corresponds to the curve sitting above the diagonal line. The ECE increases linearly with the depth of the tree.

This is adequately and efficiently compensated for by applying VS (Fig. 1, bottom row). VS exhibits low complexity in the number of classes—only two

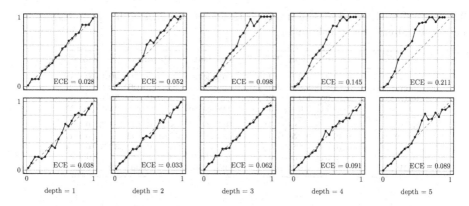

Fig. 1. Reliability plots for an ND with logistic regression base learners, cut off at increasing depth. Top: no external calibration. As the depth increases, the ND becomes increasingly under-confident because of the effect of multiplying probabilities together. Bottom: externally calibrated using vector scaling.

Table 1. Datasets used in our experiments.

Name	Instances	Features	Classes		Name	Instances	Features	Classes
optdigits[1]	5,620	64	10		RCV1[2]	15,564/518,571	47,236	53
micromass[1]	571	1,301	20		sector[2]	6,412/3,207	55,197	105
letter[1]	20,000	16	26		ALOI[2]	97,200/10,800	128	1,000
devanagari[1]	92,000	1,000	46		ILSVR2010[3]	1,111,406/150,000	1,000	1,000
					ODP-5K[4]	361,488/180,744	422,712	5,000

[1] UCI Repository [24] , [2] LIBSVM Repository [7], [3] ImageNet [34], [4] ODP [4]

parameters per class—making it suitable for problems with many classes typically handled by NDs. For externally calibrated NDs, the ECE initially increases linearly with the depth of the tree (although for $d > 1$, the ECE values are much lower than their uncalibrated counterparts). However, at $d = 5$, the ECE levels off and even begins to decrease slightly.

6   Experiments

In this section, we present experimental results of calibration of NDs using different base classifiers on a series of datasets. The datasets we used in our experiments are listed in Table 1, and were chosen to span a range of numbers of classes. Optdigits, letter and devanagari [1] are character recognition datasets for digits, latin letters, and Devanagari script respectively. Micromass [25] is for the identification of microorganisms from mass spectroscopy data. RCV1, sector and ODP-5K are text categorisation tasks, while ALOI and ILSVR2010 are object recognition tasks. We use the visual codewords representation for ILSVR2010.

In order to obtain performance estimates, we performed 10 times 10-fold cross-validation for the datasets from the UCI repository, while adopting the

Table 2. NLL (left) and classification accuracy (right) of NDs with logistic regression, before and after external calibration.

Dataset	Baseline	Ext. VS	Dataset	Baseline	Ext. VS
optdigits	0.30 (0.07)	**0.30 (0.08)**	optdigits	90.5 (0.02)	**90.6 (0.03)**
micromass	5.92 (1.83)	**1.88 (0.51)**	micromass	**80.4 (0.06)**	77.2 (0.05)
letter	1.50 (0.06)	**1.44 (0.08)**	letter	51.2 (0.03)	**53.6 (0.03)**
devanagari	2.43 (0.11)	**2.03 (0.05)**	devanagari	42.8 (0.02)	**42.8 (0.02)**
RCV1	1.00 (0.02)	**0.58 (0.01)**	RCV1	81.4 (0.01)	**85.5 (0.00)**
sector	2.86 (0.01)	**1.25 (0.02)**	sector	84.8 (0.01)	**86.7 (0.00)**
ALOI	3.60 (0.02)	**3.05 (0.03)**	ALOI	27.4 (0.01)	**33.1 (0.01)**
ILSVR2010	6.44 (0.01)	**5.78 (0.01)**	ILSVR2010	**6.3 (0.00)**	5.3 (0.00)
ODP-5K	5.80 (0.01)	**4.97 (0.01)**	ODP-5K	18.9 (0.00)	**22.8 (0.00)**

standard train/test splits for the larger datasets with a larger number of classes ($m > 50$). The number of instances stated in Table 1 for the larger datasets is split into number of training and test instances. Note that in each fold and run of 10 times 10-fold cross-validation, a different random ND structure is constructed. In the case of the larger datasets, the average of 10 randomly constructed NDs is reported. Standard deviations are given in parentheses, and the best result per row appears in bold face. The original ODP dataset contains 105,000 classes— we took the subset of the most frequent 5,000 classes to create ODP-5K for the purposes of this investigation. We also reduce the dimensionality to 1,000 when evaluating NDs with boosted trees, by using a Gaussian random projection [6].

We implemented VS [17] and NDs in Python, and used the implementations of the base learners, IR and PS available in `scikit-learn` [31]. Our implementations are available online at https://github.com/timleathart/pynd.

6.1   Well-Calibrated Base Learners

As shown in Fig. 1, overall calibration of NDs can degrade to systematically under-confident predictions as the depth of the tree increases, even if the base learners are well-calibrated. To further investigate the effects of ND depth on predictions, we performed experiments with external calibration to determine the extent to which the classification accuracy and NLL are affected as well.

Table 2 shows the NLL and accuracy of NDs with logistic regression, before and after external calibration is applied. Logistic regression models are known to be well-calibrated [30]. VS [17] is used as the external calibration model. We use a 90% sample of the training data to build the ND including the base models, and the remaining 10% to train the external calibration model.

Discussion. Table 2 shows that, for all datasets, a reduction in NLL is observed after applying external calibration with VS (Ext. VS). For some of the datasets with fewer classes (optdigits and letter), the reduction is modest, but the larger datasets see substantial improvements. Interestingly, for some datasets a large improvement in classification accuracy is also observed, especially for the datasets with more classes. A surprising finding is that the accuracy degrades for ILSVR2010 and micromass, despite a large improvement in NLL.

6.2 Poorly Calibrated Base Learners

Tables 3 and 4 show the NLL and classification accuracy respectively of NDs trained with poorly calibrated base learners, when different calibration strategies are applied. Specifically, we considered NDs with naïve Bayes and boosted decision trees as the base learners. The calibration schemes compared are internal PS (Int. PS), internal IR (Int. IR) and external VS (Ext. VS), as well as each internal calibration scheme in conjunction with external VS (Both PS and Both IR, respectively). Three-fold cross validation is used to produce the training data for the internal calibration models, rather than splitting the training data. This is to ensure that each internal calibration model has a reasonable amount of data points to train on. When external calibration is performed, 10% of the data is held out to train the external calibration model. Note that this means 10% less data is available to train the ND and (if applicable) perform internal calibration. Gaussian naïve Bayes is applied for optdigits, micromass, letter and devanagari, and multinomial naïve Bayes is used for RCV1, sector, ALOI, ILSVR2010 and ODP-5K as they have sparse features. We use 50 decision trees with AdaBoost [18], limiting the depth of the trees to three.

Discussion. Tables 3 and 4 show that applying internal calibration is very beneficial in terms of both NLL and classification accuracy. There is no combination of base learner and dataset for which the baseline gives the best results, and there are very few cases where the baseline does not perform the worst out of every scheme. When naïve Bayes is used as the base learner, applying internal calibration with IR always gives better results than the baseline, and when an ensemble of boosted trees is used as the base learner, applying internal PS always outperforms the uncalibrated case. It is well known that these calibration methods are well-suited to the respective base learners [30], and this appears to also apply when they are used in an ND.

External calibration also has a positive effect on both NLL and classification accuracy in most cases compared to the baseline. However, the best results are usually obtained when both internal and external calibration are applied together. For naïve Bayes, the smaller datasets as well as the two object recognition datasets (ALOI and ILSVR2010) generally see the best performance for both NLL and classification accuracy when applying internal IR in conjunction with external calibration. Interestingly, the best results for the three text categorisation datasets were obtained through external calibration only.

Performing both internal PS and external calibration gives the best NLL performance for NDs with boosted trees in most cases, although the improvement compared to IR is usually small. However, performance in terms of classification accuracy is less consistent, sometimes being greater when only calibrating internally.

Table 3. NLL of NDs with poorly calibrated base classifiers.

Model	Dataset	Baseline	Ext. VS	Int. PS	Both PS	Int. IR	Both IR
Naïve Bayes	optdigits	4.25 (0.92)	0.84 (0.13)	0.85 (0.11)	0.80 (0.09)	0.71 (0.12)	**0.64 (0.09)**
	micromass	8.66 (1.72)	1.62 (0.42)	0.92 (0.08)	0.75 (0.12)	0.76 (0.12)	**0.71 (0.15)**
	letter	2.33 (0.08)	2.15 (0.08)	2.16 (0.06)	2.06 (0.07)	2.05 (0.07)	**1.95 (0.06)**
	devanagari	13.14 (0.59)	3.31 (0.16)	2.98 (0.05)	2.60 (0.02)	2.75 (0.07)	**2.44 (0.05)**
	RCV1	1.69 (0.19)	**0.86 (0.01)**	1.14 (0.07)	0.94 (0.03)	0.99 (0.04)	0.91 (0.02)
	sector	3.79 (0.36)	**1.40 (0.09)**	2.07 (0.20)	1.51 (0.10)	1.91 (0.21)	1.77 (0.09)
	ALOI	32.9 (0.43)	6.84 (0.02)	5.53 (0.02)	4.33 (0.03)	4.85 (0.03)	**4.13 (0.01)**
	ILSVR2010	32.3 (0.20)	6.81 (0.00)	6.16 (0.00)	6.12 (0.01)	6.17 (0.00)	**6.11 (0.00)**
	ODP-5K	8.49 (0.31)	**5.16 (0.05)**	6.10 (0.00)	5.51 (0.02)	6.05 (0.11)	5.36 (0.01)
Boosted Trees	optdigits	3.86 (0.57)	0.63 (0.07)	0.40 (0.04)	**0.29 (0.04)**	0.39 (0.03)	0.30 (0.04)
	micromass	10.01 (2.05)	2.51 (0.52)	1.26 (0.11)	1.00 (0.14)	1.23 (0.27)	**0.95 (0.19)**
	letter	4.86 (0.27)	0.92 (0.04)	0.56 (0.02)	**0.44 (0.03)**	0.55 (0.02)	0.44 (0.03)
	devanagari	3.42 (0.28)	1.03 (0.04)	2.26 (0.17)	**0.71 (0.02)**	1.97 (0.12)	0.73 (0.02)
	RCV1	1.96 (0.02)	1.02 (0.00)	0.93 (0.01)	**0.71 (0.01)**	0.86 (0.00)	0.74 (0.01)
	sector	3.63 (0.20)	2.91 (0.11)	2.67 (0.03)	**2.03 (0.03)**	2.59 (0.05)	2.20 (0.07)
	ALOI	4.44 (0.26)	2.51 (0.05)	4.88 (0.03)	**1.05 (0.02)**	4.28 (0.04)	1.17 (0.03)
	ILSVR2010	6.55 (0.10)	5.86 (0.00)	5.64 (0.00)	5.21 (0.00)	5.45 (0.00)	**5.20 (0.00)**
	ODP-5K	7.73 (0.04)	7.19 (0.00)	7.12 (0.00)	6.60 (0.00)	6.98 (0.00)	**6.57 (0.00)**

Table 4. Accuracy of NDs with poorly calibrated base classifiers.

Model	Dataset	Baseline	Ext. VS	Int. PS	Both PS	Int. IR	Both IR
Naïve Bayes	optdigits	71.9 (0.05)	74.9 (0.04)	71.9 (0.05)	73.5 (0.04)	77.4 (0.04)	**79.5 (0.04)**
	micromass	74.9 (0.05)	72.4 (0.05)	77.0 (0.05)	76.2 (0.05)	**77.2 (0.05)**	75.6 (0.05)
	letter	32.9 (0.02)	36.4 (0.03)	31.8 (0.03)	36.5 (0.03)	37.6 (0.03)	**41.2 (0.03)**
	devanagari	20.2 (0.02)	22.4 (0.04)	16.7 (0.02)	26.9 (0.01)	26.5 (0.04)	**34.0 (0.01)**
	RCV1	64.4 (0.04)	**78.1 (0.00)**	69.1 (0.03)	75.6 (0.00)	73.4 (0.01)	76.5 (0.01)
	sector	33.7 (0.07)	**77.2 (0.01)**	63.3 (0.04)	73.7 (0.03)	69.2 (0.04)	69.0 (0.01)
	ALOI	2.4 (0.00)	2.9 (0.00)	1.9 (0.00)	12.4 (0.00)	9.4 (0.00)	**16.6 (0.01)**
	ILSVR2010	0.9 (0.00)	1.5 (0.00)	1.4 (0.00)	1.9 (0.00)	2.1 (0.00)	**2.6 (0.00)**
	ODP-5K	4.3 (0.01)	**21.0 (0.00)**	9.1 (0.00)	16.1 (0.00)	13.5 (0.01)	18.0 (0.00)
Boosted Trees	optdigits	88.8 (0.02)	88.3 (0.02)	**92.2 (0.01)**	91.7 (0.01)	92.1 (0.01)	91.5 (0.01)
	micromass	71.0 (0.06)	65.0 (0.06)	**74.1 (0.06)**	72.9 (0.06)	73.7 (0.06)	72.7 (0.05)
	letter	85.9 (0.01)	85.1 (0.01)	88.8 (0.01)	88.3 (0.01)	**89.0 (0.01)**	88.4 (0.01)
	devanagari	10.3 (0.06)	71.0 (0.01)	48.8 (0.11)	**79.3 (0.01)**	63.6 (0.04)	78.3 (0.01)
	RCV1	68.1 (0.06)	74.8 (0.01)	81.0 (0.00)	**81.4 (0.01)**	81.0 (0.00)	80.7 (0.00)
	sector	17.4 (0.08)	40.9 (0.02)	**60.3 (0.01)**	57.6 (0.01)	57.8 (0.01)	55.2 (0.01)
	ALOI	7.1 (0.03)	45.1 (0.04)	16.1 (0.04)	**74.3 (0.01)**	36.8 (0.01)	72.3 (0.00)
	ILSVR2010	1.9 (0.00)	4.0 (0.00)	9.3 (0.00)	**10.2 (0.00)**	9.7 (0.00)	10.0 (0.00)
	ODP-5K	3.2 (0.00)	3.8 (0.00)	5.5 (0.00)	6.8 (0.00)	6.2 (0.00)	**7.2 (0.00)**

7   Conclusion

In this paper, we show that the predictive performance of NDs can be substantially improved by applying calibration techniques. Calibrating the internal models increases the likelihood that the path to the leaf node corresponding to the true class is assigned high probability, while external calibration can correct for the systematic under-confidence exhibited by NDs. Both of these techniques have been empirically shown to provide large performance gains in terms of accuracy and NLL for a range of datasets when applied individually. Additionally, when both internal and external calibration are applied together, the performance often improves further, especially so when the number of classes is high.

Future work in this domain includes evaluating alternative external calibration methods. In our experiments, we applied VS as it is an efficient and scalable solution for large multiclass tasks. However, when resources are available, it is possible that employing a more complex method such as matrix scaling, or IR with one-vs-rest, could provide superior results. It would also be interesting to investigate whether such calibration measures are as effective for other methods of constructing NDs than random subset selection [13,22,23,26,35]. We expect that the calibration techniques discussed in this paper will transfer to such methods.

References

1. Acharya, S., Pant, A.K., Gyawali, P.K.: Deep learning based large scale handwritten Devanagari character recognition. In: SKIMA, pp. 1–6. IEEE (2015)
2. Agrawal, R., Gupta, A., Prabhu, Y., Varma, M.: Multi-label learning with millions of labels: recommending advertiser bid phrases for web pages. In: WWW, pp. 13–24 (2013)
3. Bengio, S., Weston, J., Grangier, D.: Label embedding trees for large multi-class tasks. In: NIPS, pp. 163–171 (2010)
4. Bennett, P.N., Nguyen, N.: Refined experts: improving classification in large taxonomies. In: SIGIR, pp. 11–18. ACM (2009)
5. Beygelzimer, A., Langford, J., Ravikumar, P.: Error-correcting tournaments. In: Gavaldà, R., Lugosi, G., Zeugmann, T., Zilles, S. (eds.) ALT 2009. LNCS (LNAI), vol. 5809, pp. 247–262. Springer, Heidelberg (2009). https://doi.org/10.1007/978-3-642-04414-4_22
6. Bingham, E., Mannila, H.: Random projection in dimensionality reduction: applications to image and text data. In: KDD, pp. 245–250. ACM (2001)
7. Chang, C.C., Lin, C.J.: LIBSVM: a library for support vector machines. ACM Trans. Intell. Syst. Technol. **2**(3), 27 (2011)
8. Choromanska, A.E., Langford, J.: Logarithmic time online multiclass prediction. In: NIPS, pp. 55–63 (2015)
9. Daumé, III, H., Karampatziakis, N., Langford, J., Mineiro, P.: Logarithmic time one-against-some. In: ICML, pp. 923–932. PMLR (2017)
10. Dekel, O., Shamir, O.: Multiclass-multilabel classification with more classes than examples. In: AISTATS, pp. 137–144. PMLR (2010)
11. Dembczyński, K., Kotłowski, W., Waegeman, W., Busa-Fekete, R., Hüllermeier, E.: Consistency of probabilistic classifier trees. In: Frasconi, P., Landwehr, N., Manco, G., Vreeken, J. (eds.) ECML PKDD 2016. LNCS (LNAI), vol. 9852, pp. 511–526. Springer, Cham (2016). https://doi.org/10.1007/978-3-319-46227-1_32
12. Dietterich, T.G., Bakiri, G.: Solving multiclass learning problems via error-correcting output codes. JAIR **2**, 263–286 (1995)
13. Dong, L., Frank, E., Kramer, S.: Ensembles of balanced nested dichotomies for multi-class problems. In: Jorge, A.M., Torgo, L., Brazdil, P., Camacho, R., Gama, J. (eds.) PKDD 2005. LNCS (LNAI), vol. 3721, pp. 84–95. Springer, Heidelberg (2005). https://doi.org/10.1007/11564126_13
14. Fox, J.: Applied Regression Analysis, Linear Models, and Related Methods. Sage, Thousand Oaks (1997)

15. Frank, E., Kramer, S.: Ensembles of nested dichotomies for multi-class problems. In: ICML, pp. 39–46. ACM (2004)
16. Friedman, J.H.: Another approach to polychotomous classification. Technical report, Statistics Department, Stanford University (1996)
17. Guo, C., Pleiss, G., Sun, Y., Weinberger, K.Q.: On calibration of modern neural networks. In: ICML, pp. 1321–1330. PMLR (2017)
18. Hastie, T., Rosset, S., Zhu, J., Zou, H.: Multi-class adaboost. Stat. Interface **2**(3), 349–360 (2009)
19. Jiang, X., Osl, M., Kim, J., Ohno-Machado, L.: Smooth isotonic regression: a new method to calibrate predictive models. In: AMIA Summits on Translational Science Proceedings, p. 16 (2011)
20. Kumar, A., Vembu, S., Menon, A.K., Elkan, C.: Beam search algorithms for multilabel learning. Mach. Learn. **92**(1), 65–89 (2013)
21. Leathart, T., Frank, E., Pfahringer, B., Holmes, G.: Probability calibration trees. In: ACML, pp. 145–160. PMLR (2017)
22. Leathart, T., Frank, E., Pfahringer, B., Holmes, G.: Ensembles of nested dichotomies with multiple subset evaluation. In: Yang, Q., et al. (eds.) PAKDD 2019. LNAI, vol. 11439, pp. xx-yy. Springer, Heidelberg (2019)
23. Leathart, T., Pfahringer, B., Frank, E.: Building ensembles of adaptive nested dichotomies with random-pair selection. In: Frasconi, P., Landwehr, N., Manco, G., Vreeken, J. (eds.) ECML PKDD 2016. LNCS (LNAI), vol. 9852, pp. 179–194. Springer, Cham (2016). https://doi.org/10.1007/978-3-319-46227-1_12
24. Lichman, M.: UCI machine learning repository (2013)
25. Mahé, P., et al.: Automatic identification of mixed bacterial species fingerprints in a MALDI-TOF mass-spectrum. Bioinformatics **30**(9), 1280–1286 (2014)
26. Melnikov, V., Hüllermeier, E.: On the effectiveness of heuristics for learning nested dichotomies: an empirical analysis. Mach. Learn. **107**(8–10), 1–24 (2018)
27. Mena, D., Montañés, E., Quevedo, J.R., Del Coz, J.J.: Using A* for inference in probabilistic classifier chains. In: IJCAI (2015)
28. Murphy, A.H., Winkler, R.L.: Reliability of subjective probability forecasts of precipitation and temperature. Appl. Stat. **26**, 41–47 (1977)
29. Naeini, M., Cooper, G., Hauskrecht, M.: Obtaining well calibrated probabilities using Bayesian binning. In: AAAI, pp. 2901–2907 (2015)
30. Niculescu-Mizil, A., Caruana, R.: Predicting good probabilities with supervised learning. In: ICML, pp. 625–632. ACM (2005)
31. Pedregosa, F., et al.: Scikit-learn: machine learning in python. JMLR **12**(Oct), 2825–2830 (2011)
32. Platt, J.: Probabilistic outputs for support vector machines and comparisons to regularized likelihood methods. Adv. Large Margin Classif. **10**(3), 61–74 (1999)
33. Rifkin, R., Klautau, A.: In defense of one-vs-all classification. JMLR **5**, 101–141 (2004)
34. Russakovsky, O., et al.: ImageNet large scale visual recognition challenge. IJCV **115**(3), 211–252 (2015)
35. Wever, M., Mohr, F., Hüllermeier, E.: Ensembles of evolved nested dichotomies for classification. In: GECCO, pp. 561–568. ACM (2018)
36. Zadrozny, B., Elkan, C.: Obtaining calibrated probability estimates from decision trees and naive Bayesian classifiers. In: ICML, pp. 609–616. ACM (2001)
37. Zadrozny, B., Elkan, C.: Transforming classifier scores into accurate multiclass probability estimates. In: KDD, pp. 694–699. ACM (2002)
38. Zhong, W., Kwok, J.T.: Accurate probability calibration for multiple classifiers. In: IJCAI, pp. 1939–1945 (2013)

Ensembles of Nested Dichotomies
with Multiple Subset Evaluation

Tim Leathart$^{(\boxtimes)}$, Eibe Frank, Bernhard Pfahringer, and Geoffrey Holmes

Department of Computer Science, University of Waikato, Hamilton, New Zealand
tml15@students.waikato.ac.nz, {eibe,bernhard,geoff}@waikato.ac.nz

Abstract. A system of nested dichotomies (NDs) is a method of decomposing a multiclass problem into a collection of binary problems. Such a system recursively applies binary splits to divide the set of classes into two subsets, and trains a binary classifier for each split. Many methods have been proposed to perform this split, each with various advantages and disadvantages. In this paper, we present a simple, general method for improving the predictive performance of NDs produced by any subset selection techniques that employ randomness to construct the subsets. We provide a theoretical expectation for performance improvements, as well as empirical results showing that our method improves the root mean squared error of NDs, regardless of whether they are employed as an individual model or in an ensemble setting.

1 Introduction

Multiclass classification problems are commonplace in real world applications. Some models, like neural networks and random forests, are inherently able to operate on multiclass data, while other models, such as classic support vector machines, can only be used for binary (two-class) problems. The standard way to bypass this limitation is to convert the multiclass problem into a series of binary problems. There exist several methods of performing this decomposition, the most well-known including one-vs-rest [26], one-vs-one [16] and error-correcting output codes [7]. Models that are directly capable of working with multiclass data may also see improved accuracy from such a decomposition [13,25].

The use of ensembles of nested dichotomies (NDs) is one such method for decomposing a multiclass problem into several binary problems. It has been shown to outperform one-vs-rest and perform competitively compared to the aforementioned methods [11]. In an ND [10], the set of classes is recursively split into two subsets in a tree structure. At each split node of the tree, a binary classifier is trained to discriminate between the two subsets of classes. Each leaf node of the tree corresponds to a particular class. To obtain probability estimates for a particular class from an ND, assuming the base learner can

Electronic supplementary material The online version of this chapter (https://doi.org/10.1007/978-3-030-16148-4_7) contains supplementary material, which is available to authorized users.

Q. Yang et al. (Eds.): PAKDD 2019, LNAI 11439, pp. 81–93, 2019.
https://doi.org/10.1007/978-3-030-16148-4_7

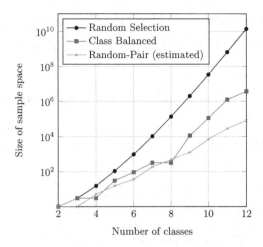

Fig. 1. Growth functions for each subset selection method discussed.

produce probability estimates, one can simply compute the product of the binary probability estimates along the path to the leaf node corresponding to the class.

For non-trivial multiclass problems, the space of potential NDs is very large. An ensemble classifier can be formed by choosing suitable decompositions from this space. In the original formulation of ensembles of NDs, decompositions are sampled with uniform probability [11], but several other more sophisticated methods for splitting the set of classes have been proposed [8,9,19]. Superior performance is achieved when ensembles of NDs are trained using common ensemble learning methods like bagging or boosting [27].

In this paper, we describe a simple method that can improve the predictive performance of NDs by considering several splits at each internal node. Our technique can be applied to NDs built with almost any subset selection method, only contributing a constant factor to the training time and no additional cost when obtaining predictions. It has a single hyperparameter λ that provides a trade-off between predictive performance and training time, making it easy to tune for a given learning problem. It is also straightforward to implement.

The paper is structured as follows. First, we describe existing methods for class subset selection in NDs. Following this, we describe our method and provide a theoretical expectation of performance improvements. We then present and discuss empirical results for our experiments. Finally, we touch on related work, before concluding and discussing future research directions.

2   Class Subset Selection Methods

At each internal node i of an ND, the set of classes present at the node $\mathcal{C}_i$ is split into two non-empty, non-overlapping subsets, $\mathcal{C}_{i1}$ and $\mathcal{C}_{i2}$. In this section, we introduce existing class subset selection methods for NDs. These techniques are designed to primarily be used in an ensemble setting, where multiple ND decompositions are generated that each form an ensemble member. Note that other methods than those listed here have been proposed for constructing NDs— these are not suitable for use with our method and are discussed later in Sect. 5.

2.1   Random Selection

The most basic class subset selection method is to split the set of classes into two subsets using a random split.[1] This approach has several attractive qualities. It is easy to compute, and does not scale with the amount of training data, making it suitable for large datasets. Furthermore, for an n-class problem, the number of possible NDs is very large, given by the recurrence relation

$$T(n) = (2n - 3) \times T(n - 1)$$

where $T(1) = 1$. This ensures that, in an ensemble of NDs, there is a high level of diversity amongst ensemble members. We refer to this function that relates the number of classes to the size of the sample space of NDs for a given subset selection method as the *growth function*. Figure 1 shows the growth functions for the three selection methods discussed in this chapter.

2.2   Balanced Selection

An issue with random selection is that it can produce very unbalanced tree structures. While the number of internal nodes (and therefore, binary models) is the same in any ND for the same number of classes, an unbalanced tree often implies that internal binary models are trained on large datasets near the leaves, which has a negative effect on the time taken to train the full model. Deeper subtrees also provide more opportunity for estimation errors to accumulate. Dong *et. al.* mitigate this effect by enforcing C_i to be split into two subsets C_{i1} and C_{i2} such that $abs(|C_{i1}| - |C_{i2}|) \leq 1$ [8]. This has been shown empirically to have little effect on the accuracy in most cases, while reducing the time taken to train NDs. Balanced selection has greater benefits for problems with many classes.

It is clear that the sample space of class balanced NDs is smaller than that of random NDs, but it is still large enough to ensure sufficient ensemble diversity. The growth function for class balanced NDs is given by

$$T_{CB}(n) = \begin{cases} \frac{1}{2}\binom{n}{n/2}T_{CB}(\frac{n}{2})T_{CB}(\frac{n}{2}), & \text{if } n \text{ is even} \\ \binom{n}{(n+1)/2}T_{CB}(\frac{n+1}{2})T_{CB}(\frac{n-1}{2}), & \text{if } n \text{ is odd} \end{cases}$$

where $T_{CB}(2) = T_{CB}(1) = 1$ [8]. Dong *et. al.* also explored a form of balancing where the amount of data in each subset is roughly equal, which gave similar results for datasets with unbalanced classes [8].

---

[1] This is a variant of the approach from [11], where each member of the space of NDs has an equal probability of being sampled.

2.3   Random-Pair Selection

Random-pair selection provides a non-deterministic method of creating C_{i1} and C_{i2} that groups similar classes together [19]. In random-pair selection, the base classifier is used directly to identify similar classes in C_i. First, a random pair of classes $c_1, c_2 \in C_i$ is selected, and a binary classifier is trained on just these two classes. Then, the remaining classes are classified with this classifier, and its predictions are stored as a confusion matrix M. C_{i1} and C_{i2} are constructed by

$$C_{i1} = \{c \in C_i \setminus \{c_1, c_2\} : M_{c,c_1} \leq M_{c,c_2}\} \cup \{c_1\}$$
$$C_{i2} = \{c \in C_i \setminus \{c_1, c_2\} : M_{c,c_1} > M_{c,c_2}\} \cup \{c_2\}$$

where $M_{j,i}$ is defined as the number of examples of class j that were classified as class i by the binary classifier. In other words, a class is assigned to C_{i1} if it is less frequently confused with c_1 than with c_2, and to C_{i2} otherwise. Finally, the binary classifier is re-trained on the new meta-classes C_{i1} and C_{i2}. This way, each split is more easily separable for the base learner than a completely random split, while exhibiting a degree of randomness, which produces diverse ensembles.

Due to the fact that the size of the sample space of NDs under random-pair selection is dependent on the dataset and base learner (different initial random pairs may lead to the same split), it is not possible to provide an exact expression for the growth function $T_{RP}(n)$; using logistic regression (LR) as the base learner, it has been empirically estimated to be

$$T_{RP}(n) = p(n) T_{RP}\left(\frac{n}{3}\right) T_{RP}\left(\frac{2n}{3}\right)$$

where $T_{RP}(2) = T_{RP}(1) = 1$ and $p(n) = 0.3812n^2 - 1.4979n + 2.9027$ [19].

3   Multiple Subset Evaluation

In class subset selection methods, for each node i, a single class split (C_{i1}, C_{i2}) of C_i is considered, produced by some splitting function $S(C_i) : \mathbb{N}^n \to \mathbb{N}^a \times \mathbb{N}^b$ where $a + b = n$. Our approach for improving the predictive power of NDs is a simple extension. We propose to, at each internal node i, consider λ subsets $\{(C_{i1}, C_{i2})_1 \ldots (C_{i1}, C_{i2})_\lambda\}$ and choose the split for which the corresponding model has the lowest training root mean squared error (RMSE). The RMSE is defined as the square root of the Brier score [5] divided by the number of classes:

$$\text{RMSE} = \sqrt{\frac{1}{nm} \sum_{i=1}^{n} \sum_{j=1}^{m} (\hat{y}_{ij} - y_{ij})^2}$$

where n is the number of instances, m is the number of classes, $\hat{y}_{ij}$ is the estimated probability that instance i is of class j, and y_{ij} is 1 if instance i actually belongs to class j, and 0 otherwise. RMSE is chosen over other measures such as classification accuracy because it is smoother and a more sensitive indicator

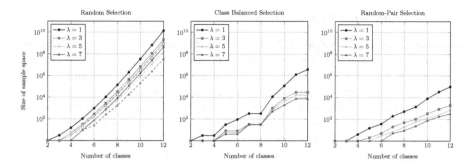

Fig. 2. Left: Growth functions for random selection with multiple subset evaluation and $\lambda \in \{1, 3, 5, 7\}$. Solid lines indicate the upper bound, and dashed lines indicate the lower bound. Middle: Considering class-balanced selection instead of random selection. Right: Growth functions for random-pair selection.

of generalisation performance. Previously proposed methods with single subset selection can be considered a special case of this method where $\lambda = 1$.

Although conceptually simple, this method has several attractive qualities. By choosing the best of a series of models at each internal node, the overall performance should improve, assuming the size of the sample space of NDs is not hindered to the point where ensemble diversity begins to suffer.

Multiple subset evaluation is also widely applicable. If a subset selection method S has some level of randomness, then multiple subset evaluation can be used to improve the performance. One nice feature is that advantages pertaining to S are retained. For example, if class-balanced selection is chosen due to a learning problem with a very high number of classes, we can boost the predictive performance of the ensemble while keeping each ND in the ensemble balanced. If random-pair selection is chosen because the computational budget for training is high, then we can improve the predictive performance further than single subset selection in conjunction with random-pair selection.

Finally, implementing multiple subset evaluation is very simple, and the computational cost for evaluating multiple subsets of classes scales linearly in the size of the tuneable hyperparameter λ, making the tradeoff between predictive performance and training time easy to navigate. Additionally, multiple subset evaluation has no effect on prediction times.

Higher values of λ give diminishing returns on predictive performance, so a value that is suitable for the computational budget should be chosen. When training an ensemble of NDs, it may be desirable to adopt a *class threshold*, where $\lambda = 1$ is used if fewer than a certain number of classes is present at an internal node. This reduces the probability that the same subtrees will appear in many ensemble members, and therefore reduce ensemble diversity. In lower levels of the tree, where the number of classes is small, the number of possible binary problems is relatively low (Fig. 2).

3.1   Effect on Growth Functions

Performance of an ensemble of NDs relies on the size of the sample space of NDs, given an n-class problem, to be relatively large. Multiple subset evaluation removes the $\lambda - 1$ class splits that correspond to the worst-performing binary models at each internal node i from being able to be used in the tree. The effect of multiple subset evaluation on the growth function is non-deterministic for random selection, as the sizes of C_{i1} and C_{i2} affect the values of the growth function for the subtrees that are children of i. The upper bound occurs when all worst-performing splits isolate a single class, and the lower bound is given when all worst-performing splits are class-balanced. Class-balanced selection, on the other hand, is affected deterministically as the size of C_{i1} and C_{i2} are the same for the same number of classes.

Growth functions for values of $\lambda \in \{1, 3, 5, 7\}$, for random, class balanced and random-pair selection methods, are plotted in Fig. 2. The growth curves for random and class balanced selection were generated using brute-force computational enumeration, while the effect on random-pair selection is estimated.

3.2   Analysis of Error

In this section, we provide a theoretical analysis showing that performance of each internal binary model is likely to be improved by adopting multiple subset evaluation. We also show empirically that the estimates of performance improvements are accurate, even when the assumptions are violated.

Let E be a random variable for the training root mean squared error (RMSE) for some classifier for a given pair of class subsets C_{i1} and C_{i2}, and assume $E \sim N(\mu, \sigma^2)$ for a given dataset under some class subset selection scheme. For a given set of λ selections of subsets $S = \{(C_{i1}, C_{i2})_1, \ldots, (C_{i1}, C_{i2})_\lambda\}$ and corresponding training RMSEs $\mathcal{E} = \{E_1, \ldots, E_\lambda\}$, let $\hat{E}_\lambda = min(\mathcal{E})$. There is no closed form expression for the expected value of $\hat{E}_\lambda$, the minimum of a set of normally distributed random variables, but an approximation is given by

$$\mathbb{E}[\hat{E}_\lambda] \approx \mu + \sigma\Phi^{-1}\left(\frac{1 - \alpha}{\lambda - 2\alpha + 1}\right) \tag{1}$$

where $\Phi^{-1}(x)$ is the inverse normal cumulative distribution function [28], and the *compromise value* α is the suggested value for λ given by Harter [15].[2]

Figure 3 illustrates how this expected value changes when increasing values of λ from 1 to 5. The first two rows show the distribution of E and estimated $\mathbb{E}[\hat{E}_\lambda]$ on the UCI dataset `mfeat-fourier`, for a LR model trained on 1,000 random splits of the class set C. These rows show the training and testing RMSE respectively, using 90% of the data for training and the rest for testing. Note that as λ increases, the distribution of the train and test error shifts to lower values and the variance decreases.

---

[2] Appropriate values for α for a given λ can be found in Table 3 of [15].

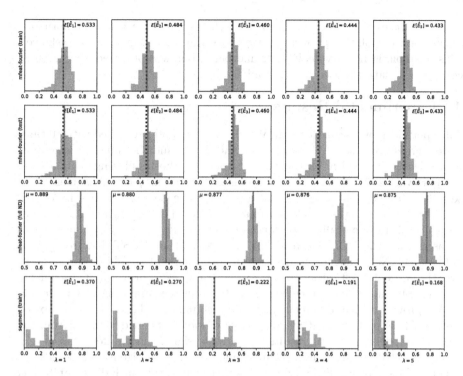

Fig. 3. Empirical distribution of RMSE of LR trained on random binary class splits, for values of λ from one to five. The shaded region indicates empirical histogram, the orange vertical line shows the empirical mean, and the black dotted vertical line is the expected value, estimated from (1). Top two rows: train and test RMSE of LR trained on random binary class splits of mfeat-fourier UCI dataset. For the test data, the approximated value of $\mathbb{E}[E_\lambda]$ is estimated from the mean and standard deviation of the train error. Third row: train RMSE of an ND built with random splits and multiple-subset evaluation, trained on mfeat-fourier for different values of λ. Bottom row: train RMSE of LR trained on random binary class splits of segment data. (Color figure online)

This reduction in error affects each binary model in the tree structure, so the effects accumulate when constructing an ND. The third row shows the distribution of RMSE of 1,000 NDs trained with multiple subset evaluation on mfeat-fourier, using LR as the base learner, considering increasing values of λ. As expected, a reduction in train error with diminishing returns is seen.

In order to show an example of how the estimate from (1) behaves when the error is not normally distributed, the distribution of E for LR trained on the segment UCI data is plotted in the bottom row. The assumption of normality is commonly violated in real datasets, as the distribution is often skewed towards zero error. As with the other examples, 1,000 different random choices for $\mathcal{C}_1$ and $\mathcal{C}_2$ were used to generate the histogram. Although the distribution in this case is not very well modelled by a Gaussian, the approximation of $\mathbb{E}[\hat{E}_\lambda]$ from (1) still closely matches the empirical mean. This shows that even when the normality assumption is violated, performance gains of the same degree can be achieved. This example is not atypical; the same behaviour was observed on the entire collection of datasets used in this study.

4   Experimental Results

All experiments were conducted in WEKA 3.9 [14], and performed with 10 times 10-fold cross validation. We use class-balanced NDs and NDs built with random-pair selection, with LR as the base learner. For both splitting methods, we compare values of $\lambda \in \{1, 3, 5, 7\}$ in a single ND structure, as well as in ensemble settings with bagging [4] and AdaBoost [12]. The default settings in WEKA were used for the `Logistic` classifier as well as for the `Bagging` and `AdaBoostM1` meta-classifiers. We evaluate performance on a collection of 15 commonly used datasets from the UCI repository [21], as well as the MNIST digit recognition dataset [20]. Note that for MNIST, we report results of 10-fold cross-validation over the entire dataset rather than the usual train/test split. Datasets used in our experiments and their characteristics are listed in the supplementary material.

We provide critical difference plots [6] to summarise the results of the experiments. These plots present average ranks of models trained with differing values of λ. Models producing results that are not significantly different from each other at the 0.05 significance level are connected with a horizontal black bar. Full results tables showing RMSE for each experimental run, including significance tests, are available in the supplementary materials.

4.1   Individual Nested Dichotomies

Restricting the sample space of NDs through multiple subset evaluation is expected to have a greater performance impact on smaller ensembles than larger ones. This is because in a larger ensemble, a poorly performing ensemble member does not have a large impact on overall performance. On the other hand, in a small ensemble, one poorly performing ensemble member can degrade ensemble performance significantly. In the extreme case, where a single ND is trained, there is no need for ensemble diversity, so a technique for improving the predictive performance of an individual ND should be effective. Therefore, we first compare the performance of single NDs for different values of λ.

Figure 4 shows critical difference plots for both subset selection methods. Class balanced selection shows a clear trend that increasing λ improves the RMSE, with the average rank for $\lambda = 1$ being exactly 4. For random-pair selection, choosing $\lambda = 3$ is shown to be statistically indistinguishable from $\lambda = 1$, while higher values of λ give superior results on average.

4.2   Ensembles of Nested Dichotomies

Typically, NDs are utilised in an ensemble, so we investigate the predictive performance of ensembles of ten NDs with multiple subset evaluation, with bagging and AdaBoost employed as the ensemble methods.

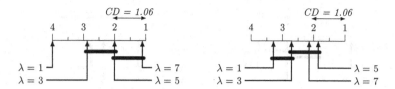

Fig. 4. Critical differences charts for individual NDs. Left: Class balanced selection. Right: Random-pair selection.

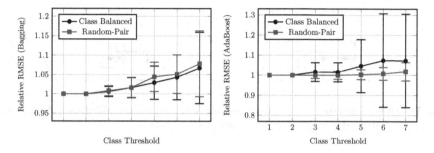

Fig. 5. Effect of changing the class threshold on RMSE for ensembles of NDs.

Class Threshold. The number of binary problems is reduced when multiple subset evaluation is applied, which can have a negative effect on ensemble diversity, potentially reducing predictive performance. To investigate this, we built ensembles of NDs with multiple subset evaluation by introducing a *class threshold*, the number of classes present at a node required to perform multiple subset evaluation, and varying its value from one to seven. We plot the test RMSE, relative to having a class threshold of one, averaged over all the datasets in Fig. 5. Interestingly, the RMSE increases monotonically, showing that the potentially reduced ensemble diversity does not have a negative effect on the RMSE for ensembles of this size. Therefore, we use a class threshold of one in our subsequent experiments. However, note that increasing the class threshold has a positive effect on training time, so it may be useful to apply it in practice.

Number of Subsets. We now investigate the effect of λ when using bagging and boosting. Figure 6 shows critical difference plots for bagging. Both subset selection methods improve when utilising multiple subset selection. When class-balanced selection is used, as was observed for single NDs, the average ranks across all datasets closely correspond to the integer values, showing that increasing the number of subsets evaluated consistently improves performance. For random-pair selection, a more constrained subset selection method, each value of $\lambda > 1$ is statistically equivalent and superior to the single subset case.

The critical difference plots in Fig. 7 (left) show boosted NDs are significantly improved by increasing the number of subsets sufficiently when class-balanced NDs are used. Results are less consistent for random-pair selection, reflected in

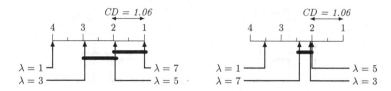

Fig. 6. Critical differences charts for ensemble of ten bagged NDs. Left: Class balanced selection. Right: Random-pair selection.

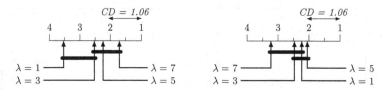

Fig. 7. Critical differences charts for ensemble of ten NDs, ensembled with AdaBoost. Left: Class balanced selection. Right: Random-pair selection.

the critical differences plot (Fig. 7, right), which shows single subset evaluation statistically equivalent to multiple subset selection for all values of λ, with $\lambda = 7$ performing markedly worse on average. As RMSE is based on probability estimates, this may be in part due to poor probability calibration, which is known to affect boosted ensembles [24] and NDs [18].

5   Related Work

Splitting a multiclass problem into several binary problems in a tree structure is a general technique that has been referred to by different names in the literature. For example, in a multiclass classification context, NDs in the broadest sense of the term have been examined as filter trees, conditional probability trees, and label trees. Beygelzimer et al. proposed algorithms which build balanced trees and demonstrate the performance on datasets with very large numbers of classes. Filter trees, with deterministic splits [3], as well as conditional probability trees, with probabilistic splits [2], were explored. Bengio et al. [1] define a tree structure and optimise all internal classifiers simultaneously to minimise the tree loss. They also propose to learn a low-dimensional embedding of the labels to improve performance, especially when many classes are present. Melnikov and Hullermeier [23] also showed that a method called best-of-k models—simply sampling k random NDs and choosing the best one based on validation error— gives competitive predictive performance to the splitting heuristics discussed so far for individual NDs. However, it is very expensive at training time, as k independent NDs must be built and tested on a held-out set.

A commonality of these techniques is that they attempt to build a single ND structure with the best performance. NDs that we consider in this paper, while conceptually similar, differ from these methods because they are intended to be trained in an ensemble setting, and as such, each individual ND is not built with optimal performance in mind. Instead, a group of NDs is built to maximise ensemble performance, so diversity amongst the ensemble members is key [17].

NDs based on clustering [9] are deterministic and used in an ensemble by resampling or reweighting the input. They are built by finding the two classes $(c_1, c_2) \in C_i$ for which the centroids are furthest from each other, and grouping the remaining classes based on the distance of their centroids from c_1 and c_2.

Wever et al. [29] utilise genetic algorithms to build NDs. In their method, a population of random NDs is sampled and is evolved for several generations. The final ND is chosen as the best performing model on a held-out validation set. An ensemble of k NDs is produced by evolving k populations independently, and taking the best-performing model from each population.

6   Conclusion

Multiple subset selection in NDs can improve predictive performance while retaining the particular advantages of the subset selection method employed. We present an analysis of the effect of multiple subset selection on expected RMSE and show empirically in our experiments that adopting our technique can improve predictive performance, at the cost of a constant factor in training time.

The results of our experiments suggest that for class-balanced selection, performance can be consistently improved significantly by utilising multiple subset evaluation. For random-pair selection, $\lambda = 3$ yields the best trade-off between predictive performance and training time, but when AdaBoost is used, multiple subset evaluation is not generally beneficial.

Avenues of future research include comparing multiple subset evaluation with base learners other than LR. It is unlikely that training RMSE of the internal models will be a reliable indicator when selecting splits based on more complex models such as decision trees or random forests, so other metrics may be needed. Also, it may be beneficial to choose subsets such that maximum ensemble diversity is achieved, possibly through information theoretic measures such as variation of information [22].

References

1. Bengio, S., Weston, J., Grangier, D.: Label embedding trees for large multi-class tasks. In: NIPS, pp. 163–171 (2010)
2. Beygelzimer, A., Langford, J., Lifshits, Y., Sorkin, G., Strehl, A.: Conditional probability tree estimation analysis and algorithms. In: UAI, pp. 51–58 (2009)
3. Beygelzimer, A., Langford, J., Ravikumar, P.: Error-correcting tournaments. In: Gavaldà, R., Lugosi, G., Zeugmann, T., Zilles, S. (eds.) ALT 2009. LNCS (LNAI), vol. 5809, pp. 247–262. Springer, Heidelberg (2009). https://doi.org/10.1007/978-3-642-04414-4_22
4. Breiman, L.: Bagging predictors. Mach. Learn. **24**(2), 123–140 (1996)

5. Brier, G.: Verification of forecasts expressed in term of probabilities. Mon. Weather Rev. **78**, 1–3 (1950)
6. Demšar, J.: Statistical comparisons of classifiers over multiple data sets. JMLR **7**(Jan), 1–30 (2006)
7. Dietterich, T.G., Bakiri, G.: Solving multiclass learning problems via error-correcting output codes. JAIR **2**, 263–286 (1995)
8. Dong, L., Frank, E., Kramer, S.: Ensembles of balanced nested dichotomies for multi-class problems. In: Jorge, A.M., Torgo, L., Brazdil, P., Camacho, R., Gama, J. (eds.) PKDD 2005. LNCS (LNAI), vol. 3721, pp. 84–95. Springer, Heidelberg (2005). https://doi.org/10.1007/11564126_13
9. Duarte-Villaseñor, M.M., Carrasco-Ochoa, J.A., Martínez-Trinidad, J.F., Flores-Garrido, M.: Nested dichotomies based on clustering. In: Alvarez, L., Mejail, M., Gomez, L., Jacobo, J. (eds.) CIARP 2012. LNCS, vol. 7441, pp. 162–169. Springer, Heidelberg (2012). https://doi.org/10.1007/978-3-642-33275-3_20
10. Fox, J.: Applied Regression Analysis, Linear Models, and Related Methods. Sage, Thousand Oaks (1997)
11. Frank, E., Kramer, S.: Ensembles of nested dichotomies for multi-class problems. In: ICML, p. 39. ACM (2004)
12. Freund, Y., Schapire, R.E.: Game theory, on-line prediction and boosting. In: COLT, pp. 325–332 (1996)
13. Fürnkranz, J.: Round robin classification. JMLR **2**(Mar), 721–747 (2002)
14. Hall, M., Frank, E., Holmes, G., Pfahringer, B., Reutemann, P., Witten, I.H.: The WEKA data mining software: an update. ACM SIGKDD Explor. Newsl. **11**(1), 10–18 (2009)
15. Harter, H.L.: Expected values of normal order statistics. Biometrika **48**(1/2), 151–165 (1961)
16. Hastie, T., Tibshirani, R., et al.: Classification by pairwise coupling. Ann. Stat. **26**(2), 451–471 (1998)
17. Kuncheva, L.I., Whitaker, C.J.: Measures of diversity in classifier ensembles and their relationship with the ensemble accuracy. Mach. Learn. **51**(2), 181–207 (2003)
18. Leathart, T., Frank, E., Holmes, G., Pfahringer, B.: On calibration of nested dichotomies. In: Yang, Q., et al. (eds.) Advances in Knowledge Discovery and Data Mining. LNAI, vol. 11439, pp. 69–80. Springer, Heidelberg (2019)
19. Leathart, T., Pfahringer, B., Frank, E.: Building ensembles of adaptive nested dichotomies with random-pair selection. In: Frasconi, P., Landwehr, N., Manco, G., Vreeken, J. (eds.) ECML PKDD 2016. LNCS (LNAI), vol. 9852, pp. 179–194. Springer, Cham (2016). https://doi.org/10.1007/978-3-319-46227-1_12
20. LeCun, Y., Bottou, L., Bengio, Y., Haffner, P.: Gradient-based learning applied to document recognition. Proc. IEEE **86**(11), 2278–2324 (1998)
21. Lichman, M.: UCI machine learning repository (2013)
22. Meilă, M.: Comparing clusterings by the variation of information. In: Schölkopf, B., Warmuth, M.K. (eds.) COLT-Kernel 2003. LNCS (LNAI), vol. 2777, pp. 173–187. Springer, Heidelberg (2003). https://doi.org/10.1007/978-3-540-45167-9_14
23. Melnikov, V., Hüllermeier, E.: On the effectiveness of heuristics for learning nested dichotomies: an empirical analysis. Mach. Learn. **107**(8–10), 1–24 (2018)
24. Niculescu-Mizil, A., Caruana, R.: Predicting good probabilities with supervised learning. In: ICML, pp. 625–632. ACM (2005)
25. Pimenta, E., Gama, J.: A study on error correcting output codes. In: Portuguese Conference on Artificial Intelligence, pp. 218–223. IEEE (2005)
26. Rifkin, R., Klautau, A.: defense of one-vs-all classification. JMLR **5**, 101–141 (2004)

27. Rodríguez, J.J., García-Osorio, C., Maudes, J.: Forests of nested dichotomies. Pattern Recognit. Lett. **31**(2), 125–132 (2010)
28. Royston, J.: Algorithm AS 177: expected normal order statistics (exact and approximate). J. R. Stat. Soc. Ser. C (Appl. Stat.) **31**(2), 161–165 (1982)
29. Wever, M., Mohr, F., Hüllermeier, E.: Ensembles of evolved nested dichotomies for classification. In: GECCO, pp. 561–568. ACM (2018)

Text and Opinion Mining

Topic-Level Bursty Study for Bursty Topic Detection in Microblogs

Yakun Wang[1], Zhongbao Zhang[1], Sen Su[1(✉)], and Muhammad Azam Zia[1,2]

[1] Beijing University of Posts and Telecommunications, Beijing, China
{wangyakun,zhongbaozb,susen,zia}@bupt.edu.cn
[2] University of Agriculture Faisalabad, Faisalabad, Pakistan

Abstract. Microblogging services, such as Twitter and Sina Weibo, have gained tremendous popularity in recent years. The huge amount of user-generated information is spread on microblogs. Such user-generated contents are a mixture of different bursty topics (e.g., breaking news) and general topics (e.g., user interests). However, it is challenging to discriminate between them due to the extremely diverse and noisy user-generated text. In this paper, we introduce a novel topic model to detect bursty topics from microblogs. Our model is based on an observation that different topics usually exhibit different bursty levels at a certain time. We propose to utilize the topic-level burstiness to differentiate bursty topics and non-bursty topics and particularly different bursty topics. Extensive experiments on a Sina Weibo Dataset show that our approach outperforms the baselines and the state-of-the-art method.

Keywords: Sina Weibo · Bursty topic detection · Topic model · Hypothesis testing

1 Introduction

Microblogs, such as Twitter and Sina Weibo, have gained an explosive growth in popularity in recent years. Users on microblogs publish short posts about various topics. Due to the real-time nature compared with traditional media, microblogs have become an important resource for reporting events. Thus, bursty topic detection in microblogs is a non-trivial work and benefits a lot of applications like crisis management and decision making.

Bursty topics usually refer to the real world happenings whose popularity goes through a burst increase at the occurring time [13,18]. The large volume of event-driven posts published on microblogs makes the bursty topic detection possible. However, people in microblog not only talk about the daily events but also their daily lives, which makes posts particularly diverse and noisy. Therefore, it is challenging to do bursty topic detection in microblogs [19].

To address this problem, many works resort to the probability topic models [2,4,6,16,17,20], where text co-occurrence at a specific time or a location is captured as the bursty topic. Another attempt is to detect bursty words first, and

© Springer Nature Switzerland AG 2019
Q. Yang et al. (Eds.): PAKDD 2019, LNAI 11439, pp. 97–109, 2019.
https://doi.org/10.1007/978-3-030-16148-4_8

then cluster them [11,15,17]. For example, [17] modeled the bursty probability of words and consider it as the prior information of bursty topics when doing clustering. However, the essential difference between topics is that they exhibit different bursty levels at a certain time. To illustrate our motivation, we give a real example regarding the popularity of three topics A, B and C in Fig. 1. Generally, more frequently talked of a topic compared with previously, more popular it is at present. From Fig. 1, we can see that different topics present different bursty levels. Notably, the popularity of topics A and B experience an increase on 3 Sept 2015, while topic C does not. Accordingly, topics A and B are bursty topics, and topic C is a general topic. In addition, topic A is more popular to people compared to topic B. The reason may be that people usually pay different attention to different topics at the specific time. This example motivates us to have the following two ideas. First, we can use the burstiness of topics to differentiate between bursty topics and non-bursty topics. Second, bursty levels can also be utilized to distinguish different bursty topics. Unfortunately, prior works only focus on the burstiness of words and did not systematically explores how to model the bursty level of topics for bursty topic detection.

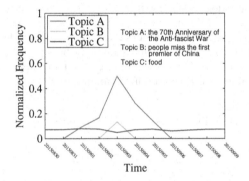

Fig. 1. Bursty topic and general topic in Sina Weibo (the normalized frequency is derived from the frequency of their keywords).

In this paper, we aim to model the topic-level bursty for bursty topic detection in microblogs. We propose a novel topic model, named Topic-level Bursty Detection model (TBD). TBD is designed to detect time-specific topics. TBD parameterizes one distribution over words associated with each temporal topic, which potentially assumes that posts published at the same time are more likely to talk about the same topic. Moreover, TBD parameterizes another distribution over words' bursty levels for each topic, which potentially assumes that words with similar bursty levels at the same time are more likely to belong to the same topic. Topics are responsible for generating both the words and words' bursty levels. When a strong co-occurrence of words and their bursty levels appears at a certain time, TBD will create a topic for them. After learning the latent parameters within the model, we identify the bursty topics based on their burstiness through a hypothesis testing method. Modeling the topic-level burstiness enjoys substantial merits for several reasons. First, burstiness is the essential difference between bursty topics and general topics and particularly different bursty topics, which makes our approach outperform the compared methods. Second, it provides a new insight to interpret the discovered topics and enable people to grasp what topics others pay more attention to, which is beneficial for applications such as crisis management and decision making. Through the experiments on a real-world dataset from Sina Weibo, we prove that TBD is a robust model that not sensitive to the number of topics in terms of precision and outperforms the baselines and the-state-of-art method in both precision and recall.

2 Bursty Topic Detection Model

In this section, we first introduce how to model the bursty level of words. Furthermore, we present our Topic-level Bursty Detection model (TBD) that models the bursty level of topics based on the bursty level of words. In what follows, we show Gibbs Sampling for parameter estimation. We give the notation used throughout this paper in Table 1.

Table 1. Notations

Notation	Description
w, b, z, d	Label for word, word's burstiness, topic and post
y	Binary indicator of topics or background topics for word and burstiness
$\alpha, \beta, \beta', \gamma, \gamma', \rho$	Parameters of the Dirichlet (Beta) priors on Multinomial (Bernoulli) distributions
θ	Per-time topic distribution
η	Per-time Bernoulli distribution over indicators
φ	Per-topic word distribution
φ'	Background word distribution
π	Per-topic burstiness distribution
π'	Background burstiness distribution
T, D, V, B, K	Number of time slices, posts, unique words, unique burstiness level, topics
N_d, N_d'	Number of words and burstiness levels of post d

2.1 Bursty Level of Words

Bursty words usually correspond to the bursty topics. Suppose a word w occurred $n_w^{(t)}$ times at time t and assume $n_w^{(t)}$ is drawn from a Gaussian distribution with mean $\hat{\mu}_w$ and standard deviation $\hat{\delta}_w$. The bursty level of w at time t can be calculated by z-score [21] as

$$b_w^t = \frac{n_w^{(t)} - \hat{\mu}_w}{\hat{\delta}_w}, \tag{1}$$

where $\hat{\mu}_w$ is the mean of n_w in the past S time slices, i.e., $\hat{\mu}_w = \frac{1}{S} \sum_{s=1}^{S} n_w^{(t-s)}$, and $\hat{\delta}_w$ is the standard variance of n_w in the past S time slices, i.e., $\hat{\delta}_w = \sqrt{\frac{1}{S} \sum_{s=1}^{S} \left(n_w^{(t-s)} - \hat{\mu}_w \right)^2}$.

The z-score is to compare an observation to a standard normal deviate and is a dimensionless quantity. The larger the z-score is, the more bursty the word is. Given the bursty level of words, in the following TBD section, we introduce how to build the connection between the burstiness of words and the burstiness of topics.

2.2   Topic-Level Bursty Detection Model

In this section, we propose the Topic-level Bursty Detection model (TBD) for bursty topic detection. The basic idea of TBD is that different topics usually present the different bursty levels at a specific time. If words with the similar bursty levels co-occur more frequently at a point, they are more likely to belong to the same topic. Unlike previous works that try to strictly divide topics into two sets (bursty topics and general topics), we detect one set of topics and associate each topic with a bursty level from which we identify the bursty topics. The plate notation for TBD is depicted in Fig. 2.

First of all, TBD is a topic model which aims to learn time-specified topical distributions θ. Each topic in TBD is viewed as a mixture of words, i.e., the multinomial distribution φ. Moreover, we associate each topic with a multinomial distribution π over the bursty levels of words. When a strong co-occurrence of words and their bursty levels appears, TBD will create an topic for them. If the topic mainly covers the high bursty levels, then it is a bursty topic, otherwise, it is a general topic.

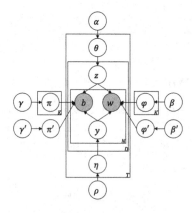

Fig. 2. Plate diagram of TBD

Besides, a binary indicator y is generated from a Bernoulli distribution η to filter out the background words that are not related to any topic. Multinomial distributions φ' and π' are responsible for capturing these background words and the corresponding bursty levels, respectively. Given that people usually focus on one topic in a post, we design TBD by associate a single hidden variable with each post to indicate its topic. Note that, in contrast to prior generative models that only model the observed data [4,6,17,20], TBD generates not only the observed words in the posts but also their bursty levels which are not directed observed. In our model, we adopt conjugate prior of Dirichlet distribution for the multinomial distribution and Beta distribution for the Bernoulli distribution. At last, each topic z will be determined by a distribution over words and a distribution over words' bursty levels.

The generative process is summarized in Algorithm 1. Consider there is a post d published at time t, we first select a topic $z_{t,d}$ by a time-topic distribution θ_t. To generate a word $w_{t,d,m}$ with the bursty level $b_{t,d,m}$ in post d, we use a Bernoulli distribution η_t to generate a binary indicator parameter $y_{t,b,m}$ to decide the resource of $w_{t,d,m}$ and $b_{t,d,m}$. When $y_{t,b,m} = 1$, the word and its bursty level are generated by a topical distribution. We use the multinomial distributions $\varphi_{z_{t,d}}$ to generate the word and $\pi_{z_{t,d}}$ to generate its bursty level. When $y_{t,b,m} = 0$, they are generated by the background distribution, where we use the multinomial distribution φ'_t and π'_t to generate the background words and the corresponding bursty levels.

Algorithm 1. Generative process for TBD

Sample $\varphi' \sim$ Dirichlet (β') and $\pi' \sim$ Dirichlet (γ)
for *each topic* $z \in [1, K]$ **do**
    Sample the distribution over words $\varphi_k \sim$ Dirichlet (β)
    Sample the distribution over words' bursty levels $\pi_k \sim$ Dirichlet (γ)

for *each time slice* $t \in [1, T]$ **do**
    Sample the distribution over topics $\theta_t \sim$ Dirichlet (α)
    for *each post* $d \in [1, D]$ *of time* t **do**
        Sample the distribution over indicators $\eta_t \sim$ Dirichlet (ρ)
        Sample the topic $z_{t,d} \sim$ Multinomial (θ_t)
        for *the* m^{th} *word and burstiness level of post* d*, where* $m \subset [1, N_d]$ **do**
            Sample the indicator $y_{t,d,m} \sim$ Bernoulli (η_t)
            if $y_{t,d,m} = 1$ **then**
                Sample the word $w_{t,d,m} \sim$ Multinomial $(\varphi_{z_{t,d}})$
                Sample the burstiness level $b_{t,d,m} \sim$ Multinomial $(\pi_{z_{t,d}})$
            if $y_{t,d,m} = 0$ **then**
                Sample the word $w_{t,d,m} \sim$ Multinomial (φ'_t)
                Sample the burstiness level $b_{t,d,m} \sim$ Multinomial (π'_t)

2.3 Parameter Estimation

Since exact inference for TBD model is intractable, we therefore utilize collapsed Gibbs sampling [8], a widely used Markov Chain Monte Carlo (MCMC) algorithm, to obtain samples of the hidden variable assignment and to estimate the model parameters from these samples. Gibbs sampling iteratively samples latent variables (i.e., z, y in TBD) from a Markov chain, whose stationary distribution is posterior.

Let $z_{\neg i}$ denote the set of all hidden variables of topics except z_i and $n_{\cdot,\neg i}^{(\cdot)}$ denote the count that the element i is excluded from the corresponding topic. We use similar symbols for other variables. Firstly, we sample the topic assignments $z_{t,d}$ for post d at time t with index $i = (t, d)$ given the observations and other assignments using a Gibbs sampling procedure as follows:

$$p(z_i = k|z_{\neg i}, w, b, \alpha, \beta, \gamma) \propto$$

$$\frac{\prod\limits_{w=1}^{W} \prod\limits_{x=1}^{n_{i,w}} (n_k^{(w)} - x + \beta) \prod\limits_{b=1}^{B} \prod\limits_{x=1}^{n_{i,b}} (n_k^{(b)} - x + \gamma)}{\prod\limits_{q=1}^{N_d} (\sum\limits_{w=1}^{W} n_k^{(w)} - q + W\beta) \prod\limits_{q=1}^{N_d'} (\sum\limits_{b=1}^{B} n_k^{(b)} - q + B\gamma)} (n_{t,\neg i}^{(k)} + \alpha) \qquad (2)$$

where $n_{i,w}$ and $n_{i,b}$ denote the number of times that word w and bursty level b occur in post d, $n_k^{(w)}$ and $n_k^{(b)}$ denote the number of times that word w and bursty level b have been observed with topic k, N_d and N_d' refer to the number of words and words' bursty levels occurs in post d, $n_{t,\neg i}^{(k)}$ denotes the number of times that topic k has been observed with a post at time t.

Then, for a word $w_{t,d,m}$ and its bursty level $b_{t,d,m}$ in post d at time t with index $j = (t, d, m)$, we sample y_j from the conditional as Eqs. 3 and 4:

$$p(y_j = 1|\beta, \gamma, \rho)$$

$$\propto \frac{n_{k,\neg j}^{(w)} + \beta}{\sum\limits_{w=1}^{W} n_{k,\neg j}^{(w)} + W\beta} \frac{n_{k,\neg j}^{(b)} + \gamma}{\sum\limits_{b=1}^{B} n_{k,\neg j}^{(b)} + B\gamma} (n_{t,\neg j}^{(y_j=1)} + \rho) \qquad (3)$$

$$p(y_j = 0|\beta', \gamma', \rho)$$

$$\propto \frac{n_{w,\neg j} + \beta'}{\sum\limits_{w=1}^{W} n_{w,\neg j} + W\beta'} \frac{n_{b,\neg j} + \gamma'}{\sum\limits_{b=1}^{B} n_{b,\neg j} + B\gamma'} (n_{t,\neg j}^{(y_j=0)} + \rho) \qquad (4)$$

where $n_{w,\neg j}$ and $n_{b,\neg j}$ denote the number of times word w and bursty level b assigned to the background distribution, $n_{t,\neg j}^{(y_j=1)}$ and $n_{t,\neg j}^{(y_j=0)}$ refer to the number of times the word and the corresponding bursty level pair (w_j, b_j) assigned to the topical distribution and the background distribution.

After a sufficient number of iterations, we can estimate the unknown parameters based on the samples as follows:

$$\theta_k = \frac{n_t^{(k)} + \alpha}{\sum\limits_{k=1}^{K} n_t^{(k)} + K\alpha}, \qquad (5)$$

$$\varphi_{k,w} = \frac{n_k^{(w)} + \beta}{\sum\limits_{w=1}^{W} n_k^{(w)} + W\beta}, \tag{6}$$

$$\varphi_w' = \frac{n_w + \beta'}{\sum\limits_{w=1}^{W} n_w + W\beta'}, \tag{7}$$

$$\pi_{k,b} = \frac{n_k^{(b)} + \gamma}{\sum\limits_{b=1}^{B} n_k^{(b)} + B\gamma}, \tag{8}$$

$$\pi_b' = \frac{n_b + \gamma'}{\sum\limits_{b=1}^{B} n_b + B\gamma'}, \tag{9}$$

$$\eta_y = \frac{n_t^{(y)} + \rho}{n_t^{(y=0)} + n_t^{(y=1)} + 2\rho}. \tag{10}$$

The time complexity of LDA model is $o(Niter * K * N_w)$, where Niter is the iteration times, K is the number of topics and N_w is the total number of words. Compared with LDA, TBD introduces an additional topic for capturing the background words. Thus it's time complexity is $o(Niter * (K+1) * N_w)$.

2.4   Hypothesis Testing

Given the estimated parameters, we can easily find out what event topic k refers to based on its top keywords in φ_k and if k is a bursty topic based on its top bursty levels in π_k. To be more scientific, we resort to hypothesis testing to determine which topics in the detected topics are bursty topics. The idea is that assuming the bursty levels in a topic are generated from a multivariate normal distribution which can be learned from the historical data. If the probability of current bursty levels in this topic is generated by the learned multivariate normal distribution over a specified confidence level, it is a bursty topic, otherwise, it is a non-bursty topic. We use $b_k^{(t)}$ to denote a bursty level set assigned to topic k at time t and assume each bursty level $b_{k,w}^{(t)} \in b_k^{(t)}$ subject to normal distribution. Thus, $b_k^{(t)}$ can be regarded as generated from the multivariate normal distribution. We introduce the null hypothesis as H_0: $b_k^{(t)}$ comes from the multivariate normal distribution, and the alternative hypothesis is H_1: $b_k^{(t)}$ do not come from the multivariate normal distribution. We set the confidence level to 0.95. If the probability of $b_k^{(t)}$ generated by the multivariate normal distribution is less than 0.95, we reject H_0 and accept H_1 so that topic k is a bursty topic.

3   Experiment Evaluation

3.1   Dataset

For comparing with the previous works, we use a dataset crawled from Sina Weibo[1], which is one of the most popular microblog platforms. The dataset is crawled by starting from a list of seed users and tracing their followers in the breadth-first traversal. The basic statistics of this dataset are given in Table 2. In the preprocess phase, we segment posts into words and remove the stop words as well as low-frequency words with a threshold of 50. We also remove the posts less than ten characters.

Table 2. Statistics of experimental dataset

Dataset	Posts	Words	Period
Sina Weibo	1.8M	71M	Aug 20 to Sept 13, 2015

3.2   Experiment Setup

Hyperparameter Setting. A significant step in such parametric method is choosing the proper hyperparameter values. We empirically set the hyperparameters $\alpha = \frac{50}{K}$, $\beta = \beta' = 0.01$, $\rho = 1$. For the number of topics K, we vary K from 5 to 20 with a step of 5 in the evaluation since the detected number of bursty topics are no more than 10 as observed. It is worth noting that we set $\gamma = \gamma' = 0$, since we intend to avoid the smoothing effect by the hyperparameter to the distribution of topics over bursty levels, i.e., π and π', and drive the learned distribution to be sparse. In this way, we get the different bursty level samples for different topics, which makes the hypothesis testing possible. The length of a time slice is set to a day, a typical setting in the literature [10]. Thus the time slice ranges from 1 to 25, where 25 denotes the most recent time slice. We use the data from previous 15 days to calculate the bursty levels of words and the following 10 days data to train different models. We run the Gibbs sampling for 500 iterations.

Compared Methods. We evaluate our approach by comparing it with the following three topic models. First, **Latent Dirichlet allocation (LDA)** [3], which is the basic topic model for documents clustering and aims to identify the latent topics within the documents. Second, **TimeLDA** [6], which is based on Twitter-LDA [19] and assumes that each post is only assigned to one topic. More than that, TimeLDA aims to learn time-specified topical distributions. Third, **Bursty Biterm Topic Model (BBTM)** [17], which is the state-of-the-art method in topic modeling for bursty topic detection in microblogs. BBTM tries to learn two kinds of topics, i.e., bursty topics and general topics, by considering the burstiness of words as the prior knowledge of topical distributions.

---

[1] http://weibo.com.

Table 3. Topics by TBD over the data of Sept 4, 2015. The third topic, the 70th Anniversary of the Victory of the World Anti-fascist War, is recognized as a non-bursty topic by TBD because it occurred 4 days ago and is not bursty any more, which is clearly illustrated in Fig. 1.

Topic	Top keywords	Top bursty levels	Hypothesis testing
People mourn the 3-year-old Syrian child whose body washed up on a beach in Turkey	Syria, photographs, Europe, refugee, war, boy, mourn, children, beach, heaven	10.7, 11.2, 5, 8.5,13.3, 2.8, 6.3,3.1, 4.1, 7	Bursty topic
Dizang Bodhisattva Festival	Bodhisattva, Dizang, all living creatures, festival, peace, vegetarian, sutras	4.3, 7.7, 2.3, 4.1, 2.5, 1.6, 6.8, 1.2, 3.8, 2.6	Bursty topic
The 70th Anniversary of the Victory of the World Anti-fascist War	Parade, victory, residence, veteran, history, commemorate, equipment, attending, formation	1.4, 0.5, 0.8, 1, −0.3, 0.9, −0.4, −0.7, −0.2, 0.6	Non-bursty topic
Work and life	Work, life, movie, feel, opportunity, way, grateful, health, hour, simple	−0.3, −0.4, −0.7, −0.2,1,0.5, −0.6, 0.9, 0.8, 1.4	Non-bursty topic

3.3   Topics Discovered from Sina Weibo Dataset

To have an intuitive idea of the detected topics, Table 3 shows some interesting results referring to bursty topics and non-bursty topics with their top keywords and bursty levels. From the bursty levels, it is clearly that the first two topics are bursty topics and the rest two topics are non-bursty topics. Hypothesis testing also gives the same results. We see that the first topic is more attractive than the second topic since its top bursty levels are greater. These results verify our idea that the bursty level of topics can be used to distinguish between bursty topics and non-bursty topics, and particularly different bursty topics.

3.4   Precision and Recall for Bursty Topic Detection

Evaluation Metrics. We detect a list of daily bursty topics and evaluate different methods in terms of precision and recall over different settings of topic number K, i.e., $P@K$ and $R@K$.

Specifically, $P@K$ is calculated as the fraction of correctly detected bursty topics over all detected bursty topics as Eq. 11.

$$P@K = \frac{\text{number of correctly detected bursty topics}}{\text{number of detected bursty topics}}. \tag{11}$$

$R@K$ is calculated as the fraction of correctly detected bursty topics over all labeled bursty topics as Eq. 12.

$$R@K = \frac{\text{number of correctly detected bursty topics}}{\text{number of labeled bursty topics}}, \tag{12}$$

where the labeled bursty topics are the union set of labeled bursty topics in all methods.

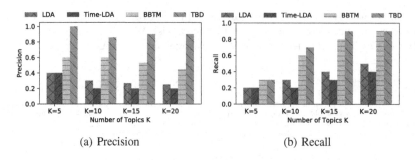

(a) Precision   (b) Recall

Fig. 3. Precision and recall at K of bursty topics detection on all data.

Ground Truth. For a quantitative evaluation in terms of precision and recall, we build up the ground truth by labeling the detected topics. Specifically, we mixed the topics detected by different methods and asked 5 students to manually label the topics by assigning a score (0: non-bursty topic, 1: bursty topic). The criterion is whether the presented topic is meaningful and bursty in its time slice. We provided the students with the top 20 keywords of each topic and time information. They are allowed to consult the external resources, like Google and Sina Weibo search, to help their judgment. A bursty topic is correctly detected if more than half of judges assigned a score of 1 to it.

Precision and Recall. Note that, all redundant topics are removed from the results. Firstly, we evaluate different methods in terms of $P@K$ and $R@K$ on all data in Fig. 3, where K is the number of topics. We see that TBD substantially outperforms the other methods in both precision and recall over all settings of K. Notably, TBD get a precision of 90% when $K = 20$, which improves LDA, Time-LDA and BBTM by 260%, 350% and 100%, respectively. Moreover, the precision of the competitors basically decreases when K increases, while TBD is not sensitive to the variance of K. This is because TBD select the topics with high bursty levels as the bursty topics from the detected K topics. However, Time-LDA and BBTM are designed to detect K bursty topics, where a bad setting of K will highly disturbance their performance.

Next, we evaluate different models in terms of mean $P@K$ and mean $R@K$ on 10 individual time slice in Fig. 4. We see that TBD still produces dramatically better results than the competitors. Note that BBTM consistently produces 0 over different settings of K in both precision and recall. The reason is that BBTM is designed to capture the co-occurrence of words within different time slices. When it works on a single time slice, it loses the basic heuristic information and falls into confusion, which leads BBTM to be ineffective on the individual time slice.

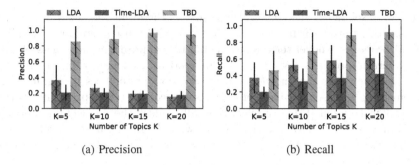

(a) Precision   (b) Recall

Fig. 4. Mean precision and recall at K of bursty topics detection on 10 single time slices. Error bars are standard errors. We omit the results of BBTM since it produces 0 over all settings of K.

4 Related Work

Bursty topic detection, such as earthquake detection [12], influenza epidemics detection [1], has attracted lots of attentions. Clustering methods are proposed to solve this problem. In work [9], a high utility pattern clustering (HUPC) framework is proposed by extracting a group of representative patterns from the microblog streams. Dirichlet-Hawkes Process is also used for bursty topic clustering in asynchronous streaming data, which is a scalable probabilistic generative model inheriting the advantages from both the Bayesian nonparametrics and the Hawkes Process [7]. Bursty topic detection problem is also considered as an anomalous subgraph detection problem. Multivariate evolving anomalous subgraphs scanning (DMGraphScan) framework is proposed in dynamic multivariate social media networks [13].

Recently, with the popularity of Latent Dirichlet Allocation (LDA) [3], many LDA-based extensions and variants have been proposed for topic discovery in microblogs [14]. Temporal information is used in LDA model to capture the bursty patterns with regards to time [18]. On the basis of temporal information, location of social messages is also captured by a location-time constrained (LTT) topic model [20] for bursty event detection. Images is also employed for generate visualized bursty topic summarization [2]. The impact of tweet images are thoroughly studied in [4], where a topic model integrating five features, i.e., text, time, position, hashtag and images, is designed to detect bursty events. The common idea of these models is to employ different features to strictly divide the topics into two sets, i.e., bursty topics and general topics. However, noise-filled microblog data seriously hurts their performance. In contrast, we detect one set of topics and associate each of them with a bursty level from which we identify the bursty topics.

Several works try to model the burstiness of words when detecting bursty topics. For example, wavelet analysis is used to build the burstiness for individual word, and then bursty words are clustered to form events with a modularity-based graph partitioning method [15]. A similar approach is employed to detect bursty tweet segments at first and then clusters the event segments into events by considering both their frequency distribution and content similarity [11]. The work [5] proposes a novel term aging model to compute the burstiness of each term, where it provides a graph-based method to retrieve the minimal set of terms, and then uses the knowledge of users' topic preferences to highlight the most emerging topics. Burstiness of words is also considered as prior information in a graphical model [17]. However, they all neglect the connections between words' bursty levels and topics, where words with the similar bursty levels usually refer to the same topic, which we have proved to be critical for the high-precision bursty topic detection.

5   Conclusion

This paper addressed the problem of bursty topic detection on microblogs. We propose to model the burstiness of topics in the microblog for bursty topic detection. For this purpose, we present a novel topic model named Topic-level Bursty Detection model, which exploits words, burstiness of words and temporal information. After that, we identify the bursty topics through the hypothesis testing method. Our qualitative and quantitative evaluations on a real dataset from Sina Weibo demonstrate that our approach is more effective than the competitors. For future work, if regarding a time slice as a data stream, our method is amenable to be a dynamic model to do bursty topic detection in microblog data streams. Meanwhile, our model can be extended to a non-parametric method such that it can automatically estimate the number of topics in each time slice.

Acknowledgements. This work was supported in part by the following funding agencies of China: National Key Research and Development Program of China under Grant 2016QY01W0200 and National Natural Science Foundation under Grant 61602050 and U1534201.

References

1. Aramaki, E., Maskawa, S., Morita, M.: Twitter catches the flu: detecting influenza epidemics using Twitter. In: Proceedings of the Conference on Empirical Methods in Natural Language Processing, pp. 1568–1576. Association for Computational Linguistics (2011)
2. Bian, J., Yang, Y., Chua, T.S.: Multimedia summarization for trending topics in microblogs. In: Proceedings of the 22nd ACM International Conference on Conference on Information and Knowledge Management, pp. 1807–1812. ACM (2013)
3. Blei, D.M., Ng, A.Y., Jordan, M.I.: Latent dirichlet allocation. J. Mach. Learn. Res. **3**, 993–1022 (2003)
4. Cai, H., Yang, Y., Li, X., Huang, Z.: What are popular: exploring Twitter features for event detection, tracking and visualization. In: Proceedings of the 23rd ACM International Conference on Multimedia, pp. 89–98. ACM (2015)
5. Cataldi, M., Caro, L.D., Schifanella, C.: Personalized emerging topic detection based on a term aging model. ACM Trans. Intell. Syst. Technol. (TIST) **5**(1), 7 (2013)

6. Diao, Q., Jiang, J., Zhu, F., Lim, E.P.: Finding bursty topics from microblogs. In: Proceedings of the 50th Annual Meeting of the Association for Computational Linguistics: Long Papers, vol. 1, pp. 536–544. Association for Computational Linguistics (2012)
7. Du, N., Farajtabar, M., Ahmed, A., Smola, A.J., Song, L.: Dirichlet-hawkes processes with applications to clustering continuous-time document streams. In: Proceedings of the 21th ACM SIGKDD International Conference on Knowledge Discovery and Data Mining, pp. 219–228. ACM (2015)
8. Griffiths, T.L., Steyvers, M.: Finding scientific topics. Proc. Nat. Acad. Sci. **101**(suppl 1), 5228–5235 (2004)
9. Huang, J., Peng, M., Wang, H.: Topic detection from large scale of microblog stream with high utility pattern clustering. In: Proceedings of the 8th Workshop on Ph.D. Workshop in Information and Knowledge Management, pp. 3–10. ACM (2015)
10. Lau, J.H., Collier, N., Baldwin, T.: On-line trend analysis with topic models: \# Twitter trends detection topic model online. In: Proceedings of COLING 2012, pp. 1519–1534 (2012)
11. Li, C., Sun, A., Datta, A.: Twevent: segment-based event detection from tweets. In: Proceedings of the 21st ACM International Conference on Information and Knowledge Management, pp. 155–164. ACM (2012)
12. Sakaki, T., Okazaki, M., Matsuo, Y.: Tweet analysis for real-time event detection and earthquake reporting system development. IEEE Trans. Knowl. Eng. **25**(4), 919–931 (2013)
13. Shao, M., Li, J., Chen, F., Huang, H., Zhang, S., Chen, X.: An efficient approach to event detection and forecasting in dynamic multivariate social media networks. In: Proceedings of the 26th International Conference on World Wide Web, pp. 1631–1639. International World Wide Web Conferences Steering Committee (2017)
14. Su, S., Wang, Y., Zhang, Z., Chang, C., Zia, M.A.: Identifying and tracking topic-level influencers in the microblog streams. Mach. Learn. **107**(3), 551–578 (2018)
15. Weng, J., Lee, B.S.: Event detection in Twitter. In: ICWSM, vol. 11, pp. 401–408 (2011)
16. Xie, W., Zhu, F., Jiang, J., Lim, E.P., Wang, K.: TopicSketch: real-time bursty topic detection from Twitter. IEEE Trans. Knowl. Data Eng. **28**(8), 2216–2229 (2016)
17. Yan, X., Guo, J., Lan, Y., Xu, J., Cheng, X.: A probabilistic model for bursty topic discovery in microblogs. In: AAAI, pp. 353–359 (2015)
18. Yin, H., Cui, B., Lu, H., Huang, Y., Yao, J.: A unified model for stable and temporal topic detection from social media data. In: 2013 IEEE 29th International Conference on Data Engineering (ICDE), pp. 661–672. IEEE (2013)
19. Zhao, W.X., et al.: Comparing Twitter and traditional media using topic models. In: Clough, P., et al. (eds.) ECIR 2011. LNCS, vol. 6611, pp. 338–349. Springer, Heidelberg (2011). https://doi.org/10.1007/978-3-642-20161-5_34
20. Zhou, X., Chen, L.: Event detection over Twitter social media streams. VLDB J. **23**(3), 381–400 (2014)
21. Zill, D., Wright, W.S., Cullen, M.R.: Advanced Engineering Mathematics. Jones & Bartlett Learning, Burlington (2011)

Adaptively Transfer Category-Classifier for Handwritten Chinese Character Recognition

Yongchun Zhu[1,2], Fuzhen Zhuang[1,2(✉)], Jingyuan Yang[3],
Xi Yang[4], and Qing He[1,2]

[1] Key Laboratory of Intelligent Information Processing of Chinese Academy of
Sciences (CAS), Institute of Computing Technology, CAS, Beijing 100190, China
`{zhuyongchun18s,zhuangfuzhen,heqing}@ict.ac.cn`
[2] University of Chinese Academy of Sciences, Beijing 100049, China
[3] George Mason University, Fairfax, USA
`jyang53@gmu.edu`
[4] Sunny Education Inc., Beijing 100102, China
`xi.yang@17zuoye.com`

Abstract. Handwritten character recognition (HCR) plays an important role in real-world applications, such as bank check recognition, automatic sorting of postal mail, the digitization of old documents, intelligence education and so on. Last decades have witnessed the vast amount of interest and research on handwritten character recognition, especially in the competition of HCR tasks on the specific data sets. However, the HCR task in real-world applications is much more complex than the one in HCR competition, since everyone has their own handwriting style, e.g., the HCR task on middle school students is much harder than the one on adults. Therefore, state-of-the-art methods proposed by the competitors may fail. Moreover, there is not enough labeled data to train a good model, since manually labelling data is usually tedious and expensive. So one question arises, is it possible to transfer the knowledge from related domain data to train a good recognition model for the target domain, e.g., from the handwritten character data of adults to the one of students? To this end, we propose a new neural network structure for handwritten Chinese character recognition (HCCR), in which we try to make full use of a large amount of labeled source domain data and a small number of target domain data to learn the model parameters. Furthermore, we make a transfer on the category-classifier level, and adaptively assign different weights to category-classifiers according to the usefulness of source domain data. Finally, experiments constructed from three data sets demonstrate the effectiveness of our model compared with several state-of-the-art baselines.

1 Introduction

The handwritten character recognition problem has attracted much interest and research for a long time, and plays an important role in various kinds of

© Springer Nature Switzerland AG 2019
Q. Yang et al. (Eds.): PAKDD 2019, LNAI 11439, pp. 110–122, 2019.
https://doi.org/10.1007/978-3-030-16148-4_9

applications [10, 20, 22], such as bank cheque recognition, automatic sorting of postal mail, the digitization of old documents, intelligence education and so on. The previous handwritten character recognition works can be grouped into different types, including the recognition tasks concerning about digits [10], English characters [11], Chinese characters [20, 22], French characters [7] etc. In this paper, we focus on the handwritten Chinese character recognition (HCCR) problem, and consider more challenging recognition scenarios which are much closer to approaching the real-world applications.

The HCCR problem has been extensively studied for more than 40 years [12], and can be further divided into two types: online and off-line recognition. The online recognizer identifies characters during the writing process using the digitised trace of the pen, while off-line recognition deals with images scanned of previously handwritten characters. Usually, the online recognition task is easier than the off-line one since there is much digitised trace information available for training the models. However, off-line recognition has broader applications, e.g., automatic sorting of postal mail and the editing of old documents. In the recent decade, there are many research works and competitions devoted to the off-line recognition of Chinese characters, especially based on the deep neural network framework [1, 18]. Convolution neural network (CNN), which was originally developed by LeCun et al. [10], provided a new end-to-end approach to handwritten Chinese character recognition with very promising results in recent years [18, 22], e.g., the extended deeper architectures of AlexNet, VGG, GoogLeNet, ResNet with dropout and nonlinear activation function ReLU. Ciresan et al. [2] proposed the multi-column deep neural network (MCDNN), which may be the first successful model based on deep neural network (DNN) used in the application of large-scale HCCR tasks. The winner of online and off-line handwritten Chinese character recognition competition in ICDAR2013 was based on MCDNN [20]. Zhong et al. [22] proposed HCCR-GoogLeNet and employed three types of directional feature maps, namely the Gabor, gradient and HoG feature maps, to enhance the performance of GoogLeNet, leading to the high accuracy of 96.74% on the ICDAR2013 off-line data set. Recently, Yang et al. [19] proposed a new training method DropSample to enhance deep convolutional neural networks for large-scale HCCR problems.

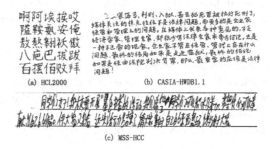

(a) HCL2000   (b) CASIA-HWDB1.1

(c) MSS-HCC

Fig. 1. The examples of three data sets, HCL2000, CASIA-HWDB1.1 and MSS-HCC.

Though there are many advanced algorithms proposed for HCCR, HCCR is still a challenging task. First, comparing English characters with only 26 basic categories, in Chinese the national standard GB2312-80 coding defined 6763 categories of commonly used Chinese characters. GB18010-2000 further expanded the Chinese character set to 27,533 categories. Even for the recently used Chinese character sets HCL2000 [21] and CASIA-HWDB1.1 [13], which both contain 3,755 categories of characters. Second, handwritten Chinese characters are very casual and diverse compared with regular printed characters. Third, there are many similar and confusing Chinese characters, which human can recognize easily but very hard for computer. For example, "己" v.s. "已", "曰" v.s. "日", "淡" v.s. "谈" and "何" v.s. "向". Moreover, there are also other technical challenges in addition to the complexity of Chinese characters. On one hand, recent HCCR competitions are targeted at improving the performance on a specific standard data set, e.g., CASIA-HWDB1.1 is adopted in ICDAR2013, but the recognition scenarios are much more complex and difficult in real-world applications. Figure 2 shows some examples from three data sets, in which HCL2000 and CASIA-HWDB1.1 are both well known off-line handwritten Chinese character data sets, while MSS-HCC is the one we collect from the middle school students writing answers and compositions in a specific exam[1]. From this figure, we can find that the recognition of MSS-HCC is much harder than HCL2000 and CASIA-HWDB1.1. On the other hand, there is usually not enough labeled data to train a satisfying model, since manually labelling is tedious, time-consuming and expensive. Also, it might fail when simply applying the model trained from one data set to another data set, e.g., the proposed model [22] trained on HCL2000 only obtains about 63% accuracy on CASIA-HWDB1.1.

Transfer learning aims to adapt the knowledge from related source domain data to the model learning in the target domain, which provides the possibility of success for HCCR tasks. Along this line, we propose a transfer handwritten Chinese character recognition model based on the successful deep network structure AlexNet [9]. Specifically, for both source and target domain data, we share the network parameters with five convolution layers and three pooling layers, and then learn the parameters of three fully connected layers separately. In addition, to adaptively transfer the category knowledge from the source domain to the target domain, we impose a regularization item with different weights to learn the similarity of category-classifiers trained from the source to the target domain. Finally, we conduct extensive experiments on three data sets to validate the effectiveness of our model.

The remainder of this paper is organized as follows. Section 2 briefly introduces the related work of handwritten character recognition and transfer learning, followed by the problem formalization and the details of the proposed model in Sect. 3. Section 4 presents the experimental results to demonstrate the effectiveness of our model, and finally Sect. 5 concludes this paper.

---

[1] This work does not consider how to segment the characters, but only focuses on the recognition of segmented isolate characters.

2   Related Work

In this section, we will briefly introduce the most related work about handwritten Chinese character recognition and transfer learning.

2.1   Handwritten Chinese Character Recognition

Ciresan et al. [4] first used CNNs to realize the classification of 1,000 types of handwritten digit characters, then based on which IDSIA Lab won the first place on off-line HCCR data set with the accuracy of 92.12% and the fourth place on online HCCR data set with 93.01% in ICDAR2011 competition [14]. The champions of off-line and online HCCR competition in ICDAR2013 [20] are based on MCDNN. Wu et al. [18] proposed to improve the performance of off-line HCCR task up to 96.06% by adopting the ensemble of four alternately trained relaxation convolutional neural networks (ATR-CNN). After that in 2015, Zhong et al. [22] proposed the HCCR-GoogLeNet with Gabor, gradient and HoG feature maps to obtain an accuracy of 96.74% accuracy on ICDAR2013 off-line data set, which is the first time computers outperformed human recognition accuracy 96.13%. Furthermore, Chen et al. [1] uses a deeper CNN network to achieve 96.79% accuracy, and Yang et al. [19] proposed a new DropSample to train ensemble CNNs model to get 97.06% accuracy on ICDAR2013 data set. However, these above methods are all designed for a specific standard data set, which might fail in more complex HCCR scenarios. Also, they don't consider to make use of the large amount of auxiliary data, i.e., HCL2000. Therefore, in this work we try to apply the knowledge from related source domain data for further improving the recognition performance in target domain.

2.2   Transfer Learning

Transfer learning targets at learning the knowledge from large amount of related source/auxiliary data to help improve the prediction performance of target domain data [16]. In recent decade, transfer learning has provoked vast amount of attention and research for variable kinds of applications, e.g., text classification [5], image classification [23], visual categorization [17] etc. To the best of our knowledge, there is little work of transfer learning for handwritten Chinese character recognition problems. The work [3] actually learnt Chinese characters by first pre-training a DNN on other data sets. Thus, we will employ transfer learning manner based on deep network to deal with HCCR. Furthermore, most previous work make transfer on the instance level [8], the model level [6], and the feature learning level [15], but we focus on transferring the knowledge on the category-classifier level.

3   Adaptively Transfer Category-Classifier for Chinese Handwriting Recognition

3.1   Problem Formulation

For clarity, the frequently used notations are listed in Table 1. Supposing we have data in both the source and the target domain $\mathcal{D}_s = \{x_i^{(s)}, y_i^{(s)}\}|_{i=1}^{n_s}$ and $\mathcal{D}_t = \mathcal{D}_t^L \cup \mathcal{D}_t^U = \{x_i^{(t)}, y_i^{(t)}\}|_{i=1}^{n_t^L} \cup \{x_i^{(t)}\}|_{i=1}^{n_t^U}$, respectively, where $x_i^{(s)}, x_i^{(t)} \in \mathbb{R}^{m \times m}$ are the data instances with image size $m \times m$, $y_i^{(s)}, y_i^{(t)} \in \{1, \cdots, c\}$ are their corresponding labels, c is the number of categories, n_s and $n_t = n_t^L + n_t^U$ are respectively the numbers of instances in source and target domains. Usually there are large number of labeled instances in source domain and only a small portion of labeled instances in target domain, i.e., $n_t^L \ll n_s$ and $n_t^L \ll n_t^U$, thus this is an challenging recognition task. Since the data distributions of source domain and target domain are different, our goal adopts transfer learning to make full use of source domain data $\mathcal{D}_s$ and a small portion of labeled target domain data $\mathcal{D}_t^L$ to train a recognition model and obtain satisfying performance on the unlabeled target domain data $\mathcal{D}_t^U$.

Table 1. The notation and denotation

$\mathcal{D}_s, \mathcal{D}_t$	The source and target domains
n_s	The number of instances in source domain
n_t	The number of instances in target domain
l	The index of layer
$n_t^L (n_t^U)$	The number of labeled (unlabeled) instances in target domain
m	The width and the height of original map
K_{ij}^l	The j-th kernel filter in the l-th layer connected to the i-th map in the $(l-1)$-th layer
k^l	The number of nodes in l-th full connected layer
κ^l	The kernel size in l-th convolutional layer
c	The number of categories
$x_i^{(s)}, x_i^{(t)}$	The i-th instance of source and target domains
$y_i^{(s)}, y_i^{(t)}$	The label of instances $x_i^{(s)}$ and $x_i^{(t)}$
a^l	The output of the l-th full connected layer
W^l, b^l	Weight matrix and bias for the l-th full connected
θ_j	$\theta_j (j \in \{1, \cdots, c\})$ is the j-th column of W^8
$\top$	The transposition of a matrix
$\xi^{(s)}, \xi^{(t)}$	The input of the softmax layer

3.2 Adaptively Transferring Category-Classifier Model

Motivated by the success of deep network structure AlexNet for image classification [9] and handwritten character recognition [22], we propose a new Adaptively Transferring Category-classifier model for HCCR (ATC-HCCR for short) based on AlexNet. The architecture of the network of ATC-HCCR is shown in Fig. 2. Particularly, this network has a total of eight layers including the first five successive convolutional layers $conv1, \cdots, conv5$ ($conv1, conv2, conv5$ are followed by pooling layers.) and three fully connected layers $fc6, fc7$, and $fc8$. In Fig. 2, two important transfer learning components are utilized to improve the performance. The first knowledge transfer is achieved when both the source and the target domain share the same parameters of the five convolutional layers and three pooling layers for the shared convolutional kernels and pooling operations. Then, the network is diverged into two branches to learn the parameters of three fully connected layers separately, where one is for the source domain and the other one is the target domain. Furthermore, we impose a regularization item with different weights as the second transfer knowledge component to adaptively transfer the category knowledge from the source domain to the target domain by learning the similarity of category-classifiers trained from the source domain to the target domain.

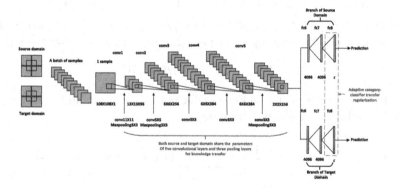

Fig. 2. The network structure of ATC-HCCR.

Given $x_i^l \in \mathbb{R}^{m^l \times m^l}$ represents the i-th map in the l-th layer and the map size is $m^l \times m^l$, j-th kernel filter in the l-th layer connected to the i-th map in the $(l-1)$-th layer denoted as $K_{ij}^l \in \mathbb{R}^{\kappa^l \times \kappa^l}$ and index maps set $M_j = \{i | i\text{-th map in the } (l-1)\text{-th layer connected to } j\text{-th map in the layer}\}$. So the convolutional operation is defined by the following equation,

$$x_j^l = f(\sum_{i \in M_j} x_i^{l-1} * K_{ij}^l + b_j^l), \tag{1}$$

where $*$ denotes convolutional operation, $f(z) = \max(0, z)$ is ReLU non-linearity activation function and b_j^l is bias.

Next, pooling layer then combines the output of the neuron cluster at one layer to single neuron in the next layer. Pooling operations are carried out to reduce the number of data points and to avoid overfitting. And pooling equation can be described as

$$x_j^l = down(x_j^l), \tag{2}$$

where $down(\cdot)$ is max pooling to computer the max value of each $p \times p$ region in x_j^l map. After the last pooling layers, pixels of pooling layers are stretched to single column vector. These vectorized and concatenated data points are fed into fully connected layers for the classification.

The output of the l-th fully connected layer of the branch of source domain is listed as follows, and the one of target domain is similar.

$$a_s^l = f(W_s^l a_s^{l-1} + b_s^l) \tag{3}$$

The $fc6$ and $fc7$ in Fig. 2 both have an output $a_s^l \in \mathbb{R}^{k^l \times 1}$ (l-th full connected layer) of k^l nodes, a weight matrix $W_s^l \in \mathbb{R}^{k^l \times k^{(l-1)}}$, and a bias vector $b_s^l \in \mathbb{R}^{k^l \times 1}$ (the l-th full connected layer). And the output of the fc7 is denoted as $\xi^{(s)} \in \mathbb{R}^{k^7 \times 1}$.

$$y^{(s)} = g(W_s^8 \xi^{(s)} + b_s^8), \tag{4}$$

$g(\cdot)$ is a softmax function, and $y^{(s)} \in \mathbb{R}^{c \times 1}, W_{(s)}^8 \in \mathbb{R}^{c \times k^7}$. Let $W_s^8 = [\theta_1^{(s)}, \theta_2^{(s)}, \cdots, \theta_c^{(s)}]^\top$, $W_t^8 = [\theta_1^{(t)}, \theta_2^{(t)}, \cdots, \theta_c^{(t)}]^\top$, $\gamma = [\gamma_1, \gamma_1, \cdots, \gamma_c]$, the objective function to be minimized in our proposed learning framework is formalized as follows:

$$\mathcal{L} = J_r(\mathcal{D}_s \cup \mathcal{D}_t^L, y) + \Omega(W_s^8, W_t^8), \tag{5}$$

where the first term is the Cross Entropy for both the source and target domain, which can be further detailed defined as

$$J_r(\mathcal{D}_s \cup \mathcal{D}_t^L, y) = -\frac{1}{n_s + n_t^L} \sum_{i=1}^{n_s+n_t^L} \sum_{j=1}^{c} 1\{y_i = j\} \log \frac{e^{\theta_j^\top \xi_i}}{\sum_{u=1}^{c} e^{\theta_u^\top \xi_i}}, \tag{6}$$

where ξ_i is the i-th instance, and $\theta_j^\top \in \mathbb{R}^{k^7 \times 1}$ ($j \in \{1, ..., c\}$) is the j-th row of W_s^8 or W_t^8. The second term of the objective is the regularization term, which is defined as

$$\Omega(W_s^8, W_t^8) = \lambda \sum_{i=1}^{c} \gamma_i \cdot |\theta_i^{(s)} - \theta_i^{(t)}|. \tag{7}$$

In the regularization term, there are two parameters, namely, the trade-off parameter λ and the weights γ. λ controls the significance of the regularization term, while γ is for transferring category-classifiers between the source and target domain. The term $|\theta_i^{(s)} - \theta_i^{(t)}|$ represents the distance of i-th category-classifier between the two domains.

We use tensorflow to implement our network and AdamOptimizer as the optimizer, and the detailed algorithm is shown in Algorithm 1. Note that, there is

---

Algorithm 1. Transfer Learning with Adaptively Transfer Category-classifier

---

Input: Given one source domain $\mathcal{D}_s = \{x_i^{(s)}, y_i^{(s)}\}|_{i=1}^{n_s}$, and one target domain $\mathcal{D}_t = \mathcal{D}_t^L \cup \mathcal{D}_t^U = \{x_i^{(t)}, y_i^{(t)}\}|_{i=1}^{n_t^L} \cup \{x_i^{(t)}\}|_{i=1}^{n_t^U}$, trade-off parameters λ and weights γ, the number of nodes in full connected layer and label layer, k and c.

Output: Results of x_i belongs to the vector of probability for each category.

1. Use both $\mathcal{D}_s$ and $\mathcal{D}_t^L$ to train AlexNet.
2. Use the parameters in Step1's model to initialize ATC-HCCR shown in Figure 2.
3. Choose a batch of instances from $\mathcal{D}_s$ or $\mathcal{D}_t^L$ as input.
4. Use AdamOptimizer with loss function Eq. (5) to update all variables.
5. Continue Step3 and Step4 until the algorithm converges.
6. Input $\mathcal{D}_t^U$ and get the vector of probability for each category that x_i belongs to.

---

only a small amount of labeled data in the target area, oversampling is required. Besides, training a randomly initialized model can waste a lot of time, so pre-training method is used. Specifically, we first use the source domain and the small amount of labeled target domain data to train a AlexNet, and then we use the parameters of this model to initialize our model. In our experiment, oversampling and pre-training are both used in ATC-HCCR.

After all the parameters are learned, we can use the classifiers to predict the target domain. That is, for any instance $x^{(t)}$ in target domain, the output of the $y^{(t)}$ can indicate $x^{(t)}$ belonging to the vector of probability for each category. We choose the maximum probability and the corresponding label as the prediction.

4   Experimental Evaluation

In this section, we conduct extensive experiments on three real-world handwritten Chinese character data sets to validate the effectiveness of the proposed framework.

4.1   Data Preparation

Two of the three data sets are standard ones, i.e., HCL2000 and CASIA-HWDB1.0[2], and the rest one MSS-HCC is collected by ourselves. The statistics of three data sets are listed in Table 2.

Table 2. The statistics of three data sets.

	HCL2000	CASIA-HWDB1.1	MSS-HCC
#category	3,755	3,755	27
#instance	3,755,000	1,126,500	5,920

---

[2] We thank the authors for providing these two data sets.

HCL2000 [21] contains 3,755 categories of frequently used simplified Chinese characters written by 1,000 different persons. All images for each character in this data set is 64×64. As shown in Fig. 1(a), the characters in the data set are neat and orderly.

CASIA-HWDB1.1 [13] is produced by 300 persons, which includes 171 categories of alphanumeric characters and symbols, and 3,755 categories of Chinese characters. The Chinese characters are used as the experimental data. As shown in Fig. 1(b), the characters are written less neat and orderly.

MSS-HCC is labeled by ourselves. We collect this data set from the middle school students writing answers and compositions in a specific exam, and after segmentation the images with size 108×108 for each Chinese character are obtained. Then we labeled about 20,000 images, and those categories of Chinese characters whose number of instances larger than 100 are selected as the experimental data. As shown in Fig. 1(c), this data set is written much in messy.

For these three data sets, we conduct some preliminary tests applying the HCCR method in [22]. We use HCL2000 as training data and CASIA-HWDB1.1 for test, then the accuracy is 63%. In contrast, using CASIA-HWDB1.1 as training data and HCL2000 for test, the accuracy is achieved at 94%. This shows that HCL2000 is more neat than CASIA-HWDB1.1. In addition, respectively using HCL2000 and CASIA-HWDB1.1 as training sets, MSS-HCC for test has accuracies of 53% and 71%. In contrast, using MSS-HCC as the training data, the accuracies on HCL2000 and CASIA-HWDB1.1 are 86% and 87%. These results reveal that MSS-HCC is more messy than HCL2000 and CASIA-HWDB1.1. In the experiments, we focus on the knowledge transfer from source domain to improve the recognition performance of much more difficult tasks. Therefore, three transfer HCCR problems are finally constructed, i.e., HCL2000 → CASIA-HWDB1.1, HCL2000 → MSS-HCC and CASIA-HWDB1.1 → MSS-HCC.

4.2 Baselines and Implementation Details

Baselines: We mainly compare our model with following two state-of-the-art baselines,

- AlexNet-HCCR [22], which uses AlexNet for HCCR task, contains five convolutional layers, three pooling layers, and three full connected layers. There is not transfer mechanism for this method.
- preDNN [3], is actually an accelerated deep neural network (DNN) model by first pretraining a DNN on a small subset of all classes and then continuing to train on all classes. As claimed in their original paper, preDNN is a transfer learning approach for handling HCCR problems.

For AlexNet-HCCR, we record three values of accuracy for each transfer learning problem. Specifically, training the models on labeled source domain data $\mathcal{D}_s$, labeled target domain data $\mathcal{D}_t^L$, labeled source and target domain data $\mathcal{D}_s \cup \mathcal{D}_t^L$, respectively, and then testing unlabeled target domain data $\mathcal{D}_t^U$, denoted as AlexNet-HCCR(s), AlexNet-HCCR(t) and AlexNet-HCCR(s+t), respectively.

For preDNN, we first pretrain the model on $\mathcal{D}_s$, then continue to train on $\mathcal{D}_t^L$, and finally make prediction on $\mathcal{D}_t^U$.

Implementation Details: There are two parameters, i.e., trade-off parameter λ and weights $\gamma_i(1 \leq i \leq c)$ for transferring category-classifiers between source and target domains. We set $\lambda = 5$ for all experiments, and for $\gamma_i(1 \leq i \leq c)$ we simply set them according to the accuracies on $\mathcal{D}_t^L$ given by the AlexNet-HCCR model trained from $\mathcal{D}_s$, i.e., the higher value of accuracy for the i-th category, the larger value is set to $\gamma_i(1 \leq i \leq c)$, and vice versa. Certainly, it would be better to study the optimum setting for $\gamma_i(1 \leq i \leq c)$, which will be our future work. The number of iterations for optimization is 50,000, and the average values of accuracy are recorded for 3 trials. Finally, a small portion of target domain data are randomly sampled as labeled ones. Specifically, we set the sampling ratio from [1.67%, 10%] with interval 1.67% for CASIA-HWDB1.1 as target domain, and from [5%, 30%] with interval 5% for MSS-HCC as target domain. The prediction accuracy is adopted as the evaluation metric.

4.3 Experimental Results

We evaluate all the approaches under different sampling ratios of labeled target domain data, and all the results are shown in Table 3. From these results, we have the following insightful observations,

- Except AlexNet-HCCR(s) only using labeled source domain data, the performance of all the other algorithms improves with the increasing values of sampling ratio of target domain data as labeled data. Generally, the performance increases significantly with the increasing of sampling ratio, and then slowly, which coincides with our expectation. Because if there are enough labeled data for training a good model, incorporating more labeled data will not take much effect.
- Our model ATC-HCCR achieves the best results over all baselines, under different sampling ratios of target domain data, which demonstrates the effectiveness of the proposed transfer learning framework for HCCR tasks. Also, we observe that ATC-HCCR beats baselines with a large margin of improvement on the problem of HCL2000 → CASIA-HWDB1.1, and much smaller margin on the problems of HCL2000 → MSS-HCC and CASIA-HWDB1.1 → MSS-HCC. This is due to the fact that the recognition of MSS-HCC data set is more challenging. On the other hand, for a challenging problem, even a small value of 0.5% improvement is remarkable.
- Both transfer learning models ATC-HCCR and preDNN outperform AlexNet-HCCR, which indicates the importance and necessity of applying transfer learning for tackling HCCR problems. ATC-HCCR is better than preDNN, since preDNN, as a simple transfer learning algorithm, only tries to adopt all network parameters from the source domain for initialization but not considers the transfer of category-classifiers.

Table 3. The performance (%) comparison on three data sets among AlexNet-HCCR, preDNN and ATC-HCCR.

	HCL2000 → CASIA-HWDB1.1						Mean
	1.67%	3.33%	5%	6.67%	8.33%	10%	
AlexNet-HCCR(s)	63.30	63.31	63.40	63.48	63.53	63.41	63.41
AlexNet-HCCR(t)	30.83	61.64	79.52	78.04	81.01	81.85	68.82
AlexNet-HCCR(s+t)	73.07	76.78	79.52	81.13	82.24	82.08	79.14
preDNN	73.01	76.89	79.37	81.47	82.47	83.56	79.46
ATC-HCCR	76.79	79.78	82.37	84.13	85.08	85.06	**82.20**
	HCL2000 → MSS-HCC						Mean
	5%	10%	15%	20%	25%	30%	
AlexNet-HCCR(s)	61.49	63.18	62.61	62.75	63.92	64.30	63.04
AlexNet-HCCR(t)	66.44	82.83	89.77	90.96	92.57	93.00	85.93
AlexNet-HCCR(s+t)	86.31	88.95	91.02	91.55	92.22	94.76	90.80
preDNN	86.93	90.69	92.61	93.45	93.90	94.61	92.03
ATC-HCCR	87.76	91.12	93.24	93.71	94.57	94.88	**92.55**
	CASIA-HWDB1.1 → MSS-HCC						Mean
	5%	10%	15%	20%	25%	30%	
AlexNet-HCCR(s)	76.01	78.38	78.27	78.12	78.38	77.87	77.84
AlexNet-HCCR(t)	66.44	82.83	89.77	90.96	92.57	93.00	85.93
AlexNet-HCCR(s+t)	89.48	91.38	92.27	93.67	94.21	94.98	92.67
preDNN	89.19	92.64	92.74	93.58	94.61	94.98	92.96
ATC-HCCR	90.98	93.14	93.80	94.55	94.68	95.29	**93.74**

Table 4. The Influence of trade-off parameter λ on the performance (%) of ATC-HCCR.

λ	HCL2000 → CASIA-HWDB1.1					
	1.67%	3.33%	5%	6.67%	8.33%	10%
0	75.77	79.58	82.03	83.83	84.65	84.90
0.05	76.30	80.14	82.56	83.83	85.05	85.10
0.5	76.51	79.76	82.41	84.08	84.84	85.29
5	76.79	79.78	82.37	84.13	85.08	85.06

4.4   The Influence of Trade-Off Parameter λ

We investigate the influence of trade-off parameter λ on the performance of ATC-HCCR over the problem HCL2000 → CASIA-HWDB1.1, and λ is sampled from {0, 0.05, 0.5, 5}. The results are shown in Table 4. $\lambda = 0$ indicates that ATC-HCCR only considers the parameters sharing of five convolutional layers and three pooling layers during the optimization, and even so ATC-HCCR can outperform preDNN. When $\lambda > 0$, the transfer category-classifier regularization is integrated in our model, and the performance of ATC-HCCR can be further improved, which shows the effectiveness of transfer category-classifier regularization. In our experiments, we simply set the weights $\gamma_i (1 \leq i \leq c)$ according to the accuracies of AlexNet-HCCR making predictions on $\mathcal{D}_t^L$. If the size of $\mathcal{D}_t^L$ is small, the estimation of γ_i may not be reliable, therefore λ is not set to a large value, i.e., $\lambda = 5$ in the experiments.

5   Conclusion

In this paper, we study the challenging handwritten Chinese character recognition (HCCR) problem in real-world applications. As there is little work about transfer learning for HCCR, based on Alexnet, we propose a new network framework by adaptively transferring category-classifier for HCCR problems. In our framework, there are actually two components for knowledge transfer. First, the parameters of five convolutional layers and three pooling operations are shared across the source and the target domain during the optimization; second, observing that the category-classifiers from two domains have different similarities, therefore different weights are imposed to regularize the category-classifier transfer. Furthermore, we also collect a small set of much more challenging HCCR data, and finally conduct experiments on three data sets to demonstrate the effectiveness of our model. In future work, we will collect more data and consider how to find the optimum values of weights.

Acknowledgments. The research work is supported by the National Key Research and Development Program of China under Grant No. 2018YFB1004300, the National Natural Science Foundation of China under Grant Nos. U1836206, U1811461, 61773361, the Project of Youth Innovation Promotion Association CAS under Grant No. 2017146.

References

1. Chen, L., Wang, S., Fan, W., Sun, J., Naoi, S.: Beyond human recognition: a CNN-based framework for handwritten character recognition. In: ACPR, pp. 695–699 (2015)
2. Ciresan, D., Meier, U., Schmidhuber, J.: Multi-column deep neural networks for image classification. In: CVPR, pp. 3642–3649 (2012)
3. Cireşan, D.C., Meier, U., Schmidhuber, J.: Transfer learning for Latin and Chinese characters with deep neural networks. In: IJCNN, pp. 1–6 (2012)
4. Ciresan, D.C., Meier, U., Gambardella, L.M., Schmidhuber, J.: Convolutional neural network committees for handwritten character classification. In: ICDAR, pp. 1135–1139 (2011)
5. Dai, W., Xue, G.R., Yang, Q., Yu, Y.: Co-clustering based classification for out-of-domain documents. In: ACM SIGKDD, pp. 210–219 (2007)
6. Gao, J., Fan, W., Jiang, J., Han, J.: Knowledge transfer via multiple model local structure mapping. In: SIGKDD, pp. 283–291 (2008)

7. Grosicki, E., El-Abed, H.: ICDAR 2011-French handwriting recognition competition. In: ICDAR, pp. 1459–1463 (2011)
8. Jiang, J., Zhai, C.: Instance weighting for domain adaptation in NLP. In: ACL, pp. 264–271 (2007)
9. Krizhevsky, A., Sutskever, I., Hinton, G.E.: Imagenet classification with deep convolutional neural networks. In: NIPS, pp. 1097–1105 (2012)
10. LeCun, Y., et al.: Handwritten digit recognition with a back-propagation network. In: NIPS, pp. 396–404 (1990)
11. Lin, Z., Wan, L.: Style-preserving English handwriting synthesis. Pattern Recognit. **40**, 2097–2109 (2007)
12. Liu, C.L., Jaeger, S., Nakagawa, M.: 'Online recognition of Chinese characters: the state-of-the-art. IEEE TPAMI **26**, 198–213 (2004)
13. Liu, C.L., Yin, F., Wang, D.H., Wang, Q.F.: Online and offline handwritten Chinese character recognition: benchmarking on new databases. Pattern Recognit. **46**, 155–162 (2013)
14. Liu, C.L., Yin, F., Wang, Q.F., Wang, D.H.: ICDAR 2011 Chinese handwriting recognition competition. In: ICDAR (2011)
15. Pan, S.J., Tsang, I.W., Kwok, J.T., Yang, Q.: Domain adaptation via transfer component analysis. IEEE TNN **22**, 199–210 (2011)
16. Pan, S.J., Yang, Q.: A survey on transfer learning. IEEE TKDE **22**, 1345–1359 (2010)
17. Shao, L., Zhu, F., Li, X.: Transfer learning for visual categorization: a survey. IEEE TNNLS **26**, 1019–1034 (2015)
18. Wu, C., Fan, W., He, Y., Sun, J., Naoi, S.: Handwritten character recognition by alternately trained relaxation convolutional neural network. In: Proceedings of 14th ICFHR, pp. 291–296 (2014)
19. Yang, W., Jin, L., Tao, D., Xie, Z., Feng, Z.: DropSample: a new training method to enhance deep convolutional neural networks for large-scale unconstrained handwritten Chinese character recognition. Pattern Recognit. **58**, 190–203 (2016)
20. Yin, F., Wang, Q.F., Zhang, X.Y., Liu, C.L.: ICDAR 2013 Chinese handwriting recognition competition. In: ICDAR, pp. 1464–1470 (2013)
21. Zhang, H., Guo, J., Chen, G., Li, C.: HCL2000-a large-scale handwritten Chinese character database for handwritten character recognition. In: ICDAR, pp. 286–290 (2009)
22. Zhong, Z., Jin, L., Xie, Z.: High performance offline handwritten Chinese character recognition using googlenet and directional feature maps. In: Proceedings of 13th ICDAR, pp. 846–850 (2015)
23. Zhu, Y., et al.: Heterogeneous transfer learning for image classification. In: AAAI (2011)

Syntax-Aware Representation for Aspect Term Extraction

Jingyuan Zhang[1,2], Guangluan Xu[1], Xinyi Wang[1,2], Xian Sun[1],
and Tinglei Huang[1(✉)]

[1] Key Laboratory of Network Information System Technology (NIST),
Institute of Electronics, Chinese Academy of Sciences,
No. 19 North 4th Ring Road West, Haidian District, Beijing, China
{zhangjingyuan16,wangxinyi16}@mails.ucas.ac.cn,
{gluanxu,sunxian,tlhuang}@mail.ie.ac.cn
[2] University of Chinese Academy of Sciences,
No.19(A) Yuquan Road, Shijingshan District, Beijing, China

Abstract. Aspect Term Extraction (ATE) plays an important role in aspect-based sentiment analysis. Syntax-based neural models that learn rich linguistic knowledge have proven their effectiveness on ATE. However, previous approaches mainly focus on modeling syntactic structure, neglecting rich interactions along dependency arcs. Besides, these methods highly rely on results of dependency parsing and are sensitive to parsing noise. In this work, we introduce a syntax-directed attention network and a contextual gating mechanism to tackle these issues. Specifically, a graphical neural network is utilized to model interactions along dependency arcs. With the help of syntax-directed self-attention, it could directly operate on syntactic graph and obtain structural information. We further introduce a gating mechanism to synthesize syntactic information with structure-free features. This gate is utilized to reduce the effects of parsing noise. Experimental results demonstrate that the proposed method achieves state-of-the-art performance on three widely used benchmark datasets.

Keywords: Aspect term extraction · Syntactic information ·
Gating mechanism

1 Introduction

Aspect term extraction plays an important role in fine-grained sentiment analysis. The goal of ATE is to identify explicit aspect terms from user generated contents such as microblogs, product reviews, etc. For example, an ATE system should extract "new Windows" and "touchscreen functions" as aspect terms from a review "I do not enjoy the new Windows and touchscreen functions".

It is a common phenomenon that there exist semantic relations between opinion indicators and aspect terms. Syntax-based models have proven to be

© Springer Nature Switzerland AG 2019
Q. Yang et al. (Eds.): PAKDD 2019, LNAI 11439, pp. 123–134, 2019.
https://doi.org/10.1007/978-3-030-16148-4_10

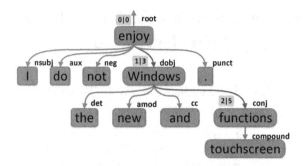

Fig. 1. Architecture of dependency tree and corresponding syntactic graph

effective in capturing such semantic relations. These methods could learn long-term syntactic dependency and semantic representation that are irrelevant to the surface form of text. For the example introduced above, we need 5 steps from "enjoy" to "functions" in sequential order, but with the help of dependency arcs as shortcut, the relative distance could be reduced to 2 as shown in Fig. 1. Apart from capturing existence evidence, the "amod" relation between "Windows" and "new" and the "compound" arc between "touchscreen" and "functions" also present the inner structure of aspect terms.

There are some prior works that utilize syntactic information for ATE. [22] use hand-crafted syntactic rules as features, however, designing these rules is labor-intensive. Recently, there are some neural network based approaches that make use of syntactic information. For example, [20] propose a syntax-based representation learning method, but the unsupervised features do not achieve satisfactory results because they could not be fine-tuned. [11,17] leverage tree-based networks to propagate syntactic information along dependency trees recursively. Besides, Graphical Convolutional Networks (GCNs) are also adopted to operate on syntactic graphs [19] to learn dependency information.

However, these approaches face several challenges. First, tree-based neural models are usually difficult to parallelize and thus computationally inefficient [21]. Second, GCNs based models could be parallelized but they directly receive information from syntactically related nodes [19] or use edge-wise gating [13] to control information flow, these models neglect the fact that there should be interactions along dependency relations. Third, previous syntax-aware methods mainly focus on modeling syntactic structure. They highly rely on correct parse trees and pay little attention to maintaining word order information. As a result, these methods suffer from error propagating along noisy dependency arcs when dealing with imperfect trees.

In this work, we introduce a novel syntax-aware representation framework to resolve these issues. Attention mechanism, which is parallelizable and computational efficient, is introduced to model syntactic information. With the constraint of syntactic structure, the attention mechanism could dynamically calculate interactions along dependency arcs and obtain information from semantically related words. Moreover, we present a novel gating mechanism to synthesize structural model with syntax-free model. This method could alleviate the effects of parsing noise. Experimental results show that both of the proposed components could improve the performance of ATE and our syntax-aware representation framework achieves state-of-the-art results on three widely used datasets.

Our main contributions can be summarized as follows:

- We propose Syntax-Directed Attention Network (SDAN) to model the interactions along dependency paths and generate high quality syntax-aware representations for ATE.
- To make the model more robust for parsing noise, we further introduce Contextual Gating Mechanism (CGM) that could synthesize syntactic information and syntax-free features.
- Experimental results show that the proposed model achieves state-of-the-art performance on three widely used datasets.

2   Related Work

Extracting explicit aspect terms has been actively investigated in the past. There are many unsupervised methods such as frequent terms mining based approach [6], rule-based method [22] and topic modeling based models [2,9]. The mainstream of supervised approach is Conditional Random Field (CRF) [4], which treats ATE as a sequence labeling problem and relies on hand-crafted features. Recently, many deep learning based methods have been explored for ATE. Most of these approaches utilize recurrent neural network [8], representation learning [20] or multi-task learning [10,18] methods to learn semantic representation. Nevertheless, these methods do not explicitly model syntactic structures and it is hard for them to capture long-term semantic dependency.

As an important linguistic resource, syntactic information has been adopted for ATE. There are some tree-based neural models [11,17] that could propagate information along parse tree, however, operating on tree structure recursively is time-consuming. As an alternative, GCNs [13,19] are introduced to learn syntactic information. GCNs could directly operate on dependency graph and obtain information from syntactic neighbors. However, these methods neglect the interactions along dependency arcs, it is hard for them to weight the importances of syntactically connected nodes.

We utilize self-attention to handle this issue. Conventional self-attention models the interactions between all word pairs in a single sequence and attends to semantically related words. In this work, instead, we use syntactic information to constrain the model and force it to only attend to syntactically connected nodes. Our SDAN also takes dependency types into consideration when calculating the scores of dependency arcs. This is different from previous works [1] because their attention values are hand-coded. Our SDAN is similar to Graph Attention Network [15], however, it is designed for node classification task and does not consider syntactic information.

Moreover, we explore a novel gating mechanism to synthesize syntactic information with syntax-free features. Previous works have shown that contextualized representation could improve syntactical models. These approaches alleviate the effects of parsing noise by simply stacking syntactic model on the top of sequential layers [13] or in the reverse order [11,17]. However, these methods could not handle noise introduced by propagating information through wrong dependency arcs. To the best of our knowledge, we are the first to introduce gating mechanism that could restrain noise propagating along dependency arcs.

3   The Proposed Method

Formally, a sentence is denoted as a sequence of tokens $X = \{x_1, x_2, \ldots, x_n\}$, where n is the length of a sentence and x_i is the index of a token in a predefined vocabulary. Following standard BIO tagging schema, the corresponding labels of a given sentence are $Y = \{y_1, y_2, \ldots, y_n\}$, where $y_i \in \{B, I, O\}$. Every token is labeled as B if the token is the beginning of an aspect term, I if the token is the inside word of a multi-word aspect term but not the first word, or O if the token does not stand for any aspect term.

3.1   Overview of the Model

The overall architecture of the proposed model consists of 3 main layers. The input sentence is first mapped into a matrix of embeddings and a Bi-LSTM layer is utilized to build initial representations. We then feed the result of Bi-LSTM layer to syntactic learning layers to generate syntax-aware representations. As shown in the right side of Fig. 2, one syntax-aware layer consists of a SDAN to learn syntactic structure and a CGM to integrate dependency information with contextual features. By stacking multiple such syntactic layers the model could learn complex dependency structure. Finally, we feed syntax-aware representations to a CRF decoding layer and generate corresponding labels.

3.2   Initial Representation

Each input sentence is firstly converted into a sequence of word embedding $W_i = \{w_1, w_2, w_3, \ldots w_n\}$ through a look-up table $W_{emb} \in \mathbb{R}^{|V| \times d_w}$, where d_w is the dimension of word vectors and $|V|$ is the vocabulary size. We utilize Bi-LSTM network to build initial token level representation. A Bi-LSTM layer consists of a forward LSTM cell that processes input embeddings from left to right sequentially and a backward LSTM cell that processes the same inputs in a reverse order. We concatenate outputs of both directions and get the initial representation $h_t = [\overrightarrow{h_t}; \overleftarrow{h_t}]$ where $h_t \in \mathbb{R}^{2*d_m}$ and d_m is the dimension of LSTM hidden unit.

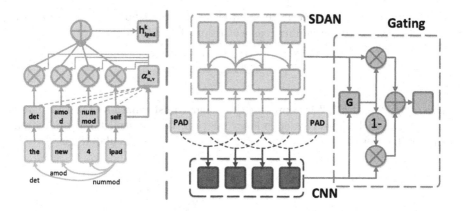

Fig. 2. Architecture of the proposed model.

3.3 Syntax-Directed Attention Network

Guided by the intuition that some dependency relations are crucial for identifying aspect terms while others contribute less, we utilize self-attention mechanism to weight syntactic relations and dynamically receive information from syntactic neighbors. we utilize an approach the same as [13] to convert dependency parse tree into syntactic graph. Syntactic graph of a sentence can be viewed as a labeled and directed graph $\mathcal{G} = \{\mathcal{V}, \mathcal{E}\}$, where $\mathcal{V}(|\mathcal{V}| = n)$ and $\mathcal{E}$ represent sets of nodes and edges in the graph $\mathcal{G}$ respectively. Each node in the graph represents a word in the sentence and each edge (u, v) in $\mathcal{E}$ is directed from one word to its syntactic governor or the inverse edge with dependency label $l(u, v)$ and dependency direction $dir(u, v)$. We also consider self-loop with a special label $l(u, u)$ and direction $dir(u, u)$.

The architecture of SDAN is shown in the left side of Fig. 2. We define the output matrix for the kth SDAN layer as $H^{sdan-k} \in \mathbb{R}^{n \times d_s}$, where d_s is the dimension of node vector and the uth row $h_u^{sdan-k} \in \mathbb{R}^{d_s}$ is the representation of a node u at the kth layer. For each node u in the graph, we firstly apply a linear transformation on each syntactic neighbor v to capture syntactic information, this is formulated as:

$$h_{u,v}^{sdan-k} = W_{d(u,v)}^k h_v^{k-1} + b_{l(u,v)}^k \tag{1}$$

Where $W_{d(u,v)}^k \in \mathbb{R}^{d_i \times d_s}$ and $b_{l(u,v)}^k \in \mathbb{R}^{d_s}$ are weight matrix and bias for the kth SDAN layer respectively, d_i is the dimension of SDAN input. This phase generates relation specific representations for nodes immediately connected to a central node in the dependency graph. Since datasets for ATE task often have moderate size with respect to the deep learning perspectives, we utilize direction specific weight matrix instead of relation type transform weights to avoid overparametrization.

The importance of syntactically related neighbors of node u is weighted by their contributions to discriminate aspect terms, formulated as:

$$e_{u,v}^k = \frac{h_{u,u}^{sdan-k} h_{u,v}^{sdan-k}}{\sqrt{d_s}} \tag{2}$$

Where $h_{u,u}^{sdan-k}$ and $h_{u,v}^{sdan-k}$ represent the linear transformation of self-loop and a nearby node with a specific dependency relation. We use scaled dot production as the similarity function. The normalized importance is calculated as:

$$\alpha_{u,v}^k = softmax(e_{u,v}^k) = \frac{\exp(e_{u,v}^k)}{\sum\limits_{k \in \mathcal{N}(u)} \exp(e_{u,k}^k)} \tag{3}$$

Where $\mathcal{N}(u)$ indicates all the direct syntactic neighbors of a node u. The output of a SDAN layer is formulated as:

$$h_u^{sdan-k} = ReLU\left(\sum_{v \in \mathcal{N}(u)} \alpha_{u,v}^k h_{u,v}^{sdan-k} \right) \tag{4}$$

We use $ReLU$ as the nonlinear activation function.

3.4   Contextual Gating Mechanism

Syntax-based models could learn rich linguistic knowledge, however, these methods highly rely on structural information and are sensitive to parsing noise. Automatically generated dependency trees are noisy in many cases, especially for ungrammatical reviews. To this end, we present a novel contextual gating mechanism to dynamically synthesize syntax-aware representation and syntax-free features to prevent information flow along noisy dependency arcs. From now on, we use sequential index i to represent a token. We firstly utilize CNN to capture syntax-free local representations. Considering a CNN component of the kth layer CNN^k with the input matrix $H^{k-1} \in \mathbb{R}^{n \times d_i}$, the output of the CNN component is formulated as:

$$h_i^{cnn-k} = W_c^k[h_{i-(w-1)/2}^{k-1}, \cdots, h_i^{k-1}, \ldots, h_{i+(w-1)/2}^{k-1}] \tag{5}$$

The convolution filter $W_c^k \in R^{w \times d_i \times d_c}$ is applied to a window of w words, where d_i and d_c are dimensions of the input and output of a CNN layer. To utilize a gating mechanism, we set the number of output channels the same as the dimension of syntax-aware representations. As shown in the right side of Fig. 2, the gate in orange block is controlled by both syntax-aware information and contextual features. Each element of the gate is influenced by both syntactic representations and context features, formulated as:

$$g_i^k = \sigma(W_{cnn}^k h_i^{cnn-k} + W_{sdan}^k h_i^{sdan-k} + b_g^k) \tag{6}$$

Where $W_{cnn}^k \in \mathbb{R}^{d_s \times d_s}$ and $W_{sdan}^k \in \mathbb{R}^{d_s \times d_s}$ are weight matrices of the kth gating mechanism and $b_g^k \in \mathbb{R}^{d_s}$ is the bias term, σ is the logistic sigmoid function. The overall output of the kth syntax-aware layer is calculated as:

$$h_i^k = ReLU(g_i^k \cdot h_i^{sdan-k} + (1 - g_i^k) \cdot h_i^{cnn-k}) \tag{7}$$

3.5 Training and Inference

We use a linear transformation to get the scores of labels. There exist strong dependencies across output labels for ATE task. Therefore, we utilize CRF [5] to model dependency of labels. Considering $P \in \mathbb{R}^{n \times (q+2)}$ is a matrix of scores output by previous network, where q is the number of distinct labels, $P_{i,j}$ corresponding to the score of the jth label of the ith token in a review; The last two columns of P are scores of start and end of a sequence. We define the score of a specific prediction sequence:

$$S(y|X) = \sum_{i=0}^{n} (A_{y_i, y_{i+1}} + P_{i, y_i}) \tag{8}$$

Where $A \in \mathbb{R}^{(q+2) \times (q+2)}$ is a label-transition matrix, $A_{i,j}$ represents the score of jumping from label i to label j; y_q and y_{q+1} are special labels indicating start and end of a sentence. The overall possibility of an output path is calculated as:

$$p(y|X) = \frac{\exp(S(y|X))}{\sum_{y' \in Y_X} \exp(S(y'|X))} \tag{9}$$

Where Y_X indicates all possible labeling sequences for a given input sentence.

The CRF layer is jointly trained with previous representation layers via Stochastic Gradient Descent (SGD). During training phase, the log-probability of a gold-standard label sequence can be calculated with dynamic programming:

$$log(p(y|X)) = S(y|X) - logadd \sum_{y' \in Y_X} S(y'|X) \tag{10}$$

While decoding, we use viterbi algorithm to decode output path with maximum score:

$$y^* = \arg\max_{y' \in Y_X} S(y'|X) \tag{11}$$

4 Experiment

4.1 Datasets and Experimental Settings

To evaluate the effectiveness of the proposed method, we conduct experiments on three widely used benchmark datasets from SemEval14[1] and SemEval15[2] ABSA challenges. Table 1 shows the statistics.

---

[1] http://alt.qcri.org/semeval2014/task4/.
[2] http://alt.qcri.org/semeval2015/task12/.

Table 1. Statistics of datasets.

Dataset	Train sentences	Train terms	Test sentences	Test terms
14-laptop	3045	2358	800	654
14-rest	3041	3693	800	1134
15-rest	1315	1192	685	542

We use Stanford CoreNLP toolkits[3] to generate parse trees. Word embeddings are initialized with 300d Glove 840B vectors[4]. Hyper-parameters are chosen by grid-search. We set the dimensions of LSTM cell, SDAN hidden units and CGM to 100. Dropout of 0.5 is applied over the input of every layer. We use 2 syntax-aware layers to learn syntactic information, and the window size for contextual features is 3. SGD is utilized to optimize the model with batch size of 20. We first train the model for 15 epochs with a fixed learning rate of 0.1, and then train another 35 epochs with a decay rate of 0.9. Following the setting of [11], we run model 20 times with the same settings and use F1 score as the evaluating metric.

4.2 Baselines

To evaluate the proposed method, we select multiple state-of-the-art models as baseline systems. (1)Top systems for SemEval14 and SemEval15 challenges such as IHS_RD [3], DLIREC(U) [14] and EliXa(U) [16]; (2)Models that only utilize sequential information such as BiLSTM+CRF [7] and BiLSTM+CNN [12]; (3)Multi-task based approaches that extract both aspect terms and opinions such as MIN [10] and CMLA [18] and (4)Previous methods that explore syntactic information such as DTBCSNN+F [19], RNCRF [17] and BiTree [11].

4.3 Main Results

The F1 scores of baselines and the proposed method are shown in Table 2. Performances on datasets from different domains and various sizes demonstrate that the proposed method consistently outperforms baseline methods.

The proposed method achieves 7% and 1.67%, 0.96% absolute gains over winning systems of SemEval challenges for three datasets respectively. This confirms the effectiveness of the proposed method. Both previous syntax-based methods and our model outperform sequential models with a large margin, which indicates that syntactic information contributes a lot to ATE. Our model also performs better than multi-task learning methods. This demonstrates that syntactic information is a better choice for ATE than leveraging a monolithic model to fit multiple tasks and share representations. The proposed model also exceeds

---

[3] https://stanfordnlp.github.io/CoreNLP/.
[4] https://nlp.stanford.edu/projects/glove/.

Table 2. Experimental results. Scores marked with '*' are copied from the paper of BiTree.

Model	14-laptop	14-rest	15-rest
IHS_RD	74.55	79.62	-
DLIREC(U)	73.78	84.01	-
EliXa(U)	-	-	70.05
BiLSTM+CRF*	76.10	82.38	65.96
BiLSTM+CNN*	78.97	83.87	69.64
MIN	77.58	73.44	-
CMLA	77.80	<u>85.29</u>	70.73
DTBCSNN+F	75.66	83.97	-
RNCRF	78.42	84.93	67.74
BiTree	<u>80.57</u>	84.83	<u>70.83</u>
OURS	**81.68**	**85.68**	**71.01**

Table 3. Effect of proposed components. Since CGM must works together with a syntactic related model, we refer replacing SDAN with SGCN [13] as "w/o SDAN".

Model	14-laptop	14-rest	15-rest
SDAN-CGM	**81.68**	**85.68**	**71.01**
w/o SDAN(SGCN-CGM)	80.52(−1.16)	85.14(−0.54)	69.66(−1.35)
w/o CGM(SDAN)	80.57(−1.11)	84.72(−0.96)	70.58(−0.43)
w/o SDAN & CGM (SGCN)	77.78(−3.90)	84.10(−1.58)	65.95(−5.06)

previous syntax-based approaches, indicating that the proposed method could learn better syntax-aware information than those models.

Further, we notice that the proposed method outperforms previous best model on 14-laptop dataset with a large margin, while the gains are less on 14-rest and 15-rest datasets. We believe that is because there are more long aspect terms in laptop domain and extracting aspect terms in laptop domain requires more long-term dependency information than that in restaurant domain. Finally, we observe that the gain on 15-rest dataset is relative small and the F1 score is lower than that on other datasets, we believe that this is caused by limited training samples of 15-rest dataset.

4.4 Analysis of the Proposed Methods

We further conduct extensive experiments to analysis the proposed components.

Computational Efficiency. We evaluate the training and inference time of BiTree model and the proposed method on a GTX 1080Ti GPU with minibatch

Table 4. Performance with different synthesize methods.

Method	14-laptop	14-rest	15-rest
w/o fusion	80.57	84.72	70.58
$STACK$	80.98	85.07	69.25
SUM	81.09	85.28	70.83
$CONCAT$	81.33	85.35	70.96
CGM	**81.68**	**85.68**	**71.01**

size of 20. The proposed framework achieves 8.4x and 9.2x speedup over Bi-Tree during training and inference. For example, inference on the test set of 15-rest dataset takes 0.45 s for the proposed method. While it requires 4.48 s for BiTree model. The speeding-up comes from that the proposed components do not require recursive operations and are convenient to parallelize.

Effect of Proposed Components. Results of different settings are shown in Table 3. As we excepted, both of the proposed components could improve the performance of ATE. Tree-based neural models tend to capture long term dependency path and ignore local information. SDAN, instead, receives informative representation from local connected neighbors and relies on hierarchy structure to learn multi-hop dependency. This method could learn long term semantic relation while focusing on local structure. Comparing with previous GCN-based method, we introduce attention mechanism instead of arc-gate. SDAN first uses an arc-specific matrix to transform representations of nearby nodes into the same semantic space and then receive relevance information with attention.

CGM could also improves the performance of both SDAN and SGCN. As a local model, CNN could effectively extracts context that is irrelevant with syntactic parsing structures. With a gating mechanism, CGM could dynamically weight the outputs of SDAN and CNN according to the confidence and information provided by these components. Notice that the F1 score of SDAN on 14-laptop dataset is comparable to previous best result, adding CGM further improves the score by 1% absolute gain. These facts clearly demonstrate that the GCM is an effective method to enhance syntactical models.

Effects of Different Fusion Methods. We further compare CGM with several fusion methods as shown in Table 4. All those fusion methods could improve the performance, demonstrating that syntax-free features do improve syntax-based models. $STACK$ indicates stacking CNN layer on SDAN, which leads to a deeper model and makes it harder to optimize. Both SUM and $CONCAT$ outperform $STACK$, however, they simply sum or concatenate features and lack the capacity to model the importance of different information channels. Our CGM utilizes gating mechanism to dynamically synthesize syntax-aware and

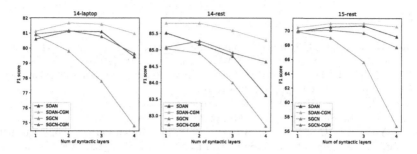

Fig. 3. Experimental results with different numbers of syntactic layers

syntax-free features, which could re-weight the contribution of these channels and reduce the effects of parsing noise, thus achieves the best performance.

Effects of Syntactic Layers. The impact of numbers of syntactic layers is shown in Fig. 3. The performance of SGCN drops rapidly with the growth of syntactic depth, while that of others do not. This is because both modeling dependency interactions and synthesizing syntax-free features contribute to reducing the effects of parsing noise. SDAN-CGM achieves best performances with two syntax-aware layers. This is reasonable because one syntactic layer could only model one-hop dependency relation, it is not sufficient for capturing long-term dependency as shown in Fig. 1. Adding more than two syntactic layers could learn longer dependency relations, however, it might introduce too much noise and even cause over-fitting considering the relative small data size.

5   Conculsion

In this work, we introduce a syntactic-based representation learning framework to obtain robust syntax-aware information for ATE. The proposed method consists of two main components, a graphical neural network SDAN to capture dependency relations and a fusion method CGM to integrate semantic information with local features and reduce the effects of parsing noise. Equipped with syntax-directed self-attention, the SDAN model could attend to syntactic neighbors that contribute a lot for the identification of current node. The CGM utilizes a gating mechanism to dynamically calculate the contribution of syntactic information and local features. Experiments results show that the proposed framework achieves state-of-the-art performances on three widely used benchmarks. We believe the proposed method could be applied to many other tasks that require syntactic information such as event extraction.

Acknowledgements. We thank Xiao Liang, Hongzhi Zhang, Yunyan Zhang, Wenkai Zhang and Hongfeng Yu, and the anonymous reviewers for their thoughtful comments and suggestions.

References

1. Chen, K., Wang, R., Utiyama, M., Sumita, E., Zhao1, T.: Syntax-directed attention for neural machine translation. In: AAAI, pp. 4792–4799 (2018)
2. Chen, Z., Mukherjee, A., Liu, B.: Aspect extraction with automated prior knowledge learning. In: ACL, pp. 347–358 (2014)
3. Chernyshevich, M.: IHS R&D belarus: cross-domain extraction of product features using CRF. In: SemEval@COLING, pp. 309–313 (2014)
4. Choi, Y., Cardie, C., Riloff, E., Patwardhan, S.: Identifying sources of opinions with conditional random fields and extraction patterns. In: HLT/EMNLP, pp. 355–362 (2005)
5. Collobert, R., Weston, J., Bottou, L., Karlen, M., Kavukcuoglu, K., Kuksa, P.P.: Natural language processing (almost) from scratch. J. Mach. Learn. Res. **12**, 2493–2537 (2011)
6. Hu, M., Liu, B.: Mining opinion features in customer reviews. In: AAAI, pp. 775–760 (2004)
7. Huang, Z., Xu, W., Yu, K.: Bidirectional LSTM-CRF models for sequence tagging. CoRR abs/1508.01991 (2015)
8. Irsoy, O., Cardie, C.: Opinion mining with deep recurrent neural networks. In: EMNLP, pp. 720–728 (2014)
9. Li, F., et al.: Structure-aware review mining and summarization. In: COLING, pp. 653–661 (2010)
10. Li, X., Lam, W.: Deep multi-task learning for aspect term extraction with memory interaction. In: EMNLP, pp. 2886–2892 (2017)
11. Luo, H., Li, T., Liu, B., Wang, B., Unger, H.: Improving aspect term extraction with bidirectional dependency tree representation. CoRR abs/1805.07889 (2018)
12. Ma, X., Hovy, E.H.: End-to-end sequence labeling via bi-directional LSTM-CNNS-CRF. CoRR abs/1603.01354 (2016)
13. Marcheggiani, D., Titov, I.: Encoding sentences with graph convolutional networks for semantic role labeling. In: EMNLP, pp. 1506–1515 (2017)
14. Toh, Z., Wang, W.: DLIREC: aspect term extraction and term polarity classification system. In: SemEval@COLING, pp. 235–240 (2014)
15. Veličković, P., Cucurull, G., Casanova, A., Romero, A., Liò, P., Bengio, Y.: Graph attention networks. In: ICLR (2018)
16. Vicente, I.S., Saralegi, X., Agerri, R.: EliXa: a modular and flexible ABSA platform. In: SemEval@NAACL-HLT, pp. 748–752 (2015)
17. Wang, W., Pan, S.J., Dahlmeier, D., Xiao, X.: Recursive neural conditional random fields for aspect-based sentiment analysis. In: EMNLP, pp. 616–626 (2016)
18. Wang, W., Pan, S.J., Dahlmeier, D., Xiao, X.: Coupled multi-layer attentions for co-extraction of aspect and opinion terms. In: AAAI, pp. 3316–3322 (2017)
19. Ye, H., Yan, Z., Luo, Z., Chao, W.: Dependency-tree based convolutional neural networks for aspect term extraction. In: Kim, J., Shim, K., Cao, L., Lee, J.-G., Lin, X., Moon, Y.-S. (eds.) PAKDD 2017. LNCS (LNAI), vol. 10235, pp. 350–362. Springer, Cham (2017). https://doi.org/10.1007/978-3-319-57529-2_28
20. Yin, Y., Wei, F., Dong, L., Xu, K., Zhang, M., Zhou, M.: Unsupervised word and dependency path embeddings for aspect term extraction. In: IJCAI, pp. 2979–2985 (2016)
21. Zhang, Y., Qi, P., Manning, C.D.: Graph convolution over pruned dependency trees improves relation extraction. CoRR abs/1809.10185 (2018)
22. Zhuang, L., Jing, F., Zhu, X.: Movie review mining and summarization. In: CIKM, pp. 43–50 (2006)

Short Text Similarity Measurement
Based on Coupled Semantic Relation
and Strong Classification Features

Huifang Ma[1,2(✉)], Wen Liu[1], Zhixin Li[2], and Xianghong Lin[1]

[1] College of Computer Science and Technology, Northwest Normal University,
Lanzhou 730000, China
{mahuifang,nwnuliuw}@yeah.net
[2] Guangxi Key Laboratory of Multi-source Information Mining and Security,
Guangxi Normal University, Guilin 541004, China
lizx@gxnu.edu.cn

Abstract. Measuring the similarity between short texts is made difficult by the fact that two texts that are semantically related may not contain any words in common. In this paper, we propose a novel short text similarity measure which aggregates coupled semantic relation (CSR) and strong classification features (SCF) to provide a richer semantic context. On the one hand, CSR considers both intra-relation (i.e. co-occurrence of terms based on the modified weighting strategy) and inter-relation (i.e. dependency of terms via paths that connect linking terms) between a pair of terms. On the other hand, Based on SCF for similarity measure is established based on the idea that the more similar two texts are, the more features of strong classification they share. Finally, we combine the above two techniques to address the semantic sparseness of short text. We carry out extensive experiments on real world short texts. The results demonstrate that our method significantly outperforms baseline methods on several evaluation metrics.

Keywords: Short text · Coupled semantic relation ·
Strong classification feature · Short text similarity

1 Introduction

Text similarity measures play a vital role in text related applications in tasks such as NLP, information retrieval, text classification, text clustering, machine translation and others. As the emergence of various social media, there are a large number of short texts, such as microblogs, and instant messages, are very prevalent on todays websites. In order to mine semantically similar information

The work is supported by the National Natural Science Foundation of China (No. 61762078, 61363058, 61663004) and Guangxi Key Lab of Multi-source Information Mining and Security, (No. MIMS18-08).

from massive data, a fast and effective similarity method for short texts has become an urgent task [1]. The challenge in measuring the similarity between short texts lies in the sparsity, i.e., there are likely to be no term co-occurrence between two texts. For example, two short texts *Dogs chase cats* and *Felines kill mice* refer to similar topics, even though there are no term co-occurrence. However, employ vector space similarity measures such as cosine similarity will still yield a value of 0. In order to get over the sparsity, enriching the semantic information of short texts using external corpus or knowledge is needed.

At present, the methods of short text similarity measure are mainly divided into two categories, namely, knowledge based and corpus based. Knowledge based approaches rely on handcrafted resources such as thesauri, taxonomies or encyclopedias, as the context of comparison. Previous methods based on lexical a taxonomy (tree), which is a hierarchical network representation consists of concepts and relations between these concepts. Most works depend on the semantic is a relations in WordNet [2,3]. Corpus based approaches work by extracting the contexts of the terms from large corpora and then inducing the distributional properties of *words* or *n-grams* [4,5]. Corpus can be anything from webpages, web search snippets to other text repositories.

However, knowledge based measures faced several challenges. First, the scale and scope of the concept or instance is not big enough, in other words, those concepts don't cover many proper nouns, or very popular senses. Second, exist methods treat concept as black and white, most of these knowledge based measures are deterministic instead of probabilistic. Nevertheless, corpus based approaches also face several serious limitations. First, corpus based measures extremely sensitivity to noise data. Second, corpus based approaches can't effectively addresses both synonymy and polysemy. Final, corpus-based methods focus on context of a term, which is more suitable to the calculation of semantic relatedness rather than similarity. For example, photo and delicacy would have high semantic relatedness because they co-occurrence very frequently.

In this paper, we propose an efficient and effective framework for computing semantic similarity (a number between 0 and 1) between two short texts. The major contributions of the paper are summarized as follows:

- We propose CSR-based method for acquiring similarity by coupling intra-relation and inter-relation to capture the richer semantic information, namely *coupled semantics relation for similarity measure* (CSRS).
- We design a strong classification feature-based similarity function, called *strong classification feature-based similarity* (SCFS), by utilizing the improved expected cross entropy to extract the strong category features of each class from labeling data set. Besides, aiming at multi-sense word, we propose a novel terms sense disambiguation by utilizing terms context similarity.
- Our approach employs a smoothing parameter to regulate the importance of the CSRS method and the SCFS method(i.e.CSRS-SCFS). This provides a complete representation of the semantic information for the document set.

- We conduct extensive experiments on the real-world text corpora. Experimental results prove the effectiveness of our proposed approach over many competitive baseline approaches.

The remainder of this paper is organized as follows. In Sect. 1 the related works are introduced. In Sect. 2 we present our proposed approach in details. In Sect. 3, we show experiments and result analysis, respectively on two real-world data sets. And the concluding remarks in Sect. 4.

2   The Proposed Methods

In our work, the whole framework consists of three phases: (i) an effective approach is presented to measure the relationship between terms by capturing both explicit and implicit semantic relations, which is implemented via utilizing modified intra-relation and inter-relation; (ii) the difference contribution of a term for a category is calculated based on the improved expected cross entropy; and (iii) the term correlation and the difference contribution of the term are aggregated to capture the comprehensive relationship between short texts.

2.1   Coupled Semantic Relation for Similarity Measure

In this section, we present the notations and define the problem of similarity measure. Let $D = \{d_1, d_2, ..., d_m\}$ is the set of m short texts in a document set D, and $T = \{t_1, t_2, ..., t_n\}$ is the set of n terms in a vocabulary set T.

We assign weight for a particular term in a certain short text by considering the co-occurrence, distance of term pairs and term discrimination (following the ideas of reference [6]). Thereafter, we can obtain the correlation weight of the term t_i ($t_i \in T$) in given text d_s ($d_s \in D$), which is defined as $w_{d_s}(t_i)$.

Intra-relation. We adapt the popular co-occurrence measure Jaccard to evaluate the intra-relation.

Definition 1 (*intra-relation*). If two terms co-occur in at least one text d_s, they are said to be **intra-related**. The intra-relation of term t_i and t_j can be described as:

$$CoR(t_i.t_j) = \frac{1}{|H|} \times \sum_{d_s \in H} \frac{w_{d_s}(t_i) \times w_{d_s}(t_j)}{w_{d_s}(t_i) + w_{d_s}(t_J) - w_{d_s}(t_i) \times w_{d_s}(t_j)} \quad (1)$$

where $w_{d_s}(t_i), w_{d_s}(t_j)$, denote the correlation weights of term t_i and t_j in d_s, respectively. $|H|$ denotes the number of elements in $H = \{d_s | w_{d_s}(t_i) \neq 0 \vee w_{d_s}(t_j) \neq 0\}$, if $H = \phi$, we define $CoR(t_i, t_j) = 0$. We further normalize $CoR(t_i, t_j)$ to [0,1] as follows:

$$UIaR(t_i|t_j) = \begin{cases} 1 & i = j \\ \frac{CoR(t_i,t_j)}{\sum_{i=1}^{n} CoR(t_i,t_j)} & i \neq j \end{cases} \quad (2)$$

where $UIaR(t_i|t_j)$ describes intra-relation of term t_i and t_j accounting for the proportion of all intra-relation of term t_i and other terms except t_j. Note that $UIaR(t_i|t_j)$ is asymmetric, we define intra-relation in a symmetric way:

$$IaR(t_i, t_j) = \frac{UIaR(t_i|t_j) + UIaR(t_j|t_i)}{2} \tag{3}$$

Inter-relation. The intra-relation introduced above only captures the explicit relatedness of two co-occurrence terms, but fails to consider the relatedness of term pairs in a global view, for the reason that the intra-relation fails to capture the semantic relatedness of term pairs by taking the interactions of other terms in the document set into consideration. In this subsection, we introduce an approach to model the implicit relatedness based on the graph, namely inter-relation [7]. Assume that the document set may be drawn as a graph which is consist of vertexes and edges to indicate the terms and their relatedness separately. If and only if term pairs co-occur in the document, they should have relation, i.e., there is an edge between the term pairs.

Definition 2 (*inter-relation path*). For given two terms t_i and t_j, there exist paths starting at t_i and ending at t_j, and linking finite terms which connect a sequence of terms. The path is defined as **inter-relation path** (IeP). Obviously, term pair may contain multiple paths. Therefore, we define IeP of any existing path as follows:

$$Path(t_i, t_j) = \{T_{i:j}^P | t_i, t_{l_1}, \cdots, t_{l_g}, t_j, t_j \in T_{i:j}^P, e_{il_1}, e_{l_1 l_2}, \cdots, e_{l_g j} \in E_{i:j}^P\} \tag{4}$$

where t_i is the initial vertex and t_j is the ending vertex, t_{l_1} stands for the terms on $Path(t_i,t_j)$, g is the number of these terms. $T_{i:j}^P$ denotes vertex set of the path. $E_{i:j}^P$ is the set of edges that passed by $Path(t_i,t_j)$. And $g \in [1, \theta]$, θ is a user-defined threshold to limit the number of t_{l_g}, i.e., the length of a path. The longer the path is, the weaker the path strength is. Nevertheless, the inter-relation is defined in fails to capture the high-order correlation among multiterms. The relations among multi-terms are too complex to be described via merely using the binary correlation. Thus, it is necessary to find a new measure to represent the high-order relations among multi-terms. We further define set of terms between two terms on a particular path as:

Definition 3 (*linking term set*). All terms between t_i and t_j on $Path(t_i, t_j)$ construct a **linking term set** T^{p-link} with length h, which is formalized as:

$$T^{p-link} = \{t_{l_h} | t_{l_h} \in T_{i:j}^P, T_{i:j}^P \in Path(t_i, t_j)\} \tag{5}$$

For any term pairs, we will consider the relation on possible paths, to discover the semantic relationship between the terms. A new measure called sharing entropy is adopted to measure the high-order correlation among multiple features [8], which is defined as:

Definition 4 (*sharing entropy*). Sharing entropy is defined as correlation degree on the p path of the term t_i and t_j from T^{p-link} as follows:

$$S_q(T^{p-link}) = (-1)^0 \sum_{t_{l_h}} J(t_{l_h}) + (-1)^1 \sum_{1 < i \le j \le h} J(t_{l_i}, t_{l_j}) + \cdots + (-1)^h J(T^{p-link}) \tag{6}$$

where $J(\bullet)$ denotes joint entropy of linking terms on the pth path, and h denotes the length of the pth path joint entropy [9] measures the amount of information contained in these multiple terms:

$$J(T^{p-link}) = - \sum_{t_i, t_{l_1}, \cdots, t_{l_h}, t_j} P(t_i, t_{l_1}, \cdots, t_{l_h}, t_j) \times \log_2 P(t_i, t_{l_1}, \cdots, t_{l_h}, t_j) \tag{7}$$

Noticeably, the sharing entropy reduces to the entropy of a single term when the path only have one linking term. If the path contains two terms, its sharing entropy degrades into information gain between the two terms. For any term pairs, the larger the shared entropy is, the closer multi-terms are correlated. In this work, we ignore the first two cases, in other words the length $h > 2$.

We normalize Eq. (6) based on the proportional heuristic that the sharing entropy of a term pair on p paths divided by the sharing entropy of all possible paths q. We define the inter-relation of the path p as follow:

$$IeR_p(t_i, t_j) = \frac{S_p(T^{p-link})}{\sum_{p=1}^{q}(T^{p-link})} \tag{8}$$

To the end, we acquire inter-relation of term pairs by selecting the max value of the sharing entropy on the all possibly paths.

$$IeR(t_i, t_j) = max\{IeR_p(t_i, t_j)\} \tag{9}$$

Given term pair (t_i, t_j), the coupled semantic relation is defined as:

$$CSR(t_i, t_j) = \begin{cases} 1 & i = j \\ \alpha IaR(t_i, t_j) + \\ (1-\alpha) IeR(t_i, t_j) & i \ne j \end{cases} \tag{10}$$

where $\alpha \in [0, 1]$ is the smoothing factor to trade off the important between intra-relation and inter-relation. Let M be a set of all term pairs with strong coupled semantic relation, which satisfies the user specified minimum threshold $CSR(t_i, t_j) \ge 0.3$. The CSRS can be defined as:

$$S_{CSRS}(d_1, d_2) = \frac{1}{\|d_1\|\|d_2\|} \sum_{t_i \in d_1} \prod w_{d_1}(t_i) \times w_{d_2} h(t_i) \times CSR(t_i, h(t_i)) \tag{11}$$

where $h(t_i) = t_j | t_j \in d_2 \wedge (t_i, t_j) \in M$. This method adopts term t_i in the text d_1 to mapping others terms that has coupled relation in the text d_2. Hence our method essentially aligns each term in d_1 with its best match term in d_2.

2.2   The Strong Classification Feature-Based Similarity

Traditional similarity methods do not take category distribution information of terms into consideration. In our work, we measure the weight of a term in a particular class, which utilize *improved expected cross entropy* (ECE'') to obtain strong classification features in given each class.

Improved Expected Cross Entropy. Let $D^l = \{d_1^l, d_2^l, \cdots, d_x^l\}$ be the set of short texts with labels. $C = \{C_1, C_2 \ldots C_y\}$ is the collection of all classes. For any term t_i that belongs to the text d_s, the correlation weight of the term t_i in the given class C_r (i.e. $C_r \in C$) is defined as:

$$cow_{C_r}(t_i) = \left(\frac{\sum_{d_s \in C_r} cow_{d_s}(t_i)}{|C_r(t_i)|} \right) \tag{12}$$

where $cow_{d_s}(t_i)$ represents the correlation weight of term t_i in the given short text. $|C_r(t_i)|$ is the number of texts that both contain term t_i and belong to C_r. In this subsection, we will discuss class-based weight by calculating the weight of a term t_i in different categories separately, i.e. *improved expected cross entropy*. The key idea is that a representative term of category A may be not of great importance in category B, therefore we should assign different weights in different categories.

The importance of term t_i in a particular class C_r (i.e.expected cross entropy) can be calculated as follows:

$$ECE_{C_r}(t_i) = \begin{cases} cow_{C_r}(t_i) P(C_r|t_i) \log \frac{P(C_r|t_i)}{P(C_r)} \\ cow_{C_r}(t_i) P(C_r|t_i) \log \frac{P(C_r)}{P(C_r|t_i)} \end{cases} \tag{13}$$

From Eq. (13), two scenarios are set up (i.e. $P(C_r) \le P(C_r|t_i)$ and $P(C_r) > P(C_r|t_i)$ respectively) to prevent value is negative. We can see that if t_i has a strong indication for C_r, it is more likely to assign a relatively higher weight with regard to category C_r for t_i. Next, the $ECE_{C_r}(t_i)$ of term t_i is defined as the average importance of term t_i on all classes except C_r, which is formulated as:

$$ECE'_{C_j}(t_i) = \frac{\sum_{j \ne r} ECE_{C_j}(t_i)}{y - 1} \tag{14}$$

The improved expected cross entropy is then defined as follows:

$$ECE''_{C_r}(t_i) = \frac{ECE_{C_r}(t_i)}{ECE'_{C_j}(t_i) + 0.01} \tag{15}$$

Equation (15) is used to measure the class-based weight of a term with regard to a category. It is clear that if t_i has a strong relationship with category C_r but weak class indication for other categories, the term t_i has higher possibility to have high weight with regard to category C_r. The value of ECE''_{C_r} reflects the importance and representativeness of term t_i in certain class C_r.

We also introduce the inverse document frequency as an external weight to enforce the global importance of the terms. The final weight of term t_i in class C_r is as follows:

$$W_{C_r}(t_i) = ECE''_{C_r}(t_i) \times idf(t_i) \tag{16}$$

The top-K distinctive terms for each category are selected to represent strong classification dictionary $S = \{s_1, s_2 \ldots s_{y \times k}\}$.

Strong Classification Feature Similarity. The basic idea of SCFS is that the more similar two texts are, the more features of strong classification they share. It is reasonable because the most representative features that are strongly related to target classes are selected from each classes. However, human language is ambiguous, so that many terms can be interpreted in multiple ways depending on the context in which they occur. For instance, consider the following short texts: "*I have apple juice*" and "*I have apple shares*" The occurrences of the term *apple* in the two short texts clearly denote different meanings: a type of fruit and a company, respectively. Therefore, we need identify terms sense, namely distinguish the term whether or not express the same meaning in different texts.

In our approach, we merely take into account the strong classification features that are contained in both texts, i.e. each terms t_i in short text should satisfy the constraint: $t_i \in s(t) = \{t_j | t_j \in d_1 \land t_j \in d_2 \land t_j \in S\}$. In general, nearby words provide strong and consistent clues to the sense of a target term, conditional on relative distance, and semantic relation. The word sense disambiguation (WSD) algorithm is performed by comparing context similarity of the term t_i ($t_i \in s(t)$) within fixed-size context windows. The context of term t_i between in two text can be defined as:

$$K_{d_1}(t_i) = \{t_j | t_j \in d_1, dis_{d_1}(t_i, t_j) \leq \varphi\} \tag{17}$$

$$K_{d_2}(t_i) = \{t_j | t_j \in d_2, dis_{d_2}(t_i, t_j) \leq \varphi\} \tag{18}$$

where $dis_{d_1}(t_i, t_j)$ is the distance between terms t_i and t_j in a given text d_1, i.e. the number of terms in-between. φ is a threshold to control the size of windows. In this paper, we set $\varphi = 2$. Then, we exploit the context of terms to calculate context similarity, which is calculated as:

$$S_{cox}(t_i) = \frac{\sum\limits_{t_1 \in K_{d_1}(t_i)} \sum\limits_{t_2 \in K_{d_2}(t_i)} CSR(t_i, t_j)}{|K_{d_1}(t_i)| \times |K_{d_2}(t_i)|} \tag{19}$$

After obtaining the context similarity of term t_i in the pair of texts, we can judge whether the strong classification features suggest the same meaning in these two short texts. An indicator function $I(t_i)$ is defined to reveal real sense for a term:

$$I(t_i) = \begin{cases} 1 & S_{cox}(t_i) \geq 0.5 \\ 0 & otherwise \end{cases} \tag{20}$$

A comprehensive weighting strategy for strong classification features in the short text is adopted combining the correlation weight(following the ideas of references [6] and ranking value of term t_i in the corresponding class (calculated by (16)). The weight of term t_i in a particular short text d_1 is redefined as:

$$w'_{d_1}(t_i) = w_{d_1}(t_i) \times w_{C_r}(t_i) \tag{21}$$

With all term senses are disambiguated and weights are updated, we can compute the *strong classification feature similarity (SCFS)* defined as:

$$S_{SCFS}(d_1, d_2) = \sum_{r=1}^{y} \sum_{t_i \in C_r} (min(w'_{d_1}(t_i), w'_{d_2}(t_i)) \times I(t_i)) \tag{22}$$

The higher the value of $S_{SCFS}(d_1, d_2)$ is, the higher probability of texts belong to same class is, the more similar the texts are. We further define the *strong classification feature similarity* via normalizing the relation between texts (i.e. $S_{SCFS}(d_1, d_2)$ to $[0, 1]$), which is calculated as follows:

$$S'_{SCFS}(d_1, d_2) = \frac{S_{SCFS}(d_1, d_2)}{\sum_{r=1}^{y} \sum_{t_i \in C_r} (max(w'_{d_1}(t_i), w'_{d_2}(t_i)) \times I(t_i))} \tag{23}$$

2.3 Combination of the Coupled Semantic Relation and Strong Classification Feature for Similarity Measure

In this paper, we propose a novel similarity measure of short text, namely *combination of the coupled semantic relation and strong classification feature similarity measure of short text* (CSRS-SCFS), which is formalized as:

$$S_{CSRS-SCFS}(d_1, d_2) = \beta \times S_{CSRS}(d_1, d_2) + (1 - \beta) \times S'_{SCFS}(d_1, d_2) \tag{24}$$

where β is a damping factor to determine relatively importance between the CSRS and the SCFS. If $\beta > 0.5$, the importance of CSRS is greater than SCFS. The value of β falls into $[0, 1]$, when $\beta = 0$ and $\beta = 1$, our approach degrade into the SCFS and the CSRS, respectively.

3 Experimental Results

In this section, we first give the experimental setup in Sects. 3.1, then we observe performance with parameter changing in Sect. 3.2. Finally, we compare the effectiveness of our approach with existing approaches in document clustering on two data sets in Sect. 3.3.

3.1   Experimental Setup

We use two data sets in the following experiments, including DBLP data set [10] and Sogou corpus data set [11]. For the experiments on the DBLP data set, we used a random sample of 10000 paper titles from the 10 domains averagely, such as Data Mining, Artificial Intelligence, Natural Language Processing and so on. Sogou corpus data set is composed of 11 categories, including car, finance, IT, health, and tourism, et al. 1000 news headlines are randomly selected in each category as experimental data. We carry out experiments on document clustering on both data sets. For a clustering algorithm, the performance of clustering depends heavily on the similarity mechanism. Namely, the performance of similarity measurements can be evaluated through the clustering results. In this paper, we adopt k-means as the clustering algorithm, and we set k equal to the number of classes in the document set for comparison. The 5-fold cross validation is employed in our experiments, and each fold composes of 80% of data for training and 20% for testing. We make use of F-measure and Rand Index (RI) as the clustering validation criteria.

3.2   Parameter Analysis

In this section, we describe some experiments on three important parameters (i.e. α, K, β) involved in our approaches. From previous analysis, we know that parameter α trades off the relative importance of intra-relation and inter-relation. The parameter K is related to the number of strong classification features, and the parameter β indicates the relative important of similarity between the two methods (CSRS and SCFS). In the following experiments, RI and F-measure are adopted to observe the effects with parameters changing on the clustering task. The obtained threshold is verified on a test data set, and, if it produces a satisfactory performance, then that value is adopted as the optimal one.

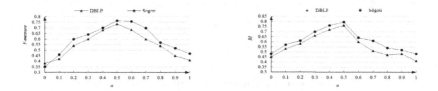

Fig. 1. The effect of the parameter α on the RI and F-measure

We report the values of RI and F-measure on two data sets varying with the values of α from 0 to 1 with a 0.1 step. The experimental results in Fig. 1 show that as the value of α increases, the value of RI and F-measure are linearly increasing on both data sets, while the value of RI and F-measure first increase to a peak value (in the case of $\alpha = 0.5$), then continuously declines. The results demonstrate that the inter-relation has great impact on the performance of document clustering, and it plays equal important role as the intra-relation, hence we set the value of $\alpha = 0.5$ as the optimal value. Fig. 2 reports the curves of the F-measure value with varying parameter K from 50 to 500 with a 50 step in SCFS on both data set. Experimental results show that as the value of K increases, the F-measure are linearly increasing. While the F-measure value first increases to a max value (in this case $K = 200$), then decreases with K value increasing. This may be due to the fact that if the value of K is too small, it will lead to the lack of strong category information. On the contrary, a large K value will incorporate some useless terms as strong category features, which may be considered as noises.

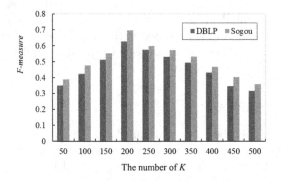

Fig. 2. The effect of varying the parameter K on both data set

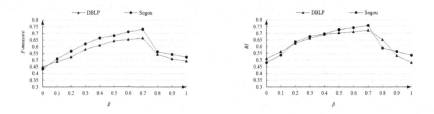

Fig. 3. The effect of the parameter α on the RI and $F - measure$

Parameter β is a damping factor to balance bias between CSRS and SCFS. For parameter β, we analyze how it affects the performance of CSRS-SCFS by fixing the remaining parameters ($\alpha = 0.5$, $K = 200$). The growth trends of RI and F-measure scores on each value of β are represented in Fig. 3, ranging from 0 to 1 with the increment of 0.1 on different data sets. From the result in Fig. 3 that we can observe that the curve keeps growing at the beginning, then starts to descend after it reaches the peak, indicating that clustering achieves best performance at a peak point with respect to a certain value of $\beta = 0.7$. Therefore we set $\beta = 0.7$ as the optimal value. We believe the reason is that CSRS comprehensively considers both explicit relation and implicit relation of terms while SCFS only relies on explicit relation (i.e. Co-occurrence relation) of terms with category information. Moreover, Fig. 3 reveals a fact that the incorporation of SCFS will improve the calculation of similarity accuracy between short texts, while richer semantic context and category knowledge between terms are revealed, leading to improve clustering performance.

3.3 Performance Comparison with Different Similarity Methods

In this section, we aim to observe the effectiveness of our approaches from two aspects. Firstly, we compare different methods proposed in this paper (namely CSRS, SCFS and CSRS-SCFS). The results in Fig. 4 demonstrate the superiority of our similarity measure on both data sets in terms of both RI and F-measure. The reason is that CSRS-SCFS not only considers the coupled relationship between the terms, but also takes the category information into account. Besides, Fig. 4 indicates that CSRS is superior to the SCFS on both evaluation metrics. It is reasonable because SCFS merely employs the terms co-occurrence relation with category information and the implicit interactions with other link terms are neglected.

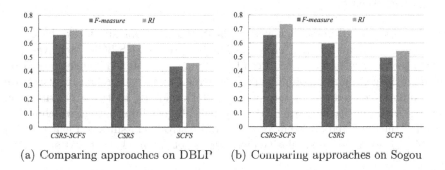

(a) Comparing approaches on DBLP   (b) Comparing approaches on Sogou

Fig. 4. Comparing the effectiveness in document clustering between our approaches

Secondly, we compare CSRS-SCFS with the three benchmark methods to verify its effectiveness. The three benchmark methods are as follows: *Co-occurrence Distance and Discrimination Based Similarity Measure on Short Text* (CDPC) [12], *Semantic Coupling Similarity* (CRM) [7], and *Strong Classification Features Affinity Propagation* (SCFAP) [13]. Table 1 compares the F-measure and RI of our approaches with that of three others. From the experimental results, we make the following observations. First, our most advanced approach, CSRS-SCFS, leads the competition against the peers by large margins in all data sets. Second, CRM performs better than CDPC, the reason is CDPC only use the co-occurrence between the terms but fails to capture the high-order correlation among terms. Finally SCFAP performs the worst, we believe the reason is that SCFAP exploits the small amount of the categories information but overlooks the semantic context and the correlation between terms. As a whole, unlike three benchmark methods which overlook the internal interactions and category information between terms, CSRS-SCFS accomplishes a comprehensive consideration that both combines coupled relation and strong classification features. Moreover, CSRS-SCFS also addresses the term sense ambiguity by utilizing context similarity.

Table 1. F-measure and RI of the k-means clustering with different similarity methods

Methods	DBLP		Sogou	
	RI	F-measure	RI	F-measure
SCFAP	0.478	0.378	0.358	0.325
CDPC	0.502	0.482	0.542	0.512
CRM	0.534	0.512	0.425	0.394
CSRS-SCFS	**0.724**	**0.665**	**0.759**	**0.732**

4   Conclusion

In this paper, we propose a novel similarity measure for short text based on coupling realton and strong classification features. CSRS-SCFS achieves this in terms of a three-step procedure: (1) Defining the coupled semantic relation based on the combination of intra- and inter-relation to comprehensively capture the semantic relatedness of term pairs. (2) Developing a strong classification features for similarity measure by considering category information, which utilize modified the expected cross entropy to acquire strong classification features. (3) Combining two kinds of similarity methods to get the final short text similarity.

References

1. Li, C.L., Wang, H.R., Zhang, Z.Q., Sun, A.: Topic modeling for short texts with auxiliary word embeddings. In: 39th International ACM SIGIR Conference on Research and Development in Information Retrieval, pp. 17–21. SIGIR, Pisa (2016)
2. Wang, C., Song, Y., Elkishky, A., Zhang, M.: Incorporating world knowledge to document clustering via heterogeneous information networks. In: 21th Knowledge Discovery and Data Mining. KDD, pp. 1215–1224. ACM, Sydney (2015)
3. Li, P.P., Wang, H., Zhu, K.Q., Wang, Z.: A large probabilistic semantic network based approach to compute term similarity. IEEE Trans. Knowl. Data Eng. **27**(10), 2604–2617 (2015)
4. Cheng, X., Miao, D., Wang, C., Cao, L.: Coupled term-term relation analysis for document clustering. In: The 2013 International Joint Conference on Neural Networks, pp. 1–8. IEEE, Dallas (2013)

5. Kusner, M.J., Sun, Y., Kolkin, N.I., Weinberger, K.Q.: From word embeddings to document distances. In: 32nd International Conference on International Conference on Machine Learning, pp. 957–966. ICML, Lille (2015)
6. Ma, H., Xing, Y., Wang, S., Li, M.: Leveraging term co-occurrence distance and strong classification features for short text feature selection. In: Li, G., Ge, Y., Zhang, Z., Jin, Z., Blumenstein, M. (eds.) KSEM 2017. LNCS (LNAI), vol. 10412, pp. 67–75. Springer, Cham (2017). https://doi.org/10.1007/978-3-319-63558-3_6
7. Chen, Q.Q., Hu, L., Xu, J., Liu, W.: Document similarity analysis via involving both explicit and implicit semantic couplings. In: 2015 International Conference on Data Science and Advanced Analytics, pp. 1–10. IEEE, Paris (2015)
8. Zhang, L., Gao, Y., Hong, C., Zhu, J.: Feature correlation hypergraph: exploiting high-order potentials for multimodal recognition. IEEE Trans. Cybern. **44**(8), 1408–1419 (2017)
9. MacKay, D.J.C.: Information Theory, Inference, and Learning Algorithms, 1st edn. Cambridge University Press, Cambridge (2003)
10. DBLP Dataset. http://dblp.uni-trier.de/xml/. Accessed 20 Apr 2016
11. Sogou Text Classification Corpus, 1 September 2012. http://www.sogou.com/labs/resource/tce.php/
12. Liu, W., Ma, H., Tuo, T., Chen, H.B.: Co-occurrence distance and discrimination based similarity measure on short text. Comput. Eng. Sci. **40**(7), 1281–1286 (2018)
13. Wen, H., Xiao, N.: A semi-supervised text clustering based on strong classification features affinity propagatione. Pattern Recognit. Artif. Intell. **27**(7), 646–654 (2014)

A Novel Hybrid Sequential Model
for Review-Based Rating Prediction

Yuanquan Lu[1], Wei Zhang[2(✉)], Pan Lu[1], and Jianyong Wang[1]

[1] Tsinghua University, Beijing, China
yuanquanlu@gmail.com, lupantech@gmail.com, jianyong@mail.tsinghua.edu.cn
[2] East China Normal University, Shanghai, China
zhangwei.thu2011@gmail.com

Abstract. Nowadays, the online interactions between users and items become diverse, and may include textual reviews as well as numerical ratings. Reviews often express various opinions and sentiments, which can alleviate the sparsity problem of recommendations to some extent. In this paper, we address the personalized review-based rating prediction problem, namely, leveraging users' historical reviews and corresponding ratings to predict their future ratings for items they have not interacted with before. While much effort has been devoted to this challenging problem mainly to investigate how to jointly model natural text and user personalization, most of them ignored sequential characteristics hidden in users' review and rating sequences. To bridge this gap, we propose a novel Hybrid Review-based Sequential Model (HRSM) to capture future trajectories of users and items. This is achieved by feeding both users' and items' review sequences to a Long Short-Term Memory (LSTM) model that captures dynamics, in addition to incorporating a more traditional low-rank factorization that captures stationary states. The experimental results on real public datasets demonstrate that our model outperforms the state-of-the-art baselines.

Keywords: Recommender systems · Rating prediction ·
Review analysis · Sequential model

1 Introduction

With the rapid development of the Internet, massive amounts of information spring up every day, posing both opportunities and challenges. Among many adopted techniques, recommender systems have been playing an increasingly vital role, being advantageous to alleviate information overload for ordinary users and increase sales for e-commerce companies. Particularly, in the field of recommender systems, rating prediction is a fundamental problem and has draw much attention since the success of Netflix Prize Competition[1]. Given historical

---

[1] https://www.netflixprize.com/.

© Springer Nature Switzerland AG 2019
Q. Yang et al. (Eds.): PAKDD 2019, LNAI 11439, pp. 148–159, 2019.
https://doi.org/10.1007/978-3-030-16148-4_12

ratings, rating prediction is required to predict users' ratings for items they have not evaluated before.

Latent factor models [10,13,19] behave well and are widely applied to the rating prediction problem. The main goal of such models is to learn low dimensional vector representations for both users and items, reflecting their proximity in the corresponding latent space. Salakhutdinov et al. [19] first formulated latent factor model from a probabilistic perspective. Beyond basic latent factor models, Koren et al. [10] introduced additional user and item rating biases as new features to improve prediction. Nowadays, the online interactions between users and items become diverse, and may include textual reviews besides ratings. According to the survey [20], reviews as a kind of side information are valuable for recommender systems because of the sentiment dimension.

Review-based rating prediction problem was well formulated in the model of Hidden Factor as Topics (HFT) [15], aiming at leveraging the knowledge from ubiquitous reviews to improve rating prediction performance. As reviews can be regarded as the interactions between users and items, they contain information related to both user and item latent factors. Previous work for solving this problem could be roughly classified into two categories. One is employing topic models to generate the latent factors for users and items based on their review texts [1,3,14,15,21,24]. Another is making use of fresh neural networks to model the semantic representation of words or sentences in the review texts [22,25,26]. However, most of the current review-based models mainly focus on learning semantic representations of reviews and ignore the sequential features among the reviews, which is the major focus of our work. Note that the task of review-based rating prediction is different from the task of sentiment classification. The difference is that our task focuses on leveraging users' **historical** reviews to predict their future ratings, but the sentiment classification task is to classify the **current** textual review' sentiment. Specifically, the method of learning semantic representation can be referred in the elementary component of our task.

To highlight the peculiarity of our proposed model, we first introduce the sequential models briefly, which take temporal dimension into consideration. Since preferences of users tend to vary along time and are influenced by the newly interacted items, sequential interaction history, as a kind of side information like reviews mentioned in [20], potentially serve as an important factor for predicting ratings. Apart from users, the characteristics of items might also be influenced by its recently interacted users. However, the existing methods based on matrix factorization [9] or deep neural networks [4,5,23], are mainly designed for mining temporal information on ratings, so that they cannot be directly employed to model the sequential features among the reviews.

From the above introduction, we can see that most of the current review-based models and sequential models only consider either review information or temporal information. To bridge this gap, we propose a novel Hybrid Review-based Sequential Model (HRSM) to capture future trajectories of users and items. The sequential information hidden in the textual reviews can help us to reveal the dynamic changes of user preferences and item characteristics. These two kinds of side information, namely reviews and temporality, are captured simultaneously in our proposed model. Furthermore, stationary latent factors of user and item generated from latent factor model potentially keep the inherent features over a long period. We integrate these stationary states with user's and item's dynamic states learned from review sequences to jointly predict ratings. The key differences between our proposed model named HRSM and the representative models for comparison in the rating prediction task, including PMF [19], BMF [10], HFT [15], DeepCoNN [26], RRN [23], are summarized in Table 1.

Table 1. Comparison of different models.

Characteristics	PMF	BMF	HFT	DeepCoNN	RRN	HRSM
Ratings	√	√	√	√	√	√
Reviews			√	√		√
Deep learning				√	√	√
Sequences					√	√

In summary, the main contributions of our work are as follows.

(1) We propose a hybrid review-based sequential model for rating prediction, which enables capturing temporal dynamics of users and items by leveraging their historical reviews.
(2) We integrate user's and item's stationary latent factors with dynamic states learned from review sequences to jointly predict ratings.
(3) Extensive experiments conducted on real public datasets demonstrate that our model outperforms the state-of-the-art baselines and obviously benefits from employing the sequential review content.

2   Related Work

Document Representation. Learning the document representation is the fundamental task of Natural Language Processing (NLP). LDA [2] as a traditional method is to learn the topic distribution from a set of documents. Based on neural networks, word2vec [17] and doc2vec [12] achieved a great success in modeling the distributed representation of words and documents, respectively. In recently years, methods employing deep learning technology outperform the previous models. Kim et al. [7] applied a convolutional layer to extract local feature among the words, and Lai et al. [11] added a recurrent structure based on it to reduce noise.

Review-Based Model for Rating Prediction. McAuley et al. [15] proposed the HFT model to use reviews to learn interpretable representation of users and items for review-based rating prediction problem. Many studies were inspired later, employing topic models as McAuley et al. did. TopicMF [1] as an extension of HFT, used non-negative matrix factorization for uncovering latent topics correlated with user and item factors simultaneously. Diao et al. [3] further designed a unified framework jointly modeling aspects, ratings and sentiments of reviews. Ling et al. [14] used mixture of Gaussian instead of matrix factorization to retain the interpretability of latent topics. Tan et al. [21] proposed a rating-boosted method to integrate review features with the sentiment orientation of the user who posted it. Recently, methods under the help of neural networks perform better in review-based rating prediction. Zhang et al. [25] combined word embedding method with biased matrix factorization, and Wang et al. [22] integrated the stacked denoising autoencoders with probabilistic matrix factorization. Zheng et al. [26] designed DeepCoNN which modeled the user and item representations using review embeddings learned by Convolutional Neural Network (CNN). However, most current review-based models fail to pay attention to the sequential features among the reviews, which is the major focus of our work.

Sequential Model for Rating Prediction. To model the dynamics, Koren et al. [9] designed a time piecewise regression to make use of dynamic information. He et al. [5] later adopted a metric space optimization method to capture additive user-item relations in transaction sequences. Recently, Recurrent Neural Networks (RNN) based models like User-based RNN [4] and RRN [23] have been shown effective in extracting temporal features from rating sequences, leading to a further improvement in prediction. However, the existing sequential models mainly focus on rating sequences. Informative review sequences ignored by them are considered in our model.

3 Preliminary

3.1 Problem Formulation

Assume the user set and item set are denoted as $\mathcal{U}$ and $\mathcal{V}$, respectively. We further represent the rating matrix as $\mathbf{R}$ and the collection of review text as $\mathcal{D}$. For $u \in \mathcal{U}$ and $v \in \mathcal{V}$, $r_{uv} \in \mathbf{R}$ indicates the rating value which the user u assigns to the item v, while $d_{uv} \in \mathcal{D}$ indicates the corresponding review text written by the user u to the item v. Given historical observed ratings and reviews, the problem of personalized review-based rating prediction is to predict the missed rating values in the rating matrix $\mathbf{R}$.

3.2 Biased Matrix Factorization

In order to verify how the temporal information and review text work, we briefly introduce a stationary model first. Biased Matrix Factorization (BMF) [10] is a collaborative filtering model for recommender systems. It is a classical and strong baseline applied in various scenes. The predicted rating $\hat{r}_{uv}$ of the user u to the item v can be computed as:

$$\hat{r}_{uv} = \mathbf{p}_u^\top \mathbf{q}_v + b_u + b_v + g, \tag{1}$$

where $\mathbf{p}_u$ and $\mathbf{q}_v$ are stationary latent vectors of the user and item, respectively. b_u and b_v correspond to their rating biases, respectively, and g is the global average rating.

4   Proposed Methodology

In this paper, we propose a novel **H**ybrid **R**eview-based **S**equential Model (**HRSM**). The overall framework of our proposed model is described in Fig. 1. Specifically, we first get each review's representation by feeding the inside words into CNN step by step. Then LSTM [6] is employed to model the sequential property of review sequences and thus we obtain the dynamic states of users and items. We further combine the dynamic states with user and item stationary latent vectors, and train them together to make the final rating prediction.

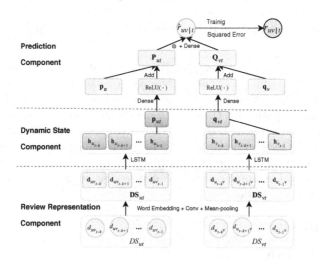

Fig. 1. The architecture of the hybrid sequential model for review-based rating prediction.

4.1   Review Representation

As we know, reviews contain abundant information. Emotional words among reviews, such as positive or negative words, indicate the preferences shown by a user to an item. Before exploring the sequential relation among reviews, we first need to obtain the representation for each given review. Each review d ($d = \{w_1, w_2, ...\}$) consists of a certain number of words, where each word $w \in \mathcal{W}$ comes from a vocabulary $\mathcal{W}$. By padding zeros in the front of review if necessary, each review could be transformed into a fixed-length matrix, with original one-hot representations for each word. After transformation by an embedding layer, each word inside the review is represented as an embedding $\mathbf{w}$. For each review, we adopt a convolutional layer to extract the local features and then adopt a mean-pooling layer to average the local features over its inner words. At last, the output vector $\mathbf{d}$ is regarded as the representation of the current input review d. The above procedures can be formulated as follows:

$$\mathbf{d} = \text{MP}(\text{CONV}(\text{EMB}(d))), \tag{2}$$

where $\text{EMB}(\cdot)$, $\text{CONV}(\cdot)$ and $\text{MP}(\cdot)$ denote the word embedding, the convolution and the mean-pooling operations, respectively.

Note that an LSTM layer can also be applied to learn the review representation from the mention in the related work part. But in the following procedure, we employ another LSTM layer to model the review sequences. The nested structures composed of these two kinds of LSTM layer will make the whole model too complicated to perform well in the attempted trial. Therefore, we choose the CNN layer as the alternative.

4.2   Review-Based States of User and Item

Because the procedures of learning dynamic state representations of users and items based on their reviews are conducted in a similar fashion, we just illustrate how to model review-based state for users in more detail. In order to model the dynamic states of users, we are supposed to take the timestamps of reviews into consideration.

Assume that the current user u already has interactions with n items. After sorting the interactions by their timestamps, we get an item id sequence denoted as VS_u ($VS_u = \{v_1, v_2, ..., v_n\}$) and a review sequence denoted as DS_u ($DS_u = \{d_{uv_1}, d_{uv_2}, ..., d_{uv_n}\}$). Different from the previous studies, we model the dynamic changes of user u from its review sequence DS_u rather than item id sequence VS_u, due to the reason that reviews denote the interactions between current user u and other items, and tend to contain both user's opinion and item's characteristic simultaneously. For user u and item v at time step t, their rating is denoted as $r_{uv|t}$. Obviously, $r_{uv|t}$ is only associated with the interactions before t. For a rating $r_{uv|t}$, the latest k interactions (v itself excluded) assigned by u, constitute a subsequence of DS_u, denoted as DS_{ut} ($DS_{ut} = \{d_{uv_{t-k}}, d_{uv_{t-k+1}}, ..., d_{uv_{t-1}}\}$). After getting review embeddings using Eq. (2), the review sequence DS_{ut} is transformed into the review embedding sequence $\mathbf{DS}_{ut}$ ($\mathbf{DS}_{ut} = \{\mathbf{d}_{uv_{t-k}}, \mathbf{d}_{uv_{t-k+1}}, ..., \mathbf{d}_{uv_{t-1}}\}) \in \mathbb{R}^{l \times k}$, where l is the dimension of review embedding and k is the sequence length. Here k also means the time window size. When the time window keeps sliding over the whole review sequence DS_u, multiple $\mathbf{DS}_{ut}$ are generated and are regarded as the input instances of the LSTM layer. To make the input sequences of LSTM have equal length, we assume that each current rating to be predicted explicitly results from its latest k interactions. The impact of parameter k will be discussed in the experiment part (see Sect. 5.5).

For the stationary model, we obtain stationary states of user and item by means of matrix factorization shown in Eq. (1). Differently, for sequential model, we apply LSTM [6] to learn the dynamic states of a user from its review sequence. At each time step τ of the sequence $\mathbf{DS}_{ut}$, the hidden state $\mathbf{h}_{u_\tau}$ of LSTM is updated based on the current review embedding $\mathbf{d}_{uv_\tau}$ and the previous hidden state $\mathbf{h}_{u_{\tau-1}}$ by a transition function f. The relationship is formulated in Eq. (3). Three gates inside the function f, namely input gate, forget gate and output gate, collaboratively control how information flows through the sequence. In this way, each review among the whole review sequence is considered together, since that current review can influence all subsequent reviews when the hidden state propagates through the sequence. When we feed sequence $\mathbf{DS}_{ut}$ into the LSTM layer, transition function f are conducted k times in total. We obtain the last hidden state as user dynamic state representation $\mathbf{p}_{ut}$ based on its recent review sequence. This procedure is formulated in Eq. (4).

$$\mathbf{h}_{u_\tau} = f(\mathbf{d}_{uv_\tau}, \mathbf{h}_{u_{\tau-1}}), \tag{3}$$
$$\mathbf{p}_{ut} = \text{LSTM}(\mathbf{DS}_{ut}). \tag{4}$$

In a similar manner, for current item v we can obtain its review sequence $DS_v(DS_v = \{d_{u_1v}, d_{u_2v}, ..., d_{u_mv}\})$ consisting of reviews written by m users. For rating $r_{uv|t}$, item v also has a subsequence $DS_{vt}(DS_{vt} = \{d_{u_{t-k}v}, d_{u_{t-k+1}v}, ..., d_{u_{t-1}v}\})$ of DS_v. Note that according to the definition, DS_{ut} and DS_{vt} have not exactly the same review documents although they have the same length. After applying the LSTM layer, we can also obtain the dynamic state $\mathbf{q}_{vt}$ of item v based on its review sequence.

4.3   Joint Rating Prediction

Up to now, we have obtained the dynamic states of user and item based on their review sequences. It is noted that both user and item have some inherent features that do not change with time. For example, user has fixed gender and item has stable appearance. Therefore, it is necessary to combine the dynamic and stationary states together for rating prediction. To be specific, we introduce a fully-connected layer consisting of a weight matrix $\mathbf{W_1}$ (biases as parameter included) and a ReLU activation function [18] to map the dynamic state into the same vector space as that of the stationary state. We formulate the final states of user and item as follows

$$\mathbf{P}_{ut} = \text{ReLU}(\mathbf{W}_1^\top \mathbf{p}_{ut}) + p_u, \tag{5}$$
$$\mathbf{Q}_{vt} = \text{ReLU}(\mathbf{W}_2^\top \mathbf{q}_{vt}) + q_v, \tag{6}$$

where $\mathbf{P}_{ut}$ denotes the joint state of user u, and $\mathbf{Q}_{vt}$ denotes the joint state of item v.

Previous work like BMF [10], mainly illustrated in Eq. (1), simply conducts the dot product of two latent vectors to produce a scalar as predicted rating. In that case, different dimensions among the latent vectors of user and item are considered to be equally important. To improve generalization, a linear transformation using a weight matrix $\mathbf{W_3}$ is added to distinguish significant dimensions. Finally, our model adopt the following equation to predict rating $\hat{r}_{uv|t}$,

$$\hat{r}_{uv|t} = \mathbf{W}_3^\top (\mathbf{P}_{ut} \odot \mathbf{Q}_{vt}) + b_u + b_v + g, \tag{7}$$

where $\odot$ is the hadamard product, representing the element-wise product of two vectors.

4.4   Inference

We define our objective function by minimizing the regularized squared error loss between the prediction and the ground truth,

$$\min_{\theta} \sum_{(u,v,t,d)\in\mathcal{K}_{\text{train}}} (r_{uv|t} - \hat{r}_{uv|t}(\theta))^2 + \text{Reg}(\boldsymbol{\theta}), \qquad (8)$$

where $\boldsymbol{\theta}$ denotes all the parameters, which can be learned using backpropagation. (u, v, t, d) means each observed tuple in the training dataset $\mathcal{K}_{\text{train}}$, and $\text{Reg}(\boldsymbol{\theta})$ denotes some optional regularizations.

5   Experiments

In this section, we describe our experimental setup and make a detailed analysis about our experimental results.

5.1   Dataset

We conduct experiments based on the Amazon dataset[2] [16]. We generally adopt two large subsets: "CDs and Vinyl" (hereinafter called **CD**) and "Movies and TV" (hereinafter called **Movie**). The CD dataset is more related to audio term while the Movie dataset is more related to video term.

To obtain enough sequence instances, we remove users and items with less than 20 occurrences in the dataset. After filtering, the total interactions (ratings or reviews) still number over 1×10^5 and 4×10^5 on CD and Movie dataset, respectively. A detailed summary is shown in Table 2. As our sequential model takes the historical reviews as the input, to ensure fair comparison with other stationary models, the test set is built with the last interacted item of each user. The remaining items form the training set. Furthermore, we partition the training set with the same strategy to obtain the validation set, which is used to tune the hyper-parameters. Mean Square Error (MSE) is employed as the evaluation metric for measuring model performance.

Table 2. Statistics of datasets.

Datasets	#ratings/reviews	#users	#items	Density	Avg text len
CD	107,518	2,230	2,672	0.018	236
Movie	441,783	7,506	7,360	0.008	242

---

[2] http://jmcauley.ucsd.edu/data/amazon/.

5.2   Baselines

Our model HRSM is compared with three traditional and three state-of-the-art models, including GloAvg, PMF, BMF, HFT, DeepCoNN, and RRN. Specifically, the first three methods use numerical ratings, and the following two methods learn review representations with topic models or neural networks, and the last one incorporates temporal information. The differences of the comparative approaches (excluding GloAvg) are summarized in Table 1.

- **GloAvg**. GloAvg simply uses the global factor g in Eq. (1) when making predictions.
- **PMF** [19]. PMF formulates matrix factorization from a probabilistic perspective with no rating biases.
- **BMF** [10]. BMF uses matrix factorization considering additional user's and item's biases on the basis of PMF.
- **HFT** [15]. HFT is the classical method that combines reviews with ratings. It integrates matrix factorization with topic models, where the former learns latent factors and the later learns review parameters.
- **DeepCoNN** [26]. This is the state-of-the-art method for review-based rating prediction problem, which indistinguishably merges all reviews of each user or item into a new large document and then employs CNN to learn review representations.
- **RRN** [23]. This is the state-of-the-art sequential model for rating prediction problem, which employs LSTM to capture the dynamics by modeling user's and item's id sequences without considering reviews.

5.3   Hyper-parameter Setting

Our model is implemented in Keras[3], a high-level neural network API framework. We employ Adam [8] to optimize parameters. To obtain the robust performance of our model and the compared baselines, we initialize each model with different seeds, and repeat the experiments five times, and report their average results.

Hyper-parameters are tuned in the validation sets using grid search. We apply 40-dimensional stationary latent vectors and 40-dimensional dynamic states based on reviews. Word embedding is 100-dimension and the LSTM layer contains 40 units. The batch size is set to 256 and the learning rate is set to 0.001. We use L2 regularization and its parameter is set to be 1×10^{-5} on CD dataset while 1×10^{-4} on Movie dataset. The hyper-parameters in baselines are also tuned in the similar method.

5.4   Results Analysis

The performances of models on two datasets are reported in Table 3. From the results, we have the following observations: (1) GloAvg is the weakest baseline,

---

[3] https://keras.io/.

since it is a non-personalized method. Compared with PMF, BMF performs better by introducing the additional rating biases. (2) Apart from using rating matrix as PMF and BMF do, the following three methods (HFT, DeepCoNN, and RRN) consider additional information like textual review or sequential property, generally achieving better results. RRN performs poorly on Movie dataset, and the reason might be that this dataset has sparser information in user's and item's id sequences. (3) RRN and DeepCoNN are the best baselines on CD and Movie datasets, respectively. It shows approaches utilizing deep neural networks usually perform better than the other baselines. (4) Our model HRSM consistently outperforms all the baselines on two datasets. Both HRSM and RRN are deep neural networks considering the sequential information, but HRSM achieves better results, which shows that review information is complementary to ratings. Although both HRSM and DeepCoNN are deep neural networks taking textual reviews into account, our model performs better due to exploiting the sequential information in addition.

Table 3. Performance comparison on two datasets.

Models	CD	Movie
GloAvg	1.4956	1.6908
PMF	1.0099	1.1694
BMF	0.9905	1.1623
HFT	0.9852	1.1618
DeepCoNN	0.9812	1.1530
RRN	0.9748	1.1793
HRSM	**0.9584**	**1.1378**

5.5 Impact of Time Window Size

In this part, we discuss how the important parameter k influences model performance. According to the definition of k, when k increases, the input review sequence of LSTM becomes longer and the number of input instances becomes less. The shortest length of user's or item's review sequence is 20 after data preprocessing (see Sect. 5.1). To obtain the best performance of our model, we examine the impact of different time window size k from 1 to 19 on both datasets. As the results are shown in Fig. 2, we have the following observations: (1) For two datasets, MSE decreases with the increase of k. In other words, when the review sequence becomes longer, the sequential information becomes more sufficient, which leads to the better performance. (2) When k increases into the later part of the range 1–19, the performance remains stable in general. Actually, our model can obtain global sequential information to some extent because the time window keeps sliding over the whole review sequence. When k is small, the input sequence of LSTM is too short to contain enough information, resulting in the

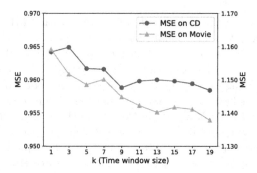

Fig. 2. Impact of the varying k on two datasets.

poor performance. But when k increases to a large value, the marginal benefit for model brought by the increment of k becomes smaller. This observation can help us determine how long a sequence should be as the input instance of LSTM.

6   Conclusion

In this paper, we propose a novel hybrid sequential model for the personalized review-based rating prediction problem. Previous models consider either review information or temporal information. But these two kinds of side information are captured simultaneously in our proposed model. Leveraging deep neural networks, our model learns the dynamic features of users and items by exploiting the sequential property contained in their review sequences. Experimental results on real public datasets demonstrate the effectiveness of our proposed model and prove that the sequential property hidden in reviews contributes a lot in the task of rating prediction.

Acknowledgement. This work was supported in part by National Natural Science Foundation of China under Grant No. 61532010 and 61521002. We also thank Yifeng Zhao and Ning Liu for helpful discussions.

References

1. Bao, Y., Fang, H., Zhang, J.: TopicMF: simultaneously exploiting ratings and reviews for recommendation. In: AAAI, pp. 2–8 (2014)
2. Blei, D.M., Ng, A.Y., Jordan, M.I.: Latent Dirichlet allocation. J. Mach. Learn. Res. **3**, 993–1022 (2003)
3. Diao, Q., Qiu, M., Wu, C.Y., Smola, A.J., Jiang, J., Wang, C.: Jointly modeling aspects, ratings and sentiments for movie recommendation (JMARS). In: SIGKDD, pp. 193–202 (2014)
4. Donkers, T., Loepp, B., Ziegler, J.: Sequential user-based recurrent neural network recommendations. In: RecSys, pp. 152–160 (2017)
5. He, R., Kang, W.C., McAuley, J.: Translation-based recommendation. In: RecSys, pp. 161–169 (2017)

6. Hochreiter, S., Schmidhuber, J.: Long short-term memory. Neural Comput. **9**(8), 1735–1780 (1997)
7. Kim, Y.: Convolutional neural networks for sentence classification. In: EMNLP, pp. 1746–1751 (2014)
8. Kingma, D.P., Ba, J.: Adam: a method for stochastic optimization. CoRR abs/1412.6980 (2014)
9. Koren, Y.: Collaborative filtering with temporal dynamics. In: SIGKDD, pp. 447–456 (2009)
10. Koren, Y., Bell, R., Volinsky, C.: Matrix factorization techniques for recommender systems. Computer **42**(8), 30–37 (2009)
11. Lai, S., Xu, L., Liu, K., Zhao, J.: Recurrent convolutional neural networks for text classification. In: AAAI, pp. 2267–2273 (2015)
12. Le, Q., Mikolov, T.: Distributed representations of sentences and documents. In: ICML, pp. II-1188–II-1196 (2014)
13. Lee, D.D., Seung, H.S.: Algorithms for non-negative matrix factorization. In: NIPS, pp. 535–541 (2000)
14. Ling, G., Lyu, M.R., King, I.: Ratings meet reviews, a combined approach to recommend. In: RecSys, pp. 105–112 (2014)
15. McAuley, J., Leskovec, J.: Hidden factors and hidden topics: understanding rating dimensions with review text. In: RecSys, pp. 165–172 (2013)
16. McAuley, J., Targett, C., Shi, Q., van den Hengel, A.: Image-based recommendations on styles and substitutes. In: SIGIR, pp. 43–52 (2015)
17. Mikolov, T., Sutskever, I., Chen, K., Corrado, G., Dean, J.: Distributed representations of words and phrases and their compositionality. In: NIPS, pp. 3111–3119 (2013)
18. Nair, V., Hinton, G.E.: Rectified linear units improve restricted Boltzmann machines. In: ICML, pp. 807–814 (2010)
19. Salakhutdinov, R., Mnih, A.: Probabilistic matrix factorization. In: NIPS, pp. 1257–1264 (2007)
20. Shi, Y., Larson, M., Hanjalic, A.: Collaborative filtering beyond the user-item matrix: a survey of the state of the art and future challenges. ACM Comput. Surv. **47**(1), 3:1–3:45 (2014)
21. Tan, Y., Zhang, M., Liu, Y., Ma, S.: Rating-boosted latent topics: understanding users and items with ratings and reviews. In: IJCAI, pp. 2640–2646 (2016)
22. Wang, H., Wang, N., Yeung, D.Y.: Collaborative deep learning for recommender systems. In: SIGKDD, pp. 1235–1244 (2015)
23. Wu, C.Y., Ahmed, A., Beutel, A., Smola, A.J., Jing, H.: Recurrent recommender networks. In: WSDM, pp. 495–503 (2017)
24. Zhang, W., Wang, J.: Integrating topic and latent factors for scalable personalized review-based rating prediction. IEEE Trans. Knowl. Data Eng. **28**(11), 3013–3027 (2016)
25. Zhang, W., Yuan, Q., Han, J., Wang, J.: Collaborative multi-level embedding learning from reviews for rating prediction. In: IJCAI, pp. 2986–2992 (2016)
26. Zheng, L., Noroozi, V., Yu, P.S.: Joint deep modeling of users and items using reviews for recommendation. In: WSDM, pp. 425–434 (2017)

Integrating Topic Model and Heterogeneous Information Network for Aspect Mining with Rating Bias

Yugang Ji[1], Chuan Shi[1(✉)], Fuzhen Zhuang[2,3], and Philip S. Yu[4]

[1] Beijing Key Laboratory of Intelligent Telecommunications Software and Multimedia, Beijing University of Posts and Telecommunications, Beijing, China
{jiyugang,shichuan}@bupt.edu.cn
[2] Key Laboratory of Intelligent Information Processing of Chinese Academy of Sciences (CAS), Institute of Computing Technology, CAS, Beijing 100190, China
zhuangfuzhen@ict.ac.cn
[3] University of Chinese Academy of Sciences, Beijing 100049, China
[4] University of Illinois at Chicago, Chicago, USA
psyu@uic.edu

Abstract. Recently, there is a surge of research on aspect mining, where the goal is to predict aspect ratings of shops with reviews and overall ratings. Traditional methods assumed that aspect ratings in a specific review text are of the same level, which equal to the corresponding overall rating. However, recent research reveals a different phenomenon: there is an obvious rating bias between aspect ratings and overall ratings. Moreover, these methods usually analyze aspect ratings of reviews with topic models at textual level, while totally ignore potentially structural information among multiple entities (users, shops, reviews), which can be captured by a Heterogeneous Information Network (HIN). In this paper, we present a novel model integrating Topic model and HIN for Aspect Mining with rating bias (called THAM). Firstly, a phrase-level LDA model is designed to extract topic distributions of reviews by using textual information. Secondly, making full use of structural information, we constructs a topic propagation network, and propagate topic distributions in this heterogeneous network. Finally, by setting review as the sharing factor, the two parts are integrated into a uniform optimization framework. Experimental results on two real datasets demonstrate that THAM achieves significant performance improvement, compared to the state of the arts.

Keywords: Aspect mining · Rating bias · Topic model ·
Topic propagation network · Heterogeneous information network

1 Introduction

With the rapid development of E-commerce, a large number of opinion reviews and ratings have been accumulated on the Web in the past decade [3,9,14].

© Springer Nature Switzerland AG 2019
Q. Yang et al. (Eds.): PAKDD 2019, LNAI 11439, pp. 160–171, 2019.
https://doi.org/10.1007/978-3-030-16148-4_13

These reviews and ratings have played an important role which can not only help people make more favorable purchase decisions, but also give valuable advice to the shops [1,10]. For instance, users may pay attention to both of overall ratings and reviews of a shop before making purchase decisions. The owner of a shop can learn the positive and negative feedback embedded in users' reviews as well.

In recent years, there is a surge of research on aspect mining and the main goal of aspect mining is tox effectively discover the aspect distribution and the aspect ratings of entities [16]. To address this problem, the earlier studies prefer to take advantage of Probabilistic Latent Semantic Analysis (PLSA). For example, both of Lu et al. [5] and Luo et al. [7] regarded reviews as several opinion phrases and respectively designed two PLSA-based models. However, these two models ignored the influence of ratings to reviews. Recently, many researchers [6,8,13,15] took the influence of ratings into consideration and utilized Latent Dirichlet Allocation (LDA) to describe the generation of reviews in details. Luo et al. [6] paid attention to the latent distribution of overall ratings and designed an LDA-based method for aspect rating prediction. Laddha et al. [2] integrated both discriminative conditional random field, regression, LDA to simultaneously extract phrases and predict ratings.

Almost all models for aspect mining usually have a basic assumption that the overall rating could be close to aspect ratings or the average score of aspect ratings. Thus, these methods preferred to directly associate review phrases or terms with the corresponding overall rating. However, recent research [4] found an insightful observation that there is an obvious rating bias between overall rating and aspect ratings. For example, in Dianping, the bias between overall rating and Environment is often +0.54, while in TripAdvisor, the bias between overall rating and Food is −0.09. This phenomenon indicates that review phrases or terms are more likely rated by latent aspect ratings rather than overall rating. Furthermore, Li et al. [4] proposed the RABI model to handle aspect rating prediction considering rating bias. Although the RABI obtains performance improvement on aspect rating prediction compared to previous models, there are several weaknesses existing in RABI. On the one hand, this model is based on PLSA without considering some other latent dependence, for example, the topic of modifier. This may restrict performance improvement. On the other hand, in the RABI model, it assumes that overall rating is on the center of the model, where it determines the reviews and aspects. Although this assumption may simplify the model, it is a little against our common sense.

Besides, contemporary aspect mining methods all focus on making use of textual information and overall rating, but ignore abundant structural information existing on review networks among the multi-typed entities, such as users, shops, and reviews. However, these structural information may be useful for aspect mining. For example, the reviews given by a user can describe his profile and the generation of a review is influenced by the quality of a shop as well as the corresponding user profile. In order to utilize the rich structural information of these multi-typed entities (e.g., users, reviews, and shops) and the various relations (e.g., writing and evaluating) among them, it is naturally to form the review network as a Heterogeneous Information Network (HIN) [11,12]. In a review HIN, both of user profile and shop profile can be easily described by propagating topic distribution of review texts to neighbour entities. Similarly, the topic distribution of review texts can be influenced by the profiles of their neighbour entities.

Motivated by these observations, we propose a novel method integrating Topic model and Heterogeneous information network for Aspect Mining with rating bias (THAM for short). To overcome the weaknesses of RABI [4] and describe the process of generating reviews more reasonably, THAM designs a LDA-based topic model at phrase-level to describe the generation of reviews and mine the aspect rating distribution of each review text. In this topic model, the modifier term of a phrase is associated with the sampled aspect rating rather than directly rated by overall rating because of the existing rating bias. Moreover, taking the abundant structural information into consideration, we propose a topic propagation network based on HIN to propagate topic distribution among users, shops and reviews for keeping the consistency of topic distributions of neighbour entities. Furthermore, in order to effectively fuse textual information and structural information, we design a uniform optimization framework through setting reviews as the sharing factor to integrate topic model and topic propagation network. An iterative optimization algorithm is proposed for this optimization framework.

2   Preliminary

Here we introduce the relevant concepts and the problem of aspect mining with rating bias.

Review: A review d is the text to express the user's opinion of a shop, and there are $|D|$ reviews in total.

Phrase: A phrase $l = < h, m >$ consists of a head term h and its modifier term m, for example, $< food, delicious >$. There are $|L|$ phrases in all.

Aspect/Topic: An aspect z is a specific topic of a shop. There are K aspects/topics. Note that, "topic" and "aspect" are used interchangeable in this paper.

Overall rating: An overall rating r is the quantified overall opinion of a review d. There are R levels of overall ratings and R is usually 5.

Aspect rating: An aspect rating $r_{s,z}$ is a numerical rating on the aspect z of the shop s. There are R levels of aspect ratings too.

Rating bias: The rating bias is the gap between the average of overall ratings and the average of aspect ratings.

Heterogeneous information network: Heterogeneous information network (HIN) is a special information network containing multiple entities and various relations [11]. For instance, the review network shown in left box of Fig. 1 is such a network, which contains three types of entities: user (u), shop (s), and review (d), and each edge represents a specific relation (e.g., "writing" for u to d).

Aspect rating prediction with rating bias: This problem is to predict ratings on each aspect for each shop with the bias prior information. Given a set of reviews D written by users U to evaluate shops S, the task is to identify the aspect of each phrase and predict the aspect ratings of shops considering rating bias.

Since aspect ratings are always missing in real applications but very valuable to users and shops, aspect rating prediction is an effective way to repair the missing information. Moreover, rating bias plays an significant role to improve the accuracy of aspect ratings [4]. Therefore, it is meaningful to study the problem of aspect mining with rating bias.

3   The THAM Model

In this section, we propose the THAM model, which makes full use of textual information and structural information for addressing the problem of aspect identification and aspect rating prediction with rating bias.

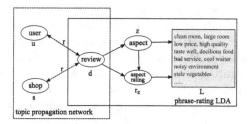

Fig. 1. The framework of THAM model. The dotted-line box is a network schema of topic propagation network, and the solid-line box describes the phrase-rating LDA.

3.1   The Phrase-Rating LDA

Here we design the phrase-rating LDA to more effectively learn topic distributions of reviews at textual level. In Fig. 2, both of aspect and aspect rating of a review are assumed as latent factor respectively sampled by aspect distributions and aspect rating distributions. Moreover, each review consists of several opinion phrases, in which head terms are generated by aspects while modifier terms are dependent on the sampled aspect ratings. Different from related methods, we consider the sampled aspect rating is associated with not only the observed overall rating but also the corresponding rating bias. Obviously, r_z, the sampled aspect rating of modifier term m, plays quite significant role in this model. Taking rating bias into consideration, we make two basic assumptions about aspect ratings. On the one hand, r_z is sampled by the aspect rating distribution of d. On the other hand, the mean of aspect rating distribution could be similar to

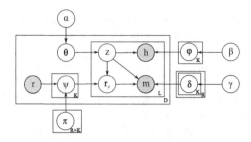

Fig. 2. The graphical model of the phrase-rating LDA.

but not equal to overall rating r because of the rating bias. Therefore, we design an aspect rating distribution ψ and regard that the Dirichlet prior parameter for aspect rating distribution, π, is related to the overall rating r. Given the observed overall rating r, the π_{r,k,r_z} is defined as follows:

$$\pi_{r,k,r_z} = B(r'_z|\omega(1-r'), \omega(r')),\tag{1}$$

where $B(\cdot)$ is the beta probability distribution, $0 < r' < 1$, $0 < r'_z < 1$ respectively represents the small scaled value of r and r_z, ω is the prior parameter. By using Eq. 1, we cleverly utilize overall rating to constrain the corresponding aspect rating distribution. Moreover, taking the rating bias into consideration, we set the aspect z's rating levels as $\{1 - b_z, 2 - b_z, ..., R - b_z\}$.

Given a set of review texts and overall ratings, both of r and $< h, m >$ are the observed variable, α, β, γ, π are the Dirichlet prior parameters, and the main latent parameters learnt are θ, ψ, ϕ, δ, z, and r_z. Given the model parameters and overall rating, the probability of observing the review text (i.e., the likelihood) is:

$$L_1 = -log(\prod_d \prod_l \sum_z \sum_{r_z} p(z|\theta_d)p(r_z|\psi_{d,z})p(h_l|\phi_z)p(m_l|\delta_{r_z,z})).\tag{2}$$

We employ Gibbs sampling to estimate the posterior probability given the observed phrases.

It is noteworthy that the phrase-rating LDA does not describe the dependence between aspect distribution θ and overall rating r, because the dependence is closely associated with user profile and shop profile.

3.2   Topic Propagation on Review Network

In order to make full use of structural information for aspect mining, we design a HIN-based topic propagation network shown in Fig. 3, to propagate topic distribution among neighbour entities so as to describe user profile and shop profile.

In the topic propagation network, the topic distribution of each entity should be related to its neighbour entities. Furthermore, we constrain the topic propagation must under the same overall rating. This constraint is reasonable because

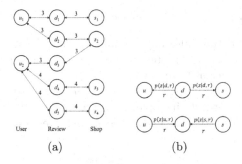

Fig. 3. An example of topic propagation network and the topic propagation strategy.

the aspect distributions under different overall ratings represent different meanings. For instance, if a user gives high overall rating to a shop, he is likely to write many positive phrases to describe the aspect he cares, and vice versa.

Therefore, given the topic distribution of a review $p(z|d,r)$ where r is the observed overall rating of the corresponding review d, we design the topic propagation strategy as shown in Fig. 3(b), and the topic distribution of a user u is constructed by his/her reviews, denoted as:

$$p(z|u,r) = \sum_{d_{u,r} \in \boldsymbol{D}_{u,r}} p(z|d_{u,r})p(d_{u,r}|\boldsymbol{D}_{u,r}) = \sum_{d_{u,r} \in \boldsymbol{D}_{u,r}} \frac{p(z|d_{u,r})}{|\boldsymbol{D}_{u,r}|}, \qquad (3)$$

where $\boldsymbol{D}_{u,r}$ is the set of reviews under overall rating r belonging to user u, $|\boldsymbol{D}_{u,r}|$ is the number of these reviews. Similarly, the topic distribution for a shop is constructed by its reviews, calculated by:

$$p(z|s,r) = \sum_{d_{s,r} \in \boldsymbol{D}_{s,r}} p(z|d_{s,r})p(d_{s,r}|\boldsymbol{D}_{s,r}) = \sum_{d_{s,r} \in \boldsymbol{D}_{s,r}} \frac{p(z|d_{s,r})}{|\boldsymbol{D}_{s,r}|}, \qquad (4)$$

where $\boldsymbol{D}_{s,r}$ is the set of reviews under overall rating r belonging to shop s, $|\boldsymbol{D}_{s,r}|$ is the number of these reviews.

Furthermore, the review d should have similar topic distribution with its author u and its shop s. Therefore, aiming at obtaining effective topic distribution of reviews, we design two functions, one of which is to calculate the similarity of d and u, and the other is to calculate the similarity of d and s. The two functions are shown as follows:

$$L_2 = \frac{1}{2} \sum_d \sum_z [p(z|d,r) - p(z|u_d,r)]^2, \qquad (5)$$

$$L_3 = \frac{1}{2} \sum_d \sum_z [p(z|d,r) - p(z|s_d,r)]^2, \qquad (6)$$

where u_d is the user who writes the review d and s_d is the shop whom the review d evaluates.

3.3   Uniform Optimization Framework

To make full use of textual information and structural information at the same time, the THAM model incorporates the topic model and the topic propagation into a uniform optimization framework.

In this framework, we consider the review d as the sharing factor, which plays significant role not only in topic propagation but also in topic modelling. To ensure the optimization process of the model, we design a combined loss function here:

$$Loss = L_1 + \frac{\lambda}{2}(L_2 + L_3), \tag{7}$$

where $\lambda \geq 0$ is to control the balance between topic modelling and topic propagation. Obviously, if $\lambda = 0$, we only take into account the loss of phrase-rating LDA. With the increase of λ, the loss from topic propagation will be paid more and more attention.

There are two main steps for learning the algorithm. In topic modelling, we sample the distribution of reviews on phrase-rating LDA to reduce L_1 where L_2 and L_3 are fixed. In topic propagation, we get rid of the Newton-Raphson updating formula, which decreases function $f(x)$ by updating $x_{t+1} = x_t - \xi \frac{f'(x_t)}{f''(x_t)}$, to decrease L_2 and L_3. $p(z|d,r)$ in topic propagation is updated by:

$$p(z|d,r)_{t+1} = (1 - \xi)p(z|d,r)_t + \frac{\xi}{2}(p(z|u_d,r)_t + p(z|s_d,r)_t), \tag{8}$$

where ξ is a step parameter. Then, the corresponding topic distribution of u_d and s_d can also be updated by in Eqs. (3) and (4) respectively. In this step, we also take L_1 into consideration because the updated $p(z|d,r)$ (i.e., θ) can also influence the value of L_1.

3.4   Aspect Identification and Rating Prediction

Based on the obtained aspect-head distribution ϕ and aspect-modifier distribution δ, we can identify the aspect which phrase $l = <h, m>$ should be assigned to by using Eq. (9):

$$g(l) = \underset{z'}{\operatorname{argmax}} \sum_{r_z} \delta_{r_z,z',m} \phi_{z',h}, \tag{9}$$

and the corresponding rating of l is:

$$r_l = \frac{\sum_{r_z} \delta_{r_z,z,m} \phi_{z,h} r_z}{\sum_{r_z} \delta_{r_z,z,m} \phi_{z,h}}, \tag{10}$$

where $z = g(l)$. And the predicted aspect rating of each shop is calculated by:

$$\widehat{r}_{s,z} = \frac{\sum_{d \in D_s} \sum_{r_z} \psi_{d,z,r_z} r_z}{\sum_{d \in D_s} \sum_{r_z} \psi_{d,z,r_z}}, \tag{11}$$

where $\widehat{r}_{s,z}$ is the rating on aspect k of shop s, D_s is the set of reviews of shop s.

4 Experiments

4.1 Dataset

There are two real datasets in different languages for conducting experiments: Dianping in Chinese [4] and TripAdvisor in English. The review information in Dianping dataset consists of a Chinese review text and three aspect ratings on Taste, Service, and Environment. Similarly, the TripAdvisor dataset crawled from the TripAdvisor website, is a set of English reviews and each review includes English comments, an overall rating and three aspect ratings on Value, Service, and Food. In addition, the range of ratings in the two datasets are in $[1, 5]$. The statistics of the two datasets are shown in Table 1. Note that, b_1, b_2, and b_3 respectively represents the rating bias of Taste, Service and Environment (on Dianping) or Value, Service, and Food (on TripAdvisor).

4.2 Preparation

To obtain phrases from reviews, the dataset is preprocessed via the process similar to that in RABI [4] Besides, the number of aspects K is set as 3 for both of Dianping and TripAdvisor. The prior parameters α, β, and γ in the phrase-rating LDA are set as $50/K, 0.01, 0.01$ respectively. ω is set as 2.5 and ξ is set as 0.1. The max iteration is 1000. The controlling parameter λ is adjusted to 9000 on both the two datasets by parameter analysis.

Since the performance of an aspect mining method may be affected by the size of the training dataset [4], we sample four subsets for Dianping and TripAdvisor with different scales of reviews (i.e., 25%, 50%, 75%, 100% of review data). To ensure that the latent aspects correspond to the given aspects, we also select several head terms as prior for each latent aspect.

We select Root Mean Square Error (RMSE) and Pearson Correlation Coefficient (PCC) as evaluation metrics. RMSE is to measure the average difference between real ratings and predicted ratings on all aspects. The smaller the value of RMSE, the better the algorithm performs. Considering that rating prediction are often used for ranking-based recommendation, we also measure the linear relation of the predicted results and the real results by the PCC metric. The larger value of PCC represents the better performance.

4.3 Comparison Methods

To demonstrate the effectiveness of THAM model, four representative methods, including QPLSA [7], SATM [13], AIR [3], and RABI [4], are adopted

Table 1. The statistic information of Dianping and TripAdvisor

Dataset	# Users	# Shops	# Reviews	# Phrases	Avg. Overall Rating	b_1	b_2	b_3
Dianping	14519	1097	216291	696608	3.97	+0.28	+0.48	+0.54
TripAdvisor	107368	5178	243186	2544148	4.10	+0.27	−0.05	−0.09

for comparisons. Since neither QPLSA and SATM nor AIR takes into account the rating bias, we adjust the results of the three baselines through subtracting/adding the rating bias for fair comparison and mark the adjusted method with "*".

Furthermore, in order to validate the effectiveness of rating bias and structure information, we also test three simplified versions of our THAM. First, we remove the assumption of rating bias from THAM, and call this version THAM\B. Second, we only use the textual information without topic propagation. We call it THAM\H. Third, we remove both of rating bias and topic propagation form THAM, and call this version THAM\HB.

Table 2. Top 10 rated phrases for different aspects of the two datasets

Datasets	Aspects	Phrases (Ratings)
Dianping	Taste	great taste (4.35), good mouth-feel (3.61), delicious dish (3.54), *suitable price (3.52)*, delicious drink (3.33), *good restaurant (3.28)*, *high taste (3.17)*, light flavour (3.04), common flavour (2.58), few dish (2.51)
	Service	enjoyable service (3.95), enthusiastic service (3.92), comfortable service (3.78), *nice shop (3.72)*, *delicious service (3.43)*, handsome waiter (3.43), good impression (3.28), good attitude (3.25), cold waiter (1.95), not enthusiastic service (1.85)
	Environment	elegant style (4.29), *cheap price (4.08)*, nice inside (3.56), *good feeling (3.32)*, suitable position (3.32), *easy to find (3.20)*, good traffic (3.09), common environment (2.76), small room (2.31), unreasonable design(2.05)
TripAdvisor	Value	unbeatable price (4.48), best quality (4.19), cheap price (3.89), reasonable price (3.57), pricey fare (3.44), *great place (3.41)*, *big price (3.29)*, *good place (3.29)*, *good selection (3.20)*, poor value (1.81),
	Service	fantastic waitress (4.12), friendly service (4.03), courteous waiter (3.89), great experience (3.53), interesting waitress (3.18), *good drink (3.31)*, *good meal (3.24)*, first experience (3.09), slowest service (1.82), disgusting service (1.57)
	Food	amazing food (4.54), wholesome food (4.20), excellent dishes (4.03), *nice location (3.98)*, rich menu (3.31), good food (3.06), *good atmosphere (3.02)*, small dish (2.91), *small restaurant (2.77)*, not fresh dish (2.59)

4.4 Aspect Identification

Since the opinion phrases are unlabelled, it is hard to quantitatively validate the effectiveness of aspect identification. Therefore, we list some representative rated phrases for each aspect on the two datasets respectively for illustration. The most possible phrases for each aspect are automatically mined and shown in Table 2. In addition, we rank these phrases by their ratings and the meaningless phrases are marked in italic type.

Here we find that most of the extracted phrases in both English and Chinese can accurately express users' feelings about specific aspects and these frequent opinion phrases are effectively assigned to the related aspects which they describe. On the one hand, the head term of a phrase can indicate the aspect which the user describes, such as "mouth-feel" for taste and "room" for environment. On the other hand, a positive modifier term can express a positive evaluation while a negative modifier term may indicate a lower rating. It is obvious that some phrases with positive modifier terms, like "great" and "unbeatable" can get high ratings while those with negative modifier terms, like "poor" and "slowest" get the lowest ratings.

4.5   Effectiveness Experiments

In this section, we present the results of predicted aspect ratings on reviews with overall ratings and measure the performances of the different methods in terms of RMSE and PCC. In addition, each method here is run ten times and the average results (RMSE and PCC) are recorded in Tables 3 and 4 respectively.

RMSE Performance. To evaluate the accuracy of these methods on predicting aspect ratings, we calculate all RMSE values of these results. As is shown in Table 3, we can clearly find the following observations:

Table 3. RMSE performances of different methods on two datasets.

Dataset	Dianping				TripAdvisor			
	25 %	50 %	75 %	100 %	25 %	50 %	75 %	100 %
QPLSA	0.5806	0.5750	0.5724	0.5705	0.5805	0.4628	0.4005	0.3876
QPLSA*	0.3639	0.3518	0.3483	0.3460	0.5798	0.4486	0.3944	0.3833
SATM	0.5783	0.5754	0.5698	0.5601	0.6101	0.4886	0.4203	0.4064
SATM*	0.3818	0.3783	0.3704	0.3642	0.6012	0.4737	0.4136	0.3822
AIR	0.5369	0.5307	0.5157	0.5112	0.6517	0.5572	0.4778	0.4408
AIR*	0.3363	0.3207	0.3055	0.3034	0.6446	0.5475	0.4546	0.4380
RABI	0.3228	0.3150	0.3024	0.2951	0.5286	0.4388	0.3771	0.3695
THAM\HB	0.5064	0.4910	0.4873	0.4855	0.5027	0.4128	0.3614	0.3247
THAM\H	0.3089	0.2897	0.2833	0.2798	0.4920	**0.4024**	0.3477	0.3191
THAM\B	0.5060	0.4906	0.4869	0.4843	0.4985	0.4160	0.3610	0.3261
THAM	**0.3078**	**0.2891**	**0.2822**	**0.2789**	**0.4889**	0.4048	**0.3475**	**0.3101**

Compared with baselines, our THAM achieves the best performances on all subsets. There are two main advantages of our THAM. On the one hand, we utilize the bias prior information more effectively by designing a reasonable LDA-based topic model, and this topic model can overcome the weaknesses of RABI. On the other hand, we take into account not only textual information but also structural information contained in the review network, while all of baselines only focus on review texts and ratings. Comparing THAM with its variant versions, we find that THAM performs the best on most situations.

We can make two main conclusions as follows: (1) Bias prior information is effectively utilized in THAM, by comparing THAM and THAM\B; (2) The topic propagation strategy can improve the performance of our method, by comparing THAM and THAM\H.

Table 4. PCC performances of different methods on two datasets.

Dataset	Dianping				TripAdvisor			
	25 %	50 %	75 %	100 %	25 %	50 %	75 %	100 %
QPLSA	0.5689	0.5766	0.5752	0.5837	0.5715	0.5855	0.5860	0.5918
SATM	0.3503	0.3656	0.3735	0.3984	0.5535	0.5919	0.6279	0.6471
AIR	0.5670	0.5707	0.5875	0.5949	0.6643	0.6667	0.6979	0.7218
RABI	0.6130	0.6245	0.6378	0.6398	0.6582	0.6621	0.6740	0.6801
THAM\H	0.6691	0.6956	0.6999	0.7060	0.7563	**0.7696**	0.7901	0.8030
THAM	**0.6721**	**0.6971**	**0.7031**	**0.7093**	**0.7594**	0.7665	**0.7907**	**0.8075**

PCC Performance. To evaluate the ability of these models to maintain relative order among shops, we also calculate all PCC performance of these models on datasets, and show the results in Table 4. Because rating bias rarely affects the order of shops, we only compare these original methods and our THAM\H, THAM.

As is shown in Table 4, obviously, the proposed THAM obtains the best performances on almost all datasets than other baselines. We can also observe that RABI performs better than AIR on Dianping dataset while AIR does better than RABI on TripAdvisor dataset. These observations once again validate that THAM is stable and robust enough to predict aspect ratings of shops. Therefore, THAM is proved as a better choice when recommending Top-N aspect ranking orders than other baselines.

5   Conclusion

In this paper, we have proposed THAM to integrate topic model and heterogeneous information network for aspect mining with rating bias. Taking advantage of both textual and structural information, THAM designs a phrase-level LDA model and the topic propagation strategy for aspect mining. In order to integrate the two parts for optimization, THAM sets the reviews as the sharing factor and proposes a uniform iterative optimization model. By comparing the performances of baselines, THAM performs better on the two datasets for aspect mining. In the future, we can make use of heterogeneous information network more effectively for aspect mining by taking the user attributes and shop attributes into consideration.

Acknowledgements. This work is supported by the National Key Research and Development Program of China (2017YFB0803304) and the National Natural Science Foundation of China (No. 61772082, U1836206, 61702296, 61806020, 61375058).

References

1. Bauman, K., Liu, B., Tuzhilin, A.: Aspect based recommendations: recommending items with the most valuable aspects based on user reviews. In: The ACM SIGKDD International Conference, pp. 717–725 (2017)
2. Laddha, A., Mukherjee, A.: Aspect opinion expression and rating prediction via LDA-CRF hybrid. Nat. Lang. Eng. **24**, 1–29 (2018)
3. Li, H., Lin, R., Hong, R., Ge, Y.: Generative models for mining latent aspects and their ratings from short reviews. In: 2015 IEEE International Conference on Data Mining, ICDM 2015, Atlantic City, NJ, USA, 14–17 November 2015, pp. 241–250 (2015)
4. Li, Y., Shi, C., Zhao, H., Zhuang, F., Wu, B.: Aspect mining with rating bias. In: Frasconi, P., Landwehr, N., Manco, G., Vreeken, J. (eds.) ECML PKDD 2016. LNCS (LNAI), vol. 9852, pp. 458–474. Springer, Cham (2016). https://doi.org/10. 1007/978-3-319-46227-1_29
5. Lu, Y., Zhai, C., Sundaresan, N.: Rated aspect summarization of short comments. In: Proceedings of the 18th International Conference on World Wide Web, WWW 2009, Madrid, Spain, 20–24 April 2009, pp. 131–140 (2009)
6. Luo, W., Zhuang, F., Cheng, X., He, Q., Shi, Z.: Ratable aspects over sentiments: predicting ratings for unrated reviews. In: 2014 IEEE International Conference on Data Mining, ICDM 2014, Shenzhen, China, 14–17 December 2014, pp. 380–389 (2014)
7. Luo, W., Zhuang, F., Zhao, W., He, Q., Shi, Z.: QPLSA: utilizing quad-tuples for aspect identification and rating. Inf. Process. Manag. **51**(1), 25–41 (2015)
8. Moghaddam, S., Ester, M.: The FLDA model for aspect-based opinion mining: addressing the cold start problem. In: International Conference on World Wide Web, pp. 909–918 (2013)
9. Pecar, S.: Towards opinion summarization of customer reviews. In: Proceedings of ACL 2018, Student Research Workshop, pp. 1–8 (2018)
10. Schouten, K., van der Weijde, O., Frasincar, F., Dekker, R.: Supervised and unsupervised aspect category detection for sentiment analysis with co-occurrence data. IEEE Trans. Cybern. **48**(4), 1263–1275 (2018)
11. Shi, C., Li, Y., Zhang, J., Sun, Y., Yu, P.S.: A survey of heterogeneous information network analysis. IEEE Trans. Knowl. Data Eng. **29**(1), 17–37 (2017)
12. Sun, Y., Han, J., Zhao, P., Yin, Z., Cheng, H., Wu, T.: RankClus: integrating clustering with ranking for heterogeneous information network analysis. In: ACM SIGKDD 2009, pp. 565–576 (2009)
13. Wang, H., Ester, M.: A sentiment-aligned topic model for product aspect rating prediction. In: Conference on Empirical Methods in Natural Language Processing, pp. 1192–1202 (2014)
14. Xiao, D., Ji, Y., Li, Y., Zhuang, F., Shi, C.: Coupled matrix factorization and topic modeling for aspect mining. Inf. Process. Manag. **54**(6), 861–873 (2018)
15. Yu, D., Mu, Y., Jin, Y.: Rating prediction using review texts with underlying sentiments. Inf. Process. Lett. **117**, 10–18 (2017)
16. Zhang, L., Liu, B.: Aspect and entity extraction for opinion mining. In: Chu, W.W. (ed.) Data Mining and Knowledge Discovery for Big Data. SBD, vol. 1, pp. 1–40. Springer, Heidelberg (2014). https://doi.org/10.1007/978-3-642-40837-3_1

Dependency-Aware Attention Model for Emotion Analysis for Online News

Xue Zhao, Ying Zhang[(⊠)], and Xiaojie Yuan

College of Computer Science, Nankai University,
38 Tongyan Road, Tianjin 300350, People's Republic of China
zhaoxue@dbis.nankai.edu.cn, {yingzhang,yuanxj}@nankai.edu.cn

Abstract. This paper studies the emotion responses evoked by the news articles. Most work focuses on extracting effective features from text for emotion classification. As a result, the valuable information contained in the emotion labels has been largely neglected. In addition, all words are potentially conveying affective meaning yet they are not equally significant. Traditional attention mechanism can be leveraged to extract important words according to the word-label co-occurrence pattern. However, words that are important to the less popular emotions are still difficult to identify. Because emotions have intrinsic correlations, by integrating such correlations into attention mechanism, emotion triggering words can be detected more accurately. In this paper, we come up with an emotion dependency-aware attention model, which makes the best use of label information and the emotion dependency prior knowledge. The experiments on two public news datasets have proved the effectiveness of the proposed model.

Keywords: Emotion analysis · Attention mechanism · Neural sentiment analysis

1 Introduction

Emotion detection is a challenging task, the complexity stems from the fact that emotions are highly sensitive to contextual and personal factors [1]. Most of the studies [2–4] in textual emotion detection rely on explicit expression of emotions using some emotion bearing words, which can be found in tweets, blog posts and product reviews. But emotion expression can also happen in an emotion provoking situation. For example, the emotions of readers evoked by news articles. Several news websites have made such affective data available by providing a mood meter widget on the news web page, as illustrated in Fig. 1.

In addition, attention mechanism [5] has shown promising results in many NLP tasks [6,7]. The attention implementations are based on context information and allow the model to focus on important words in a sentence by learning a weight vector over sentence representation. The importance of words largely depends on the co-occurrence of words and labels in the task. There also has

Q. Yang et al. (Eds.): PAKDD 2019, LNAI 11439, pp. 172–184, 2019.
https://doi.org/10.1007/978-3-030-16148-4_14

Fig. 1. The emotion votes for a *Rappler* news article.

work building attention layer over both word embedding and label embedding [8], however, the dependencies of labels are neglected. Past research [9,10] has theoretically proved that some basic emotions can coexist and blend to form more complex forms. Existing work has also reported the improvement of emotion prediction accuracy after integrating emotion dependencies into the models [12–14]. Hence, in this paper, we empower the traditional attention module with the knowledge conveyed in labels, so that the words that are important to those less popular emotions can be better captured. Specifically, we first use news headlines to guide the attention module as we observed that news headlines usually hold the important information of the news body. The attention module is followed by a CNN classifier, which can generate an emotion signal, as shown in Fig. 2. The emotion signal is then transformed by an emotion embedding layer as well as an emotion dependency matrix into emotion representation, which will guide another attention module to attend the context again. The final news representation is thus generated not only from the knowledge of words, but also from the word-label co-occurrence patterns and the label-label dependency prior knowledge, which can help improve the emotion prediction results.

In particular, instead of using symmetric emotion dependency matrix as in [12–14], we generate a "dominant emotion-centered" asymmetric emotion dependency matrix. The matrix is row-wise explicable that each row represents the correlations between the dominant emotion (top voted emotion) and the other emotions. For instance, in the matrix, the diagonal items represent the dominant emotions. In each row, the values other than the diagonal stand for the correlations between the dominant emotion and the other emotions. Therefore, the distribution of the emotions are actually different given different dominant emotions. The matrix can thus reflect the complicated emotion status more accurately.

This work validated the proposed emotion dependency-aware attention model by experimenting on two real world news datasets, Rappler[1] and Yahoo[2]. Results show the proposed model outperforms the other baselines.

---

[1] https://www.rappler.com.

[2] https://tw.news.yahoo.com.

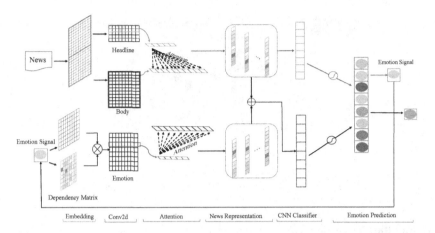

Fig. 2. Architecture of the proposed neural network. The model works in two spaces, context space (upper) and emotion space (lower), and it starts from the context space. The attention consists of two parts, headline-guided attention (upper) and emotion dependency-aware attention (lower). The emotion signal is generated and fed to an emotion embedding layer as well as an emotion dependency matrix to obtain the emotion representation, which will be used in emotion dependency-aware attention module. The news representations from two spaces will be concatenated and output to a CNN Classifier for the final prediction.

2   Approach

In the proposed model, two guided attention layers are involved. One is guided by the context information from headlines to generate emotion signals, the other is guided by the emotion information from the emotion embedding and emotion dependency matrix.

2.1   Context Guided Attentive CNN

The proposed model first generates the basic news representation with a word embedding layer and a 2-dimension CNN. In particular, it first transforms news article with a d-dimension word embedding layer into vectors $X = \{H, B\}$, where $H = \{h_1, h_2, ..., h_l\} \in R^{l \times d}$ and $B = \{b_1, b_2, ..., b_m\} \in R^{m \times d}$ are representation for news headlines and news bodies, respectively. It then extracts n-gram feature maps with a 2-dimension convolution operation with filter $W_1 \in R^{f \times n \times d}$, where f is the number of filters and (n, d) is kernel size. We denote headline feature map from filter k as $U_k = \{u_k^1, u_k^2, ..., u_k^l\}$ and body feature map as $V_k = \{v_k^1, v_k^2, ..., v_k^m\}$, where $u^i \in R^f$ and $v^i \in R^f$ represent the feature map for ith n-gram in news headlines and news bodies. We use zero paddings in all convolution operations with stride size 1, so the output size of convolution layer is the same as input.

$$u^i = \text{Elu}(W_1 \cdot H_{i:i-n+1} + b_1)$$
$$v^i = \text{Elu}(W_1 \cdot B_{i:i-n+1} + b_1) \qquad (1)$$

$b_1 \in R^f$ is the bias item. We use non-linear exponential linear unit (ELU) as activation to reduce the bias shift effect. The energy function to weight the ith n-gram feature involvement in the news bodies is denoted as $\mathbf{s}$ and defined as:

$$\mathbf{s} = \tanh(W_2 U + W_3 V + b_2) \qquad (2)$$

where $\mathbf{s} \in R^m$, W_2, W_3 are trainable weights and b_2 is bias. So the news body representation $X_b \in R^{f \times m}$ can be derived by Eq. 4.

$$\alpha_h = \text{softmax}(\mathbf{s}) \qquad (3)$$

$$X_b = \alpha_h \cdot V \qquad (4)$$

CNN Classifier. With X_b, we build a CNN classifier to generate an emotion signal. The CNN classifier consists of a 1-dimension convolution operation with a global average pooling. In particular, we apply convolution on X_b to generate a feature map for every emotion category. Let us denote $f_c \in R^m$, as the generated feature map for emotion category $c \in C$, $W_4 \in R^{C \times f}$ as filters and $b_3 \in R^C$ as bias:

$$f_c = \text{Elu}(W_4 \cdot X_b + b_3) \qquad (5)$$

We use average pooling to take the spatial average of feature map, and feed the result directly to a softmax layer.

$$p_c = \text{avgpooling}(f_c) \qquad (6)$$

$$P_1 = \text{softmax}(\mathbf{p_1}) \qquad (7)$$

$$\hat{c} = \text{argmax}(P_1) \qquad (8)$$

The average pooling result p_c can be seen as the "confidence" for predicting the emotion category c. We have $\mathbf{p_1} = \{p_c\}$, which can be interpreted into probabilities P_1 by a softmax function. Then the emotion with highest probability $\hat{c}$ will be used as an emotion signal and sent to the emotion space.

2.2   Emotion Dependency-Aware Attentive CNN

So far the features are solely from context knowledge and weighted by the closeness to the news headlines. However, some emotion triggering words and phrases that are not semantically close to headlines are also important to the task. Therefore, we design an emotion dependency-aware attentive CNN structure to extract features from emotion space, which consists of an emotion dependency matrix, an emotion embedding layer, an attention layer and a CNN classifier.

Emotion Dependency Matrix. We first divide the news from training data into C groups according to its dominant emotion label. Inside of each group, we rank the emotions for each article and then calculate the Kendall's tau, which is a more robust correlation measure. With all groups calculated, we can obtain an emotion dependency matrix $M \in R^{C \times C}$ where each row $M_{\hat{c}}$ represents the emotion distribution when $\hat{c}$ is the dominant emotion. Figures 3 and 4 display the emotion dependency of Rappler and Yahoo news dataset, from which we can tell that most emotions are correlated. By observing the first matrix by rows, we can see that when *sad* is the dominant emotion, *amused* is negatively correlated while *angry* and *afraid* are closely correlated to *sad*. From Fig. 4, we can see that *happy* can be correlated to many other emotions, even the negative ones. This demonstrates that the evoked emotion varies from person to person, which explains the complexity nature of the emotion.

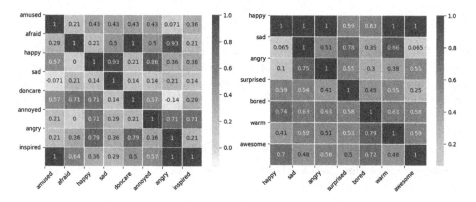

Fig. 3. Emotion dependency for *Rappler dataset.*

Fig. 4. Emotion dependency for *Yahoo dataset.*

Emotion Dependency-Aware Attention. The emotion signal $\hat{c}$ will be used to select the corresponding row $M_{\hat{c}}$ in M. On the other hand, $\hat{c}$ will be transformed into vectors $E_{\hat{c}}$ by a d'-dimensional emotion embedding layer $E \in R^{1 \times d'}$. The vectorized emotion signal $S \in R^{C \times d'}$ can be obtained by a product of two matrices.

$$S = M_{\hat{c}}^T E_{\hat{c}} \tag{9}$$

The emotion representation $R \in R^{f \times c}$ will be extracted by a 2-dimension convolution layer with kernel size $(1, d')$.

$$R = \text{Elu}(W_5 \cdot S + b_4) \tag{10}$$

$W_5 \in R^{f \times d'}$ is the trainable filter weight where f is the number of filters. b_4 is the bias. Similar to the context space, the energy function here is defined as:

$$\mathbf{s}' = \tanh(W_6 R + W_7 V + b_5) \tag{11}$$

where $W_6 \in R^{m \times c}$, $W_7 \in R^{m \times m}$ are weights and b_5 is bias. Then the news body representation $X_e \in R^{f \times m}$ attended by emotion signal can be obtained as follows:

$$\alpha_e = \text{softmax}(\mathbf{s'}) \tag{12}$$

$$X_e = \alpha_e \cdot V \tag{13}$$

We concatenate two representations from the context space and the emotion space, so the final representation can be denoted as $X = [X_b, X_e]$.

CNN Classifier. Similar to the classifier in context space, we apply another CNN classifier to predict final emotion distribution P_2.

$$f_{c'} = \text{Elu}(W_8 \cdot X + b_6) \tag{14}$$

$$p_{c'} = \text{avgpooling}(f_{c'}) \tag{15}$$

$$P_2 = \text{softmax}(\mathbf{p_2}) \tag{16}$$

$$\hat{c'} = \text{argmax}(P_2) \tag{17}$$

where $\mathbf{p_2} = \{p_{c'}\}$. Therefore, we obtain the final emotion prediction $\hat{c'}$, which is derived not only from the knowledge of words, but also the knowledge of word-emotion co-occurrence pattern and emotion dependency prior knowledge.

2.3 Composition of Networks

Noted that in our settings we use n-gram to represent the granularity of words and phrases. By simply composing different sizes of phrases, the information from both spaces will be better expressed. For example, for $3, 4, 5$-grams, there will be 3 emotion signals being predicted by using different size of kernels $3, 4, 5$. All 3 emotion signals can be sent into emotion space to set up a more reliable emotion representation E.

2.4 Training

We have two predictions in forward propagation, P_1 and P_2, which give us two different loss items. We use cross entropy $loss_{ce}$ as loss functions for both.

$$loss_{ce} = - \sum_{n \in D} \sum_{i=1}^{C} P_i^t(n) \cdot log(P_i(n)) \tag{18}$$

where D and C denote the number of news articles for training and the emotion categories, respectively. $P_i^t(n)$ denotes the true emotion label with one-hot encoding mechanism and P_i the predicted emotion distribution.

$$Loss = loss_{ce1} + loss_{ce2} \tag{19}$$

During training, batch normalization has been applied after every convolution operation, before Elu activations. Dropout is used in CNN classifier before convolution, which is set as 0.2. We set number of filters f as 128, and experiment on different sizes of kernels. All the parameters are optimized by Adam optimizer with learning rate 0.001.

3   Experiment

3.1   Datasets

Two datasets are involved, Rappler and Yahoo, both are publicly accessible. Rappler is developed by [15] with 8 emotion labels, and Yahoo is a Chinese Yahoo Kimo news dataset with 7 emotions (originally 8 emotion categories, but we excluded the label "informative" as it is not related to an emotional state). Table 1 summaries the statistic of the datasets.

Table 1. Datasets statistics: number of articles that the emotion has the highest number of votes.

Rappler	# of articles	Yahoo	# of articles
Happy	12,304	Happy	17,023
Sad	4,571	Sad	1,545
Angry	3,003	Angry	12,884
Amused	3,449	Surprised	3,001
Afraid	2,576	Bored	2,429
Annoyed	1,311	Warm	521
Inspired	2,729	Awesome	2,123
Don't care	1,164	-	-
Total	31,107	Total	49,000

We split the data into training/development/testing by 7/1/2. Because the data is highly imbalanced, stratified sampling is applied to keep the proportion of each category the same in training/development/testing.

3.2   Baselines

We classify our baselines into four groups. First group is based on machine learning methods.

– **LR**, Logsitic Regression with tf-idf weighted unigrams and bigrams.
– **SVM**, multi-class SVM classifier with unigram and bigram features.

Group 2 has RNN-based neural networks.

– **GRNN**, Gated RNN proposed in [16].
– **SelfATTN**, structured self-attentive sentence embedding proposed in [17].
– **BLSTM-ATTN**, BLSTM with an attention layer over hidden outputs.
– **HAN**, hierarchical attention networks proposed in [18].

Group 3 contains CNN-based neural networks.

- **CNN**, a CNN model for sentence classification, proposed in [19].
- **CNN-ATTN**, an attentive CNN model proposed in [8].
- **Inception**, proposed in [20] with deeper architecture of convolutions.

Group 4 includes neural networks without RNN/CNN structures.

- **Transformer** an attention-only model proposed in [5].
- **FastText**, a neural network for representation learning, proposed in [21].

We also investigate the performance of the proposed model in different settings.

- **Proposed-NT** is the proposed model with no headline attention guidance.
- **Proposed-NE** is the version without emotion dependency matrix.
- **Proposed-ngram** is the proposed model with convolution kernel size n.
- **Proposed-3, 4, 5** is the composed version having kernel sizes $3, 4, 5$.

3.3   Model Configuration

All the neural networks are using 200-dimension word embedding initialized with Glove [22] word vectors. The number of hidden outputs is 256. Dropout is set as 0.2. For uncommon parameters, such as the number of attention heads, are set according to the original papers.

4   Results

Measures. We use Accuracy and RMSE to evaluate the performance of models.

$$\text{Accuracy} = \frac{T}{D} \tag{20}$$

$$\text{RMSE} = \sqrt{\frac{\sum_{i=1}^{N} (y_i - \hat{y}_i)^2}{N}} \tag{21}$$

Accuracy is a standard metric to measure the overall emotion classification performance. T denotes the number of news that are correctly classified and D is the total number of news in test dataset. RMSE measures the divergences between predicted and the ground truth emotions, where $\hat{y}_i$ and y_i represent the predicted outputs and gold labels, respectively. We summarize the experiment results in Table 2.

Result Analysis. From Table 2, we can observe that the machine learning models are comparably weaker than deep learning models.

Table 2. Emotion classification on different models.

Models	Rappler		Yahoo	
	Accuracy %	RMSE	Accuracy %	RMSE
LR	57.93	0.234	70.04	0.187
SVM	72.99	0.234	78.43	0.234
GRNN	75.07	0.232	83.04	0.203
SelfATTN	75.01	0.229	83.86	0.223
BLSTM-ATTN	78.37	0.215	84.67	0.193
HAN	75.08	0.221	82.16	0.214
CNN	78.23	0.191	85.09	0.189
CNN-ATTN	78.18	0.214	85.27	0.225
Inception	75.38	0.222	80.45	0.227
Transformer	77.98	0.183	84.72	0.277
FastText	75.99	0.226	84.23	0.227
Proposed-NT	78.80	0.226	84.74	0.221
Proposed-NE	77.55	0.185	83.93	0.220
Proposed-3gram	77.11	**0.181**	84.48	0.228
Proposed-4gram	79.06	0.193	83.88	0.227
Proposed-5gram	**79.72**	0.211	83.33	0.217
Proposed-3, 4, 5	79.44	0.218	**85.67**	**0.175**

Compared to RNN-based methods, the proposed CNN-based method performs better. Generally, RNNs suffer from long sequence encoding. HAN does not improve the prediction results. This indicates that the information conveyed by the sentence structure is not helpful in differentiating the emotions in news articles.

From the third group we can observe that simple CNN structure outperforms all the other CNN models. The evaluation on accuracy and RMSE shows that our model can achieve better performance in most scenarios. Compared to the attention-free CNN model, such as CNN and Inception, our method performs better. It indicates that the attention modules are helpful in attending the important words and giving better news representation. Besides, our method outperforms CNN-ATTN which builds the attention module over word-label compatibility without considering the label-label dependencies.

As for the models in group 4, the proposed model outperforms these baselines on both two metrics. This indicates that CNN is more effective for feature extraction, and the combination with attention mechanism can improve the prediction results.

Additionally, as shown in Table 2, our model with 5-gram kernel performs the best on Rappler and with 3, 4, 5-gram kernel the best on Yahoo. This demonstrates that 5-gram can represent low-level information in our English dataset more properly while 3, 4, 5-gram features are better for our Chinese dataset. Furthermore, accuracy is better when the emotion dependencies are considered. We can also observe that the accuracy has a small improvement when the headline guidance is added. Besides, the composition of different kernel sizes is not always necessary. For example, on Rapper, the single emotion signal generated from 5-gram convolution outperforms the other settings, including the model compositions.

5   Discussion

5.1   Multi-task Training

In the proposed model, two outputs can be treated as two tasks and one's output is the input of the other. In order to observe how these tasks interact, we train **Proposed-3, 4, 5** and analyze the prediction accuracy for 4 outputs (3 emotion signals and 1 final prediction) at the end of each epoch, as shown in Fig. 5. *train* refers to final prediction on training dataset and *test* means on test dataset. n means the output emotion signal from the n-gram kernel. We can see from the figure that three auxiliary tasks are helping the main task to achieve good result very fast. After the 10th epoch, the model is then focusing on improving the predictions of emotion signals, which also slightly increases the accuracy of the main task.

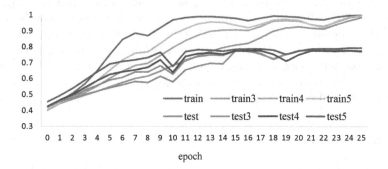

Fig. 5. The accuracy results on *Rappler* dataset.

5.2   Attention Visualization

To illustrate the results more straightforwardly, we investigate the performance of the emotion dependency-aware attention module by visualizing the attended content in a news sample, as shown in Fig. 6.

The yellow part is the context attended by the context attention module, while the red one is captured by the emotion dependency-aware attention module. We can observe that the context attention does attend useful words, yet the red part is also useful because it does trigger and exaggerate some emotional feelings.

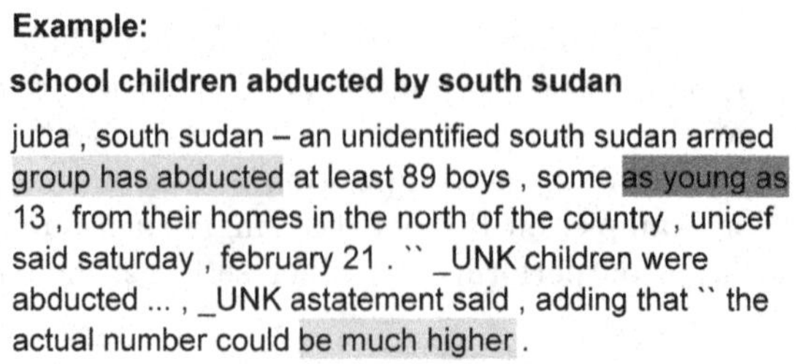

Example:

school children abducted by south sudan

juba , south sudan – an unidentified south sudan armed
group has abducted at least 89 boys , some as young as
13 , from their homes in the north of the country , unicef
said saturday , february 21 . `` _UNK children were
abducted ... , _UNK astatement said , adding that `` the
actual number could be much higher .

Fig. 6. Attention visualization on a *Rappler* news sample. 'UNK' denotes the out-of-vocabulary token. (Color figure online)

6   Related Work

There are two lines of work for emotion analysis: lexicon based and machine learning based approaches. Lexicon based methods usually depend on emotion lexicons [3, 23–27]. The lexicons are easy to use but suffer the inability of distinguishing different meanings of words in different context [2]. Machine learning based methods focus on learning the important features from the text. Emotion analysis can be cast into multi-class classification [4], multi-label classification [28], emotion distribution learning [13] or ranking problems [12].

Emotion analysis for online news originates from the SemEval task *Affective Text* [11] in 2007, which aims to explore the connection between news headlines and the evoked emotions of readers. In [2,4,28,29], authors built topic models to use a latent layer to bridge the gap between emotions and documents. Work [13] proposed a method to learn a mapping function from sentences to their emotion distributions. However, the valuable information conveyed in the emotion labels has been neglected in these methods.

7   Conclusions

In this work, we propose a framework for emotion analysis for online news. We first introduce the headline-guided attention module to learn the important words and phrases in news articles, and produce an emotion signal from the context space. Then we design a dependency-aware attention module, which utilizes an emotion embedding layer and an emotion dependency matrix to transform the emotion signal into emotion representation. The emotion representation will enable the attention module to identify emotion triggering words particularly. Our model achieved better prediction results in the experiment comparing to various of baselines.

Acknowledgement. We thank the reviewers for their constructive comments. This research is supported by National Natural Science Foundation of China (No. U1836109), Natural Science Foundation of Tianjin (No. 16JCQNJC00500) and Fundamental Research Funds for the Central Universities.

References

1. Ben-Ze'ev, A.: The Subtlety of Emotions. MIT Press, Cambridge (2001)
2. Rao, Y., Lei, J., Wenyin, L., Li, Q., Chen, M.: Building emotional dictionary for sentiment analysis of online news. WWW **17**, 723–742 (2014)
3. Staiano, J., Guerini, M.: Depeche Mood: a lexicon for emotion analysis from crowd annotated news. In: ACL (2014)
4. Bao, S., et al.: Mining social emotions from affective text. TKDE **24**, 1658–1670 (2012)
5. Vaswani, A., et al.: Attention is all you need. In: NIPS (2017)
6. Luong, T., Pham, H., Manning, C.D.: Effective approaches to attention-based neural machine translation. In: EMNLP (2015)
7. Gao, S., Ramanathan, A., Tourassi, G.: Hierarchical convolutional attention networks for text classification. In: Proceedings of The Third Workshop on Representation Learning for NLP (2018)
8. Wang, G., et al.: Joint embedding of words and labels for text classification. In: ACL (2018)
9. Ortony, A., Turner, T.J.: What's basic about basic emotions? Psychol. Rev. **97**, 315–331 (1990)
10. Ekman, P.: An argument for basic emotions. Cogn. Emot. **6**, 169–200 (1992)
11. Strapparava, C., Mihalcea, R.: Semeval-2007 task 14: affective text, June 2007
12. Zhou, D., Yang, Y., He, Y.: Relevant emotion ranking from text constrained with emotion relationships. In: NAACL (2018)
13. Deyu, Z.H.O.U., Zhang, X., Zhou, Y., Zhao, Q., Geng, X.: Emotion distribution learning from texts. In: EMNLP (2016)
14. Quan, X., Wang, Q., Zhang, Y., Si, L., Wenyin, L.: Latent discriminative models for social emotion detection with emotional dependency. TOIS **34**, 2:1–2:19 (2015)
15. Song, K., Gao, W., Chen, L., Feng, S., Wang, D., Zhang, C.: Build emotion lexicon from the mood of crowd via topic-assisted joint non-negative matrix factorization. In: SIGIR (2016)
16. Tang, D., Qin, B., Liu, T.: Document modeling with gated recurrent neural network for sentiment classification. In: EMNLP (2015)
17. Lin, Z., et al.: A structured self-attentive sentence embedding. In: ICLR (2017)
18. Yang, Z., Yang, D., Dyer, C., He, X., Smola, A., Hovy, E.: Hierarchical attention networks for document classification. In: NAACL (2016)
19. Kim, Y.: Convolutional neural networks for sentence classification. In: EMNLP (2014)
20. Szegedy, C., et al.: Going deeper with convolutions. In: ICLR (2015)
21. Joulin, A., Grave, E., Bojanowski, P., Mikolov, T.: Bag of tricks for efficient text classification. In: Proceedings of the European Chapter of the Association for Computational Linguistics: Volume 2, Short Papers (2017)
22. Pennington, J., Socher, R., Manning, C.: GloVe: global vectors for word representation. In: EMNLP (2014)
23. Mohammad, S.M., Turney, P.D.: Crowdsourcing a word-emotion association lexicon. Comput. Intell. **29**, 436–465 (2013)
24. Mohammad, S.: From once upon a time to happily ever after: tracking emotions in novels and fairy tales. In: ACL (2011)
25. Baccianella, S., Esuli, A., Sebastiani, F.: SentiWordNet 3.0: an enhanced lexical resource for sentiment analysis and opinion mining. In: LREC (2010)

26. Tausczik, Y.R., Pennebaker, J.W.: The psychological meaning of words: LIWC and computerized text analysis methods. J. Lang. Soc. Psychol. **29**(1), 24–54 (2010)
27. Strapparava, C., Valitutti, A.: WordNet affect: an affective extension of wordnet. In: LREC, vol. 4, pp. 1083–1086, May 2004
28. Li, X., et al.: Weighted multi-label classification model for sentiment analysis of online news. In: BigComp (2016)
29. Bao, S., et al.: Joint emotion-topic modeling for social affective text mining. In: ICDM (2009)

Multi-task Learning
for Target-Dependent Sentiment
Classification

Divam Gupta[1]([⊠]), Kushagra Singh[1], Soumen Chakrabarti[2],
and Tanmoy Chakraborty[1]

[1] IIIT Delhi, New Delhi, India
{divam14038,kushagra14056,tanmoy}@iiitd.ac.in
[2] IIT Bombay, Mumbai, India
soumen@cse.iitb.ac.in

Abstract. Detecting and aggregating sentiments toward people, organizations, and events expressed in unstructured social media have become critical text mining operations. Early systems detected sentiments over whole passages, whereas more recently, target-specific sentiments have been of greater interest. In this paper, we present MTTDSC, a multi-task target-dependent sentiment classification system that is informed by feature representation learnt for the related auxiliary task of passage-level sentiment classification. The auxiliary task uses a gated recurrent unit (GRU) and pools GRU states, followed by an auxiliary fully-connected layer that outputs passage-level predictions. In the main task, these GRUs contribute auxiliary per-token representations over and above word embeddings. The main task has its own, separate GRUs. The auxiliary and main GRUs send their states to a different fully connected layer, trained for the main task. Extensive experiments using two auxiliary datasets and three benchmark datasets (of which one is new, introduced by us) for the main task demonstrate that MTTDSC outperforms state-of-the-art baselines. Using word-level sensitivity analysis, we present anecdotal evidence that prior systems can make incorrect target-specific predictions because they miss sentiments expressed by words independent of target.

1   Introduction

As the volume of news, blogs [8], and social media [14] far outstrips what an individual can consume, sentiment classification (**SC**) [12,15] has become a powerful tool for understanding emotions toward politicians, celebrities, products, governance decisions, etc. Of particular interest is to identify sentiments expressed toward specific entities, i.e., *target dependent sentiment classification* (**TDSC**). Recent years have witnessed many TDSC approaches [21–23] with increasing sophistication and accuracy.

Possibly because research on passage-level and target-dependent sentiment classification were separated in time by the dramatic emergence of deep learning, TDSC systems predominantly use recurrent neural networks (RNNs) and

© Springer Nature Switzerland AG 2019
Q. Yang et al. (Eds.): PAKDD 2019, LNAI 11439, pp. 185–197, 2019.
https://doi.org/10.1007/978-3-030-16148-4_15

borrow little from passage-level tasks and trained models. From the perspective of curriculum learning [2], this seems suboptimal: representation borrowed from passage-level SC should inform TDSC well. Moreover, whole-passage labeling entails considerably lighter cognitive burden than target-specific labeling. As a result, whole-passage gold labels can be collected at larger volumes.

In this paper, we present MTTDSC, **M**ulti-**T**ask **T**arget **D**ependent **S**entiment **C**lassifier[1], a novel multi-task learning (MTL) system that uses passage-level SC as an auxiliary task and TDSC as the main task. MTL has shown significant improvements in many fields of Natural Language Processing and Computer Vision. In basic ('naive') MTL, we jointly train multiple models for multiple tasks with some shared parameters, usually in network layers closest to the inputs [13], resulting in shared input representation learning. Symmetric, uncontrolled sharing can be detrimental to some tasks.

In MTTDSC, the auxiliary SC task uses bidirectional GRUs, whose states are pooled over positions to make whole-passage predictions. This sensitizes the auxiliary GRU to target-independent expressions of sentiments in words. The main TDSC task combines the auxiliary GRUs with its own target-specific GRUs. The two tasks are jointly trained. If passages with both global and target-specific labels are available, they can be shared between the tasks. Otherwise, the two tasks can also be trained on disjoint passages. MTTDSC can be interpreted as a form of task-level curriculum learning [2], where the simpler whole-passage SC task learns to identify sentiments latent in word vectors, which then assists the more challenging TDSC task. Static sentiment lexicons, such as SentiWord-Net [6], are often inadequate for dealing with informal media.

Using two standard datasets, as well as one new dataset we introduce for the main task, we establish superiority of MTTDSC over several state-of-the-art approaches. While improved accuracy from additional training data may seem unsurprising from a learning perspective, we show that beneficial integration of the auxiliary task and data is nontrivial. Simpler multi-task approaches [13], where a common feature extraction network is used for jointly training on multiple tasks, perform poorly. We also use word-level sensitivity tests to obtain anecdotal evidence that direct TDSC approaches (that do not borrow from whole-passage SC models) make target-specific prediction errors because they misclassify the (target independent) sentiments expressed by words. Thus, MTTDSC also provides a more interpretable model, apart from accuracy gains.

The contributions of our work are summarized as follows:

- MTTDSC, a **novel neural MTL architecture designed specifically for TDSC**. We show the superiority of our model and also compare it with other state-of-the-art models of TDSC and multi-task learning.
- A **new dataset** for target dependent sentiment classification which is better for real world analysis on social media data.

---

[1] MTTDSC code and datasets are available at https://github.com/divamgupta/mttdsc.

- **Thorough investigation of the reasons behind the success** of MTTDSC. In particular, we show that current models fail to capture many emotive words owing to insufficient training data.

2   Related Work

Target Dependent Sentiment Classification: An input text passage is a sequence of words $w_1, w_2, \ldots, w_N$. We use 'tweet' and 'passage' interchangeably, given the preponderance of social media in TDSC applications. One word position or contiguous span is identified as a *target*. For simplicity, we will assume compounds like New_York to be pre-fused and consider a target as a single word position. A passage may have one or more targets. In gold labeled instances, the target is associated with one of three labels $\{-1, 0, +1\}$, corresponding to negative, neutral, and positive sentiments respectively. The passage-level task has a label associated with the whole passage [11], rather than a specific target position. E.g., in the tweet "I love listening to electronic music, however artists like Avici & Tiesta copy it from others", the overall sentiment is positive but the sentiments associated with both targets 'Avici' and 'Tiesta' are negative.

TDLSTM [20] and TDParse [22] divide the sentence into left context, right context and the target entity, and then combine their features. TDLSTM uses a left-to-right $\mathrm{LSTM_{LR}}$ on the context left of the target $(w_1, \ldots, w_i)$, a right-to-left $\mathrm{LSTM_{RL}}$ on the context right of the target $(w_i, \ldots, w_N)$, and a fully connected layer W_{TDLSTM} that combines signals from $\mathrm{LSTM_{LR}}$ and $\mathrm{LSTM_{RL}}$. If LSTM state vectors are in $\mathbb{R}^D$, then[2] $W_{\mathrm{TDLSTM}} \in \mathbb{R}^{2D \times 3}$. Given a tweet and target position i, $\mathrm{LSTM_{LR}}$ is applied to (pre-trained and pinned) input word embeddings of $w_1, \ldots, w_i$, obtaining state vectors $\mathrm{LSTM_{LR}}[1], \ldots, \mathrm{LSTM_{LR}}[i]$. Similarly, $\mathrm{LSTM_{RL}}$ is applied to (word embeddings of) $w_i, \ldots, w_N$, obtaining state vectors $\mathrm{LSTM_{RL}}[i], \ldots, \mathrm{LSTM_{RL}}[N]$. The output probability vector in Δ^3 is

$$\mathrm{SoftMax}\Big(\big[\mathrm{LSTM_{LR}}[i], \mathrm{LSTM_{RL}}[i]\big] \, W_{\mathrm{TDLSTM}} \Big), \tag{1}$$

where Δ^3 is a 3-class multinomial distribution over $\{-1, 0, 1\}$, obtained from the softmax. Standard cross-entropy against the one-hot gold label is used for training. TCLSTM [20] is a slight modification to TDLSTM, where the authors also concatenated the embedding of the target entity with each token in the given sentence. They showed that TCLSTM has a slight improvement over TDLSTM.

By pooling embeddings of words appearing on dependency paths leading to the target position, TDParse [22] improves further on TDLSTM accuracy. More details of these systems are described in Sect. 4.4, along with their performance. The major problem in TDParse is the inability to learn compositions of words. TDParse usually fails for the sentences containing a polar word which is not related to the entity.

---

[2] We elide possible scalar offsets in sigmoids and softmaxes for simplicity throughout the paper.

The "naive segmentation" (Naive-Seg) model of Wang et al. [22] concatenates word embeddings of left context, right context and sub sentences of the tweets. Various pooling functions are used to combine them and an SVM is used for labeling. Naive-Seg+ extends Naive-Seg by using sentiment lexicon based features. TDParse extends Naive-Seg by using dependency parse paths incident on the target entity to collect words whose embeddings are then pooled. TDParse+ further extends TDParse by using sentiment lexicon [6] based features. TDParse+ beats TDLSTM largely because of carefully engineered features (including SentiWord-Net based features), but may not generalize to diverse datasets.

Pooling word embeddings over dependency paths may not capture complex compositional semantics. Given enough training data, TDLSTM should capture complex compositional semantics. But in practice, neural sequence models start with word vectors that were not customized for sentiment detection, and then get limited training data.

Multi-task Learning: Multi-task learning has been used in many applications related to NLP. Peng et al. [16] showed improved results in semantic dependency parsing be learning three semantic dependency graph formalisms. Choi et al. [3] improved the performance on question answering by jointly training answer generation and answer retrieval model. Sluice networks proposed by Ruder et al. [18] claims to be a generalized model which could learn to control the sharing of information between different task models. Sluice networks do not perform well for TDSC, as the sharing of information happens at all positions of the sentence. On the other hand, our model forces the auxiliary task to learn feature representation at all positions and share them at the appropriate locations with the main task.

3   MTTDSC: A Multi-task Approach to TDSC

Recurrent models for TDSC have to solve two challenging problems in one shot: identify sentiment-bearing words in the passage, and use influences between hidden states to connect those sentiments to the target. A typical TDSC system attempts to do this as a self-contained learner, without representation support from an auxiliary learner solving the simpler task of whole-passage SC. We present anecdotes in Sect. 4.5 that reveal the limitations of such approaches. In response, we propose a multi-task learning (MTL) approach called MTTDSC. Representations trained for the auxiliary task (Sect. 3.1) inform the main task (Sect. 3.2). The combined loss objective is described in Sect. 3.3 and implementation details are presented in Sect. 3.4.

Our MTL framework is significantly different from traditional ones. In particular, we do not require auxiliary and main task gold labels to be available on the same instances. (This makes it easier to collect larger volumes of auxiliary labeled data.) As a result, in standard MTL, attempts to improve auxiliary task performance interferes with the learning of feature representations that are important for the main task. To solve this problem, we use separate RNNs for the two tasks, the output of the auxiliary RNN acting as additional features to the main model. This ensures that the gradients from the auxiliary task loss do not unduly interfere with the weights of the main task RNN.

3.1 Auxiliary Task

The network for the auxiliary task is shown at the top of Fig. 1. The auxiliary model consists of a left-to-right $\text{GRU}_{\text{LR}}^{\text{AUX}}$, a right-to-left $\text{GRU}_{\text{RL}}^{\text{AUX}}$, and a fully-connected layer $W_{\text{AUX}} \in \mathbb{R}^{2D \times 3}$. The auxiliary model is trained with tweets that are accompanied by whole-tweet sentiment labels from $\{-1, 0, 1\}$. First $\text{GRU}_{\text{LR}}^{\text{AUX}}$ and $\text{GRU}_{\text{RL}}^{\text{AUX}}$ are applied over the entire tweet (positions $1, \ldots, N$). At every token position i, we construct the concatenation

$$[\text{GRU}_{\text{LR}}^{\text{AUX}}[i-1], \text{GRU}_{\text{RL}}^{\text{AUX}}[i+1]]$$

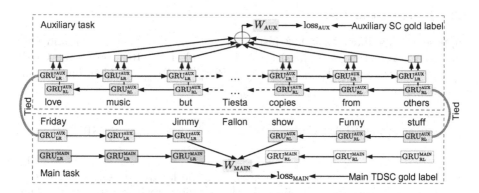

Fig. 1. MTTDSC network architecture. Passage-level gold labels are used to compute loss in the upper auxiliary network. Target-level gold labels are used to compute loss in the lower main network. These are coupled through tied parameters in auxiliary GRUs. The main task uses another set of task-specific GRUs.

These are then averaged over positions to get a fixed-size pooled representation $\bar{x} \in \mathbb{R}^{2D}$ of the whole tweet:

$$\bar{x} = \frac{1}{N} \sum_{i=1}^{N} [\text{GRU}_{\text{LR}}^{\text{AUX}}[i-1], \text{GRU}_{\text{RL}}^{\text{AUX}}[i+1]] \tag{2}$$

Average pooling lets the auxiliary model learn useful features at all positions of the tweet. This helps the primary task, as the target entity of the primary task can be at any position. The whole-tweet prediction output is $\text{SoftMax}(\bar{x} W_{\text{AUX}}) \in \Delta^3$. Again, cross-entropy loss is used.

3.2   Main Task

Beyond the auxiliary model components, the main task uses a left-to-right $\mathrm{GRU_{LR}^{MAIN}}$, a right-to-left $\mathrm{GRU_{RL}^{MAIN}}$, and a fully connected layer $W_{\mathrm{MAIN}} \in \mathbb{R}^{4D \times 3}$ as model components.

Let target entity be at token position i. $\mathrm{GRU_{LR}^{AUX}}$ and $\mathrm{GRU_{LR}^{MAIN}}$ are run over positions $1, \ldots, i-1$. $\mathrm{GRU_{RL}^{AUX}}$ and $\mathrm{GRU_{RL}^{MAIN}}$ are run over positions $i+1, \ldots, N$. The four resulting state vectors $\mathrm{GRU_{LR}^{AUX}}[i-1]$, $\mathrm{GRU_{RL}^{AUX}}[i+1]$, $\mathrm{GRU_{LR}^{MAIN}}[i-1]$, and $\mathrm{GRU_{RL}^{MAIN}}[i+1]$ are concatenated into a vector in $\mathbb{R}^{4D}$, which is input into the fully-connected layer followed by a softmax.

$$\mathrm{SoftMax}\Big([\mathrm{GRU_{LR}^{AUX}}[i-1], \mathrm{GRU_{LR}^{MAIN}}[i-1], \mathrm{GRU_{RL}^{AUX}}[i+1], \mathrm{GRU_{RL}^{MAIN}}[i+1]]\, W_{\mathrm{MAIN}} \Big). \quad (3)$$

Our network for the situation where the auxiliary and main tasks do not share instances is shown in Fig. 1.

3.3   Training the Tasks

Suppose the auxiliary task has instances $\{(x_i, y_i) : i = 1, \ldots, A\}$ and the main task has instances $\{(x_j, y_j) : j = 1, \ldots, M\}$. Let $\mathrm{GRU_*^*}$ be all the GRU model parameters in $\{\mathrm{GRU_{LR}^{AUX}}, \mathrm{GRU_{RL}^{AUX}}, \mathrm{GRU_{LR}^{MAIN}}, \mathrm{GRU_{RL}^{MAIN}}\}$. Then our overall loss objective is

$$\sum_{i=1}^{A} \mathrm{loss_{AUX}}(x_i, y_i; \mathrm{GRU_{LR}^{AUX}}, \mathrm{GRU_{RL}^{AUX}}, W_{\mathrm{AUX}}) + \alpha \sum_{j=1}^{M} \mathrm{loss_{MAIN}}(x_j, y_j; \mathrm{GRU_*^*}, W_{\mathrm{MAIN}})$$

$$(4)$$

Standard cross-entropy is used for both $\mathrm{loss_{AUX}}$ and $\mathrm{loss_{MAIN}}$. Before training the full objective above, we pre-train only the auxiliary task for one epoch. The situation where instances may be shared between the auxiliary and main tasks is similar, except that GRU cells are now directly shared between auxiliary and main tasks. We anticipate this multi-task setup to do better than, say, fine-tuning word embeddings in TDLSTM, because the auxiliary task is better related to the main task than unsupervised word embeddings. By the same token, we do not necessarily expect our auxiliary learner to outperform more direct approaches for the auxiliary task—its goal is to supply better word/span representations to the main task.

3.4   Implementation Details

GRU instead of LSTM: We used GRUs instead of LSTMs, which are more common in prior work. GRUs have fewer parameters and are less prone to overfitting. In fact, our TDGRU replacement performs better than TDLSTM (Tables 1 and 2).

Hyperparameters: We set the hidden unit size of each GRU network as 64. Recurrent dropout probability of the GRU is taken as 0.2. We also used a dropout of 0.2 before the last fully connected layer. For training the models we used the Adam optimizer with learning rate of 0.001, $\beta_1 = 0.9$, $\beta_2 = 0.999$. We used a mini-batch size of 64.

Ensemble: While reimplementing and/or running baseline system codes, we saw large variance in test accuracy scores for random initializations of the network weights. We improved the robustness of our networks by using an ensemble of the same model trained on the complete dataset with different weight initializations. The output class scores of the final model are the average of the probabilities returned by members of the ensemble. For a fair comparison, we also use the same ensembling for all our baselines.

Word Embeddings: MTTDSC, TDLSTM and TCLSTM use GloVe embeddings [17] trained on the Twitter corpus.

4   Experiments

We summarize datasets, competing approaches, and accuracy measures, followed by a detailed performance comparison and analysis.

4.1   Datasets for Auxiliary SC Task

Go [7]: This is a whole-passage SC dataset, containing 1.6M tweets automatically annotated using emoticons, highlighting that SC labeling can be easier to acquire at large scale. It has only positive and negative classes.

Sanders [19]: The second dataset is provided by Sanders Analytics and has 5,513 tweets over all 3 classes. These are manually annotated.

4.2   Datasets for Main TDSC Task

Dong [5]: Target entities are marked in tweets (one target entity per tweet), and one of three sentiment labels manually associated with each target. The training and test folds contain 6,248 and 692 instances respectively. The class distribution is 25%, 50% and 25% for negative, neutral and positive respectively.

Election1 [22]: Derived from tweets about the recent UK election, this dataset contains 3,210 training tweets that mention 9,912 target entities and 867 testing tweets that mention 2,675 target entities. The class distribution is 45.3%, 36.5%, and 17.7% for negative, neutral and positive respectively, which is highly unbalanced. There are an average of 3.16 target entities per tweet.

Election2: In this paper, we introduce a new TDSC dataset, also based on UK election tweets. We first curated a list of candidate hashtags related to the UK General Elections, such as #GE2017, #GeneralElection and #VoteLabour.

The collection was done during a period of 12 days, from June 2, 2017 through June 14, 2017. After removing retweets and duplicates (tweets with the same text), we ended up with 563,812 tweets. After running the named entity tagger, we observed that 158,978 tweets (28.19%) had at least one named entity, 38,809 tweets (6.88%) had at least two named entities and the remaining 7,992 tweets (1.42%) had three or more named entities. We took all the tweets which had at least two named entities, and randomly sampled an equal number of tweets from the set of tweets which had only one named entity.

4.3   Details of Performance Measures

Past TDSC work reports on 0/1 accuracy and macro averaged F_1 scores, and we do so too, for 3-class $\{-1, 0, 1\}$ instances and 2-class $\{-1, 1\}$ subsets. However, SC is fundamentally a regression or ordinal regression task; e.g., $\text{loss}(-1, 1) > \text{loss}(-1, 0) > \text{loss}(-1, -1) = 0$. Evaluating ordinal regression in the face of class imbalance can be tricky [1]. In addition, the system label may be discrete from $\{-1, 0, 1\}$ or continuous in $[-1, 1]$. Therefore we report on two additional performance measures. Let $(x_i, y_i, \hat{y}_i)$ be the ith of I instances, comprised of a tweet, gold label, and system-estimated label.

Mean Absolute Error (MAE): It is defined as $(1/I) \sum_i |y_i - \hat{y}_i|$. Downstream applications that use the numerical values of $\hat{y}$ will want MAE to be small.

Pair Inversion Rate (PIR): For a pair of instances (i, j), if $y_i > y_j$ but $\hat{y}_i < \hat{y}_j$, that is an *inversion*. PIR is then defined as $\left(\sum_{i \neq j} [\![(y_i - y_j)(\hat{y}_i - \hat{y}_j) < 0]\!] \right) \Big/ \binom{I}{2}$. Closely related to the area under the curve (AUC) for 2-class problems, PIR is widely used in Information Retrieval [10].

4.4   Various Methods and Their Performance

Table 1 shows aggregated and per-class accuracy for the competing methods. It has three groups of rows. The first group includes methods that use no or minimal target-specific processing. The second group includes the best-known recent target-dependent methods. The third group includes our methods and their variants, to help elucidate and justify the merits of our design.

Target Independent Baselines: In the first block, **LSTM** means a whole-tweet LSTM was applied, followed by a linear SVM on the final state. **Target-ind** [9] pools embedding features from the entire tweet. **Target-dep+** extends Target-dep. **Target-dep+** extends Target-dep by identifying sentiment-revealing context features with the help of multiple lexicons (such as SentiWordNet[3]).

---

[3] http://sentiwordnet.isti.cnr.it/.

Table 1. Performance of various methods on **Dong** dataset: (a) overall, (b) class-wise. MTTDSC beats other baselines across diverse performance measures.

		(a)							(b)		
	Model	3 class performance						2-class	Class-wise F_1		
		Accuracy	Precision	Recall	F_1	MAE	PIR%	F_1	-1	0	$+1$
Baseline	LSTM	66.5	66.0	63.3	64.7	0.367	5.39	60.6	64.4	71.8	56.9
	Target-ind	67.3	69.2	61.7	63.8	0.351	5.45	58.8	60.9	73.6	56.8
	Target-dep+	70.1	69.9	65.9	67.7	0.341	5.31	62.6	65.9	75.7	59.4
	TDLSTM	70.8	69.5	68.8	69.0	0.335	5.61	65.7	68.9	75.6	62.5
	TCLSTM	71.5	70.3	69.4	69.5	0.321	5.47	67.2	68.4	75.1	65.9
	SLUICE	69.2	67.9	67.2	67.5	0.348	5.90	64.4	67.6	73.7	61.6
	Naive-Seg+	70.7	70.6	65.9	67.7	0.332	5.54	63.2	66.0	76.3	60.0
	TDParse	71.0	70.5	67.1	68.4	0.331	5.86	64.3	65.8	76.7	62.7
	TDParse+	72.1	72.2	68.3	69.8	0.312	5.00	66.0	67.5	77.3	64.5
	TDParse+ (m)	72.5	72.6	68.9	70.3	0.308	4.95	66.6	68.3	77.6	65.0
Our	TDGRU	71.0	70.2	68.7	69.3	0.324	5.14	66.4	67.8	75.1	65.0
	TDGRU+SVM	71.7	71.4	68.7	69.7	0.315	5.16	66.5	68.9	76.2	64.1
	TD naive MTL	63.0	63.2	57.4	59.1	0.403	6.22	46.2	55.0	70.3	52.0
	TDFT	73.3	72.7	70.8	71.6	0.299	5.17	68.8	70.0	77.4	**67.7**
	MTTDSC	**74.1**	**74.0**	**71.7**	**72.7**	**0.286**	**4.22**	**70.0**	**72.8**	**77.9**	67.3

Prior Target-Dependent Baselines: The second block shows more competitive target-aware TDSC approaches. **TDLSTM** and **TCLSTM** are from Tang et al. [20]. **Naive-Seg+** segments the tweet using punctuations. Word vectors in each segment are pooled to give a segment embedding. Additional features are generated from the left and right contexts based on multiple sentiment lexicons. **TDParse** [22] uses a syntactic parse to pool embeddings of words connected to the target. **TDParse+** extends TDParse by adding features from sentiment lexicons. **TDParse+(m)** considers the presence of the same target multiple times in the tweet. Feature vectors generated from multiple target positions are merged using pooling functions.

MTTDSC and Variations: The third block shows **MTTDSC** and some variations. **TDGRU** replaces the two LSTMs of TDLSTM with two GRUs [4] which have fewer parameters but perform slightly better than LSTMs. In **TDGRU+SVM**, we first train the TDGRU model. Then, at entity position i, we extract the features $[\mathrm{GRU_{LR}}[i]; \mathrm{GRU_{RL}}[i]]$ of the two GRU models and train an SVM with RBF kernel with the extracted features. TDGRU with SVM is expected to perform better due to the non-linear nature of the features at the penultimate layer, which the SVM can then recognize without overfitting problems. **TD naive MTL** is similar to MTTDSC, but, rather than having separate GRUs for primary and auxiliary tasks, shared $\mathrm{GRU_{LR}}$ and $\mathrm{GRU_{RL}}$ are used for both tasks. The tasks are trained jointly as in MTTDSC. In **TDFT**, we first train $\mathrm{GRU_{LR}}$, $\mathrm{GRU_{RL}}$ and W_{AUX} on the auxiliary whole-passage SC task. We then use the weights of $\mathrm{GRU_{LR}}$ and $\mathrm{GRU_{RL}}$ learnt by the auxiliary task in TDGRU and train it on TDSC with a new W_{MAIN}.

Table 2. Performance of various methods on our **Election2** dataset: (a) overall, (b) class-wise. The extreme label skew makes it easy for simpler algorithms to do well at MAE and PIR, although MTTDSC still leads in traditional measures, and recognizes neutral content better.

	(a)								(b)		
	Model	3 class performance						2-class	Class-wise F_1		
		Accuracy	Precision	Recall	F_1	MAE	PIR%	F_1	-1	0	$+1$
Baseline	LSTM	55.6	53.4	43.1	42.3	0.484	7.76	29.0	47.9	57.2	56.8
	Target-ind	55.6	52.3	43.7	42.7	0.484	7.87	30.0	42.5	68.1	17.4
	Target-dep+	59.4	59.6	48.1	48.4	0.447	7.30	37.2	47.2	70.9	27.1
	TDLSTM	58.6	54.5	50.0	50.7	0.479	10.46	41.1	48.0	70.0	34.2
	TCLSTM	58.4	54.3	49.3	50.1	0.474	9.73	40.1	45.1	70.1	35.1
	SLUICE	58.8	54.8	53.0	52.9	0.489	10.9	45.2	54.9	68.3	35.4
	Naive-Seg+	60.5	60.0	51.1	52.2	**0.433**	**7.06**	43.0	51.1	70.5	34.9
	TDParse	58.9	56.5	50.8	51.6	0.460	8.67	42.9	51.3	68.8	34.5
	TDParse+	61.1	59.8	52.7	53.6	0.438	8.16	45.1	**53.4**	70.8	36.7
	TDParse+(m)	60.6	59.1	52.2	53.1	0.440	8.40	44.5	52.4	70.5	36.6
Our	TDGRU	59.5	55.4	53.0	53.8	0.475	10.67	**45.8**	50.8	69.7	**40.8**
	TDGRU+SVM	59.2	56.0	51.6	52.5	0.469	9.93	43.9	50.0	69.6	37.9
	TD naive MTL	56.5	57.1	43.2	41.4	0.473	7.04	27.3	51.2	57.5	62.7
	MTTDSC	**61.6**	**60.1**	**53.1**	**54.1**	0.439	8.79	45.3	52.6	**71.8**	38.0

Observations: Table 1 shows that MTTDSC outperforms all the baselines across all the measures on **Dong** dataset. MTTDSC achieves 2.8%, 3.41%, 7.14% and 24.5% relative improvements in accuracy, F_1, MAE and PIR respectively over TDParse+(m) (best model by Wang et al. [22]). The improvement in 2-class F_1 is also substantial (5.1%). MTTDSC maintains a better balance between precision, recall and F_1 across the three classes (Table 1(b)).

TDFT improves on TDLSTM and TDGRU because it learns important features during pre-training. TDFT is better than TD naive MTL; jointly training the latter results in auxiliary loss prevailing over primary loss. TD naive MTL also loses to MTTDSC, because, in fine tuning, the primary task training makes the model forget some auxiliary features critical for the primary task. Summarizing, MTTDSC's gains are not explained by the large volume of auxiliary data alone; good network design is critical.

Table 2 shows results for **Election2**. The trend is preserved, with our gains in macro-F_1 being more noticeable than micro-accuracy. This is expected given the label skew. Table 3 shows similar behavior on the **Election1** dataset. Although TDPWindow-12 is slightly better for 0/1 accuracy, MTTDSC achieves 4.97% and 6.77% larger F_1 score for 3-class and 2-class sentiment classification respectively.

Table 3. Performance comparison on **Election1** dataset. The results of the baselines are taken from Table 3 of Wang et al. [22] where TDPWindow-12 (which extracts features exactly like TDParse+, but limits the size of left and right contexts to 12 tokens) was reported as the best model. To save space, we report the accuracy w.r.t. only three measures. The broad trends are similar to Election2.

Model	Accuracy	3-class F_1	2-class F_1
Target-ind	52.30	42.19	40.50
Target-dep+	55.85	43.40	40.85
TDParse	56.45	46.09	43.43
TDPWindow-12	**56.82**	45.45	42.69
TDGRU	55.46	47.12	44.22
MTTDSC	56.67	**47.71**	**45.58**

4.5   Side-by-Side Diagnostics and Anecdotes

Given their related architectures, we picked TDLSTM and MTTDSC, and focused on instances where MTTDSC performed better than TDLSTM, to tease out the reasons for the improvement.

Table 4. Word sensitivity studies. (Must be viewed in color.) Green words are regarded as positive and red words are regarded as negative by the respective RNNs. Intensity of color roughly represents magnitude of sensitivity. TDLSTM makes mistakes in estimating the polarity of words independent of context, which lead to incorrect predictions. Assisted by the auxiliary task, MTTDSC avoids such mistakes.

MTTDSC Word Polarities	MTTDSC Prediction	TDLSTM Word Polarities	TDLSTM Prediction	Ground Truth
just saw stephen colbert and the roots covering friday on the jimmy fallon show. funny stuff .	0	just saw stephen colbert and the roots covering friday on the jimmy fallon show. funny stuff.	-1	0
page 12 of comedy videos. will ferrell as george bush , trunk monkey , and some hilarious pranks	0	page 12 of comedy videos. will ferrell as george bush , trunk monkey, and some hilarious pranks	-1	0
playing on the wii fit with my mum , its hilarious :p	+1	playing on the wii fit with my mum, its hilarious :p	0	+1
merry christmas! i keep seeing that a christmas carol commercial. now i feel like britney spears , randomly wishing people a merry christmas.	+1	merry christmas ! i keep seeing that a christmas carol commercial. now i feel like britney spears , randomly wishing people a merry christmas .	0	+1

Word-level sensitivity analysis: For each word in the context of the target entity, we replaced the word with UNK (unknown word) and noted the drops in scores of labels +1 and −1. A large drop in the score of label +1 means the word was regarded as strongly positive, and a large drop in the score of label −1 means the word was regarded as strongly negative. We use these scores to color-code context words in the form of a heatmap. Table 4 shows the positive and the negative words highlighted accordingly to their sensitivity scores. The words highlighted in green color contribute to the positive label and the words highlighted red contribute to the negative label. In the first row, MTTDSC correctly identifies *funny* as a positive word, whereas TDLSTM considers *funny* to be a negative word. TDLSTM also finds stronger negative polarity in neutral words like *the* and *covering*. In the second row, MTTDSC correctly identifies *hilarious* as a positive word, whereas TDLSTM finds *hilarious* strongly negative. In the third row, MTTDSC finds *hilarious* positive, whereas TDLSTM misses the signal. Although TDLSTM correctly identifies more positive words in the fourth row than MTTDSC, it also incorrectly identifies negative words like *randomly* and *people*, leading to an overall incorrect neutral prediction. The examples show that TDLSTM either misses or misclassifies crucial emotive, polarized context words.

5   Conclusion

We presented MTTDSC, a multi-task system for target-dependent sentiment classification. By exploiting the easier auxiliary task of whole-passage sentiment classification, MTTDSC improves on recent TDSC baselines. The auxiliary LSTM learns to identify corpus-specific, position-independent sentiment in words and phrases, whereas the main LSTM learns how to associate these sentiments with designated targets. We tested our model on three benchmark datasets, of which we introduce one here, and obtained clear gains in accuracy compared to many state-of-the-art models.

Acknowledgement. The project was partially supported by IBM, Early Career Research Award (SERB, India), and the Center for AI, IIIT Delhi, India.

References

1. Baccianella, S., Esuli, A., Sebastiani, F.: Evaluation measures for ordinal regression. In: Intelligent Systems Design and Applications, Pisa, Italy, pp. 283–287 (2009)
2. Bengio, Y., Louradour, J., Collobert, R., Weston, J.: Curriculum learning. In: ICML, Montreal, Canada, pp. 41–48 (2009)
3. Choi, E., Hewlett, D., Uszkoreit, J., Polosukhin, I., Lacoste, A., Berant, J.: Coarse-to-fine question answering for long documents. In: ACL, pp. 209–220 (2017)
4. Chung, J., Gulcehre, C., Cho, K., Bengio, Y.: Empirical evaluation of gated recurrent neural networks on sequence modeling. arXiv preprint arXiv:1412.3555 (2014)

5. Dong, L., Wei, F., Tan, C., Tang, D., Zhou, M., Xu, K.: Adaptive recursive neural network for target-dependent Twitter sentiment classification. In: ACL, Baltimore, Maryland, USA, pp. 49–54 (2014)
6. Esuli, A., Sebastiani, F.: SentiWordNet: a publicly available lexical resource for opinion mining. In: LREC, pp. 417–422 (2006)
7. Go, A., Bhayani, R., Huang, L.: Twitter sentiment classification using distant supervision. CS224N Project Report, Stanford, vol. 1, no. 2009, p. 12 (2009)
8. Godbole, N., Srinivasaiah, M., Skiena, S.: Large-scale sentiment analysis for news and blogs. In: ICWSM, pp. 219–222 (2007)
9. Jiang, L., Yu, M., Zhou, M., Liu, X., Zhao, T.: Target-dependent twitter sentiment classification. In: ACL, Portland, Oregon, USA, pp. 151–160 (2011)
10. Joachims, T.: Optimizing search engines using clickthrough data. In: SIGKDD Conference, pp. 133–142. ACM (2002)
11. Kim, Y.: Convolutional neural networks for sentence classification. arXiv preprint arXiv:1408.5882 (2014)
12. Liu, B.: Sentiment analysis and opinion mining. Synth. Lect. Hum. Lang. Technol. 5(1), 1–167 (2012)
13. Maurer, A., Pontil, M., Romera-Paredes, B.: The benefit of multitask representation learning. J. Mach. Learn. Res. 17(81), 1–32 (2016)
14. Pak, A., Paroubek, P.: Twitter as a corpus for sentiment analysis and opinion mining. In: LREC, Valletta, Malta, pp. 1–3 (2010)
15. Pang, B., Lee, L., et al.: Opinion mining and sentiment analysis. Found. Trends Inf. Retrieval 2(1–2), 1–135 (2008)
16. Peng, H., Thomson, S., Smith, N.A.: Deep multitask learning for semantic dependency parsing. arXiv preprint arXiv:1704.06855 (2017)
17. Pennington, J., Socher, R., Manning, C.D.: GloVe: global vectors for word representation. In: EMNLP Conference 2014, pp. 1532–1543 (2014)
18. Ruder, S., Bingel, J., Augenstein, I., Søgaard, A.: Learning what to share between loosely related tasks. arXiv preprint arXiv:1705.08142 (2017)
19. Sanders, N.: Twitter sentiment corpus (2011)
20. Tang, D., Qin, B., Feng, X., Liu, T.: Effective LSTMs for target-dependent sentiment classification. arXiv preprint arXiv:1512.01100 (2015)
21. Teng, Z., Vo, D.-T., Zhang, Y.: Context-sensitive lexicon features for neural sentiment analysis. In: EMNLP, Austin, Texas, USA, pp. 1629–1638 (2016)
22. Wang, B., Liakata, M., Zubiaga, A., Procter, R.: TDParse: multi-target-specific sentiment recognition on Twitter. In: EACL, pp. 483–493 (2017)
23. Wilson, T., Wiebe, J., Hoffmann, P.: Recognizing contextual polarity in phrase-level sentiment analysis. In: EMNLP, Vancouver, BC, Canada, pp. 347–354 (2005)

SC-NER: A Sequence-to-Sequence Model with Sentence Classification for Named Entity Recognition

Yu Wang, Yun Li[✉], Ziye Zhu, Bin Xia, and Zheng Liu

Jiangsu Key Laboratory of Big Data Security and Intelligent Processing,
School of Computer Science, Nanjing University of Posts and Telecommunications,
Nanjing 210023, Jiangsu, China
{2017070114,liyun,1015041217,bxia,zliu}@njupt.edu.cn

Abstract. Named Entity Recognition (NER) is a basic task in Natural Language Processing (NLP). Recently, the sequence-to-sequence (seq2seq) model has been widely used in NLP task. Different from the general NLP task, 60% sentences in the NER task do not contain entities. Traditional seq2seq method cannot address this issue effectively. To solve the aforementioned problem, we propose a novel seq2seq model, named SC-NER, for NER task. We construct a classifier between the encoder and decoder. In particular, the classifier's input is the last hidden state of the encoder. Moreover, we present the restricted beam search to improve the performance of the proposed SC-NER. To evaluate our proposed model, we construct the patent documents corpus in the communications field, and conduct experiments on it. Experimental results show that our SC-NER model achieves better performance than other baseline methods.

Keywords: Named Entity Recognition ·
Sequence-to-sequence model · Deep learning

1 Introduction

Named Entity Recognition (NER) [27] is a basic task in Natural Language Processing (NLP) [9], which has attracted extensive attention for a long time [3]. NER aims to identify entities (e.g., names, places, and organization names) in the text. An entity can express the core information of the sentence, which is useful for various NLP tasks [16,17].

In the last few years, deep learning has been widely used in NLP tasks such as text classification [18], language recognition [24], and machine translation [12]. In these studies, the Recurrent Neural Networks (RNNs) [22] is widely used to obtain the sequential nature of the sentence. Especially, Long Short-Term Memory network (LSTM) [24], as a popular RNN model, is adopted to extract semantic features reflecting sequential nature of the text. Moreover, the RNNs

© Springer Nature Switzerland AG 2019
Q. Yang et al. (Eds.): PAKDD 2019, LNAI 11439, pp. 198–209, 2019.
https://doi.org/10.1007/978-3-030-16148-4_16

like to model the conditional probability $P(y|x)$, where the output sequence $y = (y_1, ..., y_n)$ (n is the length of sequence.) is generated from the input sequence x. Recently, researchers leverage sequence-to-sequence (seq2seq) model [32] to solve the sequence generation [5] problems in NLP tasks. In general, the NER is considered as a sequence problem [11], where the NER model tags each word to indicate whether the word is part of any named entity. So we apply seq2seq model [13] to the NER task, which encodes a sequence as a vector and decodes the vector into a tag sequence. Both encoder and decoder are constructed based on RNN.

However, the most evident difference between the normal natural language generation task [30] (e.g., human-computer conversation) and the NER task [29] is that 60% sentences in our corpus may not have an entity. This feature is non-trivial, and we should pay attention to it when designing the seq2seq model in NER task. Moreover, the seq2seq model learns to predict an output sequence at the training time, and it chooses the best tag sequence using the beam search [34] at the test time. The beam search is a heuristic search algorithm [21], and it works well in many tasks with large search space. However, the standard beam search [14] is not suitable for NER task, since the search space of NER task is small.

In the proposed seq2seq model named SC-NER, the tag sequence is generated in two steps: (1) We use the encoder's output to construct a classifier, which identifies whether the input sequence has an entity. In particular, we choose the LSTM for both encoder and decoder. The encoder will output the cell state and the last hidden state. The last hidden state is input into the classifier. (2) The classifier's output is considered as the starting vector of the decoder. Usually, the decoder generates the tag sequence based on the starting vector and cell state. In addition, the beam search is always utilized in tag sequence generation. The standard beam search, however, is not suitable for NER task. Therefore, we present a restricted beam search. The main contributions of our work are as follows:

- To the best of our knowledge, it is the first time for the seq2seq model to be used in NER task.
- In the proposed SC-NER model, a classifier is added to determine whether sentences have entities. Moreover, the training of classifier, encoder, and decoder is seamless.
- We present a restricted beam search to adjust the search space. The restricted beam search is more suitable to the NER task than the standard beam search.

The remaining part of this paper is organized as follows. Section 2 reviews the related work. Section 3 describes the details of our model. Section 4 introduces the data set. Experiment results are reported in Sect. 5. Section 6 concludes our work.

2   Related Work

Su and Su [31] presented a Hidden Markov Model (HMM) based block marker to identify and classify names, times, and numbers. It achieves high performance on the MUC-6 and MUC-7 English NER tasks and performs better than manual rule-based methods. Hai and Ng [15] proposed a maximum entropy approach for NER task, which showed the feasibility of extracting useful features (global features) using the occurrences of each word in a document. Mccallum and Li [26] presented a feature induction method for CRFs (Conditional Random Fields), which obtained a high F1-score.

In order to improve the performance on NER with more diverse entity types, researchers tend to use the deep learning methods. Wu and Xu [35] used DNNs for the NER task, the experiments show that it outperforms the CRF's model. Chiu and Nichols [7] showed that CNN is effective in the feature engineering. However, the CNN pays more attention to learn the spatial features, while the text is mainly represented by temporal features. Instead of the CNN, researchers use the RNN [2] to learn the sequential feature in the text. However, Bengio et al. [4] presents that RNN did not consider the long-term dependencies, and Long Short-Term Memory network (LSTM), which is a particular RNN, is adopted to learn the long-term dependencies. Yao and Huang [36] used the LSTM for the word segmentation, and it has a positive impact on the NER task. Lin et al. [23] proposed a Multi-channel BiLSTM-CRF Model for NER in Social Media.

Recently, researchers tend to use the sequence-to-sequence model for the NLP task. Cho et al. 168 [8] proposed a novel neural network model called RNN Encoder-Decoder for statistical machine translation. The model can learn a semantically and syntactically meaningful representation of linguistic phrases. Shao et al. [30] used the sequence-to-sequence model for generating high-quality and informative conversation responses. Moreover, self-attention was added to the decoder to maintain coherence in longer responses. Konstas et al. [19] applied the sequence-to-sequence models for parsing and generation. It showed that sequence-based models are robust against ordering variations of graph-to-sequence conversions.

3   Model

In this section, we present our SC-NER model in detail. Section 3.1 provides an overview, Sect. 3.2 introduces the encoder and classifier, Sect. 3.3 elaborates the decoder. We describe the restricted beam search algorithm in Sect. 3.4.

3.1   Overview

Figure 1 illustrates the overall architecture of our model. The sentence classifier is added to the traditional seq2seq model. The traditional seq2seq model [10, 32] contains the encoder and the decoder. We modify the encoder and use the last hidden state h as the input of sentence classifier. Moreover, the decoder generates the final sequence based on the cell state c and the output of the classifier (i.e., the class label l).

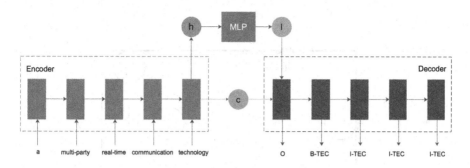

Fig. 1. An overview of our SC-NER model.

3.2   Encoder and Classifier

In the SC-NER model, we use two different LSTMs [24]: one for encoder and another for decoder. It is common to train the LSTMs on the pairs of two sequence in different features at the same time. In the NER task, the input sequence represents a natural sentence. However, the output sequence is just a set of tags without any semantic information. Moreover, attention-based LSTM [6, 33] can significantly outperform standard LSTM, then we use the LSTM with attention for encoder part.

Figure 2 shows the structure of the cell state in LSTM [2]. It uses the input gate(i_t), forget gate(f_t) and output gate(o_t) to protect and control the cell state(c_t).

$$f_t = \sigma(W_f \cdot [c_{t-1}, h_{t-1}, x_t] + b_f), \tag{1}$$
$$i_t = \sigma(W_i \cdot [c_{t-1}, h_{t-1}, x_t] + b_i), \tag{2}$$
$$\widetilde{c_t} = \tanh(W_c \cdot [c_{t-1}, h_{t-1}, x_t] + b_c), \tag{3}$$
$$c_t = f_t \cdot c_{t-1} + i_t \cdot \widetilde{c_t}, \tag{4}$$
$$o_t = \sigma(W_o \cdot [c_{t-1}, h_{t-1}, x_t] + b_o), \tag{5}$$
$$h_t = o_t \cdot \tanh(c_t), \tag{6}$$

where $\sigma(\cdot)$ is the logistic sigmoid function, W and b denote the weight matrix and the bias vector of each gate respectively, x_t presents the current input vector, h_{t-1} means the previous hidden state, and c_{t-1} denotes the previous cell state.

According to statistics, more than half of sentences in corpus have no entity. For example, the sentence "This method performs very well in many applications" contains no entity, and its tag sequence is very simple. Thus, the NER model should filter out sentences that do not contain entities. Therefore, we propose an Multi-Layer Perceptron (MLP) classifier to determine whether sentences contain entities.

Note that, the outputs of encoder are the last hidden state (h) and cell state (c) [30]. We can observe the difference between the hidden state and cell state from the above Eqs. 4 and 6. The LSTM uses the cell state to store the context

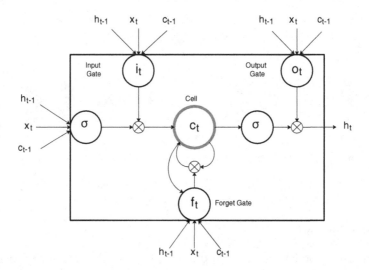

Fig. 2. The structure of the cell state in LSTM is combined with the three gates.

information. Therefore, the cell state changes slowly. The hidden state, however, changes faster than the cell state. The h_t may be very different from the h_{t-1} since Eq. 6 shows that it contains the activation function (i.e., tanh) and the dot product. Generally, the last hidden state pays more attention to the conclusion of the sentence and the cell state carries the information of the whole sentence. Therefore, the last hidden state is used as input of MLP classifier. The classifier is added after the encoder. The output of classifier based on Eq. 7 is class label l. The training objective of the classifier is to minimize the Cross-Entropy of the data.

$$l = \sigma(W_l \cdot h_t + b_l). \tag{7}$$

It is well-known that NER has the data imbalance problem. The seq2seq model with MLP classifier also can solve this problem to a certain extent.

3.3 Decoder

The decoder is also realized by LSTM. The cell state and the classifier's output are taken as the input in our SC-NER model, which is different from the traditional seq2seq model. The inputs of decoder in traditional seq2seq model are the cell state and a starting vector, and the starting vector is always initialized as the all-zero vector or the last hidden state of encoder. In other words, the output of MLP classifier is used as the starting vector in our model, which will inform the decoder whether a sentence contains entity. If there is no entity in this sentence, the decoder can almost ignore the information of the cell state.

The training method for decoder is based on SGD (Stochastic Gradient Descent) to minimize the negative logarithm likelihood.

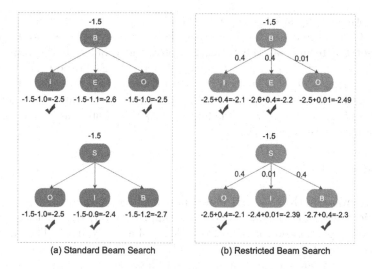

(a) Standard Beam Search   (b) Restricted Beam Search

Fig. 3. The structure of two beam search algorithm.

3.4   Restricted Beam Search

The seq2seq models are trained to generate an output sequence. At the test time, the model usually uses beam search [34] to choose the best sequence given the input. Beam Search is a heuristic graph search algorithm. It is usually used when the search space of the graph is large. As illustrated in Fig. 3(a), at each step of the depth expansion, the score of each hypothesis is calculated by standard beam search. Some low score hypotheses are cut off, and the high score hypotheses are left.

Table 1. Examples of the named entities and its tags.

Named Entity	Tag
MCU	S
Vector coding	B E
Hidden Markov models	B I E

The search space, however, is relatively small in the NER task [23], where the search space (i.e. tag set) includes five tags to represent the entity boundary information i.e. BIESO [1] (e.g., Beginning, Internal, Ending, Single and Other). 'B', 'I', and 'E' denote the beginning, the internal and the ending of an entity. 'S' denotes the entities with a single word. 'O' denotes the other words. Table 1 shows some examples of the named entities and its tags. Therefore, the beam size for NER task is small. Moreover, we add some constraints in the restricted beam search to adjust the search process. As shown in Fig. 3(b), the score of each hypothesis is updated by adding an additional term γ to the original value, where γ denotes the dependency between the current hypothesis and its parent.

We set the term γ as the transition probability between the tags in the corpus. The γ is able to reduce the error of sequence generation. For instance, the tag sequence "O O E" is never generated by the decoder since 'E' should be presented following the 'B' or 'I'. The restricted beam search thus generally favors choosing hypotheses with the higher transition probability, which leads to a more appropriate N-best list.

4   Data

We like to evaluate the SC-NER model on the patent documents corpus in the communications field. The data set contains approximately ten thousand patents and 1000 test patents downloaded from Google Patent Search[1]. The NER task is to recognize all the named entities and determine its type for each patent document. For example, given a sentence "The present invention relates to a multi-party real-time communication technology.", the NER task is to recognize the named entity "multi-party real-time communication" and determine its type. Table 2 shows the description of our corpus.

Table 2. The description of our corpus

Entity type	Total
Method	3041
Material	4827
Product	4205

4.1   Data Preprocessing

The patent documents corpus is provided in Extensible Markup Language (XML). Patent documents contain many fields. Some fields are noise in our task. Examples of such fields are country, bibliographic data, legal-status, or non-English abstracts. Therefore, in order to focus on the NER task, the following fields are chosen: Title, Abstract, Description, and Claims.

4.2   Named Entity Recognition

In the traditional area, NER is to determine whether a sentence contains a named entity and to identify its type. The entity types [27] generally include person's name, place name, and time information. However, in our task, the entity types are predefined as:

---

[1] https://patents.google.com.

Table 3. The results of different methods

Method	Type	Precision	Recall	F1-score
Passos et al. [28]	Method	87.48	81.35	84.30
	Material	88.23	83.33	85.71
	Product	88.02	81.55	84.66
Passos et al. [28] + artificial features	Method	87.36	83.42	85.34
	Material	88.63	85.85	87.22
	Product	88.25	84.87	86.53
Ma and Hovy [25]	Method	89.44	85.16	87.25
	Material	89.87	86.74	88.28
	Product	88.89	85.66	87.24
Lample et al. [20]	Method	90.30	**87.54**	88.90
	Material	91.01	**88.41**	89.69
	Product	91.36	87.33	**89.30**
SC-NER	Method	**92.30**	**87.54**	**89.86**
	Material	**92.71**	87.75	**90.16**
	Product	**91.86**	**86.30**	88.99

– Method Names:
  The meaning of a method is a theoretical solution to a certain problem.
– Material Names:
  Materials are those substances that humans use to make objects, devices, components and machines.
– Product Names:
  Products are tangible items that can be supplied to the market, used and consumed by people, and can meet people's needs, such as mobile phones, routers, processors, etc.

5   Experiments and Analysis

5.1   Experiments

In our experiment, we validate the performance of SC-NER model under different conditions. (1) The effect of the sentence classification module. (2) Whether the restricted beam search can improve the performance of the proposed model. Furthermore, we also compared the proposed model with many existing methods, such as Passos et al. [28], which is based on CRFs, and the deep learning methods based on the CNN [25] and LSTMs [20]. We evaluate all models performance using several criteria (e.g., Precision Rate, Recall Rate, and F1-score).

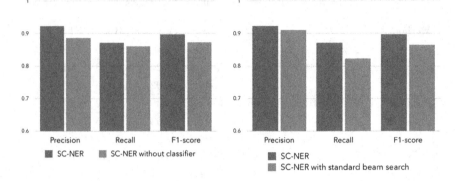

Fig. 4. The experimental results about the sentence classification module and the restricted beam search.

5.2   Results

Figure 4 shows that the performance of the SC-NER model and the SC-NER without classifier. From Fig. 4, we can observe that SC-NER achieves the better precision rate and the similar recall rate to the SC-NER without classifier. Based on the experimental result, we conclude that the sentence classifier is effective in the SC-NER model, because classifier can filter out sentences that do not contain entities.

Figure 4 also shows that the restricted beam search can improve the recall rate, and get the similar precision rate to the standard beam search. There exist two important factors that affect the recall rate. (1) Sentence contains entities, while the model can not recognize them. (2) The model tags the incorrect sequence for the sentence. The restricted beam search can alleviate the second issue above, since it can keep SC-NER model from generating the incorrect tag sequence.

Finally, Table 3 presents the results of all methods for the patent NER under precision, recall, and F1-score. The experimental result shows that the performance of our proposed model outperforms other approaches. The F1-score of the two Standford NER models are lower than SC-NER since the CRFs can not extract the effective features for patents.

As shown in Table 3, the recognition performance on different entity types are different. The performance of recognizing the Material entity is better than identifying the other types of entity, i.e., Method and Product. The reason for this result is that Material entities are usually relatively short. Method entity and Product entity appears to be compound words, which is difficult for the model to extract contextual features.

6   Conclusion

In this paper, we propose the novel seq2seq model SC-NER for the communication patent NER, and the model achieves the good performance in this NER task. We add an MLP classifier between the encoder and the decoder. The classifier can filter out sentences that do not contain entities, which can make the SC-NER model efficient and solve the data imbalance problem. Moreover, the classifier can be trained jointly with the encoder and the decoder. At the test time, we propose the restricted beam search, which is suitable for small search space in the NER task.

Experimental results show that our model makes some improvements in both the precision rate and recall rate compared with other traditional baselines such as the CRFs, CNN, and Bi-LSTMs.

In the future, we plan to enhance the generalization ability of the proposed model. The transfer learning is an option.

Acknowledgment. This work was partially supported by Natural Science Foundation of China (No. 61603197, 61772284, 61876091, 61802205), Jiangsu Provincial Natural Science Foundation of China under Grant BK20171447, Jiangsu Provincial University Natural Science Research of China under Grant 17KJB520024, the Natural Science Research Project of Jiangsu Province under Grant 18KJB520037, and Nanjing University of Posts and Telecommunications under Grant NY215045.

References

1. Deft ERE annotation guidelines: Entities. Linguistics Data Consortium (2014)
2. Graves, A., Mohamed, A., Hinton, G.: Speech recognition with deep recurrent neural networks. In: IEEE International Conference on Acoustics, Speech and Signal Processing, vol. 38, pp. 6645–6649. IEEE (2013)
3. Alotaibi, F., Lee, M.: A hybrid approach to features representation for fine-grained Arabic named entity recognition. In: Proceedings of COLING 2014, the 25th International Conference on Computational Linguistics: Technical Papers, pp. 984–995 (2014)
4. Bengio, Y., Simard, P., Frasconi, P.: Learning long-term dependencies with gradient descent is difficult. IEEE Trans. Neural Netw. **5**, 157–166 (1994)
5. Bowman, S.R., Vilnis, L., Vinyals, O., Dai, A.M., Jozefowicz, R., Bengio, S.: Generating sentences from a continuous space. arXiv preprint arXiv:1511.06349 (2015)
6. Cheng, Y., et al.: Agreement-based joint training for bidirectional attention-based neural machine translation. arXiv preprint arXiv:1512.04650 (2015)
7. Chiu, J., Nichols, E.: Named entity recognition with bidirectional LSTM-CNNs. Comput. Sci. (2015)
8. Cho, K., et al.: Learning phrase representations using RNN encoder-decoder for statistical machine translation. Comput. Sci. (2014)
9. Collobert, R., Weston, J., Karlen, M., Kavukcuoglu, K., Kuksa, P.: Natural language processing (almost) from scratch. J. Mach. Learn. Res. **12**, 2493–2537 (2011)
10. Doersch, C.: Tutorial on variational autoencoders. arXiv preprint arXiv:1606.05908 (2016)
11. Dong, C., Zhang, J., Zong, C., Hattori, M., Di, H.: Character-based LSTM-CRF with radical-level features for Chinese named entity recognition. In: Lin, C.-Y., Xue, N., Zhao, D., Huang, X., Feng, Y. (eds.) ICCPOL/NLPCC -2016. LNCS (LNAI), vol. 10102, pp. 239–250. Springer, Cham (2016). https://doi.org/10.1007/978-3-319-50496-4_20

12. Gehring, J., Auli, M., Grangier, D., Dauphin, Y.N.: A convolutional encoder model for neural machine translation. arXiv preprint arXiv:1611.02344 (2016)
13. Gehring, J., Auli, M., Grangier, D., Yarats, D., Dauphin, Y.N.: Convolutional sequence to sequence learning. arXiv preprint arXiv:1705.03122 (2017)
14. Graves, A., Mohamed, A.R., Hinton, G.: Speech recognition with deep recurrent neural networks. In: 2013 IEEE International Conference on Acoustics, Speech and Signal Processing (ICASSP), pp. 6645–6649. IEEE (2013)
15. Hai, L., Ng, H.: Named entity recognition with a maximum entropy approach. In: Conference on Natural Language Learning at HLT-NAACL, pp. 160–163 (2003)
16. Ji, Y., Tan, C., Martschat, S., Choi, Y., Smith, N.A.: Dynamic entity representations in neural language models. arXiv preprint arXiv:1708.00781 (2017)
17. Keith, K.A., Handler, A., Pinkham, M., Magliozzi, C., McDuffie, J., O'Connor, B.: Identifying civilians killed by police with distantly supervised entity-event extraction. arXiv preprint arXiv:1707.07086 (2017)
18. Kim, Y.: Convolutional neural networks for sentence classification. arXiv preprint arXiv:1408.5882 (2014)
19. Konstas, I., Iyer, S., Yatskar, M., Choi, Y., Zettlemoyer, L.: Neural AMR: Sequence-to-sequence models for parsing and generation (2017)
20. Lample, G., Ballesteros, M., Subramanian, S., Kawakami, K., Dyer, C.: Neural architectures for named entity recognition. arXiv preprint arXiv:1603.01360 (2016)
21. Li, J., Monroe, W., Jurafsky, D.: A simple, fast diverse decoding algorithm for neural generation. arXiv preprint arXiv:1611.08562 (2016)
22. Li, P.H., Dong, R.P., Wang, Y.S., Chou, J.C., Ma, W.Y.: Leveraging linguistic structures for named entity recognition with bidirectional recursive neural networks. In: Proceedings of the 2017 Conference on Empirical Methods in Natural Language Processing, pp. 2664–2669 (2017)
23. Lin, B.Y., Xu, F., Luo, Z., Zhu, K.: Multi-channel BiLSTM-CRF model for emerging named entity recognition in social media. In: Proceedings of the 3rd Workshop on Noisy User-generated Text, pp. 160–165 (2017)
24. Sundermeyer, M., Ney, H., Schluter, R.: From feedforward to recurrent LSTM neural networks for language modeling. IEEE/ACM Trans. Audio Speech Lang. Process. 23(3), 517–529 (2015)
25. Ma, X., Hovy, E.: End-to-end sequence labeling via bi-directional LSTM-CNNs-CRF. arXiv preprint arXiv:1603.01354 (2016)
26. Mccallum, A., Li, W.: Early results for named entity recognition with conditional random fields, feature induction and web-enhanced lexicons. In: Conference on Natural Language Learning at HLT-NAACL, pp. 188–191 (2003)
27. Nadeau, D., Sekine, S.: A survey of named entity recognition and classification. Lingvisticae Investig. 30(1), 3–26 (2007)
28. Passos, A., Kumar, V., McCallum, A.: Lexicon infused phrase embeddings for named entity resolution. arXiv preprint arXiv:1404.5367 (2014)
29. Peng, N., Dredze, M.: Improving named entity recognition for Chinese social media with word segmentation representation learning. In: Meeting of the Association for Computational Linguistics, Berlin, Germany, pp. 149–155, August 2016
30. Shao, L., Gouws, S., Britz, D., Goldie, A., Strope, B.: Generating high-quality and informative conversation responses with sequence-to-sequence models (2017)
31. Su, J., Su, J.: Named entity recognition using an HMM-based chunk tagger. In: Meeting on Association for Computational Linguistics, pp. 473–480 (2002)
32. Sutskever, I., Vinyals, O., Le, Q.V.: Sequence to sequence learning with neural networks. In: Advances in Neural Information Processing Systems, pp. 3104–3112 (2014)

33. Vaswani, A., et al.: Attention is all you need. In: Advances in Neural Information Processing Systems, pp. 6000–6010 (2017)
34. Wiseman, S., Rush, A.M.: Sequence-to-sequence learning as beam-search optimization. arXiv preprint arXiv:1606.02960 (2016)
35. Wu, Y., Jiang, M., Lei, J., Xu, H.: Named entity recognition in Chinese clinical text using deep neural network. Stud. Health Technol. Inform. **216**, 624–628 (2015)
36. Yao, Y., Huang, Z.: Bi-directional LSTM recurrent neural network for Chinese word segmentation. In: Hirose, A., Ozawa, S., Doya, K., Ikeda, K., Lee, M., Liu, D. (eds.) ICONIP 2016. LNCS, vol. 9950, pp. 345–353. Springer, Cham (2016). https://doi.org/10.1007/978-3-319-46681-1_42

BAB-QA: A New Neural Model
for Emotion Detection
in Multi-party Dialogue

Zilong Wang[1,2], Zhaohong Wan[1,2], and Xiaojun Wan[1,2(✉)]

[1] Institute of Computer Science and Technology, Peking University, Beijing, China
{wang_zilong,xmwzh,wanxiaojun}@pku.edu.cn
[2] The MOE Key Laboratory of Computational Linguistics, Peking University,
Beijing, China

Abstract. In this paper, we propose a new neural model BAB-QA to address the task of emotion detection in multi-party dialogues, which aims to detect emotion for each utterance in a dialogue among four label candidates: joy, sadness, anger, and neutral. A variety of models have been proposed to solve this task, but few of them manage to capture contextual information in a dialogue properly. Therefore, we adopt a Bi-directional Long Short-Term Memory network (BiLSTM) and an attention network to obtain representations of sentences and then apply a contextualization network to refine the sentence representations for classification. More importantly, we propose and incorporate a new module called QA network in our model, which is inspired by natural language inference tasks. This QA network enables our model to acquire better sentence encodings by modeling adjacent sentences in a dialogue as question-answer pairs. We evaluate our model in the benchmark EmotionX datasets provided by SocialNLP2018 and our model achieves the state-of-the-art performance.

Keywords: Emotion detection · Multi-party dialogue · Neural model

1 Introduction

With rapid development of social media, people today confront a huge amount of online information. We always express our emotional tendency through utterances in conversations with others. And the emotional tendency is of great help for social media analysis and public opinion investigation. To deal with overwhelming online information, automatic emotion detection by computer program has attracted much more attention.

The goal of this research is to develop a model to recognize emotional tendency in multi-party dialogues, i.e., to detect the emotion for each utterance in dialogues among four label candidates: joy, sadness, anger, and neutral. Unlike ordinary emotion detection task, this task is based on utterances in dialogue, which means the sentences will be very casual and arranged in a sequence. An

© Springer Nature Switzerland AG 2019
Q. Yang et al. (Eds.): PAKDD 2019, LNAI 11439, pp. 210–221, 2019.
https://doi.org/10.1007/978-3-030-16148-4_17

example of the dialogue is shown in Table 1. The last two sentences are exactly the same but, according to the context, they have different emotional tendency. So in this task, we try to take advantage of the sequential relations between sentences and enable the model to make use of more contextual information and thus obtain a better performance.

Table 1. Example of EmotionX dataset

Speaker	Sentence	Emotion
Rachel	Oh okay, I'll fix that to. What's her e-mail address?	Neutral
Ross	Rachel!	Anger
Rachel	All right, I promise. I'll fix this. I swear. I'll-I'llI'll-I'll talk to her	Non-neutral
Ross	Okay!	Anger
Rachel	Okay!	Neutral

Our proposed model BAB-QA can be divided into four parts: a word embedding layer, a sentence encoding layer, a contextualization and classification network, and a QA network. In the word embedding layer, we convert words into word vectors using pre-trained wording embeddings [18]. In the sentence encoding layer, the sequence of word vectors is taken as input of a BiLSTM network and an attention network to obtain the contextual representations of words. We then use max-pooling to get a vector representation for the sentence. For the contextualization and classification network, we use another BiLSTM to refine the sentence representations for emotion classification. The QA network is used to acquire better sentence encodings by modeling adjacent sentences in a dialogue as question-answer pairs. A joint loss considering both classification loss and QA loss is used for the optimizer.

We conduct experiments on the benchmark EmotionX datasets provided by SocialNLP2018 [4, 9][1], and our proposed BAB QA model achieves the state-of-the-art performance. We also validate the usefulness of different components in our model.

The contributions of this work are summarized as follows:

- We propose a neural model BAB-QA to address the challenging task of emotion detection in multi-party dialogues, by making full use of the context information within a sentence and between adjacent sentences.
- We propose an attention network to better capture the contextual information between words in adjacent sentences, and propose a QA network to better capture the contextual information between adjacent sentences.
- Our model achieves the state-of-the-art performance on the benchmark EmotionX datasets.

---

[1] SocialNLP2018 Workshop Challenge: http://doraemon.iis.sinica.edu.tw/emotion lines/challenge.html.

2   Related Work

As sentiment or emotional tendency is of great use for decision-making, numerous efforts have been made to obtain better performance in sentiment or emotion classification tasks. A variety of neural networks have been proposed and designed.

Convolutional neural network (CNN) [13] is used in natural language processing. With several fixed-size windows going through the sentence, the CNN model can achieve attractive results in emotion classification tasks. The Long Short-Term Memory (LSTM) [1,6,8,16] is a kind of recurrent neural network which can capture the temporary features through special gates in cells of the network Some non-recurrent networks are also proposed [10,12,14,24]. Google proposes a sequence-to-sequence model called transformer [22] for machine translation task. Instead of complex recurrent or convolutional neural network, transformer uses a crucial part called Attention to extract temporal features.

Because of the heat of social media such as twitter, emotion detection and sentiment analysis in tweets have attracted a lot of attention these years. Considering tweets usually lack interactions, there is no need to focus on context relations. Some works apply Emoji embeddings to extract information to deal with the widely-used Emoji in tweets [5]. The Transfer Learning is also adopted in this task [20]. And there are works using complex combination of RNN and Attention [2]. Early machine learning methods are adopted to solve this problem as well [7,15].

While all the models above focus on ordinary emotion detection, they do not perform well on the task of emotion detection in dialogues, because sentences in dialogues are more casual, and depend on context to express their emotional tendency. To achieve good performance in this task, SA-BiLSTM [17] and CNN-DCNN [11] from the contest have done a lot of work. The SA-BiLSTM system uses BiLSTM to deal with words relation and attention mechanism to deal with sentence relation. The CNN-DCNN system uses a CNN encoder and CNN decoder on fixed number of sentences to solve the task.

However, both models above do not pay enough attention to the relations between the sentences in the same dialogue. And the relations between sentences are of great importance to this task. Utterances in our daily conversation are dependent on their context. That's why we do a lot work on this part and our model achieves better performance than the challengers in the contest.

3   Model

The task is to recognize the emotional tendency of each utterance (typically sentences) in multi-party dialogues. The sentences are arranged in a dialogue, so the input to our model is a sequence of sentences in the same dialogue: $sent_{i1}, sent_{i2}, ..., sent_{im} \in dialogue_i$. For each sentence in the dialogue, our model will predict its emotional tendency through the words in the sentence and the contextual sentences.

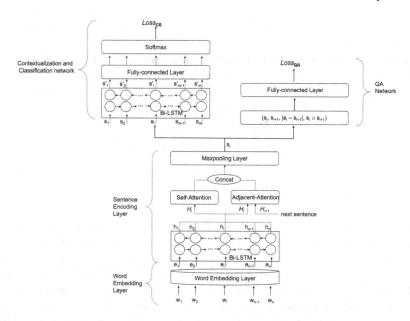

Fig. 1. The architecture of our model. w_i: word in a sentence; e_i: word embedding vector; h_i: hidden states of BiLSTM in Sentence Encoding Layer; H_i and H_{i+1} are the hidden states of sentences i and $i+1$; s_i: sentence vector; h_i': hidden states of BiLSTM in the contextualization and classification network.

Our BAB-QA model is composed of four main components: a word embedding layer, a sentence encoding layer, a contextualization and classification network, and a QA network. In the following sections, they will be discussed in detail. And the architecture of the model is depicted in Fig. 1.

3.1   Word Embedding Layer

Each sentence in a dialogue is represented in the form of a sequence of words. This layer maps the words to real-valued vectors, which captures syntactic and semantic word relations. The pre-trained word embedding vectors [18] are adopted in this study to convert the words into vectors.

For sentence $sent_i = [w_1, w_2, ..., w_n]$, this layer replaces each word with the pre-trained vector and makes a matrix out of it: $Emb_i = [e_1, e_2, ..., e_n] \in \mathbb{R}^{n \times d^w}$, where n is the length of the sentence and d^w is the dimension of the pre-trained word embedding vectors. Each row vector of the matrix is the lexical-semantic representation of the word in the sentence respectively.

3.2   Sentence Encoding Layer

This layer takes each sentence matrix Emb_i as input and produces a vector s_i to represent the sentence $sent_i$. The layer consists of two major parts: a

Bi-directional Long Short-Term Memory network (BiLSTM) and an attention network (Attention).

In the first part, the row vectors in each sentence matrix are processed by the BiLSTM network sequentially. As the BiLSTM is a kind of recurrent neural network dealing with sequential information, it outputs a sequence of hidden states $H = [h_1, h_2, ..., h_n](h \in \mathbb{R}^{d^{hs}})$ for a sentence of n words, where the d^{hs} is the dimension of the hidden states.

$$\overrightarrow{h_i} = \overrightarrow{LSTM}(\overrightarrow{h_{i-1}}, e_i) \tag{1}$$

$$\overleftarrow{h_i} = \overleftarrow{LSTM}(\overleftarrow{h_{i+1}}, e_i) \tag{2}$$

$$h_i = [\overrightarrow{h_i}; \overleftarrow{h_i}] \tag{3}$$

In the second part, we adopt attention network to capture the relation between words more precisely. Attention network used here is first introduced by Google researchers [22]. Because sentences in the same dialogue have different lengths, we pad them to the max length of sentences with zeros. The output is weighted by summing over all the hidden states. For sentence i in a dialogue, we let H_i denote the corresponding hidden states obtained in the above way, and $A_{self}(i)$ as the output of the self-attention network. We use a subscript with H and a superscript with h_i to differentiate different sentences.

$$H_i = [h_1^i, h_2^i, ..., h_n^i] \tag{4}$$

$$A_{self}(i) = softmax(\frac{H_i H_i^T}{\sqrt{d^{hs}}}) H_i \tag{5}$$

where H_i is padded to length n with zeros, and n is the max length of sentences in the dialogue.

We notice that words in each sentence and its next sentence are related as well. Another attention network is adopted here. The input to this attention network is also the hidden states vectors from BiLSTM, but the weight matrix is calculated by each sentence and its next sentence.

$$H_{i+1} = [h_1^{i+1}, h_2^{i+1}, ..., h_n^{i+1}] \tag{6}$$

$$A_{adjacent}(i) = softmax(\frac{H_{i+1} H_i^T}{\sqrt{d^{hs}}}) H_i \tag{7}$$

where H_i and H_{i+1} are the hidden states of sentences i and $i + 1$, respectively; $A_{adjacent}(i)$ is the output of the attention network with H_i, H_{i+1} as input.

We suppose that there is strong connection between each sentence and its next sentence. From the words in the next sentence, the model can attain better sentence encoding.

Because the last sentence in the dialogue does not have the next sentence, we use zero vectors to pad the space.

Then we concatenate the outputs of the two attention networks, and get the result A.

$$A(i) = [A_{self}(i); A_{adjacent}(i)] \quad \text{for } i \in [1, m - 1] \tag{8}$$

$$A(m) = [A_{self}(m); \mathbf{0}]   \tag{9}$$

where m is the number of sentences in the dialogue.

Then we conduct max-pooling to the output of the whole attention network to convert a matrix into a vector.

The final output of this layer is a vector $s_i \in \mathbb{R}^{2d^{hs}}$ encoding the word relations within a sentence and also the word relations between a sentence and its next sentence. This part is shown in Fig. 2.

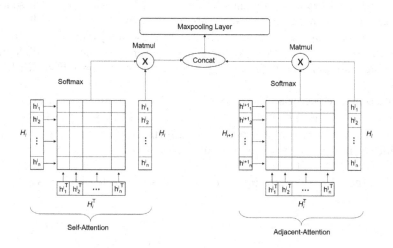

Fig. 2. The overview of our attention network

3.3   Contextualization and Classification Network

Since the sentences in this task are utterances in people's daily conversation, the relation between sentences is important for emotion detection. A same sentence may express different emotions in different dialogue contexts, so we should make full use of the sequential relations between sentences in a dialogue.

We adopt a BiLSTM network to refine the sentence representations obtained in the sentence encoding layer. The BiLSTM takes the sequence of sentence vectors s_i obtained in the previous layer as input, and acquires new sentence vectors s_i' by considering contextual information between sentences. The dimension of the vectors is decided by the parameter of new hidden size ($d^{hs'} = 2d^{hs}$), so the new vectors $s_i' \in \mathbb{R}^{d^{hs'}}$.

The new sentence vector is used to predict the emotion label of each sentence. To cater for the 4-classification task, we input the vector into a fully-connected layer and get probability scores over the four emotion labels.

$$P_{emotion} = softmax(relu(Ws' + b))   \tag{10}$$

where $P_{emotion} \in \mathbb{R}^4$, and W is weight matrix, b is bias.

The emotion label of the sentence is that corresponding to the highest probability score in $P_{emotion}$. We calculate the cross entropy loss of the prediction. We use $Loss_{CE}$ to denote it.

3.4 QA Network

As discussed before, the relation between sentences in the same dialogue is very important, and the BiLSTM is not enough for the task. Enlightened by the related task about natural language inference [3,23], we suppose each sentence in a dialogue is related to its next sentence in a similar way, just like a pair of question and answer. In this study, we propose a QA network to acquire better sentence representations by modeling adjacent sentences in a dialogue as question-answer pairs and forcing adjacent sentences to be related with each other.

For a sentence vector s_i and the vector of its next sentence s_{i+1}, we calculate $|s_i - s_{i+1}|$, $s_i \odot s_{i+1}$ and concatenate s_i, s_{i+1}, $|s_i - s_{i+1}|$, $s_i \odot s_{i+1}$ together into the concatenated vector $Cat \in \mathbb{R}^{4d^{hs}}$. We use the concatenated vector as input to a fully-connected network and obtain a relatedness score $score(i, i+1) \in \mathbb{R}$ between these two sentences.

For sentence i in dialogue $dialogue_j$, a new loss for each sentence is calculated by:

$$Loss_{QA}(s_i) = \sum_{sent_k \in dialogue_j, k \neq i, i+1} \frac{1 - score(i, i+1) + score(i, k)}{|dialogue_j| - 2} \tag{11}$$

where $|dialogue_j|$ is the number of sentences in dialogue j.

$$Loss_{QA} = \sum_{dialogue_j \in TrainingSet} \frac{\sum_{sent_i \in dialogue_j} \frac{Loss_{QA}(s_i)}{|dialogue_j|}}{|TrainingSet|} \tag{12}$$

where $|TrainingSet|$ is the number of dialogues in the training dataset.

The final loss of our model is the sum of $Loss_{CE}$ and $Loss_{QA}$.

$$Loss = Loss_{CE} + Loss_{QA} \tag{13}$$

4   Experiments and Results

4.1 Dataset

We evaluate our model on benchmark EmotionX datasets provided by SocialNLP-2018. There are two evaluation datasets, Friends and EmotionPush, where each sentence in the conversation is annotated with one label showing its emotional tendency. The Friends dataset consists of actor's lines in Friends, a famous TV series, and the EmotionPush dataset consists of chat logs on Facebook Messenger.

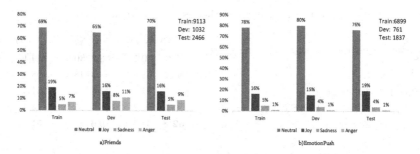

Fig. 3. The EmotionX dataset: Friends and EmotionPush

The original datasets have eight categories of emotion, but only four of them (Neutral, Joy, Sadness, Anger) are considered according to the requirement of EmotionX. Figure 3 summarizes the statistics of the two datasets. Note that each dataset has already been split into training, development and test sets.

Because the EmotionPush dataset is from the online chat logs, many words and phrases in it need to be processed beforehand. More specifically, we translate the emoji symbols to its corresponding meanings and truncate duplicated symbols. We also replace locations and names to special tokens.

4.2 Training Settings

For both datasets, test performance is assessed on the training epoch with best validation performance on the development set and the unweighted average accuracy (UWA) is reported for comparison, which is the official evaluation metric adopted by EmotionX.

$$UWA = \frac{1}{|C|} \sum_{l \in C} acc_l \qquad (14)$$

where acc_l is the accuracy of emotion category l, and C is the set of emotion categories.

The word embeddings are 300-dimensional GloVe embeddings (Global Vectors for Word Representation) [19]. They are trained on large amount of Google News and not fixed during the training in order to enhance better performance. Pack padded sequence and pad packed sequence are used to deal with the different lengths of sentences.

BiLSTMs in different layers have 300 hidden units. And the fully-connected layer in the contextualization and classification network has two hidden layers with the same size of 128. The fully-connected layer in QA network has a hidden layer with size of 100.

The model is trained using the Adam optimization method. The learning rate is initially set as 0.001 and the decay factor is set as 0.99 for every epoch. To avoid overfitting, we adopt dropout in each layer [21]. The dropout rate is set as 0.5.

Table 2. Results on the test set of Friends

Friends	UWA	Neutral	Joy	Sadness	Anger
SA-BiLSTM [17]	59.6	90.1	68.8	30.6	49.1
CNN-DCNN [11]	60.5	68.4	68.3	48.3	58.8
BiLSTM+BiLSTM (BB)	59.2	73.6	77.6	34.1	51.6
BiLSTM+Attention+BiLSTM (BAB)	60.7	66.7	66.5	35.3	74.5
BiLSTM+Attention+BiLSTM+QA (BAB-QA)	**62.8**	63.4	79.6	56.5	51.6

Table 3. Results on the test set of EmotionPush

EmotionPush	UWA	Neutral	Joy	Sadness	Anger
SA-BiLSTM	55.0	94.2	70.5	31.0	24.3
CNN-DCNN	64.9	65.2	75.6	63.2	55.6
BiLSTM+BiLSTM (BB)	62.9	79.9	79.5	43.7	48.7
BiLSTM+Attention+BiLSTM (BAB)	67.9	73.8	74.7	60.9	62.2
BiLSTM+Attention+BiLSTM+QA (BAB-QA)	**70.6**	78.1	78.2	66.7	59.5

4.3 Results and Discussion

Tables 2, 3 show the results (UWA and the accuracy for each emotion label) of our model and the published results of top models (SA-BiLSTM and CNN-DCNN) in the EmotionX Challenge.

There are three versions of our model:

- **BiLSTM+BiLSTM(BB):** The sentence encoding layer only consists of a BiLSTM layer and a max-pooling layer. Another BiLSTM is used to capture contextual information between sentences.
- **BiLSTM+Attention+BiLSTM(BAB):** The sentence encoding layer consists of a BiLSTM, the Attention network described before, and a max-pooling layer. Another BiLSTM is used to capture contextual information between sentences.
- **BiLSTM+Attention+BiLSTM+QA(BAB-QA):** The sentence encoding layer and contextualization and classification network are the same as BAB, but the QA network is added.

From Tables 2, 3, we can clearly find out that BAB has already achieved better performance than the top players attending the EmotionX contest and our overall BAB-QA model can further improve the UWA score. In summary, BAB-QA can improve the state-of-the-art performance of CNN-DCNN by 2.3 and 5.7 points on the two datasets, respectively.

Compared with the top two models in the contest, our model can get better sentence representations for emotion classification, by fully considering the relations between sentences in a dialogue. The techniques introduced in this paper enable our model to enhance better performance as well.

More specifically, with the help of pre-trained word embedding, the lexical semantic information of words is properly extracted and the BiLSTM and attention network in the sentence encoding layer can capture the subtle relation between the words. The word-wise information is thoroughly obtained.

In the part of Contextualization and Classification Network and QA network, relations between sentences is well modeled and leveraged by the BiLSTM and QA network.

From the comparison between BB and BAB, we can see the attention network is useful and it can improve UWA by 1.5 and 5 points on the two datasets, respectively. It indicates that attention network can better obtain information about strong emotions like anger. After we examine the dataset, we find out most sentences tagged as anger are highly dependent to its context. It can be explained by the Adjacent_attention's ability to catch relations between adjacent sentences.

Table 4. Sentence example with anger emotion

Speaker	Emotion	Sentence
	...	
Monica:	(neutral)	What-what was it you were gonna tell us?
Rachel:	(neutral)	Yeah. Oh! Was how you invented the cotton gin?!
Ross:	(anger)	Okay, good bye!

In Table 4, the character Ross expresses his anger emotion in the last sentence. But BB tags it as sadness, because only the BiLSTM network in BB cannot catch enough contextual information. With the help of attention network, BAB gives the right prediction.

The comparison between BAB and BAB-QA shows the improvement more clearly. The QA network can abstract relations between a sentence and its next sentence. It predicts the most possible "answer" to each sentence in the dialogue. In this way, the QA network can figure out the difference of each emotional tendency and get rid of the interference of unbalanced dataset, so BAB-QA achieves high results for most categories.

Take the sadness emotion as an example (Table 5).

Table 5. Sentence example with sadness emotion

Speaker	Emotion	Sentence
	...	
Joey:	(non-neutral)	Oh, how bad is it?
Phoebe:	(sadness)	Oh, it's bad. It's really bad. The only thing ...
Chandler:	(neutral)	How's your room Rach?

In the Phoebe's utterance, she uses word "bad" several times. The BAB just detects this feature, and predicts it as anger. However, BAB-QA considers sentences in the dialogue more thoroughly, and gives the right prediction. BAB-QA finds speakers' emotion is not that strong. Even if "bad" appears several times, the Phoebe's utterance should be tagged as sadness.

Limited by the scale of the datasets, the results of our model are not perfect. Some sentences in the dialogue are very informal and require other source of information to make the right judgment. We expect visual and audio information will be available in the future to enhance better performance on this task.

5   Conclusion

In this work, we propose a neural network model BAB-QA to detect emotional tendency in multi-party dialogues. We take advantage of unique characteristics of this task, and try to extract information both word-wise and sentence-wise. Several kinds of networks, such as BiLSTM and Attention, are combined in this model to enhance better performance. We also introduce two new attempts in the emotion detection task: Concatenated Attention and QA Network. Both attempts turn out to be useful to this task. Our model achieved the state-of-the-art UWA scores on the EmotionX datasets.

In future work, we will apply our model to other problems related to sequential emotion recognition. We will also consider visual and audio information in our model to achieve better performance on this task.

Acknowledgments. This work was supported by National Natural Science Foundation of China (61772036, 61331011) and Key Laboratory of Science, Technology and Standard in Press Industry (Key Laboratory of Intelligent Press Media Technology). We thank the anonymous reviewers for their helpful comments.

References

1. Abdul-Mageed, M., Ungar, L.: EmoNet: fine-grained emotion detection with gated recurrent neural networks. In: Proceedings of the 55th Annual Meeting of the Association for Computational Linguistics (Volume 1: Long Papers), vol. 1, pp. 718–728 (2017)
2. Baziotis, C., et al.: NTUA-SLP at SemEval-2018 task 1: predicting affective content in tweets with deep attentive RNNs and transfer learning. In: Proceedings of the 12th International Workshop on Semantic Evaluation, pp. 245–255 (2018)
3. Bowman, S.R., Angeli, G., Potts, C., Manning, C.D.: A large annotated corpus for learning natural language inference. arXiv preprint arXiv:1508.05326 (2015)
4. Chen, S.Y., Hsu, C.C., Kuo, C.C., Ku, L.W., et al.: Emotionlines: an emotion corpus of multi-party conversations. arXiv preprint arXiv:1802.08379 (2018)
5. Chen, Y., Yuan, J., You, Q., Luo, J.: Twitter sentiment analysis via bi-sense emoji embedding and attention-based LSTM. arXiv preprint arXiv:1807.07961 (2018)
6. Chung, J., Gulcehre, C., Cho, K., Bengio, Y.: Empirical evaluation of gated recurrent neural networks on sequence modeling. arXiv preprint arXiv:1412.3555 (2014)
7. Hasan, M., Rundensteiner, E., Agu, E.: EMOTEX: detecting emotions in Twitter messages (2014)

8. Hochreiter, S., Schmidhuber, J.: Long short-term memory. Neural Comput. **9**(8), 1735–1780 (1997)

9. Hsu, C.C., Ku, L.W.: SocialNLP 2018 emotionX challenge overview: recognizing emotions in dialogues. In: Proceedings of the Sixth International Workshop on Natural Language Processing for Social Media, pp. 27–31 (2018)

10. Jirayucharoensak, S., Pan-Ngum, S., Israsena, P.: EEG-based emotion recognition using deep learning network with principal component based covariate shift adaptation. Sci. World J. **2014**, Article ID 627892, 10 p. (2014). https://doi.org/10.1155/2014/627892

11. Khosla, S.: EmotionX-AR: CNN-DCNN autoencoder based emotion classifier. In: Proceedings of the Sixth International Workshop on Natural Language Processing for Social Media, pp. 37–44 (2018)

12. Kim, Y., Lee, H., Jung, K.: AttnConvnet at SemEval-2018 task 1: attention-based convolutional neural networks for multi-label emotion classification. arXiv preprint arXiv:1804.00831 (2018)

13. Kim, Y.: Convolutional neural networks for sentence classification. arXiv preprint arXiv:1408.5882 (2014)

14. Lei, Z., Yang, Y., Yang, M., Liu, Y.: A multi-sentiment-resource enhanced attention network for sentiment classification. arXiv preprint arXiv:1807.04990 (2018)

15. Liew, J.S.Y., Turtle, H.R.: Exploring fine-grained emotion detection in tweets. In: Proceedings of the NAACL Student Research Workshop, pp. 73–80 (2016)

16. Liu, P., Qiu, X., Huang, X.: Recurrent neural network for text classification with multi-task learning. arXiv preprint arXiv:1605.05101 (2016)

17. Luo, L., Yang, H., Chin, F.Y.: EmotionX-DLC: self-attentive BiLSTM for detecting sequential emotions in dialogue. arXiv preprint arXiv:1806.07039 (2018)

18. Mikolov, T., Sutskever, I., Chen, K., Corrado, G.S., Dean, J.: Distributed representations of words and phrases and their compositionality. In: Advances in neural information processing systems, pp. 3111–3119 (2013)

19. Pennington, J., Socher, R., Manning, C.: GloVe: global vectors for word representation. In: Proceedings of the 2014 Conference on Empirical Methods in Natural Language Processing (EMNLP), pp. 1532–1543 (2014)

20. Rozental, A., Fleischer, D., Kelrich, Z.: Amobee at IEST 2018: transfer learning from language models. arXiv:1808.08782

21. Srivastava, N., Hinton, G., Krizhevsky, A., Sutskever, I., Salakhutdinov, R.: Dropout: a simple way to prevent neural networks from overfitting. J. Mach. Learn. Res. **15**(1), 1929–1958 (2014)

22. Vaswani, A., et al.: Attention is all you need. In: Advances in Neural Information Processing Systems, pp. 5998–6008 (2017)

23. Yang, Y., et al.: Learning semantic textual similarity from conversations. arXiv preprint arXiv:1804.07754 (2018)

24. Zhang, Q., Chen, X., Zhan, Q., Yang, T., Xia, S.: Respiration-based emotion recognition with deep learning. Comput. Ind. **92**, 84–90 (2017)

Unsupervised User Behavior Representation for Fraud Review Detection with Cold-Start Problem

Qian Li[1(✉)], Qiang Wu[1], Chengzhang Zhu[2,3], Jian Zhang[1], and Wentao Zhao[3]

[1] Global Big Data Technologies Centre, University of Technology Sydney,
Ultimo, Australia
Qian.Li-7@student.uts.edu.au, {Qiang.Wu,Jian.Zhang}@uts.edu.au
[2] Advanced Analytics Institute, University of Technology Sydney, Ultimo, Australia
kevin.zhu.china@gmail.com
[3] College of Computer, National University of Defense Technology, Changsha, China
wtzhao@nudt.edu.cn

Abstract. Detecting fraud review is becoming extremely important in order to provide reliable information in cyberspace, in which, however, handling cold-start problem is a critical and urgent challenge since the case of cold-start fraud review rarely provides sufficient information for further assessing its authenticity. Existing work on detecting cold-start cases relies on the limited contents of the review posted by the user and a traditional classifier to make the decision. However, simply modeling review is not reliable since reviews can be easily manipulated. Also, it is hard to obtain high-quality labeled data for training the classifier. In this paper, we tackle cold-start problems by (1) using a user's behavior representation rather than review contents to measure authenticity, which further (2) consider user social relations with other existing users when posting reviews. The method is completely (3) unsupervised. Comprehensive experiments on Yelp data sets demonstrate our method significantly outperforms the state-of-the-art methods.

Keywords: Fraud review detection · Cold-start ·
Behavior representation · Unsupervised learning

1 Introduction

With the increasing popularity of E-commerce, a large number of online reviews are manipulated by fraudsters, who intend to write fraud reviews driven by strong incentives of profit and reputation. Early in 2013, it has been found that around 25% of Yelp reviews could be fake[1]. This situation becomes worse than ever recently. As reported by Forbes news[2] in 2017, Amazon is seeing a lot more

---

[1] https://www.bbc.com/news/technology-24299742.

[2] https://www.forbes.com/sites/emmawoollacott/2017/09/09/exclusive-amazons-fake-review-problem-is-now-worse-than-ever/#501eccb87c0f.

© Springer Nature Switzerland AG 2019
Q. Yang et al. (Eds.): PAKDD 2019, LNAI 11439, pp. 222–236, 2019.
https://doi.org/10.1007/978-3-030-16148-4_18

suspicious reviews than before. As a result, it has become a critical and urgent task to effectively detecting such fraudsters and fraud reviews.

Recent years have seen significant progress made in fraud detection. Current efforts mainly focused on extracting linguistic features (n-grams, POS, etc) and behavioral features [5,27]. However, linguistic features are ineffective when dealing with real-life fraud reviews [19], especially when linguistic features are easy to be imitated, a.k.a. camouflage [6]. Also, extracting behavior features require a large number of samples and usually takes months to make observations. When facing the *cold-start* problem, i.e. *a new user just posted a new review*, extracting behavior features becomes even harder because none historical information is available for a new user [28].

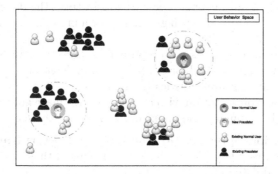

Fig. 1. Example of user behavior space. In this space, similar users will close to each other, i.e. a normal user will be majorly surrounded by normal users, and vice versa.

Recently, the cold-start problem in fraud review detection has been first studied by [25]. This method handles the cold-start problem by considering correlations between users, items, and reviews. Later, [28] makes one step further by incorporating the relations between entities (users, items, and reviews) with their attributes from different domains. Both of the above methods feed the embedded **review** representation into a classifier for cold-start fraud detection. However, two problems may arise when adopting these methods. (1) Only using review itself is ineffective as discussed in [19], and is easy to be affected by camouflage [6]. (2) Also, high-quality labeled data are required in both methods, which is really hard to obtain in real life.

We address the above problems in current cold-start fraud detection methods by focusing on **user behavior**. The rationale is similar users may result in similar behaviors when posting a review. Specifically, in a behavior representation space (Fig. 1), if a new user is closer to a group of existing fraudsters, those fraudsters will be identified as his/her similar users. Then, the new user is likely to be detected as a fraudster. Thus, the cold-start problem can be transferred to identifying the existing users who have similar behavior with a new user. Although limited information is available for a new user, existing fraudsters can be effectively detected by many methods [6,13,23].

Motivated by [25,28], we first represent user according to the relations among entities. Further, we integrate the entities relations with user social relations. The intuition is that fraud reviews are always manipulated by a group of fraudsters with close social relations [6,13]. For example, a group of users usually work together to effectively promote or demote target products. They may even know one another and copy reviews among themselves [17].

To further strength the discriminate ability of the represented behavior for fraud detection, we apply the dense subgraph mining [1,6] to generate pseudo-fraud labels to tune the representation in an iterative way. The foundation is that an end-to-end supervised training will enable the strong task-specific discriminate ability of the generated representation, as demonstrated in most of representation learning tasks. In this process, dense subgraph mining generates high-quality labels in an unsupervised way [6]. In turn, the discriminate representation adjusts the weight of each graph link for a more precise dense subgraph mining.

Based on the above analysis, we propose a socially-aware unsupervised user behavior representation method for **cold**-start fraud detection (SUPER-COLD). Our method jointly captures entities interactions and user social relations to generate behavior representation with a strong discriminate ability for cold-start fraud detection. In summary, the main contributions of this work are as follows.

- **A user behavior representation model for cold-start fraud detection**: the represented user behavior avoids camouflage and thus is more reliable for cold-start fraud detection.
- **A socially-aware user behavior representation**: the reviewing habits and social relations of a user are jointly embedded in its behavior representation to provide comprehensive evidence for fraud detection.
- **A discriminative unsupervised representation approach for cold-start user behavior**: a dense subgraph-based approach for fraudsters detection has been involved into the unsupervised representation approach, which strengths the discriminant of the representation and tackles the problem of lacking high-quality fraud labels in real life.

Comprehensive experiments on two large real-world data sets show that: (1) SUPER-COLD effectively detects cold-start fraud reviews without manual labels (improved up to 150% in terms of F-score compared with the state-of-the-art supervised detection method); (2) SUPER-COLD enjoys a significant recall gain (up to 9.23% in terms of F-score) in general review detection tasks from incorporating entities and social relations; (3) SUPER-COLD generates a user behavior representation with a strong discriminate ability.

2   Related Work

2.1   Fraud Review Detection

Fraud review detection was initially studied in [8], and has long been an attractive research topic since then. Later, more efforts were made on exploits linguistics features [7,11,20], analyzing the effectiveness of n-grams, POS, etc. However,

[19] found that linguistic features are insufficient when dealing with real-life fraud reviews. Therefore, researchers put more efforts in employing users' behavior features [3,4,10–12]. Also, [18] proved that user's behavioral features are more effective than linguistic features for fraud detection. Behavioral features were then intensively studied by introducing a set of graph-based methods. The intuition is reviews posted with similar fraud-behavior would be fraud. Wang et al. [24] first introduce review graph to capture the relationships between entities. Spotting fraudster groups were then explored by network footprints [27], community discovery with sentiment analysis [2], social interactions for sparse group [26]. In-depth, Hooi et al. [6] proposes an advanced dense subgraph mining for group fraudsters detection, targeting on detecting camouflage or hijacked accounts who manipulate their writing to look just like normal users.

2.2   Cold-Start Problem

The cold-start problem in fraud review detection was first studied in [25]. By considering the correlations between users, items, and reviews (entities relation), the review posted by a new user can be represented. Motivated by [25], the method proposed in [28] further leverages both attribute and domain knowledge for a better review representation. At last, the review representation is fed into a traditional classification model like SVM to form the fraud review classifier.

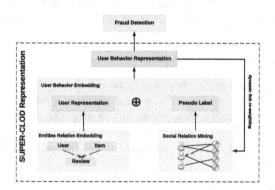

Fig. 2. The proposed SUPER-COLD Model

While both the above cold-start fraud review detection methods focus on user's review representation, we believe fraud reviews are easy to be manipulated to look like normal reviews [6] and thus may confuse existing methods. In this paper, we propose a novel user behavior representation model for fraud detection, where entities relations and user social relations are jointly embedded. In addition, we apply the dense subgraph mining to obtain pseudo-labels for existing users in an unsupervised way, which also avoids the difficulty to obtain high-quality labeled data.

3   Proposed Method

3.1   Behavior Representation Architecture

The behavior representation architecture of SUPER-COLD is shown in Fig. 2. It consists: *entities relation embedding, social relation mining* and *user behavior embedding*. It also involves a dynamic link re-weighting strategy to enable a discriminative representation for fraud detection.

SUPER-COLD first embeds the relations among users, items, and reviews in their representations (entities relation embedding), and leverages the user social relation by the dense subgraph mining based on a user-item bipartite graph (social relation mining). Then, it further learns a user behavior representation by integrating the learned entities relations, which are embedded in the user representation, and the social relations, which are involved in the pseudo fraud labels generated by the dense subgraph mining, through a neural network (user behavior embedding). A dynamic link re-weighting strategy is adopted to enhance the discriminative ability of the generated behavior representation. Specifically, after user behavior representation is learned, SUPER-COLD discovers suspicious fraudsters according to the user distribution in the user behavior representation space, and assigns a higher weight to the links corresponding to these suspicious fraudsters in the user-item bipartite graph. After re-weighting, it executes the dense subgraph mining to reveal a more accurate social relation, which is further integrated with the user representation to generate a new behavior representation. SUPER-COLD repeats this process until convergence.

3.2   Entities Relation Embedding

SUPER-COLD embeds entities relation following the method in [25]. Let's note $\mathbf{v}_u$, $\mathbf{v}_o$, $\mathbf{v}_r$ as the representations of u, o, and r, where u refers to a user, o refers to an item, and r refers to a review wrote by the user u to the item o. We further denote a tuple of $< u, o, r >$ as $\nu \in S$, where S refers to an online review data set. For a given S, SUPER-COLD embeds the entities relation by the following objective function:

$$\min_{\mathbf{V}_u, \mathbf{V}_o, \omega} \sum_{\nu \in S} \sum_{\nu' \in S} \gamma \max\{0, 1 + \|\mathbf{v}_u + \mathbf{v}_o - \mathbf{v}_r\|^2 - \|\mathbf{v}_{u'} + \mathbf{v}_{o'} - \mathbf{v}_{r'}\|^2\},$$

$$s.t. \quad \gamma = \begin{cases} 1 & u = u' \\ 0 & u \neq u', \end{cases} \tag{1}$$

$$\mathbf{v}_r = t_\omega(r),$$

where $\mathbf{V}_u$ and $\mathbf{V}_o$ is a set of the user and item representations, $t_\omega(\cdot)$ refers to a text embedding neural network with parameters ω, and $\max\{\cdots\}$ returns the maximum in a set. SUPER-COLD implements the text embedding neural network as a hierarchical bi-directional recurrent neural network.

3.3 Social Relation Mining

SUPER-COLD models users relations by a user-item bipartite graph. Motivated by [6,13], it adopts the dense subgraph mining to reveal the fraud behavior based on user social relation. The basic idea is greedily removing nodes in the bipartite graph which can maximize the subgraph density per a given density evaluation (Algorithm 1 line 1–5). The final remained subgraph (Algorithm 1 line 6) will have the largest density and thus can discover the users who may work together to manipulate reviews.

Given an online review data set S, SUPER-COLD constructs the user-item bipartite graph as $\mathcal{G} = (\mathcal{U} \cup \mathcal{I}, \mathcal{E})$, where $\mathcal{U}$ is a set of users, $\mathcal{I}$ is a set of items, and $\mathcal{E} = \{< u, o, r > \mid < u, o, r > \in S\}$ is a set of edges, i.e. links from users to items. Algorithm 1 shows the process of the dense subgraph mining for SUPER-COLD. Motivated by [6], SUPER-COLD defines the density metric $g(\cdot)$ as follows,

$$g(S) = \frac{f(S)}{|S|}, \tag{2}$$

where

$$f(S) = \sum_{<u,o,r> \in \mathcal{E}} w_{u,o}, \tag{3}$$

for link weight $w_{u,o} > 0$ between user u to item o. Initially, SUPER-COLD assigns all link weights as 1. It will further adopt a dynamic re-weighting strategy to update the link weights in an iterative process.

Algorithm 1. Dense Subgraph Mining for SUPER-COLD.

Input: Bipartite graph $\mathcal{G} = (\mathcal{U} \cup \mathcal{I}, \mathcal{E})$;
Output: The pseudo-labels set Y;
1: $\mathcal{X}_0 \leftarrow \mathcal{U} \cup \mathcal{I}$
2: **for** $t = 1, \cdots, (n_u + n_o)$ **do**
3:     $i^* \leftarrow \arg\max_{i \in \mathcal{X}_{t-1}} g(\mathcal{X}_{t-1} \backslash \{i\})$
4:     $\mathcal{X}_t \leftarrow \mathcal{X}_{t-1} \backslash \{i^*\}$
5: **end for**
6: $\mathcal{X}^* \leftarrow \arg\max_{\mathcal{X}_i \in \{\mathcal{X}_0, \cdots, \mathcal{X}_{n_u+n_o}\}} g(\mathcal{X}_i)$;
7: **for** $u = u_1, \cdots, u_{n_u}$ **do**
8:     **if** $u \in \mathcal{X}^*$ **then**
9:         $y_i = c_f$
10:     **else**
11:         $y_i = c_n$
12:     **end if**
13: **end for**
14: **return** $Y = \{y_1, \cdots, y_{n_u}\}$

SUPER-COLD assigns pseudo-labels to existing users according to the dense subgraph mining results. Specifically, it gives the candidate fraudster label (c_f) to the users in the detected dense subgraph, and sets candidate normal user label (c_n) to other users (Algorithm 1 line 7–14). These pseudo-labels inherit the social relations and will be used in the following behavior representation and cold-start fraud detection.

3.4   Integrating Entities and Social Relation for Behavior Representation

SUPER-COLD further integrates the entities and social relation for behavior representation. Specifically, it adopts a fully connected neural network, $Dense_{\mathbf{p}}(\cdot)$, to transform a user representation $\mathbf{v}_u$ to a user behavior representation $\mathbf{v}_u^*$, and minimizes the pseudo-labels prediction loss (defined as the cross-entropy between the predicted labels and pseudo-labels) based on $\mathbf{v}_u^*$ by updating the parameters $\mathbf{p}$ of $Dense_{\mathbf{p}}(\cdot)$ using a softmax function. The objective function can be formalized as follows,

$$\min_{\mathbf{p},\mathbf{w},b} \sum_{i=1}^{n_u} \sum_{y=\{c_f,c_n\}} \mathbf{1}[y_i = y] \log q_i \tag{4}$$
$$s.t. \quad q_i = \mathrm{softmax}(\mathbf{w} \cdot Dense_{\mathbf{p}}(\mathbf{v}_{u_i}) + b),$$

where y_i is the pseudo-label of the i-th user assigned by Algorithm 1, n_u is the number of existing users, $\mathbf{w}$ and b are the parameters of the softmax function.

3.5   Dynamic Link Re-weighting Strategy

The target of the user behavior representation is detecting fraudsters. To this end, the discriminative ability of the behavior representation should be strong. In SUPER-COLD, this discriminative ability is mainly obtained from the pseudo-labels generated by the dense subgraph mining. However, social relation may not comprehensively indicate all kinds of fraudsters [23]. As a result, the discriminative ability of behavior representation may not be good if only learning from the social relation.

To enhance the discriminative ability, SUPER-COLD reinforces the focusing of the dense subgraph mining on the suspicious users discovered in the behavior representation space, which reflects both the entities relation and the social relation. Specifically, SUPER-COLD clusters a set of users into two categories according to their behavior representations. It then re-weights the link of each user by the reciprocal of the number of its assigned categories. Formally, the links weight of a user u is assigned as,

$$w_{u,\cdot} = \frac{1}{|C_u|}, \tag{5}$$

where C_u refers to a set of users with the same category as user u, and $|\cdot|$ returns the size of the set. The assumption behind this re-weighting is that a user with less similar users are more suspicious as a fraudster. After re-weighting the links, SUPER-COLD conducts the dense subgraph mining again to generates new pseudo-labels, which are further integrated with the entities relation for the behavior representation. SUPER-COLD repeats this dynamic re-weighting strategy until convergence. With the dynamic link re-weighting strategy, the SUPER-COLD behavior representation procedure is summarized in Algorithm 2.

Algorithm 2. SUPER-COLD User Behavior Representation.

Input: Online review set S, convergence threshold ϵ;
Output: User behavior representation $\mathbf{V}_u^*$;
1: Entities relation embedding by Eq. (1)
2: Generating pseudo-label set Y by Alg. 1
3: Generating behavior representation $\mathbf{V}_u^*$ by Eq. (4)
4: Initializing $\Delta = +\infty$
5: **while** $\Delta > \epsilon$ **do**
6:    $Y' \leftarrow Y$
7:    Clustering $\mathbf{V}_u^*$ into two categories
8:    Re-weighting user-item graph links by Eq. (5)
9:    Generating pseudo-label set Y by Alg. 1
10:   Generating behavior representation $\mathbf{V}_u^*$ by Eq. (4)
11:   $\Delta = 1 - \dfrac{\sum\limits_{y_i \in Y, y_i' \in Y'} 1[y_i = y_i']}{|Y|}$
12: **end while**
13: **return** $\mathbf{V}_u^*$

3.6   SUPER-COLD Fraud Review Detection

SUPER-COLD detects fraud reviews according to the behavior representation in an unsupervised way. The SUPER-COLD fraud review detection procedure is shown in Algorithm 3. For a new review tuple $< u^*, o^*, r^* >$, the user behavior representation cannot be directly got from the existing model because u^* never appears and thus are not in the learned embedding layer. Alternatively, SUPER-COLD deduces the new user behavior representation by using the entities relation and the social relation. SUPER-COLD first looks up the learned item representation $\mathbf{v}_{o^*}$ and generates the review representation $\mathbf{v}_{r^*} = t_\omega(r^*)$ by the learned text embedding neural network. Then, it calculates an approximate representation of the user as $\mathbf{v}_{u^*} = \mathbf{v}_{r^*} - \mathbf{v}_{o^*}$. After that, it uses the learned fully connected neural network to generate behavior representation of the user as $\mathbf{v}_{u^*}^* = Dense_\mathbf{p}(\mathbf{v}_{u^*})$, which integrates the social relation with the approximate representation. Finally, SUPER-COLD retrieves the k-nearest existing users of the detecting user, and uses the majority voting to ensemble the retrieved users' pseudo-labels as the label assigned to the u^* (Algorithm 3 line 5). If most of the nearest users are candidate fraudsters, u^* will be treated as a fraudster and the r^* will be assigned as a fraud review (Algorithm 3 line 6–11).

4   Experiments

4.1   Data Sets

Following the literature [25,28] about cold-start fraud detection, our experiments are carried on two Yelp data sets including Yelp-hotel and Yelp-restaurant, which are also commonly used in previous fraud detection researches [16,19,22]. Tables 1 and 2 display the statistics of the data sets.

We split original data sets into two parts for cold-start fraud detection performance evaluation. The first part includes 90% earliest posted reviews. The users who posted these reviews are treated as existing users. The second part

Algorithm 3. SUPER-COLD Fraud Review Detection.

Input: An online review tuple $< u^*, o^*, r^* >$, the number of nearest users k;
Output: The fraud detection label y^*;

1: Looking up $\mathbf{v}_{o^*}$
2: Generating $\mathbf{v}_{r^*} = t_\omega(r^*)$
3: Calculating $\mathbf{v}_{u^*} = \mathbf{v}_{r^*} - \mathbf{v}_{o^*}$
4: Calculating $\mathbf{v}_{u^*}^* = Dense_\mathbf{p}(\mathbf{v}_{u^*})$
5: Retrieving $U = \arg\min\limits_{U} \sum\limits_{u \in U} \|\mathbf{v}_u - \mathbf{v}_u^*\|^2$ with $|U| = k$
6: Looking up pseudo-label set $|Y|$ of $|U|$
7: **if** $|\{y|y = c_f, y \in Y\}| > |\{y|y = c_n, y \in Y\}|$ **then**
8:     $y^* = Fraud$
9: **else**
10:     $y^* = Normal$
11: **end if**
12: **return** y^*

is the 10% latest posted reviews. From the second part, we pick up the reviews which wrote by new users at the first time as cold-start reviews. Besides, we use the whole data sets to evaluate the general fraud detection performance and do the ablation study.

4.2   Evaluation Metrics

We evaluate their performance by three metrics - *precision, recall,* and *F-score.* While precision evaluates the fraction of relevant review among detected reviews, recall reflects the fraction of relevant reviews that have been detected over the total amount of relevant reviews. The precision and recall should be jointly considered since fraud detection is an imbalance problem [14], i.e. fraud reviews are much less than normal reviews. Thus, we use F-score, which balances the precision and recall, as an averaged indicator. Higher F-score indicates a better performance for a fraud detection method. We report these three metrics per ground-truth normal and fraud classes to illustrate the performance for different categories. We further average them to show an overall performance.

We follow the literature [25] to use the results of the Yelp commercial fake review filter as the ground-truth for performance evaluation. Although its filtered (fraud reviews) and unfiltered reviews (normal reviews) are likely to be the closest to real fraud and normal reviews [19], they are not absolutely accurate [10]. The inaccuracy exists because it is hard for the commercial filer to have the same psychological state of mind as that of the users of real fraud reviews who have real businesses to promote or to demote, especially for the cold-start problem.

4.3   Parameters Settings

In our experiments, we use a hierarchical bi-directional GRU structure with 100 nodes to embed reviews, in which the pre-trained word embedding by GloVe algorithm [21] is used[3]. We train the user/item/review embedding by Adam

---

[3] The pre-trained word embedding can be downloaded from: http://nlp.stanford.edu/data/glove.6B.zip.

[9]. We set the word embedding dimension as 100 and training batch size as 32. To integrate the entities relation and the social relation, we adopt a 3-layer fully connected neural network with 100 nodes in each hidden layer. We train this fully connected neural network 10 epochs by Adam with batch size 32. All hidden nodes of the neural network used in the experiments use ReLU as their activation function. We choose k-means as the clustering method in SUPER-COLD. When inferring the behavior of a new user, we retrieve the 5 closest existing users of the user according to the distance in the embedding space. For the parameters of the compared methods, we take their recommended settings.

4.4 Effectiveness on Cold-Start Fraud Detection

Experimental Settings. SUPER-COLD is compared with the state-of-the-art method JETB [25]. This method handling the cold-start problem by considering entities (users, items and reviews) relations to represent reviews. When a new user posts a new review, this review can be represented by the trained network and classified by the classifier. JETB is also the first work that exploits the cold-start problem in review fraud detection.

In [25], support vector machine (SVM) is used as the fraud classifier based on the JETB generated review features. However, SVM is with a time complexity $O(n^3)$, where n is the number of training samples. It is not suitable for the problem with a large amount of data. In this experiments, the training data contains $619,496$ and $709,623$ reviews on Yelp-Hotel and Yelp-Restaurant data sets, respectively. To make JETB practicable, we use a 5-layer fully connected neural network instead of SVM as the fraud classifier of JETB.

Findings - SUPER-COLD Significantly Outperforming the State-of-the-art Cold-Start Fraud Detection Method. Table 1 demonstrates the SUPER-COLD fraud detection performance, compared to JETB on Yelp-Hotel and Yelp-Restaurant data sets. SUPER-COLD largely improves the fraud detection performance, i.e. 150% and 18.07% F-score increase on Yelp-Hotel and Yelp-Restaurant data sets. This averaged performance improvement is mainly contributed by the increased detection performance of fraud category (550% on

Table 1. Cold-start Fraud Detection: Precision (P), Recall (R) and F-score (F) are reported.

Data info.				SUPER-COLD			JETB			Improvement		
Name	Category	#Existing	#Cold-start	P	R	F	P	R	F	P	R	F
Hotel	Normal	376,671	60	**0.45**	0.15	0.23	0.32	**0.90**	0.47	40.63%	−83.33%	−51.06%
	Fruad	242,825	122	**0.69**	**0.91**	**0.78**	0.57	0.07	0.12	21.05%	1200.00%	550.00%
	Overall	619,496	182	**0.61**	**0.66**	**0.6**	0.49	0.34	0.24	24.49%	94.12%	150.00%
Restaurant	Normal	412,435	1,654	0.64	**0.84**	**0.73**	0.68	0.74	0.71	−5.88%	13.51%	2.32%
	Fruad	297,188	873	**0.62**	**0.68**	**0.65**	0.41	0.34	0.37	51.22%	100.00%	75.30%
	Overall	709,623	2,527	**0.63**	**0.78**	**0.70**	0.59	0.60	0.59	7.90%	30.39%	18.07%

Yelp-Hotel and 75.3% on Yelp-Restaurant). As shown in the results, SUPER-COLD "decreases" the performance of normal reviews detection. It reflects SUPER-COLD is more tough for fraud reviews. On one hand, this "decreased" performance of normal reviews detection does not decrease the averaged fraud detection performance. On the other hand, this "decreased" performance may be caused by the *noising ground-truth labels* of the cold-start fraud reviews that do not be detected by the Yelp commercial filter.

SUPER-COLD uses the represented user behavior instead of review features for cold-start fraud detection to avoid the camouflage in reviews. Because of the more reliable information, SUPER-COLD achieves significant performance improvement in cold-start fraud detection.

4.5   Effectiveness on General Fraud Detection

Experimental Settings. SUPER-COLD is further compared with three state-of-the-art competitors: Frauder [6], HoloScope [13], and SPEAGLE [23] in detecting *general fraud reviews* - all the reviews contained in a data set. These three competitors have different but relevant mechanisms compared with SUPER-COLD.

- *Fixed weighting dense subgraph mining-based method - FRAUDER* [6]. FRAUDER is a fraud detection method by dense subgraph mining. To detect camouflage and hijacked accounts, it adopts a fixed weighting strategy. Different from FRAUDER, the dense subgraph mining method used in SUPER-COLD is with a dynamic link weighting strategy to further fuse the entities relation with the social relation.
- *Dynamic weighting dense subsgraph mining-based method - HoloScope* [13]. HoloScope uses graph topology and temporal spikes to detect fraudsters groups, and employs a dynamic weighting approach to enable a more accurately fraud detection. However, the dynamic weighting is only conducted once according to the user temporal spikes. In contrast, SUPER-COLD interactively updates the dynamic weighting along the user behavior embedding process.
- *Metadata and social relation integration-based method - SPEAGLE* [23]. SPEAGLE proposes a unified framework to utilize metadata and the social relation in Markov Random Field for fraud detection. While SPEAGLE needs fraud labels, SUPER-COLD is a completely unsupervised method which jointly considers the entities relation and the social relations for user behavior representation.

While FRAUDER and HoloScope directly predict fraud reviews, SPEAGLE gives a probability of a review may be fake. To make a fair comparison, we only report the averaged precision of SPEAGLE but ignore its the recall and F-score.

Table 2. General Fraud Detection: Precision (P), Recall (R) and F-score (F) are reported.

Data Info.			SUPER-COLD			HoloScope			FRUADER			SPEAGLE			Improvement		
Name	Category	#Review	P	R	F	P	R	F	P	R	F	P	R	F	P	R	F
Hotel	Normal	420,785	**0.69**	0.96	**0.8**	0.64	0.6	0.62	0.64	**0.98**	0.77	0.53	-	-	7.81%	-2.04%	3.90%
	Fraud	267,544	**0.82**	0.31	**0.45**	0.42	**0.46**	0.44	**0.82**	0.11	0.31	0.72	-	-	0.00%	-32.61%	2.27%
	Overall	888,329	**0.74**	**0.71**	**0.66**	0.55	0.55	0.55	0.71	0.65	0.55	0.60	-	-	4.23%	9.23%	20.00%
Restaurant	Normal	461,190	**0.67**	0.9	**0.77**	0.51	**0.95**	0.66	0.63	**0.95**	0.76	0.42	-	-	6.35%	-5.26%	1.32%
	Fraud	326,981	0.73	**0.37**	0.49	**0.74**	0.12	0.21	**0.74**	0.21	0.33	0.58	-	-	-1.35%	76.19%	48.48%
	Overall	788,741	**0.69**	**0.68**	**0.65**	0.63	0.52	0.43	0.68	0.64	0.58	0.49	-	-	1.47%	6.25%	12.07%

Findings - SUPER-COLD Significantly Improving General Fraud Detection Performance, Especially Recall. The precision, recall, and F-score of SUPER-COLD, Frauder, HoloScope, and SPEAGLE are reported in Table 2. Overall, SUPER-COLD significantly outperforms the competitors. It improves 20% and 12.07% compared with the best-performing method in terms of F-score on two data sets.

Unlike FRAUDER and HoloScope that ignore the entities relation when they perform dense subgraph mining based on the social relation, SUPER-COLD couples these two independent relations to iteratively refine their performance by the dynamic link weighting. This enables SUPER-COLD to avoid camouflage by the social relation and effectively detect personalized fraud by the entities relation. Consequently, SUPER-COLD obtains up to 76.19% recall improvement compared with the competitors.

4.6   Quality of Behavior Representation

Experimental Settings. We visualize the behavior representation in a two-dimensional space trough TSNE [15]. To evaluate the representation quality, we plot pseudo-labels of each user according to the dense subgraph mining-based fraud detection results. A high-quality behavior representation will enable a separate location of users with different pseudo-labels. The behavior representation generated by SUPER-COLD is compared with that generated by JETB.

Findings - SUPER-COLD Generated Behavior Representation Is with Strong Discriminate Ability. The behavior representations generated by SUPER-COLD and JETB are visualized in Fig. 3. In the JETB generated representation space, there is a large overlap between users with different pseudo-labels, especially on Yelp-Hotel data set. In contrast, the SUPER-COLD generated representation is with a stronger discriminative ability that separates users with pseudo-labels well. These qualitative illustrations are consistent with the quantitative results in Table 2 that the improvement brought by SUPER-COLD on Yelp-Hotel data set is much larger than that on Yelp-Restaurant data set.

Based on the JETB generated user behavior representation, SUPER-COLD moves one step further. It adopts the pseudo-labels generated by the dense subgraph mining-based fraud detection to fine-tune the representation learning from the entities relation. Since the dense subgraph mining-based fraud detection involves much more effective patterns (e.g. group manipulation) about fraudsters, the fine-tuned representation thus has a stronger discriminative ability for fraud and normal users.

4.7   Ablation Study

Experimental Settings. We further study the contribution from each SUPER-COLD components: entities relation learning, social relation learning, dynamic graph link re-weighting, and behavior-based cold-start fraud detection. This contribution can be analyzed from the Tables 1, 2 and Fig. 3. Here, we assume FRAUDER reflects the performance of social relation learning, HoloScope demonstrates the performance of dynamic weighting, SPEAGLE stands for the performance of combining the entities and social relations, and JETB implies the performance of review-based cold-start fraud detection.

      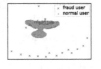

| (a) JETB: Hotel. | (b) SUPER-COLD: Hotel. | (c) JETB: Restaurant. | (d) SUPER-COLD: Restaurant. |

Fig. 3. User behavior embedding of different methods.

Findings - SUPER-COLD Is Contributed by Learning Entities and Social-Relation and Dynamically Re-weighting Graph Links, Especially by Social Relation. As shown in Table 1, SUPER-COLD outperforms FRAUDER at least 12.07% in terms of F-score. Meanwhile, SPAEGLE also achieves much better performance compared with FRAUDER. This demonstrates that incorporating entities relation with social-relation gains a large performance improvement, which reflects the contribution of entities relation.

Social relation also makes a significant contribution, which is much greater than that made by the entities relation. Compared with JETB (with entities relation but without social relation), SUPER-COLD (with entities and social relation) gains 150% F-score improvement on Yelp-Hotel data set because of integrating social relation. In contrast, on that data set, the improvement brought by entities relation is only 20% according to the comparison of SUPER-COLD (with entities and social relation) and FRAUDER (with social relation but without entities relation) shown in Table 2.

Dynamically re-weighting graph links make the dynamic weighting strategy more reliable. As shown in Table 2, the performance of HoloScope (with dynamic weighting) is similar to FRAUDER (without dynamic weighting) on

Yelp-Hotel data set but slightly lower than FRAUDER on Yelp-Restaurant data set. However, SUPER-COLD (with dynamic re-weighting) consistently outperforms FRAUDER on both data set. The reason may lie in the re-weighting mechanism that iteratively enhances weighting quality.

5   Conclusion

This paper proposes a socially-aware unsupervised user behavior representation method to tackle the cold-start problem in fraud review detection. The proposed unsupervised method integrates both entities and social relations for user behavior representation, and further strengths the discriminative ability of the behavior representation by a dynamic link re-weighting strategy. It can effectively detect fraud reviews with the cold-start problem as demonstrated by comprehensive experiments.

References

1. Chen, J., Saad, Y.: Dense subgraph extraction with application to community detection. TKDE **24**(7), 1216–1230 (2012)
2. Choo, E., Yu, T., Chi, M.: Detecting opinion spammer groups through community discovery and sentiment analysis. In: Samarati, P. (ed.) DBSec 2015. LNCS, vol. 9149, pp. 170–187. Springer, Cham (2015). https://doi.org/10.1007/978-3-319-20810-7_11
3. Fei, G., Mukherjee, A., Liu, B., Hsu, M., Castellanos, M., Ghosh, R.: Exploiting burstiness in reviews for review spammer detection. In: ICWSM 2013, pp. 175–184 (2013)
4. Feng, S., Xing, L., Gogar, A., Choi, Y.: Distributional footprints of deceptive product reviews. In: ICWSM 2012, pp. 98–105 (2012)
5. Hooi, B., Shin, K., Song, H.A., Beutel, A., Shah, N., Faloutsos, C.: Graph-based fraud detection in the face of camouflage. TKDD **11**(4), 44 (2017)
6. Hooi, B., Song, H.A., Beutel, A., Shah, N., Shin, K., Faloutsos, C.: FRAUDAR: bounding graph fraud in the face of camouflage. In: ACM SIGKDD, pp. 895–904. ACM (2016)
7. Hovy, D.: The enemy in your own camp: how well can we detect statistically-generated fake reviews-an adversarial study. In: ACL, vol. 2, pp. 351–356 (2016)
8. Jindal, N., Liu, B.: Opinion spam and analysis. In: WSDM, pp. 219–230. ACM (2008)
9. Kingma, D.P., Ba, J.: Adam: a method for stochastic optimization. arXiv preprint arXiv:1412.6980 (2014)
10. Li, H., Chen, Z., Liu, B., Wei, X., Shao, J.: Spotting fake reviews via collective positive-unlabeled learning. In: ICDM, pp. 899–904. IEEE (2014)
11. Li, H., Chen, Z., Mukherjee, A., Liu, B., Shao, J.: Analyzing and detecting opinion spam on a large-scale dataset via temporal and spatial patterns. In: ICWSM, pp. 634–637 (2015)
12. Li, H., et al.: Modeling review spam using temporal patterns and co-bursting behaviors. arXiv preprint arXiv:1611.06625 (2016)
13. Liu, S., Hooi, B., Faloutsos, C.: HoloScope: topology-and-spike aware fraud detection. In: CIKM, pp. 1539–1548. ACM (2017)

14. Luca, M., Zervas, G.: Fake it till you make it: reputation, competition, and yelp review fraud. Manag. Sci. **62**(12), 3412–3427 (2016)
15. van der Maaten, L., Hinton, G.: Visualizing data using t-SNE. JMLR **9**(Nov), 2579–2605 (2008)
16. Mukherjee, A., et al.: Spotting opinion spammers using behavioral footprints. In: ACM SIGKDD, pp. 632–640. ACM (2013)
17. Mukherjee, A., Liu, B., Wang, J., Glance, N., Jindal, N.: Detecting group review spam. In: WWW, pp. 93–94. ACM (2011)
18. Mukherjee, A., Venkataraman, V., Liu, B., Glance, N.: Fake review detection: classification and analysis of real and pseudo reviews. Technical report UIC-CS-2013-03, University of Illinois at Chicago (2013)
19. Mukherjee, A., Venkataraman, V., Liu, B., Glance, N.S.: What yelp fake review filter might be doing? In: ICWSM (2013)
20. Ott, M., Choi, Y., Cardie, C., Hancock, J.T.: Finding deceptive opinion spam by any stretch of the imagination. In: ACL HLT, pp. 309–319. Association for Computational Linguistics (2011)
21. Pennington, J., Socher, R., Manning, C.: GloVe: global vectors for word representation. In: EMNLP, pp. 1532–1543 (2014)
22. Rayana, S., Akoglu, L.: Collective opinion spam detection: bridging review networks and metadata. In: ACM SIGKDD, pp. 985–994. ACM (2015)
23. Rayana, S., Akoglu, L.: Collective opinion spam detection using active inference. In: ICDM, pp. 630–638. SIAM (2016)
24. Wang, G., Xie, S., Liu, B., Philip, S.Y.: Review graph based online store review spammer detection. In: ICDM, pp. 1242–1247. IEEE (2011)
25. Wang, X., Liu, K., Zhao, J.: Handling cold-start problem in review spam detection by jointly embedding texts and behaviors. In: ACL, vol. 1, pp. 366–376 (2017)
26. Wu, L., Hu, X., Morstatter, F., Liu, H.: Adaptive spammer detection with sparse group modeling. In: ICWSM, pp. 319–326 (2017)
27. Ye, J., Akoglu, L.: Discovering opinion spammer groups by network footprints. In: Appice, A., Rodrigues, P.P., Santos Costa, V., Soares, C., Gama, J., Jorge, A. (eds.) ECML PKDD 2015. LNCS (LNAI), vol. 9284, pp. 267–282. Springer, Cham (2015). https://doi.org/10.1007/978-3-319-23528-8_17
28. You, Z., Qian, T., Liu, B.: An attribute enhanced domain adaptive model for cold-start spam review detection. In: COLING, pp. 1884–1895 (2018)

Gated Convolutional Encoder-Decoder for Semi-supervised Affect Prediction

Kushal Chawla$^{(\boxtimes)}$, Sopan Khosla, and Niyati Chhaya

Big Data Experience Lab, Adobe Research, Bangalore, India
{kchawla,skhosla,nchhaya}@adobe.com

Abstract. Analyzing human reactions from text is an important step towards automated modeling of affective content. The variance in human perceptions and experiences leads to a lack of uniform, well-labeled, ground-truth datasets, hence, limiting the scope of neural supervised learning approaches. Recurrent and convolutional networks are popular for text classification and generation tasks, specifically, where large datasets are available; but are inefficient when dealing with unlabeled corpora. We propose a gated sequence-to-sequence, convolutional-deconvolutional autoencoding (GCNN-DCNN) framework for affect classification with limited labeled data. We show that compared to a vanilla CNN-DCNN network, gated networks improve performance for affect prediction as well as text reconstruction. We present a regression analysis comparing outputs of traditional learning models with information captured by hidden variables in the proposed network. Quantitative evaluation with joint, pre-trained networks, augmented with psycholinguistic features, reports highest accuracies for affect prediction, namely frustration, formality, and politeness in text.

1 Introduction

Affect refers to the experience of a feeling or emotion [22]. The importance of affect analysis in human communications and interactions has been well discussed. The study of human affect from text and other published content is an important topic in language understanding. Word correlation with social and psychological processes is discussed by [21]. Personality and psycho-demographic preferences through social media content have also been studied [23]. Human communication, especially through language, reflects psychological and emotional states. Examples include the use of opinion and emotion words [8]. The analysis of affect in interpersonal communication such as emails, chats, and longer articles is necessary for applications including consumer behavior and psychology, understanding audiences and opinions in computational social science, and more recently, for dialogue systems and conversational agents. Interpersonal communication illustrates fine-grained affect categories. Frustration is one such dominant affect that is expressed in human interactions [1].

K. Chawla and S. Khosla—denotes equal contribution.

© Springer Nature Switzerland AG 2019
Q. Yang et al. (Eds.): PAKDD 2019, LNAI 11439, pp. 237–250, 2019.
https://doi.org/10.1007/978-3-030-16148-4_19

We present a **computational approach for affect analysis in text; namely, Frustration, Formality, and Politeness**. Human perceptions on the same content vary across individuals. This introduces a challenge in creating and gathering labeled datasets for affect modeling tasks. The lack of labeled data makes it challenging to build supervised models, and in particular, neural models for the affect prediction tasks. Semi-supervised or unsupervised models for affect understanding and modeling are also unexplored.

RNNs are a popular choice for neural language modeling. However, they are unable to model longer sequences and suffer drawbacks such as exposure bias [30]. Convolutional networks (CNN) on the other hand are good at extracting position invariant features [25] while RNN trump at modeling units in a sequence. Our work focuses on affect modeling, which is similar to sentiment classification, where global cues are important. We leverage an encoder-decoder framework with a CNN as encoder and a deconvolutional (transposed convolutional) neural network as decoder proposed by [32] as the base framework for learning language representations for this work.

Gating mechanisms control the information flow in a network and have been useful for RNNs. In [6], gating was introduced for a convolutional language modeling setup. In this work, we introduce a Gated CNN-DCNN architecture that combines the encoder-decoder setup with advantages of the gating mechanism. The lack of labeled data introduces a need for the network to learn from unlabeled datasets, if available. Semi-supervised methods such as joint training [29] and self-learning using pre-training, dropouts, and error forgetting provide techniques for this training. We focus on a large email dataset with limited labeled subset in this work. We introduce shared embeddings and pre-training in the network to exploit the availability of the large unlabeled corpus in the proposed network.

This paper introduces a semi-supervised deep neural network for affect prediction in text by enhancing a gated convolutional architecture with pre-training and data-relevant psycholinguistic features. The proposed approach outperforms the vanilla neural networks. We show that pre-training leads to improved performance with minimal labeled data. Further, gated networks combined with shared embeddings (joint-training) out perform standard networks. The key contributions of this paper include:

- A **joint semi-supervised** framework with **pre-training** and **psycholinguistic features** for affect prediction tasks. We present the first deep neural network built on a **Gated CNN-DCNN** architecture, for computational modeling of Frustration in text (email) data. The proposed approach reports highest accuracies for affect dimensions including Formality and Politeness.
- A **Gated CNN-DCNN** auto-encoder for text reconstruction. Experiments show significant improvements over state-of-the-art approaches for standard datasets.

We also present a regression study to better understand the features captured by the hidden layers of the network. This analysis leads towards an interpretability study for this work.

Paper Structure: The prior work is discussed in Sect. 2 along with the ENRON-FFP dataset. Our proposed approach is described in Sect. 3. This is followed by the experiments, a discussion section with error analysis, and finally the conclusion.

2   Related Work

Encoder-decoder models are used to learn language and sentence representations [17]. RNN and LSTM-based approaches have gained recent popularity, given their high performance for language related tasks [2]. Other sets of approaches for similar tasks are built on convolutional networks. An extensive analysis of the advantage of convolutional methods over LSTMs is presented in [30]. We leverage the convolutional models in this work. RNNs have benefited from gating mechanisms to reach state-of-the-art performances. LSTMs use *input* and *forget* gates to solve the vanishing gradient problem. Similar gating mechanisms have been proposed for convolutional modeling of images [18] which were later extended to language modeling [9]. In contrast to the *tanh* based mechanism proposed in [18], a gated linear unit (GLU) is proposed in [6] which outperforms the former as it allows the gradient to propagate through the linear unit without scaling as shown by the equation, $\nabla[\boldsymbol{X} \otimes \sigma(\boldsymbol{X})] = \nabla\boldsymbol{X} \otimes \sigma(\boldsymbol{X}) + \boldsymbol{X} \otimes \sigma'(\boldsymbol{X})\nabla\boldsymbol{X}$.

While gating mechanisms have shown promise in language modeling and machine translation, they remain unexplored in an auto-encoder setup. Most CNN-based architectures contain a single convolution layer followed by a pooling layer which captures n-gram features. More recently, a deep architecture was proposed [32] to capture higher level linguistic features at every step in this hierarchy. A deconvolutional decoder is used in the model to overcome the shortcomings of a RNN-based decoder. This has been shown to improve text classification and summarization as it accounts for distant dependencies in long sentences. Our approach is inspired from this work. The availability of large well-labeled data is not assured, more so for tasks where human perception plays an important role. Authors in [12] introduce a joint learning framework; the model is trained using labeled and unlabeled data simultaneously, hence generating pseudo-labels for the unlabeled training instances. Pre-training was employed in [4] using two unsupervised learning approaches: language modeling and auto-encoders. They use pre-trained weights to initialize the model for various classification tasks in NLP and computer vision domain. We leverage both **joint training** and **pre-training**.

Neural Methods for Affect Analysis. The lack of labeled datasets for affect prediction in text makes it difficult to build supervised neural models for classification. A CNN-based model was presented in [16] for personality detection from essays. They work on a dataset with 3000 text essays and predict the author personality using a combination of the word embeddings and Mairesse features [15]. We use a similar feature enhancement technique in this work, but the proposed architecture focuses on leveraging a combination of large unlabeled and limited labeled data for text classification and reconstruction tasks. Affects such as formality and politeness have been explored using standard machine learning models [5,20], but neural models for these are absent.

In the related domain of sentiment analysis, works based on deep learning are prevalent [26]. Most of these methods rely on large labeled datasets. Note that affect dimensions such as frustration are fine-grained and complex as against sentiment. To the best of our knowledge, this is the first attempt at a semi-supervised deep learning approach for fine-grained affect prediction, specifically, for frustration detection in text.

ENRON-FFP Dataset. We leverage the ENRON-FFP dataset[1] [11] which is a subset of 960 emails from the ENRON-email dataset [3] annotated with formality, politeness, and frustration scores using a crowd-sourced experiment. The scores are converted to binary labels for this work. After conversion, we have 418 emails tagged as Formal, 389 as Frustrated and 404 as Polite. This dataset is referred to as **ENRON-FFP**, and is used as the limited labeled dataset, while the complete dataset is addressed as **ENRON** for rest of this paper.

3   Method

We propose a semi-supervised, joint learning framework for affect prediction, built on a Gated convolutional encoder (GCNN). Figure 1 shows the network architecture. Our joint network learns to represent the input text sequence while simultaneously capturing the affect information from the labeled data (**ENRON–FFP**). GCNN with a deconvolutional decoder (DCNN) provides the mechanism for text reconstruction, i.e. to reproduce the text sequences from latent representations. The learned encoding, augmented with linguistic features, acts as the input feature space for a fully connected classification network, trained to predict affect labels in text.   The lack of large labeled dataset and the availability of the unlabeled **ENRON** data enables us to pre-train the network with the unlabeled samples. The proposed network aims at affect classification and in turn also learns the reconstruction objective.

Architecture Overview. Consider a text input d. Each word w_d^t in d is embedded into a k-dimensional representation, $e_t = \boldsymbol{E}[w_d^t]$ where $\boldsymbol{E}$ is a learned matrix. The embedding layer is passed through a GCNN to create a fixed-length vector $\boldsymbol{h}_L$. This latent representation, appended with linguistic features is sent to a fully connected layer with a softmax classifier. Simultaneously, $\boldsymbol{h}_L$ is also fed to a deconvolutional decoder, which attempts to reconstruct d from the latent vector. Hence, the final loss function:

$$\alpha_{ae}L_{ae} + (1 - \alpha_{ae})L_c \tag{1}$$

for the model is defined as the combination of the classification error L_c and the reconstruction error L_{ae} explained in the following subsections. Here, α_{ae} controls the weight for reconstruction loss L_{ae}. The dearth of labeled data in

---

[1] Link to the annotated ENRON-FFP dataset: https://bit.ly/2IAxPab.

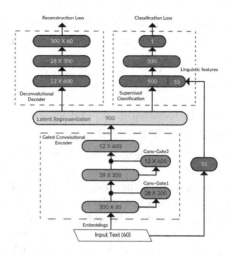

Fig. 1. Architecture diagram: Gated CNN encoder for affect prediction

the **ENRON-FFP** dataset, motivates **unsupervised pre-training** using the **ENRON** data ($\alpha_{ae} = 1$) and **joint-training** with the limited labeled samples ($0 < \alpha_{ae} < 1$).

Gated Convolutional Autoencoder (GCNN-DCNN). A sequence-to-sequence 3-layer convolutional encoder followed by a 3-layer deconvolutional decoder (CNN-DCNN) framework for learning latent representations from text data was introduced in [32]. Their proposed framework outperforms RNN-based networks for text reconstruction and semi-supervised classification tasks. We extend their network in our work.

Gated Convolutional Encoder. A CNN with L layers, inspired from [24] is used to encode the document into a latent representation vector, h_L. The first $L - 1$ convolutional layers create a feature map $C^{(L-1)}$ which is fed into a fully-connected layer implemented as a convolutional layer. This final layer produces the latent representation h_L which acts as a fixed-dimensional summarization of the document [32].

In order to control the information that propagates through the convolutional layers, we add **output gates** to the encoder. The gating function follows from [6], such that the hidden layers $h_0, h_1..., h_L$ are computed as

$$h_l(X) = (X * W_l + b_l) \otimes \sigma(X * W_g + b_g) \tag{2}$$

where X is the input and W_l, W_g, b_l, b_g are learned parameters for layer l. $\sigma(X * W_g + b_g)$ modulates the information transferred by each element in the output, $X * W_l + b_l$ to the next layer. Gates are added only to the encoder layers as we would like the encoded representation to be self-sufficient for decoding.

Deconvolutional Decoder. We leverage the deconvolutional decoder introduced by [32] for our model. The reconstruction loss is defined as,

$$L_{ae} = \sum_{d \in D} \sum_t \log p(\hat{w}_d^t = w_d^t), \tag{3}$$

where D is the set of observed sentences. w_d^t and $\hat{w}_d^t$ correspond to the words in the input and output sequences respectively.

Language Features. Using features to enrich neural architectures is a popular technique in NLP tasks [10,27]. Affective language studies focus on analyzing various features including lexical, syntactic, and psycho-linguistic features to detect affect dimensions. We augment the latent vector produced by the GCNN encoder with these features to capture human reactions.

Let h' denote the representation vector for linguistic features extracted from the input data d. h' is normalized and concatenated with h_L to derive $h'' = h_L \frown h'$. h'' is used as an input to a fully connected network, producing a probability p corresponding to the positive class with the ground-truth label y. Here, we use the cross entropy loss for binary classification which is defined as:

$$L_c = -(y \log(p) + (1 - y) \log(1 - p)) \tag{4}$$

Joint Learning. The GCNN-DCNN network learns the text sequences while the linguistic features augment it with affect information. Joint learning introduces a mechanism to learn shared representations during the network training for both text reconstruction and affect prediction. We implement joint learning using simultaneous optimization for both these tasks. The loss function is hence given by,

$$L = \alpha_{ae} L_{ae} + (1 - \alpha_{ae}) L_c. \tag{5}$$

where α_{ae} is a balancing hyperparameter with $0 \leq \alpha_{ae} \leq 1$. Higher the value of α_{ae}, higher is the importance given to the reconstruction loss L_{ae} while training and vice versa. While training, given a dataset D of word sequences and $\alpha_{ae} = 1$, we first **pre-train** the model for text reconstruction. The model, initialized with the learned weight parameters, is then **jointly-trained** using the limited labeled dataset (**ENRON–FFP**) for both the prediction and the reconstruction task simultaneously $(0 < \alpha_{ae} < 1)$.

4   Experiments

Three kinds of experiments are presented here:

- Evaluation of the gated CNN architecture, i.e. does introduction of gates to a CNN-DCNN autoencoder improve the network performance? We present the performance of GCNN-DCNN for text reconstruction (on Hotel Review dataset and ENRON) and affect prediction (ENRON).

Table 1. Text reconstruction using hotel reviews dataset [13].

(a) Reconstruction evaluation on Hotel Reviews and ENRON. GCNN-DCNN-12 having gates on first two layers gives best results.

Model	BLEU4	ROUGE1	ROUGE2
Hotel Reviews [13]			
LSTM-LSTM [13]	24.1	57.1	30.2
Hier. LSTM-LSTM [13]	26.7	59.0	33.0
Hier. + att. LSTM-LSTM [13]	28.5	62.4	35.5
CNN-LSTM [32]	18.3	56.6	28.2
CNN-DCNN [32]	89.7	95.5	91.5
GCNN-DCNN-1	92.7	96.7	93.9
GCNN-DCNN-12	**92.8**	**96.8**	**94.1**
GCNN-DCNN-123	**92.8**	**96.8**	94.0
ENRON [3]			
CNN-DCNN [32]	88.8	91.8	90.3
GCNN-DCNN-1	90.7	**94.6**	93.5
GCNN-DCNN-12	**90.8**	**94.6**	**93.6**
GCNN-DCNN-123	90.4	94.4	93.2

(b) Text Reconstruction Example.

Ground-truth:	on every visit to nyc , the hotel beacon is the place we love to stay . so conveniently located to central park , lincoln center and great local restaurants . the rooms are lovely . beds so comfortable , a great little kitchen and new wizz bang coffee maker . the staff are so accommodating and just love walking across the street to the fairway super-market with every imaginable goodies to eat .
Hier. LSTM	every time in new york , lighthouse hotel is our favorite place to stay . very convenient , central park , lincoln center, and great restaurants . the room is wonderful , very comfortable bed , a kitchenette and a large explosion of coffee maker . the staff is so inclusive , just across the street to walk to the supermarket channel love with all kinds of what to eat .
CNN-DCNN	on every visit to nyc , the hotel beacon is the place we love to stay . so **closely** located to central park , lincoln center and great local restaurants . the rooms are lovely . beds so comfortable , a great little kitchen and new **UNK james** coffee maker . the staff **turned** so accommodating and just love walking across the street to the fairway supermarket with every **pasta receptions** to eat .
GCNN-DCNN-12	on every visit to nyc , the hotel beacon is the place we love to stay . so conveniently located to central park , lincoln center and great local restaurants . the rooms are lovely . beds so comfortable , a great little kitchen and new **UNK** bang coffee maker . the staff are so accommodating and just love walking across the street to the fairway super-market with every **flower** goodies to eat .

- Evaluation of pre-training and joint learning. We evaluate the effect of introducing semi-supervised learning to overcome the lack of ground truth data. Extensive experiments for Frustration, Formality, and Politeness prediction are discussed.
- Evaluating the use of psycholinguistic features. Experiments showing the change in network performance with introduction of linguistic features are discussed.

Experimental Setup. Identifying Predictive Psycholinguistic Features. A Logistic Regression based analysis on the various language features shows the significant contribution of the psycholinguistic features for affect prediction. These features include emotion, personality features, and affect-lexica based features. The features identified through this analysis are appended to the encoded vector as discussed in the Method section when augmenting the network with language features.

Baselines. We use the experimental setup from [32] to define baselines for the evaluation of the Gated architecture for reconstruction task. Affect prediction is modeled as a binary classification task. We compare our approach to a baseline which directly predicts the majority class as the output. We also build standard ML models such as Decision Tree, Support Vector Machine (SVM), Random Forest, and Nearest Neighbors using the language features. These models are trained on 55 linguistic features including lexical, syntactic, and psycholinguistic features. A vanilla CNN model is defined as another baseline. This model contains a convolutional layer with 5 filters each with 10X5 size, followed by a pooling layer with size $5X5$ and stride as 5, and lastly a dense layer of size 50 (dropout rate: 0.2). Rectified Linear Unit (ReLU) is used as the activation throughout with a cross entropy loss function.

Results: Text Reconstruction. Text reconstruction tasks [32] have gained popularity along with identification of related topics or sentiments, and abstracting (short) summaries from user generated content [7]. First, an investigation on the performance of the proposed auto-encoder in terms of learning representations that can preserve text information is discussed. The evaluation criteria from [32], i.e. BLEU score [19] and ROUGE score [14] are used to measure the closeness of the reconstructed text (model output) to the input text. BLEU and ROUGE scores measure the n-gram recall and precision between the model outputs and the ground-truth references. We use BLEU-4, ROUGE-1, 2 in our evaluation, in alignment with [32]. Along with CNN-DCNN and GCNN-DCNN (our model), a comparison with Hierarchical LSTM encoder [13] and LSTM-LSTM encoder [32] is also reported. The experimental setup by [32] on the Hotel Reviews dataset [13] is used for the experiment. Latent representation dimension is set to h = 500 and training is done for 50 epochs. This dataset consists of $348,544$ training samples and $39,023$ testing samples. Results are also reported for the ENRON email dataset [3]. The dataset is split into $467,401$ training and $50,000$ test samples. We compare the CNN-DCNN [32] with variants of our proposed network GCNN-DCNN (h = 900). Training converges in 20 epochs.

Table 1a shows the performance of text reconstruction. Gates significantly improve results for both datasets across all metrics. For the Hotel Reviews dataset, the best performing architecture is with gates added on first and second encoder layers (GCNN-DCNN-12). The addition of a gate to third layer (GCNN-DCNN-123) yields no further improvement. Similarly for ENRON, adding a gate to first and second convolutional layers is sufficient to capture the underlying complexity of data. Table 1b shows the reconstruction outputs for the example paragraph from Hotel Reviews dataset used by [13]. GCNN-DCNN-12 corrects most word-level errors made by vanilla CNN-DCNN architecture. An observation of the training loss profiles show that the gated versions converge faster.

Results: Affect Prediction. The gated architecture is used to predict Formality, Frustration, and Politeness on the **ENRON-FFP** dataset. We use a CNN encoder and a fully connected network (with one hidden-layer of 300 dimensions) for binary classification. This corresponds to training the architecture in Fig. 1 using only the labeled **ENRON-FFP** dataset i.e. setting α_{ae} as 0. This approach is termed 'Supervised'(S). Extending this model, we introduce Gates (G, on first and second convolutional layers), Joint learning (J), Pre-training (P), and the augmentation with Language features (L). These approaches are also compared against standard ML models trained using various language features.

To pre-train the model, we run the reconstruction task on **ENRON**. **ENRON-FFP** is held out during this training. Hyper-parameters are set to $\alpha_{ae} = 1$, hidden layer representation $h = 900$, and a batch size of 32. Empirical experiments show that convergence is achieved when training both gated and non-gated networks for 15 epochs. For the labeled data **ENRON–FFP**, the batch size is 16 with 5 epochs. For joint learning, $\alpha_{ae} = 0.5$ reports best results. We use a 300 dimensional pre-trained word vectors trained on a part of Google News Dataset[2] for this task.

Figure 2a shows accuracy across various network configurations. Figure 2b shows the change in F1 scores of the minority class (i.e. positive affect class) across different models. Supervised approaches have moderate accuracy but poor F1 scores, especially for frustration. The models fail to predict the minority class given the lack of data. For the non-gated version, combining reconstruction task with affect prediction using Joint learning and Pretraining, we achieve an improvement of 4.7, 0.5 and 3.3 in accuracy for formality, frustration and politeness respectively. For the gated version, these numbers are 2.4, 1.7 and 2.6 respectively.

Introducing gates improves performance over simple CNN-DCNN in most cases. Combining GCNN-DCNN with pre-training, joint learning, and language features reports best performance for all three dimensions. The best performing model beats the traditional SVM, the simplified CNN model and other baselines by a significant margin for all three affects. Using Joint learning with the supervised model performs poorly, but improves when combined with pre-training. This is because the joint learning objective is difficult to optimize as against the affect prediction objective, it demands for further training or more data. Pre-training the model with unlabeled data, helps address this issue and generalizes better.

Sentiment Analysis. Table 2 shows the performance of our model on Yelp Review Polarity dataset [31] for the Sentiment analysis task. We follow the setup defined by [32]. To establish the utility of gates in a semi-supervised setup (large unlabeled, small labeled dataset), the dataset is split into two partitions: one simulating the unlabeled corpus and the other corresponding to the limited labeled dataset. Experiments with different amounts of labeled data (1%, 10% and 100%), with 100% unlabeled data used for pre-training, are shown. As seen in the table, the proposed model GCNN-DCNN-12 outperforms the baseline in the semi-supervised setup for Sentiment classification: 1% and 10%, hence supporting the hypothesis that introducing gates creates a robust model when there is limited labeled data.

Table 2. Percentage accuracy for Sentiment analysis on the Yelp review dataset. Results show that the proposed model GCNN-DCNN-12 out performs the baseline

Proportion	1%	10%	100%
CNN-DCNN [32]	88.2	93.3	**96.0**
GCNN-DCNN-12	**88.5**	**93.5**	95.9

---

[2] https://code.google.com/archive/p/word2vec/.

Model	Formality	Frustration	Politeness
Majority Baseline	56.5	59.5	57.9
Decision Tree	73.8	59.5	63.8
SVM	69.6	51.6	56.7
Random Forest	79.3	64.9	69.5
Nearest Neighbor	76.9	60.5	65.4
Vanilla CNN	71.8	58.6	60.0
Supervised(S)	77.6	62.9	66.1
S + Joint(J)	75.5	60.3	63.9
S + Pretraining(P)	80.8	63.6	70.9
S + J + P	82.3	63.4	69.4
S + Gates(G)	80.4	62.4	66.9
S + G + J	77.5	60.8	63.4
S + G + P	81.9	63.9	70.3
S + G + J + P	82.8	64.1	69.5
S + J + P + L	82.4	63.1	70.5
S + G + J + P + L	**83.8**	**67.6**	**72.1**

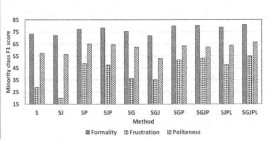

(a) Percentage Accuracy for affect prediction. All values show mean accuracy over 10-fold CV on **ENRON-FFP**.

(b) F1-scores across different models (Minority class)

Fig. 2. Performance on affect prediction. **S**: Supervised learning using CNN encoder trained on labeled data only, **J**: Joint learning with reconstruction task using DCNN decoder, **P**: Pre-training model for reconstruction, **G**: GCNN encoder with gates on first and second encoder layer (corresponding to GCNN-DCNN-12 in Table 1a), **L**: Linguistic features. $\alpha_{ae} = 0.5$.

Model	Formality	Frustration	Politeness
SGJP	(82.8, 0.047)	(64.1, 0.035)	(69.5, 0.055)
+ Lex + Syn	(81.5, 0.027)	(66.4, 0.024)	(70.0, 0.059)
+ Lex + Syn + Deriv	(82.9, 0.04)	(64.9, 0.044)	(70.0, 0.067)
+ Psycho	(82.5, 0.027)	(66.2, 0.056)	(71.0, 0.067)
+ Lex + Syn + Psych	(83.5, 0.035)	(65.8, 0.041)	(**72.4, 0.045**)
+ All	(**83.8, 0.035**)	(**67.6, 0.034**)	(72.1, 0.057)

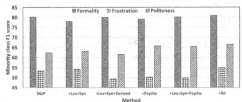

(a) Percentage accuracy across feature combinations. All values show (Mean accuracy, Std. dev) over 10-fold CV on **ENRON-FFP** Dataset.

(b) F1-scores for minority class (positive affect)

Fig. 3. Performance of **SGJP** (best performing) using linguistic features for Affect prediction. The features are based on **Lex**ical, **Syn**tactic, **Deriv**ed, and **Psycho**linguistic sets.

5   Discussion

An analysis on the contribution of the linguistic features and how it reflects in the hidden parameters is presented here.

Error Analysis: Model Predictions. The contribution of gates in the architecture is analyzed by creating a set of emails that are correctly classified by the Gated architecture but are mis-classified by the non-gated counterpart. Let that set be G_c. For the emails in G_c, we investigate the distribution of various lexical, syntactic, psycholinguistic, and derived features. The average feature values for the set G_c are observed to be higher as compared to averages computed across all the emails in **ENRON-FFP** for various structural features such as Difficult Words, NumChars, TextDensity, and NumSentences, whereas lower for

FleschReadingEase. These structural differences are consistent across Formality, Frustration, and Politeness (p-value< 0.01 using Student's T-test for Formality), showing that the addition of gates to the architecture allows the model to correctly classify these structurally more complex inputs, where the non-gated model fails.

Table 3. Dependency of language features against the ENRON-FFP labels and hidden features. "***" corresponds to $p < 0.001$, "**" for $p < 0.01$, and "*" for $p < 0.05$.

Features	p_f	p_fr	p_p	c_f	c_fr	c_p	g_f	g_fr	g_p
Derived features									
Contractions							***		*
NER_PERSON			*				***	***	***
NER_ORG			***	***	***	***	***	***	***
PersonLength			***	***	***	***	***	***	***
FleschReadingEase	***	***		***	***	***	*	*	*
Lexical & syntactic features									
AvgWordLength	***	***	*	***	***	***	*	*	*
NumWords	***	***	***	***	***	***	***	***	***
NumSyllables	***		***	***	***	***	***	***	***
NonAlphaChars	*	*	***	***	***	***			*
PunctuationChars		*	***				***		
POS_N	***	***	***	***					
POS_A	***	***	***						
POS_V	***	***	***	***	*		*		
Exclamation	***		***						
LowerCase	***	***	***						
Psycho-linguistic features									
EmoloxIntensity_sad		***	*	***	***	***	***	***	***
EmolexIntensity_anger	***	*		***	***		*	***	*
EmolexIntensity_joy	***		*	***	*	***	***	*	***
Perma_NEG_M	***	***	***		*	***			
Perma_NEG_R	***	***	***		*				
Perma_NEG_P	***	***	***			***			
Sentiment	***		***	***					
EmolexIntensity_fear	***		***		*				
Emolex_surprise								*	
Emolex_anticipation	***	***	*		*				
ANEW_dominance	***	*	***						
Emolex_fear	***	***	***						

Figure 3 reports experiments for various combinations of the language features. The experiments are reported on the best performing model (**S+G+J+P**) from Fig. 2. Augmenting the network with linguistic features improves performance in almost all cases. In case of formality and frustration, the best performance is reported with all features, while for politeness, inclusion of lexical, syntactic, and psycholinguistic features outperforms other combinations.

Analyzing Hidden Representations. In order to quantify the contribution of language features, we analyze their linear dependency using a regression analysis on the hidden representation created by the auto-encoder models. We follow the experimental setup presented by [28] for this analysis. Table 3 shows the significance of these dependencies (i.e. p-value of regression) over the **ENRON–FFP** dataset for a selected set of features. c_f (formality), c_{fr} (frustration), and c_p (politeness) correspond to the CNN-DCNN architecture i.e. no gates, and the extreme right columns (g_f, g_{fr}, g_p) correspond to the proposed GCNN-DCNN-12 architecture. Dependency of language features against the **ENRON-FFP** labels is also reported (p_f, p_{fr}, and p_p). The p-values indicate that the learned latent vector representation captures most syntactic and some lexical (e.g. Average Word Length, NumWords, NumNonAlphaCharacters) features. Part-of-Speech (POS) features are not captured by either the gated or the non-gated model. On the other hand, Named entity related features such as NER_PERSON or NER_ORGANIZATION are well represented by the CNN models, especially the gated version, but are less significant for the affect-prediction task. Note that the psycholinguistic features are not well represented by the deep learning models. This indicates that the information captured by hand-engineered psycholinguistic features indeed complements the signals in the hidden features.

Further, language features (e.g. POS_*, EmolexIntensity_*, PERMA_NEG_* and Sentiment) are in general more dependent on the latent-representation of the non-gated model than its gated counterpart. This could partly explain as to why augmenting language features substantially improves accuracy for the Gated variant ($SGJP$ in Fig. 2a) but not for the non-Gated counterpart (SJP in Fig. 2a).

6   Conclusion

We propose the first neural model, a Gated CNN autoencoder with joint learning for affect prediction: predicting frustration, formality, and politeness in email data. The proposed GCNN-DCNN outperforms the state-of-the-art for text reconstruction on the Hotel review dataset as well as the ENRON email data. Introduction of joint learning, with pre-training and data-relevant language features improves the performance of the model for affect prediction. We show that the introduction of gates leads to an improved performance. Our evaluation shows that subjective tasks such as affect prediction and computational models for frustration can be addressed using semi-supervised neural approaches. The analysis in the Discussion section supports the use of the language features. We plan to extend this work towards non-parallel text generation.

References

1. Burgoon, J.K., Hale, J.L.: The fundamental topoi of relational communication. Commun. Monogr. **51**(3), 193–214 (1984). https://doi.org/10.1080/03637758409390195
2. Chung, J., Gülçehre, Ç., Cho, K., Bengio, Y.: Empirical evaluation of gated recurrent neural networks on sequence modeling. CoRR abs/1412.3555 (2014)
3. Cohen, W.W.: Enron email dataset (2009)
4. Dai, A.M., Le, Q.V.: Semi-supervised sequence learning. In: Advances in Neural Information Processing Systems, pp. 3079–3087 (2015)
5. Danescu-Niculescu-Mizil, C., Sudhof, M., Jurafsky, D., Leskovec, J., Potts, C.: A computational approach to politeness with application to social factors. In: Proceedings of the 51st Annual Meeting of the Association for Computational Linguistics (2013)
6. Dauphin, Y.N., Fan, A., Auli, M., Grangier, D.: Language modeling with gated convolutional networks. CoRR abs/1612.08083 (2016)
7. Dieng, A.B., Wang, C., Gao, J., Paisley, J.W.: TopicRNN: a recurrent neural network with long-range semantic dependency. CoRR abs/1611.01702 (2016)
8. Ghosh, S., Chollet, M., Laksana, E., Morency, L.P., Scherer, S.: Affect-LM: a neural language model for customizable affective text generation. In: Proceedings of the 55th Annual Meeting of the Association for Computational Linguistics (2017)
9. Kalchbrenner, N., Espeholt, L., Simonyan, K., van den Oord, A., Graves, A., Kavukcuoglu, K.: Neural machine translation in linear time. arXiv preprint arXiv:1610.10099 (2016)
10. Ke, Y., Hagiwara, M.: Alleviating overfitting for polysemous words for word representation estimation using lexicons. In: International Joint Conference on Neural Networks (2017)
11. Khosla, S., Chhaya, N., Chawla, K.: Aff2Vec: affect-enriched distributional word representations. In: Proceedings of the 27th International Conference on Computational Linguistics, pp. 2204–2218. Association for Computational Linguistics (2018)
12. Lee, D.H.: Pseudo-label: the simple and efficient semi-supervised learning method for deep neural networks. In: Workshop on Challenges in Representation Learning, ICML (2013)
13. Li, J., Luong, M., Jurafsky, D.: A hierarchical neural autoencoder for paragraphs and documents. CoRR abs/1506.01057 (2015)
14. Lin, C.Y.: Rouge: a package for automatic evaluation of summaries. In: Proceedings of ACL Workshop on Text Summarization Branches Out (2004)
15. Mairesse, F., Walker, M.A.: Trainable generation of big-five personality styles through data-driven parameter estimation. In: ACL, pp. 165–173 (2008)
16. Majumder, N., Poria, S., Gelbukh, A., Cambria, E.: Deep learning-based document modeling for personality detection from text. IEEE Intell. Syst. **32**(2), 74–79 (2017)
17. Mikolov, T., Karafiát, M., Burget, L., Černocký, J., Khudanpur, S.: Recurrent neural network based language model. In: Proceedings of the 11th Annual Conference of the International Speech Communication Association, vol. 2010, pp. 1045–1048 (2010)

18. Van den Oord, A., Kalchbrenner, N., Espeholt, L., Vinyals, O., Graves, A., et al.: Conditional image generation with pixelCNN decoders. In: NIPS (2016)
19. Papineni, K., Roukos, S., Ward, T., Zhu, W.J.: BLEU: a method for automatic evaluation of machine translation. In: Proceedings of the 40th Annual Meeting on Association for Computational Linguistics, ACL 2002, pp. 311–318 (2002)
20. Pavlick, E., Tetreault, J.: An empirical analysis of formality in online communication. Trans. Assoc. Comput. Linguist. **4**, 61–74 (2016)
21. Pennebaker, J.W.: The secret life of pronouns. New Scientist **211**(2828), 42–45 (2011)
22. Picard, R.W.: Affective Computing. MIT Press, Cambridge (1997)
23. Preotiuc-Pietro, D., Liu, Y., Hopkins, D.J., Ungar, L.: Personality driven differences in paraphrase preference. In: Proceedings of the Workshop on Natural Language Processing and Computational Social Science (NLP+CSS). ACL (2017)
24. Radford, A., Metz, L., Chintala, S.: Unsupervised representation learning with deep convolutional generative adversarial networks. arXiv preprint arXiv:1511.06434 (2015)
25. dos Santos, C., Gatti, M.: Deep convolutional neural networks for sentiment analysis of short texts. In: Proceedings of COLING 2014, the 25th International Conference on Computational Linguistics: Technical Papers, pp. 69–78 (2014)
26. Severyn, A., Moschitti, A.: Twitter sentiment analysis with deep convolutional neural networks. In: Proceedings of the 38th International ACM SIGIR Conference on Research and Development in Information Retrieval, SIGIR 2015, pp. 959–962 (2015)
27. Shen, Y., Lin, Z., Huang, C., Courville, A.C.: Neural language modeling by jointly learning syntax and lexicon. CoRR abs/1711.02013 (2017)
28. Subramanian, S., Baldwin, T., Cohn, T.: Content-based popularity prediction of online petitions using a deep regression model. In: Proceedings of the 56th Annual Meeting of the Association for Computational Linguistics, pp. 182–188 (2018)
29. Titov, I., Klementiev, A.: Semi-supervised semantic role labeling: approaching from an unsupervised perspective. In: Proceedings of COLING 2012, pp. 2635–2652 (2012)
30. Yin, W., Kann, K., Yu, M., Schütze, H.: Comparative study of CNN and RNN for natural language processing. CoRR abs/1702.01923 (2017)
31. Zhang, X., Zhao, J., LeCun, Y.: Character-level convolutional networks for text classification. In: Advances in Neural Information Processing Systems, pp. 649–657 (2015)
32. Zhang, Y., Shen, D., Wang, G., Gan, Z., Henao, R., Carin, L.: Deconvolutional paragraph representation learning. CoRR abs/1708.04729 (2017)

Complaint Classification Using Hybrid-Attention GRU Neural Network

Shuyang Wang, Bin Wu[✉], Bai Wang, and Xuesong Tong

Beijing Key Laboratory of Intelligent Telecommunications
Software and Multimedia, Beijing University of Posts and Telecommunications,
Beijing 100876, China
{shalnark, wubin, wangbai, txsllfuji}@bupt.edu.cn

Abstract. Recently, a growing number of customers tend to complain about the services of different enterprises on the Internet to express their dissatisfaction. The correct classification of complaint texts is fairly important for enterprises to improve the efficiency of transaction processing. However, the existing literature lacks research on complaint texts. Most previous approaches of text classification fail to take advantage of the information of specific characters and negative emotions in complaint texts. Besides, some grammatical and semantic errors caused by violent mood swings of customers are another challenge. To address the problems, a novel model based on hybrid-attention GRU neural network (HATT-GRU) is proposed for complaint classification. The model constructs text vectors at character level, and it is able to extract sentiment features in complaint texts. Then a hybrid-attention mechanism is proposed to learn the importance of each character and sentiment feature, so that the model can focus on the features that contribute more to text classification. Finally, experiments are conducted on two complaint datasets from different industries. Experiments show that our model can achieve state-of-the-art results on both Chinese and English datasets compared to several text classification baselines.

Keywords: Text classification · Recurrent neural network · Attention

1 Introduction

Nowadays, there is a large volume of users complaining about services in various industries on the Internet to express dissatisfaction, which consequently generates lots of complaint texts. The classification of complaint texts is helpful for companies to deal with relevant issues promptly and efficiently according to users' feedback, thus having great commercial value.

Text classification is a classic topic in natural language processing (NLP). Recently, deep learning methods have demonstrated powerful ability for text classification. Convolutional neural network (CNN) and recurrent neural network (RNN) are the two most commonly used network structures. CNN uses convolution kernels to extract local, position-independent features [1], while RNN is to model the whole sequence and capture long-term dependencies [2]. Two typical variants of RNN are LSTM [3] and GRU [4].

© Springer Nature Switzerland AG 2019
Q. Yang et al. (Eds.): PAKDD 2019, LNAI 11439, pp. 251–262, 2019.
https://doi.org/10.1007/978-3-030-16148-4_20

Although various texts are well studied such as reviews [5] and questions [6], analysis on complaint texts is rare in the current literature. For the existing text classification methods, there are two difficulties in classifying complaint texts:

- Complaint texts usually contain obvious negative emotions, as few users remain happy when making complaints. Effective use of negative sentiment features can have a positive impact on classification. However, As opposed to sentiment classification, the category of each complaint text does not simply rely on sentiment polarities. Therefore, it is necessary to design an appropriate feature fusion strategy.
- Users are more likely to generate spelling and grammatical errors in complaint texts due to mood fluctuations, and it leads to poor segmentation results in Chinese. Besides, people with highly varying backgrounds express their feelings in distinct ways, which also brings more difficulties to unified modeling.

In response to these challenges, we proposes a novel hybrid-attention GRU neural network (HATT-GRU) for complaint classification. The model constructs text vectors based on character level, which works without any knowledge on the syntactic or semantic structures of a language [7]. Meanwhile, negative sentiment features are extracted to fully utilize the characteristics of complaint texts. Besides, our model assigns attention scores for salient character and sentiment features and aggregates those into a document representation. We summarize our main contributions as follows:

- We propose a method to capture deeper information in complaint texts by extracting both character features and negative sentiment features, and integrate them to improve text representation;
- We introduce a novel neural network model which combines hybrid attention of features, so that the network can focus on features that contribute more to complaint classification;
- The experimental results on two complaint datasets in different languages show that our model outperforms several state-of-the-art baselines. HATT-GRU has been proved to better solve the problem of complaint classification.

2   Related Work

Deep learning was initially applied in computer vision [8] and image analysis [9], and has since spread to NLP tasks. Sentiment analysis is a widely studied one, and sentiment classification on datasets like product reviews and twitter is also common in the existing literature. Some prior studies have incorporated sentiment features into classification models [10, 11]. However, unlike review texts, almost all complaint texts merely have negative emotions, so the classification of complaint texts does not depend on sentiment orientation. Besides, instead of paying attention to the position information or target information of sentiment features, complaint classification attaches more importance to the different contribution of each sentiment feature.

In recent years, nearly most of deep learning methods are based on CNN or RNN. Kim et al. [12] first utilized convolutional neural network for sentence-level classification task. Xu et al. [13] presented cached long short-term memory neural networks for document-level sentiment classification. Yang et al. [14] proposed a hierarchical attention network to better capture the important information of documents. Zhang *et al.* [7] introduced character-level convolutional networks for text classification, which can be directly applied to texts without the need for semantic knowledge of specific languages. Conneau *et al.* [15] proposed very deep convolutional networks using up to 29 convolutional layers. Some researchers have also attempted to combine the advantages of CNN and RNN. Lai *et al.* [16] introduced RCNN, which applied RNN to learn word representation and CNN to get final representation. Zhou *et al.* [17] utilized CNN to model phrases and finally fed them to LSTM to obtain the sentence representation.

In terms of applications, the work of Rios *et al.* [18] showed the great potential of CNN in the classification of biomedical texts. Seo *et al.* [19] proved the good effect of character-level convolutional networks for offensive sentence classification. Shirai *et al.* [20] used the naive bi-directional recurrent neural network to classify Thai complaint texts, but without any adaptive changes to the complaint classification problem.

In fact, research on analyzing complaint texts is still rare in current works. Complaint texts have great commercial value, and companies all need to continuously improve their services from users' feedback. Considering the characteristics of complaint texts and the shortcomings of existing methods in dealing with complaint classification, we propose a complaint classification model based on hybrid-attention GRU neural network. Experiments on two complaint datasets demonstrate the superior performance of our model over compared baselines on both Chinese and English datasets.

3 Our Approach

In this section, we introduce our HATT-GRU model. We first explain the character embedding and sentiment embedding strategy, followed by the introduction of hybrid-attention mechanism. At last, we present the overall network structure of our model.

3.1 Character Embedding

Clients from highly varying backgrounds in general use different words and syntax, and semantic and grammatical errors caused by negative emotions are also common in complaint texts. Traditional methods are prone to problems such as sparse features and unclear semantics, which brings difficulties to text modeling. Therefore, we generate text representation at character level to avoid grammar restrictions, which directly models texts of different languages without additional semantic or syntactic knowledge.

The character embedding module is used to convert the raw characters of a text into encoded character vectors. First, an alphabet of size n is built for the input language. Specifically, an alphabet with 70 characters is constructed for English dataset, which covers all English letters, numbers and punctuation marks that may appear in English texts. For Chinese datasets, the alphabet is created by selecting the top 5000 characters with the highest frequency. Then, each character in the alphabet is mapped into a multi-dimensional and continuous vector, thus obtaining a character vector matrix $\mathbf{E} \in \mathbf{R}^{m \times |V|}$, where m is the vector dimension of each character, and $|V|$ is the size of the alphabet. That means each character is converted to an m-dimensional vector $c_i \in \mathbf{R}^m$, and this encoding is done by Xavier Initialization [21].

Thus, by searching for a character vector corresponding to each character, the original input of the characters in a text is transformed to a sequence of such m-sized vectors with fixed length l:

$$\mathbf{t}_j = \mathbf{c}_{j1} \oplus \mathbf{c}_{j2} \oplus \mathbf{c}_{j3} \oplus \ldots \oplus \mathbf{c}_{jl} \tag{1}$$

where c_{jk} is the vector of the k-th character in the j-th text. Characters that exceed length l are ignored, and for character sequences with length less than l, all-zero vectors are used for padding. In addition, any characters not in the alphabet are also encoded as all-zero vectors. Hence, due to the character embedding method, the complaint classification model we proposed can be directly applied to multiple languages.

3.2 Sentiment Embedding

Complaint texts differ from other texts in many aspects, especially in being emotional. Customers tend to have strong negative emotions while expressing their complaints, and their judgement on relevant personnel will also be subjective. These emotions and feelings can reflect the category information to some extent. However, the information contained in specific sentiment words will be ignored because of the character embedding mechanism. In order to make up for this deficiency and better learn the features of complaint texts, we introduce a sentiment embedding method.

We use Hownet sentiment lexicons, including minus feeling lexicon and minus sentiment lexicon, to identify all negative sentiment words and phrases in Chinese and English respectively. Some of the words in the Hownet sentiment lexicons are not commonly used in complaint texts and, therefore, cannot represent the sentiment characteristics of complaint texts. Hence, top 300 negative sentiment words are selected according to the word frequency to construct the sentiment dictionary. Similar to character embedding, each sentiment word is mapped to an m-dimensional vector $s_i \in \mathbf{R}^m$. Since sentiment distribution takes precedence over sentiment polarities in complaint classification, so sentiment features of a complaint text are incorporated by forming a sequence of sentiment feature vectors of length p for the input text, according to the occurrence frequency of the sentiment words in the text:

$$\mathbf{y}_j = \mathbf{s}_{j1} \oplus \mathbf{s}_{j2} \oplus \mathbf{s}_{j3} \oplus \ldots \oplus \mathbf{s}_{jp} \tag{2}$$

where s_{jk} is the vector of sentiment feature with the k-th frequency in the j-th text. Equally, sentiment features that exceed length p are ignored, and sequences shorter than p are filled with all-zero vectors.

Different from sentiment classification, sentiment features do not play a key role in complaint classification. Therefore, when constructing the input matrix, we make sure that $p < l$ to prevent the influence of sentiment features from being too large. The weight coefficient is used for adjustment as well:

$$\mathbf{x}_j = (\alpha \times \mathbf{t}_j) \oplus (\beta \times \mathbf{y}_j) \tag{3}$$

where α and β are weight coefficients, which are 1.2 and 0.8 respectively in the experiment.

3.3   Attention Mechanism

Not all features contribute equally to the representation of text meaning. Therefore, we utilize attention mechanism to learn a weight matrix for each feature, and assign more attention to the more informative ones. The attention scores of character features and sentiment features are separately calculated and aggregated to form a text vector. Specifically, the characters are fed into a bi-directional GRU:

$$\overrightarrow{\mathbf{h}}_i^c, \overleftarrow{\mathbf{h}}_i^c = bi\_GRU(\mathbf{c}_i), \quad i \in [1, l] \tag{4}$$

where $\overrightarrow{\mathbf{h}}_i^c$ and $\overleftarrow{\mathbf{h}}_i^c$ respectively represent the forward and backward hidden states of the i-th input character vector. The representation of the hidden state is then calculated by a one-layer MLP:

$$\mathbf{e}_i^c = tanh(\mathbf{W}_c[\overrightarrow{\mathbf{h}}_i^c, \overleftarrow{\mathbf{h}}_i^c] + \mathbf{b}_c), \quad i \in [1, l] \tag{5}$$

Next, calculate the similarity of $\mathbf{e}_i^c$ and a character-level context vector $\mathbf{u}_c$ to evaluate the importance of the character, and get the character attention distribution through a softmax function:

$$\alpha_i^c = \frac{exp(\mathbf{e}_i^{cT}\mathbf{u}_c)}{\sum_{k=1}^{l} exp(\mathbf{e}_k^{cT}\mathbf{u}_c)} \tag{6}$$

where $\mathbf{W}_c, \mathbf{b}_c$ and $\mathbf{u}_c$ are learnable parameters. Based on the attention scores, the final representation of character features is calculated by a weight pooling:

$$\mathbf{v}_c = \sum_{i=1}^{l} [\overrightarrow{\mathbf{h}}_i^c, \overleftarrow{\mathbf{h}}_i^c] \alpha_i^c \tag{7}$$

The representation of sentiment features is formed likewise:

$$\mathbf{e}_i^s = tanh(\mathbf{W}_s[\overrightarrow{\mathbf{h}}_i^s, \overleftarrow{\mathbf{h}}_i^s] + \mathbf{b}_s), \quad i \in [1, p] \tag{8}$$

$$\alpha_i^s = \frac{exp(\mathbf{e}_i^{sT}\mathbf{u}_s)}{\sum_{k=1}^{p} exp(\mathbf{e}_k^{sT}\mathbf{u}_s)} \tag{9}$$

$$\mathbf{v}_s = \sum_{i=1}^{p} [\overrightarrow{\mathbf{h}}_i^s, \overleftarrow{\mathbf{h}}_i^s] \alpha_i^s \tag{10}$$

Finally, a text vector containing all the information is obtained by concatenating the character vector and the sentiment vector:

$$\mathbf{v} = \mathbf{v}_c \oplus \mathbf{v}_s \tag{11}$$

Each feature provides different information for the final prediction. Through the attention mechanism, the model can focus on the most salient features, thus further improving the text representation.

3.4 Network Structure

In this section, we introduce the network structure of our hybrid-attention GRU neural network. The overall framework is in Fig. 1.

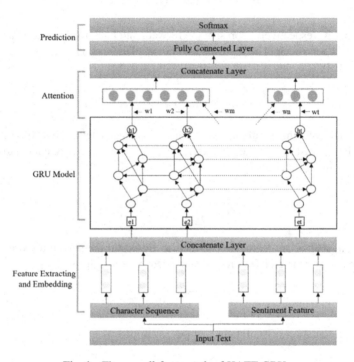

Fig. 1. The overall framework of HATT-GRU

Feature Extraction and Embedding. We model the characteristics of complaint texts from two aspects, namely character sequence features and sentiment features. The two kind of features are converted into multi-dimensional and continuous vectors, which are concatenated and serve as input to the bi-directional GRU neural network.

GRU Neural Network. In contrast to other deep learning networks, bi-directional recurrent neural networks can fully capture sequential information with varying lengths and learn the associations between features. Among them, GRU achieves promising results in multiple tasks with less computation and performance decay [2], which is selected as the basic recurrent unit in the proposed model. Besides, we use two hidden layers to learn deeper information and output the hidden state step by step.

Attention. The attention module assigns attention scores for character representation and sentiment representation, and aggregates all the embedding vectors to get the final text representation. The attention mechanism allows the model to focus on the most salient features.

Prediction. The prediction module completes the prediction task according to the final text representation. It consists of a fully connected layer and a softmax layer. The fully connected layer maps the representation vector into the category space of the sample, followed by a dropout [22] layer to avoid overfitting. The activation function used for non-linearity is rectified linear units (ReLUs). Ultimately, the classification result is determined through the softmax layer.

We uses Adam algorithm [23] to perform optimization through backpropagation, and cross entropy is selected as the loss function:

$$loss = -\sum_{c=1}^{M} q_{oc} log(p_{oc}) \tag{12}$$

where M is the number of categories, q_{oc} is a binary indicator with a value of 0 or 1, indicating whether category c is the correct label for text o. And p_{oc} is the predicted probability that text o belongs to category c. The implementation is done using tensorflow. The hyperparameter settings of HATT-GRU are shown in Table 1.

Table 1. Hyperparameters of our model

Hyperparameter	Values
Embedding dimension	64
Number of hidden layers	2
Hidden layer dimension	128
Learning rate	1e−3
Dropout rate	0.5
Mini-batch size	32
Epoch size	10

4 Experiments

4.1 Evaluation

For the whole dataset, we choose accuracy to evaluate the performance of each model. And for each class, precision, recall and F1-score are used. F1-score is the harmonic average of the precision rate and the recall rate. It takes into account the precision and recall rate of the model. The calculation formula is:

$$F1 - score = \frac{2 \times Precision \times Recall}{Precision + Recall} \tag{13}$$

4.2   Datasets

Our model is evaluated on two complaint datasets from different industries, one in Chinese and the other in English. The details of each dataset are described as below:

- **Express Complaint Dataset.** This is a collection of consumer complaints about express delivery issues and is a Chinese dataset. It consists of 8 categories and 73,458 Chinese samples, but some categories contain only a few hundred texts. Hence, we select the top 3 classes with the largest number of samples for experiments, including package loss, package delay and delivery service.
- **Finance Complaint Dataset.** This dataset is a collection of consumer complaints about financial issues, and is an English dataset. The complete dataset has 11 classes and 555,958 samples, of which only 66,806 have complaint narratives. For the experiment, similarly, 3 largest classes are chosen, namely credit reporting, debt collection and mortgage. This dataset can be obtained from Kaggle[1].

For the two datasets, we both use 30,000 samples for training, 3,000 samples for validation and 6,000 samples for testing.

4.3   Comparison Models

A series of experiments has been conducted to prove the good performance of our model. Brief descriptions of the classification models are listed as follows:

- **Decision Tree.** A popular machine learning classification algorithm and we use TF-IDF method to generate text vectors.
- **SVM.** TF-IDF is also used as textual features for classification.
- **Bi-LSTM.** A basic bi-directional LSTM network [3] with one hidden layer;
- **Bi-GRU.** A basic bi-directional GRU network with one hidden layer based on [4];
- **Word-level CNN.** A convolutional neural network [12] is used to extract features based on embedding word vectors.
- **Char-level CNN.** A character-level convolutional neural network proposed in [7]. Similarly, we use Pinyin to replace Chinese characters in complaint texts as well.
- **RCNN.** The recurrent convolutional neural network proposed in [16], which uses bi-directional LSTM to capture context information and then introduces a pooling layer to extract key features.
- **HAN.** Based on the hierarchical attention network proposed in [13]. It adopts the bi-directional GRU model and applies the attention mechanism to the word level and sentence level respectively.
- **FastText.** An effective and fast method for generating word vectors for classification in [24]. It has 10 hidden units and is evaluated with bigram.
- **CATT-GRU.** We introduce a comparative experiment based on character-attention GRU neural network to verify the effect of sentiment features.
- **HATT-GRU.** The hybrid-attention GRU neural network proposed in this paper.

---

[1] https://www.kaggle.com/dushyantv/consumer_complaints.

The above models are all based on words in addition to Char-level CNN, CATT-RNN and HATT-GRU. For Chinese datasets, we use *jieba* for word segmentation.

4.4   Results and Discussion

We test the performance of each classification models on two complaint datasets. Table 2 shows the overall classification accuracy of different models on each dataset.

Table 2. The total accuracy rate of different models on each dataset

Model	Accuracy	
	Express	Finance
Decision tree	71.10	80.96
SVM	71.38	86.43
Bi-LSTM	84.23	87.12
Bi-GRU	84.65	87.58
Word-level CNN	82.69	87.85
Char-level CNN	85.85	89.28
RCNN	85.78	88.78
HAN	86.92	90.90
FastText	85.17	88.93
CATT-GRU	86.50	91.18
HATT-GRU	**87.68**	**91.80**

As shown in Table 2, HATT-GRU model has improved the accuracy of complaint classification greatly on both express complaint dataset and finance complaint dataset. The results demonstrate the superiority of the proposed architecture.

Two traditional methods, decision tree and support vector machine, do not work well on the two datasets, especially on the Chinese complaint dataset. Complaint data is a kind of user-generated data which varies in the degree of how users of different backgrounds organize their narratives. It can be concluded that deep learning methods generally perform better on extracting the important features than traditional ones.

Remarkably, HATT-GRU outperforms HAN and CATT-GRU. It also confirms the effectiveness of character embedding and integrating sentiment features to the network. And HATT-GRU is superior to models without attention, which shows that hybrid attention mechanism can indeed help to improve complaint classification results.

In addition, the results of English dataset is better than that of Chinese dataset. One the one hand, there's no need of word segmentation for English texts, which avoids the negative impact of poor segmentation results. On the other hand, Chinese is more diverse in terms of colloquial expression, and this adds to difficulty of feature extraction. However, our HATT-GRU has still achieved the best results in both Chinese and English complaint datasets.

To further understand the results, we present details of precision, recall and F1-score on each class of the two complaint datasets. Tables 3 and 4 illustrate that our

Table 3. Details for each class in the Chinese dataset compared with other models. "P" stands for precision, "R" stands for recall and "F" stands for F1-Score.

Model	Package loss			Package delay			Delivery service		
	P	R	F	P	R	F	P	R	F
Decision tree	77.6	75.9	76.7	76.2	70.9	73.4	60.9	66.5	63.5
SVM	78.5	77.0	77.7	71.9	70.4	71.1	64.2	66.7	65.4
Bi-LSTM	87.7	91.2	89.4	84.0	77.3	80.5	80.9	83.1	82.0
Bi-GRU	87.2	91.6	89.3	82.4	79.9	81.1	83.0	81.3	82.1
Word-level CNN	90.0	88.4	89.2	77.8	80.2	79.0	80.5	78.3	79.4
Char-level CNN	90.5	90.7	90.6	84.8	80.6	82.7	82.4	86.2	84.2
RCNN	90.0	91.0	90.5	85.6	79.1	82.2	82.0	87.2	84.6
HAN	88.5	**93.1**	90.8	**90.6**	79.2	84.5	82.3	88.3	85.2
FastText	88.0	91.9	89.9	84.9	80.8	82.8	**83.8**	81.7	82.7
CATT-GRU	90.1	91.5	90.8	86.7	80.4	83.4	82.8	87.5	85.1
HATT-GRU	**90.3**	92.5	**91.4**	**90.6**	**82.0**	**86.1**	82.7	**88.5**	**85.5**

Table 4. Details for each class in the English dataset compared with other models. "P" stands for precision, "R" stands for recall and "F" stands for F1-Score.

Model	Credit reporting			Debt collection			Mortgage		
	P	R	F	P	R	F	P	R	F
Decision tree	76.5	79.5	77.9	78.6	76.6	77.5	88.0	86.7	87.3
SVM	79.9	88.2	84.0	87.6	80.2	83.7	92.7	90.8	91.7
Bi-LSTM	85.9	85.6	85.8	84.4	84.7	84.6	90.9	90.9	90.9
Bi-GRU	90.5	82.0	86.0	82.6	88.2	85.8	90.2	92.1	91.5
Word-level CNN	87.1	85.6	86.4	85.9	85.5	85.7	90.4	92.3	91.4
Char-level CNN	89.8	86.2	88.0	85.4	89.1	87.2	92.8	92.5	92.7
RCNN	88.9	86.5	87.7	87.4	86.6	87.0	90.0	93.2	91.6
HAN	89.6	90.0	89.8	90.3	87.3	88.8	92.6	95.3	93.9
FastText	88.6	87.7	88.2	87.6	85.9	86.7	90.5	93.1	91.8
CATT-GRU	**90.7**	89.3	90.0	89.2	**89.3**	89.3	93.4	94.9	94.2
HATT-GRU	89.0	**92.6**	**90.7**	**92.3**	87.1	**89.7**	**94.3**	**95.7**	**95.0**

model achieves competitive results on every class. For both Chinese and English datasets, our HATT-GRU works well in classifying complaint texts.

To visualize the effectiveness of the proposed model, we show the relative errors of F1-score with regard to each comparison model in Fig. 2. Each column is computed by taking the difference of F1-score between the comparison model and our HATT-GRU. F1-score is an indicator which comprehensively takes into account precision and recall. It can be concluded that our model reaches the best result in every class of both Chinese and English complaint dataset.

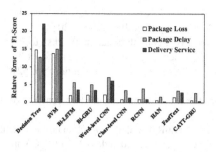 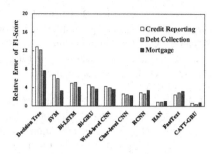

(a) comparison results of express dataset   (b) comparison results of finance dataset

Fig. 2. Relative error of F1-Score of each model compared with HATT-GRU

5   Conclusion

In this paper, we propose a hybrid-attention GRU neural network for complaint classification. We generate representation vectors at character level, thus avoiding the negative impact of semantic and syntactic errors. Meanwhile, negative sentiment features are integrated to further improve text representation. Hybrid attention is then incorporated into a bi-directional GRU network to effectively extract the most salient features in complaint texts. Experiments show that our model outperforms several state-of-the-art baselines on both Chinese and English complaint datasets.

For future work, we want to explore other more powerful network structure to encode complaint texts, and extend our model to a wider range of texts in different fields.

Acknowledgments. This work is supported by the National Key R&D Program of China (No. 2018YFC0831500) and the National Social Science Foundation of China under Grant 16ZDA055.

References

1. Collobert, R., Weston, J., Bottou, L., Karlen, M., Kavukcuoglu, K., Kuksa, P.: Natural language processing (almost) from scratch. J. Mach. Learn. Res. **12**(8), 2493–2537 (2011)
2. Chung, J., Gulcehre, C., Cho, K., Bengio, Y.: Empirical evaluation of gated recurrent neural networks on sequence modeling. arXiv preprint arXiv:1412.3555 (2014)
3. Cho, K., et al.: Learning phrase representations using RNN encoder-decoder for statistical machine translation. arXiv preprint arXiv:1406.1078 (2014)
4. Tang, D., Qin, B., Liu, T.: Document modeling with gated recurrent neural network for sentiment classification. In: Proceedings of the 2015 Conference on Empirical Methods in Natural Language Processing, pp. 1422–1432 (2015)
5. Wang, M., Chen, S., He, L.: Sentiment classification using neural networks with sentiment centroids. In: Pacific-Asia Conference on Knowledge Discovery and Data Mining, pp. 56–67 (2018)

6. Xia, W., Zhu, W., Liao, B., Chen, M., Cai, L., Huang, L.: Novel architecture for long short-term memory used in question classification. Neurocomputing **299**, 20–31 (2018)
7. Zhang, X., Zhao, J., LeCun, Y.: Character-level convolutional networks for text classification. In: Advances in Neural Information Processing Systems, pp. 649–657 (2015)
8. Yang, J., et al.: IEEE Conference on Computer Vision and Pattern Recognition, pp. 5216–5225 (2017)
9. Krizhevsky, A., Sutskever, I., Hinton, G.E.: ImageNet classification with deep convolutional neural networks. In: NIPS, pp. 1097–1105 (2012)
10. Wang, L., Niu, J., Song, H., Atiquzzaman, M.: SentiRelated: a cross-domain sentiment classification algorithm for short texts through sentiment related index. J. Netw. Comput. Appl. **101**, 111–119 (2018)
11. Yang, M., Qu, Q., Chen, X., Guo, C., Shen, Y., Lei, K.: Feature-enhanced attention network for target-dependent sentiment classification. Neurocomputing **307**, 91–97 (2018)
12. Kim, Y.: Convolutional neural networks for sentence classification. arXiv preprint arXiv: 1408.5882 (2014)
13. Xu, J., Chen, D., Qiu, X., Huang, X.: Cached long short-term memory neural networks for document-level sentiment classification. arXiv preprint arXiv:1610.04989 (2016)
14. Yang, Z., Yang, D., Dyer, C., He, X., Smola, A., Hovy, E.: Hierarchical attention networks for document classification. In: NAACL-HLT, pp. 1480–1489 (2016)
15. Conneau, A., Schwenk, H., Barrault, L., Lecun, Y.: Very deep convolutional networks for text classification. arXiv preprint arXiv:1606.01781 (2016)
16. Lai, S., Xu, L., Liu, K., Zhao, J.: Recurrent convolutional neural networks for text classification. In: AAAI, pp. 2267–2273 (2015)
17. Zhou, C., Sun, C., Liu, Z., Lau, F.: A C-LSTM neural network for text classification. arXiv preprint arXiv:1511.08630 (2015)
18. Rios, A., Kavuluru, R.: Convolutional neural networks for biomedical text classification: application in indexing biomedical articles. In: Proceedings of the 6th ACM Conference on Bioinformatics, Computational Biology and Health Informatics, pp. 258–267 (2015)
19. Seo, S., Cho, S.B.: Offensive sentence classification using character-level CNN and transfer learning with fake sentences. In: Liu, D., Xie, S., Li, Y., Zhao, D., El-Alfy, E.S. (eds.) ICONIP 2017, vol. 10635, pp. 532–539. LNCS. Springer, Cham (2017). https://doi.org/10.1007/978-3-319-70096-0_55
20. Shirai, K., Sornlertlamvanich, V., Marukata, S.: Recurrent neural network with word embedding for complaint classification. In: Proceedings of the Third International Workshop on Worldwide Language Service Infrastructure and Second Workshop on Open Infrastructures and Analysis Frameworks for Human Language Technologies, pp. 36–43 (2016)
21. Glorot, X., Bengio, Y.: Understanding the difficulty of training deep feedforward neural networks. In: Proceedings of the Thirteenth International Conference on Artificial Intelligence and Statistics, pp. 249–256 (2010)
22. Srivastava, N., Hinton, G., Krizhevsky, A., Sutskever, I., Salakhutdinov, R.: Dropout: a simple way to prevent neural networks from overfitting. J. Mach. Learn. Res. **15**(1), 1929–1958 (2014)
23. Kinga, D., Adam, J.B.: A method for stochastic optimization. In: International Conference on Learning Representations (2015)
24. Joulin, A., Grave, E., Bojanowski, P., Mikolov, T.: Bag of tricks for efficient text classification. arXiv preprint arXiv:1607.01759 (2016)

Spatio-Temporal and Stream Data Mining

FGST: Fine-Grained Spatial-Temporal Based Regression for Stationless Bike Traffic Prediction

Hao Chen[1], Senzhang Wang[2], Zengde Deng[3], Xiaoming Zhang[1],
and Zhoujun Li[1(✉)]

[1] Beihang University, Beijing, China
{chh,yolixs,lizj}@buaa.edu.cn
[2] Nanjing University of Aeronautics and Astronautics, Nanjing, China
szwang@nuaa.edu.cn
[3] The Chinese University of Hong Kong, Hong Kong, China
zddeng@se.cuhk.edu.hk

Abstract. Currently, fully stationless bike sharing systems, such as Mobike and Ofo are becoming increasingly popular in both China and some big cities in the world. Different from traditional bike sharing systems that have to build a set of bike stations at different locations of a city and each station is associated with a fixed number of bike docks, there are no stations in stationless bike sharing systems. Thus users can flexibly check-out/return the bikes at arbitrary locations. Such a brand new bike-sharing mode better meets people's short travel demand, but also poses new challenges for performing effective system management due to the extremely unbalanced bike usage demand in different areas and time intervals. Therefore, it is crucial to accurately predict the future bike traffic for helping the service provider rebalance the bikes timely. In this paper, we propose a Fine-Grained Spatial-Temporal based regression model named FGST to predict the future bike traffic in a stationless bike sharing system. We motivate the method via discovering the spatial-temporal correlation and the localized conservative rules of the bike check-out and check-in patterns. Our model also makes use of external factors like Point-Of-Interest(POI) informations to improve the prediction. Extensive experiments on a large Mobike trip dataset demonstrate that our approach outperforms baseline methods by a significant margin.

Keywords: Traffic prediction · Spatial-temporal data · Sharing-bikes

1 Introduction

Bike sharing systems provide an eco-friendly solution for *the last-mile problem*, which refers to the gap from the transportation hub to the final destination [13]. Traditional bike sharing systems, such as *Citi Bike* in New York and *Ubike* in

© Springer Nature Switzerland AG 2019
Q. Yang et al. (Eds.): PAKDD 2019, LNAI 11439, pp. 265–279, 2019.
https://doi.org/10.1007/978-3-030-16148-4_21

Taipei, require users to return the bikes to vacant docks of the bike stations after riding. Currently, stationless bike sharing systems such as Mobike and Ofo are becoming increasingly popular in China as well as some big cities in the world, enables users to check out and return the bikes at arbitrary locations, and thus better meets people's travel demand. But on the other side of the coin, it also poses new challenges for efficient system management. As there are no bike stations, the number of idle bikes at hot spots like subway stations can increase sharply, causing a severely unbalanced bike distribution problem. This will lead to negative effects on system resource usage and user experience. Thus, it is crucially important to predict the future bike traffic flow accurately for helping the service provider make a reasonable system management strategy in advance.

Existing works on traffic prediction can be broadly divided into three classes: time-series learning based methods, statistical learning based methods, and deep learning based methods. Vogel et al. [11,12] and Yoon et al. [21] adopted time series analysis methods to predict the bike demand and the available bike supply for each bike station. Li et al. [3] and Liu et al. [7] used statistical learning methods and considered the external factors like weather to predict the future check-in/check-out numbers in each bike station. Recently, there are also some works using deep models for urban traffic prediction. For example, Ma et al. [8], Zhang et al. [22,23] and Yao et al. [19,20] considered city-wide traffic as heat-map images, where the value of each pixel represented the traffic volume in the corresponding region. However, the assumption of CNN and RNN may not perfectly model the complex spatial-temporal correlations in traffic prediction problems. In addition, different from previous traffic prediction tasks which mainly focus on predicting one single traffic volume value, in our study, the bike check-outs and check-ins are highly correlated and follow the conservative rules. This also makes the previous deep models challenging to be applied.

Compared with traditional bike sharing systems, forecasting the future bike traffic in fully station-less bike sharing systems faces the following two major challenges. First, the destination of a bike trip is more uncertain. In station-based bike sharing systems, users must return their bikes to a bike station with vacant docks. While in station-less bike sharing systems, a user can return the bike just at or very close to the final destination which is more flexible and reflects the real travel demand of the user. Second, the spatial correlation of the bike traffic in station-less bike sharing systems is more complex and harder to capture. For station-based bikes, the spatial correlations among neighbor bike stations are usually high [3]. However, this may not be the case in the station-less bike sharing systems. The bike traffic patterns of two neighbor regions may be quite different due to their different functions reflected by their different Point-Of-Interest(POI) distributions.

To address the challenges mentioned above, in this paper, we propose a Fine-Grained Spatial-Temporal based regression model named FGST for traffic prediction in stationless bike sharing systems. First, we build up the basic fine-grained regression model. For each region and each time slot, the regression model aims to learn two projection vectors to project the feature vector to the corresponding check-in/check-out bike numbers, respectively. Based on the basic model, we further propose to incorporate the spatial-temporal correlations into the model by solving a joint optimization problem. The idea is that, for the regions in some time slots which present similar check-in/check-out patterns, their corresponding projection vectors should be similar. We also propose to cluster the regions into localized groups based on the traffic flows among them. By constraining the check-in bike number should be close to the check-out bike number in each cluster, the conservation rules are also integrated into our model. Finally, we propose a regression model to infer the projection vector of the next time slot, which is used to predict the future traffic.[1] Our main contributions can be summarized as follows:

- To the best of our knowledge, this is the first work that studies the traffic prediction problem in fully station-less bike sharing systems, which is a more challenging problem compared with it in traditional bike sharing systems.
- We propose a fine-grained approach to extract and utilize the spatial-temporal correlations and the conservation rules while considering the region-specific features.
- The proposed model is evaluated on a large public dataset, and the results demonstrate its superior performance compared with baseline models.

2   Problem Definition and Framework

2.1   Problem Definition

In this paper, we use calligraphic uppercase characters for high-order tensors (e.g. $\mathcal{A}$), bold uppercase characters for matrices (e.g. $\mathbf{A}$), bold lowercase characters for vectors (e.g. $\boldsymbol{a}$), and lowercase characters for scalars (e.g. a). The i-th row of matrix $\mathbf{A}$ is represented as $\boldsymbol{a}_{i,:}$, the j-th column of $\mathbf{A}$ as $\boldsymbol{a}_{:,j}$, and the (i,j)-th entry of $\mathbf{A}$ as $a_{i,j}$.

Definition 1 *Regions.* We divide a city into a set of equal sized grid regions as $R = \{r_1, r_2, \ldots, r_N\}$ based on the latitude and longitude of the regions, where N denotes the total number of the regions.

Definition 2 *Feature Tensor.* Given regions R and time slots $\{t_1, \ldots t_K\}$, we use a 3-dimensional tensor $\mathcal{X} \in \mathbb{R}^{N \times K \times M}$ to denote the features of all the regions in all the time slots. M denotes the dimension of the feature vector. $\boldsymbol{x}_{n,k}$ denotes the feature vector of region r_n in time slot t_k. $\boldsymbol{x}_{n,:}$ denotes all the feature vectors of region r_n and $\boldsymbol{x}_{:,k}$ denotes all the feature vectors in time slot t_k. Detailed information is given in Sect. 4.1.

---

[1] The data and code of this work is publicly available at https://github.com/coderhaohao/FGST.

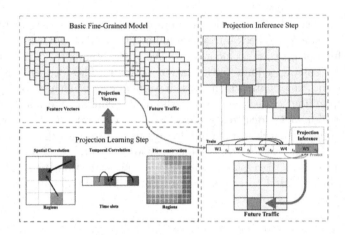

Fig. 1. Framework of FGST

Definition 3 *Bike Traffic Tensor.* We use a 3-dimensional tensor $\mathcal{Y} \in \mathbb{R}^{N \times K \times 2}$ to denote the bike traffic data. $y_{n,k}^{in}$ denotes the check-in number of region r_n at time slot t_k, and $y_{n,k}^{out}$ denotes the check-out number of region r_n at time slot t_k.

Definition 4 *Projection Tensor.* We use two 3-dimensional tensors $\mathcal{W}^{in}, \mathcal{W}^{out} \in \mathbb{R}^{N \times K \times M}$ to denote the projection tensors that project the features to the prediction of the future bike check-out and check-in respectively. $\boldsymbol{w}_{n,k}^{in}$ and $\boldsymbol{w}_{n,k}^{out}$ denote the projection vectors in region r_n and time slot t_k. For simplicity, we will use $\mathcal{W}$ if there is no confusion, and have $\mathbf{W}_{:,k} = [\boldsymbol{w}_{1,k}, \dots, \boldsymbol{w}_{n,k}] \in \mathbb{R}^{M \times N}$, $\mathbf{W}_{n,:} = [\boldsymbol{w}_{n,1}, \dots, \boldsymbol{w}_{n,K}] \in \mathbb{R}^{M \times K}$.

Problem Definition. Given the feature tensor of the previous K time slots, and the history bike traffic tensor $\mathcal{Y}$, our goal is to predict the bike check-in and check-out numbers $y_{n,k+1}^{in}$ and $y_{n,k+1}^{out}$ for all the regions in the next time slot $K + 1$.

2.2 Framework

Figure 1 presents the framework of the proposed approach, which contains three major steps. We first build up the ***basic multi-model framework***. For each region and each time slot, there will be two projection vectors that project the feature vector to the corresponding check-in/check-out bike numbers, respectively. Then we introduce the ***projection learning step*** to learn the projection vectors by solving an optimization problem while preserving the spatial-temporal correlations and the flow conservation rules. Finally, a ***projection inference step*** is introduced to predict the future traffic by inferring the projection vector of the next time slot.

To capture the spatial-temporal correlations, we make the following three assumptions. First, in the spatial view, if two regions present similar bike traffic patterns and have similar POI features, their projection vectors should be similar. In the temporal view, a region in the same time slot of different weekdays or weekends should also have similar projection vectors due to the periodic bike usage patterns. Second, the projection vector should follow the global conservation rules, which means the total number of check-out bikes should be the same as the number of check-in bikes. As showed in Fig. 1, we cluster the regions based on the bike traffic flows among them, and relax the constraint to let the bike trips in each region cluster follow the localized conservation rules. Third, the future projection vectors of a region in a future time slot can be inferred from the history projection vectors of this region.

3   Methodology

In this section, we start with the basic fine-grained regression model. Then we introduce how to incorporate the spatial-temporal correlation component and the flow conservation constraints. Finally, we introduce the inference method to predict the future bike traffic flow for each region.

3.1   The Basic Model

We introduce the basic optimization function to minimize the projection loss for the projection vectors of each region and each time slot. Given the observed feature vectors x and the corresponding bike check-in/out number y^{in} and y^{out}, we first propose to solve the following optimization problem to learn the projection vectors w^{in} and w^{out} for each region and each time slot,

$$\min_{\mathbf{W}} \sum_{k=1}^{K} \sum_{n=1}^{N} (\mathcal{L}_{n,k}^{in} + \mathcal{L}_{n,k}^{out} + \theta(\|w_{n,k}^{in}\|^2 + \|w_{n,k}^{out}\|^2)) \tag{1}$$

where $\mathcal{L}_{n,k}^{in}$ and $\mathcal{L}_{n,k}^{out}$ denote the projection errors which are $(\mathbf{X}_{i,k} w_{i,k}^{in} - y_{n,k}^{in})^2$ and $(\mathbf{X}_{i,k} w_{i,k}^{out} - y_{n,k}^{out})^2$, $\ell 2$-norm is imported to avoid over-fitting, controlled by a non-negative parameter θ.

3.2   Spatial Correlation

As described in the assumption, we first calculate the spatial similarity d_{ij} of the region r_i and r_j through the *Euclidean distance* of the historical bike traffic flow data and POI distributions of the two regions. Then we capture the spatial correlation of two regions r_i and r_j at time slot t_k by minimizing the following formula:

$$\sum_{i=1}^{N} \sum_{j=1}^{N} g(d_{ij}) \| w_{i,k} - w_{j,k} \|_2^2 \tag{2}$$

where $\| \cdot \|_2^2$ denotes the 2-norm distance. $g(d_{ij})$ is a non-increase function of d_{ij}. When the distance d_{ij} is small, $g(d_{ij})$ is large to make $\boldsymbol{w}_{i,k}$ and $\boldsymbol{w}_{j,k}$ more similar. We empirically set $g(d_{ij}) = d_{ij}^{-h}$, where h is set as 1. By using Laplacian matrix, the spatial correlation over all the time slots can be written as the matrix form as follows. For simplicity, we also employ an abbreviation by the following derivation.

$$\sum_{k=1}^{K} (\sum_{i=1}^{N} \sum_{j=1}^{N} g(d_{ij}) \| \boldsymbol{w}_{i,k} - \boldsymbol{w}_{j,k} \|^2) = \sum_{k=1}^{K} 2tr(\mathbf{W}_{:,k} \mathbf{L}^{sp} \mathbf{W}_{:,k}^T) \triangleq \sum_{k=1}^{K} 2tr(\mathbf{S}_k) \quad (3)$$

where $\mathbf{D}^{sp}, \mathbf{Z}^{sp}, \mathbf{L}^{sp} \in \mathbb{R}^{N \times N}$. $\mathbf{Z}^{sp}$ denotes the spatial correlation matrix with each entry $z_{i,j}^{sp} = g(d_{ij})$. $\mathbf{D}^{sp}$ is a diagonal matrix with diagonal entry $d_{i,i}^{sp} = \sum_j z_{i,j}^{sp}$. The matrix $\mathbf{L}^{sp} = \mathbf{D}^{sp} - \mathbf{Z}^{sp}$ is the Laplacian matrix of the spatial correlation matrix $\mathbf{Z}^{sp}$. $\mathbf{S}_k^{in}$ denotes $\mathbf{W}_{:,k}^{in} \mathbf{L}^{sp} \mathbf{W}_{:,k}^{in T}$, and $\mathbf{S}_k^{out}$ denotes $\mathbf{W}_{:,k}^{out} \mathbf{L}^{sp} \mathbf{W}_{:,k}^{out T}$ for brevity.

3.3 Temporal Correlation

The bike traffic data presents high temporal correlations such as periodicity [19]. That means in the same time slot of different days, the traffic data tend to be similar. Assume that d denotes the day, h denotes the hour of the day and Δd denotes the number of days between two time slots. For the same hour of two different days t_{d*24+h} and $t_{(d+\Delta d)*24+h}$ in region r_n, we propose to minimize the following term to capture the periodical temporal correlation in region r_n,

$$f(\Delta d) \| \boldsymbol{w}_{n,d*24+h} - \boldsymbol{w}_{n,(d+\Delta d)*24+h} \|_2^2 \quad (4)$$

where $f(\Delta d)$ is a decay function on Δd. When Δd is small, $f(\Delta d)$ is large to make $\boldsymbol{w}_n^{(d+\Delta d)*24+h}$ and $\boldsymbol{w}_n^{d*24+h}$ similar. $f(\Delta d)$ tends to be zero while Δd keeps increasing. For simplicity, we define $f(\Delta d)$ as follows: $f(\Delta d) = 1$ if $\Delta d = 1$; $f(\Delta d) = 0$, otherwise. With this definition, only the same hour of two consecutive days will be taken into consideration. Then the temporal correlation component of region r_n can be rewritten as follows,

$$\sum_{h=1}^{24} \sum_{d=1}^{D-1} \| \boldsymbol{w}_{n,d*24+h} - \boldsymbol{w}_{n,(d+1)*24+h} \|_2^2 \quad (5)$$

where D denotes the total number of days. The temporal correlation over all the regions can be rewritten as the matrix form with the help of Laplacian matrix and abbreviated as follows,

$$\sum_{n=1}^{N} 2tr(\mathbf{W}_{n,:} \mathbf{L}^{sp} \mathbf{W}_{n,:}^T) \triangleq \sum_{n=1}^{N} 2tr(\mathbf{T}_n) \quad (6)$$

where $\mathbf{D}^{te}, \mathbf{Z}^{te}, \mathbf{L}^{te} \in \mathbb{R}^{K \times K}$. We use $\mathbf{Z}^{te}$ to denote the temporal correlation matrix and $\mathbf{Z}_{(d+1)*24+h,d*24+h}^{te} = 1$, $\mathbf{Z}_{d*24+h,(d+1)*24+h}^{te} = 1$ for $d = 1, \cdots, D-1$.

$\mathbf{D}^{te}$ is a diagonal matrix with diagonal entries $\mathbf{D}_{i,i}^{te} = \sum_j \mathbf{Z}_{i,j}^{te}$. The matrix $\mathbf{L}^{te} = \mathbf{D}^{te} - \mathbf{Z}^{te}$ is the Laplacian matrix of the temporal correlation matrix $\mathbf{Z}_{te}$. $\mathbf{T}_n^{in}$ denotes $\mathbf{W}_{n,:}^{in} \mathbf{L}^{te} {\mathbf{W}_{n,:}^{in}}^T$, and $\mathbf{T}_n^{out}$ denotes $\mathbf{W}_{n,:}^{out} \mathbf{L}^{te} {\mathbf{W}_{n,:}^{out}}^T$ for brevity.

3.4   Flow Conservation Constraint

Sharing bikes are usually used for short-trip transportation, so there will exist some region clusters that the inter-cluster flow is small while the intra-cluster flow is large. Thus the bike usage in each region cluster should be balanced which means the total check-out bike number should be close to the total check-in number. To utilize this constraint, we cluster the grid regions into several groups whose inter-group bike flows will be of a relatively low proportion and can be neglected compared with the large volume of intra-group bike flows. Thus, we introduce an agglomerative hierarchical clustering method, and the similarity between two region clusters is defined as follows,

$$ sim\,(c_i, c_j) = \frac{\mathbf{F}_{c_i->c_j}}{\mathbf{F}_{c_i->\bullet}} + \frac{\mathbf{F}_{c_i->c_j}}{\mathbf{F}_{\bullet->c_j}} + \frac{\mathbf{F}_{c_j->c_i}}{\mathbf{F}_{c_j->\bullet}} + \frac{\mathbf{F}_{c_j->c_i}}{\mathbf{F}_{\bullet->c_i}} \tag{7}$$

where c_i and c_j represent two region clusters, and $\mathbf{F}_{c_i->c_j}$ denotes the number of the trips starting from the regions in c_i and ending at the regions in c_j. $\bullet$ denotes all the regions, and $\mathbf{F}_{c_i->\bullet}$ is the number of the trips starting from the regions in c_i and ending at any regions. $\mathbf{F}_{\bullet->c_i}$ denotes the number of the trips starting from any regions and ending at the regions in c_i.

Then for each given cluster C_o in time slot t_k, we minimize the summation of the absolute unbalance value between the inflow and outflow in cluster C_o at time slot t_k.

$$ |\sum_{i \in C_o} \mathbf{X}_{i,k}(\boldsymbol{w}_{i,k}^{in} - \boldsymbol{w}_{i,k}^{out})| \tag{8}$$

We sum over all the time slots and region clusters to get the global flow conservation constraint component.

$$ \sum_{k=1}^{K}\sum_{o=1}^{O}|\sum_{i \in C_o} \mathbf{X}_{i,k}(\boldsymbol{w}_{i,k}^{in} - \boldsymbol{w}_{i,k}^{out})| \tag{9}$$

3.5   The Unified Optimization Model

By incorporating the temporal correlation, spatial correlation and bike traffic flow conservation constraint to the basic model, the final objective function can be written as follows.

$$ \min_{\mathbf{W}} L = \sum_{k=1}^{K}\sum_{n=1}^{N}(\mathcal{L}_{n,k}^{in} + \mathcal{L}_{n,k}^{out} + \theta(\|\boldsymbol{w}_{n,k}^{in}\|^2 + \|\boldsymbol{w}_{n,k}^{out}\|^2)) + \lambda_1 \sum_{k=1}^{K} tr(\mathbf{S}_k^{in} + \mathbf{S}_k^{out}) $$
$$ + \lambda_2 \sum_{n=1}^{N} tr(\mathbf{T}_n^{in} + \mathbf{T}_n^{out}) + \lambda_3 \sum_{k=1}^{K}\sum_{o=1}^{O}|\sum_{i \in C_o} \boldsymbol{x}_{i,k}(\boldsymbol{w}_{i,k}^{in} - \boldsymbol{w}_{i,k}^{out})| $$
$$ \tag{10}$$

The parameters λ_1, λ_2, λ_3 are used to control the importance of the temporal correlation, spatial correlation and flow conservation constraint terms, respectively. Formula (10) can be solved by existing optimization methods like ADMM. More details of the derivation can also be found at my GitHub.

3.6 Projection Inference for Traffic Prediction

After the optimization, we obtain the trained historical projection vectors that preserve the spatial-temporal correlation. We assume that the future projection vector $w_{n,k+1}$ of region r_n can be estimated by the linear combination of the historical projection vectors of r_n from t_k to t_{k-g} as follows.

$$w_{n,k+1} = \alpha_1 w_{n,k} + \cdots + \alpha_g w_{n,k-g} \tag{11}$$

The coefficients $\alpha = \{\alpha_1, \cdots, \alpha_g\}$ are introduced to control the contributions of the history projection vectors $\{w_{n,k}, \cdots, w_{n,k-g}\}$. Then we can predict the bike traffic flow of region r_n in time slot t_{k+1} by using $w_{n,k+1}$.

$$
\begin{aligned}
y_{n,k+1} &= x_{n,k+1} \left(\alpha_1 w_{n,k} + \cdots + \alpha_g w_{n,k-g} \right) \\
&= \alpha_1 x_{n,k+1} w_{n,k} + \cdots + \alpha_g x_{n,k+1} w_{n,k-g}
\end{aligned} \tag{12}
$$

Based on formula (12), we define a regression problem to learn the coefficients α and then make the prediction. The regression problem can be rewritten as follows.

$$\{x_{n,k+1} w_{n,k}, x_{n,k+1} w_{n,k-1}, \cdots, x_{n,k+1} w_{n,k-g}\} \rightarrow y_{n,k+1} \tag{13}$$

Considering the different bike check-in/out patterns in different hours of a day, it is necessary to train different coefficient vectors α for different hours. To reduce the number of parameters and speed up the model training process, we cluster the regions based on their similarity in POI feature and historical bike traffic patterns. The regions in the same cluster share the same coefficient vector. It means that, we only estimate one group of parameters α, for the same region cluster and the same hour of a day.

4 Experiment

In this section, we first introduce the dataset, baseline methods, and experiment setup. Then we compare the performance of FGST with baseline methods. We also compare FGST with its variants to show the impacts of different components on the model performance. Finally, we give a case study to further concretely demonstrate the superior performance of FGST.

4.1 Dataset and Settings

In this paper, we use the publicly available Mobike sharing-bike dataset, released in 2017 for Mobike Cup, containing 3.2 million bike trip data in Beijing. For each trip, we extract its start time, end time and check-in/check-out locations. Then we partition Beijing into 32×32 grid regions and count the bike checking numbers in each region at each time slot. Besides, we also collect the POI data, which is aggregated and counted by different categories for each region.

The feature tensor contains two parts: historical bike checking data and POI data. In our evaluation, we set the length of each time slot as one hour. The FGST framework is first trained on the first two days of the dataset and iteratively evaluated on the later days. While for baseline models, we use 70% data for training and the remaining 30% for testing. The values of the λs are set with both considering the relative sizes of both the loss function and the gradient of the loss.

4.2 Evaluation Metrics

In our experiment, we use Mean Average Error (MAE) and Rooted Mean Square Error (RMSE) as the evaluation metrics, which are widely used in previous literatures [7,19] for traffic prediction. The two metrics are defined as follows:

$$MAE = \frac{1}{N}\frac{1}{K_s}\sum_{n=1}^{N}\sum_{k=1}^{K_S}|\mathbf{Y}_n^k - \hat{\mathbf{Y}}_n^k|; RMSE = \sqrt{\frac{1}{N}\frac{1}{K_s}\sum_{n=1}^{N}\sum_{k=1}^{K_S}(\mathbf{Y}_n^k - \hat{\mathbf{Y}}_n^k)^2}$$

(14)

where K_S denotes the number of time slots for evaluation, $\hat{\mathbf{Y}}_n^k$ denotes the predicted bike traffic volume of region n at time slot k, and $\mathbf{Y}_n^k$ denotes the ground truth value.

4.3 Baseline

We compare FGST with three types of prediction methods: time series based prediction models HA and ARMA, statistical learning based prediction models LR and RF, and deep learning based prediction models ConvLSTM and DMVST.

- **Historical average (HA)** uses the historical average traffic flow of a region in a time slot as the prediction of the future traffic flow of the region in the same time slot in a day.
- **ARMA** [9] is a traditional time series data analysis method, which analyzes and predicts the temporal tendency of the sequence data with moving average and autoregressive components.
- **Linear Regression (LR)** [10] is a very regression method, which ignores the spatial-temporal correlations and the conservation constraint. ℓ_1 regularization is used to avoid over-fitting.

- **Random Forest (RF)** [4] is an ensemble learning method for classification and regression, which usually presents stable performance in the prediction tasks.
- **RF-hour** is an variation of RF. The difference is that RF-hour builds separated models for different hours of a day.
- **ConvLSTM** [18] extends the LSTM model with convolution structures. ConvLSTM extracts the temporal information with LSTM and extracts the spatial information with CNN.
- **DMVST** [20] is a deep learning based model for taxi demand prediction. The method consists of three views: temporal view (LSTM structure), spatial view (CNN structure), and semantic view (graph embedding).

Table 1. Overall performance comparison

	MAE (in)	MAE (out)	RMSE (in)	RMSE (out)
HA	8.6195	8.6241	12.2538	14.2175
ARMA	6.2738	6.3166	8.6068	8.6368
LR	9.0145	11.159	13.058	17.734
RF	5.7162	5.6288	8.9739	8.7611
RF-hour	5.5721	5.4617	8.6142	8.2856
ConvLSTM	6.2839	6.4121	9.0726	9.1700
DMVST	5.6134	5.6297	8.1190	8.0202
FGST	**4.8591**	**4.7750**	**6.9512**	**6.8322**

4.4 Experiment Results

Table 1 shows the experiment result of various methods over the two evaluation metrics. We show the prediction results for check-out and check-in bikes separately. The best performance is highlighted with bold font. One can see that our proposed FGST outperforms all the baselines in all the cases. The traditional linear regression method LR does not perform well while the time series based prediction methods like HA and ARMA perform better than LR. This is mainly because the bike traffic patterns in different regions and different hours can be quiet different, while LR model cannot distinguish the difference and thus achieves the worst preference. RF achieves relatively better performance in general, while RF-hour model also performs rather well and outperforms most of the other baseline methods including the two deep learning based methods. It indicates that deep learning methods like DMVST and ConvLSTM cannot get a good performance on our task. DMVST has a more complex model and performs better than ConvLSTM. Our proposed method significantly outperforms ConvLSTM and DMVST and other baselines by training individual models for each region by considering the spatial and temporal correlations dynamically in a fined-grained manner.

4.5   Evaluation on Model Components

Model Components Analysis. To evaluate whether each component of FGST can contribute to a better performance, we compare FGST with the following four variants.

- FGST-b: It is the basic model. We can derive it by setting $\lambda_1, \lambda_2, \lambda_3 = 0$.
- FGST-s: This model only uses the spatial correlation and the basic model. We can derive it by setting $\lambda_2, \lambda_3 = 0$.
- FGST-t: This model only considers the temporal correlation and the basic model in the FGST model. We can derive it by setting $\lambda_1, \lambda_3 = 0$.
- FGST-f: This model is used to evaluate the importance of the flow conservation constraint. We can derive it by setting $\lambda_1, \lambda_2 = 0$.

Table 2. Performance comparison under different optimization settings

	MAE (in)	MAE (out)	RMSE (in)	RMSE (out)
FGST-b	8.3861	8.3125	12.3024	13.0324
FGST-s	5.2866	5.3894	8.8917	9.0618
FGST-t	5.1013	5.0633	7.5889	7.5284
FGST-f	7.2689	7.1876	11.9924	10.7618
FGST	**4.8591**	**4.7750**	**6.9512**	**6.8322**

The experimental results are shown in Table 2. One can see than FGST-b achieves the worst performance, which indicates that the correlations and constraints play an important role in this model. Moreover, FGST-f gets the second-worst performance, which indicates the flow conservation constraint is helpful. However, it does not work well by only considering the flow conservation constraint. The FGST-s and FCST-t outperform FGST-f and FGST-b, which verifies that the spatial and temporal correlations are both meaningful in the prediction. FGST outperforms all the other variations which verifies that the spatial and temporal correlations and the flow conservation constraint are complementary rather than conflicting. Aggregating these components can provide a better model to predict the future bike traffic.

Different Regression Settings Analysis. As described in the traffic prediction part, we apply a regression method to predict the future traffic flow with the fitted projection vectors. There are several variations of the regression method, we list several variations of the regression model below and test the performance for each variation.

- FGST-*equal*: This variant treats each region and time slot the same, and trains the same α for all the regions and time slots.

- FGST-*hour*: This variant treats each region the same, and the training set is divided by the different hours of the day. For each hour of a day, we train a regression model.
- FGST-*cluster*: This variant treats each hour the same, and the training set is divided by the different clusters of regions. For each region cluster, we train a regression model.
- FGST-*h&c*: This variant divides the training set by both the different clusters of regions and the hours. For each region cluster and in each hour of the day, we train a regression model.

The experimental results are shown in Table 3, one can see that FGST-*h&c* achieves the best performance while the FGST-*cluster* achieves the worst. FGST-*equal* performs much better than FGST-*cluster*. From this result, one can draw the following conclusion. (1) Different hours of a day have different traffic flow patterns, and thus the regression parameters should not be shared in different hours. (2) Training the models for each hour and region cluster separately can get a better result, but may also lead to over-fitting. When training the model, one should limit the strength of the regression model to avoid over-fitting. (3) Ignoring the time information and the regions in a cluster sharing one model cannot get a desirable performance, which indicates that the time information is important in bike traffic flow prediction.

Table 3. Performance comparison under different regression settings

	MAE (in)	MAE (out)	RMSE (in)	RMSE (out)
FGST-equal	5.2357	5.1650	8.0429	7.7560
FGST-hour	5.1634	5.0257	7.8964	7.4835
FGST-cluster	5.3575	5.2804	8.4179	8.0182
FGST-h& c	**4.8591**	**4.7750**	**6.9512**	**6.8322**

4.6   Case Study

Figure 2(a) depicts the prediction of the bike check-out numbers of FGST and RF-hour against the ground truth in two regions of Beijing. The plot on the left shows the prediction result of a region near to a subway station. At morning, people ride to the station and check-in bikes, while in the afternoon, people check out the bikes from the station. The plot in the middle shows the prediction result of a central business district, whose bike check-out number has a main-peak at around 8:00 am, and two sub-peaks at around 12:00 am and 6:00 pm. The prediction result of FGST can accurately fit the ground truth curves (including the peak and the sub-peak), which demonstrates the effectiveness of FGST. From the experiment, one can see that the prediction accuracy is highly relative to the hour of the day. Thus, in Fig. 2(b), we plot the mean average loss at different

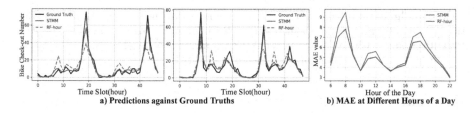

Fig. 2. Prediction Visualization

hours of our model and the best baseline model RF-hour. The MAE value of FGST is significantly lower than RF-hours at peak hours.

5   Related Work

The studies in system prediction can be classified into three classes of methods: (1) time series analysis, (2) statistical learning and (3) deep learning. In time series communities, researchers give the predictions by using ARMA, Kalman filtering and their variants for each region [6,11,12]. Statistical learning models provide several ways to predict future situations by extracting features from raw data. Forehlich et al. [1] and Kaltenbrunner et al. [2] adopted simple statistical models to predict the available number of bikes and docks for each station. Li et al. [3] proposed a multi-factor hierarchical prediction model to predict the future bike demands based on the clusters of stations. Liu et al. [7] proposed a network-based method to predict the bike pick-up and drop-off demand of each station. Nowadays, deep learning models show their abilities in several fields. And they also be used in the prediction of traffic flows. Lin et al. [5] proposed a deep learning model to predict the station-level hourly demands of sharing bikes. Zhang et al. [23] and Zhang et al. [22] treated the whole city as image and constructed three same convolutional structures to capture the trend, period and closeness information. Besides [14–17] used social media data to enhance the traffic prediction.

6   Conclusion

In this paper, we studied the bike traffic prediction problem in a fully stationless bike sharing system. We proposed a Fine-Grained Spatial-Temporal based regression model prediction framework. Our model extracted the spatial-temporal correlations by minimizing the distance between the projection vectors of similar regions and periodic time slots. Besides, we also made use of the localized conservation rules to enhance our prediction. Moreover, the extensive evaluations on the Mobike bike sharing dataset of Beijing demonstrated the superiority of FGST.

Acknowledgement. This work was supported in part by the Natural Science Foundation of China (Grand Nos. U1636211,61672081,61370126,61602237), the National Key R&D Program of China (No.2016QY04W0802), and the Natural Science Foundation of Jiangsu Province of China under Grant BK20171420.

References

1. Froehlich, J., Neumann, J., Oliver, N., et al.: Sensing and predicting the pulse of the city through shared bicycling. In: IJCAI (2009)
2. Kaltenbrunner, A., Meza, R., Grivolla, J., Codina, J., Banchs, R.: Urban cycles and mobility patterns: exploring and predicting trends in a bicycle-based public transport system. Pervasive Mob. Comput. **6**(4), 455–466 (2010)
3. Li, Y., Zheng, Y., Zhang, H., Chen, L.: Traffic prediction in a bike-sharing system. In: SIGSPATIAL (2015)
4. Liaw, A., Wiener, M., et al.: Classification and regression by randomforest. R News **2**(3), 18–22 (2002)
5. Lin, L., He, Z., Peeta, S., Wen, X.: Predicting station-level hourly demands in a large-scale bike-sharing network: a graph convolutional neural network approach. arXiv (2017)
6. Lippi, M., Bertini, M., Frasconi, P.: Short-term traffic flow forecasting: an experimental comparison of time-series analysis and supervised learning. IEEE Trans. Intell. Transp. Syst. **14**(2), 871–882 (2013)
7. Liu, J., Sun, L., Chen, W., Xiong, H.: Rebalancing bike sharing systems: a multi-source data smart optimization. In: SIGKDD (2016)
8. Ma, X., Dai, Z., He, Z., Ma, J., Wang, Y., Wang, Y.: Learning traffic as images: a deep convolutional neural network for large-scale transportation network speed prediction. Sensors **17**(4), 818 (2017)
9. Marple, S.L., Marple, S.L.: Digital Spectral Analysis: With Applications, vol. 5. Prentice-Hall, Englewood Cliffs (1987)
10. Seber, G.A., Lee, A.J.: Linear Regression Analysis, vol. 329. Wiley, Hoboken (2012)
11. Vogel, P., Greiser, T., Mattfeld, D.C.: Understanding bike-sharing systems using data mining: exploring activity patterns. Procedia - Soc. Behav. Sci. **20**, 514–523 (2011)
12. Vogel, P., Mattfeld, D.C.: Strategic and operational planning of bike-sharing systems by data mining – a case study. In: Böse, J.W., Hu, H., Jahn, C., Shi, X., Stahlbock, R., Voß, S. (eds.) ICCL 2011. LNCS, vol. 6971, pp. 127–141. Springer, Heidelberg (2011). https://doi.org/10.1007/978-3-642-24264-9_10
13. Wang, H., Odoni, A.: Approximating the performance of a last mile transportation system. Transp. Sci. **50**(2), 659–675 (2014)
14. Wang, S., He, L., Stenneth, L., Philip, S.Y., Li, Z., Huang, Z.: Estimating urban traffic congestions with multi-sourced data. In: MDM (2016)
15. Wang, S., He, L., Stenneth, L., Yu, P.S., Li, Z.: Citywide traffic congestion estimation with social media. In: SIGSPATIAL (2015)
16. Wang, S., Li, F., Stenneth, L., Yu, P.S.: Enhancing traffic congestion estimation with social media by coupled hidden Markov model. In: Frasconi, P., Landwehr, N., Manco, G., Vreeken, J. (eds.) ECML PKDD 2016. LNCS (LNAI), vol. 9852, pp. 247–264. Springer, Cham (2016). https://doi.org/10.1007/978-3-319-46227-1_16
17. Wang, S., et al.: Computing urban traffic congestions by incorporating sparse gps probe data and social media data. ACM Trans. Inf. Syst. (TOIS) **35**(4), 40 (2017)

18. Xingjian, S., Chen, Z., Wang, H., Yeung, D.Y., Wong, W.K., Woo, W.C.: Convolutional LSTM network: a machine learning approach for precipitation nowcasting. In: NIPS (2015)
19. Yao, H., Tang, X., Wei, H., Zheng, G., Yu, Y., Li, Z.: Modeling spatial-temporal dynamics for traffic prediction. arXiv (2018)
20. Yao, H., et al.: Deep multi-view spatial-temporal network for taxi demand prediction. In: AAAI (2018)
21. Yoon, J.W., Pinelli, F., Calabrese, F.: Cityride: a predictive bike sharing journey advisor. In: MDM (2012)
22. Zhang, J., Zheng, Y., Qi, D.: Deep spatio-temporal residual networks for citywide crowd flows prediction. In: AAAI (2017)
23. Zhang, J., Zheng, Y., Qi, D., Li, R., Yi, X.: DNN-based prediction model for spatio-temporal data. In: SIGSPATIAL (2016)

Customer Segmentation Based on Transactional Data Using Stream Clustering

Matthias Carnein$^{(\boxtimes)}$ and Heike Trautmann

University of Münster, Münster, Germany
matthias.carnein@uni-muenster.de, trautmann@wi.uni-muenster.de

Abstract. Customer Segmentation aims to identify groups of customers that share similar interest or behaviour. It is an essential tool in marketing and can be used to target customer segments with tailored marketing strategies. Customer segmentation is often based on clustering techniques. This analysis is typically performed as a snapshot analysis where segments are identified at a specific point in time. However, this ignores the fact that customer segments are highly volatile and segments change over time. Once segments change, the entire analysis needs to be repeated and strategies adapted. In this paper we explore stream clustering as a tool to alleviate this problem. We propose a new stream clustering algorithm which allows to identify and track customer segments over time. The biggest challenge is that customer segmentation often relies on the transaction history of a customer. Since this data changes over time, it is necessary to update customers which have already been incorporated into the clustering. We show how to perform this step incrementally, without the need for periodic re-computations. As a result, customer segmentation can be performed continuously, faster and is more scalable. We demonstrate the performance of our algorithm using a large real-life case study.

Keywords: Customer segmentation · Market segmentation ·
Stream clustering · Data streams · Machine learning

1 Introduction

Customer Segmentation is one of the most important tools in marketing. It divides a market of potential customers into distinct subsets with common characteristics. This allows to select the most attractive and profitable segments and target them with tailored marketing strategies [13]. Virtually every marketing department uses some form of customer segmentation, either based on intuition and experience or based on cluster analysis. The biggest drawback of these methods is that they require that segments are stable over time. Once segments change, the entire process needs to be repeated. Stream clustering [3,7] is an extension of traditional clustering which handles a continuous stream of new

© Springer Nature Switzerland AG 2019
Q. Yang et al. (Eds.): PAKDD 2019, LNAI 11439, pp. 280–292, 2019.
https://doi.org/10.1007/978-3-030-16148-4_22

observations. It updates the underlying clustering over time without the need to recompute the entire model.

While it seems promising to apply stream clustering for customer segmentation, there are several challenges that come with it. Most importantly, customer segmentation often uses aggregated customer data such as the number of purchases or revenue of customers. When processing a stream, these values change over time, e.g. when a customer makes another purchase. This makes it necessary to update existing observations and adjust the clustering accordingly, even when the observation has already been incorporated into the model. In this paper we show how to apply stream clustering for customer segmentation. Our strategy first removes the existing observations from the clustering before re-adding it with the updated values. Neither of these steps requires a recomputation of the model or repeated analysis of all data points. We term our algorithm userStream since we apply it for the segmentation of users.

The remainder of this paper is structured as follows: Sect. 2 introduces the concepts of Customer Segmentation and Stream Clustering. In Sect. 3, we propose a new stream clustering algorithm which allows to identify and track customer segments over time. Section 4 evaluates the proposed algorithm on more than 1.7 million real-life transactions. Finally, Sect. 5 concludes with a summary of the results and gives and outlook on future research.

2   Background

2.1   Customer Segmentation

The early days of marketing were characterised by mass-marketing strategies which use a single marketing strategy to address all customers [13]. However, this is hardly applicable for today's businesses who address customers globally with diverse preferences, values and behaviour. For example, younger customers are far more susceptible to social media marketing rather than television advertisements. Therefore, it is necessary to acknowledge and identify the different preferences and sub-markets and cater suitable products to them accordingly.

Customer Segmentation aims to divide a market into distinct groups of customers such that individuals within a group share similar behaviour or interest. The acknowledgement of customer segments allows to target each group of customers individually with distinct marketing and communication strategies, products, prices, packaging or method of distribution [13]. Nowadays, customer segmentation exists in virtually every marketing department. At the very least, it is performed informally and marketers identify segments based on experience and intuition [4]. More mature marketing departments will rely on data-driven approaches for better results, often based on clustering techniques, in order to identify segments.

Desirable segments should have diverse characteristics, be easily identifiable and have sufficient size [13,14]. In addition, segments are typically required to be stable:

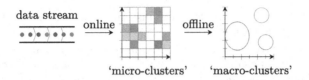

'micro-clusters'　　　'macro-clusters'

Fig. 1. Two-phase grid-based stream clustering approach [5].

'Stability is necessary, at least for a period long enough for identification of the segments, implementation of the segmented marketing strategies and the strategy to produce results' [5,14].

This is a technical necessity which demonstrates an important problem of traditional customer segmentation. If the segments change over time, all efforts and results of the marketing strategy can be rendered unusable. In practice, however, segments are highly volatile and are bound to change as trends come and go and preferences change. Throughout the remainder of this paper, we introduce a more adaptive approach for customer segmentation which handles changes in segments over time more gracefully.

2.2　Stream Clustering

Clustering is a popular tool for customer segmentation which aims to identify groups of similar observations. Most clustering algorithms are designed to work on a fixed dataset which does not change over time. When it does, the entire model needs to be recomputed. An extension to this are stream clustering algorithms which aim to find and maintain clusters over time in an endless stream of new observations. These algorithms process data points in real time and avoid expensive recomputations by incrementally updating the model.

To do so, clusters need to contain enough information to update them with new information. For example, merely storing the centres of clusters allows to merge clusters. However, it does not allow to split clusters again if necessary. To overcome this, stream clustering algorithms usually rely on an online and offline phase [1]. The online component evaluates the stream in real time and captures relevant summary statistics. As an example, it is possible to split the data space into grid-cells and simply count the number of observations per cell as shown in Fig. 1. This results in a large number of *micro-clusters* that summarize the data stream. When desired, an offline component 'reclusters' the micro-clusters to derive a final set of *macro-clusters*. This step often uses a variant of traditional clustering algorithm.

A popular concept in stream clustering and most suited for our application scenario was first proposed by the BIRCH algorithm [16] and has since received considerable attention [1,10]. The algorithm reduces the information maintained about a micro-cluster to only a few summary statistics stored in a Clustering Feature (CF): $(n, \boldsymbol{LS}, \boldsymbol{SS})$. Here, n is the number of data points in the cluster and $\boldsymbol{LS}, \boldsymbol{SS}$ are the linear and squared sum of data points for each dimension, respectively. This information is sufficient to calculate measures of location and variance for a CF such as the mean and radius:

$$\mu = \boldsymbol{LS}/n \qquad\qquad r = \sum_i \boldsymbol{SS}_i/n - (\boldsymbol{LS}_i/n)^2. \qquad (1)$$

Another important characteristic is that two CFs can be merged by summing their respective components:

$$\boldsymbol{CF}_1 + \boldsymbol{CF}_2 = (n_1 + n_2, \quad \boldsymbol{LS}_1 + \boldsymbol{LS}_2, \quad \boldsymbol{SS}_1 + \boldsymbol{SS}_2). \qquad (2)$$

An important aspect in stream clustering is to identify emerging clusters while simultaneously forgetting outdated ones. This can be implemented by a fading function which decays the importance of micro-clusters over time if no observation is assigned to them [8]. Commonly, a cluster is faded exponentially by multiplying it with $2^{-\lambda}$ in every time step, where λ is a fading factor. In practice, clusters do not need to be continuously updated but can be faded on demand using $2^{-\lambda \Delta t}$ where Δt is the time since the last update [8]. For CFs, this strategy requires to store the time of last update alongside every CF:

$$CF = (n, \boldsymbol{LS}, \boldsymbol{SS}, t). \qquad (3)$$

This allows to fade a CF as follows:

$$CF = (n \cdot 2^{-\lambda(t_{now}-t)}, \quad \boldsymbol{LS} \cdot 2^{-\lambda(t_{now}-t)}, \quad \boldsymbol{SS} \cdot 2^{-\lambda(t_{now}-t)}, \quad t_{now}), \qquad (4)$$

where t_{now} is the time of update. Note that by fading all components, the mean and variance can still be computed accurately [1,16]. Once the weight n of a cluster, i.e. the number of points, has decayed below a threshold it can be removed since its information is outdated.

A common problem in clustering is that dimensions are usually of vastly different range and variance. This is particularly true in the streaming case where the data cannot be easily normalized beforehand. A lesser known property of CFs to alleviate this is that they allow to incrementally normalize the different dimensions [2]. In particular, it is possible to use the standard deviation σ_i per dimension i as a scaling factor:

$$\boldsymbol{LS}'_i = \boldsymbol{LS}_i/\sigma_i \qquad\qquad \boldsymbol{SS}'_i = \boldsymbol{SS}_i/\sigma_i^2. \qquad (5)$$

Since the standard deviation changes over time, it is necessary to periodically update this scaling. For this, the current scaling can be removed from the CFs again in order to re-scale them with the updated standard deviation:

$$\boldsymbol{LS}_i = \boldsymbol{LS}'_i \cdot \sigma_i \qquad\qquad \boldsymbol{SS}_i = \boldsymbol{SS}'_i \cdot \sigma_i^2. \qquad (6)$$

This requires to incrementally maintain the standard deviation for each dimension. Generally, it is possible to use the information contained in the sum of all CFs for this. However, this approach suffers from numerical instability which is why we use Welford's method for computing variance as a more robust approach [15].

3   Customer Segmentation Using Stream Clustering

Features in customer segmentation can range from demographic and geographical attributes (e.g. age or location) to psychological attributes (e.g. personality) or usage-related characeristics [13,14]. In practice, usage-related segmentation is widely popular since it is easier to measure and yields more actionable results. As an example, in the retailing business users can be segmented according to their number of purchases and return rate of products. The segmentation is therefore based on the past transactions of customers. It is trivial to use these transactions as input for a (stream) clustering algorithm, but this can only identify similar *transactions* not *customers*. In order to segment customers, the transactions need to be aggregated per customer. Therefore, a stream clustering algorithm for customer segmentation must be able to process new observations but also update observations which have already been incorporated into the model and adjust the clustering accordingly. In traditional clustering, this is achieved by recomputing the entire model periodically. However, this is generally undesirable, time consuming and does not allow to track the development of clusters properly.

In this section we propose a new stream clustering algorithm called userStream which is applicable to customer segmentation. For our algorithm we utilize the concept of a time-faded Clustering Feature (CF) as well as the two-phase clustering approach as described in Sect. 2.2. Therefore, we use an online-component which summarizes the stream in a large number of micro-clusters. We then recluster this summary on demand in order to derive the final macro-clusters.

Generally, the additivity property of CFs already allows to transfer changes of an observation to its corresponding cluster, i.e. by adding the changes to the cluster. However, this does not account for changes in the clustering structure, e.g. because the updated customer does no longer fit to its current cluster. Instead, the core idea of our algorithm is to remove the existing observation from the clustering altogether before re-inserting it as a new observation. We will demonstrate that neither of these steps requires a recomputation of the entire model and both can be performed incrementally.

Algorithm 1. userStream algorithm

Require: radius threshold T, fading factor λ, minimum fading interval t_{gap}, max scaling interval m

Initialize: clusters $C = \emptyset$, scaling factor $\sigma_i' = 1$ per dimension i

1: **while** stream is active **do**
2:    Read new observation x of user u at time t_{now} from stream
3:    **if** user u already known **then**   ▷ if user known
4:        $cf \leftarrow$ previous CF of user
5:        REMOVEUSER($cf, \cdot$)   ▷ see Algorithm 2
6:    $cf \leftarrow (1, x, x^2, t_{now})$ ▷ new CF using Eq. (3)
7:    INSERTUSER($cf, \cdot$)   ▷ see Algorithm 3

This principle is core to our online component as shown in Algorithm 1. In a first step, we continuously read new observations from the stream as they become available (Line 2). For each observation x, we then evaluate whether we have previously incorporated the same customer into the model (Line 3). If that is the case, we take the customer's CF and remove it from its corresponding cluster, effectively removing the customer from the current model (Lines 4–5). Finally, we take the new observation x and initialize a new CF with it (Line 6). The new CF is then inserted into an appropriate cluster, regardless of whether it describes a new customer or simply updates an existing one (Line 7).

The main challenge of this approach is how to remove observations from the model. The pseudo-code of our removal strategy is outlined in Algorithm 2. In a first step, we identify the cluster that the customer is currently assigned to (Line 3). In order to remove the observation from this cluster, we note that the idea of the additivity property of CFs is equally applicable when subtracting CFs [1].

$$CF_1 - CF_2 = (n_1 - n_2, \quad LS_1 - LS_2, \quad SS_1 - SS_2, \quad t_{now}). \tag{7}$$

As usual, we need to make sure that both CFs are properly faded to the current time t_{now}. For this reason, we first fade both CFs (Lines 4–5) before subtracting the user CF from its cluster (Line 6). This removes the user from the current clustering but will also cause the cluster to lose some of its weight since fewer customers are assigned to it. Therefore, we need to evaluate whether the cluster still as sufficient or whether it can be removed (Line 8). As a threshold value we use $2^{-\lambda t_{gap}}$, where t_{gap} is the minimum time it takes for a cluster to become obsolete [9]. If we also want to normalize the CFs, we need to do this first by dividing each dimension i of the CF by the current scaling factor σ'_i (Line 2).

Algorithm 2. Remove a user from its cluster

1: **procedure** REMOVEUSER($cf, \cdot$)
2:     $cf \leftarrow$ NORMALIZE(cf, σ')   ▷ Using Eq. (5)
3:     $C_i \leftarrow$ current cluster of user u
4:     FADE(C_i)   ▷ Using Eq. (4)
5:     FADE(cf)   ▷ Using Eq. (4)
6:     $C_i \leftarrow C_i - cf$ ▷ Remove using Eq. (7)
7:     **if** WEIGHT(C_i) $\leq 2^{-\lambda t_{gap}}$ **then**
8:         Remove C_i from C

Once the customer is removed, we can re-insert the updated observation into the model. The pseudo-code of the insertion strategy is given in Algorithm 3. As described in Sect. 2.2, we need to periodically adjust the scaling parameters to account for changes in variance. Since the variance will change more drastically at the beginning of the stream, we perform this adjustment step in exponentially increasing intervals, i.e. after $2^1, 2^2, \ldots$ but at least every 2^m observations.

To adjust the scaling, we retrieve the used scaling factors σ' and the current standard deviations σ. The scaling with σ' is then removed from every cluster in order to normalize it with the new factor σ instead (Lines 2–7).

Once the adjustment step is finished, the main insertion step begins. Based on the current scaling factor, we start by scaling the CF that we wish to insert (Line 8). Afterwards we search for an appropriate cluster to insert the customer into. If there are currently no clusters, we initialize a new cluster at the location of the CF (Lines 9–10). Otherwise we select the cluster whose centroid is closest the customer, e.g. based on the squared Euclidean distance (Line 12). Before inserting the customer into the cluster, we need to make sure the cluster is not obsolete. To do so, we fade the cluster and check its weight. If its weight is below the weight threshold, we remove the cluster and restart the search for the closest cluster (Lines 13–16). Once we have found a suitable candidate, we check whether it can absorb the new observation. For this, we test whether the cluster can absorb it without its radius increasing beyond a threshold T. If the radius increases beyond the threshold, the customer does not fit into the cluster and we initialize a new cluster at its location instead. However, if the cluster can absorb the customer, it is inserted and the cluster updated accordingly. This is the same strategy as already used in BIRCH [16].

Algorithm 3. Insert a user into its closest cluster

1: **procedure** INSERTUSER($cf, \cdot$)
2:     **if** $t \mod \{2^1, \ldots, 2^m\} = 0$ **then**   ▷ Periodically adjust scaling
3:         $\sigma \leftarrow$ current standard deviation   ▷ using Welford's method
4:         **for each** $C_i \in C$ **do**
5:             $C_i \leftarrow$ DENORMALIZE(C_i, σ')   ▷ Using Eq. (6)
6:             $C_i \leftarrow$ NORMALIZE(C_i, σ)   ▷ Using Eq. (5)
7:         $\sigma' \leftarrow \sigma$ ▷ Update scaling factor
8:     $cf \leftarrow$ NORMALIZE(cf, σ')   ▷ Using Eq. (5)
9:     **if** $|C| = 0$ **then**   ▷ If no clusters
10:         append cf to C ▷ Create new cluster
11:     **else**
12:         $C_i \leftarrow$ closest micro-cluster to cf
13:         FADE(C_i)   ▷ Using Eq. (4)
14:         **if** WEIGHT(C_i) $\leq 2^{-\lambda t_{gap}}$ **then**   ▷ Check candidate cluster
15:             Remove C_i from C
16:             **go to** Line 9
17:         **if** RADIUS($C_i + cf$) $< T$ **then**   ▷ Radius from Eq. (1)
18:             $C_i \leftarrow C_i + cf$ ▷ Merge using Eq. (2)
19:         **else**
20:             append cf to C ▷ Create new cluster

The above online component maintains micro-clusters at dense areas in the stream. In order to derive the final clusters, the centres of these micro-clusters can be used as virtual points. This allows to apply any traditional clustering algorithm in order to recluster the micro-clusters [16]. The result is a final set of macro-clusters. In our setup we chose to use a (weighted) k-means [11] due to its simplicity and the possibility to incorporate weights of micro-clusters if necessary. Note that for reclustering, it is also advisable to scale the range of values of the micro-clusters, e.g. by subtracting the mean and dividing by the standard deviation. Alternatively, the sacling factors from the online component can be used. While the online component is working in real-time, the reclustering step can be performed on-demand, whenever macro-clusters are requested by the user.

4   Evaluation

4.1   Experimental Setup

We implemented our algorithm in C++ with interfaces to the statistical programming language R[1]. For our analysis we use real transactions from a retailer in the home furnishings and textiles sector. The retailer operates a total of 120 brick-and-mortar stores as well as an online shop. Additionally, a loyalty program is available which customers can join for easier access to the online shop, newsletters and discounts. Almost $500,000$ customers joined the loyalty programme and more than 1.7 Million transactions between June 2014 and December 2017 are available. The loyalty program assigns a unique identifier to customers which allows us to aggregate usage-related attributes per customer. Unfortunately, no information about product prices or order value is available. For our analysis we derive a handful of usage-related features. In particular we use the number of purchases, the percentage of returned products, the median order size, the time since last purchase (recency) as well as the ratio of online purchases.

Evaluating cluster quality is generally difficult since no ground-truth exists. For this reason, intrinsic properties are often used as quality indicators. As an example, the Silhouette Width measures how similar an observation i is to its own cluster, compared to other clusters:

$$s(i) = \frac{b(i) - a(i)}{\max\{a(i), b(i)\}},\tag{8}$$

where $a(i)$ is the average distance of i to other observations in its cluster and $b(i)$ is the lowest average distance of i to observations in another cluster. It can take values between -1 and $+1$ where larger values indicate a better fit. As a rule-of-thumb, Average Silhouette Widths larger than 0.7 indicate strong structures and Widths larger than 0.5 still indicate reasonable structures [12].

Since our algorithm operates on a stream of data, we aim to evaluate the clustering result over time. Therefore, we process the stream in chunks with a horizon of $10,000$ and calculate the Silhouette Width for each horizon. Throughout our

---

[1] Implementation available at: http://www.matthias-carnein.de/userStream. For reproducability, we also show how to apply the algorithm on a public dataset.

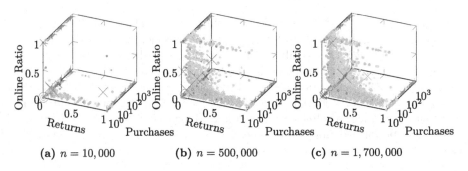

(a) $n = 10,000$ **(b)** $n = 500,000$ **(c)** $n = 1,700,000$

Fig. 2. Clustering result along three dimensions at different time steps. Grey points denote all previous customers, circles mark the centres of micro-clusters and crosses the centres of macro-clusters. Customers are scaled relative to the recency of their last purchase and clusters are scaled relative to their size. The colours of micro-clusters indicate the macro-clusters that they have been assigned to. (Color figure online)

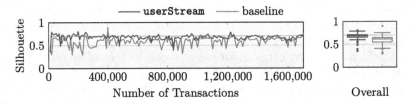

Fig. 3. Average Silhouette Width of new observations over all time horizons. The background color highlights common thresholds for strong, reasonable and weak clusterings. The boxplot summarizes the range of values. (Color figure online)

analysis we set the radius threshold to $T = 2.5$, the number of macro-clusters to $k = 5$, choose the decay rate to be $\lambda = 0.01$ and set t_{gap} to 30 days. Additionally, the largest interval to adjust the scaling is set to $m = 10$.

4.2   Results

As a simple illustration, Fig. 2 visualizes the clustering result for three of our five features at different times throughout the stream. We can see that the clusters nicely adapt over time as new customers are added and existing ones make transactions. The micro-clusters (circles) nicely summarize the observations (grey points) and the macro-clusters (crosses) use the information from the micro-clusters to create the final segments. In this example, we can see that the majority of customers only make few purchases and have no returns or online purchases. Throughout the entire stream, these one-time buyers form one of the segments. However, we also observe an increasing number of repeat buyers over time. Some customers in particular accumulate hundreds of purchases. The algorithm identifies these loyal customers nicely. Additionally, online sales seem to grow in importance over time and the segments adapt to this trend.

In order to evaluate the quality of the clustering result, Fig. 3 shows the Average Silhouette Width of new observations across all horizons. The results show that the algorithm is able to maintain a very good Silhouette Width of roughly 0.7 throughout the entire stream, indicating reasonable to strong structures. The quality only seldomly drops below 0.5 and quickly recovers afterwards. For comparison, we also show the results of a baseline approach. In this baseline approach, we do not remove customers before insertion but treat each user as a new customer. It is obvious that the baseline results are consistently worse because repeat-buyers are not assigned to their appropriate cluster.

After establishing the high quality of the clustering result, we can start profiling the selected segments. To do so, we can look at their common characteristics, i.e. what distinguishes observations in the cluster from other observations. In the following we analyse the macro-clusters at the end of the stream, i.e. after $n = 1,700,000$ transactions. Specifically, we compare the divergence of the within-cluster mean from the global mean as shown in Fig. 4. It is obvious that Cluster 1 describes loyal customers with frequent purchases. With over $300,000$ customers this is the largest segment and describes the main target group of the company. Cluster 2 identified online-shoppers where 85% of purchases are made in the online shop. This is 26 times larger than usual for customers of the company. Online-shoppers also seem to purchase more items per order. This segment might be the most accessible for digital marketing strategies and could be informed about new products in the online shop. Similarly, Cluster 3 also has an increased online-ratio. However, the return rate is also 13 times higher than usual and often products are purchased individually. This indicates that these customers make excessive use of a free shipping and free return policy of the online shop. It is important to evaluate the profitability for customers of this segment. Unprofitable customers could loose these free shipping options. Cluster 4 on the other hand represents customers with very traditional behaviour. In general, purchases are less frequent and the last purchase is further in the past. In addition, they do not use the online-shop and return products rarely. Despite this, the cluster represents a large user base of more than $135,000$ customers. Interestingly, this traditional behaviour does not seem to be related to age, since customers in this segment are of similar age as in the remaining segments. Cluster 5 represents customers that place very large orders but make purchases less frequent. It is possible that these customers are commerical resellers with bulk orders. This cluster is smallest of the identified segments.

All of the above experiments confirm the good performance of userStream. This performance is largely dependent on the number of micro-clusters that are summarizing the stream. With more micro-clusters, the stream can also be represented more accurately. The development of the number of micro-clusters throughout our analysis is shown in Fig. 5. While we do see some variation, the number of micro-clusters is relatively stable over time. In our setup, the algorithm requires at most 71 micro-clusters in order to summarize almost $500,000$ customers.

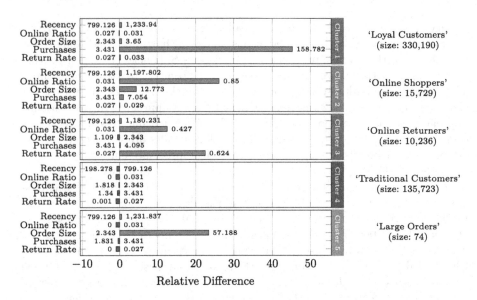

Fig. 4. Relative difference between within-cluster mean and global mean at $n = 1,700,000$. The absolute values of the global mean and within-cluster mean are shown at the origin (black) and bars (green/red) respectively. (Color figure online)

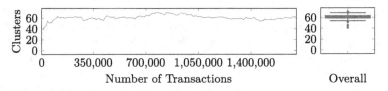

Fig. 5. Number of micro-clusters across all time horizons

5   Conclusion

Customer Segmentation is a common tool in marketing which helps to identify groups of similar customers. This knowledge can help to improve marketing strategies or product development. Unfortunately, customer segmentation is generally a slow and static process where segments are only identified at a specific point in time. In this paper we proposed userStream, a stream clustering algorithm applicable to customer segmentation. The algorithm is able to identify and track customer segments over time without expensive recalculations.

In particular userStream can handle usage-related features such as the number of purchases or return rate of products. This kind of information can be easily extracted from the transaction history. However, the biggest challenge is that the transaction history changes over time as customers make more purchases. This makes it necessary to update customers which have already been incorporated in the clustering model. In order to solve this challenge, we propose to remove the corresponding customer from its cluster when a new transaction occurs.

Afterwards, the updated values can be re-inserted into the model as a new observation in order to adjust the clustering accordingly. By carefully choosing the underlying data structure, neither of these operations require a recalculation of the entire model.

Our experiments on a large real-life case have shown that the strategy is able to identify and track meaningful clusters over time. Specifically, the silhouette width indicates very strong structures throughout most of the data stream. Additionally, we profiled the segments and gave advice how marketing strategies could be improved. In future work we will further explore the applicability of stream clustering to customer segmentation. In particular, we aim to combine the presented approach with a recently proposed stream clustering algorithm which makes uses of idle times within a stream ensuring very efficient macro-cluster generation [6]. Additionally, we only performed little parameter tuning for the parameters of our algorithm. This is a generally difficult in the context of stream clustering and automatic and adaptive approaches for choosing threshold values are desirable.

References

1. Aggarwal, C.C., Han, J., Wang, J., Yu, P.S.: A framework for clustering evolving data streams. In: Proceedings of the 29th International Conference on Very Large Data Bases, VLDB 2003, Berlin, Germany, vol. 29, pp. 81–92. VLDB Endowment (2003)
2. Aggarwal, C.C., Han, J., Wang, J., Yu, P.S.: A framework for projected clustering of high dimensional data streams. In: Proceedings of the Thirtieth International Conference on Very Large Data Bases, VLDB 2004, Toronto, Canada, vol. 30, pp. 852–863. VLDB Endowment (2004)
3. Bifet, A., Gavalda, R., Holmes, G., Pfahringer, B.: Machine Learning for Data Streams with Practical Examples in MOA. MIT Press, Cambridge (2018)
4. Buttle, F.: Customer Relationship Management: Concepts and Technologies. Elsevier Butterworth-Heinemann, Oxford (2009)
5. Carnein, M., Assenmacher, D., Trautmann, H.: An empirical comparison of stream clustering algorithms. In: Proceedings of the ACM International Conference on Computing Frontiers (CF 2017), pp. 361–365. ACM (2017). https://doi.org/10.1145/3075564.3078887
6. Carnein, M., Trautmann, H.: Evostream - evolutionary stream clustering utilizing idle times. Big Data Res. (2018). https://doi.org/10.1016/j.bdr.2018.05.005
7. Carnein, M., Trautmann, H.: Optimizing data stream representation: an extensive survey on stream clustering algorithms. Bus. Inf. Syst. Eng. (BISE) (2019). https://doi.org/10.1007/s12599-019-00576-5
8. Chen, Y., Tu, L.: Density-based clustering for real-time stream data. In: Proceedings of the 13th ACM SIGKDD International Conference on Knowledge Discovery and Data Mining, KDD 2007, San Jose, California, USA, pp. 133–142. ACM (2007). https://doi.org/10.1145/1281192.1281210
9. Hahsler, M., Bolaños, M.: Clustering data streams based on shared density between micro-clusters. IEEE Trans. Knowl. Data Eng. 28(6), 1449–1461 (2016). https://doi.org/10.1109/TKDE.2016.2522412

10. Kranen, P., Assent, I., Baldauf, C., Seidl, T.: Self-adaptive anytime stream clustering. In: 9th IEEE International Conference on Data Mining (ICDM 2009), pp. 249–258, December 2009. https://doi.org/10.1109/ICDM.2009.47
11. Lloyd, S.: Least squares quantization in PCM. IEEE Trans. Inf. Theor. **28**(2), 129–137 (2006). https://doi.org/10.1109/TIT.1982.1056489
12. Rousseeuw, P.J., Kaufman, L.: Finding Groups in Data. Wiley, Hoboken (1990)
13. Schiffman, L.G., Hansen, H., Kanuk, L.L.: Consumer Behaviour: A European Outlook. Pearson Education, London (2008)
14. Wedel, M., Kamakura, W.A.: Market Segmentation, 2nd edn. Springer, USA (2000). https://doi.org/10.1007/978-1-4615-4651-1
15. Welford, B.P.: Note on a method for calculating corrected sums of squares and products. Technometrics **4**(3), 419–420 (1962)
16. Zhang, T., Ramakrishnan, R., Livny, M.: BIRCH: a new data clustering algorithm and its applications. Data Min. Knowl. Discov. **1**(2), 141–182 (1997)

Spatio-Temporal Event Detection
from Multiple Data Sources

Aman Ahuja[1(✉)], Ashish Baghudana[2], Wei Lu[3], Edward A. Fox[2],
and Chandan K. Reddy[1]

[1] Virginia Tech, Arlington, VA, USA
aahuja@vt.edu, reddy@cs.vt.edu
[2] Virginia Tech, Blacksburg, VA, USA
{ashishb,fox}@vt.edu
[3] Singapore University of Technology and Design, Singapore, Singapore
luwei@sutd.edu.sg

Abstract. The proliferation of Internet-enabled smartphones has ushered in an era where events are reported on social media websites such as Twitter and Facebook. However, the short text nature of social media posts, combined with a large volume of noise present in such datasets makes event detection challenging. This problem can be alleviated by using other sources of information, such as news articles, that employ a precise and factual vocabulary, and are more descriptive in nature. In this paper, we propose Spatio-Temporal Event Detection (STED), a probabilistic model to discover events, their associated topics, time of occurrence, and the geospatial distribution from multiple data sources, such as news and Twitter. The joint modeling of news and Twitter enables our model to distinguish events from other noisy topics present in Twitter data. Furthermore, the presence of geocoordinates and timestamps in tweets helps find the spatio-temporal distribution of the events. We evaluate our model on a large corpus of Twitter and news data, and our experimental results show that STED can effectively discover events, and outperforms state-of-the-art techniques.

Keywords: Topic modeling · Probabilistic models · Event detection

1 Introduction

Social media platforms such as Twitter and Facebook have become a central mode of communication in people's lives. They are regularly used to discuss and debate current events, ranging from natural calamities such as *Hurricane Harvey*, to political incidents like the *U.S. Elections*. These events span different locations and time periods. With strong Internet penetration and the ubiquity of location-enabled smartphones, a large number of social media posts also have the geographical coordinates of the users. These are rich sources of information, aiding location-specific event detection and analysis.

© Springer Nature Switzerland AG 2019
Q. Yang et al. (Eds.): PAKDD 2019, LNAI 11439, pp. 293–305, 2019.
https://doi.org/10.1007/978-3-030-16148-4_23

Fig. 1. Tweets and news related to *Brexit*, originating from different parts of the world, and containing several aspects of the event.

Event detection aims to discover content describing an important occurrence. Applications of event detection include the modeling of a disease outbreak, such as an epidemic of influenza, based on Twitter data [4], and reactions to sporting events [20]. Hence, significant research has been conducted on mining topics from Twitter data [1,19]. However events are not merely topics, and have aspects of time and location. Researchers have previously defined events with an approximate geolocation, temporal range and a set of words [18]. However, these definitions do not account for events that span across multiple regions, such as *Hurricane Harvey*, nor do they identify sub-themes of an event, such as *destruction and damage*, and *help and relief*.

In this paper, we propose **S**patio **T**emporal **E**vent **D**etection (STED), a probabilistic model that discovers events using information from various data sources, such as news and Twitter. It combines the location, time, and the text, from tweets, aided with textual information from news articles, to discover the various parameters associated with an event. An event is characterized by the following three attributes:

- The *time* of occurrence. For instance, most tweets about *Rio Olympics* occur in August 2016. Each event, therefore, has a temporal mean and variance.
- A *regional distribution* describing where the event occurs. Global events such as *Brexit* have tweets from several countries, whereas tweets about the *Burning Man Festival* are concentrated in Nevada, US. Hence, an event can occur in one or more regions. Regions are defined by their geographical center and the corresponding covariance.
- A *set of topics* describing the event, where each topic is a facet of the event. *U.S. Elections 2016* contain several topics such as the Republican and Democratic campaigns, as well as the FBI investigation into Russian meddling.

We use timestamp and geolocation information from tweets to estimate the temporal and regional distributions of events, respectively. We supplement the vocabulary in tweets with news text to provide larger context about the facts surrounding the event. This is summarized in Fig. 1. This ensures that noisy tweets are eliminated and do not contribute to aspects of an event, while news articles provide more factual information about the event.

2   Related Work

2.1   Topic Modeling

Topic modeling has been widely studied in the domain of text mining to discover latent topics. One of the earliest methods to discover topics in text documents was probabilistic Latent Semantic Indexing (pLSI) [9]. However, since pLSI was based on the likelihood principle and did not have a generative process, it cannot assign probabilities to new documents. This was alleviated by Latent Dirichlet Allocation (LDA) [6], which models each document as a mixture over topics, and topics as a mixture over words. Inspired by its success, LDA has been extended and applied to various corpora including microblogs such as tweets [19], as well as news documents [14].

2.2   Event Extraction from Text

The most common data-driven approach to event extraction uses text clustering. Within text clustering, the two major paradigms are similarity-based methods and statistical techniques. Similarity-based efforts generally use cosine similarity [11]. These techniques are fast and efficient, however they ignore all the statistical dependencies between variables. Graphical models bring more insight to event detection by modeling dependencies and hierarchies [5]. Another class of event detection models uses spikes in activity as an indication of an event. These bursts change the distribution of the existing data and are detected as new events [12,13]. These models rely on detecting words that have a sudden increase in activity, while trying to penalize terms that occur consistently in the data. Thus, events are defined only by a subset of terms that have increased co-occurrence.

2.3   Geospatial and Temporal Models

With social media platforms like Twitter and Facebook allowing users to embed their locations in posts, there has been an increase in the availability of data with geospatial and temporal information. As a result, several researchers have incorporated this information in event detection systems. [16] built an earthquake detection system by correlating Twitter messages during a disaster event in Japan. A sudden increase in the volume of tweets in a specific region within a timeframe indicated an event. [7] introduced the Geographical Topic Model where they aimed to discover variation of different topics in latent regions. However, it does not assume a dependency between the latent topics and regions. [2,10] proposed probabilistic models that address the problem of modeling geographical topical patterns on Twitter. This improved upon prior models that used predefined region labels instead of actual latitude and longitude. However, their focus was more on geographical topics, rather than events. The model proposed in [18] explicitly uses geospatial and temporal information to discover events. It assumes that every event has a single temporal and regional distribution. Furthermore, the authors use data only from Twitter. We improve upon this approach by allowing an event to be spatially distributed by creating a joint model for news and social media. While social media provides quick and short details about an event, the text often contains personal opinion. When combined with news data, event summaries are both factual, and provide views of the people about an event.

3 The Proposed STED Model

In this section, we introduce STED, a probabilistic graphical model that discovers events, and their aspects, across different geographical regions and temporal ranges, from a multimodal corpus of geo-tagged microblogs, such as tweets, and news. Our model is built on the following observations:

- An event refers to an incident that is discussed widely in news and social media, such as "U.S. Elections 2016" and "Rio Olympics 2016". Events have a definite geospatial and temporal distribution. Thus, a particular event is more likely to be discussed within a specific period of time.
- An event can be discussed in multiple geographical regions. Each region can be represented using a bivariate Gaussian distribution, with a geographical center μ_r^l, and variance determined by a diagonal covariance matrix Σ_r^l. For example, "U.S. Elections 2016" is discussed in New York, California, and Texas – each of which belongs to a different region – but not as much in Asia.
- Similarly, an event has a temporal distribution given by its mean time of occurrence, μ_e^t, and variance, σ_e^t. "Brexit Vote" and "Rio Olympics" may have similar regional distributions but occurred during different months – June 2016 and August 2016, respectively.
- News articles and tweets (now with an increased character limit of 280) cover several topical aspects within an event. "Trump Campaign" and "Clinton Campaign" form two topics in the event "U.S. Elections 2016".

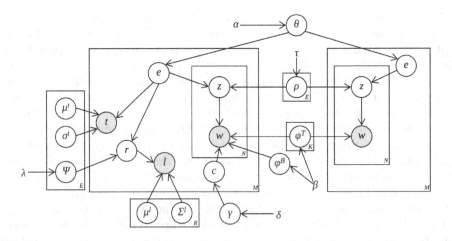

Fig. 2. Plate notation for the model.

Table 1. Notations used in this paper.

Symbol	Description	Symbol	Description	Symbol	Description
M	number of documents	c_m	category of tweet m	ψ	event-region distribution
N_m	number of words in document m	α	Dirichlet prior for θ	γ	tweet-category distribution
E	number of events	β	Dirichlet prior for ϕ^B, ϕ^T	ϕ^T	topic-word distribution
Z	number of topics	δ	Beta prior for γ	ϕ^B	background word distribution
R	number of regions	λ	Dirichlet prior for ψ	$n_{e,g}$	number of documents with event e
V	vocabulary	τ	Dirichlet prior for ρ		and region g
e	event	μ_r^l	geographical center of region r	$p_{e,k}$	number of words with event e
z	topic	Σ_r^l	regional variance of region r		and topic k
r	geographical region	μ_e^t	temporal mean of event e	$s_{k,v}$	number of times term v is used
l_m	latitude, longitude of tweet m	σ_e^t	temporal variance of event e		with topic k
t_m	time of tweet m	θ	corpus-event distribution	$q_{c,v}$	number of times term v is used
w	word	ρ	event-topic distribution		in category c

- Finally, different events can have recurring subthemes – both "Hurricane Irma" and "Hurricane Harvey" speak about wind speeds, damage, and loss of life, despite having different geospatial and temporal distributions.

3.1   Problem Statement

Given a set of news articles $D^n = \{d_1^n, \ldots, d_{|D^n|}^n\}$, a set of tweets $D^m = \{d_1^m, \ldots, d_{|D^m|}^m\}$, their geolocations $L^m = \{l_1^m, \ldots, l_{|L^m|}^m\}$, their timestamps $T^m = \{t_1^m, \ldots, t_{|T^m|}^m\}$, the goal of STED is to find for each event e, a ranked list of topics and regions, temporal mean μ_e^t and variance σ_e^t, as well as a ranked list of words for each topic k. For each geographical location $r \in R$, STED also finds it's geographical mean μ_r^l and variance Σ_r^l (Table 1).

3.2   Model Definition

STED is a generative model that incorporates the key characteristics described above. It discovers latent events and their corresponding latent topics from a corpus of long documents, such as news, and short geotagged documents, such as social media posts. Figure 2 illustrates the plate notation of our model.

- The model assumes there are E events, K topics, and R regions, the values of which are fixed.
- It models each event e as a mixture of topics and regions, along with a definite temporal distribution.
- For each news article, an event e is drawn from the corpus event distribution θ. Subsequently, for each word in the document, a topic z is drawn from the event topic distribution ρ_e.
- Since tweets are often noisy, they may or may not be related to an event. Hence, for every tweet, a category c is sampled from the category distribution.
  - If $c = 1$, an event e is drawn from the corpus event distribution θ, and the region r, geolocation (latitude, longitude) l, and time t of the tweet are drawn. Subsequently, for each word, a topic is sampled from the event topic distribution ρ_e.

- If $c = 0$, the tweet is regarded as a noisy tweet and each word in the document is sampled from the background word distribution ϕ^B.

The detailed generative process of STED is described in Algorithm 1.

Algorithm 1. Generative Process of STED.

Draw event distribution $\theta \sim Dir(\alpha)$
for each event e **do**
    Draw region distribution $\psi_e \sim Dir(\lambda)$
    Draw topic distribution $\rho_e \sim Dir(\tau)$
end for
for each topic z **do**
    Draw word distribution $\phi_z^T \sim Dir(\beta)$
end for
Draw background word distribution $\phi^B \sim Dir(\beta)$
for each news article m **do**
    Draw event $e_m \sim Mult(\theta)$
    for each word n **do**
        Draw topic $z_{m,n} \sim Mult(\rho_{e_m})$
        Draw word $w_{m,n} \sim Mult(\phi_{z_{m,n}}^T)$
    end for
end for
For tweets, draw category distribution $\gamma \sim Beta(\delta)$

for each tweet m **do**
    Draw category $c_m \sim Bin(\gamma)$
    if $c_m = 1$ **then**
        Draw event $e_m \sim Mult(\theta)$
        Draw timestamp $t_m \sim N(\mu_{e_m}^t, \sigma_{e_m}^t)$
        Draw region $r_m \sim Mult(\psi_{e_m})$
        Draw geolocation $l_m \sim N(\mu_{r_m}^l, \Sigma_{r_m}^l)$
        for each word n **do**
            Draw topic $z_{m,n} \sim Mult(\rho_{e_m})$
            Draw word $w_{m,n} \sim Mult(\phi_{z_{m,n}}^T)$
        end for
    else if $c_m = 0$ **then**
        for each word n **do**
            draw word $w_{m,n} \sim Mult(\phi^B)$
        end for
    end if
end for

3.3 Model Inference

We use the Gibbs-EM algorithm [3,8] for inference in the STED model. We first integrate out the model parameters θ, ρ, ψ, γ, ϕ^T, and ϕ^B using Dirichlet-Multinomial conjugacy. After this, the latent variables left in the model are e, r, c, z, μ^l, Σ^l, μ^t, and σ^t. We sample the latent variables e, z, r, and c in the E-step of the algorithm using the following equations:

Sampling Event. For a news article m, the event e_m can be sampled by:

$$P(e_m = x|*) \propto (n_{x,*}^{-m} + \alpha_x) \times \frac{\prod_{j \in Z_m} \prod_{y=0}^{p_{x,j}^m - 1} (p_{x,j}^{-m} + \tau_{x,j} + y)}{\prod_{y=0}^{N_m - 1} (\sum_{j=1}^{K} (p_{x,j}^{-m} + \tau_{x,j}) + y)} \tag{1}$$

For tweets, given the region is g and the timestamp is t,

$$P(e_m = x|*) \propto (n_{x,*}^{-m} + \alpha_x) \times \frac{n_{x,g}^{-m} + \lambda_g}{\sum_{i=1}^{R} n_{x,i}^{-m} + \lambda_i} \times \frac{\prod_{j \in Z_m} \prod_{y=0}^{p_{x,j}^m - 1} (p_{x,j}^{-m} + \tau_{x,j} + y)}{\prod_{y=0}^{N_m - 1} (\sum_{j=1}^{K} (p_{x,j}^{-m} + \tau_{x,j}) + y)} \times N(t_m | \mu_x^t, \sigma_x^t) \tag{2}$$

Sampling Topic. For a news article or tweet m, the topic z_{mn} for the word n with vocabulary index t can be sampled by:

$$P(z_{mn} = k|e = x) \propto \frac{s_{k,t}^{-mn} + \beta_t}{\sum_{r=1}^{V} s_{k,r}^{-mn} + \beta_r} \times \frac{p_{x,k}^{-mn} + \tau_{x,k}}{\sum_{j=1}^{K} p_{x,j}^{-mn} + \tau_{x,j}} \tag{3}$$

Sampling Region. Geographical region is sampled only for tweets with category $c = 1$, given their corresponding event is e and geo-coordinates is l_m, as follows:

$$P(r_m = g|*) \propto \frac{n_{e,g}^{-m} + \lambda_g}{\sum_{j=1}^{R} n_{e,j}^{-m} + \lambda_j} \times \mathcal{N}(l_m|\mu_g^l, \Sigma_g^l) \tag{4}$$

Sampling Category. The category (background or event-related) is sampled only for tweets, using the following equation:

$$P(c_m = d) \propto \frac{q_{d,*}^{-m} + \delta}{\sum_{i=0}^{1} q_{i,*}^{-m} + \delta} \times \frac{\prod_{r=1}^{V} \Gamma(q_{d,r}^{-m} + \beta_r)}{\Gamma(\sum_{r=1}^{V} q_{d,r}^{-m} + \beta_r)} \times \frac{\prod_{r \in V_m} \prod_{y=0}^{N_m - 1}(q_{d,r}^{-m} + \beta_r + y)}{\prod_{y=0}^{N_m - 1}(\sum_{r=1}^{V}(q_{d,r}^{-m} + \beta_r) + y)} \tag{5}$$

After sampling the latent variables e, z, c, and r, the geographical center μ_r^l, and covariance matrix Σ_r^l, is updated for each region r. The temporal mean μ_e^t and variance σ_e^t, is also updated for each event e.

3.4 Priors for Model Initialization

The STED model uses a bivariate Normal distribution on the location variable l. The mean μ_r^l and covariance Σ_r^l for the regions in R serve as the prior for this Normal distribution. To initialize these parameters, we run K-means clustering on the tweet geo-coordinates. The values of the mean and average co-variance obtained for the clusters are used as the prior μ_r^l and Σ_r^l for latent regions. The latent variables e, z, and c for all the tweets and news articles are randomly initialized, and all the distribution parameters are set using the initialized values of variables they use.

4   Experiments

4.1   Dataset Description and Preprocessing

For empirical evaluation of STED, we estimate its performance on a large real-world data, composed of tweets and news articles from the year 2016.

1. *Tweet Data:* This dataset consists of tweets with geolocations collected through 2016 using the Twitter Streaming API for a period of 7 months from June 2016 to December 2016. The Twitter Streaming API is believed to give a 1% random sample of tweets streaming on Twitter. We further performed a random sampling and obtained 1 million tweets from the collected data. Subsequently, we filtered out all the tweets that had less than 90% English characters and encoded the remaining tweets with an ASCII codec. This final dataset contained 715,262 tweets.
2. *News Data:* We collected news data from the articles published in Washington Post for the time period mentioned above. This dataset contained 148,769 news articles.

All the documents in both the datasets were lowercased and preprocessed to remove common stop words and punctuation marks. Tweets were further processed to remove all usernames and URLs. However, we retained all hashtags as they contain valuable information about events.

4.2   Performance Evaluation

For quantitative comparison of STED against baseline techniques, we use the following two metrics:

- *Perplexity:* This is a measure of the degree of uncertainty in fitting test documents to a language model. It is defined as the negative log-likelihood of test documents using the trained model.

$$Perp(D) = exp\left\{ \frac{-\sum_{d \in D} \log p(w_d)}{\sum_{d \in D} N_d} \right\} \tag{6}$$

A lower perplexity indicates better predictive performance. $p(w_d)$ is the joint probability of the word w_d occurring in an event-related and non-event related document d, and N_d is the number of words in document d.

- *Topic Coherence:* It is measured using Pointwise Mutual Information (PMI), which is a measure of information overlap between two variables. Prior research indicates that PMI is well correlated with human judgment of topic coherence [15].

$$\text{PMI-Score} = \frac{1}{EZ} \sum_{e=1}^{E} \sum_{z=1}^{Z} \sum_{i<j} \log\left\{ \frac{P(w_i, w_j)}{P(w_i)P(w_j)} \right\} \tag{7}$$

where E = number of events, and Z = number of topics. $P(w_i)$ indicates the proportion of documents containing word w_i. Consequently, $P(w_i, w_j)$ indicates the proportion of documents containing words w_i and w_j. A higher PMI score shows better topic coherence.

4.3   Baseline Methods

We compare the aforementioned metrics on the following models:

1. **LDA** [6]: An implementation of LDA using collapsed Gibbs sampling.
2. **GeoFolk** [17]: A spatial topic model that aims to discover topics and their geographical centers.
3. **BGM** [18]: A Bayesian Graphical Model to discover latent events from Twitter, that models events with geographical and temporal centers, and their associated variances. We refer to this model as **BGM**.
4. **STED-T:** A variant of our model which uses only tweets to discover events.

4.4   Parameter Setting

To initialize STED, the following hyperparameters are required: α, β, τ, λ, and δ. These hyperparameters serve as priors for each of the distributions. We used symmetric values for these hyperparameters, all of which were derived empirically. Specifically, we set $\alpha = 1$, $\beta = 0.01$, $\tau = 0.1$, $\delta = 0.1$, and $\lambda = 10$. The priors μ_e^t and σ_e^t for temporal mean and variance, as well as μ_r^l, and Σ_r^l were set as specified in Sect. 3.4.

We ran our model, its variant, and BGM, for 50 EM iterations, with 10 Gibbs sampling steps in each E-step of the iteration. We varied both the number of events and the number of topics. The other baseline models were run for 500 Gibbs sampling iterations.

4.5   Experimental Results

Quantitative Results. In this section, we discuss the quantitative metrics of the STED model. We compare the perplexity and topic coherence of our model against baselines, and also show how the addition of news articles improves the performance of the model. Since our model is hierarchical, we measure these metrics by first varying the number of events, fixing the number of topics to 50, and then varying the number of topics, fixing the number of events to 10.

a. Perplexity: We observe that the perplexity of STED is consistently better than that of all baseline models (Fig. 3(a)). This shows that event detection is not merely dependent on words in each document, but also on the spatial and temporal distribution of the documents. We further observe that STED outperforms STED-T, indicating that the addition of news data improves the predictive power of the model.

We also notice that even though BGM performs worse than STED, it's performance is at-par with STED-T. Therefore, for a dataset such as tweets, that contains geolocation information, it is better to consider latent regions as a mixture of Gaussian distributions, rather than using predefined regions based on the coordinates. The rise in perplexity beyond $e = 10$ can be explained by overfitting of the BGM model. This trend remains the same even when we vary the number of topics while keeping the number of events constant (Fig. 3(b)).

It is also interesting to see that the perplexity plots are uniformly flat for most of the baselines, indicating that the dataset was noisy. Despite the noise, the qualitative results show that STED correctly identified events of world importance, that occurred during the timeframe that the dataset was collected.

b. Topic Coherence: As described in Sect. 4.2, we use PMI as an indicator of topic coherence. We compare the normalized PMI score of our model to those of LDA, GeoFolk, and BGM, for the top twenty words in each event (or topic).

Figures 3(c) and (d) show that STED has the highest PMI score. GeoFolk and LDA perform comparably in topic coherence, i.e., the topics are equally interpretable in both of these models. This is expected since GeoFolk only accounts for geographical topics and does not consider temporal information. Moreover, it is trained only on tweets, and not news data. For the same reason, STED outperforms STED-T. This demonstrates that vocabulary from news articles improves the readability of topics generated from the model. BGM makes the implicit assumption that events are concentrated in a specific region, which fails for events with more distributed geolocations, such as U.S. Elections or Brexit. The joint modeling makes STED's PMI score higher than BGM and GeoFolk.

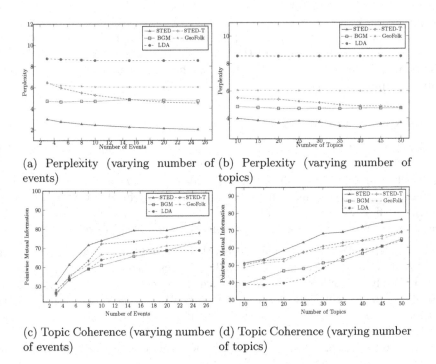

(a) Perplexity (varying number of events)

(b) Perplexity (varying number of topics)

(c) Topic Coherence (varying number of events)

(d) Topic Coherence (varying number of topics)

Fig. 3. Performance comparison of STED with other baseline methods.

Qualitative Results. For qualitative evaluation of our model, we identify two events, *U.S. Elections 2016* and *Brexit.* We characterize these events across three features – latent regions in which these events were prevalent, their temporal distributions, and their topics.

Tweets about *U.S. Elections 2016* were largely localized to North America, while those about *Brexit* were concentrated in Europe (Fig. 4a). Since we model temporal distributions as Gaussian (Fig. 4b), we observe that the event *U.S. Elections 2016* was centered at Oct. 31, 2016 (elections were held on Nov. 8, 2016), and *Brexit* was centered at June 30, 2016 (actual vote happened on June 23, 2016).

In Fig. 5, we show the top two topics for each of these events, generated from STED. Each topic is described by its corresponding top-ranking words. The first topic in Fig. 5(a) describes the Republican campaign with references to Donald Trump, Jeb Bush, and Ted Cruz, while the second topic details the Democratic campaign focusing on Hillary Clinton, and Bernie Sanders. The first topic in Fig. 5(b) mentions the Prime Minister of Britain, David Cameron, and the second topic illustrates the anti-immigration sentiment prevalent at the time of the vote.

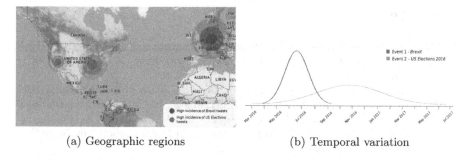

(a) Geographic regions   (b) Temporal variation

Fig. 4. Geographic regions and temporal variation for events *Brexit* and *U.S. Elections 2016*.

(a) U.S. Elections 2016   (b) Brexit

Fig. 5. Word clouds representing the top two topics generated from STED.

5 Conclusion

In this paper, we presented STED, a novel probabilistic topic model to extract latent events, from a heterogenous corpus of documents from multiple data sources, such as long and short documents. Because of the growing importance of social media, which also has location and time information, but limited textual information, we used Twitter data as one of the data sources in STED. To overcome the sparsity of textual information available in social media data, we use a much more elaborate form of data, such as news articles, as other source. This improves the predictive power of the model, by providing relevant vocabulary, along with spatial and temporal information. Furthermore, the use of latent regions helps define events more naturally – geospatially distributed, but temporally centered. The results obtained on Twitter and news data from 2016 show that the model obtains meaningful results and outperforms state-of-the-art techniques on several quantitative metrics. We hope that STED can be an important tool in detecting the different aspects of events, such as disasters, and help government agencies better plan and mitigate such events.

Acknowledgments. This work was supported in part by the US National Science Foundation grants IIS-1619028, IIS-1707498, and IIS-1838730.

References

1. Ahuja, A., Wei, W., Carley, K.M.: Microblog sentiment topic model. In: 2016 IEEE 16th International Conference on Data Mining Workshops (ICDMW), pp. 1031–1038. IEEE (2016)
2. Ahuja, A., Wei, W., Lu, W., Carley, K.M., Reddy, C.K.: A probabilistic geographical aspect-opinion model for geo-tagged microblogs. In: 2017 IEEE International Conference on Data Mining (ICDM), pp. 721–726. IEEE (2017)
3. Andrieu, C., De Freitas, N., Doucet, A., Jordan, M.I.: An introduction to MCMC for machine learning. Mach. Learn. **50**(1–2), 5–43 (2003)
4. Aramaki, E., Maskawa, S., Morita, M.: Twitter catches the flu: detecting influenza epidemics using Twitter. In: Proceedings of the Conference on Empirical methods in Natural Language Processing, pp. 1568–1576. Association for Computational Linguistics (2011)
5. Benson, E., Haghighi, A., Barzilay, R.: Event discovery in social media feeds. In: Proceedings of the 49th Annual Meeting of the Association for Computational Linguistics: Human Language Technologies-Volume 1, pp. 389–398. Association for Computational Linguistics (2011)
6. Blei, D.M., Ng, A.Y., Jordan, M.I.: Latent dirichlet allocation. J. Mach. Learn. Res. **3**(Jan), 993–1022 (2003)
7. Eisenstein, J., O'Connor, B., Smith, N.A., Xing, E.P.: A latent variable model for geographic lexical variation. In: Proceedings of the 2010 Conference on Empirical Methods in Natural Language Processing, pp. 1277–1287. Association for Computational Linguistics (2010)
8. Griffiths, T.L., Steyvers, M.: Finding scientific topics. Proc. Natl. Acad. Sci. **101**(suppl 1), 5228–5235 (2004)
9. Hofmann, T.: Probabilistic latent semantic indexing. In: Proceedings of the 22nd annual international ACM SIGIR Conference on Research and Development in Information Retrieval, pp. 50–57. ACM (1999)
10. Hong, L., Ahmed, A., Gurumurthy, S., Smola, A.J., Tsioutsiouliklis, K.: Discovering geographical topics in the Twitter stream. In: Proceedings of the 21st International Conference on World Wide Web, pp. 769–778. ACM (2012)
11. Kumaran, G., Allan, J.: Text classification and named entities for new event detection. In: Proceedings of the 27th Annual International ACM SIGIR Conference on Research and Development in Information Retrieval, pp. 297–304. ACM (2004)
12. Mathioudakis, M., Koudas, N.: TwitterMonitor: trend detection over the Twitter stream. In: Proceedings of the 2010 ACM SIGMOD International Conference on Management of Data, pp. 1155–1158. ACM (2010)
13. Matuszka, T., Vinceller, Z., Laki, S.: On a keyword-lifecycle model for real-time event detection in social network data. In: 2013 IEEE 4th International Conference on Cognitive Infocommunications (CogInfoCom), pp. 453–458. IEEE (2013)
14. Newman, D., Chemudugunta, C., Smyth, P., Steyvers, M.: Analyzing entities and topics in news articles using statistical topic models. In: Mehrotra, S., Zeng, D.D., Chen, H., Thuraisingham, B., Wang, F.-Y. (eds.) ISI 2006. LNCS, vol. 3975, pp. 93–104. Springer, Heidelberg (2006). https://doi.org/10.1007/11760146_9
15. Newman, D., Lau, J.H., Grieser, K., Baldwin, T.: Automatic evaluation of topic coherence. In: Human Language Technologies: The 2010 Annual Conference of the North American Chapter of the Association for Computational Linguistics, pp. 100–108. Association for Computational Linguistics (2010)

16. Sakaki, T., Okazaki, M., Matsuo, Y.: Earthquake shakes Twitter users: real-time event detection by social sensors. In: Proceedings of the 19th International Conference on World Wide Web, pp. 851–860. ACM (2010)

17. Sizov, S.: GeoFolk: latent spatial semantics in web 2.0 social media. In: Proceedings of the Third ACM International Conference on Web Search and Data Mining, pp. 281–290. ACM (2010)

18. Wei, W., Joseph, K., Lo, W., Carley, K.M.: A Bayesian graphical model to discover latent events from Twitter. In: ICWSM, pp. 503–512 (2015)

19. Zhao, W.X., et al.: Comparing Twitter and traditional media using topic models. In: Clough, P., et al. (eds.) ECIR 2011. LNCS, vol. 6611, pp. 338–349. Springer, Heidelberg (2011). https://doi.org/10.1007/978-3-642-20161-5_34

20. Zubiaga, A., Spina, D., Amigó, E., Gonzalo, J.: Towards real-time summarization of scheduled events from Twitter streams. In: Proceedings of the 23rd ACM Conference on Hypertext and Social Media, pp. 319–320. ACM (2012)

Discovering All-Chain Set in Streaming Time Series

Shaopeng Wang[1(✉)], Ye Yuan[2], and Hua Li[1]

[1] Inner Mongolia University, Hohhot 010021, China
wangshaopeng1984@163.com
[2] Northeastern University, Shenyang 110819, China

Abstract. Time series chains discovery is an increasingly popular research area in time series mining. Previous studies on this topic process fixed-length time series. In this work, we focus on the issue of all-chain set mining over the streaming time series, where the all-chain set is a very important kind of the time series chains. We propose a novel all-chain set mining algorithm about streaming time series (ASMSTS) to solve this problem. The main idea behind the ASMSTS is to obtain the mining results at current time-tick based on the ones at the last one. This makes the method more efficiency in time and space than the Naïve. Our experiments illustrate that ASMSTS does indeed detect the all-chain set correctly and can offer dramatic improvements in speed and space cost over the Naive method.

Keywords: Streaming time series · Time series chains · All-chain set

1 Introduction

Time series motifs are repeated subsequences in a longer time series [1]. While time series motifs have been in literature for 15 years [2–4], they recently have been a topic of great interest in the time series data mining community, and have been used as sub-routine in higher-level analytics, including classification, clustering, visualization [5] and rule discovery [6]. Moreover, they also have been applied to domains as diverse as severe weather prediction, medicine [7] and seismology [8]. However, from a knowledge discovery viewpoint, a more interesting problem is the description on the evolving of motifs. We call such descriptions time series chains (TSCs) [9–11]. TSCs are introduced by Zhu et al. in [9] firstly, and defined as the temporally ordered set of subsequence patterns. TSCs can capture the evolution of systems, and help predict the future. As such, they potentially have implications for prognostics. In [9], Zhu et al. show the differences between TSCs and other related researches, including story chain in text domain [12], periodicity in time series [13], and concept drift [14]. In [10], Zhu et al. further discuss the issue of uniform scaling time series chain mining.

The current TSCs mining algorithms work in a *one-time fashion*: mine the current entire time series and obtain results set, because the length of time series is pre-determined. However, in many applications, time series arrive continuously, and the TSCs of time series at each time-tick need to be obtained. The most fundamental support needed in these applications is to require an efficient TSCs mining mechanism

© Springer Nature Switzerland AG 2019
Q. Yang et al. (Eds.): PAKDD 2019, LNAI 11439, pp. 306–318, 2019.
https://doi.org/10.1007/978-3-030-16148-4_24

to monitor the streaming time series. We can take the current TSCs mining algorithm to handle the time series at each time-tick to solve this problem. Unfortunately, the time complexity of this method is high. Thus, the problem of TSCs mining over streaming time series still remains unsolved.

In this paper, we address the problem of efficiently monitoring the streaming time series based on the all-chain set mining, where the all-chain set is a very important kind of the TSCs, and defined as a set of all anchored TSCs within time series that are not subsumed by another chain in [9–11]. To the best of our knowledge, this is the first study that investigates all-chain set mining mechanism for monitoring streaming time series. Intuitively, this problem is equivalent to all-chain set mining in an online fashion. We design all-chain set mining algorithm about streaming time series (ASMSTS), by which the all-chain set of streaming time series can be obtained rapidly and accurately. The ASMSTS is developed from the main idea that obtaining the mining results at current time-tick based on the ones at last one. Our experiments show that ASMSTS is more efficiency than the Naive in time and space cost.

The rest of the paper is organized as follows. Related work on the all-chain set is provided in Sect. 2. Notations and problem definition are provided in Sect. 3. The detail of ASMSTS algorithm is explained in Sect. 4. Experiments are demonstrated in Sect. 5. Finally, the conclusion and future work are discussed in Sect. 6.

2   Related Work

Our review of related work is brief. To the best of our knowledge, there are only three known works [9–11] related to our research. Zhu et al. [9, 11] introduced TSCs for the first time. They proposed ATSC algorithm to compute the anchored time series, and ALLC algorithm to compute both the all-chain set and unanchored time series chain. Zhu et al. also introduced uniform scaling time series chain in [10]. They proposed UniformScaleChain algorithm to compute the scaling time series chain. However, none of these works consider the TSCs mining over the streaming time series, including the all-chain set on-line mining. Thus, all-chain set mining over the streaming time series problem still remains unsolved.

3   Notations and Problem Definition

3.1   Notations

In this section, we review some related definitions (Definitions 3–12) in [9], and create some new ones (Definitions 1–2 and 13–16).

Definition 1: A *streaming time series X* is a discrete sequence of real-valued numbers $x_i : X = x_1, x_2, \ldots, x_n, \ldots$, where x_n is the most recent value. Notice that n increases with each new time-tick, and $S_{TSC,n}$ is the all-chain set mining results on X of length n.

Definition 2: A *new (old) streaming time series at time-tick n* is the time series after (before) the arrival of the x_n.

Definition 3: A *subsequence* $X_{i,m}$ of X is a continuous subset of the values from X of length m starting from position i. Formally, $X_{i,m} = x_i, x_{i+1}, \ldots, x_{i+m-1}$, where $1 \leq i \leq n - m + 1$.

Definition 4: A *distance profile* D_i of time series X is a vector of the Euclidean distance between a given $X_{i,m}$ and each subsequence in X. Formally, $D_i = [d_{i,1}, d_{i,2}, \ldots, d_{i,n-m+1}]$, where $d_{i,j}$ is the distance between $X_{i,m}$ and $X_{j,m}$.

Definition 5: A *left distance profile* DL_i of X is a vector of the Euclidean distances between a given $X_{i,m}$ and each subsequence that appears before $X_{i,m}$ in X. Formally, $DL_i = [d_{i,1}, d_{i,2}, \ldots, d_{i,i-m/4}]$.

Definition 6: A *right distance profile* DR_i of time series X is a vector of the Euclidean distances between a given query subsequence $X_{i,m}$ and each subsequence that appears after $X_{i,m}$ in X. Formally, $DR_i = [d_{i,i+m/4}, d_{i,i+m/4+1}, \ldots, d_{i,n-m+1}]$.

Definition 7: A *left (or right) nearest neighbor of* $X_{i,m}$, $LNN(X_{i,m})$ (or $RNN(X_{i,m})$) is a subsequence that appears before (or after $X_{i,m}$) in X, and is most similar to $X_{i,m}$. Formally, $LNN(X_{i,m})$ (or $RNN(X_{i,m})) = X_{j,m}$ if $d_{i,j} = min(DL_i)$ (or $min(DR_i)$).

Definition 8: A *left (or right) matrix profile PL of* X is a vector of the z-normalized Euclidean distance between each $X_{i,m}$ and its left (or right) nearest neighbor in X. Formally, $PL = [min(DL_1), min(DL_2), \ldots, min(DL_{n-m+1})]$ (or $PR = [min(DR_1), min(DR_2), \ldots, min(DR_{n-m+1})]$) where DL_i (or DR_i) is a left (or right) distance profile of X.

Definition 9: A *left (or right) matrix profile index IL of* X is a vector of integers: $IL = [IL_1, IL_2, \ldots, IL_{n-m+1}]$ (or $IR = [IR_1, IR_2, \ldots, IR_{n-m+1}]$), where IL_i (or $IR_i) = j$ if $LNN(X_{i,m})$ (or $RNN(X_{i,m})) = X_{j,m}$.

Definition 10: A *time series chain* of X is an ordered set of subsequences: $TSC = \{X_{c1,m}, X_{c2,m}, \ldots, X_{ck,m}\}$ ($c1 \leq c2 \leq \ldots \leq ck$), such that for any $1 \leq i \leq k - 1$, we have $RNN(X_{ci,m}) = X_{c(i+1),m}$ and $LNN(X_{c(i+1),m}) = X_{ci,m}$. We denote k the length of TSC. Note that for facilitating analysis, the minimum value of k is set to 1 in our work.

Definition 11: An *anchored time series chain* of X starting from $X_{j,m}$ is an ordered set of subsequences: $TSC_{j,m} = \{X_{c1,m}, X_{c2,m}, \ldots, X_{ck,m}\}$ ($c1 \leq c2 \leq \ldots \leq ck$, $c1 = j$), such that for any $1 \leq i \leq k - 1$, we have $RNN(X_{ci,m}) = X_{c(i+1),m}$, and $LNN(X_{c(i+1),m}) = X_{ci,m}$.

Definition 12: An *all-chain set* S_{TSC} of X of length n is a set of all anchored time series chains within X that are not subsumed by another chain.

Assume that there are k old subsequences (subsequences of old streaming time series) whose *RNNs* change after x_n comes, and these subsequences form $S_{tem,n}$. Given the i-th element of $S_{TSC,n}$ $S_{TSC,n}(i)$, the length of $S_{TSC,n}(i)$ $Slen_{n,i}$, and starting position of j-th element of $S_{TSC,n}(i)$ $sp(S_{TSC,n}(i,j))$, we have following definitions.

Definition 13: A *pre-result position vector PPV_i* of X of length n at current time- tick is used to record starting position of each element of $S_{TSC,n}(i)$. Formally, $PPV_i = \left[sp\left(S_{TSC,n}(i,1)\right), sp\left(S_{TSC,n}(i,2)\right), \ldots, sp\left(S_{TSC,n}\left(i, Slen_{n,i}\right)\right)\right]$, where $sp\left(S_{TSC,n}(i,j)\right) < sp\left(S_{TSC,n}(i,j+1)\right)$.

Definition 14: An *update position vector UPV* of X of length n at current time-tick is used to record starting position of each element of $S_{tem,n}$. Let $S_{tem,n}(i)$ be the i-th element of $S_{tem,n}$, $STlen_n$ be the number of elements of $S_{tem,n}$, $sp\left(S_{tem,n}(i)\right)$ be the starting position of the i-th element of $S_{tem,n}$. Formally, $UPV = \left[sp\left(S_{tem,n}(1)\right), sp\left(S_{tem,n}(2)\right), \ldots, sp\left(S_{tem,n}(STlen_n)\right)\right]$, where $sp\left(S_{tem,n}(i)\right) < sp\left(S_{tem,n}(i+1)\right)$.

Assume that the length of streaming time series X is n at current time-tick.

Definition 15: A *temporary vector of PPV_i TVP_i* is a vector of length $n-m$. We initial it with all zeros firstly, and then set all elements whose indexes are the values of PPV_i elements to 1.

Definition 16: A *temporary vector of UPV TVU* is a vector of length $n-m$. We initial it with all zeros firstly, and then for the vector, we set all elements whose indexes are the values of UPV elements to 1.

3.2 Problem Definition

Problem Statement (*all-chain set mining over streaming time series*). Given a streaming time series X, where x_n is the most recent value, the subsequence length m, process the new element x_n and report all-chain set in the X as early as possible. In other words, our work is to get all-chain set mining results of current streaming time series after each new time series element comes.

4 Proposed Method

4.1 Naive Algorithm

For the issue of all-chain set mining over X, the most straightforward solution would be to apply LRSTROM and ALLC algorithms in [9] to X when new element x_n arrives. Note that after each x_n comes, the *PL, PR, IL* and *IR* of the new streaming time series are all needed to be rebuilded in the LRSTROM. We refer to this method as **Super-Naive**. Obviously, not only is this method expensive, but it also cannot be extended to the streaming case. A better solution is after x_n comes, we can get the *PL, PR, IL* and *IR* of the new streaming time series by updating the ones of the old streaming time series incrementally based on *DL* about $X_{n-m+1,m}$ firstly, then the ALLC is taken to handle the new streaming time series. This is because by Definitions 4–9, we have after x_n comes, the *PL* and *IL* of old streaming time series remain unchanged. We refer to this method as **Naive.** Overall, the Super-Naive and Naive are all extremely expensive, and cannot be extended to the streaming case.

4.2   All-Chain Set Mining Algorithm About Streaming Time Series (ASMSTS)

Basic Idea. The ASMSTS is based on the idea that obtaining mining results at current time-tick based on the ones at last one. Specifically, we begin by stating some properties from the analysis for all-chain set mining on the X of length n.

Property 1. Given streaming time series X of length n, subsequence $X_{i,m}$, we have that the $RNN(X_{i,m})$ and $LNN(X_{i,m})$ are both unique if these two values exist.

Property 2. Given any two anchored time series chains $ATS_i, ATS_j \in S_{TSC,n}$, we have that there no exists any common subsequence in ATS_i and ATS_j.

Property 3. Given streaming time series X, all-chain set of X $S_{TSC,n}$, subsequence $X_{i,m}$, we have $X_{i,m}$ exists and only exists in one anchored time series chain of $S_{TSC,n}$.

Theorem 1. Assume that there are k old subsequences whose $RNNs$ change after x_n comes, and these subsequences form $S_{tem,n}$. We can get $S_{TSC,n}$ by following steps:

1. if $k = 0$, after adding $X_{n-m+1,m}$ into the $S_{TSC,n-1}$, the $S_{TSC,n-1}$ is exactly the $S_{TSC,n}$.
2. if $k > 0$, For each subsequence $X_{q,m}$ of the $S_{tem,n}$,

Firstly, *Searching*. Find the anchored time series chain where $X_{q,m}$ is in the $S_{TSC,n-1}$;

Secondly, *Splitting*. Split the anchored time series chain into two halves, the first half ends with $X_{q,m}$, while the rest of the anchored time series chain is the second half.

Thirdly, *Appending*. For the new subsequence $X_{n-m+1,m}$ which caused by the arrival of x_n, if LNN of it is $X_{q,m}$, append $X_{n-m+1,m}$ at end of the first half after the splitting step.

Finally, *Adding*. After all elements of $S_{tem,n}$ are handled according to the previous three steps, if $X_{n-m+1,m}$ is still not added at end of any subsequence, then add it into the $S_{TSC,n-1}$. After these four steps above, the $S_{TSC,n-1}$ is exactly the $S_{TSC,n}$.

Theorem 1 can be inferred easily based on Definitions 10–12. Using Theorem 1, we can obtain the all-chain set mining results of streaming time series at current time-tick based on the ones at last time-tick.

Optimal Strategy. Theorem 1 needs to scan the $S_{TSC,n-1}$ $STlen_n$ times in all at each time-tick. By Properties 1–3 and Definitions 11–12, we have that for each element of original $S_{TSC,n-1}$, if we have learned which elements of $S_{tem,n}$ are in it, we just need to scan $S_{TSC,n-1}$ once to get $S_{TSC,n}$. Theorem 2 shows specific steps.

Theorem 2. After the new streaming time series element x_n comes in Theorem 1, if there exist k $(k > 0)$ elements in $S_{tem,n}$, a more simple method is as follows:

For each anchored time series chain of the original $S_{TSC,n-1}$, $S_{TSC,n-1}(i)$, assume that there exist r_i subsequences of $S_{tem,n}$ in it,

1. If $r_i = 0$, the $S_{TSC,n-1}(i)$ remains unchanged;
2. If $0 < r_i \leq k$, completing following steps:

Firstly, *Splitting*. Split the $S_{TSC,n-1}(i)$ into $r_i + 1$ parts. The front r_i parts end with these r_i subsequences respectively, while the rest of the $S_{TSC,n-1}(i)$ is the last part.

Secondly, *Appending*. If *LNN* of $X_{n-m+1,m}$ is any one of these r_i subsequences, then add $X_{n-m+1,m}$ at the end of the subsequence;

Thirdly, *Adding*. After all elements of the original $S_{TSC,n-1}$ are handled according to the previous two steps, if $X_{n-m+1,m}$ is still not added at the end of any subsequence, then one additional new element $X_{n-m+1,m}$ should be added into the current $S_{TSC,n-1}$.

After these three steps, the current $S_{TSC,n-1}$ is exactly the original $S_{TSC,n}$.

Theorem 2 starts with the premise that for each $S_{TSC,n-1}(i)$, we have learned which subsequences of $S_{tem,n}$ exist in it. However, it is still a rather time consuming work to finish this premise. Theorem 3 can be used to facilitate this process.

Theorem 3. Let dot product of TVP_i and TVU be RV_i. For each element of RV_i $RV_i(j)$, if $RV_i(j) = 1$, then in $S_{TSC,n-1}(i)$, we can find the subsequence starting with the element whose index is j in old streaming time series, and the subsequence also exists in $S_{tem,n}$. These subsequences are all common elements between $S_{TSC,n-1}(i)$ and $S_{tem,n}$.

Theorem 3 can be inferred easily by Definitions 15–16. We take the example of Ref. 9 to illustrate the Definitions 13–16 and Theorem 3.

Example 1: Given the snippet of streaming time series: 47, 32, 1, 22, 2, 58, 3, which is the old one. The new coming element is 36. Figure 1 shows the illustration.

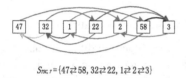

$S_{TSC,7} = \{47 \rightleftarrows 58, 32 \rightleftarrows 22, 1 \rightleftarrows 2 \rightleftarrows 3\}$

(a) Old streaming time series and $S_{TSC,7}$

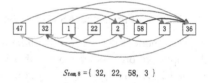

$S_{tem,8} = \{ 32, 22, 58, 3 \}$

(b) New streaming time series and $S_{tem,8}$

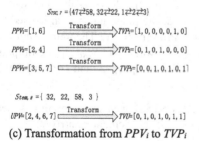

(c) Transformation from PPV_i to TVP_i

(d) Construction of RV and construction of TVU

Fig. 1. Illustration of Definitions 13–16 and Theorem 3.

As shown in Fig. 1(a)(b), when 36 comes, the old streaming time series is: 47, 32, 1, 22, 2, 58, 3, and $S_{TSC,7} = \{47 \rightleftarrows 58, 32 \rightleftarrows 22, 1 \rightleftarrows 2 \rightleftarrows 3\}$. The arrival of 36 changes RNNs of 32, 22, 58 and 3, so $S_{tem,8} = \{32,22,58,3\}$. Figure 1(c) shows the transformation from *PPV* to *TVP*, and the construction of *TVU*. Consider $47 \rightleftarrows 58$ of $S_{TSC,7}$, by Definitions 13 and 15, we have $PPV_1 = [1, 6]$, $TVP_1 = [1,0,0,0,0,1,0]$. Similarly, TVP_2 and TVP_3 can both be obtained. Consider $S_{tem,8}$, by Definition 14, we have $UPV = [2, 4, 6, 7]$.

Similarly, we can get *TVU* based on *UPV*. Figure 1(d) shows Theorem 3. For example, by RV_1, we have 58 is the common subsequence between $47 \rightleftharpoons 58$ and $S_{tem,8}$

all-chain Set Mining Algorithm about Streaming Time Series (ASMSTS)
Architecture and Workflows of the ASMSTS. The ASMSTS algorithm is proposed based on Theorems 1–3. Figure 2 illustrates the architecture of the ASMSTS.

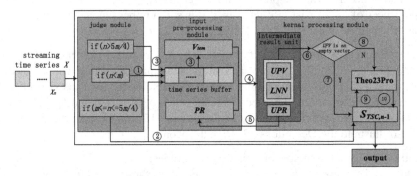

Fig. 2. Architecture and workflows of the ASMSTS

As shown in Fig. 2, the ASMSTS consists of three components:

1. A judge module that performs different operations based on different conditions.
2. An input pre-processing module that buffers streaming time series, and maintains *PR* and V_{tem}. The *PR* is right matrix profile on the old streaming time series, and the V_{tem} consists of z-normalized Euclidean distance between $X_{n-m+1,m}$ and each subsequence of new streaming time series.
3. A kernel processing module that consists of the intermediate results unit where outputs of input pre-processing module are stored, the judge unit which is used to determine whether the *UPV* is null, the Theo23Pro unit where Theorem 2 and the *splitting*, *appending* and *adding* steps of Theorem 3 work, and $S_{TSC,n-1}$. The outputs of input pre-processing module include *UPV*, *LNN* of $X_{n-m+1,m}$, and *UPR* (updated *PR*).

Figure 2 also illustrates the procedure of the ASMSTS. Given streaming time series *X*, subsequence length *m*, when new streaming time series element x_n arrives:

Step 1: if $n < m$, Put x_n into time series buffer, and move on to x_{n+1};
Step 2: if $m \leq n < 5m/4$, Put $X_{n-m+1,m}$ and x_n into $S_{TSC,n-1}$ and the time series buffer respectively, then move on to x_{n+1} after outputting contents of current $S_{TSC,n-1}$. Note that the current $S_{TSC,n-1}$ is exactly the $S_{TSC,n}$;
Step 3: if $n \geq 5m/4$, perform **Steps** 3–11. Put x_n into time series buffer firstly, and then calculate z-normalized Euclidean distance between $X_{n-m+1,m}$ and all subsequence of new streaming time series. These distances form a vector V_{tem} of length $n - m + 1$. Let all elements of V_{tem} whose indexes are in the exclusion zone of $X_{n-m+1,m}$ be *infs* [10, 15];

Step 4: after adding one element whose value is *infs* at the end of *PR* which is on the old streaming time series, the *UPR* can be obtained by comparing *PR* with V_{tem} as the LRSTROM does. During this progress, $S_{tem,n}$ and the *LNN* of $X_{n-m+1,m}$ can also be obtained. Furthermore, by Definition 14 and $S_{tem,n}$, the *UPV* can be obtained;

Step 5: update *PR* of the input pre-processing module with the *UPR* to obtain the *PR* of new streaming time series;

Step 6: determine whether the *UPV* is null;

Step 7: if *UPV* is an empty vector, after adding a new element $X_{n-m+1,m}$ into the $S_{TSC,n-1}$, output contents of the current $S_{TSC,n-1}$ firstly, and then move on to t x_{n+1}.

Steps 8–10: if *UPV* is not empty, for each element of $S_{TSC,n-1}$, process it by Theo23Pro firstly, then replace the element with results of Theo23Pro.

Step 11: output contents of the current $S_{TSC,n-1}$, and then move on to the x_{n+1}. Note that the current $S_{TSC,n-1}$ is exactly the $S_{TSC,n}$.

all-chain Set Mining Algorithm about Streaming Time Series (ASMSTS). The ASMSTS algorithm is given in Fig. 3.

In lines 2 –6, the length of current streaming time series is less than *m*, so we just need to put x_n into time series buffer. In lines 7–14, any two subsequences of *X* at current time-tick are trivial match, and each subsequence is a time series chain of length 1. All these subsequences form all-chain set of *X*. We store each subsequence in the form of vector of length 1, and initialize the vector with starting position of the subsequence. In lines 15–34, after x_k comes, there are subsequences which are not trivial match with $X_{n-m+1,m}$ in *X*, so we can construct a vector of length $k-m + 1$ V_{tem} with z-normalized Euclidean distances between $X_{n-m+1,m}$ and all subsequences of *X*; then, as LRSTROM does, we can get updated *PR*, *UPV*, $LNN_{n-m+1,m}$ based on V_{tem} and *PR*; thirdly, deal with each element of $S_{TSC,n-1}$ based on **Steps 8–10** of the ASMSTS workflows (lines 28–33). After these operations, $S_{TSC,n-1}$ is exactly the $S_{TSC,n}$ (line 34).

We still take the example of Ref. 9 to illustrate how the ASMSTS works.

Example 2: Given the snippet of streaming time series: 47, 32, 1, 22, when 47 comes, the streaming time series buffer is {47}, and the $S_{TSC,1}$ is {47}, *PR* = [*inf*].

When 32 comes, the time series buffer = {47,32}, initial $PR = [inf, inf], V_{tem} = [15, inf]$. Comparing initial *PR* with V_{tem} as LRSTROM does, we have $UPR = [15, inf], UPV = [1], LNN_{2,1} = 47$. After updating *PR* with $UPR, PR = [15, inf]$. As $S_{TSC,1}$ = {47}, so PPV_1 = [1]. By Definitions 15–16 and Theorems 2–3, we have $RV_1 = TVP_1 . * TVU = [1]$, and $S_{TSC,2} = \{47 \rightleftarrows 32\}$. Similarly, when 1 comes, the time series buffer = {47, 32, 1}, $S_{TSC,3} = \{47 \rightleftarrows 32 \rightleftarrows 1\}$, and *PR* = [15, 31, *inf*].

When 22 comes, the time series buffer = {47, 32, 1, 22}, initial *PR* = [15, 31, *inf*, *inf*], $V_{tem} = [25, 10, 21, inf]$. Comparing initial *PR* with V_{tem} as LRSTROM does, we have $UPR = [15, 10, 21, inf], UPV = [2, 3], LNN_{4,1} = 32$. After updating *PR* with *UPR*, the *PR* = [15, 10, 21, *inf*]. As $S_{TSC,3} = \{47 \rightleftarrows 32 \rightleftarrows 1\}$, PPV_1 = [1–3]. By Definitions 15–16 and Theorems 2–3, we have $RV_1 = TVP_1 . * TVU = [0, 1, 1]$, and $S_{TSC,4} = \{47 \rightleftarrows 32 \rightleftarrows 22, 1\}$.

Algorithm ASMSTS

Input: a new value x_n at time-tick n, subsequence length m

Output: the all-chain set mining results of streaming time series at time-tick n

1. $k=len+1$;　　// len is the number of elements of the time series buffer before x_n comes
2. **If** $(k<m)$
3.　　Put x_n into the streaming time series buffer;
4.　　$S_{TSC,n}$=null;
5.　　Output contents of the current $S_{TSC,n}$, and then process the x_{n+1};
6. **End**
7. **If** $(m \leq k \leq 5m/4)$
8.　　Put x_n into the streaming time series buffer;
9.　　$S_{TSC,n} \leftarrow$ ConsVec(1, $k-m+1$) //Construct vector of length 1, initialize the value of
　　　　　　　　　　　　　　　　　　　vector with $k-m+1$, then put it into $S_{TSC,n}$;
10.　　**If** $(k==5m/4)$
11.　　　Construct PR vector of length $k-m+1$, and initial each element of PR with $infs$;
12.　　**End**
13　　Output contents of the current $S_{TSC,n}$, and then process the x_{n+1};
14. **End**
15. **If** $(k>5m/4)$
16.　　Put x_n into the streaming time series buffer;
17.　　Add one element whose value is $infs$ at the end of PR;
18.　　**For** (each subsequence of X in the time series buffer $X_{s,m}$, where $n-k+1 \leq s \leq n-m+1$)
19.　　$V_{tem}(s) = $ Calz-Euc$(X_{n-m+1,m}, X_{s,m})$
　　　　　　　　　　　　　　// calculate the z-normalized Euclidean distance between
　　　　　　　　　　　　　　$X_{n-m+1,m}$ and $X_{s,m}$, and V_{tem} is a vector of length $k-m+1$;
20.　　　Initial elements of V_{tem} whose indexes are in the exclusion zone of length $m/4$
　　　　before the location of $X_{n-m+1,m}$ with $infs$;　　　　　　　// avoid trivial match
21.　　**End**
22.　　$(UPR, UPV, LNN_{n-m+1,m})$ = LRSTROM-COMP(V_{tem}, PR);
　　　　　　　　　　　　　// compare PR with V_{tem} as the LRSTROM does, and
　　　　　　　　　　　　　obtain the UPR, UPV, $LNN_{n-m+1,m}$.
23.　　$PR \leftarrow UPR$;　　//update the PR with the UPR
24.　　**If** UPV is empty vector
25.　　　$S_{TSC,n-1} \leftarrow$ ConsVec(1, $k-m+1$);
26.　　　Output contents of the current $S_{TSC,n-1}$, then process the x_{n+1};
27.　　**End**
28.　　**For** (each element of $S_{TSC,n-1}$ $S_{TSC,n-1}(i)$)
29.　　ppv_i=GetPPV$(S_{TSC,n-1}(i))$;　　// get ppv about $S_{TSC,n-1}(i)$ based on Definition 8
30.　　　Construct TVU and TVP_i based on the UPV and PPV_i respectively;
31.　　　$RV_i = TVP_i .* TVU$;　　　　//dot product of TVP_i and TVU
32.　　　RUN-Theorem 2-3 $(RV_i, LNN_{n-m+1,m}, ppv_i, S_{TSC,n-1}(i))$;　　//run Theorem 2 and 3
33.　　**End**
34.　　Output contents of the current $S_{TSC,n-1}$, then process the x_{n+1};
35. **End**

Fig. 3. ASMSTS algorithm

Complexity Analysis. For each new x_n, the ASMSTS just needs to traverse the $S_{TSC,n-1}$ once, so the time overhead of ASMSTS at each time-tick is $2*O(n)$, where the first and second $O(n)$ are time costs of V_{tem} construction and $S_{TSC,n-1}$ traversing respectively. The space cost of ASMSTS is $3*O(n)$, as X, PR and $S_{TSC,n-1}$ are all need to be stored.

5   Experiments

Our experiments were conducted on an Intel i7-4710MQ 2.5 GHz with 4 GB of memory, running windows 7. The experiments were designed to answer the following questions: 1. how successful is ASMSTS in detecting all-chain set of streaming time series? 2. how does ASMSTS scale with the lengths of streaming time series and subsequence in terms of the computational time and memory space?

5.1   Dataset

Penguinshort and TiltABP are both the data sets of [9]. The Penguinshort is a part of Penguin dataset collected by attaching a small multi-channel data-logging device to the Magellanic penguin, and consists of 900 data points. TiltABP is the patient's arterial blood pressure that is a response to changes in position induced by a tilt table. The full data consists of 40000 data points. All these data set are available in [16].

5.2   Discovery of All-Chain Set

The ALLC proposed in [9] can detect all-chain set of time series, so the Super-Naive can be taken as a criterion. Given any dataset, if ASMSTS has the same results as the Super-Naive, we have the ASMSTS is effective in discovering all-chain set of streaming time series. Table 1 shows the details of the experiment results. Given a data set DS, we define the SR(similarity rate) of the algorithm A relative to algorithm B on DS as $SR_{A,B} = |Result_A(DS) \cap Result_B(DS)|/|Result_A(DS)|$. The $Result_A(DS)$ and $Result_B(DS)$ are the results of algorithm A and B on DS respectively, and the $|Result_A(DS)|$ and $|Result_B(DS)|$ are the elements number of $Result_A(DS)$ and $Result_B(DS)$ respectively. Obviously, if $SR_{A,B} = SR_{B,A}$ holds, we have the results of A and B algorithms on DS are the same. Note that $SR_{AS,SN}$, $SR_{SN,AS}$, $SR_{N,SN}$, and $SR_{SN,N}$ of Table 1 are the abbreviated form of $SR_{ASMSTS,Super-Naive}$, $SR_{Super-Naive,ASMSTS}$, $SR_{Naive,Super-Naive}$ and $SR_{Super-Naive,Naive}$ respectively.

Table 1. Results of all-chain set mining results

Dataset	Size of dataset	m	$SR_{AS,SN}$	$SR_{SN,AS}$	$SR_{N,SN}$	$SR_{SN,N}$
penguinshort	900	28	1	1	1	1
TiltABP	40000	200	1	1	1	1

As we can see in Table 1, the results of these three algorithms are all the same. So we have ASMSTS can discover all-chain set of the streaming time series.

5.3   Performance

We did experiments to evaluate the efficiency and to verify the complexity of the ASMSTS. The space cost and wall clock time are the average space cost and processing time needed to carry out the algorithm for each time-tick respectively.

Figure 4 compares ASMSTS and other two implementations in terms of computation time based on TiltABP. The default value of the dataset sizes and m are 40000 and 200 respectively.

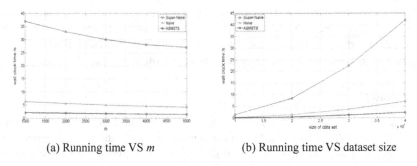

(a) Running time VS m (b) Running time VS dataset size

Fig. 4. Performance study based on computation time

From Fig. 4(a), we can observe the time costs of these three methods decrease as m increases. This is because the larger subsequence length is, the less subsequences needed to be handled are. In addition, we can observe the running time of ASMSTS is less than the ones of other two algorithms, which also demonstrates the ASMSTS is efficiency in terms of execution time. From Fig. 4(b), we can also observe ASMSTS can identify all-chain set much faster than other two methods, which also demonstrates the ASMSTS is more efficiency in running time than other two ones.

Figure 5 compares ASMSTS and other implementations in terms of space cost based on TiltABP. The default value of the dataset sizes and m are 40000 and 200 respectively in this experiment.

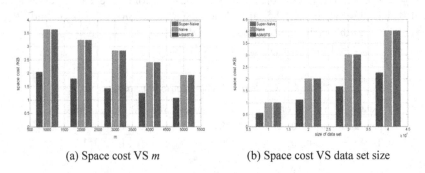

(a) Space cost VS m (b) Space cost VS data set size

Fig. 5. Performance study based on space cost

Figure 5(a) shows the space costs of each algorithm decreases as m increases. This is because the larger m is, the fewer elements of PL, IL, PR, and IR are. Figure 5(b) shows the space cost of each algorithm increases monotonically with the sample size increasing. This is because the larger sample size is, the larger sizes of PL, IL, PR, and IR are. Figure 5(a)(b) also show the space cost of ASMSTS is less than the ones of other two ones, which demonstrates ASMSTS is more efficiency in space cost than them.

6 Conclusion and Future Work

In this paper, we introduced the problem of all-chain set mining over streaming time series, and proposed ASMSTS, a new fast method to solve this problem. The ASMSTS outperforms the two Naives by a wide margin in time and space, and guarantees the same results as the ones of the Naive. The experiments show ASMSTS works as expected. There are many interesting research problems related to all-chain set mining over streaming time series that should be pursued further. For example, further improving the ASMSTS, the all-chain set mining based on the damped window are all interesting problems for future research.

Acknowledgement. This research is supported by: the Natural Science Foundation of Inner Mongolia in China (Grant nos. 2018BS06001), the National Natural Science Foundation of China (Grant nos. 61862047, 61572119, 61622202), and the Fundamental Research Funds for the Central universities (Grant No.N150402005).

References

1. Begum, N., Keogh, E.: Rare time series motif discovery from unbounded streams. In: VLDB 2015, pp. 149–160. Association for Computing Machinery, USA (2015)
2. Patel, P., Keogh, E., Lin, J., Lonardi, S.: Mining motifs in massive time series databases. In: ICDM 2002, pp. 370–377. IEEE Computer Society, Piscataway (2002)
3. Zhu, Y., Zimmerman, Z., Senobari, N.S., et al.: Matrix profile ll: exploiting a novel algorithm and GPUs to break the one hundred million barrier for time series motifs and joins. In: ICDM 2016, pp. 739–748. IEEE Computer Society, Piscataway (2016)
4. Yeh, C.C.M., Zhu, Y., Ulanova, L., et al.: Matrix profile I: all pairs similarity joins for time series: a unifying view that includes motifs, discords and shapelets. In: ICDM 2016, pp. 1317–1322, IEEE Computer Society, Piscataway (2016)
5. Hao, M.C., Marwah, M., Janetzko, H., et al.: Visualexploration of frequent patterns in multivariate time series. Inf. Vis. **11**(1), 71–83 (2012)
6. Shokoohi-Yekta, M., Chen, Y.P., Campana, B., et al.: Discovery of meaningful rules in time series. In: Proceedings of the 21th ACM SIGKDD, Philadelphia, PA, USA, pp. 1085–1094 (2015)
7. Syed, Z., Stultz, C., Kellis, M., et al.: Motif discovery in physiological datasets: a methodology for inferring predictive elements. TKDD **4**(1), 2 (2010)
8. Zhu, X., Oates, T.: Finding story chains in newswire articles. In: Proceedings of the 2012 IEEE 13th International Conference on Information Reuse and Integration, pp. 93–100. IEEE Computer Society, Piscataway (2012)

9. Zhu, Y., Imamura, M., Nikovski, D.: Matrix profile VII: time series chains: a new primitive for time series data mining. In: ICDM 2017, pp. 695–704. IEEE, Computer Society, Piscataway (2017)
10. Zhu, Y., Imamura, M., Nikovski, D.: Introducing time series chains: a new primitive for time series data mining. Knowl. Inf. Syst. (2018). https://doi.org/10.1007/s10115-018-1224-8
11. Zhu, Y., Imamura, M., Nikovski, D., Keogh, E.: Time series chain: A Novel tool for time series data mining. In: Proceedings of the Twenty-Seventh International Joint Conference on Artificial Intelligence 2018, pp. 5414–5418, Springer Verlag, Heidelberg (2018)
12. Bögel, T., Gertz, M.: Time will tell: temporal linking of news stories. In: Proceedings of the ACM/IEEE Joint Conference on Digital Libraries 2015, pp. 195–204. IEEE Computer Society, Piscataway (2015)
13. Li, Z., Han, J., Ding, B., Kays, R.: Mining periodic behaviors of object movements for animal and biological sustainability studies. Data Min. Knowl. Discov. **24**(2), 355–386 (2012)
14. Gama, J., Žliobaitė, I., Bifet, A., Pechenizkiy, M., Bouchachia, A.: A survey on concept drift adaptation. ACM Comput. Surv. (CSUR) **46**(4), 1–37 (2014)
15. Chiu, B., Keogh, E., Lonardi, S.: Probabilistic discovery of time series motifs. In: ACM SIGKDD 2003, Philadelphia, PA, USA, pp. 493–498 (2003)
16. Supporting webpage. https://sites.google.com/site/timeserieschain/

Hawkes Process with Stochastic Triggering Kernel

Feng Zhou[1,4(✉)], Yixuan Zhang[2], Zhidong Li[3,4], Xuhui Fan[1], Yang Wang[3,4], Arcot Sowmya[1], and Fang Chen[3,4]

[1] University of New South Wales, Sydney, Australia
[2] The University of Sydney, Sydney, Australia
[3] University of Technology Sydney, Sydney, Australia
[4] CSIRO DATA61, Sydney, Australia
feng.zhou@data61.csiro.au

Abstract. The impact from past to future is a vital feature in modelling time series data, which has been described by many point processes, e.g. the Hawkes process. In classical Hawkes process, the triggering kernel is assumed to be a deterministic function. However, the triggering kernel can vary with time due to the system uncertainty in real applications. To model this kind of variance, we propose a Hawkes process variant with stochastic triggering kernel, which incorporates the variation of triggering kernel over time. In this model, the triggering kernel is considered to be an independent multivariate Gaussian distribution. We derive and implement a tractable inference algorithm based on variational auto-encoder. Results from synthetic and real data experiments show that the underlying mean triggering kernel and variance band can be recovered, and using the stochastic triggering kernel is more accurate than the vanilla Hawkes process in capacity planning.

Keywords: Hawkes process · Stochastic triggering kernel

1 Introduction

Point process is a common statistical model in describing the pattern of event occurrence in many real world applications, such as a series of earthquakes and the order book in finance. Mutual dependence between events is an important factor in describing the clustering effect in point process. A variety of models are proposed for the dependence, such as Hawkes process (HP) [10] and correcting model [16]. Among those models, HP is the most extensively used one for modelling the self-exciting phenomenon where the influence decays over time.

HP has been used to estimate the intensity (rate of event occurrence) by accumulating the triggering effect from past events. As an intensity estimator, it has been used widely in social networks [18], crime [14] and financial engineering [8]. The triggering kernel in most HP implementations [8] is modelled as a deterministic function. In the real world, however, the actual triggering effect

© Springer Nature Switzerland AG 2019
Q. Yang et al. (Eds.): PAKDD 2019, LNAI 11439, pp. 319–330, 2019.
https://doi.org/10.1007/978-3-030-16148-4_25

from each event can vary because of the system uncertainty and the deterministic triggering kernel is rather limited in capability to model the variation. To model this phenomenon, we introduce variance into the triggering kernel to enable the triggering kernel of HP to be stochastic. We visualize it as a band addition to the triggering kernel (see the example in Fig. 1a).

The importance of the band may be ignored in real applications, because the learned average triggering kernel usually has the largest likelihood to fit the observed data. As a result, when we do prediction, the vanilla HP would eventually be used. However, as we can see later, this band is meaningful for the risk-based planning. For example, when capacity planning is performed in the taxi allocation problem with HP [7], the arriving rate of pickup events is predicted from historic pickups. Based on the prediction, vehicles can be allocated to an area to cover the pickup need (i.e. #pickups $\leq$ #vehicles). If the taxi company uses the intensity $\lceil \lambda \rceil$ learned from vanilla HP as the expected rate of pickups to satisfy, about 50% probability that the pickup need can be satisfied. To plan for a higher probability, more vehicles need to be sent, e.g. for extra probability $P_m = Poisson(x \leq M|\lambda) - Poisson(x \leq \lambda|\lambda)$, extra $m = \lceil M - \lambda \rceil$ vehicles need to be sent. However, in Sect. 6, when there is a significant variance on the triggering effect, sending m vehicles can only satisfy pickup need with extra probability less than P_m, which will lead to a decision with insufficient capacity. Using our stochastic triggering kernel, one can obtain extra information about the distribution of the triggering effect, so the insufficient capacity could be compensated. The similar issue could happen in other HP-based capacity planning applications, as long as there is a significant variance on the triggering kernel.

We propose a HP variant with stochastic triggering kernel (HP-STK), aimed at quantifying the variance of triggering kernel so as to overcome the problem mentioned above. Based on Gaussian white noise, we consider two cases for the variance: homoscedasticity (i.e. constant variance) and heteroscedasticity (i.e. time-varying variance). Then we propose a tractable inference method to replace the original maximum likelihood estimation (MLE) and apply the inference of both cases to the variational auto-encoder (VAE) [11] framework.

To our best knowledge, no work has been done before to model the variance of triggering kernel in HP. Specifically, our work makes the following contributions: (1) we propose a new HP variant named HP-STK, in which the variance of triggering kernel is incorporated to overcome the underestimation problem in capacity planning; (2) two special cases are considered: homoscedasticity and heteroscedasticity; (3) the uniform-trigger-kernel-based MLE is proposed to replace the original MLE and a VAE-based algorithm is used for inference.

2   Related Work

The model proposed in this paper is motivated by the Cox process [2]. The Cox process, also known as the doubly stochastic Poisson process, is a stochastic process which is an extension of a Poisson process where the intensity function is itself a stochastic process. It has been widely used in many applications, such as astronomy [9] and neuroscience [3]. A common version of Cox process is the Gaussian Cox process [15], where the intensity function is modeled as a Gaussian process. However, the inference is intractable because of non-conjugacy and integration over infinite-dimensional random function. Different inference algorithms based on Markov chain Monte Carlo (MCMC) or Laplace approximation have been proposed in [1,4]. In Cox process, the randomness is added to the intensity, but in this paper the randomness is on triggering kernel to reduce dimensions.

There are also HP extensions to model the randomness of triggering kernel. For example, Dassios [5] proposed a stochastic HP, where jumps in the intensity function are considered to be independent and identically distributed (i.i.d.) random variables. Lee [12] extended all jumps to a stochastic process and solved it using stochastic differential equation. Both works focus on stochastic jumps, but our proposed model considers the whole triggering kernel as a stochastic process which is more generalized.

Another related direction is VAE [11]. VAE has a similar architecture with auto-encoder, but makes an assumption about the distribution of latent variables. VAE is a generative model, which combines ideas from neural network with statistical inference. It can be used to learn a low dimensional representation Z of high dimensional data X. It assumes that the data is generated by a decoder $P(X|Z)$ and the encoder is learning an approximation $Q(Z|X)$ to the posterior distribution $P(Z|X)$. It uses the variational method for latent representation learning, which results in a specific loss function. In this paper we apply the loss of VAE into our model.

3   Proposed Model

3.1   Hawkes Process

A Hawkes process is a stochastic process, whose realization is a sequence of timestamps $\{t_i\} \in [0, T]$. Here, t_i stands for the time of occurrence for the i-th event and T is the observation duration for this process. An important way to characterize a HP is through the definition of a conditional intensity function that captures the temporal dynamics. The conditional intensity function is defined as the probability of event occurring in an infinitesimal time interval $[t, t + dt)$ given the history:

$$\lambda(t) = \lim_{\Delta t \to 0} \frac{P(event\ occurring\ in\ [t, t + \Delta t)|\mathcal{H}_t)}{\Delta t} \tag{1}$$

where $\mathcal{H}_t = \{t_i | t_i < t\}$ are the historical timestamps before time t. Then the specific form of intensity for HP is:

$$\lambda(t) = \mu + \sum_{t_i < t} \gamma^*(t - t_i) \tag{2}$$

where $\mu > 0$ is the baseline intensity which is a constant, and $\gamma^*(\cdot)$ is the triggering kernel. In most cases, the triggering kernel is assumed to be an exponential decay function. The summation of triggering kernels explains the nature of self-excitation, which is the occurrence of events in the past will intensify the intensity of events occurring in the future. Then the log-likelihood function can be expressed using the above conditional intensity as:

$$\log \mathcal{L} = \sum_{i=1}^{n} \log \lambda(t_i) - \int_0^T \lambda(t)dt \tag{3}$$

3.2   HP with Stochastic Triggering Kernel

In HP-STK, we target to introduce variance into the triggering kernel of HP. We define the HP-STK model and see what is the variance of triggering kernel.

Definition 1. *HP-STK is a Hawkes process whose triggering kernel after event t_i can be written as a sample drawn from a stochastic process with $\Delta t \in \mathbb{R}^+$ as:*

$$\gamma_i(\Delta t) = \bar{\gamma}(\Delta t | \boldsymbol{\xi}) + \epsilon_i(\Delta t), where\ \epsilon_i(\Delta t) \sim P(\epsilon(\Delta t) | \boldsymbol{\theta}) \tag{4}$$

where $\gamma_i(\Delta t)$ is the triggering kernel after event t_i, $\bar{\gamma}(\Delta t | \boldsymbol{\xi})$ is a deterministic triggering kernel with parameters $\boldsymbol{\xi}$, $\epsilon_i(\Delta t)$ is a noise function for $\gamma_i(\Delta t)$ and $P(\cdot)$ is a distribution over function with parameters $\boldsymbol{\theta}$.

Naturally, $P(\epsilon(\Delta t)|\boldsymbol{\theta})$ can be defined as a Gaussian process. Here for simplicity $P(\epsilon(\Delta t)|\boldsymbol{\theta})$ is defined as an independent multivariate Gaussian distribution $\boldsymbol{N}(\epsilon(\Delta t)|\boldsymbol{0}, \sigma^2(\Delta t) \cdot \boldsymbol{I})$ (expressed in finite dimensions) where $\boldsymbol{I}$ is the identity matrix which means there is no covariance. $\bar{\gamma}(\Delta t|\boldsymbol{\xi})$ and $\sigma^2(\Delta t)$ are both defined to be in parametric form. In conclusion, we define $\bar{\gamma}(\Delta t|\boldsymbol{\xi}) = \alpha \exp(-\beta \Delta t)$, $\sigma^2(\Delta t) = \sigma_c^2$ in homoscedastic case and $\sigma^2(\Delta t) = (\alpha_\sigma \exp(-\beta_\sigma \Delta t))^2$ in heteroscedastic case. Here we define the $\sigma(\Delta t)$ to be an exponential decay function because in many scenarios it would be common to have a high variance just after a triggering event and have a lower variance afterwards, but in fact $\sigma(\Delta t)$ can be extended to other cases, e.g. linear decreasing variance or periodic variance. It can be seen that the homoscedasticity is just a special case of heteroscedasticity by setting: $\alpha_\sigma = \sigma_c$ and $\beta_\sigma = 0$.

The intensity of HP-STK can be written as:

$$\lambda(t) = \mu + \sum_{t_i < t}(\alpha \exp(-\beta(t - t_i)) + \epsilon_i(t - t_i)) \tag{5}$$

To avoid the superposition of $\epsilon_i(t - t_i)$ to explode, $\epsilon_i(\Delta t)$ and $\bar{\gamma}(\Delta t)$ are both defined on the support of $[0, T_\gamma]$ and 0 afterwards. In the theory of point process the intensity has to be positive, so $\lambda(t)$ is restricted to $(\lambda(t))_+$ (i.e. $\lambda(t) = 0$ if $\lambda(t) < 0$). Because the $\gamma_i(\Delta t)$ is subject to Gaussian distribution: see (4), so the $\lambda(t)$ is also subject to Gaussian distribution: see (5) and $(\lambda(t))_+$ is subject to a truncated Gaussian distribution. In real applications, the triggering kernel variance $\sigma^2(\Delta t)$ is always small compared with the intensity, so the truncated Gaussian distribution can be seen as a Gaussian distribution approximately. As we can see later, using Gaussian to describe $\gamma_i(\Delta t)$ introduces computation convenience to the inference.

3.3 Stability Condition

The stability condition of Hawkes process has been proposed in [10]: the Hawkes process P_t is stable if and only if $\int_0^\infty \gamma^*(\cdot) < 1$.

Because $\gamma_i(\cdot)$ is subject to Gaussian distribution in our model, the $\int_0^\infty \gamma_i(\cdot)$ is also subject to Gaussian distribution:

$$\int_0^\infty \gamma_i(\cdot) \sim N(x| \int_0^\infty \alpha\exp(-\beta t)dt, \int_0^\infty \sigma^2(t)dt) \tag{6}$$

where Δt is replaced by t and x is the integral value.

Definition 2. *HP-STK is probabilistically stable with $P(\int_0^\infty \gamma_i(\cdot) < 1)$.*

For homoscedasticity, in order to avoid the $\int_0^\infty \sigma_c^2 dt$ to explode, $\gamma_i(\Delta t)$ is defined on the support of $[0, T_\gamma]$. Given $\exp(-\beta T_\gamma) \approx 0$, we have the stability probability:

$$P_{homo} = \int_{-\infty}^1 N(x|\frac{\alpha}{\beta}, \sigma_c^2 T_\gamma)dx \tag{7}$$

For heteroscedasticity, the stability probability is:

$$P_{hetero} = \int_{-\infty}^1 N(x|\frac{\alpha}{\beta}, \frac{\alpha_\sigma^2}{2\beta_\sigma})dx \tag{8}$$

The probabilistic stability of homoscedastic HP-STK is constrained by T_γ. Therefore, when T_γ is undetermined, heteroscedastic HP-STK is recommended.

4 Inference

4.1 Inference with Uniform Triggering Kernel

Given a set of observed data, the goal of inference is to evaluate these parameters: μ, α, β, σ_c for homoscedastic case, and μ, α, β, α_σ, β_σ for the heteroscedastic case. We use MLE to infer parameters where the log-likelihood is:

$$\log \mathcal{L}(\{t_i\}_{i=1}^n | \mu, \boldsymbol{\Theta})$$
$$= \log \int_{\gamma_n} \cdots \int_{\gamma_2} \int_{\gamma_1} \mathcal{L}(\{t_i\}_{i=1}^n | \mu, \gamma_1, \gamma_2, \cdots, \gamma_n) \tag{9}$$
$$P(\gamma_1|\boldsymbol{\Theta})P(\gamma_2|\boldsymbol{\Theta})\cdots P(\gamma_n|\boldsymbol{\Theta})d\gamma_1 d\gamma_2 \cdots d\gamma_n$$

where the $\boldsymbol{\Theta}$ stands for θ and $\boldsymbol{\xi}$ in (4). However, this log marginal likelihood is complicated to work out because of multiple integrals. To solve this problem, an intuitive way is to assume $\gamma_1 = \gamma_2 = \cdots = \gamma_n = \gamma$ (i.e. the uniform triggering kernel). Then the log-likelihood can be rewritten as:

$$\log \mathcal{L}(\{t_i\}|\mu, \boldsymbol{\Theta}) = \log \int_{\gamma} \mathcal{L}(\{t_i\}|\mu, \gamma) \cdot P(\gamma|\boldsymbol{\Theta})d\gamma \qquad (10)$$

It is worth noting that, given a set of observed data, the estimation with the uniform triggering kernel is equivalent to the original one. This is proved by (9) and (10), because we get the same $\log \mathcal{L}(\{t_i\})$ with respect to $\mu, \boldsymbol{\Theta}$.

After transforming (9) to (10), we can directly infer the parameters using Monte Carlo integration. However, we still need to calculate the likelihood which is not numerically stable. To solve this problem, we propose an inference method based on VAE.

4.2   Inference with VAE

In fact, our proposed model can be considered as a VAE to some extent. So the loss function [11] of VAE can be applied to our model for inference. The loss function of VAE is the negative log-likelihood with a regularizer:

$$L = -\mathbb{E}[\log \mathcal{L}(\{t_i\}|\mu, \gamma(\cdot))] + \kappa \cdot D_{KL}[P(\gamma(\cdot)|\boldsymbol{\Theta})\|Q(\gamma(\cdot))] \qquad (11)$$

where the first term is the expectation of log-likelihood of $\{t_i\}$ given $\gamma(\cdot)$. The expectation is taken with respect to the encoder's distribution over $\gamma(\cdot)$. This term encourages the decoder to learn to construct the observed sequence data. If the decoder's output does not fit the data well, it will incur a large cost in the loss function. The second term is a regularizer with a weight parameter κ. It is the Kullback-Leibler (KL) divergence between the encoder's distribution $P(\gamma(\cdot)|\boldsymbol{\Theta})$[1] and $Q(\gamma(\cdot))$. $Q(\gamma(\cdot))$ is a benchmark distribution and it describes a priori about $\gamma(\cdot)$. This divergence measures how close $P(\gamma(\cdot)|\boldsymbol{\Theta})$ is to $Q(\gamma(\cdot))$.

In the loss function, the first term can be rewritten as $-\int_{\gamma} \log \mathcal{L}(\{t_i\}|\mu, \gamma) \cdot P(\gamma|\boldsymbol{\Theta})d\gamma$. It is an integral over an infinite-dimensional stochastic function and it has no analytical solution because of non-conjugacy. To solve these problems, we use discretization and Monte Carlo integration to transform the integral into an average of log-likelihood. By putting log into the integration, we avoid the calculation of likelihood by log-likelihood which is more numerically stable. The Monte Carlo integration will produce a volatile loss function which is not differentiable because of the sampling process, and we can use the reparameterization trick [6] in VAE to make it differentiable. The reparameterization trick is as follows: if we have $x \sim \mathcal{N}(m, \sigma^2)$ and then standaridize it to $\mathcal{N}(0, 1)$, we could revert it back to the original distribution by $x = m + x' \cdot \sigma$ where $x' \sim \mathcal{N}(0, 1)$. Now the sampling process is outside the loss function, so the gradient of loss function will not be affected by sampling.

---

[1] Customarily, $Q(\cdot)$ is used for encoder's distribution in VAE, but here to be consistent with the previous discussion $P(\cdot)$ is used.

The second term is a KL divergence. In VAE, a popular choice of $Q(\cdot)$ is $\mathcal{N}(0, 1)$ to express the prior knowledge [6]. In our setting, we select $Q(\gamma(\Delta t)) = \mathcal{N}(0, I)$ for the homoscedastic case, and $Q(\gamma(\Delta t)) = \mathcal{N}(0, \exp(-\phi\cdot\Delta t)\cdot I)$ for the heteroscedastic case, where ϕ is a constant which can be set manually in experiment. Having $Q(\gamma(\Delta t))$ to be a Gaussian distribution also introduces another benefit. Because the $P(\gamma(\cdot)|\Theta)$ in our model is also assumed to be Gaussian: see (4), the KL divergence between $P(\gamma(\cdot)|\Theta)$ and $Q(\gamma(\Delta t))$ could be computed in closed form. The KL divergence between two Gaussian distributions is:

$$D_{KL}[\mathcal{N}(m_1, \Sigma_1)\|\mathcal{N}(m_2, \Sigma_2)] = \frac{1}{2}[\log|\Sigma_2| - \log|\Sigma_1|$$
$$- k + \mathrm{Tr}\{\Sigma_2^{-1}\Sigma_1\} + (m_2 - m_1)^{\mathrm{T}}\Sigma_2^{-1}(m_2 - m_1)]$$

(12)

where k is the dimension of Gaussian, $\mathrm{Tr}\{\}$ is the trace of matrix, $|\cdot|$ is the determinant. Both Gaussian distributions in our model are assumed to be independent which means the covariance Σ_1 and Σ_2 are both diagonal matrices. This independence assumption improves the computational efficiency further.

After getting the loss function, we can train the model using the generic gradient descent method to optimize the loss with respect to the parameters $\mu, \alpha, \beta, \sigma_c$ in homoscedastic case or $\mu, \alpha, \beta, \alpha_\sigma, \beta_\sigma$ in heteroscedastic case.

5   Synthetic Data Experiment

In synthetic data experiments, we prove that the underlying mean triggering kernel and the corresponding variance parameters can be recovered.

5.1   Homoscedastic Stochastic Triggering Kernel

Based on the thinning algorithm [17], we generate data by setting $\mu = 10$, $\bar{\gamma}(\Delta t) = 1 \cdot \exp(-2 \cdot \Delta t)$, $\sigma_c = 0.5$ and $T_\gamma = 3$. We sampled 10 sets of synthetic data and each of them is a sequence of timestamps in $[0, T]$ where $T = 20$, with a realization of about 400 events.

We use both of the vanilla HP and the homoscedastic HP-STK to recover the parameters for each set of the synthetic data. For both models, the evaluation of parameters is the average of 10 results. For vanilla HP, the final estimations are $\hat{\mu} = 11.04$, $\hat{\alpha} = 0.88$, $\hat{\beta} = 2.71$; for homoscedastic HP-STK, with the configuration of $\kappa = 0.015$, $Q(\gamma(\Delta t)) = \mathcal{N}(0, I)$ and 300 samples from $\mathcal{N}(0, I)$ to perform Monte Carlo integration, the final estimations are $\hat{\mu} = 10.98$, $\hat{\alpha} = 0.88$, $\hat{\beta} = 2.51$, $\hat{\sigma}_c = 0.36$. The learned triggering kernel is shown in Fig. 1a. We can see that the vanilla HP only gives out a deterministic function, while the homoscedastic version gives out an additional variance band.

5.2   Heteroscedastic Stochastic Triggering Kernel

Similarly, in heteroscedastic case, we set $\mu = 2$, $\bar{\gamma}(\Delta t) = 1\cdot\exp(-2\cdot\Delta t)$, $\sigma(\Delta t) = 0.5 \cdot \exp(-2 \cdot \Delta t)$ and $T_\gamma = 3$. We generate timestamps in $[0, T]$ where $T = 100$, resulting in a realization of about 400 events. 10 synthetic datasets are generated.

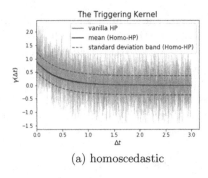
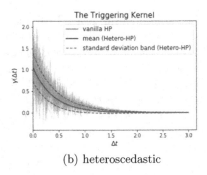

(a) homoscedastic   (b) heteroscedastic

Fig. 1. The triggering kernel from vanilla HP and HP-STK (black and red lines overlap), the shade region is the $\gamma_i(\Delta t)$s of 10 sets of synthetic data. (Color figure online)

We use the similar setting for this experiment, except that homoscedastic HP-STK is replaced by heteroscedastic HP-STK. For vanilla HP, the final estimations are $\hat{\mu} = 2.32$, $\hat{\alpha} = 1.05$, $\hat{\beta} = 2.60$; for heteroscedastic HP-STK, with the configuration of $\kappa = 0.015$, $Q(\gamma(\Delta t)) = \mathcal{N}(0, \exp(-4 \cdot \Delta t) \cdot I)$ and 300 samples from $\mathcal{N}(0, I)$ for Monte Carlo integration, the final estimations are $\hat{\mu} = 2.33$, $\hat{\alpha} = 1.06$, $\hat{\beta} = 2.60$, $\hat{\alpha_\sigma} = 0.33$, $\hat{\beta_\sigma} = 1.52$. The learned triggering kernel is shown in Fig. 1b. The vanilla HP only gives out a deterministic function, while the heteroscedastic version gives out an additional time-decreasing variance band.

6   Applications

To evaluate the effectiveness of our model, we conduct experiments on two real datasets, taxi pickup and crime. We discuss the results and show how HP-STK outperforms vanilla HP in the application of decision on capacity planning.

6.1   Datasets and Experiment Setting

Green Taxi Pickup in New York City: This dataset includes trip records from all trips completed in green taxis in New York City from January to June in 2016. In the experiment, the data from January 1st to 15th is used. We filter out pick-up dates and times for all long-distance trips (>15 miles), since the long distance trips usually have different patterns with short ones [13]. In addition, we pre-process the data by adding a small time interval to separate all the simultaneous records. As a result, we obtain 6223 pickups for 15 days, and the observed variance is 50.39 given 1 h as time interval. This means the actual number of pickups in short periods can be very unstable, so we model it with the homoscedastic HP-STK.

We apply both of the vanilla HP and homoscedastic HP-STK to model the triggering effect of pickups. We assume the triggering kernels are independent for different days. The support of $\gamma(\Delta t)$ is $[0, 3]$ and the time unit is 1 h.

The evaluation of parameters is the average of 15 training results. For vanilla HP, the final estimations are $\hat{\mu} = 5.23$, $\hat{\alpha} = 0.78$, $\hat{\beta} = 1.09$; for homoscedastic HP-STK, with $\kappa = 0.015$, $Q(\gamma(\Delta t)) = \mathcal{N}(0, I)$ which is consistent with the synthetic experiment, the final estimations are $\hat{\mu} = 5.25$, $\hat{\alpha} = 0.78$, $\hat{\beta} = 0.98$, $\hat{\sigma}_c = 0.34$. The learned $\gamma(\cdot)$ is shown in Fig. 2a. It can be seen that the mean $\gamma(\cdot)$ from homoscedastic version is close to the vanilla result, but it gives out an additional variance band. The corresponding intensity of January 4th is plotted in Fig. 3a. The black solid line is the intensity learned from vanilla HP, the gray band corresponds to the variance band of intensity with $\pm\sigma_\lambda(t)$.

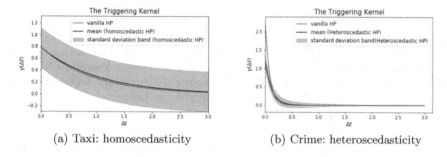

(a) Taxi: homoscedasticity   (b) Crime: heteroscedasticity

Fig. 2. Trigger kernels learned from two real datasets for vanilla HP and HP-STK.

Theft of Vehicle in Vancouver: The data of crimes in Vancouver comes from the Vancouver Open Data Catalogue. It is extracted on 2017-07-18 and it includes all valid felony, misdemeanour and violation crimes from 2003-01-01 to 2017-07-13. We filter out the records of which the crime type is 'Theft of Vehicle' from 2012 to 2016. As a result, we obtain 6320 records for 5 years and the observed variance is 4.29 given 1 day as the time interval. This is stabler than taxi pickups, therefore we model it with the heteroscedastic HP-STK.

We apply both the vanilla HP and heteroscedastic HP-STK to model the triggering effect in crime. We assume the triggering kernels are independent for different years. The support of $\gamma(\Delta t)$ is $[0, 3]$ and the time unit is 1 day.

The evaluation of parameters is the average of 5 training results. For vanilla HP, the final estimations are $\hat{\mu} = 2.80$, $\hat{\alpha} = 2.29$, $\hat{\beta} = 12.21$; for heteroscedastic HP-STK, with the configuration of $\kappa = 0.015$, $Q(\gamma(\Delta t)) = \mathcal{N}(0, \exp(-4 \cdot \Delta t) \cdot I)$ which is consistent with the synthetic experiment, the final estimations are $\hat{\mu} = 2.95$, $\hat{\alpha} = 1.21$, $\hat{\beta} = 8.31$, $\hat{\alpha}_\sigma = 0.17$, $\hat{\beta}_\sigma = 1.25$. The learned $\gamma(\cdot)$ is shown in Fig. 2b. It can be seen that the mean $\gamma(\cdot)$ from heteroscedastic version is close to the vanilla result, but it gives out an additional variance band. The corresponding intensity of 2016-year crime is plotted in Fig. 3b. The black solid line is the intensity learned from vanilla HP, the gray band corresponds to the variance band of intensity with $\pm\sigma_\lambda(t)$.

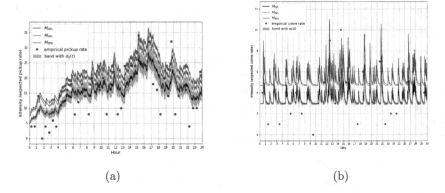

(a)   (b)

Fig. 3. (a): M_{HP_1}, M_{HP_2} and M_{STK} of 4th Jan. in taxi dataset based on vanilla HP and homoscedastic HP-STK. Blue points are empirically estimated pickup rates in every 30 min. (b): M_{HP_1}, M_{HP_2} and M_{STK} of 2016 year crime in Vancouver based on vanilla HP and heteroscedastic HP-STK. Blue points are empirically estimated crime rates in each day. (Only 30 days are shown). (Color figure online)

6.2   Use Case for HP-STK

We examine the use case based on the variance of triggering kernel for HP-STK. It is discussed with the comparison with vanilla HP and applied to both datasets.

Decision for Capacity Planning: In the taxi dataset, the taxi company needs to decide the number ($M(t)$) of taxis to meet the pickup need on time t. We omit t in the following discussion for simplicity. If the company uses intensity λ_{HP} learned from vanilla HP, and send $M_{HP_1} = \lambda_{HP}$ (black line in Fig. 3a) taxis to satisfy the pickup need, about 50% probability[2] that all pickups can be satisfied. To plan for a higher probability, the planner needs to send more taxis. So if the variance of Poisson distribution is taken into consideration, we let $M_{HP_2} = \lambda_{HP} + \sqrt{\lambda_{HP}}$ (green line in Fig. 3a), then theoretically extra 29.7% ($Poisson(x < \lambda_{HP} + \sqrt{\lambda_{HP}}) - Poisson(x < \lambda_{HP})$ where x is real pickup need) probability should be added to satisfy the need, given that the average intensity of 15 days is about $\overline{\lambda}_{HP} = 17$ pickups per hour. To empirically estimate this probability, we compute pickup rates in each 0.5 h (blue points in Fig. 3a). The probability is defined as the number of blue points under the corresponding intensity line divided by the total number, which is shown in Table 1. However, in Table 1, only about 23.2% probability has been added using M_{HP_2} comparing with using M_{HP_1} by averaging the probabilities of 15 days. The difference between theoretical and empirical results means that using vanilla HP underestimates the uncertainty while our method can provide more accurate one.

To demonstrate the superiority of our model, here we also show the M_{STK} got from homoscedastic HP-STK. After we learn the variance of triggering kernel σ_c, we can get the variance of intensity σ_λ (gray band in Fig. 3a) using (5). Then we sample $\{\lambda_{STK}^i\}_{i=1}^{100}$ from the Gaussian distribution $N(\overline{\lambda}_{STK}, \sigma_\lambda^2)$,

---

[2] Based on the Poisson process, the probability could be larger when intensity is low.

Table 1. The probability of satisfying pickup need for M_{HP_1}, M_{HP_2} and M_{STK} of Jan. 1st to 15th taxi pickup in NYC.

Probability	Jan. 1st	Jan. 2nd	Jan. 3th	Jan. 4th	Jan. 5th	Jan. 6th	Jan. 7th	Jan. 8th	Jan. 9th	Jan. 10th	Jan. 11th	Jan. 12th	Jan. 13th	Jan. 14th	Jan. 15th	Average
M_{HP_1}	50.0%	62.5%	47.9%	54.2%	56.3%	72.9%	66.7%	52.1%	56.3%	52.1%	64.6%	72.9%	62.5%	64.6%	52.1%	59.2%
M_{HP_2}	64.6%	83.3%	83.3%	83.3%	89.6%	87.5%	85.4%	81.3%	83.3%	85.4%	83.3%	91.7%	77.1%	81.3%	75.0%	82.4%
M_{STK}	72.9%	85.4%	89.6%	89.6%	93.8%	89.6%	89.6%	87.5%	85.4%	89.6%	89.6%	93.8%	89.6%	91.7%	81.3%	87.9%

Table 2. The probability of satisfying security need for M_{HP_1}, M_{HP_2} and M_{STK} of 2012 to 2016 crime in Vancouver.

Probability	2012	2013	2014	2015	2016	Average
M_{HP_1}	64.2%	69.6%	55.1%	53.7%	44.3%	57.4%
M_{HP_2}	88.3%	91.2%	87.7%	79.2%	75.7%	84.4%
M_{STK}	91.0%	92.6%	88.2%	82.7%	79.8%	86.9%

which are samples larger than the mean. We set the expected rate of pickups as $M_{STK} = \frac{1}{100} \sum_{i=1}^{100} (\lambda_{STK}^i + \sqrt{\lambda_{STK}^i})$ (magenta line in Fig. 3a). We also compute the probability of satisfying pickup need of M_{STK} which is shown in Table 1. It can be seen that about 28.7% probability has been added using M_{STK} comparing with using M_{HP_1} by averaging the probabilities of 15 days. This result is close to the theoretical result 29.7%, which means HP-STK is more accurate.

Similarly, the capacity planning decision task is also performed in crime dataset using same definition for M_{HP_1}, M_{HP_2} and M_{STK} (black, green and magenta lines in Fig. 3b, respectively). In the dataset, $\overline{\lambda}_{HP} = 4$, therefore theoretically we should observe 26.05% difference between M_{HP_2} and M_{HP_1}. To empirically estimate this probability, we compute crime occurrence rates in each day which are shown as blue points in Fig. 3b. The probability result is shown in Table 2 which shows that the difference between M_{HP_2} and M_{HP_1} is 27.0% that is close to the theoretical one. This means the variance of triggering kernel is quite small, which is consistent with the result in Fig. 2b. In such case, the vanilla HP is good enough for capacity planning and there is no need to use HP-STK because the magenta line is very close to the green line (see Fig. 3b).

7   Conclusion

We extended HP with stochastic triggering kernel and considered both the homoscedastic and heteroscedastic cases. Our proposed model can provide the variance of triggering kernel, so allow us to overcome the underestimation problem in capacity planning. Along with the model, we also propose a tractable inference based on VAE loss function. Results from synthetic data show that the HP-STK model can recover the underlying mean triggering kernel and the corresponding variance. The usage of HP-STK in taxi allocation discloses that the taxi pickup has a highly stochastic triggering kernel. Vanilla HP will underestimate the expected pickup rate. Without misleading the taxi dispatcher, HP-STK could provide a more accurate rate. Furthermore, another case in crime with a stabler triggering kernel is used to test that HP-STK could disclose the data stability as expected. There is also freedom to maneuver the stochastic triggering kernel to adapt to other real-life applications or to invent nonparametric stochastic triggering kernels.

References

1. Adams, R.P., Murray, I., MacKay, D.J.: Tractable nonparametric Bayesian inference in Poisson processes with Gaussian process intensities. In: Proceedings of the 26th Annual International Conference on Machine Learning, pp. 9–16. ACM (2009)
2. Cox, D.R.: Some statistical methods connected with series of events. J. Roy. Stat. Soc. Ser. B (Methodol.) **17**, 129–164 (1955)
3. Cunningham, J.P., Byron, M.Y., Shenoy, K.V., Sahani, M.: Inferring neural firing rates from spike trains using Gaussian processes. In: Advances in Neural Information Processing Systems, pp. 329–336 (2008)
4. Cunningham, J.P., Shenoy, K.V., Sahani, M.: Fast Gaussian process methods for point process intensity estimation. In: Proceedings of the 25th International Conference on Machine Learning, pp. 192–199. ACM (2008)
5. Dassios, A., Zhao, H., et al.: Exact simulation of Hawkes process with exponentially decaying intensity. Electron. Commun. Probab. **18**, 1–13 (2013)
6. Doersch, C.: Tutorial on variational autoencoders. arXiv preprint arXiv:1606.05908 (2016)
7. Du, N., Dai, H., Trivedi, R., Upadhyay, U., Gomez-Rodriguez, M., Song, L.: Recurrent marked temporal point processes: embedding event history to vector. In: Proceedings of the 22nd ACM SIGKDD International Conference on Knowledge Discovery and Data Mining, pp. 1555–1564. ACM (2016)
8. Embrechts, P., Liniger, T., Lin, L.: Multivariate Hawkes processes: an application to financial data. J. Appl. Probab. **48**(A), 367–378 (2011)
9. Gregory, P., Loredo, T.J.: A new method for the detection of a periodic signal of unknown shape and period. Astrophys. J. **398**, 146–168 (1992)
10. Hawkes, A.G.: Spectra of some self-exciting and mutually exciting point processes. Biometrika **58**(1), 83–90 (1971)
11. Kingma, D.P., Welling, M.: Auto-encoding variational bayes. arXiv preprint arXiv:1312.6114 (2013)
12. Lee, Y., Lim, K.W., Ong, C.S.: Hawkes processes with stochastic excitations. In: International Conference on Machine Learning, pp. 79–88 (2016)
13. Menon, A.K., Lee, Y.: Predicting short-term public transport demand via inhomogeneous Poisson processes. In: Proceedings of the 2017 ACM on Conference on Information and Knowledge Management, pp. 2207–2210. ACM (2017)
14. Mohler, G.O., Short, M.B., Brantingham, P.J., Schoenberg, F.P., Tita, G.E.: Self-exciting point process modeling of crime. J. Am. Stat. Assoc. **106**(493), 100–108 (2011)
15. Møller, J., Syversveen, A.R., Waagepetersen, R.P.: Log gaussian cox processes. Scand. J. Stat. **25**(3), 451–482 (1998)
16. Ogata, Y., Vere-Jones, D.: Inference for earthquake models: a self-correcting model. Stoch. Process. Appl. **17**(2), 337–347 (1984)
17. Ogata, Y.: On lewis' simulation method for point processes. IEEE Trans. Inf. Theory **27**(1), 23–31 (1981)
18. Rodriguez, M.G., Balduzzi, D., Schölkopf, B.: Uncovering the temporal dynamics of diffusion networks. arXiv preprint arXiv:1105.0697 (2011)

Concept Drift Based Multi-dimensional Data Streams Sampling Method

Ling Lin[1,2], Xiaolong Qi[1,2], Zhirui Zhu[1], and Yang Gao[1(✉)]

[1] State Key Laboratory for Novel Software Technology, Collaborative Innovation
Center of Novel Software Technology and Industrialization, Nanjing University,
Nanjing, China
linling_xj@163.com, qxl_0712@sina.com, julie1996@foxmail.com,
gaoy@nju.edu.cn
[2] Electronics and Information Engineering College, Yili Normal University,
Yining, Xinjiang, China

Abstract. A summary can immensely reduce the time and space complexity of an algorithm. This concept is considered a research hotspot
in the field of data stream mining. Data streams are characterized as
having continuous data arrival, rapid speed, large scale, and cannot be
completely stored in memory simultaneously. A summary is often formed
in the memory to approximate the database query or data mining task.
A sampling technique is a commonly used method for constructing data
stream summaries. Traditional simple random sampling algorithms do
not consider the conceptual drift of data distributions that change over
time. Therefore, a challenging task is sampling the summary of the data
distribution in multi-dimensional data streams of a concept drift. This
study proposes a sampling algorithm that ensures the consistency of
the data distribution with the data streams of the concept drift. First,
probability statistics is used on the data stream cells in the reference
window to obtain data distribution. A probability sampling is performed
on the basis of this distribution. Second, the sliding window is used to
continuously detect whether the data distribution has changed. If the
data distribution does not change, then the original sampling data are
maintained. Otherwise, the data distribution in the statistical window
is restated to form a new sampling probability. The proposed algorithm
ensures that the data distribution in the data profile is continually consistent with the population distribution. We compare our algorithm with
the state-of-the-art algorithms on synthetic and real data sets. Experimental results demonstrate the effectiveness of our algorithm.

Keywords: Data stream clustering · Sampling · Summary

1 Introduction

The rapid development of mobile Internet, e-commerce, Internet of things, communication, and spatial positioning technology has generated extensive streaming data. However, Data streams make storing data completely in a database

© Springer Nature Switzerland AG 2019
Q. Yang et al. (Eds.): PAKDD 2019, LNAI 11439, pp. 331–342, 2019.
https://doi.org/10.1007/978-3-030-16148-4_26

impossible. Therefore, selecting part of the data temporarily stored in the memory is necessary to form a summary and perform data mining and query analysis on the summary to obtain an approximate but sufficiently reliable result [1]. One of the main characteristics of the summary is the capability to reduce the amount of data. Nonetheless, the data are remain extensive and the approximate data mining results can be obtained from the compressed data (occasionally identical to the overall data mining effect).

A research hotspot in data stream mining is the method of designing a summary that is smaller than the data set size based on a rapid, continuously arriving data streams to make the data processed in the memory [2]. Sampling is one of the main methods used to generate a summary [3]. A small percentage of data from the data set is taken as a sample to represent the entire data set and obtain data mining or database query results on the basis of the sample set. Simple random sampling is a basic sampling technique, in which the sampling probability of each sample is equal. This method performs well on traditional data but is unsuitable for streaming data. The streaming data are dynamic (i.e., continuously arrive), while the number of samples is uncertain (i.e., the sample size N is uncertain). The reservoir sampling algorithm [4] solves the random selection of K samples from N samples ($K \ll N$), where N is large (such that N samples cannot be placed in memory at the same time) or N is an unknown number. Reservoir sampling is a generalization of random sampling, which is one of the classic streaming data sampling schemes, such as adaptive size reservoir [5] and chain sampling [6]. The existing data stream sampling is generally a simple random sampling of discrete attribute data streams. For continuous attribute data streams, the data obey a Gaussian distribution and sampled on the basis of this distribution. Simple random sampling cannot describe the population distribution and easily loses key change data. Data distribution in the data stream environment changes over time. Moreover, the assumption that the data obey the Gaussian distribution is impractical and fails to consider the phenomenon of drift of data stream distribution over time. Hence, the problem to be solved sampling the summary that is consistent with population distribution in the multi-dimensional continuous attribute data streams of the concept drift.

A challenging task is to detect whether data distribution has changed in unsupervised data streams without a class label [7]. The existing literature [8] has proposed a change detection framework for data streams. This framework can immediately detect the distribution change over multidimensional data streams by projecting data into low dimensional space using principal component analysis. Inspired by this algorithm, we propose a multi-dimensional continuous attribute data stream sampling algorithm based on consistent distribution. This algorithm opens a buffer in the memory for storing stream summary. We first divide the data into cells in a sliding window and compute its distribution. Thereafter, we apply probability sampling on the data on the basis of the distribution. Lastly, we observe whether the distribution of the data has been changed by continuously comparing the reference and detected windows. Thus, we can decide if a reestimation of the distribution is needed to obtain new probability statistics for the following sampling. Our algorithm consistently ensures that the summary distribution is consistent with the population distribution. In addition, our algorithm guarantees that the most recent and recently arrived valuable data streams are stored in the summary by discarding the historical data that first entered the buffer.

The main contributions of this study are as follows.

- We consider the true distribution of data instead of simple random sampling and the assumption that data obey the Gaussian distribution. Subsequently, the summary formed after sampling will continually be consistent with the population distribution.
- The majority of the existing data stream sampling algorithms are discrete attribute data stream sampling and rarely consider continuous attribute data stream sampling. Our algorithm is suitable for sampling continuous data streams and discrete data stream sampling.
- We consider the drift of the data stream concept. When the data distribution changes, the algorithm will sample based on the actual data distribution and the original summary will be maintained when the data distribution does not change.

We compare our algorithm with random sampling and hypothesis data with Gaussian distribution sampling. This comparison utilizes the KL divergence between the summary distribution generated by the three sampling algorithms to compare the similarities of the data distribution. The original population distribution is used as the evaluation index. The experimental results show that the statistical probability sampling based on concept drift outperforms the random and Gaussian-assumption samplings.

The remainder of this paper is organized as follows. Section 2 presents the related studies. Sections 3 and 4 present our proposed sampling algorithm for concept drift data streams and evaluation results, respectively. Lastly, Sect. 5 provides the conclusion.

2 Related Work

As an important step in reducing data size and accelerating data mining tasks, the summary is a beneficial and effective technique for supporting big data analytics [9, 10] and extensively used in such fields as medical informatics, astronomy and earth science, and social and sensor networks. Figure 1 shows how the summary of data streams can reduce the amount of data and accelerate the process of completing data mining tasks. However, [11–13] indicated that no uniform definition of a summary is available and its definition or role depends on the application purpose.

The existing literature [11] has proposed a summary generation method based on concise and counting sampling for the database query. Concise sampling solves the problem of inefficient data expression in reservoir sampling. Each sampling element in the reservoir sampling occupies a single position. Even if the elements

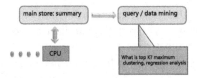

Fig. 1. The process of knowledge discovery in data streams

have the same value, a storage unit is allocated. To improve storage efficiency, the concise sampling algorithm is represented by a *<value, count>* structure for elements that appear multiple times, where value represents the element code and count represents the number of elements in the sample set. The counting sampling method is a variant of the concise sampling method. The difference between the two is the method of dealing with the sample set overflow.

Certain in-vehicle devices with GPS and other communication technologies have limited storage capacity. Therefore, a summary should be generated for continuously arriving data streams. In [12], the infinite marginal vehicle-to-vehicle (V2V) traffic data streams are sampled to generate the summary to provide users with rapid traffic information services. Three sampling techniques, namely, sliding window, reservoir, and extended biased, are used for the summary generation of the V2V traffic data streams to provide real-time data analysis for users.

Feature preserved sampling (FPS) is another sampling technique for data streams [13]. This method uses a sliding window model and aims to maintain a similar distribution of attribute values before and after the sampling. FPS can immediately decide whether to retain or discard the current instance. However, FPS is only suitable for discrete data stream sampling and unsuitable for continuous attribute value data streams.

With the exception of the definition and application role, the conceptual drift phenomenon of a data set should be considered in the summary generation. However, existing random sampling algorithms [3,14,15] do not consider the conceptual drift phenomenon of streaming data. Moreover, only a few sampling algorithms for continuous attribute data streams are available, even though continuous attribute value data streams are common. The objective of the current study is to generate a summary that is consistent with the population distribution by sampling the continuous attribute data streams, in which the summary is stored in the memory buffer pool for online data mining or database query tasks at any time. When new data mining tasks or query requirements are requested, the summary can be used directly to immediately calculate results.

3   Summary for Concept Drift Data Steams

Sampling algorithms based on concept drift should address two key issues. The first issue is how to estimate the probability density of high-dimensional unclassified data (we can use the closeness of the data distribution to sample after determining the probability density of the data). The second issue is how to determine whether the data distribution has changed (the time point when the distribution is determined to be changed) and how to sample when the data distribution changes (in the case of conceptual drift). This section focuses on the two issues.

3.1   Probability Sampling for Concept Drift Data Streams

The application value of streaming data will decrease rapidly over time. In data streams, we are concerned with data that have arrived recently because the latest data are critical to our decision-making. The sliding window model is ideal for applications that process data in the most recent period. The sliding window on the data streams refers to an interval set on the data streams, which includes only the partial data of the data streams. If the size of the window is W, at any time point t_c, then we only consider the W data in the $w[t_c - w + 1, t_c]$ sliding window. Moreover, moving the window forward indicates that new data have arrived and replaced the previous data. The size of window W is often determined by the actual application problem. Different window sizes may lead to different results and we need multiple experiments to determine a suitable window size W.

We use the sampling method of cell division to obtain an accurate sampling. For the data in the window, we first divide the cells (each cell is composed of bins in k dimension), count the data stream samples falling in each cell, and eventually perform probability sampling on the statistics in the cells. The sampling equation is $\frac{n}{w} * p$, where W is the window size, n is the statistics of the cells (i.e., data density), and p is the sampling rate. In each detection window, we need to detect whether the data distribution has changed. If the data distribution does not change, then the original sampling data are maintained. Otherwise, the data distribution in the statistical window is restated to form a new sampling probability. Figure 2 shows the probability sampling process based on concept drift. The first reference window data is divided into cells and probability sampling is performed thereafter. The sampled data are placed in the first area of the buffer for data summary. In the second window, no distribution changes are determined. Thus, no samples are taken. Accordingly, the original summary is maintained. In the third detected window, the algorithm divides the detection window into cells, probability sampling is performed, and the sampled data are placed in the second buffer. When the buffer is full, the algorithm moves the samples of the earliest entry buffer out of the memory and constantly ensures that the data in the summary are consistent with the most recently arrived data streams. The size of the buffer can be set on the basis of the size of the memory and the amount of summary that needs to be retained. Algorithm 2 provides the specific implementation process.

3.2   Change Detection

For distribution change detection, we use the KDE density estimation method [8]. After receiving the data of the first batch of the W window size in the

---

ALGORITHM 1. Change Detection

Parameters: window size w, δ, ζ;
Online flow in: streaming data $S = (x_1, x_2, \ldots x_t)$;
Online output: time t when detecting a change;
Procedure:
1. Initialize $t_c = 0$, $setp = \min(0.05w, 100)$;

2. Initialize $\bar{S}_c, m, M$ to NULL ;
3. Set reference window $S_1 = (x_{t_c+1}, \ldots x_{t_c+w})$;
4. Extract principal components by applying PCA on to obtain $Z_1, Z_2 \ldots Z_k$;

5. Project S_1 on $Z_1, Z_2 \ldots Z_k$ to obtain $\hat{S}_1$;

6. $\forall i \, (1 \le i \le k)$ estimate $\hat{f}_i$ using data of the $i - th$ component of $\hat{S}_1$;

7. Clear S_1 and $\hat{S}_1$;
8. Set test window $S_2 = (x_{t_c+w+1}, \ldots x_{t_c+2w})$;

9. Project S_2 on $Z_1, Z_2 \ldots Z_k$ to obtain $\hat{S}_2$;
10. clear S_2;

11. Estimate $\hat{g}_i$ using data of the $i - th$ component of $\hat{S}_2$;
12. **while** a new sample x_t arrives in the stream **do**;

13.   Project x_t on $Z_1, Z_2 \ldots Z_k$ to obtain $\hat{x}_t$;

14.   Remove $\hat{x}_{t-w}$ from $\hat{S}_2$;

15. $\forall i \, (1 \le i \le k)$ update $\hat{g}_i$ using $\hat{x}_t^{(i)}$ and $\hat{x}_{t-w}^{(i)}$;
16.   **if** $\mod(t, step) = 0$ **then** ;

17. $curScore = \max_i \left(D_{MKL} \left(\hat{g}_i \| \hat{f}_i \right) \right)$;

18.   **if** Change $(curScore, Sc, m, M, \xi, \delta)$ **then** ;
19.   Report a change at time t and set $t_c = t$;

20.   Clear $\hat{S}_2$ and GOTO step 2;
21. **Subprocedure: Change** $(curScore, Sc, m, M, \xi, \delta)$

22. Update $\bar{Sc}$ Sc to include the curScore in the average.;

23. $new\_m = m + \bar{S}_c - curScoure + \delta$;
24. **if** $|new\_m| \ge M$ **then** ;
25. $new\_M = new\_m$.;

26. $\tau_t = \xi * \bar{S}_c$;
27. **if** $curScore > \tau_t$ **then** ;
28.   return True;
29. **else**;
30. $M = new\_M, m = new\_m$;
31.   return False;

---

reference window $S1$, PCA is performed on the data in the window, while the first k feature values to satisfy the $\frac{\sum_{i=1}^{k} \lambda_i}{\sum_{i=1}^{d} \lambda_i} \ge 0.999$ conditions extracted. The same operation is performed on the detection window $S2$, while the data of

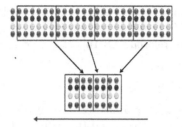

Fig. 2. Example of the sampling process based on concept drift

---

ALGORITHM 2. Sampling for Concept Drift

Parameters: bins size l, b, k;
Online flow in: streaming data $S = (x_1, x_2, \ldots x_t)$;
Online output: sampled data;
Procedure:
1. Initialize Initialize bins size l, b, k ;
2. $Cells = l^k$;
3. Saved_window $= 0$;
4. Save sampled data in summary from first window ;
5. Saved_window $=$ Saved_window $+ 1$;
6. **While** a new sample x_t arrives in the stream **do**;
7.   Change detection;
8.   if find change **then** ;
9.   Divide the cells and count the number of samples per cell in windows;
10.   Sampling from the cells;
11.   if saved_window $>$ b **then** ;
12.   Delete the block that first entered the summary;
13.   Repeat 9,10;
14.   **Else**;
15.   **Next**;

---

the two windows $S1$ and $S2$ are respectively projected onto the feature vectors composed of the k principal components. The change scores of the two windows are compared and recorded. The maximum of the k change values is considered the change point. [8] proposed a dynamic threshold setting method to change-score (procedure change lines 21–31). m_t is a cumulative variable, which is used to store the cumulated difference between the mean of the previously observed values and presently observed values, and defined as follows:

$$m_t = \sum_{t'=1}^{t} \left(\bar{s}_{t'} - s_{t'} + \delta \right) \tag{1}$$

where $\bar{s}_t^i = \frac{1}{t'} \sum_{i=1}^{t'} s_i$, $s_{t'}$ is the observed value at time t', δ is the allowed change's magnitude, which is often set near to zero. When the $m_t > \theta$ (θ is a threshold) and the difference between $M_t = \max\{m_1, \ldots, m_t\}$, a change is reported. Algorithm 1 presents the specific implementation process of the change detection. For additional details, refer to reference [8,16].

3.3 Complexity Analysis

The probability sampling of cells indicates that the data streams in the window should be divided into cells and sampled thereafter. The time complexity of the algorithm is $O(\text{bin}s \wedge k)$, where bins are the number of cells in one dimension and k is the largest variance dimension. If statistical sampling is performed every time a window is swept, then the time complexity increases as the data dimension increases. The method we propose is to sample when the data distribution changes, refrain from sampling when the distribution is unchanged, and retain the original summary. The time complexity of the change detection algorithm is $O\left(qdw \log\left(\frac{1}{v}R_c\right)\right)$, where R_c is the number of reported changes, v is the minimum cell width, w is the window size, d is the data dimensionality, and q is the number of bootstrap samples.

4   Experimental Evaluation

The performance of our proposed sampling algorithm is compared with that of the random sampling and data hypothesis that obeys the Gaussian distribution sampling on synthetic data and real data sets. The KL divergence measures the similarity between the raw and sampled data distributions.

4.1   Parameter Settings

The parameters setting for the change detection method follows the settings in [7,8] with $\delta = 0.05$, and the window size W is set to 10^4. For concept drift sampling algorithm, the parameters are set as $l = 100$, sampling rate $p = 0.2$, and buffer length $b = 10$.

4.2   Experiments on Synthetic Data

We use two-dimensional synthetic data sets, the distribution of which changes to evaluate our proposed sampling algorithm for the concept drift data streams. Each data set contains $5 * 10^6$ samples and the data distribution changes once per $5 * 10^4$ data points. The construction of synthetic data refers to the data generation method provided by the literature [7,8]. The symbols given to the data sets indicate the different types of change. ($M(\Delta)$ represents the varying mean value, $D(\Delta)$ represents the varying standard deviation, and $C(\Delta)$ represents the varying correlation). At each change point, a set of random numbers in the interval $[-\Delta, -\Delta/2] \cup [\Delta/2, \Delta]$ is generated and added to the distributions parameters that will be changed. The parameter Δ controls the degree of change, in which large values for Δ make changes easy to detect and vice versa.

Table 1. KL divergence between the population and summary distributions generated by the three sampling algorithm on synthetic data

Datasets	SRS	GS	CDSS
M(0.01)	0.0716	0.31125	**0.0503**
M(0.02)	0.0523	0.0998	**0.0320**
M(0.05)	0.1011	0.2156	**0.0839**
D(0.01)	0.0965	0.4103	**0.0246**
D(0.02)	**0.0712**	0.0913	0.0831
D(0.05)	0.7691	1.2141	**0.4415**
C(0.01)	0.0062	0.0721	**0.0004**
C(0.02)	0.0821	0.7131	**0.0125**
C(0.05)	0.0931	0.0551	**0.0021**

The generation parameters of the three data sets M(0.01), M(0.02), and M(0.05) for the first group are μ_1, μ_2 are changing by θ_1, θ_2 selected randomly from the interval $[-\Delta, -\Delta/2] \cup [\Delta/2, \Delta]$. The standard deviation and correlation coefficient are set to a fixed value of ($\sigma_1 = \sigma_2 = 0.2$) and ($\rho = 0.5$), respectively. The three data sets D(0.01), D(0.02), and D(0.05) of the second group are generated by fixing the correlation coefficient value ($\rho = 0.5$) and mean value $\mu_1 = \mu_2 = 0.5$. The standard deviation $\sigma_1 \sigma_2$ in the interval $[-\Delta, -\Delta/2] \cup [\Delta/2, \Delta]$ adds random values. In the three data sets C(0.01), C(0.02), and C(0.05) of the last group, the coefficient ρ makes random walks in the interval $(-1, 1)$ with random steps selected from $[-\Delta, -\Delta/2] \cup [\Delta/2, \Delta]$. The standard deviation and mean values are set to fixed values $\sigma_1 = \sigma_2 = 0.2$ and $\mu_1 = \mu_2 = 0.5$.

If we suppose that $P(x)$ is the original data probability density, then $C(x)$ is the probability density of the concept drift sampling, $R(x)$ is the probability density of the random sampling, and $G(x)$ is the probability density of the Gaussian distribution sampling. $D_{KL}(P \parallel R)$, $D_{KL}(P \parallel G)$, and $D_{KL}(P \parallel C)$ are used to calculate the similarity between real data distribution and each sampled data distribution. Table 1 shows the KL divergence between the distributions of the population and summary generated by three sampling algorithms. Accordingly, SRS represents random sampling, GS represents the data obeying Gaussian distribution sampling, and CDSS represents the probability sampling for concept drift data streams. The best results are presented in boldface. Table 1 shows that the probability sampling has the smallest KL divergence value in the majority of the cases. The distribution of the probability sampling is nearest to the original data distribution. Figure 3 is the visualized form of the experimental results on the M(0.02) data set. Subfigure(b) is most similar to the subfigure(a). In the experimental results, we note that the value of the KL divergence is near 0 because our synthetic data are a value of between 0 and 1.

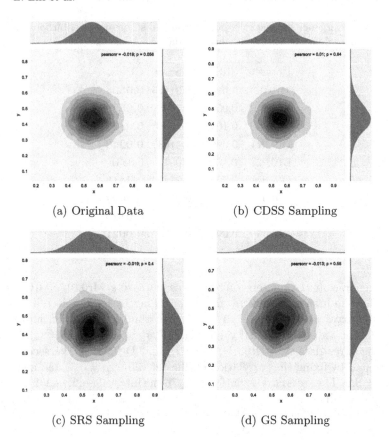

(a) Original Data   (b) CDSS Sampling

(c) SRS Sampling   (d) GS Sampling

Fig. 3. (a) Original data distribution vs: (b) Probability sampling for concept drift, (c) Random sampling, (d) Hypothesis data with Gaussian distribution sampling

4.3   Experiments on Real Data

We evaluated our method with the SRS and GS methods on five real data sets Walking (3D)[1], Jogging (3D)[2], El Nino (5D)[3], Spruce (10D)[4], Ascending Stairs(30D)[5] that were obtained from machine learning repositories. The experiments required that the data sets are sufficiently large and have change points at which concept drifts occur. To comply with the experimental requirements, we follow the method in [7,8]. First, the data set is extended by interpolation without changing the data distribution until the data set reaches the experiments required size. Second, we sample a batch of original data every $2*10^4$ samples and

---

[1] http://www.cis.fordham.edu/wisdm/dataset.php.

[2] http://www.cis.fordham.edu/wisdm/dataset.php.

[3] http://kdd.ics.uci.edu/databases/el_nino/el_nino.html.

[4] http://kdd.ics.uci.edu/databases/covertype/covertype.html.

[5] http://www.pamap.org/demo.html.

replace the batch with one of the following two techniques to change the distribution of the data. (1) Scale-1D (S1D): changes the batch of data by multiplying the random selected dimension to 2. (2) Gauss-1D (GID) changes the batch of data by randomly selecting one dimension and adding a standard Gaussian variable. Table 2 shows the experiment results on the real data set. The results confirm that the probability sampling distribution is considerably similar to the original data distribution. The experimental results on synthetic and real data set show that the proposed method is effective.

Table 2. KL divergence between the population and summary distributions generated by the three sampling algorithm on real data

Datasets	SRS	GS	CDSS
D1: Walking (3D)	0.6768	1.4122	**0.4414**
D2: Jogging (3D)	0.0965	0.1238	**0.0153**
D3: El Nino (5D)	2.0468	3.1224	**1.5656**
D4: Spruce (10D)	**2.2214**	3.6612	2.2411
D5: Ascending Stairs (30D)	5.0987	5.9753	**4.1614**

5   Conclusion

This study proposes a novel sampling method for concept drift data streams. Our method selects representative samples consistent with the population distribution and chooses k dimensions with the largest variance for statistical sampling. Unlike continuously random sampling, the concept drift stream sampling algorithm can generate a sample when concept drift occurs (i.e., data distribution changes) in data streams. However, we do not assume that the data obey any distribution. Instead, statistical sampling is performed on the basis of the actual distribution of the data streams. The use of the FIFO queue to eliminate outdated data makes our method suitable for data stream applications with limited memory resources. The experimental results on synthetic and real data show that the proposed algorithm is feasible and efficient.

Acknowledgments. This work is supported by the National Key R&D Program of China (2017YFB0702600, 2017YFB0702601), the National Natural Science Foundation of China (61432008, U1435214, 61503178) and Yili Normal University Project (No. 2016WXDZD001).

References

1. Agarwal, P.K., Cormode, G., Huang, Z., Phillips, J.M., Wei, Z., Yi, K.: Mergeable summaries. ACM Trans. Database Syst. (TODS) **38**(4), 26 (2013)
2. Rivetti, N., Busnel, Y., Mostefaoui, A.: Efficiently summarizing data streams over sliding windows. In: 2015 IEEE 14th International Symposium on Network Computing and Applications (NCA), pp. 151–158. IEEE (2015)
3. Cormode, G., Duffield, N.: Sampling for big data: a tutorial. In: Proceedings of the 20th ACM SIGKDD International Conference on Knowledge Discovery and Data Mining, pp. 1975–1975. ACM (2014)
4. Vitter, J.S.: Random sampling with a reservoir. ACM Trans. Math. Softw. (TOMS) **11**(1), 37–57 (1985)
5. Al-Kateb, M., Lee, B.S., Wang, X.S.: Adaptive-size reservoir sampling over data streams. In: 19th International Conference on Scientific and Statistical Database Management, p. 22. IEEE (2007)
6. Babcock, B., Datar, M., Motwani, R.: Sampling from a moving window over streaming data. In: Proceedings of the Thirteenth Annual ACM-SIAM Symposium on Discrete Algorithms, pp. 633–634. Society for Industrial and Applied Mathematics (2002)
7. Song, X., Wu, M., Jermaine, C., Ranka, S.: Statistical change detection for multi-dimensional data. In: Proceedings of the 13th ACM SIGKDD International Conference on Knowledge Discovery and Data Mining, pp. 667–676. ACM (2007)
8. Qahtan, A.A., Alharbi, B., Wang, S., Zhang, X.: A PCA-based change detection framework for multidimensional data streams: change detection in multidimensional data streams. In: Proceedings of the 21th ACM SIGKDD International Conference on Knowledge Discovery and Data Mining, pp. 935–944. ACM (2015)
9. Ahmed, M.: Data summarization: a survey. Knowl. Inf. Syst. **58**, 1–25 (2018)
10. Hesabi, Z.R., Tari, Z., Goscinski, A., Fahad, A., Khalil, I., Queiroz, C.: Data summarization techniques for big data—a survey. In: Khan, S.U., Zomaya, A.Y. (eds.) Handbook on Data Centers, pp. 1109–1152. Springer, New York (2015). https://doi.org/10.1007/978-1-4939-2092-1_38
11. Gibbons, P.B., Matias, Y.: New sampling-based summary statistics for improving approximate query answers. In: ACM SIGMOD Record, vol. 27, no. 2, pp. 331–342. ACM (1998)
12. Zhang, J., Xu, J., Liao, S.S.: Sampling methods for summarizing unordered vehicle-to-vehicle data streams. Transp. Res. Part C: Emerg. Technol. **23**, 56–67 (2012)
13. Chuang, K.-T., Chen, H.-L., Chen, M.-S.: Feature-preserved sampling over streaming data. ACM Trans. Knowl. Discov. Data (TKDD) **2**(4), 15 (2009)
14. Tillé, Y.: Sampling algorithms. In: Lovric, M. (ed.) International Encyclopedia of Statistical Science, pp. 1273–1274. Springer, Heidelberg (2011)
15. Al-Kateb, M., Lee, B.S.: Adaptive stratified reservoir sampling over heterogeneous data streams. Inf. Syst. **39**, 199–216 (2014)
16. Zhang, X., Furtlehner, C., Germain-Renaud, C., Sebag, M.: Data stream clustering with affinity propagation. IEEE Trans. Knowl. Data Eng. **26**(7), 1644–1656 (2014)

Spatial-Temporal Multi-Task Learning for Within-Field Cotton Yield Prediction

Long H. Nguyen[1(✉)], Jiazhen Zhu[2], Zhe Lin[3], Hanxiang Du[1], Zhou Yang[1], Wenxuan Guo[3], and Fang Jin[1]

[1] Department of Computer Science, Texas Tech University, Lubbock, USA
{long.nguyen,hanxiang.du,zhou.yang,fang.jin}@ttu.edu
[2] Department of Computer Science, George Washington University, Washington, D.C., USA
jiazhen_zhu@gwmail.gwu.edu
[3] Department of Plant and Soil Science, Texas Tech University, Lubbock, USA
{zhe.lin,wenxuan.guo}@ttu.edu

Abstract. Understanding and accurately predicting within-field spatial variability of crop yield play a key role in site-specific management of crop inputs such as irrigation water and fertilizer for optimized crop production. However, such a task is challenged by the complex interaction between crop growth and environmental and managerial factors, such as climate, soil conditions, tillage, and irrigation. In this paper, we present a novel Spatial-temporal Multi-Task Learning algorithm for within-field crop yield prediction in west Texas from 2001 to 2003. This algorithm integrates multiple heterogeneous data sources to learn different features simultaneously, and to aggregate spatial-temporal features by introducing a weighted regularizer to the loss functions. Our comprehensive experimental results consistently outperform the results of other conventional methods, and suggest a promising approach, which improves the landscape of crop prediction research fields.

1 Introduction

Cotton is an important cash crop native to tropical and subtropical regions in the world. Accurate yield prediction not only provides valuable information to cotton producers for effective management of the crop for optimized production, but also is important to policymakers, as well as consumers of agricultural products. However, cotton yield prediction is challenging due to complex interactions between crop growth and weather factors, soil conditions, as well as management factors, such as irrigation, tillage, rotation, etc. Moreover, simply applying other crop yield prediction models on cotton may lead to nothing but disappointment: a prediction model that works on other crops like wheat, rice, and sugarcane, however, fails on predicting cotton yield [1].

The existing approaches estimate crop yield based either on the crop sown areas, crop-cutting experiments or market arrivals show wide variability because of their inability to capture the indeterminate nature of the crop and its

© Springer Nature Switzerland AG 2019
Q. Yang et al. (Eds.): PAKDD 2019, LNAI 11439, pp. 343–354, 2019.
https://doi.org/10.1007/978-3-030-16148-4_27

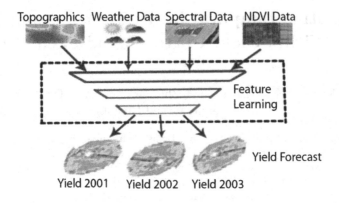

Fig. 1. Data sources in our prediction model. (Color figure online)

response to environmental conditions [2]. One of the attempts is to apply Grey model [3] on production prediction, utilizing short-term forecasting with exponential growth. The regression-based method such as time series analysis can also be applied to production prediction [4], but it suffers from great variation when the external environment is under significant variation. Another frequently used method for prediction is the differential equation model [5], which demands the system to be stable and requires extra work to solve the equation.

Conventional yield prediction treats the field uniformly despite inherent variability. Uniform assumption may result in over- or under-application of resources in specific locations within a field, which may have a negative impact on the environment and profitability [6]. However, consistent and accurate within-field yield prediction is challenging due to the high accuracy requirement under the complex interactions between yield-influencing factors, such as soil, weather, water, and spatial correlations.

With the introduction of the global positioning system (GPS), geographic information systems (GIS), and yield monitors along with other new technologies, we can quantify spatial variability in soil properties and crop yield in small areas of a field. As satellite and drone technologies develop, we are able to collect remote sensing images at fine resolutions to support within-field yield forecast. Within-field scale crop yield prediction provides valuable information for producers to site-specifically manages their crop, which can optimize crop production for maximum profitability. In the within-field prediction procedure, we use a 30-m grid to represent a continuous surface.

The advancement of machine learning offers a different approach compared with the traditional ways for yield forecasting. The rapid advances in sensing technologies, the use of fully automated data recording, unmanned systems, remote sensing through satellites or aircraft, and real-time non-invasive computer vision, are additional boosts for enabling the new yield forecasting model. Due to the capability of machine learning based systems to process a large number of inputs and handle non-linear tasks, people have attempted to predict county-level soybean yield in the United States [7]. However, using deep learning for within-field cotton forecast remains as an untouched ground. In our work, the within-field forecasting is based on each grid for one field in West Texas area across three years (2001, 2002, 2003) in order to predict the cotton yield before harvest.

On this account, we propose a Multi-Task learning model to predict within-field cotton yield. As shown in Fig. 1, this model ingests many sources of data which contain features for different learning tasks, including soil topographic attributes (elevation, slope, curvature, etc.), spectral data (Blue, Green, Red, and NIR bands denoted as BAND1, BAND2, BAND3, and BAND4, respectively), normalized difference vegetation index (NDVI) during the crop seasons; and weather (temperature, rainfall, etc.) data. These multiple data sources are aggregated in the shared layer before transferring to task-specific layers. This type of design in a Multi-Task learning model makes it capable of enhancing specific learning task by utilizing all sources of information of other related tasks. In other words, this allows us to take various factors and variables into consideration to achieve a more accurate yield prediction. On the other hand, crop yield within a field is typically autocorrelated, meaning yield values close together are likely more similar than those farther apart. Hence, to incorporate the spatial relationship, we propose a spatial regularization term to minimize the yield difference between one region and the weighted average of neighboring regions. Therefore, we termed this technique as Spatial-temporal Multi-Task Learning. The main contributions of this paper are summarized as below:

- We design an innovative multi-task learning approach to predict within-field cotton yield across several years. Different from other machine learning models, to predict the cotton yield for a specific year is one of the tasks in our model; each task is enhanced through its access to all available data from prior years.
- This work provides an entirely new vision for grid-scale crop yield prediction. To the best of our knowledge, this is among the first attempts to predict fine-grain cotton yield with the Multi-Task Learning approach, as existing work focus more on county-level or country level.
- We introduce a spatial weight regularizer to overcome the effects of geographical distance on yield prediction. Each grid is trained to minimize not only the difference between the prediction and the actual value, but also the difference between its yield and its neighbors.
- We perform a comprehensive set of experiments using the real-world dataset that produced results consistently outperformed other competitive methods, which could provide guidance for achieving higher crop production.

2   Related Work

Crop Yield Prediction. Crop yield prediction is challenging. Many studies have been conducted based on NDVI derived from the new moderate resolution imaging spectroradiometer (MODIS) sensor [8], MODIS two-band Enhanced

Vegetation Index [9], even future weather variables [10]. Various methodologies are employed, such as statistical models [11], fuzzy systems and Artificial Neural Networks [12], deep long short-term memory model [7] and deep neural network [13]. You et al. [14] provided an end to end soybean yield prediction using remote sensing images as input. Ji et al. [15] investigated the effectiveness of machine learning methods and found, unfortunately, most of the academic endeavors centered on Artificial Neural Networks with one or a few data sources undermine their predict performance. Most previous studies assume that the crop yield uniformly distributed over space ignoring the spatial variations.

Multi-Task Learning. Multi-Task learning (MTL) is implemented to predict spatial events due to its competence to exploit dynamic features and scalability [16]. Through learning multiple related tasks simultaneously and treating prediction at each time point as a single task, MTL captures the progression trend of Alzheimer's Disease better [17]. MTL also has outstanding performance in event forecasting across cities [16] and fine-grain sentiment analysis [18], as well as in distance speech recognition [19]. Lu et al. [20] proposed a principled approach for designing compact MTL architecture by starting with a thin network and dynamically widening it in a greedy manner. Xu et al. [21] designed an online learning framework that can be used to solve online multi-task regression problems, although it is not a memory-efficient solution for data intensive application. To the best of our knowledge, usage of MTL in crop yield forecast in within-field practice is untouched. We propose a Multi-Task Learning model which targets at predicting each grid in the field for a crop season as an individual task. Meanwhile, we incorporate the spatial correlations as a regularization term to minimize the prediction errors.

3   Proposed Model

3.1   Overview

Figure 2 presents the framework of our prediction model. The cotton field is split into 475 grids for fine-grain prediction. We utilized the Dense and Dropout layers in the network. A shared Dense layer is used to extract latent features from all data dimensions, which are aggregated and fed into multiple sub-networks. Each sub-network represents the architecture of forecasting task in one year for all the grids. In other words, cotton yield prediction for all grids of each year are achieved in parallel via the separated sub-networks.

This aggregation is shared among all task-specific sub-networks. Therefore, it helps the task-specific sub-network to learn features from other tasks and to enhance its own prediction performance. Dense layer helps receive input from all the neurons in the previous layer with the intuition that all factors contribute each layer output neurons. Mathematically, the latent feature $\hat{p}^t$ learned after a fully connected layer is computed as:

$$\hat{p}^t = \sigma(\sum_{i=1}^{N} x_i * w_i + b), \tag{1}$$

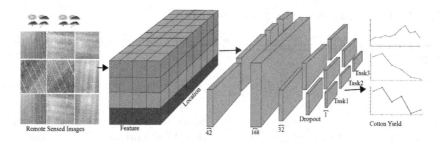

Fig. 2. The cotton yield prediction framework.

where N is the number of neurons in a layer, x_i represents input feature, w_i is a weight element, b is a bias and σ is the activation function. In our setting, $\sigma(x)$ is a Sigmoid function defined as $\sigma(x) = 1/(1+e^{-x})$. Moreover, the dropout layer or dropout regularization is also used to randomly exclude some neurons (20% in ours) to avoid over-fitting.

3.2 Cotton Yield Prediction

Our feature set is enriched by concatenating the latent features and feeding the output into a shared Dense layer. Suppose $p^{t_1}, p^{t_2}, p^{t_3}$ and p^{t_4} are features from our sources, the joint feature v_{fc} is the concatenation (denote as $\oplus$) of those features [22]:

$$v^{fc} = p^{t_1} \oplus p^{t_2} \oplus p^{t_3} \oplus p^{t_4}. \tag{2}$$

Stacked on the top of the shared Dense layer are three separate sub-networks, and each is used for one yearly cotton yield forecasting task, as shown in Fig. 2. After this layer, the latent feature is learned at time j following the equation: $h_j = \sigma(W_j * v_j^{fc} + b)$. Because cotton is usually planted by the end of May and harvest at the end of September or early October, we cut cotton's life cycle into several pieces, and each piece represents 2 weeks. Instead of taking it as time series data, we treat it as a couple of separate temporal features and utilize fully connected Dense layers and Dropout layers behind the shared Dense layer. We define the regression function for task t as:

$$\hat{y}_j^t = \sigma(W_j^t * h_j^t + b_j^t), \tag{3}$$

where W_j^t and b_j^t are learnable parameters, h_j^t is the output of the last hidden layer, and σ is a linear activation function. The model output lies in the interval $[0, 1]$ after value normalization. We will recover them to the original values when doing performance evaluation.

3.3 Spatial Feature in the Loss Function

A loss function is defined as the mean square error between the observation and the prediction:

$$\mathcal{L}(\theta) = \sum_{k=1}^{N} (y_k - \hat{y}_k)^2, \tag{4}$$

where θ means all learnable parameters, N is the number of regions in the field, y_k represents actual yield value, and $\hat{y}_k$ represents the predicted yield value. To train θ by minimizing the loss function may introduce overfitting. Therefore, for the grid-scale crop forecasting within a field, the spatial correlations depend heavily on the factor of distance. This drives us to define the spatial influence via a regularization term that the yield difference between the predicting region and the weighted average yield of the neighboring regions should be minimized. In particular, suppose $G(k)$ is the set of neighbors of region k (determined by all regions whose distance to region k less than a threshold), and $w(k, j)$ is the inverse distance weight between region k and region j, the loss function now becomes:

$$\mathcal{L}(\theta) = \sum_{k=1}^{N} [(y_k - \hat{y}_k)^2 + \lambda \sum_{j \in G(k)} \frac{w(k,j)}{|G(k)|} * |\hat{y}_k - y_j|^2], \tag{5}$$

where λ is the hyper parameter, $d(k, j)$ is the Euclidean distance between these regions $k \neq j$. The spatial weight $w(k, j)$ is computed as:

$$w(k, j) = \frac{1}{d(k, j)^p}, \tag{6}$$

where p is the power parameter (which equals to 2 in our experiment).

4   Experiments

4.1 Dataset and Feature Extraction

Our dataset includes weather data, soil properties, spectral data, and NDVI. Spectral data and NDVI are extracted from Lantsat 5 and Landsat 7 remote sensing images. The multi-spectral images were collected from 2001 to 2003 of a cotton field in west Texas. The total area is approximately 48 ha. The sensed images spatial resolution is 30 m. Hence, there are 475 grid cells under investigation. Figure 3 shows the distribution of some features over the field.

Weather Data. Weather data includes the daily temperature and rainfall level. For simplicity, we use the average of every two weeks' weather data as features to match the sensed images.

Soil Properties. Topographic variation is a common characteristic of large agricultural fields that has effects on spatial variability of soil water and ultimately on crop yield [23]. Besides, soil electrical conductivity (ECa) is also a reliable

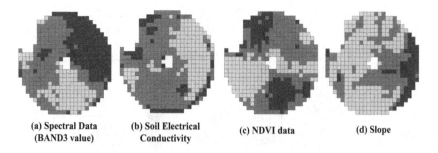

**(a) Spectral Data
(BAND3 value)** **(b) Soil Electrical
Conductivity** **(c) NDVI data** **(d) Slope**

Fig. 3. (a) to (d) Within-field feature value distributions. Darker colors indicate higher values. Each feature value is normalized into [0, 1].

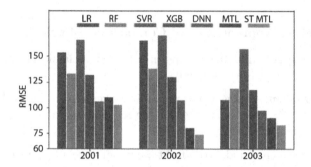

Fig. 4. Performance comparison in each year measured in RMSE, using whole dataset until September.

measurement of field variability. The relationship between ECa and crop yield depends on climate, crop type and other specific field conditions [24]. The variables that are considered in this paper include elevation, slope, curvature, the average electrical conductivity of soil, etc.

Field Spectral Data Before Planting. Field spectrum before planting may influence the entire crop yield. This data has four spectral bands extracted from the sensed images with 30-m spatial resolution. Band1, Band2, Band3, and Band4 represent blue, green, red and near infrared value, respectively.

NDVI Data. NDVI represents Normalized Difference Vegetation Index. It *is typically related to amount or density of vegetation, which is calculated as the difference between the reflectance in near-infrared (which vegetation strongly reflects) and red wavelengths divided by the sum of these two.* NDVI is computed as: $NDVI = \frac{NIR-RED}{NIR+RED}$, where NIR represents the spectral reflectance in near-infrared wavelength and RED is the spectral reflectance in the red wavelength.

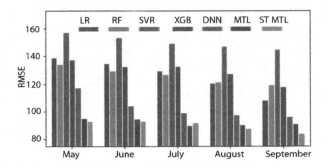

Fig. 5. Model performance in each month of 2003 measured in RMSE.

4.2   Competing Approaches and Comparison Metrics

Competing Approaches. In our experiment, a list of classical forecasting models are used for comparison and analysis:

Linear Regression. This is a traditional linear regression model. Its standard formula is: $y = \sum_{i=1}^{n} \alpha_i x_i + \epsilon$ where y is the response variable, x_i is the feature and ϵ is the deviation.

Random Forest. This model tries to fit a number of regression trees on various sub-samples of the dataset and uses averaging to improve the predictive accuracy and control overfitting. Each leaf of the tree contains a distribution for the continuous output variable.

Support Vector Regression (SVR). SVR is a nonparametric technique that aims to find a function $f(x)$ that produces output deviated from observed response values y_n by a value no greater than ϵ for each training point x, and meanwhile, as flat as possible.

XGBoost. This is an extension of gradient boosting machine (GBM) algorithm that tries to divide the optimization problem into two parts by first determining the direction of the step and then optimizing the step length.

Deep Neural Network (DNN). A fully connected deep neural network composed of three Dense layers, connecting to a Dropout and followed by another Dense layer is developed and used as another baseline for comparison.

Evaluation Metrics. Let N be the number of grids under forecast. We denote A_i as the actual crop yield and F_i is the forecast yield for grid i. A set of classical metrics such as Mean square error (MSE), Root mean square error (RMSE), Mean absolute error (MAE), Mean absolute percentage error (MAPE) and Max error (ME) are used to elaborate the performance. These measures are computed as: $MSE = \frac{1}{N} \sum_{i=1}^{N} (A_i - F_i)^2$, $RMSE = \sqrt{MSE}$, $MAE = \frac{1}{N} \sum_{i=1}^{N} |A_i - F_i|$, $MAPE = \frac{100\%}{N} \sum_{i=1}^{N} \frac{|A_i - F_i|}{A_i}$ and $ME = max(|A_i - F_i|)$ where $i = 1, \ldots, N$.

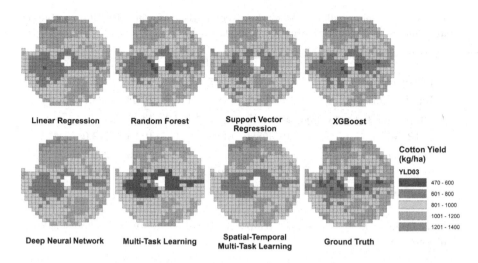

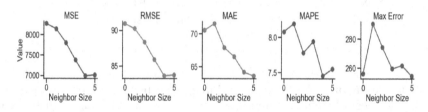

Fig. 6. Within-field cotton yield prediction on different algorithms versus ground truth yield monitor data in 2003.

Fig. 7. Impact of neighborhood size on model performance

4.3 Experimental Results

Average Performance. Figure 4 shows the performance of our proposed model compared with other baselines in term of RMSE metric. The Multi-Task learning and our proposed Spatial-Temporal Multi-Task learning model have the least error in all three years. Figure 6 and Table 1 take yield prediction performance of 2003 as an example. The Multi-Task Learning and our Spatial-Temporal Multi-Task learning methods show significant superiority than all the other approaches. With the whole data package, our model achieves the smallest error metrics (MSE, RMSE, MAE, MAPE, ME) which are 7,013.5, 83.7, 63.6, 7.55 and 254.4, respectively. The overall performance of Multi-Task Learning is second and close to our proposed approach, while Support Vector Regression shows the worst performance.

Real-Time Prediction Throughout the Year. Considering the life cycle of cotton in the U.S., we train the model using partially available input features and predict the cotton yield in each month in an online manner. Figure 5 shows the

Table 1. Cotton yield prediction performance comparison of year 2003 while combining all data sources. Bold values represent the best results.

	MSE	RMSE	MAE	MAPE	Max error
Linear regression	17,875.6	133.7	109.0	10.96	276.7
Random forest	19,295.3	138.9	112.0	10.68	321.7
Support vector regression	33,588.5	183.3	142.2	13.35	475.4
XGBoost	19,498.4	139.6	111.4	10.05	353.0
Deep neural network	9,504.8	97.5	77.8	9.96	269.4
Multi-task learning	8,267.5	90.9	70.5	8.08	256.2
Spatial-temporal M.T.L.	**7,013.5**	**83.7**	**63.6**	**7.55**	**254.4**

Table 2. Performance of the spatial-temporal Multi-task Learning model using individual source of data.

Input source	MSE	RMSE	MAE	MAPE	Max Error
Soil properties	15,791.9	125.7	98.2	12.04	**372.7**
Spectral data	21,292.5	145.9	114.4	13.62	450.1
NDVI	**15,134.5**	**123.0**	**94.7**	**11.80**	470.4

performance when we try to make a prediction in May, June, July, August and September, using only the data available up to that point. As more information is available, most of the models improve. The improvement during the first three months is less than that of later two months. All models perform better in August and reach the best in September.

Spatial Correlations: We also vary the neighborhood distance of each region from 1 to 5 to verify the impact of spatial correlation among regions under prediction. As shown in Fig. 7, MSE, RMSE and MAE gradually decrease when the distance increases. This trend stops when neighborhood size equals to 4. The performance becomes more stable afterwards. Even though we see a random value in MAPE and ME metrics with respect to the neighborhood distance, there is also a decreasing trend on MAPE and ME when neighborhood size increases. Therefore, in our experiment, we set the neighbor distance as 5.

Understanding the Importance of the Features: Since weather data is shared in all regions under prediction, we do not evaluate its impact. Instead, we explore the impacts of soil properties, spectral conditions before planting and NDVI on the cotton yield prediction. We split the data by dimensions and conduct two experiments: in Table 2, we use one data source at a time to compare the importance of this single source, in Table 3 we remove one data source and use the rest input each time to compare the performance.

Table 3. Discover the importance of sources of data by removing one source at a time.

Removed source	MSE	RMSE	MAE	MAPE	Max error
Soil properties	13,383.0	115.7	**88.8**	**10.79**	481.6
Spectral data	**13,189.6**	**114.8**	93.0	11.54	**276.5**
NDVI	15,642.0	125.1	100.3	12.39	330.0

Table 2 shows that NDVI data contributes most significantly to better precision. It gets MSE and RMSE values at 15,134.5 and 123.0 while the spectral data produces the worst results, whose MSE and RMSE are 21,292.5 and 145.9, respectively. Table 3 indicates that if we ignore the spectral feature, the model achieves the best results compared with ignoring the soil properties or NDVI features. These results demonstrate that the NDVI impacts the prediction the most, then soil properties, while spectral data before planting has the minimal impact.

5   Conclusion

This paper proposes a novel Multi-task Learning framework for within-field scale cotton yield prediction, which ingests multiple heterogeneous data sources, such as soil type, weather, topographic, and remote sensing, and is capable of predicting within-field cotton yield throughout the growing season. By aggregating these multiple data sources in the shared layer before transferring to task-specific layers, this creative strategy is able to enhance specific learning task by utilizing sources from other related tasks. To minimize the spatial errors in prediction, this work introduces a spatial regularization to measure the correlations between a certain grid and its neighboring grids. The experimental results show the proposed approach consistently outperforms other competing approaches, and has a promising future in the crop yield prediction research field.

Acknowledgement. This work was supported by the U.S. National Science Foundation under the Grant CNS-1737634.

References

1. Bastiaanssen, W.G., Ali, S.: A new crop yield forecasting model based on satellite measurements applied across the Indus Basin, Pakistan. Agric. Ecosyst. Environ. **94**(3), 321–340 (2003)
2. Hebbar, K., et al.: Predicting cotton production using infocrop-cotton simulation model, remote sensing and spatial agro-climatic data. Curr. Sci. **95**, 1570–1579 (2008)
3. Akay, D., Atak, M.: Grey prediction with rolling mechanism for electricity demand forecasting of Turkey. Energy **32**(9), 1670–1675 (2007)
4. Sugihara, G., May, R.M.: Nonlinear forecasting as a way of distinguishing chaos from measurement error in time series. Nature **344**(6268), 734 (1990)

5. Garcia, O.: A stochastic differential equation model for the height growth of forest stands. Biometrics **39**, 1059–1072 (1983)
6. Guo, W., Cui, S., Torrion, J., Rajan, N.: Data-driven precision agriculture: opportunities and challenges. In: Soil-Specific Farming, pp. 366–385. CRC Press (2015)
7. Jiang, Z., Liu, C., Hendricks, N.P., Ganapathysubramanian, B., Hayes, D.J., Sarkar, S.: Predicting county level corn yields using deep long short term memory models, arXiv preprint arXiv:1805.12044 (2018)
8. Mkhabela, M., Bullock, P., Raj, S., Wang, S., Yang, Y.: Crop yield forecasting on the canadian prairies using MODIS NDVI data. Agric. Forest Meteorol. **151**(3), 385–393 (2011)
9. Bolton, D.K., Friedl, M.A.: Forecasting crop yield using remotely sensed vegetation indices and crop phenology metrics. Agric. Forest Meteorol. **173**, 74–84 (2013)
10. Schlenker, W., Roberts, M.J.: Nonlinear temperature effects indicate severe damages to US crop yields under climate change. Proc. Natl. Acad. Sci. **106**(37), 15594–15598 (2009)
11. Lobell, D.B., Burke, M.B.: On the use of statistical models to predict crop yield responses to climate change. Agric. Forest Meteorol. **150**(11), 1443–1452 (2010)
12. Dahikar, S.S., Rode, S.V.: Agricultural crop yield prediction using artificial neural network approach. IJIREEICE **2**(1), 683–686 (2014)
13. Oliveira, I., Cunha, R.L., Silva, B., Netto, M.A.: A scalable machine learning system for pre-season agriculture yield forecast, arXiv preprint arXiv:1806.09244 (2018)
14. You, J., Li, X., Low, M., Lobell, D., Ermon, S.: Deep Gaussian process for crop yield prediction based on remote sensing data. In: AAAI, pp. 4559–4566 (2017)
15. Ji, B., Sun, Y., Yang, S., Wan, J.: Artificial neural networks for rice yield prediction in mountainous regions. J. Agric. Sci. **145**(3), 249–261 (2007)
16. Zhao, L., Sun, Q., Ye, J., Chen, F., Lu, C.-T., Ramakrishnan, N.: Multi-task learning for spatio-temporal event forecasting. In: Proceedings of KDD 2015, pp. 1503–1512. ACM (2015)
17. Zhou, J., Yuan, L., Liu, J., Ye, J.: A multi-task learning formulation for predicting disease progression. In: Proceedings of KDD 2011, pp. 814–822. ACM (2011)
18. Balikas, G., Moura, S., Amini, M.-R.: Multitask learning for fine-grained Twitter sentiment analysis. In: Proceedings of SIGIR 2017, pp. 1005–1008. ACM (2017)
19. Zhang, Y., Zhang, P., Yan, Y.: Attention-based LSTM with multi-task learning for distant speech recognition. In: Interspeech 2017, pp. 3857–3861 (2017)
20. Lu, Y., Kumar, A., Zhai, S., Cheng, Y., Javidi, T., Feris, R.S.: Fully-adaptive feature sharing in multi-task networks with applications in person attribute classification. In: CVPR, vol. 1, no. 2, p. 6 (2017)
21. Xu, J., Tan, P.-N., Zhou, J., Luo, L.: Online multi-task learning framework for ensemble forecasting. IEEE Trans. Knowl. Data Eng. **29**(6), 1268–1280 (2017)
22. Yao, H., et al.: Deep multi-view spatial-temporal network for taxi demand prediction, arXiv preprint arXiv:1802.08714 (2018)
23. Hanna, A., Harlan, P., Lewis, D.: Soil available water as influenced by landscape position and aspect 1. Agron. J. **74**(6), 999–1004 (1982)
24. Kitchen, N., Sudduth, K., Drummond, S.: Soil electrical conductivity as a crop productivity measure for claypan soils. J. Prod. Agric. **12**(4), 607–617 (1999)

Factor and Tensor Analysis

Online Data Fusion Using Incremental Tensor Learning

Nguyen Lu Dang Khoa[1(✉)], Hongda Tian[2], Yang Wang[2], and Fang Chen[2]

[1] Data61, CSIRO, Sydney, NSW, Australia
khoa.nguyen@data61.csiro.au
[2] University of Technology Sydney, Sydney, NSW, Australia
{hongda.tian,yang.wang,fang.chen}@uts.edu.au

Abstract. Despite the advances in Structural Health Monitoring (SHM) which provides actionable information on the current and future states of infrastructures, it is still challenging to fuse data properly from heterogeneous sources for robust damage identification. To address this challenge, the sensor data fusion in SHM is formulated as an incremental tensor learning problem in this paper. A novel method for online data fusion from heterogeneous sources based on incrementally-coupled tensor learning has been proposed. When new data are available, decomposed component matrices from multiple tensors are updated collectively and incrementally. A case study in SHM has been developed for sensor data fusion and online damage identification, where the SHM data are formed as multiple tensors to which the proposed data fusion method is applied, followed by a one-class support vector machine for damage detection. The effectiveness of the proposed method has been validated through experiments using synthetic data and data obtained from a real-life bridge. The results have demonstrated that the proposed fusion method is more robust to noise, and able to detect, assess and localize damage better than the use of individual data sources.

Keywords: Data fusion · Incrementally-coupled tensor learning ·
Online learning · Anomaly detection

1 Introduction

Civil infrastructures are critical to our society as they support the flows of people and goods within cities. Any problem on such a structure from small damage to catastrophic failures would result in certain economic and potential life losses. Currently most of structural maintenances are time-based, e.g. visual inspections at predefined regular schedules. Structural Health Monitoring (SHM) is a condition-based monitoring using sensing system which provides actionable information on the current and future states of infrastructures. SHM systems built on advanced sensing technologies and data analytics allow the shift from time-based to condition-based maintenance [3].

© Springer Nature Switzerland AG 2019
Q. Yang et al. (Eds.): PAKDD 2019, LNAI 11439, pp. 357–369, 2019.
https://doi.org/10.1007/978-3-030-16148-4_28

In SHM, measured data are often in a multi-way form, i.e. multiple sensors at different locations simultaneously collect data over time. These data are highly redundant and correlated, which are suitable to be analyzed using tensor analysis [6,7]. There were efforts to apply tensor CANDECOMP/PARAFAC (CP) [7] decomposition in SHM for damage identification [5,6,10]. However, these approaches are confined to a fusion from sensors of the same type to guarantee the data can be formed in a single tensor. In many SHM systems, data come from heterogeneous sources due to an availability of different types of sensors (e.g. accelerometers, strain gauges and thermometers). Additionally, existing methods for data fusion from heterogeneous sources using tensor analysis [1,12] mainly work in an offline manner, which is not practical for SHM applications. In this paper, we propose a method to fuse data online from heterogeneous sources based on incremental tensor learning, which is then used for damage identification in SHM. Our contributions are summarized as follows:

- We propose a method for online data fusion from heterogeneous sources using incrementally-coupled tensor learning. Specifically, our method collectively and incrementally updates component matrices for CP decomposition from multiple tensors when new data arrive.
- We develop a case study used in SHM for sensor data fusion and online damage identification. In the case study, the SHM data are formed as multiple tensors to which an incremental tensor fusion is applied, followed by a one-class support vector machine (OCSVM) for damage detection.
- We demonstrate the effectiveness of the proposed method through experiments using synthetic data and real data obtained from a bridge in Sydney.

In this paper, we represent a tensor as a three-way array, which is a typical case in SHM. However, all the theories could be generalized for a n-way array. The remainder of the paper is organized as follows. Section 2 summarizes the related work. Section 3 describes our novel method to incrementally update component matrices from multiple tensors at the same time and its uses for online damage identification in SHM. Section 4 presents the experimental results. We conclude our work in Sect. 5.

2   Related Work

Incremental tensor analysis, which is used for online applications, mainly focuses on Tucker decomposition [8,13] since it makes use an extensive literature of incremental singular value decomposition (SVD). There are a few works [5,9,15] on an incremental learning for CP decomposition. Nion and Sidiropoulos [9] proposed a method to incrementally track the SVD of the unfolded tensor for CP. However, this technique scales linearly with time, which is impractical to use for large datasets. Zhou et al. [15] discussed a method to incrementally track CP decomposition over time. It follows an alternating least square (ALS) style: update a component matrix while fixing all the others. However, the update only occurs once instead of an iterative process, which makes the approximation sometimes ineffective. Khoa et al. [5] extended this method in a proper ALS style, resulting in more accurate updated component matrices.

Data fusion using coupled matrix/tensor decomposition has become popular recently [1,12,14]. Instead of using ALS algorithms, Acar et al. [1] proposed an all-at-once optimization approach for coupled matrix and tensor factorization. Sorber et al. [12] presented a framework where the type of tensor decomposition, the coupling between factorizations and the structure imposed on the factors can all be chosen freely without any changes to the solver. In [14], the authors proposed a method to learn a clustered low-rank representation for multiview spectral clustering using structured matrix factorization. However, these methods all work offline which limits their applications.

3   Online Damage Identification Using Incrementally-Coupled Tensor Learning

The proposed method to identify damage online using incrementally-coupled tensor learning is depicted in Fig. 1. In SHM, vibration responses of a structure are measured over time by different types of sensors (e.g. accelerometers and strain gauges). The data from each type of sensors when the structure is in a healthy condition can be considered as a three-way tensor ($feature \times location \times time$). Thus we have tensor $\mathcal{X}_1$ for accelerometers and $\mathcal{X}_2$ for strain gauges. Feature is the information extracted from the raw signals; location represents sensor positions; and time indicates data snapshots at different timestamps. Each slice along the time axis shown in Fig. 1 is a frontal slice representing all features across all locations at a particular time.

Training tensors $\mathcal{X}_1$ and $\mathcal{X}_2$ are jointly decomposed into matrices of different modes using coupled tensor-tensor decomposition as described in Sect. 3.1. When new data arrive, these matrices are jointly updated using our proposed incremental tensor analysis as in Sect. 3.2. A monitoring of these factor matrices over time will help identify the damage in the structure (Sect. 3.3).

3.1   Data Fusion Using Coupled Tensor-Tensor Decomposition

Two typical approaches for tensor decomposition are CP decomposition and Tucker decomposition [7]. After a decomposition of a three-way tensor, three component matrices can be obtained representing information in each mode. In the case of SHM data as in Fig. 1, they are associated with feature (matrix A_1), location (matrix B_1) and time modes (matrix C) (for tensor $\mathcal{X}_1$). We also obtain component matrices A_2, B_2 and C (for tensor $\mathcal{X}_2$). Note that C is the same for $\mathcal{X}_1$ and $\mathcal{X}_2$ since time information is shared between these two types of sensors.

In CP method, the decomposed matrices are unique provided that we permute the rank-one components [7]. Therefore it is easy to interpret the artifact in each mode separately using its corresponding component matrix. Thus, CP method is used in this paper for our SHM applications.

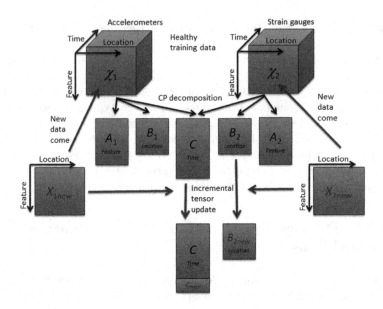

Fig. 1. A flowchart of incremental tensor fusion for online damage identification.

The problem to jointly decompose $\mathcal{X}_1$ and $\mathcal{X}_2$ using CP can be formulated as

$$f(A_1, B_1, C, A_2, B_2) = \frac{1}{2}\left\|\mathcal{X}_1 - [A_1, B_1, C]\right\|^2 + \frac{1}{2}\left\|\mathcal{X}_2 - [A_2, B_2, C]\right\|^2 \quad (1)$$

where $\mathcal{X}_i = [A_i, B_i, C]$ represents the CP decomposition and can be formulated as $X_{i(1)} = A_i(C \odot B_i)^{\top}$, $X_{i(2)} = B_i(C \odot A_i)^{\top}$ and $X_{i(3)} = C(B_i \odot A_i)^{\top}$ ($X_{i(j)}$ is an unfolding matrix of $\mathcal{X}_i$ in mode j and $\odot$ is the Khatri-Rao product) [7]. Equation 1 can be solved using ALS and it is summarized in Algorithm 1. In our SHM application (as in Sect. 4.2), the time matrix C is used for damage detection in time mode while location matrix B_2 is used for damage localization.

3.2   Incremental Tensor Update

OnlineCP-ALS [5] was proposed to incrementally update the component matrices of a tensor when new data arrive, which was shown to be better than other incremental CP decomposition methods. Using similar ideas, we propose a technique to jointly and incrementally update component matrices from different tensors over time as follows.

Update Temporal Mode C. Due to an arrival of new information (new frontal slices in time mode), additional rows will be added to component matrix C.

Algorithm 1. Coupled Tensor-Tensor Decomposition

Input: Tensors $\mathcal{X}_1$, $\mathcal{X}_2$, number of components R
Output: Component matrices A_1, B_1, C, A_2 and B_2

1: Initialize A_1, B_1, C, A_2 and B_2
2: **repeat**
3:    $A_1 = \arg\min_{A_1} \frac{1}{2} \left\| X_{1(1)} - A_1(C \odot B_1)^\top \right\|^2$ (fixing B_1 and C)
4:    $B_1 = \arg\min_{B_1} \frac{1}{2} \left\| X_{1(2)} - B_1(C \odot A_1)^\top \right\|^2$ (fixing A_1 and C)
5:    $A_2 = \arg\min_{A_2} \frac{1}{2} \left\| X_{2(1)} - A_2(C \odot B_2)^\top \right\|^2$ (fixing B_2 and C)
6:    $B_2 = \arg\min_{B_2} \frac{1}{2} \left\| X_{2(2)} - B_2(C \odot A_2)^\top \right\|^2$ (fixing A_2 and C)
7:    $C = \arg\min_{C} \frac{1}{2} \left\| [X_{1(3)} \; X_{2(3)}] - C \left[(B_1 \odot A_1)^\top \; (B_2 \odot A_2)^\top\right] \right\|^2$ (fixing A_1, B_1, A_2 and B_2)
8: **until** convergence

By fixing A_1, B_1, A_2 and B_2, we can solve C from Eq. 1 as:

$$C = \arg\min_{C} \frac{1}{2} \left\| [X_{1(3)} \; X_{2(3)}] - C \left[(B_1 \odot A_1)^\top \; (B_2 \odot A_2)^\top\right] \right\|$$

$$= \arg\min_{C} \frac{1}{2} \left\| \begin{bmatrix} [X_{1old(3)} \; X_{2old(3)}] - C_{old} \left[(B_1 \odot A_1)^\top \; (B_2 \odot A_2)^\top\right] \\ [X_{1new(3)} \; X_{2new(3)}] - C_{new} \left[(B_1 \odot A_1)^\top \; (B_2 \odot A_2)^\top\right] \end{bmatrix} \right\|.$$

Thus,

$$C = \begin{bmatrix} C_{old} \\ C_{new} \end{bmatrix} = \begin{bmatrix} C_{old} \\ [X_{1new(3)} \; X_{2new(3)}] \left[(B_1 \odot A_1)^\top \; (B_2 \odot A_2)^\top\right]^\dagger \end{bmatrix}, \quad (2)$$

where $\dagger$ is a matrix pseudo-inverse. Therefore, new rows added to C can be estimated using only new information appending in time mode.

Update Non-temporal Mode A_1, B_1, A_2 and B_2. By fixing B_1 and C for updating A_1, the Eq. 1 can be written as $\frac{1}{2} \left\| X_{1(1)} - A_1(C \odot B_1)^\top \right\|^2$. Using the approach as in [15], by taking the derivative of this function with regard to A_1 and setting it to zero, we have:

$$A_1 = \frac{X_{1(1)}(C \odot B_1)}{(C \odot B_1)^\top(C \odot B_1)} = P_1 Q_1^{-1},$$

where $P_1 = X_{1(1)}(C \odot B_1)$ and $Q_1 = (C \odot B_1)^\top(C \odot B_1)$.

Directly calculating P_1 and Q_1 is costly since $(C \odot B_1)$ is a big matrix. By representing $X_{1(1)}$ and C with *old* and *new* information, we can have $P_1 = P_{1old} + X_{1new(1)}(C_{new} \odot B_1)$ and $Q_1 = Q_{1old} + C_{new}^\top C_{new} \circ B_1^\top B_1$ ($\circ$ is the Hadamard product). Therefore, A_1 can be computed as:

$$A_1 = P_1 Q_1^{-1} = \frac{P_{1old} + X_{1new(1)}(C_{new} \odot B_1)}{Q_{1old} + C_{new}^\top C_{new} \circ B_1^\top B_1}. \quad (3)$$

Similarly we can derive the update for B_1 as:

$$B_1 = U_1 V_1^{-1} = \frac{U_{1old} + X_{1new(2)}(C_{new} \odot A_1)}{V_{1old} + C_{new}^\top C_{new} \circ A_1^\top A_1}, \tag{4}$$

where $U_1 = X_{1(2)}(C \odot A_1)$ and $V_1 = C^\top C \circ A_1^\top A_1$ [15].

Likewise, A_2 and B_2 can be updated as:

$$A_2 = P_2 Q_2^{-1} = \frac{P_{2old} + X_{2new(1)}(C_{new} \odot B_2)}{Q_{2old} + C_{new}^\top C_{new} \circ B_2^\top B_2}; \tag{5}$$

$$B_2 = U_2 V_2^{-1} = \frac{U_{2old} + X_{2new(2)}(C_{new} \odot A_2)}{V_{2old} + C_{new}^\top C_{new} \circ A_2^\top A_2}. \tag{6}$$

We can see that by storing information from previous decomposition (i.e. P_1, Q_1, U_1, V_1, P_2, Q_2, U_2 and V_2), component matrices A_1, B_1, A_2 and B_2 are updated using only new information arriving in time mode.

OnlineCP-Fusion. For two three-way tensors that grow with time (shared C mode), a two-staged procedure is proposed to jointly incrementally update tensor component matrices. The technique, which is called onlineCP-Fusion, is described in Algorithm 2.

Algorithm 2. Incrementally-Coupled Tensor Update: onlineCP-Fusion

---

Input: Training tensors $\mathcal{X}_{1train}$ and $\mathcal{X}_{2train}$
Output: Component matrices C, A_1, B_1, A_2 and B_2 when new data arrive

1: **Initialization/training stage:**
   C, A_1, B_1, A_2 and B_2 are obtained using Algorithm 1 on training tensors
   $P_1 = X_{1train(1)}(C \odot B_1)$ and $P_2 = X_{2train(1)}(C \odot B_2)$
   $Q_1 = C^\top C \circ B_1^\top B_1$ and $Q_2 = C^\top C \circ B_2^\top B_2$
   $U_1 = X_{1train(2)}(C \odot A_1)$ and $U_2 = X_{2train(2)}(C \odot A_2)$
   $V_1 = C^\top C \circ A_1^\top A_1$ and $V_2 = C^\top C \circ A_2^\top A_2$
2: **Update/test stage:** when new data arrive as new slices appended to time mode
   Repeat
      Update P_1, Q_1, U_1, V_1, P_2, Q_2, U_2 and V_2 using '*old*' information
      C is updated using Eq. (2) (fixing A_1, B_1, A_2 and B_2)
      A_1 is updated using Eq. (3) (fixing B_1 and C)
      B_1 is updated using Eq. (4) (fixing A_1 and C)
      A_2 is updated using Eq. (5) (fixing B_2 and C)
      B_2 is updated using Eq. (6) (fixing A_2 and C)
   Until convergence

Complexity Analysis. As in [15], the time complexity of onlineCP for a n-way tensor $\mathcal{X}_1(I_1 \times ... \times I_{n-1} \times K)$ and $\mathcal{X}_2(J_1 \times ... \times J_{n-1} \times K)$ are $O(nR \prod_{i=1}^{n-1} I_i)$ and $O(nR \prod_{i=1}^{n-1} J_i)$ where I_i and J_i are sizes of non-temporal modes. So it takes $O(nR(\prod_{i=1}^{n-1} I_i + \prod_{i=1}^{n-1} J_i)t)$ for onlineCP-Fusion, where t is the number of iterations for an ALS update. Since in a tensor growing in time mode normally $I_i, J_i \ll K$ (the size in time mode) and n, R, t are very small, the complexity for an update from onlineCP-Fusion can be consider as constant.

3.3   Online Damage Identification

Building a Benchmark Model. In practice, events corresponding to damaged states of structures are often unavailable for a supervised learning approach. In this work, we use OCSVM [11] with Gaussian kernel as an anomaly detection method. The technique in [4] is adopted to tune σ in the kernel.

In this step, C, A_1, B_1, A_2 and B_2 are obtained using Algorithm 1 on training tensors. Each row of the component matrix C represents an event in time mode. We build a benchmark model using healthy training events which are represented by rows of C by means of OCSVM.

Damage Identification. Due to an arrival of a new time event, an additional row C_{new} is added to the component matrix C, and matrices A_1, B_1, A_2 and B_2 are incrementally updated as described in Algorithm 2. The new row C_{new} will be checked if it agrees with the benchmark model built at the training stage, answering the condition of the structure. In the case of OCSVM, a negative decision value indicates that the new event is likely a damaged event.

Location matrices B_1 and B_2, where each row captures meaningful information for each sensor location, could be used for damage localization. By analyzing these matrices when each new data instance arrives, it is able to find anomalies, which correspond to damaged locations. In this work we only use B_2 (which represents sensor locations for one type of sensors) for damage localization due to the specific sensor instrumentation for the bridge tested in the experiment. An average distance from a sensing location (a row in B_2) to k nearest neighboring locations ($k = 2$) is regarded as an anomaly score to localize damage.

To estimate the extent of damage, we analyze the decision values by the OCSVM model. The rationale is that a structure with a more severe damage (e.g. a longer crack) will behave more differently from a normal situation. Different ranges of the decision values may imply different severity levels of damage.

4   Experimental Results

Experiments were conducted using both synthetic data and data collected from a real bridge in operation. For all experiments, we compare our onlineCP-Fusion (for fusion $\mathcal{X}_1$ and $\mathcal{X}_2$) with onlineCP-ALS [5] (as baselines which learn from $\mathcal{X}_1$ and $\mathcal{X}_2$ separately). Another baseline called naive-Fusion to fuse data by concatenating the features from all strain gauge and cable sensors in each time instance as a feature vector, followed by random projection (with dimension size $k = 50$) to reduce the feature dimension. Then self-tuning OCSVM (with $\nu = 5\%$) was used for anomaly detection on feature spaces obtained by all these methods. About 80% of healthy data used for training and the rest for testing. All reported results were averaged over 10 trials.

4.1   Synthetic Data

We generated 5 matrices randomly from standard normal distribution with different means and variances: $A_1(50 \times R), B_1(20 \times R), A_2(25 \times R), B_2(10 \times R)$ and $C(500 \times R)$. They were considered as latent factors decomposed from two tensors as in Fig. 1. Then 5% of data instances in C were replaced with data generated randomly from a uniform distribution (as damage/anomaly). All the matrices were then normalized to have unity norm for all their columns. Then a tensor $\mathcal{X}_1$ was constructed from A_1, B_1, C and a tensor $\mathcal{X}_2$ was constructed from A_2, B_2, C using CP. Next Gaussian noise was randomly added to 35% of data along third dimension of tensor $\mathcal{X}_1$ and to 50% of data along third dimension of tensor $\mathcal{X}_2$. The purpose is to check if a data fusion of two tensors can eliminate the adverse effects of noise from individual tensors.

$R = 5$ was selected for tensor construction and decomposition in the experiment. $F1$-score was adopted to measure the accuracy of OCSVM for anomaly detection in the learned time matrix C. Factor similarity was used to estimate the similarity between decomposed latent factors and the real ones we generated (this is not applicable to naive-Fusion). The similarity score for each column of each component matrix is computed as $\frac{|\hat{a}_r^T a_r|}{\|\hat{a}_r\|\|a_r\|}$ after finding the best matching permutation of the columns ($\hat{a}_r$ and a_r are a real latent factor column and its decomposed one respectively). The product of all these scores for all columns of all component matrices represents the final similarity score.

Factor similarities and $F1$-scores based on different methods are shown in Table 1, indicating that fusing data from two noisy tensors by means of onlineCP-Fusion overall yields better result than all other baselines.

Table 1. Factor similarities and $F1$-scores based on different methods.

	Factor similarity	$F1$-score
OnlineCP-ALS for $\mathcal{X}_1$	0.86	0.85
OnlineCP-ALS for $\mathcal{X}_2$	0.68	0.72
Naive-Fusion	N/A	0.54
OnlineCP-Fusion for $\mathcal{X}_1$ and $\mathcal{X}_2$	0.86	0.87

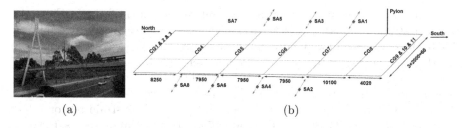

(a)   (b)

Fig. 2. The cable-stayed bridge and the longitudinal and lateral girders under the deck.

4.2   Real Bridge Data

A cable-stayed bridge (Fig. 2a) in Sydney, Australia was considered as a case study in this work. It has a steel tower with a composite steel-concrete deck. The deck is supported by four I-beam steel girders, which are internally attached by a set of equally-spaced cross girders (CG). A dense array of sensing system has been deployed on the bridge since 2016. All the sensors are timely synchronized and are continuously measuring the dynamic response of the bridge under normal operation at 600 Hz. Each cable has been instrumented with a full axial Wheatstone bridge to measure the dynamic strain response of the cables (namely SA1 to SA8 which are, respectively, installed on cables 1 to 8 as in Fig. 2b. They are also aligned with CGs 4–7). After a test, it was realised that sensor SA4 was not operational and thus it was eliminated from the analysis. In our work, we used two sets of sensors for data fusion experiments: a set of 6 strain gauge sensors mounted to the bridge deck and a set of aforementioned 7 cable sensors.

We emulated damage by locating stationary mass on the bridge at different locations as real damage was not available. This additional mass can be treated as a damaged event for evaluation purpose since the increment of mass results in a similar effect on the bridge dynamic properties as the decrement of stiffness caused by an actual damage. Two extensive field experiments were conducted on this bridge which are referred to as Bus Damage Test and Car Damage Test. The Bus Damage Test was conducted in a way that a 13t three-axle bus was placed at a stationary location at mid-span of the bridge. Due to the distributed effect of mass in this case, this dataset is not suitable for damage localization and it will be solely adopted for detection and assessment of damage. In the Car Damage Test, a 2.4t car was utilized. In each damage case (i.e. Car Damage 1 to Car Damage 4), the vehicle was placed in each cross girder (CGs 4–7 respectively, where the 4 pairs of cable sensors are placed) and the dynamic response of the bridge was recorded under ambient excitation. The Car Damage Test could be used to verify whether the proposed method is capable of locating damage.

Feature Extraction. The change in the cable-forces was adopted for damage identification as any damage in the structure changes the distribution of the cable-forces. Ambient strain responses from each cable sensor in both healthy and damaged cases were split into events of 2 s for analysis. Then the following steps were applied to extract features for our damage identification. First, the dynamic strain responses due to the live load effects from each cable (except SA4 due to the sensor issue) were normalized by subtracting the average strain of the healthy training data from the same cable. Then the absolute normalized strain was transformed into an unique direction by taking into account the orientation of each cable. This resulted in seven time series responses for seven cable sensors (i.e. SA1, SA2, SA3, SA5, SA6, SA7 and SA8). Since each strain response had 1200 samples (2 s at 600 Hz) and there were 7 locations of cable sensors, the data formed a cable tensor of 1200 *features* × 7 *locations* × 187 *events* where 187 indicates the total number of healthy events and damaged events (including 4 Car Damage cases and 1 Bus Damage case). For 6 strain gauges on the bridge deck, the feature extraction is the same except their orientations were not used as in the cable sensors. Similarly, we have a strain gauge tensor of 1200 *features* × 6 *locations* × 187 *events*.

Damage Detection and Severity Assessment. Similar to experiments for the synthetic data, Gaussian noise was randomly added to 35% of data along third dimension of strain gauge tensor and to 50% of data along third dimension of cable tensor. Anomaly detection using self-tuning OCSVM was applied on the feature spaces learned by onlineCP-Fusion and all baselines. The number of latent factors R was selected as 2 using core consistency diagnostic technique (CORCONDIA) [2]. $F1$-scores of 0.99, 1 and 0.87 were achieved by onlineCP-ALS for the strain gauge tensor and cable tensor, and naiveFusion, respectively. Data fusion from two tensors using the proposed onlineCP-Fusion led to an $F1$-score of 1, which improved the overall performance from the approach without tensor data fusion. The results in Fig. 3 (obtained C matrices, $R = 2$) indicate the proposed method (Fig. 3c) is not only more capable to distinguish between healthy and damaged data for damage detection, but also between Bus Damage and Car Damage cases for severity assessment (i.e. Bus Damage samples were further away from the healthy data compared to those of Car Damage).

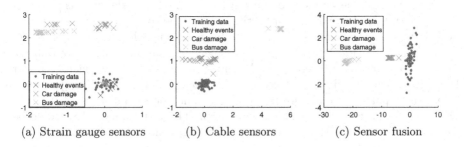

(a) Strain gauge sensors   (b) Cable sensors   (c) Sensor fusion

Fig. 3. Damage detection and severity assessment.

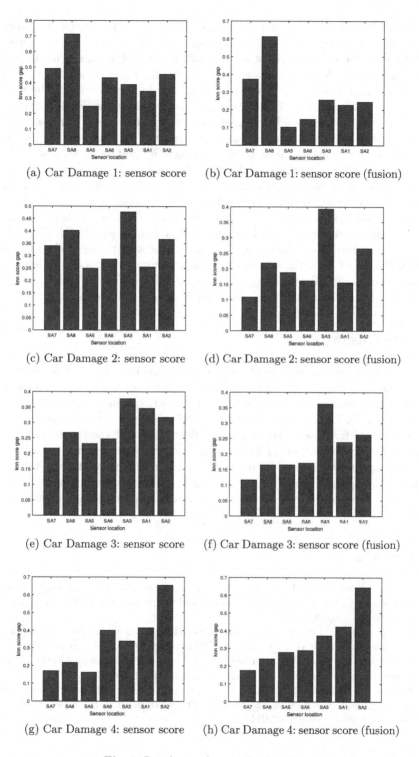

(a) Car Damage 1: sensor score   (b) Car Damage 1: sensor score (fusion)

(c) Car Damage 2: sensor score   (d) Car Damage 2: sensor score (fusion)

(e) Car Damage 3: sensor score   (f) Car Damage 3: sensor score (fusion)

(g) Car Damage 4: sensor score   (h) Car Damage 4: sensor score (fusion)

Fig. 4. Results on damage localization.

Damage Localization. Using the techniques in Sect. 3.2, component matrix B_{2new} was incrementally updated for every new test event. Sensor scores at each location were computed as in Sect. 3.3. We used the change of this score as an indicator to localize damage. Specifically the sensor which has the most change of this score is likely located near to the damage location. The results on damage localization for 4 Car Damage cases are shown in Fig. 4. As noticed, each damage case corresponds to a pair of figures for comparison purpose. Taking Figs. 4a and b as an example, Fig. 4a measures score changes (the difference or gap between the average sensor score of the healthy and damaged test data) of cable sensors when we only considered cable tensor. Figure 4b measures score changes of cable sensors when a data fusion for cable and strain gauge sensors was used. It is shown in Figs. 4a and b that the damage location was close to SA7/SA8 (in CG4), which is true as the car was in CG4 in Car Damage 1. We achieved similar results for Car Damage 3 (CG6, SA3) and 4 (CG7, SA1/SA2), except in Car Damage 2 where SA3 was the sensor with the most change. Even though the use of data fusion achieved similar results with the use of only cable sensors, the score changes of localized sensors using onlineCP-Fusion are more pronounced compared with those obtained from only cable sensors.

5   Conclusion

This paper has proposed a novel method for online data fusion from heterogeneous sources based on incrementally-coupled tensor learning, where component matrices from multiple tensors are updated collectively and incrementally when new data arrive. The method has been applied to a developed case study in SHM for sensor data fusion and online damage identification. Experiments using synthetic data and data obtained from a real-life bridge have verified the effectiveness of the proposed method for data fusion. The results show that the proposed data fusion approach is more robust to noise than the approach using individual data sources and has a potential for data fusion for damage identification in SHM.

References

1. Acar, E., Kolda, T.G., Dunlavy, D.M.: All-at-once optimization for coupled matrix and tensor factorizations. In: Proceedings of Mining and Learning with Graphs, MLG 2011, August 2011
2. Bro, R., Kiers, H.A.L.: A new efficient method for determining the number of components in PARAFAC models. J. Chemometr. **17**(5), 274–286 (2003)
3. Farrar, C.R., Worden, K.: An introduction to structural health monitoring. Philos. Trans. Roy. Soc. A: Math. Phys. Eng. Sci. **365**(1851), 303–315 (2007)
4. Khazai, S., Homayouni, S., Safari, A., Mojaradi, B.: Anomaly detection in hyperspectral images based on an adaptive support vector method. IEEE Geosci. Remote Sens. Lett. **8**(4), 646–650 (2011)

5. Khoa, N.L.D., Anaissi, A., Wang, Y.: Smart infrastructure maintenance using incremental tensor analysis: extended abstract. In: Proceedings of the 2017 ACM on Conference on Information and Knowledge Management, CIKM 2017, pp. 959–967. ACM, New York (2017)
6. Khoa, N.L.D., et al.: On damage identification in civil structures using tensor analysis. In: Cao, T., Lim, E.-P., Zhou, Z.-H., Ho, T.-B., Cheung, D., Motoda, H. (eds.) PAKDD 2015, Part I. LNCS (LNAI), vol. 9077, pp. 459–471. Springer, Cham (2015). https://doi.org/10.1007/978-3-319-18038-0_36
7. Kolda, T.G., Bader, B.W.: Tensor decompositions and applications. SIAM Rev. **51**(3), 455–500 (2009)
8. Liu, W., Chan, J., Bailey, J., Leckie, C., Kotagiri, R.: Utilizing common substructures to speedup tensor factorization for mining dynamic graphs. In: Proceedings of the 21st ACM International Conference on Information and Knowledge Management, CIKM 2012, pp. 435–444. ACM, New York (2012)
9. Nion, D., Sidiropoulos, N.D.: Adaptive algorithms to track the PARAFAC decomposition of a third-order tensor. Trans. Sig. Process. **57**(6), 2299–2310 (2009)
10. Prada, M.A., Toivola, J., Kullaa, J., Hollmén, J.: Three-way analysis of structural health monitoring data. Neurocomputing **80**, 119–128 (2012). Special Issue on Machine Learning for Signal Processing 2010
11. Schölkopf, B., Williamson, R.C., Smola, A.J., Shawe-Taylor, J., Platt, J.C.: Support vector method for novelty detection. In: NIPS, pp. 582–588 (1999)
12. Sorber, L., Barel, M.V., Lathauwer, L.D.: Structured data fusion. IEEE J. Sel. Top. Sig. Process. **9**(4), 586–600 (2015)
13. Sun, J., Tao, D., Papadimitriou, S., Yu, P.S., Faloutsos, C.: Incremental tensor analysis: theory and applications. ACM Trans. Knowl. Discov. Data **2**(3), 11:1–11:37 (2008)
14. Wang, Y., Wu, L., Lin, X., Gao, J.: Multiview spectral clustering via structured low-rank matrix factorization. IEEE Trans. Neural Netw. Learn. Syst. **29**(10), 4833–4843 (2018)
15. Zhou, S., Vinh, N.X., Bailey, J., Jia, Y., Davidson, I.: Accelerating online cp decompositions for higher order tensors. In: Proceedings of the 22nd ACM SIGKDD International Conference on Knowledge Discovery and Data Mining, KDD 2016, pp. 1375–1384. ACM, New York (2016)

Co-clustering from Tensor Data

Rafika Boutalbi[1,2(✉)], Lazhar Labiod[1], and Mohamed Nadif[1]

[1] LIPADE, University of Paris Descartes, 45 rue des Saints Pères, 75006 Paris, France
{rafika.boutalbi,lazhar.labiod,mohamed.nadif}@parisdescartes.com
[2] TRINOV, 196 rue Saint Honoré, 75001 Paris, France

Abstract. With the exponential growth of collected data in different fields like recommender system (user, items), text mining (document, term), bioinformatics (individual, gene), co-clustering which is a simultaneous clustering of both dimensions of a data matrix, has become a popular technique. Co-clustering aims to obtain homogeneous blocks leading to an easy simultaneous interpretation of row clusters and column clusters. Many approaches exist, in this paper we rely on the latent block model (LBM) which is flexible allowing to model different types of data matrices. We extend its use to the case of a tensor (3D matrix) data in proposing a Tensor LBM (TLBM) allowing different relations between entities. To show the interest of TLBM, we consider continuous and binary datasets. To estimate the parameters, a variational EM algorithm is developed. Its performances are evaluated on synthetic and real datasets to highlight different possible applications.

Keywords: Co-clustering · Tensor · Data science

1 Introduction

Co-clustering addresses the problem of simultaneous clustering of both dimensions of a data matrix. Many of the datasets encountered in data science are two-dimensional in nature and can be represented by a matrix. Classical clustering procedures seek to construct separately an optimal partition of rows (individuals) or, sometimes (features), of columns. In contrast, co-clustering methods cluster the rows and the columns simultaneously and organize the data into homogeneous blocks (after suitable permutations); see for instance [3]. Methods of this kind have practical importance in a wide variety of applications where data are typically organized in two-way tables. However, in modern datasets, instead of collecting data on every individual-feature pair, we may collect supplementary individual or item information leading to tensor representation. This kind of data has emerged in many fields such as recommender systems where the data are collected on multiple items rated by multiple users, information about users and items is also available yielding as a tensor rather than a data matrix.

Despite the great interest for co-clustering and the tensor representation, few works tackles the co-clustering from tensor data. We mention the work of

Q. Yang et al. (Eds.): PAKDD 2019, LNAI 11439, pp. 370–383, 2019.
https://doi.org/10.1007/978-3-030-16148-4_29

[1] based on Minimum Bregman information (MBI) to find co-clustering of a tensor. Most recently, in [14] the General Tensor Spectral Co-clustering (GTSC) method for co-clustering the modes of non-negative tensor has been developed. In [4] the authors proposed a tensor biclustering algorithm able to compute a subset of tensor rows and columns whose corresponding trajectories form a low-dimensional subspace. However, the majority of authors consider the same entities for the row and columns or do not consider the tensor co-clustering under a probabilistic approach. To the best of our knowledge, this is the first attempt to formulate our objective when both sets -row and column- are different and with model-based co-clustering. To this end, we rely on the latent block model [8] for its flexibility to consider any type of data matrices.

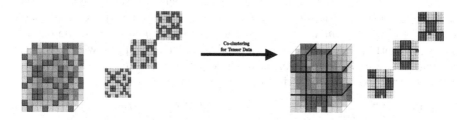

Fig. 1. Goal of co-clustering for binary tensor data.

In this paper, we propose a co-clustering model for tensor data, where clustering of row and column entities is done not only on principal relation matrix but on tensor including multiple covariates and/or relations between entities. The proposed model can also be viewed as multi-way clustering approach where each slice of the third dimension of the tensor represents a relation or covariate (see Fig. 1). The goal is to simultaneously discover the row and columns clusters and the relationship between these clusters for all slices. To achieve this, we propose to extend Latent block model (LBM) to tensor data referred to as TLBM. This model is suitable for several applications.

The main contributions of this paper are summarized as follows: (i) we propose an extension of latent block model for tensor data (TLBM) (ii) we show its flexibility to be applied with different types of data (iii) we derive a variational EM for co-clustering. The remainder of this paper is organized as follows. Section 2 describes classical latent block model and presents its extension TLBM. Section 3 details the proposed algorithm variational EM for co-clustering of tensor data. Section 4 presents experimental results on the synthetic and real-world data set and comparisons with several algorithms. Section 5 concludes this paper and provides some directions for future work.

2   From Latent Block Model for 2D Data Matrix to Tensor Data

2.1   Latent Block Model

The latent block model [8] in $g \times m$ blocks is defined as follows. Given a matrix $\mathbf{X}$ of size $n \times d$, we assume that there is a couple of partitions $(\mathbf{z}, \mathbf{w})$ where $\mathbf{z}$ is partitioned in g clusters on the set of rows I and $\mathbf{w}$ is partitioned in m clusters on the set of columns J, such that each element x_{ij} belonging to the block $k\ell$ is generated according to a probability distribution, where k represents the class of the line i, while ℓ represents the class of the column j. The $\mathbf{z}$ partition can be represented by a vector of labels or by $\mathbf{z} = (z_{ik})$ of size $n \times g$ where $z_{ik} = 1$ if the line i belongs to the class k, and $z_{ik} = 0$ otherwise. In the same way, the $\mathbf{w}$ partition can be represented by a label vector or by a column classification matrix $\mathbf{w} = (w_{j\ell})$ of size $d \times m$ where $w_{j\ell} = 1$ if the column j belongs to the class ℓ, and $w_{j\ell} = 0$ otherwise. Under the independence assumption $p(\mathbf{z}, \mathbf{w}) = p(\mathbf{z})p(\mathbf{w})$ and noting $\mathcal{Z}$ and $\mathcal{W}$ the sets of all possible partitions $\mathbf{z}$ and $\mathbf{w}$, the likelihood of the observed data can be written as follows:

$$f(\mathbf{X}; \mathbf{\Omega}) = \sum_{(\mathbf{z},\mathbf{w}) \in \mathcal{Z} \times \mathcal{W}} \prod_{i,k} \pi_k^{z_{ik}} \prod_{j,\ell} \rho_\ell^{w_{j\ell}} \prod_{i,j,k,\ell} \left(\Phi(x_{ij}; \lambda_{k\ell}) \right)^{z_{ik} w_{j\ell}} \tag{1}$$

where $\mathbf{\Omega} = (\boldsymbol{\pi}, \boldsymbol{\rho}, \boldsymbol{\lambda})$ are the unknown parameters of LBM with $\boldsymbol{\pi} = (\pi_1, \dots, \pi_g)$ and $\boldsymbol{\rho} = (\rho_1, \dots, \rho_m)$ where $(\pi_k = p(z_{ik} = 1), k = 1, \dots, g)$, $(\rho_\ell = p(w_{j\ell} = 1), \ell = 1, \dots, m)$ are the proportions of clusters and $\lambda_{k\ell}$ represents the parameters of $k\ell$ block distribution. The classification log-Likelihood takes the following form:

$$L_C(\mathbf{z}, \mathbf{w}, \mathbf{\Omega}) = \sum_{i,k} z_{ik} \log \pi_k + \sum_{j,\ell} w_{j\ell} \log \rho_\ell + \sum_{i,j,k,\ell} z_{ik} w_{j\ell} \log(\Phi(x_{ij}; \lambda_{k\ell})) \tag{2}$$

2.2   Latent Block Model for Tensor Data (TLBM)

Hereafter, we propose a novel Latent Block model for tensor data (TLBM). Few studies have addressed the issue of co-clustering for tensor data [4,14]. Unlike classical LBM which considers data matrix $\mathbf{X} = [x_{ij}] \in \mathbb{R}^{n \times d}$, TLBM considers 3D data matrix $\mathbf{X} = [\mathbf{x}_{ij}] \in \mathbb{R}^{n \times d \times v}$ where n is the number of rows, d the number of columns, and v the number of covariates. Figure 2a presents the data structure and Fig. 2b the probabilistic graphical model TLBM. The generative process is described in Algorithm 1; TLBM is flexible and can be used with different types of data.

Binary Data. In this case, we can consider an extension of the Bernoulli LBM (Bernoulli TLBM), thereby $\boldsymbol{\mu}_{k\ell}$ is a probability vector. Specifically, assuming the conditional independence (independence per block), Φ is given by $\Phi(\mathbf{x}_{ij}; \lambda_{k\ell})$ defined as follows $\prod_{a=1}^{v} (\boldsymbol{\mu}_{k\ell}^a)^{\mathbf{x}_{ij}^a} (1 - \boldsymbol{\mu}_{k\ell}^a)^{1 - \mathbf{x}_{ij}^a}$ and the classification log-likelihood can be written as

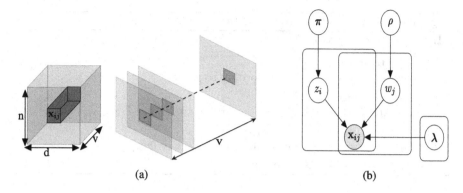

Fig. 2. (a) Data structure, (b) Graphical model of TLBM.

$$L_C(\mathbf{z}, \mathbf{w}, \mathbf{\Omega}) = \sum_{i,k} z_{ik} \log \pi_k + \sum_{j,\ell} w_{j\ell} \log \rho_\ell + \sum_{k\ell} z_{.k} w_{.\ell} \sum_a \log(1 - \mu_{k\ell}^a)$$
$$+ \sum_{i,j,k,\ell} z_{ik} w_{j\ell} \left(\sum_{a=1}^{v} \mathbf{x}_{ij}^a \log \frac{\mu_{k\ell}^a}{1 - \mu_{k\ell}^a} \right) \tag{3}$$

with $z_{.k} = \sum_i z_{ik}$ and $w_{.\ell} = \sum_j w_{j\ell}$.

Continuous Data. In this case, we can assume $\Phi(\mathbf{x}_{ij}; \lambda_{k\ell})$ as a multivariate normal distribution with $\boldsymbol{\mu}_{k\ell}$ the v-dimensional mean vector and $\boldsymbol{\Sigma}_{k\ell}$ its $v \times v$ covariance matrix. Hence, the parameter $\boldsymbol{\Omega}$ is formed by π, $\boldsymbol{\rho}$ and $\boldsymbol{\lambda} = (\lambda_{11}, \ldots, \lambda_{gm})$ where $\boldsymbol{\lambda}_{k\ell} = (\boldsymbol{\mu}_{k\ell}, \boldsymbol{\Sigma}_{k\ell})$ with $\boldsymbol{\mu}_{k\ell}^\top = (\mu_{k\ell}^1, \ldots, \mu_{k\ell}^v)$. Hence, $\Phi(\mathbf{x}_{ij}; \lambda_{k\ell})$ takes the following form.

$$\Phi(\mathbf{x}_{ij}; \lambda_{k\ell}) = \frac{1}{(2\pi)^{n/2} |\Sigma_{k\ell}|^{0.5}} \exp \left\{ -\tfrac{1}{2} (\mathbf{x}_{ij} - \boldsymbol{\mu}_{k\ell})^\top \boldsymbol{\Sigma}_{k\ell}^{-1} (\mathbf{x}_{ij} - \boldsymbol{\mu}_{k\ell}) \right\} \tag{4}$$

and,

$$L_C(\mathbf{z}, \mathbf{w}, \mathbf{\Omega}) = \sum_{i,k} z_{ik} \log \pi_k + \sum_{j,\ell} w_{j\ell} \log \rho_\ell - \frac{1}{2} \sum_{k,\ell} z_{.k} w_{.\ell} \log |\Sigma_{k\ell}|$$
$$- \frac{1}{2} \sum_{i,j,k,\ell} z_{ik} w_{j\ell} (\mathbf{x}_{ij} - \boldsymbol{\mu}_{k\ell})^\top \boldsymbol{\Sigma}_{k\ell}^{-1} (\mathbf{x}_{ij} - \boldsymbol{\mu}_{k\ell}). \tag{5}$$

Algorithm 1. Generative process of `Tensor LBM` model

Input: n, d, g, m, $\boldsymbol{\pi}$, $\boldsymbol{\rho}$, $\boldsymbol{\lambda}$

for $i \leftarrow 1$ **to** n **do**
$\quad$⌊ Generate the row label z_i according to $\mathcal{M}(\pi_1, \ldots, \pi_g)$
for $j \leftarrow 1$ **to** d **do**
$\quad$⌊ Generate the column label w_j according to $\mathcal{M}(\rho_1, \ldots, \rho_m)$
for $i \leftarrow 1$ **to** n **and** $j \leftarrow 1$ **to** d **do**
$\quad$⌊ Generate a vector $\mathbf{x}_{ij}$ according to the density $\Phi(\mathbf{x}_{ij}; \lambda_{k\ell})$.
return Tensor matrix $\mathbf{X}$, $\mathbf{z}$ and $\mathbf{w}$

---

3   Variational EM Algorithm

To estimate $\boldsymbol{\Omega}$, the EM algorithm [2] is a candidate for this task. It maximizes the log-likelihood $f(\mathbf{X}, \boldsymbol{\Omega})$ w.r. to $\boldsymbol{\Omega}$ iteratively by maximizing the conditional expectation of the complete data log-likelihood $L_C(\mathbf{z}, \mathbf{w}; \boldsymbol{\Omega})$ w.r. to $\boldsymbol{\Omega}$, given a previous current estimate $\boldsymbol{\Omega}^{(c)}$ and the observed data $\mathbf{x}$. Unfortunately, difficulties arise owing to the dependence structure among the variables x_{ij} of the model. To solve this problem an approximation using the interpretation of the EM algorithm can be proposed; see, e.g., [6]. More precisely, the authors rely on the variational approach which consists in approximating the true likelihood by another expression using the following independence assumption: $P(z_{ik} = 1, w_{j\ell} = 1|\mathbf{X}) = P(z_{ik} = 1|\mathbf{X})P(w_{j\ell} = 1|\mathbf{X})$. Hence, the aim is to maximize the following lower bound of the log-likelihood criterion:

$$F_C(\tilde{\mathbf{z}}, \tilde{\mathbf{w}}; \boldsymbol{\Omega}) = L_C(\tilde{\mathbf{z}}, \tilde{\mathbf{w}}, \boldsymbol{\Omega}) + H(\tilde{\mathbf{z}}) + H(\tilde{\mathbf{w}}) \tag{6}$$

where $H(\tilde{\mathbf{z}}) = -\sum_{i,k} \tilde{z}_{ik} \log \tilde{z}_{ik}$ with $\tilde{z}_{ik} = P(z_{ik} = 1|\mathbf{X})$, $H(\tilde{\mathbf{w}}) = -\sum_{j,\ell} \tilde{w}_{j\ell} \log \tilde{w}_{j\ell}$ with $\tilde{w}_{j\ell} = P(w_{j\ell} = 1|\mathbf{X})$, and $L_C(\tilde{\mathbf{z}}, \tilde{\mathbf{w}}; \boldsymbol{\Omega})$ is the fuzzy complete data log-likelihood (up to a constant). With the Bernoulli LBM for tensor data, $L_C(\tilde{\mathbf{z}}, \tilde{\mathbf{w}}; \boldsymbol{\Omega})$ is given by

$$
\begin{aligned}
L_C(\tilde{\mathbf{z}}, \tilde{\mathbf{w}}, \boldsymbol{\Omega}) = {} & \sum_{i,k} \tilde{z}_{ik} \log \pi_k + \sum_{j,\ell} \tilde{w}_{j\ell} \log \rho_\ell + \sum_{k,\ell} \tilde{z}_{.k} \tilde{w}_{.\ell} \sum_a \log(1 - \boldsymbol{\mu}_{k\ell}^a) \\
& + \sum_{i,j,k,\ell} \tilde{z}_{ik} \tilde{w}_{j\ell} \left(\sum_{a=1}^{v} \mathbf{x}_{ij}^a \log \frac{\mu_{k\ell}^a}{1 - \mu_{k\ell}^a} \right),
\end{aligned} \tag{7}
$$

where $\tilde{z}_{.k} = \sum_i \tilde{z}_{ik}$ et $\tilde{w}_{.\ell} = \sum_j \tilde{w}_{j\ell}$. Similarly, it takes the following form with the Gaussian TLBM.

$$
\begin{aligned}
L_C(\tilde{\mathbf{z}}, \tilde{\mathbf{w}}, \boldsymbol{\Omega}) = {} & \sum_{i,k} \tilde{z}_{ik} \log \pi_k + \sum_{j,\ell} \tilde{w}_{j\ell} \log \rho_\ell - \frac{1}{2} \sum_{k,\ell} \tilde{z}_{.k} \tilde{w}_{.\ell} \log |\Sigma_{k\ell}| \\
& - \frac{1}{2} \sum_{i,j,k,\ell} \tilde{z}_{ik} \tilde{w}_{j\ell} (\mathbf{x}_{ij} - \boldsymbol{\mu}_{k\ell})^\top \boldsymbol{\Sigma}_{k\ell}^{-1} (\mathbf{x}_{ij} - \boldsymbol{\mu}_{k\ell}).
\end{aligned} \tag{8}
$$

The maximization of $F_C(\tilde{\mathbf{z}}, \tilde{\mathbf{w}}, \mathbf{\Omega})$ can be reached by realizing the three following optimization: update $\tilde{\mathbf{z}}$ by $\underset{\tilde{\mathbf{z}}}{\arg\max}\ F_C(\tilde{\mathbf{z}}, \tilde{\mathbf{w}}, \mathbf{\Omega})$, update $\tilde{\mathbf{w}}$ by $\underset{\tilde{\mathbf{w}}}{\arg\max}\ F_C(\tilde{\mathbf{z}}, \tilde{\mathbf{w}}, \mathbf{\Omega})$ and update $\mathbf{\Omega}$ by $\underset{\mathbf{\Omega}}{\arg\max}\ F_C(\tilde{\mathbf{z}}, \tilde{\mathbf{w}}, \mathbf{\Omega})$. In what follows, we detail the Expectation (E) and Maximization (M) step of the Variational EM algorithm for tensor data.

3.1 E-step

The E-step consists in computing, for all i, k, j, ℓ the posterior probabilities $\tilde{z}_{ik}$ and $\tilde{w}_{j\ell}$ maximizing $F_C(\tilde{\mathbf{z}}, \tilde{\mathbf{w}}, \mathbf{\Omega})$ given the estimated parameters $\mathbf{\Omega}_{k\ell}$. It is easy to show that, the posterior probability $\tilde{z}_{ik}$ maximizing $F_C(\tilde{\mathbf{z}}, \tilde{\mathbf{w}}, \mathbf{\Omega})$ (See Appendix A) is given by: $\tilde{z}_{ik} \propto \pi_k \exp\left(\sum_{j,\ell} \tilde{w}_{j\ell} \log\left(\Phi(\mathbf{x}_{ij}; \boldsymbol{\lambda}_{k\ell})\right)\right)$. In the same manner, the posterior probability $\tilde{w}_{j\ell}$ is given by: $\tilde{w}_{j\ell} \propto \rho_\ell \exp\left(\sum_{i,k} \tilde{z}_{ik} \log\left(\Phi(\mathbf{x}_{ij}; \boldsymbol{\lambda}_{k\ell})\right)\right)$.

3.2 M-step

Given the previously computed posterior probabilities $\tilde{\mathbf{z}}$ and $\tilde{\mathbf{w}}$, the M-step consists in updating , $\forall k, \ell$, the parameters π_k, ρ_ℓ, $\boldsymbol{\mu}_{k\ell}$ and $\boldsymbol{\lambda}_{k\ell}$ maximizing $F_C(\tilde{\mathbf{z}}, \tilde{\mathbf{w}}, \mathbf{\Omega})$. The estimated parameters are defined as follows. First, taking into account the constraints $\sum_k z_{ik} = 1$ and $\sum_\ell w_{j\ell} = 1$, it is easy to show that $\pi_k = \frac{\sum_i \tilde{z}_{ik}}{n} = \frac{\tilde{z}_{.k}}{n}$ and $\rho_\ell = \frac{\sum_j \tilde{w}_{j\ell}}{d} = \frac{\tilde{w}_{.\ell}}{d}$. Secondly, the update of $\boldsymbol{\lambda}_{k\ell}$ depends on the choice of Φ. For Bernoulli TLBM, it easy to show that $\boldsymbol{\lambda}_{k\ell}$ which is a probability vector is given by $\boldsymbol{\lambda}_{k\ell} = \frac{\sum_{i,j} \tilde{z}_{ik} \tilde{w}_{j\ell} \mathbf{x}_{ij}}{\sum_{i,j} \tilde{z}_{ik} \tilde{w}_{j\ell}}$. For Gaussian TLBM, $\boldsymbol{\lambda}_{k\ell}$ is formed by $(\boldsymbol{\mu}_{k\ell}, \boldsymbol{\Sigma}_{k\ell})$ where $\boldsymbol{\mu}_{k\ell}$ is the mean vector and $\boldsymbol{\Sigma}_{k\ell} = \frac{\sum_{i,j} \tilde{z}_{ik} \tilde{w}_{j\ell} (\mathbf{x}_{ij} - \mu_{k\ell})(\mathbf{x}_{ij} - \mu_{k\ell})^\top}{\sum_{i,j} \tilde{z}_{ik} \tilde{w}_{j\ell}}$. The proposed algorithm for tensor data, referred to as VEM-T in Algorithm 2, alternates the two previously described steps Expectation-Maximization. At the convergence, a hard co-clustering is deduced from the posterior probabilities.

Algorithm 2. VEM-T

Input: $\mathbf{X}$, g, m.

Initialization $(\mathbf{z}, \mathbf{w})$ randomly, compute $\mathbf{\Omega}$

repeat

    E-Step

    – **Compute** $\tilde{z}_{ik}$ **using**

      $\tilde{z}_{ik} \propto \pi_k \exp\left(\sum_{j,\ell} \tilde{w}_{j\ell} \log\left(\Phi(\mathbf{x}_{ij}; \boldsymbol{\lambda}_{k\ell})\right)\right)$

    – **Compute** $\tilde{w}_{j\ell}$ **using**

      $\tilde{w}_{j\ell} \propto \rho_\ell \exp\left(\sum_{i,k} \tilde{z}_{ik} \log\left(\Phi(\mathbf{x}_{ij}; \boldsymbol{\lambda}_{k\ell})\right)\right)$

    M-Step

      Update $\mathbf{\Omega}$

until *convergence*;

return $\mathbf{z}$, $\mathbf{w}$, $\mathbf{\Omega}$

---

4   Experimental Results

First we evaluate VEM-T on binary and continuous synthetic datasets in terms of (Co)-clustering. We compare VEM-T with multiple clustering methods. We retain two widely used measures to assess the quality of clustering, namely the Normalized Mutual Information (NMI) [12] and the Adjusted Rand Index (ARI) [11]. Intuitively, NMI quantifies how much the estimated clustering is informative about the true clustering. The ARI is related to the clustering accuracy and measures the degree of agreement between an estimated clustering and a reference clustering. Both NMI and ARI are equal to 1 if the resulting clustering is identical to the true one. Secondly, we present results on real datasets for two different areas namely recommender systems and multi-spectral images clustering. Through this evaluation, we aim to demonstrate the impact of covariate information on interpretation and improvement of clustering results.

4.1   Synthetic Datasets

We generated tensor data $\mathbf{X}$ according to the Bernoulli and Gaussian TLBM (Algorithm 1) with $v = 3$. Following each model, we considered two scenarios by varying the centers $\boldsymbol{\mu}_{k\ell}$'s; an example where the co-clusters are well separated and another where the co-clusters are not. The size of each tensor, number of co-clusters and their proportions are reported in Tables 1 and 2. Herein other characteristics of each tensor dataset. For continuous data we take the same covariance matrix for all blocks $\begin{bmatrix} 0.2 & 0 & 0 \\ 0 & 0.2 & 0 \\ 0 & 0 & 0.2 \end{bmatrix}$ for example 3 and $\begin{bmatrix} 1 & 0.8 & 0.8 \\ 0.8 & 1 & 0.8 \\ 0.8 & 0.8 & 1 \end{bmatrix}$

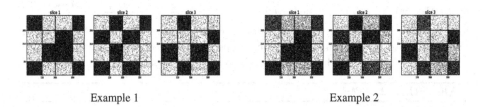

Example 1   Example 2

Fig. 3. Simulated binary datasets

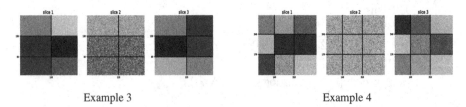

Example 3   Example 4

Fig. 4. Simulated continuous datasets

Table 1. Evaluation of co-clustering in terms of NMI and ARI for binary datasets

Algorithm	Metrics NMI-ARI		Example 1 400 × 400 × 3 $(g, m) = (4, 4)$ $\pi = [0.23, 0.3, 0.23, 0.24]$ $\rho = [0.27, 0.23, 0.3, 0.2]$			Example 2 400 × 400 × 3 $(g, m) = (4, 4)$ $\pi = [0.23, 0.3, 0.23, 0.24]$ $\rho = [0.27, 0.23, 0.3, 0.2]$		
			Slice 1	Slice 2	Slice 3	Slice 1	Slice 2	Slice 3
K-means	NMI	Row	0.801 ± 0.0	0.817 ± 0.001	0.98 ± 0.004	0.831 ± 0.004	0.82 ± 0.001	0.946 ± 0.007
		Column	0.83 ± 0.014	0.76 ± 0.003	0.76 ± 0.002	0.83 ± 0.012	0.8 ± 0.001	0.8 ± 0.005
	ARI	Row	0.688 ± 0.001	0.706 ± 0.001	0.962 ± 0.001	0.73 ± 0.014	0.713 ± 0.001	0.902 ± 0.022
		Column	0.7 ± 0.001	0.6 ± 0.001	0.59 ± 0.011	0.75 ± 0.002	0.64 ± 0.002	0.66 ± 0.003
GMM	NMI	Row	0.815 ± 0.001	0.814 ± 0.001	1.0 ± 0.0	0.825 ± 0.003	0.826 ± 0.001	0.903 ± 0.021
		Column	0.79 ± 0.021	0.77 ± 0.018	0.76 ± 0.017	0.83 ± 0.01	0.88 ± 0.003	0.83 ± 0.01
	ARI	Row	0.684 ± 0.002	0.72 ± 0.001	1.0 ± 0.0	0.733 ± 0.009	0.729 ± 0.001	0.831 ± 0.058
		Column	0.71 ± 0.001	0.6 ± 0.002	0.59 ± 0.001	0.72 ± 0.002	0.82 ± 0.004	0.71 ± 0.005
VEM	NMI	Row	0.66 ± 0.005	0.72 ± 0.005	0.78 ± 0.019	0.71 ± 0.001	0.73 ± 0.003	0.86 ± 0.017
		Column	0.7 ± 0.001	0.71 ± 0.003	0.71 ± 0.007	0.72 ± 0.001	0.73 ± 0.003	0.8 ± 0.006
	ARI	Row	0.53 ± 0.001	0.57 ± 0.009	0.7 ± 0.031	0.52 ± 0.002	0.57 ± 0.015	0.78 ± 0.024
		Column	0.51 ± 0.0	0.51 ± 0.002	0.51 ± 0.002	0.54 ± 0.003	0.56 ± 0.01	0.56 ± 0.01
VEM-T	NMI	Row		**0.94 ± 0.004**			**0.901 ± 0.005**	
		Column		**0.935 ± 0.004**			**0.977 ± 0.002**	
	ARI	Row		**0.896 ± 0.016**			**0.81 ± 0.016**	
		Column		**0.873 ± 0.018**			**0.957 ± 0.007**	

Table 2. Evaluation of co-clustering in terms of NMI and ARI for continuous datasets

Algorithm	Metrics NMI-ARI		Example 3 200 × 200 × 3 $(g, m) = (3, 2)$ $\pi = [0.3, 0.35, 0.35]$ $\rho = [0.55, 0.45]$			Example 4 500 × 500 × 3 $(g, m) = (3, 3)$ $\pi = [0.34, 0.34, 0.32]$ $\rho = [0.28, 0.34, 0.38]$		
			Slice 1	Slice 2	Slice 3	Slice 1	Slice 2	Slice 3
K-means	NMI	Row	1.0 ± 0.0	0.62 ± 0.0	1.0 ± 0.0	1.0 ± 0.0	0.09 ± 0.0	1.0 ± 0.0
		Column	1.0 ± 0.0	0.0 ± 0.0	1.0 ± 0.0	1.0 ± 0.0	0.29 ± 0.01	1.0 ± 0.0
	ARI	Row	1.0 ± 0.0	0.54 ± 0.0	1.0 ± 0.0	1.0 ± 0.0	0.09 ± 0.0	1.0 ± 0.0
		Column	1.0 ± 0.0	0.0 ± 0.0	1.0 ± 0.0	1.0 ± 0.0	0.28 ± 0.01	1.0 ± 0.0
GMM	NMI	Row	1.0 ± 0.0	0.62 ± 0.0	1.0 ± 0.0	1.0 ± 0.0	0.15 ± 0.0	1.0 ± 0.0
		Column	1.0 ± 0.0	0.0 ± 0.0	1.0 ± 0.0	1.0 ± 0.0	0.24 ± 0.01	1.0 ± 0.0
	ARI	Row	1.0 ± 0.0	0.54 ± 0.0	1.0 ± 0.0	1.0 ± 0.0	0.16 ± 0.0	1.0 + 0.0
		Column	1.0 ± 0.0	0.0 ± 0.0	1.0 ± 0.0	1.0 ± 0.0	0.23 ± 0.01	1.0 ± 0.0
VEM	NMI	Row	0.98 ± 0.005	0.77 ± 0.0	0.98 ± 0.005	0.95 ± 0.01	0.5 ± 0.0	1.0 ± 0.0
		Column	1.0 ± 0.0	0.0 ± 0.0	1.0 ± 0.0	0.95 ± 0.011	0.6 ± 0.001	0.95 ± 0.011
	ARI	Row	0.96 ± 0.015	0.59 ± 0.0	0.96 ± 0.015	0.91 ± 0.032	0.52 ± 0.0	1.0 ± 0.0
		Column	1.0 ± 0.0	0.01 ± 0.0	0.01 ± 0.0	0.9 ± 0.038	0.52 ± 0.001	0.52 ± 0.001
VEM-T	NMI	Row		**1.0 ± 0.0**			**0.95 ± 0.01**	
		Column		**1.0 ± 0.0**			**0.95 ± 0.009**	
	ARI	Row		**1.0 ± 0.0**			**0.91 ± 0.028**	
		Column		**1.0 ± 0.0**			**0.92 ± 0.027**	

for example 4. All variables (slice) are standardized to have values between zero and one. In Figs. 4 and 3 are depicted the true simulated tensor data into $v = 3$ slices.

4.2   Competitive Methods

In our experiments, we compare VEM-T with K-means, Gaussian Mixture Model (GMM: EM with the full model, see for instance [5]) and VEM for co-clustering applied on each slice [7]. The ARI and NMI metrics for rows and columns are computed by averaging on ten random initialization. Thereby, in Tables 1 and 2 are reported the performances for the three slices obtained by K-means, GMM, VEM for data matrix and by VEM-T for tensor data. From these comparisons, we observe that whether the block structure is easy to identify (Examples 1, 3) or not (Examples 2, 4), the ability of VEM-T to outperform other algorithms that, it should be recalled, act on each slice separately.

4.3   Recommender System Application

To show the benefits of our approach, we use the binary model on Movielens100K which is one of the more popular datasets on the recommender system field. The objective of this study is identifying patterns according to users and movies characteristics. The Movielens100K[1] database consists of 100,000 ratings of 943 users and 1682 movies, where each user has rated at least 20 movies. We convert the users-movies rating matrix (943×1682) to binary matrix by assigning 0 to the movie without rating and 1 to rated movies. This binary matrix can be considered as viewing matrix, in fact most users rates movies after watching them. Furthermore, Movielens includes 22 user covariates including age, gender, and 21 employment status. The age covariate is used to analyze clustering results and does not take into account in co-clustering. There are also 19 movie covariates related to movie genres, considering that movie may belong to one or more genres. The data structure can be represented as tensor with size $943 \times 1682 \times 42$. The objective of this work is not being to select the number of clusters, then we fixed the number of row clusters $g = 2$ and the number of column clusters $m = 3$, based on the works of [13]. Figures 5 and 6a represent the mean vectors $\mu_{k\ell}$ and co-clustering of rating matrix respectively. We observe two row clusters, a smaller cluster of 202 users which is more active in reviewing than a second large cluster. On the other hand, we obtain three movies clusters of different sizes 232, 355 and 1,095 respectively. The first cluster represents the most attractive movies.

The first row cluster includes three blocks $(1, 1)$, $(1, 2)$ and $(1, 3)$. The two first ones represent the more active users with a higher proportion of rating. The MovieLens100K dataset includes 29% of female reviews, an important part of them (64%) belong to a first row cluster. In addition, we notice that the top 3 of occupations for users of the first row cluster are a student, educator, and administrator. Thereby, Fig. 6b shows that 65% of them are quite young and under 31 years of age. However, the two blocks $(1, 1)$ and $(1, 2)$ are distinguished by movie genres, since the top 3 ones for first and second column clusters are Action-Thriller-Sci-Fi and Comedy-Drama-Romance respectively. Consequently, we can

---

[1] http://grouplens.org/datasets/movielens/.

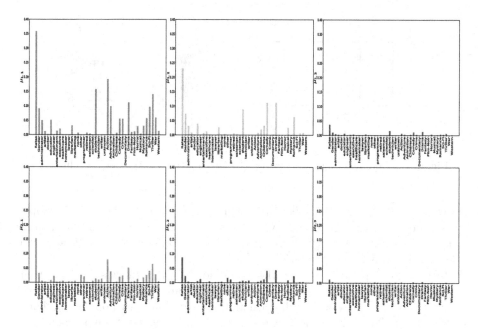

Fig. 5. Distribution of the centers $\boldsymbol{\mu}_{k\ell}$ for all co-clusters

identify two profiles of young active users; they are attracted by both categories of movies namely Action-Thriller-Sci-Fi for the first profile and Comedy-Drama-Romance for the second. The second row cluster regroups the users of different ranges of age with almost equal proportions (see Fig. 6b) and different occupations since the top three occupations include engineer, student, and another employment status. Finally the third column cluster seems representing movies with different genres Action-Drama-comedy. The block (2, 3) represents the less attractive movies watched by the less active users.

4.4 Multi-spectral Images Analysis

The used dataset is composed by 37 multispectral images of prostate cells with 16 bands which have size 512×512 pixels. Several studies showed that clustering accuracy increases according bands number [10]. The four types of multispectral images cells are: Normal cells (Stroma), Benign Hyperplasia (BHp), Interpithelial Neoplasy (PIN) which is a cancer precursory state, and the Carcinoma (CA) which corresponds to a cancer of the abnormal tissue proliferation. Figure 7a presents cell's types and the example of 16 bands of Stroma cells type are showed in Fig. 7b. Some elements allow to differentiate the cell's types, among those morphological and textural features. In this way, we limited ourselves to textural characteristics for clustering. Haralick [9] defined several metrics computed from the gray level co-occurrence matrix (GLCM). The Haralick's parameters showed their efficiency in the literature for the textures analysis [9,10]. The 14

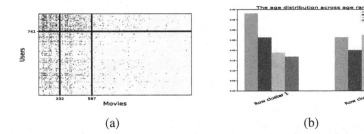

(a)   (b)

Fig. 6. (a) Co-clustering data matrix, (b) Distribution of Age per row clusters

Haralick's features are the following: Energy, Correlation, Contrast, Entropy, Homogeneity, Inverse Difference Moment, Sum Average, Sum Variance, Sum Entropy, Difference Average, Difference Variance, Difference Entropy and two Information measure of correlation.

In the most previous studies, the extraction of 14 Haralick's features from all bands are performed, and the 14×16 features are extracted for each image involving features selection or dimensionality reduction with popular methods such as PCA. These operations can provide interesting results but leads to a loss of information. To overcome this drawback, we propose to construct tensor data *Images* × *Bands* × *Features* in order to exploit all available data without requiring dimensionality reduction. The objective of this study is improving clustering results of multispectral images which highly used on biomedical and geology fields.

As we known the true number of image clusters, we take $g = 4$ and as we have no information about column clusters we postulate $m = g = 4$. As shown in Fig. 8, the Stroma cells are characterized by higher values of entropy, contrast and difference variance on the first three column clusters, and low values of inverse difference moment feature on two first band clusters. The PIN type is characterized by low values of information measure correlation 1 on bands cluster 2, 3 and 4. The cell type with the closer values of features is BHP. The CA type is characterized by higher values of information measure correlation 1 on the third

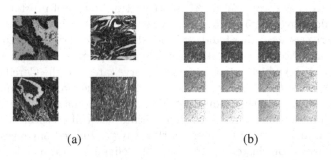

(a)   (b)

Fig. 7. (a) The four cells type, (b) Example of multispectral image from dataset

and fourth band clusters and the lower values of information measure correlation 2 on all bands. Finally, BHP cells are characterized by the lowest values of sum average on two last bands clusters.

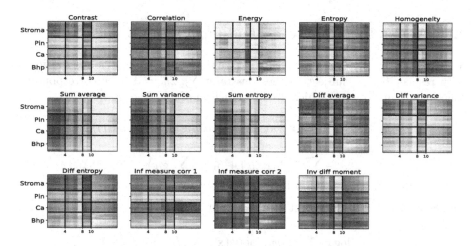

Fig. 8. Co-clustering matrix of different slice of features

Table 3. Evaluation of K-means, GMM, VEM and VEM-T in terms of NMI, ARI and ACC

Algorithms	NMI	ARI	ACC
K-means	0.67	0.56	0.78
GMM	0.7	0.59	0.78
VEM	0.61	0.49	0.7
VEM-T	**0.9**	**0.87**	**0.95**

The VEM-T algorithm is compared with K-means, GMM and EM. For this, a reduced matrix of tensor data by averaging all bands for each feature provides a *Images × Features* data matrix used to perform classical clustering. Table 3 summarizes the obtained results. For each algorithm, the best result rather than 100 random initial runs are used. Clearly the proposed algorithm achieves best results as regards NMI, ARI and ACC (Accuracy).

5   Conclusion

Inspired by the flexibility of the latent block model (LBM) for data matrix, we proposed to extend it to tensor data (TLBM). This gives rise to new variational EM algorithm for co-clustering. Empirical results on synthetic and real-world datasets –binary and continuous– show that VEM-T for tensor data does a better job than original VEM applied on each slice of tensor data. More interestingly, our findings open up good opportunities for future research such as deal with temporal data or challenges such as the assessing the number of co-clusters.

A   Appendix: Update $\tilde{z}_{ik}$ and $\tilde{w}_{j\ell}$ $\forall i, k, j, \ell$

To obtain the expression of $\tilde{z}_{ik}$, we maximize the above soft criterion $F_C(\tilde{\mathbf{z}}, \tilde{\mathbf{w}}; \Omega)$ with respect to $\tilde{z}_{ik}$, subject to the constraint $\sum_k \tilde{z}_{ik} = 1$. The corresponding Lagrangian, up to terms which are not function of $\tilde{z}_{ik}$, is given by:

$$L(\tilde{\mathbf{z}}, \beta) = \sum_{i,k} \tilde{z}_{ik} \log \pi_k + \sum_{i,j,k,\ell} \tilde{z}_{ik} \tilde{w}_{jk} \log(\Phi(\mathbf{x}_{ij}, \boldsymbol{\lambda}_{k\ell}))$$
$$- \sum_{i,k} \tilde{z}_{ik} \log(\tilde{z}_{ik}) + \beta(1 - \sum_k \tilde{z}_{ik}).$$

Taking derivatives with respect to $\tilde{z}_{ik}$, we obtain:

$$\frac{\partial L(\tilde{\mathbf{z}}, \beta)}{\partial \tilde{z}_{ik}} = \log \pi_k + \sum_{j,\ell} w_{j\ell} \log(\Phi(\mathbf{x}_{ij}, \boldsymbol{\lambda}_{k\ell})) - \log \tilde{z}_{ik} - 1 - \beta.$$

Setting this derivative to zero yields: $\tilde{z}_{ik} = \frac{\pi_k \exp(\sum_{j,\ell} w_{j\ell} \log(\Phi(\mathbf{x}_{ij}, \boldsymbol{\lambda}_{k\ell})))}{\exp(\beta+1)}$. Summing both sides over all k' yields $\exp(\beta + 1) = \sum_{k'} \pi_{k'} \exp(\sum_{j,\ell} w_{j\ell} \log(\Phi(\mathbf{x}_{ij}, \boldsymbol{\lambda}_{k'\ell})))$. Plugging $\exp(\beta)$ in $\tilde{z}_{ik}$ leads to: $\tilde{z}_{ik} \propto \pi_k \exp(\sum_{j,\ell} w_{j\ell} \log(\Phi(\mathbf{x}_{ij}, \boldsymbol{\lambda}_{k\ell})))$. In the same way, we can estimate $\tilde{w}_{jk}$ maximizing $F_C(\tilde{\mathbf{z}}, \tilde{\mathbf{w}}; \Omega)$ with respect to $\tilde{w}_{j\ell}$, subject to the constraint $\sum_\ell \tilde{w}_{j\ell} = 1$; we obtain $\tilde{w}_{j\ell} \propto \rho_k \exp(\sum_{i,k} \tilde{z}_{ik} \log(\Phi(\mathbf{x}_{ij}, \boldsymbol{\lambda}_{k\ell})))$.

References

1. Banerjee, A., Krumpelman, C., Ghosh, J., Basu, S., Mooney, R.J.: Model-based overlapping clustering. In: Proceedings of the Eleventh ACM SIGKDD, pp. 532–537 (2005)
2. Dempster, A.P., Laird, N.M., Rubin, D.B.: Maximum likelihood from incomplete data via the EM algorithm. J. Roy. Stat. Soc.: Ser. B **39**, 1–38 (1977)
3. Dhillon, I.S., Mallela, S., Modha, D.S.: Information-theoretic co-clustering. In: Proceedings of the Ninth ACM SIGKDD, pp. 89–98 (2003)
4. Feizi, S., Javadi, H., Tse, D.: Tensor biclustering. In: Advances in Neural Information Processing Systems 30, pp. 1311–1320. Curran Associates, Inc. (2017)
5. Fraley, C., Raftery, A.E.: How many clusters? Which clustering method? Answers via model-based cluster analysis. Comput. J. **41**(8), 578–588 (1998)
6. Govaert, G., Nadif, M.: An EM algorithm for the block mixture model. IEEE Trans. Pattern Anal. Mach. Intell. **27**(4), 643–647 (2005)
7. Govaert, G., Nadif, M.: Fuzzy clustering to estimate the parameters of block mixture models. Soft Comput. **10**(5), 415–422 (2006)
8. Govaert, G., Nadif, M.: Co-clustering. Wiley-IEEE Press (2013)

9. Haralick, R., Shanmugam, K., Dinstein, I.: Textural features for image classification. IEEE Trans. Syst. Man Cybern. **3**(6), 610–621 (1973)

10. Kumar, R.M., Sreekumar, K.: A survey on image feature descriptors. Int. J. Comput. Sci. Inf. Technol. (IJCSIT) **5**(1), 7668–7673 (2014)

11. Steinley, D.: Properties of the hubert-arabie adjusted rand index. Psychol. Methods **9**(3), 386 (2004)

12. Strehl, A., Ghosh, J.: Cluster ensembles - a knowledge reuse framework for combining multiple partitions. J. Mach. Learn. Res. **3**, 583–617 (2002)

13. Vu, D., Aitkin, M.: Variational algorithms for biclustering models. Comput. Stat. Data Anal. **89**, 12–24 (2015)

14. Wu, T., Benson, A.R., Gleich, D.F.: General tensor spectral co-clustering for higher-order data. In: Lee, D.D., Sugiyama, M., Luxburg, U.V., Guyon, I., Garnett, R. (eds.) Advances in Neural Information Processing Systems 29, pp. 2559–2567. Curran Associates, Inc. (2016)

A Data-Aware Latent Factor Model for Web Service QoS Prediction

Di Wu[1], Xin Luo[1(✉)], Mingsheng Shang[1], Yi He[3],
Guoyin Wang[2(✉)], and Xindong Wu[3]

[1] Chongqing Key Laboratory of Big Data and Intelligent Computing,
Chongqing Institute of Green and Intelligent Technology,
Chinese Academy of Sciences, Chongqing 400714, China
{wudi,luoxin21,msshang}@cigit.ac.cn
[2] Chongqing Key Laboratory of Computational Intelligence,
Chongqing University of Posts and Telecommunications,
Chongqing 400065, China
wanggy@ieee.org
[3] University of Louisiana at Lafayette, Lafayette 70503, USA
{yi.he1,xwu}@louisiana.edu

Abstract. Accurately predicting unknown quality-of-service (QoS) data based on historical QoS records is vital in web service recommendation or selection. Recently, latent factor (LF) model has been widely and successfully applied to QoS prediction because it is accurate and scalable under many circumstances. Hence, state-of-the-art methods in QoS prediction are primarily based on LF. They improve the basic LF-based models by identifying the neighborhoods of QoS data based on some additional geographical information. However, the additional geographical information may be difficult to collect in considering information security, identity privacy, and commercial interests in real-world applications. Besides, they ignore the reliability of QoS data while unreliable ones are often mixed in. To address these issues, this paper proposes a data-aware latent factor (DALF) model to achieve highly accurate QoS prediction, where 'data-aware' means DALF can easily implement the predictions according to the characteristics of QoS data. The main idea is to incorporate a density peaks based clustering method into an LF model to discover the neighborhoods and unreliable ones of QoS data. Experimental results on two benchmark real-world web service QoS datasets demonstrate that DALF has better performance than the state-of-the-art models.

1 Introduction

Web services are software components used to exchange data between two software systems over a network [1]. In this era of the Internet, there are numerous online web services [2]. How to select optimal ones from a large candidate set and recommend them to potential users becomes a hot yet thorny issue [3].

Quality-of-Service (QoS) is essential for addressing such an issue because it is a significant factor to evaluate the performance of web services [1, 2, 4]. Once QoS data of candidate web services are obtained, reliable ones can be selected and recommended

© Springer Nature Switzerland AG 2019
Q. Yang et al. (Eds.): PAKDD 2019, LNAI 11439, pp. 384–399, 2019.
https://doi.org/10.1007/978-3-030-16148-4_30

to potential users accordingly. Conducting warming-up tests is an important way to acquire QoS data. However, it is economically expensive [3, 5, 6].

Alternatively, QoS prediction is another widely used way to acquire QoS data [5–9]. Its principle is to predict unknown QoS data based on historical records and/or other information. Collaborative filtering (CF), which has been successfully applied to e-commerce recommendation systems [10, 11], is frequently adopted to implement QoS prediction [5–9, 12–17]. CF-based QoS prediction is developed based on a user-service QoS matrix [5–9, 12–17], where each column denotes a specified web service, each row denotes a specified user, and each entry stands for a historical QoS record produced by a specified user invoking a specified web service. Such a matrix is sparse [5–9, 12–17]. Thus, how to accurately predict the missing data of the sparse user-service QoS matrix based on its known ones is the key to achieve CF-based QoS prediction.

Among CF-based QoS prediction methods, latent factor (LF)-based models are more widely adopted [8, 9, 12–15, 17]. Originated from matrix factorization (MF) techniques [3, 10], an LF-based model works by building a low-rank approximation to the given user-service QoS matrix based on its known data only. It maps both users and services into the same low-dimensional LF space, trains desired LFs on the known data, and then predicts the missing data heavily relying on these resultant LFs [18].

Since LF-based model has the powerful ability on QoS prediction, the state-of-the-art methods in this area are primarily based on LF [8, 9, 12, 17]. They improve the basic LF-based models by identifying the neighborhoods of QoS data based on historical QoS records plus some additional geographical information. However, these geography-LF-based models have the following drawbacks:

(a) They adopt a common set on raw QoS data to identify the neighborhoods. Since the raw user-service QoS matrix can be very sparse, resultant common sets of users/services are commonly too small to identify the neighborhoods precisely. For example, Fig. 1 shows that many known data (red entries) are abandoned in finding the common sets among users, making the resultant neighborhoods lack reliability.

(b) They ignore the data reliability. Unreliable QoS data or called noises collected from malicious users (e.g., badmouthing a specific service) are often mixed up with the reliable ones [15]. Their QoS prediction accuracy would be impaired instead of being improved if they employ the unreliable QoS data.

(c) Additional geographical information can be difficult to gather in considering identity privacy, information security, and commercial interest. Moreover, geographical similarities can be influenced by unexpected factors like information facilities, routing policies, network throughput, and time of invocation.

To address the above drawbacks, this paper proposes a data-aware latent factor (DALF) model to achieve highly accurate QoS prediction. The main idea is to incorporate a density peaks based clustering method (DPClust) [19] into an LF model to discover the characteristics of QoS data, which can guide DALF to implement QoS prediction appropriately. The main contributions of this work include:

(a) We propose a method that can simultaneously identify a neighborhood for a user or a web service and detect the unreliable QoS data existed in the known ones.

(b) We theoretically analyze DALF and design its algorithm.

(c) We conduct detailed experiments on two benchmark real-world web service QoS datasets to evaluate DALF and compare it with the state-of-the-art models.

To the best of our knowledge, this work is never encountered in any previous works because (i) it can not only identify the neighborhoods but also detect the unreliable QoS data, (ii) it builds reliable neighborhoods solely based on a given QoS matrix but considering its full information, and (iii) it does not require any additional information.

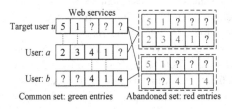

Fig. 1. The dilemma in building neighborhoods based on common sets defined on raw QoS data. (Color figure online)

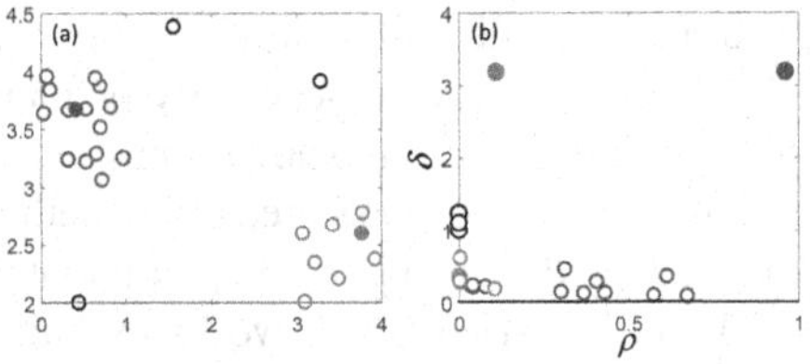

Fig. 2. The example of DPClust: (a) data distribution; (b) decision graph for data in (a); different colors correspond to different clusters. (Color figure online)

2 Preliminaries

2.1 LF Model

The QoS data is a user-service QoS matrix R defined as *Definition* 1 [9–11].

Definition 1. Given a user set U and a web service set S; let R be a $|U| \times |S|$ matrix where each element $r_{u,s}$ describes a user u's ($u \in U$) experience on a web service s ($s \in S$). R_K and R_U indicate the known and unknown entry sets of R respectively. R usually is a sparse matrix with $|R_K| \ll |R_U|$.

Definition 2. Given R, U, S, and f; given a $|U| \times f$ matrix P for U and an $f \times |S|$ matrix Q for S; $\hat{R}$ is R's rank-f approximation based on R_K under the condition of $f \ll \min(|U|, |S|)$. An LF model is to seek for P and Q to obtain $\hat{R}$ and error $\sum_{(u,s) \in RK} (r_{u,s} - \hat{r}_{u,s})^2$ is minimized. $\hat{R}$ is given by $\hat{R} = PQ$ where each element $\hat{r}_{u,s}$ is the prediction for each $r_{u,s}$ of R, $u \in U$ and $s \in S$. f is the dimension of LF space. P and Q are the LF matrices for users and web services respectively.

According to *Definition* 2, the loss function for LF model is [9–11]:

$$\arg \min_{P,Q} \varepsilon(P, Q) = \frac{1}{2} \sum_{(u,s) \in RK} \left(r_{u,s} - \sum_{k=1}^{f} p_{u,k} q_{k,s} \right)^2. \tag{1}$$

As analyzed in [9, 10, 18], it is important to integrate the Tikhonov regularization into (1) to improve its generality as follow:

$$\arg\min_{P,Q} \varepsilon(P, Q) = \frac{1}{2} \sum_{(u,s) \in R_K} \left(r_{u,s} - \sum_{k=1}^{f} p_{u,k} q_{k,s} \right)^2 + \frac{\lambda}{2} \sum_{(u,s) \in R_K} \left(\sum_{k=1}^{f} p_{u,k}^2 + \sum_{k=1}^{f} q_{k,s}^2 \right),$$

(2)

where λ is the regularization controlling coefficient. By minimizing (2) with an optimizer, e.g., stochastic gradient descent (SGD), P and Q are extracted from R.

2.2   DPClust Algorithm

DPClust is a clustering algorithm based on the idea that cluster centers are characterized by a higher density than their neighbors and by a relatively large distance from data points with higher densities [19]. We employ DPClust to develop DALF because it can not only find the characteristics of data but also spotted the outliers.

Given a dataset $X = \{x_1, x_2, \ldots, x_G\}$, for each data point $x_i, i \in \{1, 2, \ldots, G\}$, its local density ρ_i is computed via cut-off kernel or Gaussian kernel. Cut-off kernel is as follow:

$$\rho_i = \sum_{j=1, j \neq i}^{N} \Phi(d_{ij} - d_c), \quad \Phi(t) = \begin{cases} 1 & t < 0 \\ 0, & others \end{cases}$$

(3)

where $d_{i,j}$ is the distance between data points x_i and x_j and the number of all the $d_{i,j}$ is $G \times (G-1)/2$, and d_c is a cutoff distance with a fixed value. Gaussian kernel is as follow:

$$\rho_i = \sum_{j=1, j \neq i}^{G} e^{-\left(\frac{d_{ij}}{d_c}\right)^2}$$

(4)

For a robust computing of ρ_i, d_c can be set as [19, 20]:

$$Vec = sort(d_{ij}), \quad d_c = Vec(\lfloor P_{Vec} \times G \times (G-1)/2 \rfloor)$$

(5)

where Vec is a vector obtained by sorting all the $d_{i,j}$ in ascending order, P_{Vec} is a percentage denoting the average percentage of neighbors of all the data points. According to [19, 20], P_{Vec} is usually set around 1% to 2% as a rule of thumb.

For each data point x_i, δ_i is the minimum distance between x_i and any other data point with higher local density:

$$\delta_i = \begin{cases} \min_{j:\rho_i < \rho_j}(d_{ij}), & others \\ \max_j(d_{ij}), & \forall j, \ \rho_i \geq \rho_j \end{cases}$$

(6)

Then, cluster centers are recognized as data points for which the value of ρ and δ are anomalously large.

Figure 2 is a simple example for illustrating DPClust. Figure 2(a) shows 26 data points embedded in the two-dimensional space. After computing ρ and δ for all the data points, the decision graph can be drawn in Fig. 2(b). Then, we can easily recognize the blue and pink solid data points as the cluster centers. Note that the three black hollow data points are the outliers and have a relatively small ρ and a large δ, which means that DPClust can also detect the outliers by computing outlier factor γ_i for each x_i as follow:

$$\gamma_i = \rho_i / \delta_i. \tag{7}$$

Formula (7) indicates that an outlier has an anomalously small value of γ.

3   The Proposed DALF Model

Figure 3 depicts the flowchart of DALF that has three parts. Part 1 is extracting LF matrices P for users and Q for services. Part 2 is identifying neighborhoods of QoS data and detecting unreliable QoS data by employing DPClust algorithm. Concretely, P is used to identify neighborhoods of users and detect unreliable users, and Q is used to identify neighborhoods of services and detect unreliable services. Part 3 is predicting the unknown entries in R based on Part 2. There are four prediction strategies in Part 3. The characteristics of QoS data will determine which one or more are appropriate to implement predictions. Next, we give the detailed descriptions on the three parts.

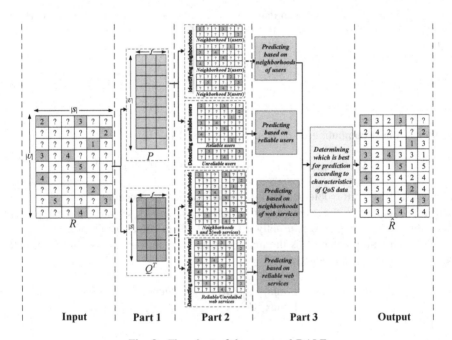

Fig. 3. Flowchart of the proposed DALF

3.1   Extracting LF Matrices for Users and Services

This part aims to extract LF matrices P for users and Q for services from R based on an LF model. We apply SGD to (2) to consider instant loss on a single element $r_{u,s}$:

$$\varepsilon_{u,s} = \frac{1}{2}\left(r_{u,s} - \sum_{k=1}^{f} p_{u,k} q_{k,s}\right)^2 + \frac{\lambda}{2}\left(\sum_{k=1}^{f} p_{u,k}^2 + \sum_{k=1}^{f} q_{k,s}^2\right) \tag{8}$$

Then, LFs involved in (8) are trained by moving them along the opposite of the stochastic gradient of (8) with respect to each single LF, i.e., we make

$$On\ r_{u,s},\ for\ k = 1 \sim f: \begin{cases} p_{u,k} \leftarrow p_{u,k} + \eta q_{k,s}\left(r_{u,s} - \sum_{k=1}^{f} p_{u,k} q_{k,s}\right) - \lambda\eta p_{u,k}, \\ q_{k,s} \leftarrow q_{k,s} + \eta p_{u,k}\left(r_{u,s} - \sum_{k=1}^{f} p_{u,k} q_{k,s}\right) - \lambda\eta q_{k,s}. \end{cases} \tag{9}$$

After LFs are trained on all the elements in R_K by computing (9), P and Q are extracted. For ease of formulation, we use the function (10) to represent extracting P and Q from R based on an LF model as follow.

$$\{P, Q\} = F^{LF}(P, Q|R) \tag{10}$$

3.2   Identifying Neighborhoods of QoS Data and Detecting Unreliable QoS Data

Since LF matrices P and Q respectively reflect the users and services characteristics hidden in R, we can identify neighborhoods of QoS data and detect unreliable QoS data based on them. Here, we use parameter α to denote the ratio of unreliable QoS data.

A.   With respect to users

This section explains how to identify a neighborhood for a user and detect unreliable users based on P. Here P is seen as the dataset of users. For each user u, its local density ρ_u is computed via cut-off kernel as:

$$\rho_u = \sum_{u'=1, u' \neq u}^{|U|} \Phi(d_{u,u'} - d_U), \ \Phi(t) = \begin{cases} 1 & t < 0 \\ 0, & others \end{cases} \tag{11}$$

where d_U is the cutoff distance with respect to users, u' denotes another user that is different from user u, $d_{u,u'}$ denotes the distance between users u and u'. Here we compute $d_{u,u'}$ with Euclidean distance as:

$$d_{u,u'} = \sqrt{\sum_{k=1}^{f} \left(p_{u,k} - p_{u',k}\right)^2} \tag{12}$$

ρ_u also can be computed via Gaussian kernel as:

$$\rho_u = \sum_{u'=1,u'\neq u}^{|U|} e^{-\left(\frac{d_{u,u'}}{d_U}\right)^2} \tag{13}$$

According to (5), d_U is computed as:

$$Vec = sort\left(d_{u,u'}\right), \; d_U = Vec(\lfloor P_{Vec} \times |U| \times (|U|-1)/2 \rfloor) \tag{14}$$

Then, the minimum distance δ_u of user u between itself and any other user with higher local density is computed as:

$$\delta_u = \begin{cases} \min_{u':\rho_u < \rho_{u'}}(d_{u,u'}), & others \\ \max_{u'}(d_{u,u'}), & \forall u', \; \rho_u \geq \rho_{u'} \end{cases} \tag{15}$$

Finally, the outlier factor γ_u of user u can be computed as:

$$\gamma_u = \rho_u/\delta_u \tag{16}$$

Based on all the ρ_u, δ_u, and γ_u, we can discover the user dataset P's clusters and outliers. Here clusters represent neighborhoods of users, outliers represent unreliable users. By computing (11)–(16), the original R can be separated into N matrices $\{R_1^U, R_2^U, \ldots, R_N^U\}$, where each matrix $R_n^U, n \in \{1, 2, \ldots, N\}$, denotes a neighborhood of users; or separated into two matrices $\{R_r^U, R_u^U\}$, where R_r^U denotes the reliable users and R_u^U denotes the unreliable users. Here the ratio of unreliable QoS data α is computed by:

$$\alpha = |R_u^U|/(|R_r^U| + |R_u^U|). \tag{17}$$

B. With respect to services

This section explains how to identify a neighborhood for a service and detect unreliable services based on Q. Here Q is seen as the dataset of services. For each service s, its local density ρ_s, minimum distance δ_s between itself and any other service with higher local density, and outlier factor γ_s can be computed by (18)–(23).

$$\rho_s = \sum_{s'=1,s'\neq s}^{|S|} \Phi(d_{s,s'} - d_S), \; \Phi(t) = \begin{cases} 1 & t < 0 \\ 0, & others \end{cases} \tag{18}$$

$$d_{s,s'} = \sqrt{\sum_{k=1}^{f} \left(q_{k,s} - q_{k,s'}\right)^2} \tag{19}$$

$$\rho_s = \sum_{s'=1,s'\neq s}^{|S|} e^{-\left(\frac{d_{s,s'}}{d_S}\right)^2} \tag{20}$$

$$Vec = sort\left(d_{s,s'}\right), \ d_S = Vec(\lfloor P_{Vec} \times |S| \times (|S|-1)/2 \rfloor) \tag{21}$$

$$\delta_s = \begin{cases} \min_{s':\rho_s < \rho_{s'}}(d_{s,s'}), & others \\ \max_{s'}(d_{s,s'}), & \forall s', \ \rho_s \geq \rho_{s'} \end{cases} \tag{22}$$

$$\gamma_s = \rho_s / \delta_s \tag{23}$$

where s' is another service that is different from service s, $d_{s,s'}$ is the distance between services s and s', d_S is the cutoff distance for services. Similarly, the original R also can be separated into N matrices $\{R_1^S, R_2^S, \ldots, R_N^S\}$, where each matrix $R_n^S, n \in \{1, 2, \ldots, N\}$, denotes a neighborhood of services; or separated into two matrices $\{R_r^S, R_u^S\}$, where R_r^S and R_u^S denote the reliable and unreliable services respectively. Here, α is computed by

$$\alpha = |R_u^S|/(|R_r^S| + |R_u^S|). \tag{24}$$

3.3 Prediction

After Sect. 3.2, we can accurately predict the missing data in R by employing the four matrices sets of $\{R_1^U, R_2^U, \ldots, R_N^U\}$, $\{R_r^U, R_u^U\}$, $\{R_1^S, R_2^S, \ldots, R_N^S\}$, and $\{R_r^S, R_u^S\}$ respectively. Each matrices set can be used to implement the prediction, but which one is the best? This is determined by the characteristics of QoS data and please refer to Sect. 4.3 to see an example. Next, we respectively explain how to implement the prediction based on the four matrices sets and formula (10).

First, if matrices set $\{R_1^U, R_2^U, \ldots, R_N^U\}$ is used to predict, the computing formulas are

$$for \ n = 1 \sim N: \ \{P_n^U, \ Q_n^U\} = F^{LF}\left(P_n^U, \ Q_n^U \big| R_n^U\right), \tag{25}$$

$$for \ n = 1 \sim N: \ \hat{R}_n^U = P_n^U Q_n^U, \tag{26}$$

$$\hat{R} = \hat{R}_1^U \cup \hat{R}_2^U \cup \ldots \cup \hat{R}_N^U. \tag{27}$$

Second, if matrices set $\{R_r^U, R_u^U\}$ is employed to predict, the computing formulas are

$$\{P^U_r, Q^U_r\} = F^{LF}\left(P^U_r, Q^U_r \big| R^U_r\right),\tag{28}$$

$$\hat{R}^U_r = P^U_r Q^U_r,\tag{29}$$

$$\hat{R} = \hat{R}^U_r \oplus PQ\big|_{row}.\tag{30}$$

where $\hat{R}^U_r \oplus PQ\big|_{row}$ indicates that using $\hat{R}^U_r$ to replace the corresponding rows of the matrix product of PQ.

Third, if matrices set $\{R^S_1, R^S_2, \ldots, R^S_N\}$ is used to predict, the computing formulas are

$$for\ n = 1 \sim N:\ \{P^S_n, Q^S_n\} = F^{LF}\left(P^S_n, Q^S_n \big| R^S_n\right),\tag{31}$$

$$for\ n = 1 \sim N:\ \hat{R}^S_n = P^S_n Q^S_n,\tag{32}$$

$$\hat{R} = \hat{R}^S_1 \cup \hat{R}^S_2 \cup \ldots \cup \hat{R}^S_N.\tag{33}$$

Fourth, if matrices set $\{R^S_r, R^S_u\}$ is employed to predict, the computing formulas are

$$\{P^S_r, Q^S_r\} = F^{LF}\left(P^S_r, Q^S_r \big| R^S_r\right),\tag{34}$$

$$\hat{R}^S_r = P^S_r Q^S_r,\tag{35}$$

$$\hat{R} = \hat{R}^S_r \oplus PQ\big|_{column}.\tag{36}$$

where $\hat{R}^S_r \oplus PQ\big|_{column}$ indicates that using $\hat{R}^S_r$ to replace the corresponding columns of the matrix product of PQ.

3.4 Algorithm Design and Analysis

DALF relies on four algorithms. Algorithm 1 is extracting LF matrices (ELFM), Algorithm 2 is computing QoS data with respect to users (U-QoS), Algorithm 3 is computing QoS data with respect to services (S-QoS), and Algorithm 4 is Prediction. Their pseudo codes and time cost of each step are given in Algorithms 1–4. For Algorithms 1–3, their computational complexities are $\Theta(N_{mtr} \times |R_K| \times f)$, $\Theta\left(|U|^2 \times f\right)$, and $\Theta\left(|S|^2 \times f\right)$, respectively, where N_{mtr} is the maximum training round. For Algorithm 4, its computational complexity is $\Theta\left(\left(|U|^2 + |S|^2\right) \times f\right) + \Theta(N_{mtr} \times |R_K| \times f)$, which is the total computational complexity of DALF.

4   Experimental Results

4.1   Datasets

Two benchmark datasets, which are real-world web service QoS data collected by the WS-Dream system (https://github.com/wsdream/wsdream-dataset) and frequently used in prior researches [2, 3, 7–9, 12–15, 17], are selected to conduct the experiments. First dataset (D1) is the Response Time that contains 1,873,838 records and second dataset (D2) is the Throughput that contains 1,831,253 records. These records are generated by 339 users on 5,825 web services. For both two datasets, different test cases are designed to evaluate the performance of DALF. Table 1 summarizes the properties of all the test cases, where column 'Density' denotes the density of the training matrix.

Algorithm 1. ELFM

Input: R; Output: P, Q		Cost
1	initializing $f, \lambda, \eta, N_{mtr}$=*max-training-round*	$\Theta(1)$
2	**while** $t \leq N_{mtr}$ && not converge	$\times N_{mtr}$
3	**for** each known entry $r_{u,s}$ in R // $r_{u,s} \in R_K$	$\times |R_K|$
4	**for** k=1 to f	$\times f$
5	computing $p_{u,k}$ according to (9)	$\Theta(1)$
6	computing $q_{k,s}$ according to (9)	$\Theta(1)$
7	**end for**	--
8	**end for**	--
9	$t=t+1$	$\Theta(1)$
10	**end while**	--
11	**return** P, Q	$\Theta(1)$

Algorithm 2. U-QoS

Input: P, R; Output: $\{R_1^U, R_2^U, ..., R_N^U\}, \{R_r^U, R_u^U\}$		Cost
1	**for** u=1 to $|U|$	$\times |U|$
2	**for** u'=u+1 to $|U|$	$\times(|U|-1)/$
3	computing $d_{u,u'}$ according to (12)	$\Theta(f)$
4	**end for**	--
5	**end for**	--
6	computing d_U according to (14)	$\Theta(|U|^2)$
7	**for** u=1 to $|U|$	$\times |U|$
8	computing ρ_u according to (11) or (13)	$\Theta(|U|-1$
9	**end for**	--
10	**for** u=1 to $|U|$	$\times |U|$
11	computing δ_u according to (15)	$\Theta(|U|)$
12	computing γ_u according to (16)	$\Theta(1)$
13	**end for**	--
14	clustering P according to all the ρ_u and δ_u	$\Theta(1)$
15	separating R into $\{R_1^U, R_2^U, ..., R_N^U\}$ according to step 14	$\Theta(1)$
16	detecting unreliable users according to all γ_u	$\Theta(1)$
17	separating R into $\{R_r^U, R_u^U\}$ according to step 16	$\Theta(1)$
18	**return:** $\{R_1^U, R_2^U, ..., R_N^U\}, \{R_r^U, R_u^U\}$	$\Theta(1)$

Algorithm 3. S-QoS

Input: Q, R; Output: $\{R_1^s, R_2^s, ..., R_N^s\}, \{R_r^s, R_u^s\}$		Cost
1	**for** s=1 to $|S|$	$\times |S|$
2	**for** s'= s+1 to $|S|$	$\times(|S|-1)/2$
3	computing $d_{s,s'}$ according to (19)	$\Theta(f)$
4	**end for**	--
5	**end for**	--
6	computing d_S according to (21)	$\Theta(|S|^2)$
7	**for** s=1 to $|S|$	$\times |S|$
8	computing ρ_s according to (18) or (20)	$\Theta(|S|-1)$
9	**end for**	--
10	**for** s=1 to $|S|$	$\times |S|$
11	computing δ_s according to (22)	$\Theta(|J|)$
12	computing γ_s according to (23)	$\Theta(1)$
13	**end for**	--
14	clustering Q according to all the ρ_s and δ_s	$\Theta(1)$
15	separating R into $\{R_1^s, R_2^s, ..., R_N^s\}$ according to step 14	$\Theta(1)$
16	detecting unreliable users according to all the γ_s	$\Theta(1)$
17	separating R into $\{R_r^s, R_u^s\}$ according to step 16	$\Theta(1)$
18	**return:** $\{R_1^s, R_2^s, ..., R_N^s\}, \{R_r^s, R_u^s\}$	$\Theta(1)$

Algorithm 4. Prediction

Input:R; Output: $\hat{R}$		Cost
1	Calling **Algorithm 1**	$\Theta(N_{mtr}\times|R_K|\times f)$
2	Calling **Algorithm 2**	$\Theta(|U|^2 \times f)$
3	Calling **Algorithm 3**	$\Theta(|S|^2 \times f)$
4	determining which matrices set is best for prediction	$\Theta(1)$
5	**if** $\{R_1^U, R_2^U, ..., R_N^U\}$ is best for prediction	
6	**for** n=1 to N	
7	computing $\hat{R}_n^u$ according to (25) and (26)	$\Theta(N_{mtr}\times|R_K|\times f)$
8	**end for**	
9	computing $\hat{R}$ according to (27)	
10	**else if** $\{R_r^U, R_u^U\}$ is best for prediction	
11	computing $\hat{R}$ according to (28)—(30)	$\Theta(N_{mtr}\times|R_K|\times f)$
12	**else if** $\{R_1^s, R_2^s, ..., R_N^s\}$ is best for prediction	
13	**for** n=1 to N	
14	computing $\hat{R}_n^s$ according to (31) and (32)	$\Theta(N_{mtr}\times|R_K|\times f)$
15	**end for**	
16	computing $\hat{R}$ according to (33)	
17	**else** computing $\hat{R}$ according to (34)—(36)	$\Theta(N_{mtr}\times|R_K|\times f)$
18	**end if**	
19	**return:** $\hat{R}$	$\Theta(1)$

Table 1. Properties of all the designed test cases.

Dataset	No.	Density	Training data	Testing data
D1	D1.1	5%	93,692	1,780,146
	D1.2	10%	187,384	1,686,454
	D1.3	15%	281,076	1,592,762
	D1.4	20%	374,768	1,499,070
D2	D2.1	5%	91,563	1,739,690
	D2.2	10%	183,125	1,648,128
	D2.3	15%	274,689	1,556,564
	D2.4	20%	366,251	1,465,002

4.2 Evaluation Protocol

To evaluate the prediction quality of DALF, mean absolute error (MAE) is computed by $MAE = \left(\sum_{(w,j) \in \Gamma} \left| r_{w,j} - \hat{r}_{w,j} \right|_{abs} \right) / |\Gamma|$, where Γ denotes the testing set.

4.3 Prediction According to the Characteristics of QoS Data

This section illustrates how to predict the unknown QoS data according to its characteristics. First, extracting LF matrices P for users and Q for services on D1.4 and D2.4 respectively. Then, computing ρ, δ, and γ for each user or service. After that, the decision graphs for D1.4 and D2.4 can be drawn in Figs. 4 and 5. With respect to users, we observe that there are two cluster centers on D1.4 and three cluster centers on D2.4, which means that users of D1.4 and D2.4 could be separated into two and three neighborhoods respectively. With respect to services, there is only one cluster center on both D1.4 and D2.4, which means that there are no neighborhoods in services. Besides, we can find that there are many outliers (red rectangles) in both D1.4 and D2.4 with respect to services, which means that there are many unreliable services. Similar results are obtained on the other test cases. Thus, we conclude that predicting based on neighborhoods of users or reliable services is the best strategy for the eight test cases. In addition, we have conducted some experiments to verify that these two prediction strategies have better performance than the other two. For saving space, we never show their results.

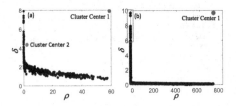

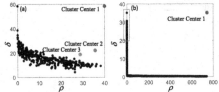

Fig. 4. The decision graph for D1.4 with respect to: (a) users, (b) services. (Color figure online)

Fig. 5. The decision graph for D2.4 with respect to: (a) users, (b) services. (Color figure online)

4.4 Predicting Based on Neighborhoods of Users

A. Impacts of f

The parameters are set as $\eta = 0.01$ for D1, $\eta = 0.0001$ for D2, $\lambda = 0.01$, and $P_{Vec} = 2\%$, uniformly. Figure 6 shows the experimental results when f increases from 10 to 320. Since the higher dimension of LF space has better representation learning ability, DALF has a lower MAE as f increases. However, as f increases over 80, the MAE tends to decrease slightly or even increase. One reason is that with $f = 80$, DALF's representation learning ability is strong enough to precisely represent the test cases. As a result, continuous increase of f after 80 cannot bring significant improvement in prediction accuracy.

B. Impacts of λ

This set of experiments sets the parameters as $\eta = 0.01$ for D1, $\eta = 0.0001$ for D2, $f = 20$, and $P_{Vec} = 2\%$, uniformly. Figure 7 records the MAE as λ increases. We test a larger range of λ on D2 than on D1 because D2 has a much larger range of value than D1. The MAE decreases at first as λ increases in general on all the test cases. However, it then increases when λ grows over the optimal threshold, which means that DALF may be greatly impacted by the regularization terms.

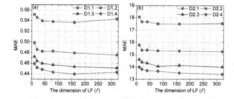
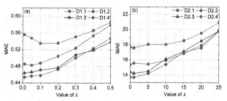

Fig. 6. MAE of DALF with different f predicting based on neighborhoods of users: (a) D1, (b) D2.

Fig. 7. MAE of DALF as λ increases predicting based on neighborhoods of users: (a) D1, (b) D2.

4.5 Predicting Based on Reliable Services

A. Impacts of α

The parameters are set as $\lambda = 0.01$, $\eta = 0.01$ for D1, $\eta = 0.0001$ for D2, $f = 20$, and $P_{Vec} = 2\%$, uniformly. Figure 8 is the measured MAE as α increases. On D1, the MAE decreases at first and then increases in general as α increases. The lowest MAE is obtained when α around to 0.3. On D2, the results are more complicated. Concretely, DALF has the lowest MAE when $\alpha = 0.05$ or 0.1. According to these observations, it seems that more services on D1 are detected as unreliable ones than that on D2. Overall, these results validate that the prediction accuracy of DALF can be improved by employing reliable services to train.

B. Impacts of f and λ

Since these results are very similar to that in Sects. 4.4(A) and (B), they are not described in detail for saving space. Please refer to Sects. 4.4(A) and (B).

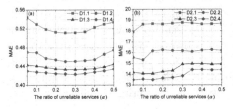

Fig. 8. MAE of DALF as α increases predicting based on reliable services: (a) D1, (b) D2.

Table 2. Descriptions of all the compared models.

Models	Descriptions
BLF	Basic LF model proposed in 2009 [18]
RSNMF	Improved LF-based model proposed in 2016 [3]
NIMF	Improved LF-based model proposed in 2013 [21]
NAMF	Geography-LF-based model proposed in 2016 [9]
GeoMF	Geography-LF-based model proposed in 2017 [8]
LMF-PP	Geography-LF-based model proposed in 2018 [12]
AutoRec	The DNN-based model proposed in 2015 [22]
DALF-1	Predicting based on neighborhoods of users
DALF-2	Predicting based on reliable services

Table 3. MAE of all the compared models on each test case.

TestCases	BLF	RSNMF	NIMF	NAMF	GeoMF	LMF-PP	AutoRec	DALF	
								DALF-1	DALF-2
D1.1	0.5561	0.5438	0.5502	0.5465	0.5305	0.5285	0.5467	0.5457	**0.5114**
D1.2	0.4944	0.4868	0.4842	0.4976	0.4827	0.4725	0.5055	0.4857	**0.451**
D1.3	0.4691	0.4492	0.4508	0.4625	0.4495	0.4472	0.4598	0.4642	**0.4331**
D1.4	0.4531	0.4371	0.4346	0.436	0.4366	0.426	0.4482	0.452	**0.4232**
D2.1	18.9776	21.4302	17.7153	22.736	24.7465	18.3091	21.3118	**17.6576**	17.9117
D2.2	16.1924	17.2305	15.5264	17.9148	22.4728	15.9125	17.031	**15.3595**	15.5734
D2.3	14.9278	14.6879	14.2146	15.9876	17.7908	14.745	15.0156	14.3836	**14.1739**
D2.4	14.3061	14.3654	13.5638	14.7462	16.2852	14.1033	14.2265	13.6697	**13.4772**

Table 4. Statistical results of prediction accuracy with a significance level of 0.05.

Comparison	DALF vs. BLF	DALF vs. RSNMF	DALF vs. NIMF	DALF vs. NAMF	DALF vs. GeoMF	DALF vs. LMF-PP	DALF vs. AutoRec
p-value	**0.0039**	**0.0039**	**0.0039**	**0.0039**	**0.0039**	**0.0039**	**0.0039**

Note that the best one between DALF-1 and DALF-2 for each test case is selected to conduct statistical analysis

4.6 Comparisons Between DALF and Other Models

We compare DALF with state-of-the-art models on prediction accuracy and computational complexity. They are three LF-based models, three geography-LF-based models, and one deep neural network (DNN)-based model, and described in Table 2.

On prediction accuracy, the dimension of LF is set as $f = 20$ for all models except for AutoRec (because AutoRec is a DNN-based model) to conduct the fair comparisons. Besides, all other parameters of the compared models are set according to their original papers. There are two situations for DALF to conduct the comparisons. They are also marked in Table 2. Meanwhile, the other parameters for DALF are set as: $\alpha = 0.3$ and $\eta = 0.01$ for D1, $\alpha = 0.05$ and $\eta = 0.0001$ for D2, $\lambda = 0.01$, and $P_{Vec} = 2\%$, uniformly.

The compared results are shown in Table 3. Besides, the Wilcoxon signed-ranks test [23], as a nonparametric pairwise comparison procedure, is adopted to conduct statistical test. The results are recorded in Table 4. From them, we have two findings:

(a) DALF has significantly better prediction accuracy than the other models. For example, it has around 5.27%–17.15% lower MAE than AutoRec on all test cases.
(b) DALF-1 has much higher MAE than DALF-2 on D1, while they have similar MAE on D2. Figure 4 shows that neighborhoods of users are not very clear on D1 but clear on D2, and unreliable services can be easily detected on both D1 and D2. These findings mean that predicting based on neighborhoods of users is better for D2 than for D1, and predicting based on reliable services are appropriate for both D1 and D2.

On computational complexity, AutoRec is not compared because it is DNN-based model with extremely high computational cost [24]. Table 5 concludes the computational complexities for all the models, where K_1 and K_2 are the number of nearest neighbors for a user and for a service respectively. From it, we have two conclusions:

(a) BLF and RSNMF have lowest computational complexity because they are basic LF-based models and never consider neighborhood or unreliable factors of QoS data.
(b) DALF's computational complexity is lower than or at least comparable to that of the geography-LF-based models because f is much smaller than $|U|$ and $|S|$.

Table 5. The computational complexities of all the compared models.

Model	Complexity
BLF [18]	$\Theta(N_{mtr} \times |R_K| \times f)$
RSNMF [3]	$\Theta(N_{mtr} \times |R_K| \times f)$
NIMF [21]	$\Theta(|U|^2 \times |S|) + \Theta(N_{mtr} \times |R_K| \times f \times K_1^2)$
NAMF [9]	$\Theta(|U|^2) + \Theta(N_{mtr} \times |R_K| \times f)$
GeoMF [8]	$\Theta(|U|^2 \times |S| + |S|^2 \times |U|) + \Theta(N_{mtr} \times |R_K| \times f^2 \times (K_1 + K_2))$
LMF-PP [12]	$\Theta(|U|^2 \times |S| + |S|^2 \times |U|) + \Theta(N_{mtr} \times |R_K| \times f)$
DALF	$\Theta((|U|^2 + |S|^2) \times f) + \Theta(N_{mtr} \times |R_K| \times f)$

5   Conclusions

We propose a data-aware latent-factor (DALF) model to achieve highly accurate QoS prediction. The main idea is to incorporate a density peaks based clustering method (DPClust) into a latent factor (LF)-based model to improve the prediction accuracy. Empirical studies on two benchmark real-world web service QoS datasets demonstrate that: (i) DALF can discover the characteristics of QoS data only based on the user-service QoS matrix, (ii) DALF is a data-aware model because it can easily choose the appropriate strategies to implement prediction according to the characteristics of QoS data, and (iii) DALF has better performance than state-of-the-art models in QoS prediction. Finally, an open challenge for DALF is how to combine the two respects of users and services to further improve it. We plan to address this challenge in our future work.

Acknowledgments. This work was supported in part by the National Key Research and Development Program of China under Grant 2016YFB1000900, and in part by the National Natural Science Foundation of China under Grants 61702475, 91646114 and 61772096.

References

1. Zheng, Z., Ma, H., Lyu, M.R., King, I.: WSRec: a collaborative filtering based web service recommender system. In: Proceeding of 2009 IEEE International Conference on Web Services, pp. 437–444. IEEE (2009)
2. Zheng, Z., Zhang, Y., Lyu, M.R.: Distributed QoS evaluation for real-world web services. In: Proceeding of 2010 IEEE International Conference on Web Services, pp. 83–90. IEEE (2010)
3. Luo, X., Zhou, M., Xia, Y., Zhu, Q., Ammari, A.C., Alabdulwahab, A.: Generating highly accurate predictions for missing QoS data via aggregating nonnegative latent factor models. IEEE Trans. Neural Netw. Learn. Syst. **27**(3), 524–537 (2016)
4. Geebelen, D., et al.: QoS prediction for web service compositions using kernel-based quantile estimation with online adaptation of the constant offset. Inf. Sci. **268**, 397–424 (2014)
5. Chen, X., Liu, X., Huang, Z., Sun, H.: RegionKNN: a scalable hybrid collaborative filtering algorithm for personalized web service recommendation. In: Proceeding of 2010 IEEE International Conference on Web Services, pp. 9–16. IEEE (2010)
6. Zheng, Z., Ma, H., Lyu, M.R., King, I.: Qos-aware web service recommendation by collaborative filtering. IEEE Trans. Serv. Comput. **4**(2), 140–152 (2011)
7. Lee, K., Park, J., Baik, J.: Location-based web service QoS prediction via preference propagation for improving cold start problem. In: Proceeding of 2015 IEEE International Conference on Web Services, pp. 177–184. IEEE (2015)
8. Chen, Z., Shen, L., Li, F., You, D.: Your neighbors alleviate cold-start: on geographical neighborhood influence to collaborative web service QoS prediction. Knowl.-Based Syst. **138**, 188–201 (2017)
9. Tang, M., Zheng, Z., Kang, G., Liu, J., Yang, Y., Zhang, T.: Collaborative web service quality prediction via exploiting matrix factorization and network map. IEEE Trans. Netw. Serv. Manag. **13**(1), 126–137 (2016)

10. Luo, X., Zhou, M., Li, S., You, Z., Xia, Y., Zhu, Q.: A nonnegative latent factor model for large-scale sparse matrices in recommender systems via alternating direction method. IEEE Trans. Neural Netw. Learn. Syst. **27**(3), 579–592 (2016)

11. Shi, Y., Larson, M., Hanjalic, A.: Collaborative filtering beyond the user-item matrix: a survey of the state of the art and future challenges. ACM Comput. Surv. **47**(1), 1–45 (2014)

12. Ryu, D., Lee, K., Baik, J.: Location-based web service QoS prediction via preference propagation to address cold start problem. IEEE Trans. Serv. Comput. (2018)

13. Wu, H., Yue, K., Li, B., Zhang, B., Hsu, C.-H.: Collaborative QoS prediction with context-sensitive matrix factorization. Future Gener. Comput. Syst. **82**, 669–678 (2018)

14. Zhu, J., He, P., Zheng, Z., Lyu, M.R.: Online QoS prediction for runtime service adaptation via adaptive matrix factorization. IEEE Trans. Parallel Distributed Syst. **28**(10), 2911–2924 (2017)

15. Wu, C., Qiu, W., Zheng, Z., Wang, X., Yang, X.: QoS prediction of web services based on two-phase k-means clustering. In: Proceeding of 2015 IEEE International Conference on Web Services, pp. 161–168. IEEE (2015)

16. Liu, A., et al.: Differential private collaborative Web services QoS prediction. World Wide Web 1–24 (2018, in Press)

17. Feng, Y., Huang, B.: Cloud manufacturing service QoS prediction based on neighbourhood enhanced matrix factorization. J. Intell. Manuf. 1–12 (2018)

18. Koren, Y., Bell, R., Volinsky, C.: Matrix factorization techniques for recommender systems. Computer **42**(8), 30–37 (2009)

19. Rodriguez, A., Laio, A.: Clustering by fast search and find of density peaks. Science **344** (6191), 1492–1496 (2014)

20. Wu, D., et al.: Self-training semi-supervised classification based on density peaks of data. Neurocomputing **275**, 180–191 (2018)

21. Zheng, Z., Ma, H., Lyu, M.R., King, I.: Collaborative web service QoS prediction via neighborhood integrated matrix factorization. IEEE Trans. Serv. Comput. **6**(3), 289–299 (2013)

22. Sedhain, S., Menon, A.K., Sanner, S., Xie, L.: AutoRec: autoencoders meet collaborative filtering. In: Proceedings of the 24th International Conference on World Wide Web, pp. 111–112 (2015)

23. Wu, D., Luo, X., Wang, G., Shang, M., Yuan, Y., Yan, H.: A highly accurate framework for self-labeled semi supervised classification in industrial applications. IEEE Trans. Ind. Inf. **14** (3), 909–920 (2018)

24. Zhou, Z.-H., Feng, J.: Deep forest: towards an alternative to deep neural networks. In: Proceedings of the 26th International Joint Conference on Artificial Intelligence (2017)

Keyword Extraction with Character-Level Convolutional Neural Tensor Networks

Zhe-Li Lin and Chuan-Ju Wang[✉]

Research Center for Information Technology Innovation, Academia Sinica,
Taipei, Taiwan
{joli79122,cjwang}@citi.sinica.edu.tw

Abstract. Keyword extraction is a critical technique in natural language processing. For this essential task we present a simple yet efficient architecture involving character-level convolutional neural tensor networks. The proposed architecture learns the relations between a document and each word within the document and treats keyword extraction as a supervised binary classification problem. In contrast to traditional supervised approaches, our model learns the distributional vector representations for both documents and words, which directly embeds semantic information and background knowledge without the need for handcrafted features. Most importantly, we model semantics down to the character level to capture morphological information about words, which although ignored in related literature effectively mitigates the unknown word problem in supervised learning approaches for keyword extraction. In the experiments, we compare the proposed model with several state-of-the-art supervised and unsupervised approaches for keyword extraction. Experiments conducted on two datasets attest the effectiveness of the proposed deep learning framework in significantly outperforming several baseline methods.

1 Introduction

Keyword extraction is the automatic identification of important, representative terms which accurately and concisely capture the main topics of a given document. Such keywords or keyphrases provide rich information about the content and help improve the performance of natural language processing (NLP) and information retrieval (IR) tasks such as text summarization [25] and text categorization [7]. Due to its importance and its connections to other NLP tasks, various approaches for keyword or keyphrase extraction have been proposed in the literature.

Typically, methods for automatic keyword extraction can be divided into two major categories: supervised and unsupervised learning. Supervised approaches often treat keyword extraction as a binary classification problem [4,6,20] in which the terms in a given document are classified into positive and negative examples of keywords. A feature vector is created to illustrate the instance from aspects such as statistical information (e.g., term frequency), lexical features, and

© Springer Nature Switzerland AG 2019
Q. Yang et al. (Eds.): PAKDD 2019, LNAI 11439, pp. 400–413, 2019.
https://doi.org/10.1007/978-3-030-16148-4_31

syntactic patterns; with these handcrafted features, the positive and negative instances are then trained with learning algorithms including naive Bayes [20], decision trees [18], and support vector machines [23], in building the classifier. However, traditional supervised learning methods require an empirical process of feature engineering driven by intuition, experience, and domain expertise. Furthermore, since supervised machine learning approaches use a set of labeled documents to train the model, the problem of *unknown words* – words in testing documents that do not appear in the training corpus – is usually not handled explicitly in traditional supervised learning approaches.

Existing unsupervised learning methods approach the problem using a wide variety of techniques, including graph-based ranking algorithms [13,16], clustering-based approaches [1,12], and language modeling [17]. Of these techniques, graph-based approaches are the most popular: they transform the words in a document into a graph-of-words representation [13] and then rank the importance of each word recursively using the random walk algorithm [2]. Compared with state-of-the-art supervised methods that capture only local vertex (word)-specific information on the candidate terms, graph-based algorithms incorporate into the model global information from co-occurrence relations between each word. However, for unsupervised approaches, it is difficult to incorporate deeper background knowledge extracted from external databases [5] or leverage the information and relations of other documents and keywords in the given dataset into their learning phase due to the nature of such unsupervised methods that in general involve only the given document.

Recently, deep learning approaches have been the focus of much research on a wide variety of NLP and IR problems. The main feature of these approaches is that during the learning process, they discover not only the mapping from representation to output but also the representation itself, which thus removes the empirical process of manual feature engineering characteristic of traditional supervised methods, greatly reducing the need to adapt to new tasks and domains. Moreover, due to the efficiency of convolutional neural networks (CNNs), many recent studies have demonstrated the superior performance of word-level CNN-based algorithms in various NLP and IR tasks [8,10,15]. For example, in [8,10], the authors use word-based CNNs for modeling sentences; in [15], the authors propose a word-based CNN architecture for reranking pairs of short texts. However, since such approaches learn the embeddings of words in the training phase, the unknown word problem becomes critical if the training data is rather small and there are many words in testing documents that do not appear in the training corpus. To take this into account, several recent studies have narrowed down the semantics by incorporating character-level representations to deal with several fundamental NLP tasks, such as part-of-speech tagging [14], text classification [24], and other applications [19].

In this article we treat text as a raw signal at the character level and propose character-level convolutional neural tensor networks (here after abbreviated as charCNTN), a simple yet efficient architecture for keyword extraction. To our best knowledge, this paper is the first to apply convolutional neural networks

(ConvNets) solely on characters for this essential NLP problem. By following the widely used TextRank [13] and the state-of-the-art graph degeneracy-based approach [16], we focus on keyword extraction only and build keyphrases with a post-processing step. The proposed model consists of two main building blocks: one word model and one document model based on ConvNets; these two underlying models operate in parallel, mapping documents and candidate keywords to their distributional vector representations. Most importantly, by introducing a sparse document representation – the document tensor – we model semantics down to the character level to capture morphological information in words, which can be of great help in mitigating the unknown word problem in many supervised learning approaches.

In the experiments, we compare the proposed model with several state-of-the-art supervised and unsupervised approaches for keyword extraction. With the post-processing step of reconstructing keyphrases from the extracted keywords [13], we also compare our performance of keyphrase extraction with TextRank on both datasets. Experiments conducted on two datasets attest the effectiveness of the proposed deep learning framework in providing significantly better performance than several baseline methods.

2   Keyword Extraction

2.1   Supervised Approaches

Given a set of documents $D = \{d_1, d_2, \ldots, d_n\}$ and a word sequence $[w_1 w_2 \ldots w_{|d_i|}]$ for each document d_i, each word $w \in W_{d_i}$ comes with a judgment y_w, where W_{d_i} denotes the set of all words in d_i, and words that are keywords are assigned labels equal to 1 and 0 otherwise. The goal of a supervised task for keyword extraction is to build a model that generates keywords for each given document taking into consideration the given document set D and the corresponding labels. This task can be formally defined as a binary classification problem with model f as

$$f(g(d_i, w)) = \begin{cases} 1 & w \text{ is a keyword,} \\ 0 & \text{otherwise,} \end{cases} \tag{1}$$

where $g(\cdot)$ denotes the function to generate features from the given d_i and w. In the literature, several common features are term frequency, term frequency-inverse document frequency, first occurrence, and part-of-speech tags [20,23]. Support vector machines and logistic regression are commonly adopted to learn the classifier.

2.2   Unsupervised Approaches

In contrast to the above supervised approaches, unsupervised approaches generate keywords mainly by learning the structure of the given documents, such as their word co-occurrences. Among the literature, graph-based ranking methods

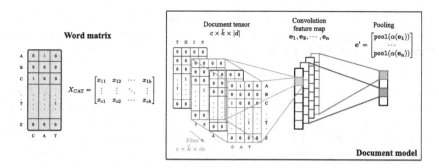

Fig. 1. Document model

such as TextRank [13] and TopicRank [1] are the most widely used unsupervised methods for keyword extraction. In this type of approach, the keyword extraction problem is defined as follows. Given a document d_i with word sequence $[w_1 w_2 \ldots w_{|d_i|}]$, a graph $G = (V, E)$ is constructed to represent the structure of document d_i. For example, in TextRank, vertices are words and the edge between two words is weighted according to their co-occurrence relation. The goal is to build a scoring function for each of the vertices on graph G and rank the words based on the scores.

2.3   Problem Formulation

In this paper, keyword extraction is formulated as a binary classification problem. However, unlike traditional supervised approaches in which the features for each document are manually determined, in the proposed deep neural network approach, the two inputs d_i and w in Eq. (1) are directly represented as a third-order tensor and a second-order tensor (a matrix), respectively. Therefore, given a tensor T_{d_i} for document $d_i \in D$ and a matrix M_w for word $w \in W_{d_i}$, Eq. (1) is rewritten as

$$f(T_{d_i}, M_w) = \begin{cases} 1 \ w \text{ is a keyword,} \\ 0 \text{ otherwise.} \end{cases}$$

In the next section, we show how documents are represented as tensors and how words are represented as matrices based on a character-level sparse representation.

3   Proposed CharCNTN Architecture

3.1   Document Model and Word Model

Shown in Fig. 1 is the architecture of ConvNet for mapping documents to distributed feature vectors, as inspired by the architecture used in [15] for document ranking and the idea of the character-level CNN model for text classification [24]. The document model consists of a single narrow convolutional layer followed by a non-linearity and simple max pooling. In the following subsections, we briefly introduce the main parts of our ConvNet: the document tensor, convolution feature maps, and pooling layers.

Document Tensor. In our model, the input is a document-word pair in which each document is represented as a third-order tensor and each word is represented as a second-order tensor (a matrix), as illustrated in Fig. 1. For simplicity, below we focus on the structure of the third-order tensor for documents; note that the input for each word is a degenerated version of a document. Each document d is treated as a word sequence: $[w_1, .., w_{|d|}]$, where $|d|$ denotes the document length of d. As illustrated in the left panel of Fig. 1, each word is represented as a character-level one-hot encoding matrix.

Let $X_j \in \mathbb{R}^{c \times k_j}$ be the $c \times k_j$ word matrix corresponding to the j-th word in the document d_i, where k_j denotes the length of the j-th word (i.e., the number of characters of this word) and c denotes the size of the character vocabulary set C. A word w is then converted to the matrix X,[1] where the $x_{ab} \in X$ is set to 1 if the b-th character of w is the a-th element in C and 0 otherwise.

For each input document d, a document tensor $T_d \in \mathbb{R}^{c \times \bar{k} \times |d|}$ is constructed as follows (see the document tensor in Fig. 1): $T_d = X_1 \oplus X_2 \oplus \cdots \oplus x_{|d|}$, where $\oplus$ is the concatenation operator, X_j denotes the matrix of the j-th word in document d, and $\bar{k} = \max(k_j)$; each matrix X_j is padded with $c \times (\bar{k} - k_j)$ zeros.

It is worth mentioning that while our model allows for learning the distributed character embeddings directly for the given task, in this paper we keep the document tensor parameter static, resulting in much fewer parameters to be learned during the training process. This characteristic is especially beneficial to the task of keyword extraction since it is usually hard to obtain a large amount of label data for such a task. In addition, different from word embedding, there is no standard procedure or commonly used algorithm to obtain the pre-trained character embedding for constructing our static input tensor. We empirically find that using sparse one-hot or a random vector to represent a character obtains similar performance, but the former one is better in practice since it requires much less memory usage than the later one.

Convolution Feature Maps. To extract patterns from the input, a narrow convolution has been used. Given that the inputs of our model are document tensors $T_d = (t_{ijk}) \in \mathbb{R}^{c \times \bar{k} \times |d|}$, a convolution filter is also a tensor of weights, $F = (f_{ijk}) \in \mathbb{R}^{c \times \bar{k} \times m}$. The filter slides along the third dimension of T_d and results in a vector $\mathbf{e} \in \mathbb{R}^{|d|-m+1}$, each component of which is

$$e_i = (T_d * F)_i = \sum_{x=1}^{c} \sum_{y=1}^{\bar{k}} \sum_{z=i}^{i+m-1} t_{xyz} f_{xyz} + b,$$

---

[1] Note that for simplicity, we omit the j subscript of each word matrix X_j.

where $*$ denotes the convolution operator and $b \in \mathbb{R}$ is the bias added to the result of the convolution. Note that each e_i is the result of calculating an element-wise product between a slice of tensor T_d and the filter tensor F.

Above, we describe the process of computing a convolution between an input document tensor and a single filter, resulting in a single feature map. In practice, deep ConvNets apply a set of n filters, each of which works in parallel to generate multiple feature maps. The resulting filter bank $F \in \mathbb{R}^{n \times c \times \bar{k} \times m}$ then yields a set of feature maps of dimension $n \times (|d| - m + 1)$.

Pooling Operation. The output of the convolution layer is then passed to the pooling layer to gather information and reduce the representations. After the pooling layer, the resulting representation vectors are $\mathbf{e}' = [\text{pool}(\alpha(\mathbf{e_1})), \cdots, \text{pool}(\alpha(\mathbf{e_n}))]$, where $\mathbf{e_i}$ denotes the i-th convolutional feature map ($\mathbf{e_i} \in \mathbb{R}^{|d|-m+1}$).

There are two conventional choices for the function $\text{pool}(\cdot)$: average pooling and max pooling. Both operations map each feature map $\mathbf{e}$ to to a single value: $\text{pool}(\mathbf{e}) : \mathbb{R}^{|d|+m-1} \rightarrow \mathbb{R}$, as shown in Fig. 1. In this paper, the simple max-pooling method is considered, the goal of which is to capture the most representative feature – that with the highest value – for each feature map; by nature, this pooling scheme is suitable for documents of varying lengths.

3.2   Learning for Keyword Extraction

In the proposed model, one document model and one word model are built to map documents and candidate words to their distributional vector representations. Below, we briefly explain the concatenation layer that joins these intermediate representations in a single representation vector and then introduce the fully connected layer and the output layer.

Concatenation Layer. Since our architecture comprises two ConvNets, a document model, and a word model, two representation vectors $\mathbf{x}_{\text{doc}}$ and $\mathbf{x}_{\text{word}}$ are produced after the pooling operation; the concatenation layer then joins these vectors from both models end-to-end. In addition, in this layer, it is natural to add supplementary features $\mathbf{x}_{\text{sup}}$ such as the term frequency of the candidate word to the model; thus we have a single joined vector

$$\mathbf{x} = \mathbf{x}_{\text{doc}} \frown \mathbf{x}_{\text{word}} \frown \mathbf{x}_{\text{sup}}, \tag{2}$$

where $\frown$ denotes concatenation between two vectors, i.e., $\mathbf{a} \frown \mathbf{b} = (a_1, \ldots, a_k, b_1, \ldots, b_\ell)$.

Fully Connected Layer and Output Layer. The resulting joined vector from the above concatenation layer is then passed through a fully connected hidden layer that models the interactions between the components in the vector. The following transformation is then performed via the fully connected layer:

$$\mathbf{h} = \alpha(\mathbf{w_h} \cdot \mathbf{x} + b), \tag{3}$$

where $\mathbf{w_h}$ denotes the weight matrix of the layer, b is the bias term, and $\alpha(\cdot)$ is a non-linear activation function.

Finally the output layer is then fed to a softmax classification layer, the output layer. Since our problem is previously defined as a binary classification problem, the output layer of our classification model contains only two units and generates the probability distribution over the two class labels:

$$p(y = \ell | \mathbf{h}) = \frac{e^{\mathbf{h}^\mathsf{T} \delta_\ell}}{e^{\mathbf{h}^\mathsf{T} \delta_0} + e^{\mathbf{h}^\mathsf{T} \delta_1}}, \tag{4}$$

where $\ell \in \{0, 1\}$ and δ_ℓ denotes the weight vector of the ℓ class. Thus, the two output units of the network are the probabilities that the given word is a keyword of the document and that the word is *not* a keyword of the document.

3.3 Optimization

To recognize the keywords, charCNTN minimizes the objective C by:

$$C = -\log \prod_{i=1}^{N} p\left(y_i | d_i \in D, w \in W_{d_i}\right) = -\sum_{i=1}^{N} \left[y_i \log q_i + (1 - y_i) \log(1 - q_i)\right],$$

where q_i is the output of the softmax function defined in Eq. (4), and N denotes the number of training examples.

The above parameters can be optimized using stochastic gradient descent (SGD) with the backpropagation algorithm, by far the most common algorithm to optimize neural networks, central to many machine learning success stories. In order to improve the convergence rate of SGD, various efficient stochastic optimization techniques have been proposed, such as AdaGrad [3] and Adadelta [22]. In this paper, Adam [11], an algorithm for first-order gradient-based optimization of stochastic objective functions, is adopted to optimize the parameters.

4 Experiments

4.1 Datasets and Preprocessing

To conduct our experiments, we used two standard, publicly available datasets with different document sizes and types. **Inspec** [6], the first dataset, is a collection of 2,000 English abstracts of journal papers including their paper titles; the dataset is partitioned into three subsets: a training set, a validation set, and a testing set, each of which contains 1,000 abstracts, 500 abstracts, and the remaining 500 abstracts, respectively. In addition, the dataset provides two sets of keyphrases assigned by professional indexers: the controlled keyphrases and the uncontrolled keyphrases; following many previous studies (e.g., [1,13]), we treat the uncontrolled keyphrases as our ground truth in the following evaluation. **SemEval**, the second dataset, was compiled by [9] for the SemEval 2010 keyphrase extraction evaluation campaign. This dataset consists of 284 scientific articles from the ACM Digital Libraries, including conference and workshop papers; the total 284 documents are divided into 40 documents for trial, 144 for training, and the remaining 100 for test. Note that since the labeled keyphrases in the testing set are in their stemmed forms, for this dataset, we use trial data instead of the testing data as our testing set. Table 1 shows the statistics of the two data collections.

Table 1. Datasets

	Inspec		SemEval	
	Train	Test	Train	Test
Documents	1,000	500	144	40
Avg. length of documents	150	156	67	69
Unique words	9,258	5,660	2,047	1,047
Unique words (stemmed)	6,804	4,219	1,564	846
Unique keywords	6,377	3,792	1,310	680
Unique keywords (stemmed)	4,954	3,010	1,150	617
Unique keyphrases	-	4,913	-	621
Training instances	30,231	-	2,112	-

As the labeled answers of each document in the above two datasets are keyphrases, we divide the keyphrases into keywords and use these keywords as our ground truth for keyword extraction. Recall that in this paper, we focus on keyword extraction only and build keyphrases with a post-processing step. Furthermore, for both datasets, we use only the abstract parts in our experiments.

It is also worth mentioning that that the performance on these two datasets varies among many of the previous studies (e.g., [1,5,13,16]) due to the different text preprocessing steps adopted in their experiments. For fair comparison, all baselines and our methods preprocess the data per [16]: (1) Stop words from the SMART information retrieval system are removed;[2] (2) Only nouns and adjectives are kept; (3) The Porter stemmer is applied to both documents and ground truth keywords. Detailed statistics for the data after preprocessing are listed in Table 1.

---

[2] http://www.ai.mit.edu/projects/jmlr/papers/volume5/lewis04a/a11-smart-stop-list/english.stop.

4.2   Experimental Setup

Baseline Methods. Due to the commonly occurred "unknown word problem" in supervised learning methods, unsupervised learning is still considered as the main stream technique for the task of keyword extraction. As for the selection of our baseline methods, we consider the unsupervised method proposed by [16] as the state-of-the-art algorithm, which yields a set of strong baselines (i.e., CoreRank, dens, inf). Moreover, in this paper, we focus on CNN-based models due to its advantages in terms of simplicity and efficiency; therefore, in this stage, we do not compare our method with RNN-based models in the experiments. In sum, we compare our results with seven algorithms, including five unsupervised approaches (1)–(5) and two supervised approaches (6)–(7):

(1) **TF-IDF:** The first baseline is term frequency-inverse document frequency, which is a numerical statistic that reflects the importance of a word with respect to a document in a collection. In the experiments, we compute the TF-IDF of all candidate words, rank them with their TF-IDF values, and select the top one-third candidate words as our keywords.

(2) **TextRank:** This baseline is an unsupervised graph-based ranking model [13], one of the most commonly adopted keyword extraction methods. We apply TextRank to a directed, weighted text graph, and use co-occurrence links to express the relations between two words, which are created based on the co-occurrence of words within a window of size 2. As with the setting of the above TF-IDF, the top one-third highest-scoring candidate words are used as our keywords.

(3)–(5) **CoreRank, dens, inf:** The three baselines are the state-of-the-art unsupervised graph-based methods proposed by [16]. In this paper, we adopt as baselines their three best-performing settings: CoreRank, dens, and inf. And the settings are following the original paper.[3]

(6) **SVM:** The supervised approach proposed in [23] is chosen as a baseline in which keyword extraction is treated as a binary classification problem. A support vector machines (SVMs) are used to build the binary classifier with several handcrafted features are used to represent the data: TF-IDF, position, and first occurrence are global features, and part of speech and linkage are local features.

(7) **CNN:** We also compare the proposed charCNTN with a traditional word-level CNN model with a set of 300-dimensional word embeddings trained on Google News.[4] The convolution network architecture is the same as the proposed charCNTN except that the input document and word are represented by a matrix and a vector, respectively, and the size of the convolution filter is set to 300×3.

---

[3] These are the best models among different window sizes; we used the code at https://github.com/Tixierae/EMNLP_2016 to reproduce the experiments.

[4] https://code.google.com/archive/p/word2vec/.

Training Details. The parameters of the proposed charCNTN are listed as follows. The width of the convolution filters m is set to 3; the characters we consider are case-insensitive and are from a to z, resulting in $c = 26$. As the maximum length of characters in each word in Inspec and SemEval is 48 and 23, respectively; in our experiments, each word is padded to length 15 since for both datasets.[5] On the other hand, the maximum length of documents in Inspec and SemEval is 254 and 274, respectively, and each document is padded to length 300. The number of convolutional feature maps n is set to 50, which yields the best performance resulting from a grid search ranging from $n = 2$ to $n = 60$. A simple pooling layer is adopted, after which we apply a ReLU activation function. The size of the fully connected layer equals that of the joined vector $\mathbf{x}$ in Eq. (2) obtained after concatenating the word and document representation vectors from the distributional models plus additional features (if used).

To train the network we use stochastic gradient descent with mini-batches and a batch size of 64; the Adam algorithm is adopted to optimize the neural networks. Moreover, we randomly select 10% of the training data as our development set and stop early to avoid model overfitting; specifically, the training process stops when the loss on the development set no longer decreases. Also, to ensure balanced training data, we randomly select the non-keywords from each training set to equal the numbers of positive and negative examples. The numbers of training instances of Inspec and SemEval are 30,231 and 2,112, respectively (also listed in Table 1).

4.3 Keyphrase Extraction

We follow the method in [13] to reconstruct the keyphrases. All words selected as keywords using the TextRank algorithm or our charCNTN model are marked in the document, and sequences of adjacent keywords are collapsed into a multi-word keyphrase. For example, if both machine and learning are selected as keywords and the two words are adjacent in the given document, they are combined as the single keyphrase "machine learning". Note that in the following experiments, we only compare the results of keyphrase extraction with TextRank because only the TextRank paper deals with the keyphrase reconstruction; the CoreRank, dens, inf, and SVM papers all focus solely on keyword extraction.

4.4 Experimental Results

In the following experiments, we first compare the performance of the proposed charCNTN model with the seven aforementioned baseline algorithms, including five unsupervised approaches and two supervised approaches. In all experiments, the standard precision, recall, and F1 measures for each document are computed and are averaged at the dataset level (macro-averaging). As shown in the table, the simple TF-IDF method serves as a strong baseline, similar to the results

---

[5] Over 96%–97% words are with less or equal than 15 characters.

reported in [1]; also, the results of the other four unsupervised baselines are similar to the results in [16], which attests the validity of the results of the baselines.

Table 2. Keyword extraction

		Unsupervised methods					Supervised methods			
		TD-IDF[†]	TextRank[†]	CoreRank[†]	dens[†]	inf[†]	SVM[†]	CNN[†]	charCNTN	charCNTN+
Inspec	Precision	43.18	63.45	63.59	48.06	48.18	60.10	54.35	57.32	69.31
	Recall	84.90	39.96	39.98	72.29	72.51	56.96	55.21	69.25	68.08
	F1-measure	55.45	47.20	47.25	55.35	55.62	56.18	50.12	60.80*	**61.45***
SemEval	Precision	33.12	50.89	48.86	36.66	37.21	49.61	51.27	41.09	42.63
	Recall	50.81	25.21	24.28	43.20	42.41	30.49	28.79	42.27	39.72
	F1-measure	38.68	32.75	31.49	37.96	37.72	36.71	33.84	**40.02***	39.61*

† denotes baseline methods; * denotes the statistical significance at $p < 0.05$ with respect to all baselines.

For the proposed charCNTN, we also include the term frequency as a supplementary feature x_{sup} in the proposed model. From Table 2, we observe that the proposed method both without and with the additional term frequency feature (denoted as charCNTN and charCNTN+, respectively) outperforms all seven baseline methods in terms of the F1 metric. A closer look at the table reveals that our method yields greater improvements on Inspec than SemEval as the training set of Inspec is much bigger than that of SemEval.

It is also worth mentioning that performing stemming on the corpus and ground truth indeed makes the problem easier than the original one and thus in general results in better performance, though this is a commonly adopted approach in the related literature. As our method captures morphological information about words without the need for stemming, our model achieves comparable or even better performance than several baseline approaches with stemming. To attest this claim, we also conducted experiments on the two datasets without stemming; the F-measure scores of Inspec and SemEval for the proposed charCNTN were 59.02 and 37.34, respectively, which are still superior to most of the baseline methods with stemming, as shown in Table 2.

Table 3. Performance on unknown words and keyphrase extraction

		Unknown words				Keyphrase		
		SVM	CNN	charCNTN	charCNTN+	TextRank	charCNTN	charCNTN+
Inspec	Precision	38.52	35.86	42.78	42.41	30.40	24.43	23.43
	Recall	71.24	52.37	78.20	74.67	16.02	35.81	36.01
	F1-measure	50.01	42.57	**55.29**	54.05	19.91	**28.03**	27.37
SemEval	Precision	38.13	31.95	37.05	35.95	19.38	14.25	14.40
	Recall	41.22	27.84	71.16	61.31	6.05	17.15	18.77
	F1-measure	39.62	29.76	**48.62**	45.24	8.71	14.94	**15.62**

Additionally, to demonstrate the ability of our model to better handle unknown words, we designed an experiment involving only those words that appear in the testing documents but not in the training corpus; that is, we keep only the predicted keywords and target keywords that are unknown words when calculating the three evaluation metrics. The results listed in Table 3 show that the character-level information used in our model indeed greatly improves the F1 performance as compared with the other two supervised methods: SVM and the word-based CNN model.

Finally, in Table 3 we briefly compare our performance on keyphrase extraction with TextRank. Observe that our method yields much better performance than the commonly-adopted TextRank because both charCNTN and charCNTN+ retrieve more target keywords than TextRank (higher recall) but with low precision loss.

Remarks. Before designing the proposed architecture based on character-level convolutional neural tensor networks, we attempted to tackle the problem of keyword extraction by adopting an architecture similar to [15], using word embeddings to represent a document as an initial input of the networks. However, after experiments and network tweaking (the results of which are denoted as CNN and shown in Table 2), we realize that the poor performance of the traditional CNN method is due to the large number of unknown words in the testing data, many of which lack word embeddings even in the pretrained dataset. This inspired us to turn to the character-level based method, which captures morphological information about words and thus greatly mitigates the unknown word problem in supervised learning approaches for keyword extraction.

5   Conclusion and Future Work

We present a simple yet efficient supervised deep neural network architecture for keyword extraction, in which we model semantics down to the character level to capture morphological information about words; therefore, the proposed approach effectively mitigates the unknown word problem in supervised learning approaches. The experimental results show that the proposed charCNTN architecture significantly outperforms both state-of-the-art supervised and unsupervised approaches.

Although in this paper we use CNN rather than the recurrent neural network (RNN) due to its advantages in terms of simplicity and efficiency, there have been many attempts to adopt RNN models or combine RNN with CNN models to accomplish several other NLP and IR tasks [21]. In the future, we plan to investigate how to integrate the proposed charCNTN architecture with RNN models to better capture word sequence behavior and to incorporate attention mechanism to analyze for potential changes of word-document tensor interaction. Additionally, how to extend the proposed architecture to directly handle keyphrase extraction is also a challenging problem. The source codes are available at https://github.com/cnclabs/CNTN.

References

1. Bougouin, A., Boudin, F., Daille, B.: TopicRank: graph-based topic ranking for keyphrase extraction. In: Proceedings of the 6th International Joint Conference on Natural Language Processing, pp. 543–551 (2013)
2. Brin, S., Page, L.: The anatomy of a large-scale hypertextual web search engine. In: Proceedings of the 7th International Conference on World Wide Web, pp. 107–117 (1998)
3. Duchi, J., Hazan, E., Singer, Y.: Adaptive subgradient methods for online learning and stochastic optimization. J. Mach. Learn. Res. **12**(Jul), 2121–2159 (2011)
4. Gutwin, C., Paynter, G., Witten, I., Nevill-Manning, C., Frank, E.: Improving browsing in digital libraries with keyphrase indexes. Decis. Support Syst. **27**(1–2), 81–104 (1999)
5. Hasan, K.S., Ng, V.: Automatic keyphrase extraction: a survey of the state of the art. In: Proceedings of the 52nd Annual Meeting of the Association for Computational Linguistics (vol. 1: Long Papers), pp. 1262–1273 (2014)
6. Hulth, A.: Improved automatic keyword extraction given more linguistic knowledge. In: Proceedings of the 2003 Conference on Empirical Methods in Natural Language Processing, pp. 216–223 (2003)
7. Hulth, A., Megyesi, B.B.: A study on automatically extracted keywords in text categorization. In: Proceedings of the 21st International Conference on Computational Linguistics and the 44th Annual Meeting of the Association for Computational Linguistics, pp. 537–544 (2006)
8. Kalchbrenner, N., Grefenstette, E., Blunsom, P.: A convolutional neural network for modelling sentences. In: Proceedings of the 52nd Annual Meeting of the Association for Computational Linguistics (vol. 1: Long Papers), pp. 655–665 (2014)
9. Kim, S.N., Medelyan, O., Kan, M.Y., Baldwin, T.: Semeval-2010 task 5: automatic keyphrase extraction from scientific articles. In: Proceedings of the 5th International Workshop on Semantic Evaluation, pp. 21–26 (2010)
10. Kim, Y.: Convolutional neural networks for sentence classification. In: Proceedings of the 2014 Conference on Empirical Methods in Natural Language Processing, pp. 1746–1751 (2014)
11. Kingma, D., Ba, J.: Adam: a method for stochastic optimization. arXiv preprint arXiv:1412.6980 (2014)
12. Liu, Z., Li, P., Zheng, Y., Sun, M.: Clustering to find exemplar terms for keyphrase extraction. In: Proceedings of the 2009 Conference on Empirical Methods in Natural Language Processing, vol. 1, pp. 257–266 (2009)
13. Mihalcea, R., Tarau, P.: TextRank: bringing order into texts. In: Proceedings of the 2004 Conference on Empirical Methods in Natural Language Processing, pp. 404–411 (2004)
14. Santos, C.D., Zadrozny, B.: Learning character-level representations for part-of-speech tagging. In: Proceedings of the 31st International Conference on Machine Learning, pp. 1818–1826 (2014)
15. Severyn, A., Moschitti, A.: Learning to rank short text pairs with convolutional deep neural networks. In: Proceedings of the 38th International ACM SIGIR Conference on Research and Development in Information Retrieval, pp. 373–382 (2015)
16. Tixier, A.J.P., Malliaros, F.D., Vazirgiannis, M.: A graph degeneracy-based approach to keyword extraction. In: Proceedings of the 2016 Conference on Empirical Methods in Natural Language Processing, pp. 1860–1870 (2016)

17. Tomokiyo, T., Hurst, M.: A language model approach to keyphrase extraction. In: Proceedings of the ACL 2003 Workshop on Multiword Expressions: Analysis, Acquisition and Treatment, vol. 18, pp. 33–40 (2003)

18. Turney, P.D.: Learning algorithms for keyphrase extraction. Inf. Retr. **2**(4), 303–336 (2000)

19. Vosoughi, S., Vijayaraghavan, P., Roy, D.: Tweet2Vec: learning tweet embeddings using character-level CNN-LSTM encoder-decoder. In: Proceedings of the 39th International ACM SIGIR conference on Research and Development in Information Retrieval, pp. 1041–1044 (2016)

20. Witten, I.H., Paynter, G.W., Frank, E., Gutwin, C., Nevill-Manning, C.G.: KEA: practical automatic keyphrase extraction. In: Proceedings of the 4th ACM Conference on Digital Libraries, pp. 254–255 (1999)

21. Yin, W., Kann, K., Yu, M., Schütze, H.: Comparative study of CNN and RNN for natural language processing. arXiv preprint arXiv:1702.01923 (2017)

22. Zeiler, M.D.: Adadelta: an adaptive learning rate method. arXiv preprint arXiv:1212.5701 (2012)

23. Zhang, K., Xu, H., Tang, J., Li, J.: Keyword extraction using support vector machine. In: Yu, J.X., Kitsuregawa, M., Leong, H.V. (eds.) WAIM 2006. LNCS, vol. 4016, pp. 85–96. Springer, Heidelberg (2006). https://doi.org/10.1007/11775300_8

24. Zhang, X., Zhao, J., LeCun, Y.: Character-level convolutional networks for text classification. In: Advances in Neural Information Processing Systems, pp. 649–657 (2015)

25. Zhang, Y., Zincir-Heywood, N., Milios, E.: World wide web site summarization. Web Intell. Agent Syst. **2**(1), 39–53 (2004)

Neural Variational Matrix Factorization with Side Information for Collaborative Filtering

Teng Xiao[1] and Hong Shen[1,2](✉)

[1] School of Data and Computer Science, Sun Yat-sen University, Guangzhou, China
[2] School of Computer Science, The University of Adelaide, Adelaide, Australia
hong@cs.adelaide.edu.au

Abstract. Probabilistic Matrix Factorization (PMF) is a popular technique for collaborative filtering (CF) in recommendation systems. The purpose of PMF is to find the latent factors for users and items by decomposing a user-item rating matrix. Most methods based on PMF suffer from data sparsity and result in poor latent representations of users and items. To alleviate this problem, we propose the neural variational matrix factorization (NVMF) model, a novel deep generative model that incorporates side information (features) of both users and items, to capture better latent representations of users and items for the task of CF recommendation. Our NVMF consists of two end-to-end variational autoencoder neural networks, namely user neural network and item neural network respectively, which are capable of learning complex nonlinear distributed representations of users and items through our proposed variational inference. We derive a Stochastic Gradient Variational Bayes (SGVB) algorithm to approximate the intractable posterior distributions. Experiments conducted on three publicly available datasets show that our NVMF significantly outperforms the state-of-the-art methods.

Keywords: Collaborative filtering · Neural network · Matrix factorization · Deep generative process · Variational inference

1 Introduction

Recommendation system (RS) is of paramount importance in social networks and e-commerce platforms. RS aims at inferring users' preferences over items by utilizing their previous interactions. Traditional methods for RS can be categorized into two classes [27]: content-based methods and collaborative filtering (CF) ones. Content-based methods make use of users' and items' features and recommend items that are similar to the items that the users have liked before. CF methods utilize previous rating information to recommend items to users. Matrix factorization (MF) is the one of the most successful and popular CF

© Springer Nature Switzerland AG 2019
Q. Yang et al. (Eds.): PAKDD 2019, LNAI 11439, pp. 414–425, 2019.
https://doi.org/10.1007/978-3-030-16148-4_32

approaches that first infers users' and items' latent factors from a user-item rating matrix and then recommends items to users who share similar latent factors to the items [12]. However, traditional MF suffers from data sparsity problem. In order to alleviate the data sparsity problems in MF, hybrid methods that enjoy advantages of different categories of methods for CF have bee proposed. Some hybrid methods [1,10,18] take the advantages of both content-based and MF methods, and incorporate side information, such as demographics of a user, types of a item, etc., into MF. Regardless of their better performance compared to the traditional MF, we find most of them apply Gaussian or Poisson distributions to model the generative process of users' ratings, resulting in learning poor representations of users and items from complex data.

Recently, deep generative model has drawn significant attention and research efforts since the deep generative model have both non-linearity of neural network, and can get more robust and subtle latent representations of users and items due to it's Bayesian nature. Li et al. [15] proposed Collaborative Variational Autoencoder (CVAE), which utilizes VAE to extract latent item information and incorporates it into matrix factorization. Liang et al. [17] proposed VAE-CF model to the CF task. However, these VAE-based methods don't consider side information of both users and items. The choose the same Gaussian priors for different users and items, which is not unrealistic for different users and items. Accordingly, we solve the problems existing in the previous hybrid methods by proposing a novel neural variational matrix factorization (NVMF) for CF. We model the relationship of latent factor and side information by a novel deep generative process so as to enable it to effectively alleviate data sparsity problem and can learn better latent representations. In addition, our proposed model is a variant variational autoencoder for MF but differs itself from existing neural network hybrid methods by treating neural network as a full Bayesian probabilistic framework, resulting in the fact that it is able to enjoy the advantages of both deep learning and probabilistic matrix factorization to capture more subtle and complex latent representations of users and items. To sum up, our main contributions are as follows:

(1) We proposed a novel neural variational matrix factorization (NVMF) model to effectively learn nonlinear latent representations of users and items with side information for collaborative filtering.
(2) We derived tractable variational evidence lower bounds for our proposed model and devised a neural network to infer latent factors of users and items.
(3) We systematically conducted experiments to show that the proposed NVMF model outperforms state-of-the-art CF methods.

2   Notation and Problem Definition

Similar to recent recommendation methods [15,25], we focus on implicit feedback in our paper. Let $R \in \{0,1\}^{M \times N}$ be the user-item feedback matrix with M and

N being the total number of users and items, respectively. $\boldsymbol{F} \in \mathbb{R}^{P \times M}$ and $\boldsymbol{G} \in \mathbb{R}^{Q \times N}$ are the side information matrix of all users and items, respectively, with P and Q being the dimensions of each user's and item's side information, respectively; $\boldsymbol{U} = [\boldsymbol{u}_1, \ldots \boldsymbol{u}_M] \in \mathbb{R}^{D \times M}$ and $\boldsymbol{V} = [\boldsymbol{v}_1, \ldots \boldsymbol{v}_N] \in \mathbb{R}^{D \times N}$ are the two rank matrices serving for users and items, respectively, with D denoting the dimensions of latent factor space. For convenient discussion, we represent each user i's rating scores including the missing/unobserved ones over all items as $\boldsymbol{s}_i^u = [R_{i1}, \ldots, R_{iN}] \in \mathbb{R}^{N \times 1}$, where R_{ij} is an element in $\boldsymbol{R}$. Similarly, we represent each item j's rating scores from all users including those who do not provide rating for j as $\boldsymbol{s}_j^v = [R_{1j}, \ldots, R_{Mj}] \in \mathbb{R}^{M \times 1}$. We call $\boldsymbol{s}_i^u$ and $\boldsymbol{s}_j^v$ as the collaborative information of user i and item j, respectively. The problem we address in this paper is to infer the posterior latent factors of users and items and predict the missing value in user-item feedback matrix $\boldsymbol{R}$ given $\boldsymbol{R}$, $\boldsymbol{F}$ and $\boldsymbol{G}$.

3   Neural Variational Matrix Factorization

3.1   Neural Variational Matrix Factorization

Unlike any other probabilistic matrix factorization that directly utilizes the rating value R_{ij} in user-item matrix $\boldsymbol{R}$ to infer users' and items' latent factors, we assume the i-th user latent factor $\boldsymbol{u}_i$ can be jointly inferred by both the user's rating history on all items $\boldsymbol{s}_i^u$ (the collaborative information of user i) and the user's features $\boldsymbol{f}_i$. Similarly, the j-th item latent factor $\boldsymbol{v}_j$ can be jointly inferred by both all users' rating history on the item $\boldsymbol{s}_j^v$ (the collaborative information of item j) and its own features $\boldsymbol{g}_j$. Under these assumptions, we first model the features of users and items through latent variable model. Although we do not know the real distributions of user features and item features, we know that any distribution can be generated by mapping the standard Gaussian through a sufficiently complicated function [5]. Thus for user i, given a standard Gaussian latent variable $\boldsymbol{b}_i$ assigned to the user, his features $\boldsymbol{f}_i$ are generated from its latent variable $\boldsymbol{b}_i$ through a neural network, which is called "user generative network" (see Fig. 1), and are governed by the parameter θ in the network such that we have:

$$\boldsymbol{b}_i \sim \mathcal{N}(0, \boldsymbol{I}_{K_b}), \quad \boldsymbol{f}_i \sim p_\theta(\boldsymbol{f}_i | \boldsymbol{b}_i), \tag{1}$$

where $\boldsymbol{I}_{K_b}$ is the covariance matrix, K_b is the dimension of $\boldsymbol{b}_i$, and the specific form of the probability of generating $\boldsymbol{f}_i$ given $\boldsymbol{b}_i$, $p_\theta(\boldsymbol{f}_i | \boldsymbol{b}_i)$, depends on the type of data. For instance, if $\boldsymbol{f}_i$ is binary vector, $p_\theta(\boldsymbol{f}_i | \boldsymbol{b}_i)$ can be a multivariate Bernoulli distribution $Ber(F_\theta(\boldsymbol{b}_i))$ with $F_\theta(\cdot)$ being the highly no-linear function parameterized by the parameter θ in the network. Similarly, for item j, it's features $\boldsymbol{g}_j$ are modeled to be generated from a standard Gaussian latent variable $\boldsymbol{d}_j$ through another generation network, which is called "item generative network" (see Fig. 1), and are governed by a parameter τ in the network such that we have:

$$\boldsymbol{d}_j \sim \mathcal{N}(0, \boldsymbol{I}_{K_d}), \quad \boldsymbol{g}_j \sim p_\tau(\boldsymbol{g}_j | \boldsymbol{d}_j), \tag{2}$$

where I_{K_d} is the covariance matrix and K_d is the dimension of d_j.

Traditional PMF assumes the prior distributions of user latent factor u_i and item latent factor v_j are standard Gaussian distributions and predict rating only through collaborative information such as the user-item feedback matrix. In our model, to further enhance the performance, besides the collaborative information, we believe the user' features f_i can also positively contribute to the inference of his latent factor u_i. Similarly, for better inferring the j-th item's latent factor v_j, we also fully utilize user's features g_j. Unlike most MF methods [10,18] that incorporate side information via linear regression, in order to get more subtle latent relations, we consider the conditional prior $p(u_i|f_i)$ and $p(v_j|g_j)$ are Gaussian distributions such that we have $p(u_i|f_i) = \mathcal{N}(\mu_u(f_i), \Sigma_u(f_i))$ and $p(v_j|g_j) = \mathcal{N}(\mu_v(g_j), \Sigma_v(g_j))$, where

$$\mu_u(f_i) = F_{\mu_u}(f_i), \qquad \Sigma_u(f_i) = \mathrm{diag}(\exp(F_{\delta_u}(f_i))), \qquad (3)$$
$$\mu_v(g_j) = G_{\mu_v}(g_j), \qquad \Sigma_v(g_j) = \mathrm{diag}(\exp(G_{\delta_v}(g_j))), \qquad (4)$$

where $F_{\mu_u}(\cdot)$, $F_{\delta_u}(\cdot)$, are the two highly non-linear functions parameterized by μ_u and δ_u in the neural network, i.e., the user prior network, serving for all users, and $G_{\mu_v}(\cdot)$ and $G_{\delta_v}(\cdot)$ are the two non-linear ones parameterized by μ_v and δ_v in another neural network, i.e., the item prior network, serving for all items, respectively. For simplicity, note that we set $\gamma = \{\mu_u, \delta_u\}$ and $\psi = \{\mu_v, \delta_v\}$.

For the collaborative information of user i (s_i^u), we assign a standard Gaussian latent variable a_i to it and believe user latent factor u_i can potentially affect user collaborative information. Then we consider s_i^u is generated from both a standard Gaussian latent variable a_i and user latent factor u_i, and is governed by the parameter α in the generative network (see Fig. 1 and the caption in the figure) such that we have:

$$a_i \sim \mathcal{N}(0, I_{K_a}), \quad s_i^u \sim p_\alpha(s_i^u|a_i, u_i), \qquad (5)$$

where I_{K_a} is the covariance matrix and K_a is the dimension of a_i. Similarly, the j-th item's collaborative information, s_j^v, is generated from it's standard Gaussian latent variable c_j and item latent factor v_j, and is governed by the parameter β in the generative network such that we have:

$$c_j \sim \mathcal{N}(0, I_{K_c}), \quad s_j^v \sim p_\beta(s_j^v|c_j, v_j), \qquad (6)$$

where I_{K_c} is the covariance matrix and K_c is the dimension of c_j. Similar to the form of the probability distribution, $p_\theta(f_i|b_i)$, in Eq. 1, the specific forms of the probability distributions in Eqs. 2, 5 and 6 depend on the type of data. The rating R_{ij} is drawn from the Gaussian distribution whose mean is the inner product of the user i and item j latent factor representations such that we have:

$$p(R_{ij}|u_i, v_j) = \mathcal{N}(u_i^\top v_j, C_{ij}^{-1}). \qquad (7)$$

where C_{ij}^{-1} is the precision of Gaussian distribution, and similar to the collaborative topic modeling [25], C_{ij} serves as a confidence parameter for rating R_{ij}, which is defined as:

$$C_{ij} = \begin{cases} \varphi_1 & \text{if } R_{ij} \neq 0, \\ \varphi_2 & \text{if } R_{ij} = 0, \end{cases} \tag{8}$$

where φ_1 and φ_2 are the parameters satisfying $\varphi_1 > \varphi_2 > 0$, the basic reason behind which is that if $R_{ij} = 0$ it means the user i is not interested in the item j or the user i is unware of it. According to the generative process of NVMF, the joint distribution of NVMF can be factorized as:

$$p(\boldsymbol{R}, \boldsymbol{F}, \boldsymbol{G}, \mathcal{Z}) = \prod_{i=1}^{M} \prod_{j=1}^{N} \underbrace{p(\boldsymbol{a}_i)p(\boldsymbol{b}_i)p(\boldsymbol{f}_i|\boldsymbol{b}_i)p(\boldsymbol{u}_i|\boldsymbol{f}_i)p(\boldsymbol{s}_i^u|\boldsymbol{a}_i, \boldsymbol{u}_i)}_{\text{for users}} \cdot$$

$$\underbrace{p(\boldsymbol{c}_j)p(\boldsymbol{d}_j)p(\boldsymbol{g}_j|\boldsymbol{d}_j)p(\boldsymbol{v}_j|\boldsymbol{g}_j)p(\boldsymbol{s}_j^v|\boldsymbol{c}_j, \boldsymbol{v}_j)}_{\text{for items}} \, p(R_{ij}|\boldsymbol{u}_i, \boldsymbol{v}_j). \tag{9}$$

Instead of inferring the joint distribution, i.e., Eq. 9, we are more interested in approximately inferring its posterior distributions over users' and items' factor matrices, $\boldsymbol{U}$, $\boldsymbol{V}$. Let $\mathcal{Z} = \{\boldsymbol{U}, \boldsymbol{V}, \boldsymbol{B}, \boldsymbol{C}, \boldsymbol{D}\}$ be a set of all latent variables in Eq. 9 that need to be inferred, and $\mathcal{Z}_{ij} = \{\boldsymbol{u}_i, \boldsymbol{v}_j, \boldsymbol{a}_i, \boldsymbol{b}_i, \boldsymbol{c}_j, \boldsymbol{d}_j\}$. However, it is difficult to infer $\mathcal{Z}$ by using traditional mean-field approximation since we do not have any conjugate probability distribution in our model which requires by traditional mean-field approachs. Inspired by VAE [11], we use Stochastic Gradient Variational Bayes (SGVB) estimator to approximate posteriors of the latent variables related to user ($\boldsymbol{a}_i$, $\boldsymbol{b}_i$, $\boldsymbol{u}_i$) and latent variables related to item ($\boldsymbol{c}_j$, $\boldsymbol{d}_j$, $\boldsymbol{v}_j$) by introducing two inference networks, i.e., the user inference network and the item inference network (see Fig. 1), parameterized by ϕ and λ, respectively. To do this, we first decompose the variational distribution q into two categories of variational distributions used in the two networks in our NVMF model—user inference network and item inference network (see Fig. 1), q_ϕ and q_λ, by assuming the conditional independence:

$$q(\mathcal{Z}_{ij}|\mathcal{X}_i, \mathcal{Y}_j, R_{ij}) = \underbrace{q_\phi(\boldsymbol{u}_i|\mathcal{X}_i)q_\phi(\boldsymbol{a}_i|\mathcal{X}_i)q_\phi(\boldsymbol{b}_i|\mathcal{X}_i)}_{\text{for users}} \cdot \underbrace{q_\lambda(\boldsymbol{v}_j|\mathcal{Y}_j)q_\lambda(\boldsymbol{c}_j|\mathcal{Y}_j)q_\lambda(\boldsymbol{d}_j|\mathcal{Y}_j)}_{\text{for items}},$$

$$\tag{10}$$

where $\mathcal{X}_i = (\boldsymbol{s}_i^u, \boldsymbol{f}_i)$ represents the set of user observed variables, and $\mathcal{Y}_j = (\boldsymbol{s}_j^v, \boldsymbol{g}_j)$ represents the set of item observed variables.

Like VAE [11], the variation distributions are chosen to be a Gaussian distribution $\mathcal{N}(\boldsymbol{\mu}, \boldsymbol{\Sigma})$, whose mean $\boldsymbol{\mu}$ and covariance matrix $\boldsymbol{\Sigma}$ are the output of the inference network. Thus, in our NVMF, for latent variables related to the i-th user, we set:

$$q_\phi(\boldsymbol{u}_i|\mathcal{X}_i) = \mathcal{N}(\mu_{\phi_{u_i}}(\mathcal{X}_i), \text{diag}(\exp(\delta_{\phi_{u_i}}(\mathcal{X}_i)))), \tag{11}$$

$$q_\phi(\boldsymbol{a}_i|\mathcal{X}_i) = \mathcal{N}(\mu_{\phi_{a_i}}(\mathcal{X}_i), \text{diag}(\exp(\delta_{\phi_{a_i}}(\mathcal{X}_i)))), \tag{12}$$

$$q_\phi(\boldsymbol{b}_i|\mathcal{X}_i) = \mathcal{N}(\mu_{\phi_{b_i}}(\mathcal{X}_i), \text{diag}(\exp(\delta_{\phi_{b_i}}(\mathcal{X}_i)))), \tag{13}$$

where the subscripts of μ and δ indicate the parameters in our user inference network corresponding to u_i, a_i and b_i, respectively. Similarly, for j-th item:

$$q_\lambda(v_j|\mathcal{Y}_j) = \mathcal{N}(\mu_{\lambda_{v_j}}(\mathcal{Y}_j), \text{diag}(\exp(\delta_{\lambda_{v_j}}(\mathcal{Y}_j)))), \tag{14}$$

$$q_\lambda(c_j|\mathcal{Y}_j) = \mathcal{N}(\mu_{\lambda_{c_j}}(\mathcal{Y}_j), \text{diag}(\exp(\delta_{\lambda_{c_j}}(\mathcal{Y}_j)))), \tag{15}$$

$$q_\lambda(d_j|\mathcal{Y}_j) = \mathcal{N}(\mu_{\lambda_{d_j}}(\mathcal{Y}_j), \text{diag}(\exp(\delta_{\lambda_{d_j}}(\mathcal{Y}_j)))), \tag{16}$$

where the subscripts of μ and δ indicate the parameters in item inference network corresponding to v_j, c_j and d_j, respectively.

Thus, the tractable standard evidence lower bound (ELBO) for the inference can be computed as follows:

$$\mathcal{L}(q) = \mathbb{E}_q[\log p(\mathcal{O}, \mathcal{Z}) - \log q(\mathcal{Z}|\mathcal{O})]$$

$$= \sum_{i=1}^{M} \sum_{j=1}^{N} (\mathcal{L}_i(q_\phi) + \mathcal{L}_j(q_\lambda) + \mathbb{E}_q[\log p(R_{ij}|u_i, v_j)]), \tag{17}$$

where $\mathcal{O} = (F, G, R)$ is a set of all observed variables. q_ϕ and q_λ are user term and item term in Eq. 10, respectively. For user i and item j, we have:

$$\mathcal{L}_i(q_\phi) = \mathcal{L}(\phi, \alpha, \theta, \gamma; \mathcal{X}_i) = \mathbb{E}_{q_\phi(a_i, u_i|\mathcal{X}_i)}[\log p_\alpha(s_i^u|a_i, u_i)]$$

$$+ \mathbb{E}_{q_\phi(b_i|\mathcal{X}_i)}[\log p_\theta(f_i|b_i)] - \text{KL}(q_\phi(a_i|\mathcal{X}_i)\|p(a_i)) \tag{18}$$

$$- \text{KL}(q_\phi(b_i|\mathcal{X}_i)\|p(b_i)) - \omega_1 \text{KL}(q_\phi(u_i|\mathcal{X}_i)\|p_\gamma(u_i|f_i)),$$

$$\mathcal{L}_j(q_\lambda) = \mathcal{L}(\lambda, \beta, \tau, \psi; \mathcal{Y}_j) = \mathbb{E}_{q_\lambda(c_j, v_j|\mathcal{Y}_j)}[\log p_\beta(s_j^v|c_j, v_j)]$$

$$+ \mathbb{E}_{q_\lambda(d_j|\mathcal{Y}_j)}[\log p_\tau(g_j|d_j)] - \text{KL}(q_\lambda(c_j|\mathcal{Y}_j)\|p(c_j)) \tag{19}$$

$$- \text{KL}(q_\lambda(d_j|\mathcal{Y}_j)\|p(d_j)) - \omega_2 \text{KL}(q_\lambda(v_j|\mathcal{Y}_j)\|p_\psi(v_j|g_j)),$$

where in the standard ELBO, the free parameters ω_1 and ω_2 are 1, and $\text{KL}(q_\phi(u_i|\mathcal{X}_i)\|p_\gamma(u_i|f_i))$ is the Kullback-Leibler divergence between the approximate posterior distribution $q_\phi(u_i|\mathcal{X}_i)$ and the prior $p_\gamma(u_i|f_i)$. The variational distribution $q_\phi(u_i|\mathcal{X}_i)$ acts as an approximation to the true posterior $p(u_i|\mathcal{O})$ when maximizing Eq. 18. Similarly, $q_\lambda(v_j|\mathcal{Y}_j)$ acts as an approximation to the true posterior $p(v_j|\mathcal{O})$ when maximizing Eq. 19. Inspired by β-VAE and the previous work [9,17], we use two free trade-off parameters ω_1 and ω_2 for the last terms in Eqs. 18 and 19, respectively, in the ELBO to control the KL regularization instead of directly applying $\omega_1 = \omega_2 = 1$ (other KL terms in Eqs. 18 and 19 do not need to apply the trade-off parameters as they are not related to our final latent factors u_i and v_j). Since we assume the posterior is Gaussian distribution, the KL terms in Eqs. (18) and (19) have analytical forms. However, for the expectations terms, we can not to compute them analytically. To handle this problem, we use Monte Carlo method [20] to approximate the expectations by drawing samples from the posterior distribution. By using the reparameterization trick

[20], the ELBO for user network is given:

$$\mathcal{L}(\phi, \alpha, \theta, \gamma; \mathcal{X}_i) \approx \frac{1}{K} \sum_{k=1}^{K} (\log p_\alpha(\boldsymbol{s}_i^u | \boldsymbol{a}_i^k, \boldsymbol{u}_i^k) + \log p_\theta(\boldsymbol{f}_i | \boldsymbol{b}_i^k)) \tag{20}$$

$$- \text{KL}(q_\phi(\boldsymbol{a}_i | \mathcal{X}_i) || p(\boldsymbol{a}_i)) - \text{KL}(q_\phi(\boldsymbol{b}_i | \mathcal{X}_i) || p(\boldsymbol{b}_i)) - \omega_1 \text{KL}(q_\phi(\boldsymbol{u}_i | \mathcal{X}_i) || p_\gamma(\boldsymbol{u}_i | \boldsymbol{f}_i)),$$

where K is the size of the samplings, $\boldsymbol{a}_i^k = \boldsymbol{\mu}_a + \boldsymbol{\delta}_a \odot \boldsymbol{\epsilon}_a^k$, $\boldsymbol{b}_i^k = \boldsymbol{\mu}_b + \boldsymbol{\delta}_b \odot \boldsymbol{\epsilon}_b^k$, $\boldsymbol{u}_i^k = \boldsymbol{\mu}_u + \boldsymbol{\delta}_u \odot \boldsymbol{\epsilon}_u^k$, $\odot$ is an element-wise multiplication and $\boldsymbol{\epsilon}_a^k, \boldsymbol{\epsilon}_b^k, \boldsymbol{\epsilon}_u^k$ are samples drawn from standard multivariate normal distribution. The superscript k denotes the k-th sample. The ELBO for item network, $\mathcal{L}(\lambda, \beta, \tau, \psi; \mathcal{Y}_j)$, can be derived similarly, and thus we omit it here.

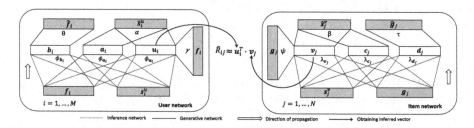

Fig. 1. The architecture of NVMF (shaded rectangles indicate observed vectors). The left part and right part are user network and item network which extract latent factors for users and items, respectively.

3.2   Optimization

Since minimizing the objection function is equivalent to maximizing the log likelihood of the observed data. Based on $\mathcal{L}(\phi, \alpha, \theta, \gamma; \mathcal{X}_i)$ in Eq. 21 and $\mathcal{L}(\lambda, \beta, \tau, \psi; \mathcal{Y}_j)$, the objective function is:

$$\mathcal{L} = -\sum_{i=1}^{M} \sum_{j=1}^{N} (\mathcal{L}(\phi, \alpha, \theta, \gamma; \mathcal{X}_i) + \mathcal{L}(\lambda, \beta, \tau, \psi; \mathcal{Y}_j)$$

$$+ \frac{C_{ij}}{2} \mathbb{E}_{q_\phi(\boldsymbol{u}_i | \mathcal{X}_i) q_\lambda(\boldsymbol{v}_j | \mathcal{Y}_j)} [(R_{ij} - \boldsymbol{u}_i^\top \boldsymbol{v}_j)^2]), \tag{21}$$

where the expectation term is given by:

$$\mathbb{E}_{q_\phi(\boldsymbol{u}_i | \mathcal{X}_i) q_\lambda(\boldsymbol{v}_j | \mathcal{Y}_j)} [(R_{ij} - \boldsymbol{u}_i^\top \boldsymbol{v}_j)^2] = R_{ij}^2 - 2R_{ij} \mathbb{E}[\boldsymbol{u}_i]^\top \mathbb{E}[\boldsymbol{v}_j]$$

$$+ \text{tr}((\mathbb{E}[\boldsymbol{v}_j] \mathbb{E}[\boldsymbol{v}_j]^\top + \Sigma_v) \Sigma_u) + \mathbb{E}[\boldsymbol{u}_i]^\top (\mathbb{E}[\boldsymbol{v}_j] \mathbb{E}[\boldsymbol{v}_j]^\top + \Sigma_v) \mathbb{E}[\boldsymbol{u}_i], \tag{22}$$

where $\text{tr}(\cdot)$ denotes the trace of a matrix. Since NVMF is a fully end-to-end neural network, the whole parameters of the model are the weight matrix of entire network, we can use back-propagation algorithm to optimize the weights of the user network and the item network.

3.3 Prediction

After the training is converged, we can get the posterior distributions of u_i and v_j through the user and item inference networks, respectively. So the prediction R_{ij} can be made by:

$$\mathbb{E}[R_{ij}|\mathcal{O}] = \mathbb{E}[u_i|\mathcal{O}]^\top \mathbb{E}[v_j|\mathcal{O}] \tag{23}$$

4 Experimental Setup

4.1 Dataset

We use three benchmark datasets in our experiments which are commonly used to previous recommendation algorithms.

MovieLens-100K[1] (ML-100K): Similar to [6,14], we extract the features of users and movies provided by the dataset to construct our side information matrices F and G respectively. The user's feature contains user's id, age, gender, occupation and zipcode, correspondingly the movie's feature contains movie's title, release data and 18 categories of movie genre.

MovieLens-1M[2] (ML-1M): This dataset is also a movie rating dataset. Similar to ML-100K, we can get side information of users and items.

Bookcrossing[3]: For this dataset, we also extract the features of users and book provided by the dataset. We encoded the user and book feature into binary vector of length 30 and 30 respectively.

Since we will evaluate our model performance on implicit feedback. Thus, following to [6,14], we interpret three datasets above as implicit feedback.

4.2 Baselines and Experimental Settings

For implicit feedback, as demonstrated in [15,26], the hybrid collaborative filtering model incorporating side information outperforms the other method without side information. So the most baselines we choose are hybrid models. The baselines we use to compare our proposed method are listed as follows:

(1) **CMF** [23]: This model is MF model which decompose the user-item matrix R, user's side information matrix F, and item's side information matrix G to get the consistent latent factor of user and item.
(2) **CDL** [26]: This model is a hierarchical Bayesian model for joint learning of stacked denoising autoencoder and collaborative filtering.
(3) **CVAE** [15]: This model is a Bayesian probabilistic model which unifies item feature and collaborative filtering through stochastic deep learning and probabilistic model.

---

[1] https://grouplens.org/datasets/movielens/100k/.
[2] https://grouplens.org/datasets/movielens/1m/.
[3] https://grouplens.org/datasets/book-crossing/.

(4) **aSDAE** [6]: This model is a deep learning model which can integrate the side information into matrix factorization via additional stack denoising autoencoder.

(5) **NeuMF** [8]: This model is a state-of-the-art collaborative filtering method for implicit feedback.

Since the implicit feedback matrix $R \in \{0, 1\}^{M \times N}$, we set $p_\theta(f_i|b_i)$ and $p_\theta(g_j|c_j)$ as multivariate Bernoulli distribution. We set ω_1, ω_2 and the learning rate η to 0.05, 0.05 and 0.0001. The value of φ_1, φ_2 and λ_w are set to 1, 0.01 and 0.0001, respectively. The dimensions K_a, K_b, K_c, K_d are all chosen to be 20. The inference and generative networks are both two-layer network architectures and the last layer of generative network is a softmax layer. Similar to explicit feedback, the prior network of user and item are both set to one layer. We also set the dimensions of use and item latent factor D as 30. We use 80% ratings of dataset to train each method, the 10% as validation, the rest 10% for testing. We repeat this procedure five times and report the average performance.

4.3   Evaluation Metrics

For evaluation, we use the Hit Ratio (HR) [16] and the Normalized Discounted Cumulative Gain (NDCG) [8] as our evaluation as metrics. For each user, we sort the top-K items based on the predicted ratings. We report the average recall scores over all users in our experimental analysis.

5   Results and Analysis

Table 1 shows the experiment result that compare CMF, CDL, CVAE, aSDAE and NeuMF using three datasets. As we can see, our proposed methods NVMF significantly outperform the compared methods in all cases on both ML-100K, ML-1M and BookCrossing datasets. Compared with other methods that have a deep learning structure, CMF achieves the worst performance. This demonstrates the deep learning structure can learn more subtle and complex representation than the tradition MF method. We also observe that although CDL and CVAE both have a deep learning structure, CVAE achieves better performance than CDL. This is because CDL is based on denoising autoencoder which can be seen as point estimation, however CVAE is fully deep probabilistic model which make it hard to overfit the data. From Table 1 We observe the strongest baseline in our experiment is aSDAE which outperforms than other baselines. Although aSDAE is no a probabilistic model, it incorporates user's side information (user's features) into matrix factorization which CDL and CVAE don not. By incorporating both user's feature and item's feature and applying deep generative model, our model NVMF outperform all the baselines. Specifically, the average improvements of NVMF over the state-of-the-art method. Compared to other datasets, our model has the greatest performance improvement on Bookcrossing which is most sparse matrix among three datasets. This shows

NVMF can more effectively alleviate the sparsity problem on implicit feedback than aSDAE. Figure 2 shows the contours of NDCG@5 for NVMF on three dataset. When the parameters equals 1, i.e $\omega_1 = 1$ and $\omega_2 = 1$, it means we directly optimize the standard ELBO -which has a degraded performance and this confirmed in Fig. 2(b). When we decrease ω_1 to 0.1 at fixed ω_2, we find NVMF's performance improves(small ω_1 implies we want condense more user's collaborative information into user's latent factor). Similar observation can be made for varying ω_2 at fixed ω_1. Moreover, It can be observed that there is a region of values of ω_1 and ω_2 (near (0.1,0.1)), around which NVMF provides the best performance in terms recall. Altogether, Fig. 2 shows treating ω_1 and ω_2 as trade-off parameters can yields significant improvement in performance of the recommendation.

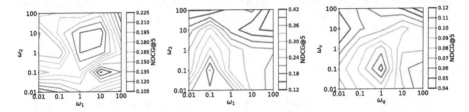

Fig. 2. The NDCG@5 for NVMF by varying ω_1 and ω_2 on three datasets (ML-100K, ML-1M and BookCrossing).

Table 1. Recommendation performance comparison between our NVMF and baselines.

Datasets	Metrics	CMF	CDL	CVAE	aSDAE	NeuMF	NVMF (ours)
ML-100K	HR@5	0.4121	0.4564	0.4721	0.4981	0.4942	**0.5083**
	NDCG@5	0.2124	0.2991	0.3012	0.3156	0.3351	**0.3417**
	HR@10	0.5587	0.6123	0.6421	0.6871	0.6692	**0.6982**
	NDCG@10	0.3387	0.3654	0.3871	0.4231	0.4103	**0.4358**
ML-1M	HR@5	0.4237	0.5011	0.5141	0.5411	0.5211	**0.5681**
	NDCG@5	0.2578	0.3362	0.3621	0.4124	0.4011	**0.4325**
	HR@10	0.5921	0.6557	0.6874	0.7321	0.7202	**0.7412**
	NDCG@10	0.3328	0.3547	0.3864	0.4121	0.4025	**0.4205**
Bookcrossing	HR@5	0.1565	0.1714	0.1921	0.2234	0.2123	**0.2354**
	NDCG@5	0.0523	0.0717	0.0921	0.1121	0.1024	**0.1244**
	HR@10	0.2347	0.2654	0.2876	0.3097	0.3012	**0.3142**
	NDCG@10	0.1024	0.1451	0.1612	0.1911	0.1876	**0.2087**

6   Conclusion

In this paper, we study the problem of how to learn subtle and robust latent factors from feedback matrix with side information for collaborative filtering. We propose neural variational matrix factorization—a novel deep generative model to learn the latent factors of users and items. NVMF incorporates features of users and items into matrix factorization through a novel generative process, which enables it to effectively handle data sparsity and cold start problems. To infer our model, we derived a variational lower bound and devised fully end-to-end network architectures so that back-propagation can be applied for efficient parameter estimation. Experiments conducted with implicit feedbacks have demonstrated the effectiveness of the learned latent factors by NVMF.

Acknowledgement. This work is supported by the National Key Research and Development Program of China (No. #2017YFB0203201) and Australian Research Council Discovery Project DP150104871.

References

1. Adams, R.P., Dahl, G.E., Murray, I.: Incorporating side information in probabilistic matrix factorization with Gaussian processes. arXiv (2010)
2. Agarwal, D., Chen, B.C.: Regression-based latent factor models. In: KDD, pp. 19–28 (2009)
3. Bowman, S.R., Vinis, L., Vinyals, O., Dai, A., Jozefowicz, R., Bengio, S.: Generating sentences from a continuous space. In: Proceedings of The 20th SIGNLL Conference on Computational Natural Language Learning, pp. 10–21 (2016)
4. Chen, Y., de Rijke, M.: A collective variational autoencoder for top-n recommendation with side information. In: Proceedings of the 3rd Workshop on Deep Learning for Recommender Systems, pp. 3–9. ACM (2018)
5. Doersch, C.: Tutorial on variational autoencoders (2016)
6. Dong, X., Yu, L., Wu, Z., Sun, Y., Yuan, L., Zhang, F.: A hybrid collaborative filtering model with deep structure for recommender systems. In: AAAI, pp. 1309–1315 (2017)
7. Goodfellow, I., et al.: Generative adversarial nets. In: NIPS, pp. 2672–2680 (2014)
8. He, X., Liao, L., Zhang, H., Nie, L., Hu, X., Chua, T.S.: Neural collaborative filtering. In: WWW, pp. 173–182 (2017)
9. Higgins, I., et al.: beta-VAE: learning basic visual concepts with a constrained variational framework (2016)
10. Kim, Y.D., Choi, S.: Scalable variational Bayesian matrix factorization with side information, pp. 493–502 (2014)
11. Kingma, D.P., Welling, M.: Auto-encoding variational bayes. arXiv preprint arXiv:1312.6114 (2013)
12. Koren, Y., Bell, R., Volinsky, C.: Matrix factorization techniques for recommender systems. Computer (2009)
13. Lee, D.D., Seung, H.S.: Algorithms for non-negative matrix factorization. In: NIPS, pp. 556–562 (2001)
14. Li, S., Kawale, J., Fu, Y.: Deep collaborative filtering via marginalized denoising auto-encoder. In: CIKM, pp. 811–820 (2015)

15. Li, X., She, J.: Collaborative variational autoencoder for recommender systems. In: KDD, pp. 305–314 (2017)
16. Liang, D., Charlin, L., McInerney, J., Blei, D.M.: Modeling user exposure in recommendation. In: Proceedings of the 25th International Conference on World Wide Web, pp. 951–961 (2016)
17. Liang, D., Krishnan, R.G., Hoffman, M.D., Jebara, T.: Variational autoencoders for collaborative filtering. arXiv (2018)
18. Park, S., Kim, Y.D., Choi, S.: Hierarchical Bayesian matrix factorization with side information. In: IJCAI, pp. 1593–1599 (2013)
19. Porteous, I., Asuncion, A., Welling, M.: Bayesian matrix factorization with side information and Dirichlet process mixtures. In: AAAI, pp. 563–568 (2010)
20. Rezende, D.J., Mohamed, S., Wierstra, D.: Stochastic backpropagation and approximate inference in deep generative models. arXiv (2014)
21. Salakhutdinov, R., Mnih, A.: Bayesian probabilistic matrix factorization using Markov chain Monte Carlo. In: ICML, pp. 880–887 (2008)
22. Singh, A., Gordon, G.J.: A Bayesian matrix factorization model for relational data. In: UAI, pp. 556–563 (2010)
23. Singh, A.P., Gordon, G.J.: Relational learning via collective matrix factorization. In: KDD, pp. 650–658 (2008)
24. Wainwright, M.J., Jordan, M.I., et al.: Graphical Models, Exponential Families, and Variational Inference. Foundations and Trends® in Machine Learning, pp. 1–305 (2008)
25. Wang, C., Blei, D.M.: Collaborative topic modeling for recommending scientific articles. In: KDD, pp. 448–456 (2011)
26. Wang, H., Wang, N., Yeung, D.Y.: Collaborative deep learning for recommender systems, pp. 1235–1244 (2014)
27. Zhang, S., Yao, L., Sun, A.: Deep learning based recommender system: a survey and new perspectives (2017)

Variational Deep Collaborative Matrix Factorization for Social Recommendation

Teng Xiao[1], Hui Tian[2], and Hong Shen[1,3(✉)]

[1] School of Data and Computer Science, Sun Yat-sen University, Guangzhou, China
[2] School of Information and Communication Technology, Griffith University,
Gold Coast, Australia
[3] School of Computer Science, The University of Adelaide, Adelaide, Australia
hong@cs.adelaide.edu.au

Abstract. In this paper, we propose a **V**ariational **D**eep **C**ollaborative **M**atrix **F**actorization (VDCMF) algorithm for social recommendation that infers latent factors more effectively than existing methods by incorporating users' social trust information and items' content information into a unified generative framework. Unlike neural network-based algorithms, our model is not only effective in capturing the non-linearity among correlated variables but also powerful in predicting missing values under the robust collaborative inference. Specifically, we use variational auto-encoder to extract the latent representations of content and then incorporate them into traditional social trust factorization. We propose an efficient expectation-maximization inference algorithm to learn the model's parameters and approximate the posteriors of latent factors. Experiments on two sparse datasets show that our VDCMF significantly outperforms major state-of-the-art CF methods for recommendation accuracy on common metrics.

Keywords: Recommender System · Matrix Factorization ·
Deep Learning · Generative model

1 Introduction

Recommender System (RS) has been attracting great interests recently. The most commonly used technology for RS is Collaborative Filtering (CF). The goal of CF is to learn user preference from historical user-item interactions, which can be recorded by a user-item feedback matrix. Among CF-based methods, matrix factorization (MF) [17] is the most commonly used one. The purpose of MF is to find the latent factors for users and items by decomposing the user-item feedback matrix. However, the feedback matrix is usually sparse, which would result in the poor performance of MF. To track this problem, many hybrid methods such as those in [12,21–24,26], called *content MF* methods, incorporate auxiliary information, e.g., content of items, into MF . The content of items can be their tags and descriptions etc. These methods all first utilize some models (e.g., Latent

© Springer Nature Switzerland AG 2019
Q. Yang et al. (Eds.): PAKDD 2019, LNAI 11439, pp. 426–437, 2019.
https://doi.org/10.1007/978-3-030-16148-4_33

Dirichlet Allocation (LDA) [1], Stack Denoising AutoEncoders (SDAE) [20] or marginal Denoising AutoEncoders (mDAE) [3]) to extract items' content latent representations and then input them into probabilistic matrix factorization [17] framework. However, these methods demonstrate a number of major drawbacks: (a) They assume that users are independent and identically distributed, and neglect the social information of users, which can be used to improve recommendation performance [15,16]. (b) those methods [21,22] which are based on LDA only can handle text content information which is very limited in current multimedia scenario in real world. The learned latent representations by LDA are often not effective enough especially when the auxiliary information is very sparse [24] (c) For those methods [12,23,24,26] which utilize SDAE or aSDAE. The SDAE and mDAE are in fact not probabilistic models, which limits them to effectively combine probabilistic matrix factorization into a unified framework. They first corrupt the input content, and then use neural networks to reconstruct the original input. So those model also need manually choose various noise (masking noise, Gaussian noise, salt-and-peper noise, etc), which hinders them to expand to different datasets. Although some hybrid recommendation methods [2,10,18,25] that consider user social information have been proposed, they still suffer from problem (b) and (c) mentioned above. Recently, the deep generative model such Variational AutoEncoder (VAE) [11] has been utilized to the recommendation task and achieve promising performance due to it's full Bayesian nature and non-linearity power. Liang et al. proposed VAE-CF [14] which directly utilize VAE to the CF task. To incorporate item content information into VAE-CF, chen et al. proposed a collective VAE model [4] and Li et al. proposed Collaborative Variational Autoencoder [13]. However those methods all don't consider users' social information. To tackle the above problems, we propose a **V**ariational **D**eep **C**ollaborative **M**atrix **F**actorization algorithm for social recommendation, abbreviated as VDCMF, for social recommendation, which integrates item contents and user social information into a unified generative process, and jointly learns latent representations of users and items. Specifically, we first use VAE to extract items' latent representation and consider users' preferences effect by the personal tastes and their friends' tastes. We then combine these information into probabilistic matrix factorization framework. Unlike SDAE based methods, our model needs not to corrupt the input content, but instead to directly model the content's generative process. Due to the full Bayesian nature and non-linearity of deep neural networks, our model can learn more effective and better latent representations of users and items than LDA-based and SDAE-based methods and can capture the uncertainty of latent space [11]. In addition, with both item content and social information, VDCMF can effectively tackle the matrix sparsity problem. In our VDCMF, to infer latent factors of users and items, we propose a EM-algorithm to learn model parameters. To sum up, our main contributions are: (1) We propose a novel recommendation model called VDCMF for recommendation, which incorporates rich item content and user social information into MF. VDCMF can effectively learn latent factors of users and items in matrix sparsity cases. (2) Due to the full Bayesian nature and non-linearity of deep neural networks, our VDCMF is able to capture more effective latent representations of users and items than state-of-the-art methods and can capture the uncertainty of latent content representation. (3) We derive an efficient parallel variational EM-style algorithm to infer latent factors of users and items. (4) Comprehensive experiments conducted on two large real-world datasets show VDCMF can significantly outperform state-of-the-art hybrid MF methods for CF.

2   Notations and Problem Definition

Let $R \in \{0,1\}^{N \times M}$ be a user-item matrix, where N and M are the number of users and items, respectively. $R_{ij} = 1$ denotes the implicit feedback from user i over item j is observed and $R_{ij} = 0$ otherwise. Let $\mathcal{G} = (\mathcal{U}, \mathcal{E})$ denote a trust network graph, where the vertex set $\mathcal{U}$ represents users and $\mathcal{E}$ represents the relations among them. Let $T = \{T_{ik}\}^{N \times N}$ denote the trust matrix of a social network graph $\mathcal{G}$. We also use N_i to represent user i's direct friends and U_{N_i} as their latent representations. Let $X = [x_1, x_2, \ldots, x_M] \in \mathbb{R}^{L \times M}$ represent item content matrix, where L denotes the dimension of content vector x_j, and x_j be the content information of item j. For example, if item j is a product or a music, the content x_j can be bag-of-words of its tags. We use $U = [u_1, u_2, \ldots, u_N] \in \mathbb{R}^{D \times N}$ and $V = [v_1, v_2, \ldots, v_M] \in \mathbb{R}^{D \times M}$ to denote user and item latent matrices, respectively, where D denotes the dimension. I_D represents identity matrix with dimension D.

3   Variational Deep Collaborative Matrix Factorization

In this section, we propose a **V**ariational **D**eep **C**ollaborative **M**atrix **F**actorization for social recommendation, the goal of which is to infer user latent matrix U and item latent matrix V given item content matrix X, user trust matrix T and user-item rating matrix R.

3.1   The Proposed Model

To incorporate users' social information and item content information in to probabilistic matrix, we consider a users' feedback or rating on items are a balance between item content, user's taste and friend's taste. For example, users's rating on movies is effected by the movie's content information (e.g., the genre and the actors) and his friend advices from their tastes. For items' content information, since it can be very complex and various, we do not know its real distribution. However, we know any distribution can be generated by mapping simple Gaussian through a sufficiently complicated function [7]. In our proposed model, we consider item contents to be generated by their latent content vectors through a generative network. The generative process of VDCMF is as follows:

1. For each user i, draw user latent vector $u_i \sim \mathcal{N}(0, \lambda_u^{-1}) \prod_{f \in N_i} \mathcal{N}(u_f, \lambda_f^{-1} T_{if}^{-1} I_D)$.

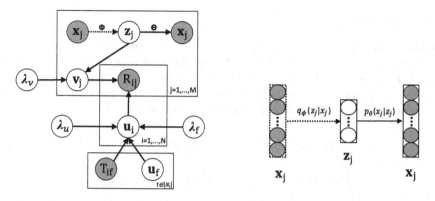

Fig. 1. Graphical model of VDCMF: generative network (left) and inference network (right), where solid and dashed lines represent generative and inference process, shaded nodes are observed variables.

2. For each item j:
   (a) Draw item content latent vector $z_j \sim \mathcal{N}(0, I_D)$.
   (b) Draw item content vector $p_\theta(x_j | z_j)$.
   (c) Draw item latent offset $k_j \sim \mathcal{N}(0, \lambda_v I_D)$ and set the item latent vector as $v_j = z_j + k_j$.
3. For each user-item pair (i, j) in R, draw R_{ij}:

$$R_{ij} \sim \mathcal{N}(u_i^\top v_j, c_{ij}^{-1}). \tag{1}$$

In the process, λ_v, λ_u and λ_g are the free parameters, respectively. Similar to [21,24], c_{ij} in Eq. 1 serves as confident parameters for R_{ij} and S_{ik}, respectively:

$$c_{ij} = \begin{cases} \varphi_1 & \text{if} \quad R_{ij} = 1, \\ \varphi_2 & \text{if} \quad R_{ij} = 0, \end{cases} \tag{2}$$

where $\varphi_1 > \varphi_2 > 0$ is the free parameters. In our model, we follow [18,24] to set $\varphi_1 = 1$ and $\varphi_2 = 0.1$. $p_\theta(x_j | z_j)$ represents item content information and x_j is generated from latent content vector z_i through a generative neural network parameterized by θ. It should be noted that the specific form of the probability $p_\theta(x_j | z_j)$ depends on the type of the item content vector. For instance, if x_j is binary vector, $p_\theta(x_j | z_j)$ can be a multivariate Bernoulli distribution $\text{Ber}(F_\theta(z_j))$ with $F_\theta(z_j)$ being the highly no-linear function parameterized by θ.

According to the graphic model in Fig. 1, the joint probability of R, X, U, V, Z and T can be represented as:

$$p(\mathcal{O}, \mathcal{Z}) = \prod_{i=1}^{N} \prod_{j=1}^{M} \prod_{k=1}^{N} p(\mathcal{O}_{ijk}, \mathcal{Z}_{ijk}) = \prod_{i=1}^{N} \prod_{j=1}^{M}$$
$$\prod_{k=1}^{N} p(z_j) p(u_i | U_{N_i}, T) p_\theta(x_j | z_j) p(v_j | z_j) p(R_{ij} | u_i, v_j), \tag{3}$$

where $\mathcal{O} = \{R, S, X\}$ is the set of all observed variables, $\mathcal{Z} = \{U, V, Z\}$ is the set of all latent variables needed to be inferred, and $\mathcal{O}_{ijk} = \{R_{ij}, T_{ik}, x_j\}$ and $\mathcal{Z}_{ijk} = \{u_i, v_j, z_j\}$ for short.

3.2   Inference

Previous work [21,24] has shown that using an expectation-maximization (EM) algorithm enables recommendation methods that integrate them to obtain high-quality latent vectors (in our case, $\boldsymbol{U}$ and $\boldsymbol{V}$). Inspired by these work, in this section, we derive an EM algorithm called VDCMF from the view of Bayesian point estimation. The marginal log likelihood can be given by:

$$\log p(\mathcal{O}) = \log \int p(\mathcal{O}, \mathcal{Z}) \mathrm{d}\mathcal{Z} \geq \int q(\mathcal{Z}) \log \frac{p(\mathcal{O}, \mathcal{Z})}{q(\mathcal{Z})} \mathrm{d}\mathcal{Z}$$
$$= \int q(\mathcal{Z}) \log p(\mathcal{O}, \mathcal{Z}) - \int q(\mathcal{Z}) \log q(\mathcal{Z}) \equiv \mathcal{L}(q), \tag{4}$$

where we apply Jensen's inequality, and $q(\mathcal{Z})$ and $\mathcal{F}(q)$ are variational distribution and the evidence lower bound (ELBO), respectively. For variational distribution $q(\mathcal{Z})$, we consider variational distributions in it to be matrix-wise independent:

$$q(\mathcal{Z}) = q(\boldsymbol{U}) q(\boldsymbol{V}) q(\boldsymbol{Z}) \tag{5}$$
$$= \prod_{i=1}^{N} q(\boldsymbol{u}_i) \prod_{j=1}^{M} q(\boldsymbol{v}_j) \prod_{j=1}^{M} q(\boldsymbol{z}_j).$$

For Bayesian point estimation, we assume the variational distribution of $\boldsymbol{u}_i$ is:

$$q(\boldsymbol{u}_i) = \prod_{d=1}^{D} \delta(U_{id} - \hat{U}_{id}). \tag{6}$$

$$q(\boldsymbol{v}_j) = \prod_{d=1}^{D} \delta(V_{jd} - \hat{V}_{jd}). \tag{7}$$

where $\{\hat{U}_{id}\}_{d=1}^{D}$ are variational parameters and δ is a Dirac delta function. Variational distributions of $\boldsymbol{v}_j$ and $\boldsymbol{g}_k$ are defined similarly. When U_{id} are discrete, the entropy of $\boldsymbol{u}_i$ is:

$$H(\boldsymbol{u}_i) = - \int q(\boldsymbol{u}_i) \log q(\boldsymbol{u}_i) = \sum_{d=1}^{D} \sum_{U_{id}} \delta(U_{id} - \hat{U}_{id}) \log \delta(U_{id} - \hat{U}_{id}) = 0. \tag{8}$$

Similarly, $H(\boldsymbol{v}_j)$ is 0 when the elements are discrete. Then the evidence lower bound $\mathcal{L}(q)$ (Eq. 4) can be written as:

$$\mathcal{L}_{\text{point}}(\hat{\boldsymbol{U}}, \hat{\boldsymbol{V}}, \boldsymbol{\theta}, \boldsymbol{\phi}) = \langle \log p(\boldsymbol{U}|\boldsymbol{T}, \boldsymbol{U}_{N_i}) p(\boldsymbol{V}|\boldsymbol{Z}) \tag{9}$$
$$p(\boldsymbol{X}|\boldsymbol{Z}) p(\boldsymbol{R}|\boldsymbol{U}, \boldsymbol{V}) \rangle_q - \text{KL}(q_\phi(\boldsymbol{Z}|\boldsymbol{X}) \| p(\boldsymbol{Z})),$$

where $\langle \cdot \rangle$ is the statistical expectation with respect to the corresponding variational distribution. $\hat{\boldsymbol{U}} = \{\hat{U}_{id}\}$ and $\hat{\boldsymbol{V}} = \{\hat{V}_{jd}\}$ are variational parameters corresponding to the variational distribution $q(\boldsymbol{U})$ and $q(\boldsymbol{V})$, respectively.

For latent variables Z, However, it is intractable to infer Z by using traditional mean-field approximation since we do not have any conjugate probability distribution in our model which requires by traditional mean-field approaches. To track this problem, we use amortized inference [6,8], it consider a shared structure for every variational distributions, instead. Consequently, similar to VAE [11], we also introduce a variational distribution $q_\phi(Z|X)$ to approximate the true posterior distribution $p(Z|\mathcal{O})$. $q_\phi(Z|X)$ is implemented by a inference neural network parameterized by ϕ (see Fig. 1). Specifically, for z_j we have:

$$q(z_j) = q_\phi(z_j|x_j) = \mathcal{N}(\mu_j, \operatorname{diag}(\delta_j^2)), \tag{10}$$

where the mean μ_j and variance δ_j are the outputs of the inference neural network.

Directly maximizing the ELBO (Eq. 9) involves solving parameters $\hat{U}$, $\hat{V}$, θ and ϕ, which is intractable. Thus, we derive an iterative variational-EM (VEM) algorithm to maximize $\mathcal{L}_{\text{point}}(\hat{U}, \hat{V}, \theta, \phi)$, abbreviated $\mathcal{L}_{\text{point}}$.

Variational E-step. We first keep θ and ϕ fixed, then optimize evidence lower bound $\mathcal{L}_{\text{point}}$ with respect to $\hat{U}$ and $\hat{V}$. We take the gradient of $\mathcal{L}$ with respect to u_i and v_j and set it to zero. We will get the updating rules of $\hat{u}_i$ and $\hat{v}_j$:

$$\hat{u}_i \leftarrow (VC_iV^\top + \lambda_u I_D + \lambda_f T_i 1_I I_D)^{-1}(\lambda_f UT_i^\top + VC_iR_i), \tag{11}$$

$$\hat{v}_j \leftarrow (\hat{U}C_j\hat{U}^\top + \lambda_v I_D)^{-1}(\hat{U}C_jR_j + \lambda_v \langle z_j \rangle), \tag{12}$$

where $C_i = \operatorname{diag}(c_{i1}, ...c_{iM})$, $T_i = \operatorname{diag}(T_{i1}, ...T_{iM})$, $R_i = [R_{i1}, ...R_{iM}]$. I_N is a N dimensional column vector with all elements elements to 1. For item latent vector v_j, C_j and R_j are defined similarly. $\hat{u}_i = [\hat{U}_{i1}, ...\hat{U}_{iD}]$ and $\hat{v}_j = [\hat{V}_{j1}, ...\hat{V}_{jD}]$. For z_j, its expectation is $\langle z_j \rangle = \mu_j$, which is the output of the inference network.

It can be observed that λ_v governs how much the latent item vector z_j affects item latent vector v_j. For example, if $\lambda_v = \infty$, it indicate we direct use latent item vector to represent item latent vector v_j; if $\lambda_v = 0$, it means we do not embed any item content information into item latent vector. λ_f serves as a balance parameter between social trust matrix and user-item matrix on user latent vector u_i. For example, if $\lambda_f = \infty$, it means we only use the social network information to model user's preference; if $\lambda_f = 0$, we only use user-item matrix and item content information for prediction. So λ_v and λ_f are regarded as collaborative parameters for item content, user-item matrix and social matrix.

Variational M-step. Keep $\hat{U}$ and $\hat{V}$ fixed, we optimize $\mathcal{L}_{\text{point}}$ w.r.t. ϕ and θ (we only focus on terms containing ϕ and θ).

$$\mathcal{L}_{\text{point}} = \text{constant} + \sum_{j=1}^{M} \mathcal{L}(\theta, \phi; x_j, v_j) = \text{constant} + \sum_{j=1}^{M} \tag{13}$$

$$-\frac{\lambda_v}{2}\langle (v_j - z_j)^\top(v_j - z_j)\rangle_{q(\mathcal{Z})} + \langle \log p_\theta(x_j|z_j)\rangle_{q_\phi(z_j|x_j)} - \text{KL}(q_\phi(z_j|x_j)||p(z_j)),$$

where M is the number of items and the constant term represents terms which don't contain θ and ϕ. For the expectation term $\langle p_\theta(x_j|z_j)\rangle_{q_\phi(z_j|x_j)}$, we can not solve it analytically. To handle this problem, we approximate it by the Monte Carlo sampling as follows:

$$\langle \log p_\theta(\boldsymbol{x}_j|\boldsymbol{z}_j)\rangle_{q_\phi(\boldsymbol{z}_j|\boldsymbol{x}_j)} = \frac{1}{L}\sum\nolimits_{l=1}^{L} p_\theta(\boldsymbol{x}_j|\boldsymbol{z}_j^l), \tag{14}$$

where L is the size of samplings, and $\boldsymbol{z}_j^l$ denotes the l-th sample, which is reparameterized to $\boldsymbol{z}_j^l = \boldsymbol{\epsilon}_j^l \odot \mathrm{diag}(\boldsymbol{\delta}_j^2) + \boldsymbol{\mu}_j$. Here $\boldsymbol{\epsilon}_j^l$ is drawn from $\mathcal{N}(0, \boldsymbol{I}_D)$ and $\odot$ is an element-wise multiplication. By using this reparameterization trick and Eq. 10, $\mathcal{L}(\boldsymbol{\theta}, \boldsymbol{\phi}; \boldsymbol{x}_j, \boldsymbol{v}_j)$ in Eq. 13 can be estimated by:

$$\mathcal{L}(\boldsymbol{\theta}, \boldsymbol{\phi}; \boldsymbol{x}_j, \boldsymbol{v}_j) \simeq \tilde{\mathcal{L}}^j(\boldsymbol{\theta}, \boldsymbol{\phi}) = -\frac{\lambda_v}{2}(-2\boldsymbol{\mu}_j^\top \hat{\boldsymbol{v}}_j + \boldsymbol{\mu}_j^\top \boldsymbol{\mu}_j$$
$$+ \mathrm{tr}(\mathrm{diag}(\boldsymbol{\delta}_j^2))) + \frac{1}{L}\sum\nolimits_{l=1}^{L} p_\theta(\boldsymbol{x}_j|\boldsymbol{z}_j^l) - \mathrm{KL}(q_\phi(\boldsymbol{z}_j|\boldsymbol{x}_j)||p(\boldsymbol{z}_j)) + \mathrm{constant}. \tag{15}$$

We can construct an estimator of $\mathcal{L}_{\mathrm{point}}(\boldsymbol{\phi}, \boldsymbol{\theta}; \boldsymbol{X}, \boldsymbol{V})$, based on minibatches:

$$\mathcal{L}_{\mathrm{point}}(\boldsymbol{\theta}, \boldsymbol{\phi}) \simeq \tilde{\mathcal{L}}^P(\boldsymbol{\theta}, \boldsymbol{\phi}) = \frac{M}{P}\sum\nolimits_{j=1}^{P} \tilde{\mathcal{L}}^j(\boldsymbol{\theta}, \boldsymbol{\phi}). \tag{16}$$

As discussed in [11], the number of samplings L per item j can be set to 1 as long as the minibatch size P is large enough, e.g., $P = 128$. We can update $\boldsymbol{\theta}$ and $\boldsymbol{\phi}$ by using the gradient $\nabla_{\boldsymbol{\theta}, \boldsymbol{\phi}}\tilde{\mathcal{L}}^P(\boldsymbol{\theta}, \boldsymbol{\phi})$.

We iteratively update $\boldsymbol{U}, \boldsymbol{V}, \boldsymbol{G}, \boldsymbol{\theta}$, and $\boldsymbol{\phi}$ until it converges.

3.3 Prediction

After we get the approximate posteriors of $\boldsymbol{u}_i$ and $\boldsymbol{v}_j$. We predict the missing value R_{ij} in $\boldsymbol{R}$ by using the learned latent features $\boldsymbol{u}_i$ and $\boldsymbol{v}_j$:

$$R_{ij}^* = \langle R_{ij}\rangle = (\langle \boldsymbol{z}_j\rangle + \langle \boldsymbol{k}_j\rangle)^\top \langle \boldsymbol{u}_i\rangle = \langle \boldsymbol{v}_j\rangle^\top \langle \boldsymbol{u}_i\rangle \tag{17}$$

For a new item that is not rated by any other users, the offset $\boldsymbol{\epsilon}_j$ is zero, and we can predict R_{ij} by:

$$R_{ij}^* = \langle R_{ij}\rangle = \langle \boldsymbol{z}_j\rangle^\top \langle \boldsymbol{u}_i\rangle \tag{18}$$

4 Experiments

4.1 Experimental Setup

Datasets. In order to evaluate performance of our model, we conduct experiments on two real-world datasets from Lastfm[1] (*lastfm-2k*) and Epinions[2] (*Epinions*) datasets:

---

[1] http://www.lastfm.com.
[2] http://www.Epinions.com.

Lastfm. This dataset contains user-item, user-user, and user-tag-item relations. We first transform this dataset as implicit feedback. For Lastfm dataset, we consider the user-item feedback is 1 if the user has listened to the artist (item); otherwise, it is 0. Lastfm only contains 0.27% observed feedbacks. We use items bag-of-word tag representations as their items content information. We direct use user social matrix as trust matrix.

Epinions. This dataset contains rating , user trust and review information. We transform this dataset as implicit feedback. For those >3 ratings, we transform it as '1'; otherwise, it is 0. We use item's review as its content information. Epinions contains 0.08% observed feedbacks.

Baselines. For fair comparisons, like that in our VDCMF, the baselines we used also incorporate user social information or item content information into matrix factorization. **(1) PMF**. This model [17] is a famous MF method, and only uses user-item feedback matrix. **(2) SoRec**. This model [15] jointly decomposes user-user social matrix and user-item feedback matrix to learn user and item latent representations. **(3) Collaborative topic regression (CTR)**. This model [21] utilizes topic model and matrix factorization to learn latent representations of users and items. **(4) Collaborative deep learning (CDL)**. This model [24] utilizes stack denoising autoencoder to learn latent items' content representations, and incorporates them into probabilistic matrix factorization. **(5) CTR-SMF**. This model [18] incorporates topic modeling and probabilistic MF of social networks. **(6) PoissonMF-CS**. This model [19], jointly models use social trust, item content and users preference using Poisson matrix factorization framework. It is a state-of-the-art MF method for Top-N recommendation on the Lastfm dataset. **(7)Neural Matrix Factorization (NeuMF)**. This model is a state-of-the-art collaborative filtering method, which utilizes neural network to model the interaction between user model [9] is a state-of-the-art collaborative filtering method, which utilizes neural network to model the interaction between users and items.

Settings. For fair comparisons, We first set the parameters for PMF, SoRec, CTR, CTR-SMF, CDL, NeuMF via five-fold cross validation. For our model, we set $\lambda_u = 0.1$, $D = 25$ for Lastfm and $D = 50$ for Epinions. Without special mention, we set $\lambda_v = 0.1$ and $\lambda_f = 0.1$. We will further study the impact of the key hyper-parameters for the recommendation performance.

Evaluation Metrics. The metrics we used are Recall@K, NDCG@K and MAP@K [5] which are common metrics for recommendation.

4.2   Experimental Results and Discussions

Overall Performance. To evaluate our model in top-K recommendation task, we evaluate our model and baselines in two datasets in terms of Recall@20, Recall@50, ND CG@20 and MAP@20. Table 1 shows the performance of our VDCMF and the baselines using the two datasets. According to Table 1, we have following findings: (a) VDCMF outperforms the baselines in terms of all

Table 1. Recommendation performance of VDCMF and baselines. The best baseline method is highlighted with underline.

	Lastfm dataset				Epinions dataset			
	Recall@20	Recall@50	NDCG@20	MAP@20	Recall@20	Recall@50	NDCG@20	MAP@20
PMF	0.0923	0.1328	0.0703	0.1083	0.4012	0.5121	0.3019	0.3365
SoRec	0.1088	0.1524	0.0721	0.1128	0.4341	0.5547	0.3254	0.3621
CTR	0.1192	0.1624	0.0799	0.1334	0.5024	0.6125	0.3786	0.4197
CTR-SMF	0.1232	0.1832	0.0823	0.1386	0.5213	0.6217	0.3942	0.4437
CDL	0.1346	0.2287	0.0928	0.1553	0.5978	0.6597	0.4502	0.4792
NeuMF	<u>0.1517</u>	0.2584	0.1036	<u>0.1678</u>	<u>0.6043</u>	<u>0.6732</u>	0.4611	<u>0.4987</u>
PoissonMF-CS	0.1482	<u>0.2730</u>	<u>0.1089</u>	0.1621	0.5876	0.6533	<u>0.4628</u>	0.4876
VDCMF (ours)	**0.1613**	**0.3006**	**0.1114**	**0.1695**	**0.6212**	**0.6875**	**0.4782**	**0.5123**

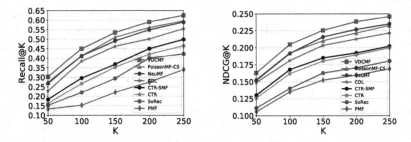

Fig. 2. Evaluation of Top-K item recommendation where K ranges from 50 to 250 on Lastfm

matrices on Lastfm and Epinions, which demonstrates the effectiveness of our method of inferring the latent factors of users and items, and leading to better recommendation performance. (b) For more sparse dataset, Epinions, VDCMF also achieves the best performance, which demonstrates our model can effectively handle matrix sparsity problem. We attribute this improvement to the incorporated item content and social trust information. (c) We can see methods which both utilizes content and social information (VDCMF, NeuMF and PoissonMF-CS) outperform others (CDL,CTR, CTR-SMF, SoRec and PMF), which demonstrates incorporating content and social information can effectively alleviate matrix sparse problem. (d) Our VDCMF outperforms the strong baseline PoissonMF-CS, though they are both Bayesian generative model. The reason VDCMF is that our VDCMF incorporates neural network into Bayesian generative model, which makes it have powerful non-linearity to model item content's latent representation. To further evaluate our VDCMF robustness, we evaluate the empirical performance of large recommendation list for our VDCMF on Lastfm and report results in Fig. 2. We can find our VDCMF significantly and consistently outperforms other baselines. This, again, demonstrates the effectiveness of our model. All of these findings demonstrates that our VDCMF is robust and it is able to achieve significant improvements of top-k recommendation over the state-of-the-art.

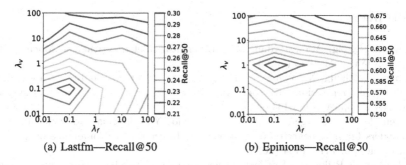

(a) Lastfm—Recall@50   (b) Epinions—Recall@50

Fig. 3. The effect of λ_v and λ_f of the proposed VDCMF with Recall@50 on Lastfm and Epinions.

Impact of Parameters. In this section, we study the effect of the key hyper-parameters of the proposed model. We first study the parameters of λ_f and λ_v. We use Recall@50 as an example, the plot the contours on Lastfm and Epinions datasets. Figure 3(a) and (b) show the contour of Recall@50. As we can see, VDCMF achieves the best recommendation performance when $\lambda_v = 0.1$ and $\lambda_f = 0.1$ on Lastfm, and $\lambda_v = 1$ and $\lambda_f = 0.1$ on Epinions. From Fig. 3(a) and (b), we can find our model is sensitive to λ_v and λ_f. The reason is that λ_v can control how much item content information is incorporated into item latent vector, λ_q can control how much social information is incorporated into user latent vector. Figure 3(a) and (b) show that we can balance the content information and social information by varying λ_v and λ_q, leading to better recommendation performance.

5   Conclusion

In this paper, we studied the problem of inferring effective latent factors of users and items for social recommendation. We have proposed a novel Variational Deep Collaborative Matrix Factorization algorithm, VDCMF, which incorporates rich item content and user social trust information into a full Bayesian deep generative framework. Due to the full Bayesian nature and non-linearity of deep neural networks, our proposed model is able to learn more effective latent representations of users and items than those generated by state-of-the-art neural networks based recommendation algorithms. To effectively infer latent factors of users and items, we derived an efficient expectation-maximization algorithm. We have conducted experiments on two publicly available datasets. We evaluated the performance of our VDCMF and baselines methods based on Recall, NDCG and MAP metrics. Experimental results demonstrate that our VDCMF can effectively infer latent factors of users and items.

Acknowledgement. This work is supported by the National Key Research and Development Program of China (No. #2017YFB0203201) and Australian Research Council Discovery Project DP150104871.

References

1. Blei, D.M., Ng, A.Y., Jordan, M.I.: Latent dirichlet allocation. J. Mach. Learn. Res. **3**(Jan), 993–1022 (2003)
2. Chen, C., Zheng, X., Wang, Y., Hong, F., Lin, Z., et al.: Context-aware collaborative topic regression with social matrix factorization for recommender systems. In: AAAI, pp. 9–15 (2014)
3. Chen, M., Weinberger, K., Sha, F., Bengio, Y.: Marginalized denoising auto-encoders for nonlinear representations. In: International Conference on Machine Learning, pp. 1476–1484 (2014)
4. Chen, Y., de Rijke, M.: A collective variational autoencoder for top-n recommendation with side information. In: Proceedings of the 3rd Workshop on Deep Learning for Recommender Systems, pp. 3–9. ACM (2018)
5. Croft, W.B., Metzler, D., Strohman, T.: Search Engines: Information Retrieval in Practice. Addison-Wesley, Reading (2015)
6. Dayan, P., Hinton, G.E., Neal, R.M., Zemel, R.S.: The Helmholtz machine. Neural Comput. **7**(5), 889–904 (1995)
7. Doersch, C.: Tutorial on variational autoencoders. CoRR abs/1606.05908 (2016)
8. Gershman, S., Goodman, N.: Amortized inference in probabilistic reasoning. In: Proceedings of the Annual Meeting of the Cognitive Science Society, vol. 36 (2014)
9. He, X., Liao, L., Zhang, H., Nie, L., Hu, X., Chua, T.S.: Neural collaborative filtering. In: WWW, pp. 173–182 (2017)
10. Hu, G.N., et al.: Collaborative filtering with topic and social latent factors incorporating implicit feedback. ACM Trans. Knowl. Discov. Data (TKDD) **12**(2), 23 (2018)
11. Kingma, D.P., Welling, M.: Auto-encoding variational bayes. In: ICLR (2014)
12. Li, S., Kawale, J., Fu, Y.: Deep collaborative filtering via marginalized denoising auto-encoder. In: CIKM, pp. 811–820 (2015)
13. Li, X., She, J.: Collaborative variational autoencoder for recommender systems. In: KDD, pp. 305–314 (2017)
14. Liang, D., Krishnan, R.G., Hoffman, M.D., Jebara, T.: Variational autoencoders for collaborative filtering. In: Proceedings of the 2018 World Wide Web Conference on World Wide Web, pp. 689–698. International World Wide Web Conferences Steering Committee (2018)
15. Ma, H., Yang, H., Lyu, M.R., King, I.: Sorec: social recommendation using probabilistic matrix factorization. In: CIKM, pp. 931–940 (2008)
16. Ma, H., Zhou, D., Liu, C., Lyu, M.R., King, I.: Recommender systems with social regularization. In: WSDM, pp. 287–296 (2011)
17. Mnih, A., Salakhutdinov, R.R.: Probabilistic matrix factorization. In: NIPS, pp. 1257–1264 (2008)
18. Purushotham, S., Liu, Y., Kuo, C.C.J.: Collaborative topic regression with social matrix factorization for recommendation systems. In: ICML, pp. 691–698 (2012)
19. da Silva, E.S., Langseth, H., Ramampiaro, H.: Content-based social recommendation with poisson matrix factorization. In: Ceci, M., Hollmén, J., Todorovski, L., Vens, C., Džeroski, S. (eds.) ECML PKDD 2017. LNCS (LNAI), vol. 10534, pp. 530–546. Springer, Cham (2017). https://doi.org/10.1007/978-3-319-71249-9_32
20. Vincent, P., Larochelle, H., Lajoie, I., Bengio, Y., Manzagol, P.A.: Stacked denoising autoencoders: learning useful representations in a deep network with a local denoising criterion. J. Mach. Learn. Res. **11**(Dec), 3371–3408 (2010)

21. Wang, C., Blei, D.M.: Collaborative topic modeling for recommending scientific articles. In: KDD, pp. 448–456 (2011)
22. Wang, H., Chen, B., Li, W.J.: Collaborative topic regression with social regularization for tag recommendation. In: IJCAI, pp. 2719–2725 (2013)
23. Wang, H., Shi, X., Yeung, D.Y.: Relational stacked denoising autoencoder for tag recommendation. In: Twenty-Ninth AAAI Conference on Artificial Intelligence (2015)
24. Wang, H., Wang, N., Yeung, D.Y.: Collaborative deep learning for recommender systems. In: KDD, pp. 1235–1244 (2015)
25. Wu, H., Yue, K., Pei, Y., Li, B., Zhao, Y., Dong, F.: Collaborative topic regression with social trust ensemble for recommendation in social media systems. Knowl.-Based Syst. **97**, 111–122 (2016)
26. Zhang, F., Yuan, N.J., Lian, D., Xie, X., Ma, W.Y.: Collaborative knowledge base embedding for recommender systems. In: KDD, pp. 353–362 (2016)

Healthcare, Bioinformatics and Related Topics

Time-Dependent Survival Neural Network for Remaining Useful Life Prediction

Jianfei Zhang[1,2], Shengrui Wang[1,2(✉)], Lifei Chen[1], Gongde Guo[1],
Rongbo Chen[2], and Alain Vanasse[3,4]

[1] College of Mathematics and Informatics, Fujian Normal University, Fuzhou, China
{clfei,ggd}@fjnu.edu.cn
[2] Département d'Informatique, Université de Sherbrooke, Sherbrooke, Canada
{jianfei.zhang,shengrui.wang,rongbo.chen}@usherbrooke.ca
[3] Département de Médecine de Famille et de Médecine d'Urgence,
Université de Sherbrooke, Sherbrooke, Canada
alain.vanasse@usherbrooke.ca
[4] Centre de Recherche du Centre Hospitalier Universitaire de Sherbrooke,
Sherbrooke, Canada

Abstract. Remaining useful life (RUL) prediction has been a topic
of practical interest in many fields involving preventive intervention,
including manufacturing, medicine and healthcare. While most of the
conventional approaches suffer from censored failures arising and statistically circumscribed assumptions, few attempts have been made to
predict RUL by developing a survival learning machine that explores
the underlying relationship between time-varying prognostic variables
and failure-free survival probability. This requires a purely data-driven
prediction approach, devoid of any a survival model and all statistical
assumptions. To this end, we propose a time-dependent survival neural
network that additively estimates a latent failure risk and performs multiple binary classifications to generate prognostics of RUL-specific probability. We train the neural network by a new survival learning criterion
that minimizes the censoring Kullback-Leibler divergence and guarantees
monotonicity of the resulting probability. Experiments on four datasets
demonstrate the great promise of our approach in real applications.

Keywords: RUL prediction · Neural network · Survival learning ·
Failure risk · Time-varying data

1 Introduction

This paper is about RUL predictive analytics. Let's think about all the in-
service machines we use daily, and organisms under pharmaceutical care, from
an engine-propelled vehicle on the way to work or a lift going up and down,
to an in-patient in the early stages of breast cancer. Imagine that one of these

© Springer Nature Switzerland AG 2019
Q. Yang et al. (Eds.): PAKDD 2019, LNAI 11439, pp. 441–452, 2019.
https://doi.org/10.1007/978-3-030-16148-4_34

should fail (e.g., break down, worsen, die) every day from now on. What impact would that have? The truth is that some failures are just an inconvenience or a financial loss, while others could mean life or death. Therefore, preventive intervention (e.g., predictive maintenance and health care) to defer failures has recently been of great practical interest [23]. But how can we find the right moment for intervention? Providing an answer to this question is the aim of RUL prediction, which seeks to build models for accurate prognostics of RUL in engines, patients and other life entities in machine manufacturing, medicine, epidemiology, economics, etc.

Predicting RUL with great accuracy in the distant future is very challenging and indeed almost impossible in most practical situations. Rather, we turn our attention to the easier and more meaningful prognostic of *how long and how probably to remain failure-free (aka survival before failure)*. RUL prediction here thus refers exclusively to the prognostic of RUL-specific probability, i.e., failure-free survival probability at a specific time. By failure, we refer in particular to a non-recurring, single, adverse incident. With a prediction model in hand, decision makers can be provided with information about when a mechanical fault that can lead to whole system failure might take place. For instance, the probability of fault-free steering in a 15-year-old vehicle engine up to 50,000 km is 80% but for up to 80,000 km the probability drops to 10%; this knowledge allows for predictive maintenance (before 50,000 km) which may prolong engine usage and holds out the promise of considerable cost savings.

In this paper, we propose a time-dependent survival neural network (TSNN) which additively estimates a latent failure risk and performs multiple classifications to generate prognostics of RUL-specific probability. We provide a new censoring Kullback-Leibler divergence for evaluating the dissimilarity between the binary classification probabilities and the actual survival process. A generalized survival learning approach is developed to minimize such divergence, running under a constraint that guarantees monotonicity of the resulting probability. Experimental results on four real-world datasets from the fields of engineering, medicine and healthcare demonstrate the promise of TSNN in real applications. The paper makes the following primary contributions: (1) It develops a purely data-driven prediction approach free of any existing survival model and all statistical assumptions. (2) It transforms prediction into multiple classifications that potentially relate to each other. (3) It makes full use of time-varying prognostic variables by exploring latent failure risk in an additive manner. (4) It provides a learning criterion that allows automatic exploitation of data with censoring.

2   Motivation

To build a model for RUL prediction, training data should allow us to capture information regarding prognostic variables leading to failure. In the observational world, however, we need to know whether failure, dropout or study cutoff comes first. Thus, the outcome of interest in data is not only whether or not a failure occurred, but also when that failure occurred. Traditional regression methods are not able to include both the failure and time aspects as the outcome in the model, though they are used to perform a time prediction on most time series data to answer the questions like *"How many days are left before failure?"*. In contrast, a considerable number of survival models have long been developed to utilize the partial information on each entity with censored failure and provide unbiased survival estimates. They incorporate data from multiple time points across entities for prediction of failure probability over time and thus can answer the question like *"How does the risk of failure change over time?"*.

These statements naturally lead one to consider using survival models for predicting RUL-specific probability. However, the three prominent survival modeling approaches developed primarily for retrospective cohort studies are characterized by their inherent disadvantages [20]. (1) Models utilizing the non-parametric approach, an analysis intended to generate unbiased descriptive statistics, cannot generally be used to assess the effect of multiple prognostic variables on failure. (2) The parametric approach suffers from an even more critical weakness, relying as it does on the assumption that the underlying failure distribution (i.e., how the probability of failure changes over time) has been correctly specified. (3) The semi-parametric approach requires an assumption on how the variables influence the risk of failure, which is often violated in practical use.

The increasing availability of complex lifetime data with time-varying prognostic variables poses more challenges to these approaches and is stimulating numerous research efforts that use data mining and machine learning methods in conjunction with survival models. Typical examples include multi-task learning [9,10,13,19], active learning [18], neural networks [4,7], transfer learning [11], Bayesian inference [16] and feature engineering [12,24] that extended to the semi-parametric Cox proportional-hazards model [3], as well as a random forest technique [6] that employed a non-parametric Nelson-Aalen estimator to predict the time to censored failures for establishing terminal nodes of forest. These approaches still suffer from the implausibility of the survival study hypothesis and prior knowledge and therefore cannot be selected as prediction models for our desired output. For example, although the feed-forward network proposed in [4] preserved most of the advantages of a typical Cox proportional-hazards hypothesis, it was still not the optimal way to model the baseline variations [7]. In addition, these time-to-failure prediction methods are not specifically designed to handle time-varying prognostic variables. The common approach employed is to predict the survival probability at a certain time (i.e., RUL-specific probability in this paper) using only the values of variables at that moment. The historical values are discarded in prediction but have been proven to latently affect the survival probability [15,25,26]. These arguments in turn demonstrate a need for a prediction model that releases priori statistical assumptions, explores latent risk and makes full use of time-varying prognostic variables.

3   Proposed Approach

Imagine a binary classification performed to predict failure of a machine in a given t-day time window; i.e., to answer the question *"Will a machine remain failure-free over the next t days?"*. This allows us to transform the original RUL prediction problem into a series of binary classification problems, as long as each has an RUL-specific output probability that the actual RUL, say T, is not earlier than t, denoted $\Pr(T > t)$. In this section, we provide a neural network that allows data to drive the survival learning inference, i.e., devoid of any a survival model and all statistical assumptions, to perform the binary classifications.

3.1   Time-Dependent Survival Neural Network

Survival Neural Network Classifier Architecture. We concentrate our attention on a one-hidden-layer neural network, i.e., three-layer networks with V input neurons, K output neurons and H hidden neurons, as shown in Fig. 1. The input layer's role is solely to distribute the inputs to the hidden layer, where the neuron $v = 1, 2, \ldots, V$ takes value x_v and the hidden neuron $h = 1, 2, \ldots, H$ computes a sum of all the inputs weighted by $\mathbf{w}_h^{\text{hide}} \in \mathbb{R}^V$, adds a bias b_h^{hide}, and applies an activation function to obtain its output. The outputs of the hidden layer subsequently become the inputs of the output layer, in which the output neuron $k = 1, 2, \ldots, K$ computes a sum of these inputs weighted by $\mathbf{w}_k^{\text{out}} \in \mathbb{R}^H$, adds a bias b_k^{out}, and then applies the activation function to obtain $S_k(\mathbf{x})$.

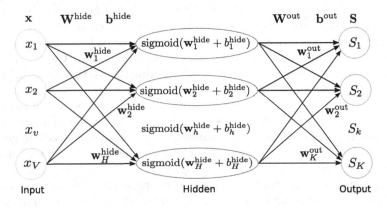

Fig. 1. A survival neural network

Survival neural network, in principle, is a combination of multiple classifiers, each performing a binary classification on every entity that is or is not still failure-free. Hence, we interpret $S_k(\mathbf{x}) \in (0, 1)$ as the classification probability that the entity with variables $\mathbf{x}$ remains failure-free by τ_k. In doing so, with the K classification outputs over disjoint time snapshots $\tau_1 < \tau_2 < \cdots < \tau_K$ in hand, we are able to estimate an RUL-specific probability curve which depicts *how long and how probably the entity will remain failure-free*. Hence, given the weights $\mathbf{W}^{\text{hide}} \in \mathbb{R}^{H \times V}$, $\mathbf{W}^{\text{out}} \in \mathbb{R}^{K \times H}$ and the biases $\mathbf{b}^{\text{hide}} \in \mathbb{R}^H$, $\mathbf{b}^{\text{out}} \in \mathbb{R}^K$ for computing the hidden and output layers, respectively, we scale the outputs $\mathbf{S}(\mathbf{x}) \in \mathbb{R}^K$ to the range of the logistic sigmoid function that is applied component-wise to the vector, i.e.,

$$\mathbf{S}(\mathbf{x}) = \mathtt{sigmoid}\left(\mathbf{W}^{\mathrm{out}} \cdot \mathtt{sigmoid}\left(\mathbf{W}^{\mathrm{hide}} \cdot \mathbf{x} + \mathbf{b}^{\mathrm{hide}}\right) + \mathbf{b}^{\mathrm{out}}\right). \tag{1}$$

For the time-varying variables, the output is yielded with input values observed at corresponding time snapshots, that is, $S_k(\mathbf{x}) = S(\mathbf{x}(\tau_k))$, where $\mathbf{x}(\tau_k)$ consists of V observational values at τ_k. Nevertheless, this approach does not take the historical variable values into account in estimating the failure risk.

TSNN with Latent Failure Risk Estimation. Note that the exponential component in Eq. 1 can serve as the failure risk, like the conventional cumulative risk in the Cox [3] and accelerated failure time (AFT) models [21]. Obviously, the risk is not dependent on any historical values at all. To address this issue, we propose the form $\gamma(*, t)$ to stand for the decay ratio of the failure risk. By such decay, we can model the amount of the latent risk produced by the values at time $*$ remaining at time $t(\geq *)$. This can be an exponential function of time in the form $\gamma(*, t) = \exp\{G(* - t)\}$. Simply, we make the decay coefficient G take a positive value and thus $0 < \gamma \leq 1$. Note that such a positive decay ratio indicates that the risk will shrink over time but not vanish. Given all historical values observed at time points $j \in R(t)$ before t, we estimate the failure risk in an additive manner and compute the TSNN output at τ_k as follows:

$$S_k(\mathbf{x}) = \left(1 + \frac{1}{|R(\tau_k)|} \sum_{j \in R(\tau_k)} \exp\{G(j - t)\} \exp\{-\mathbf{W}^{\mathrm{out}}\phi(\mathbf{x}(j)) - \mathbf{b}^{\mathrm{out}}\}\right)^{-1}$$

$$\phi(\mathbf{x}(j)) = \left(1 + \frac{1}{|R(j)|} \sum_{u \in R(j)} \exp\{G(u - t)\} \exp\{-\mathbf{W}^{\mathrm{hide}}\mathbf{x}(u) - \mathbf{b}^{\mathrm{hide}}\}\right)^{-1}.$$

Our approach can be thought of as a generalization of multi-task classification, which enables flexible modeling of RUL-specific probability in parallel. Each task executes on all training entities but has an individual variable input. As was discussed in [10], such multi-task transformation will further reduce the prediction error on each task and hence provide a more accurate estimate than models which aim at modeling the probabilities at once.

3.2 RUL-Specific Probability Evaluation

Survival Process. Given N_{tr} training entities, the actual survival process for entity i can be modulated as $\varepsilon_i(\tau_1) \varepsilon_i(\tau_2) \cdots \varepsilon_i(\tau_K)$. Each survival status $\varepsilon_i(\tau_k)$ indicates whether or not the failure occurs by time τ_k, taking a value of 1 up to τ_k, and 0 thereafter, and -1 for unknown cases. Once $\varepsilon_i(t)$ becomes "0" it will not turnover to"1", there are thus $K + 1$ possible legal sequences of the form $(1, 1, \ldots, 0, 0, \ldots)$, including the sequences of all "1"s and all"0"s. Supposing $\mathcal{K}_i^\epsilon = \{k : \varepsilon_i(\tau_k) = \epsilon_i\}$, the observed statuses are greater than or equal to unknown statues if the failure is (right-)censored, i.e.,

$\epsilon_i(\tau_k) \geq \varepsilon_i(\tau_{k'})$, $\forall k \in \mathcal{K}_i^1$ and $\forall k' \in \mathcal{K}_i^{-1}$. For an uncensored case, the survival statuses during lifetime are strictly greater than those after failure, i.e., $\varepsilon_i(\tau_k) > \varepsilon_i(\tau_{k'})$, $\forall k \in \mathcal{K}_i^1$ and $\forall k' \in \mathcal{K}_i^0$.

Censoring Kullback-Leibler Divergence. The TSNN cannot be an effective prediction model unless it achieves the objective that *the predicted RUL-specific probabilities approach the actual survival process*. In order to qualify such approachability, we define the censoring Kullback-Leibler (KL) divergence, an alternative to the relative error [17], between the distributions of the RUL-specific probability $S_k \in (0,1)$ and the survival status $\varepsilon(\tau_k) \in \{0,1\}$, as follows:

$$D_i(k) = \varepsilon_i(\tau_k) \ln \frac{\varepsilon_i(\tau_k)}{S_k(\mathbf{x}_i)} + (1 - \varepsilon_i(\tau_k)) \ln \frac{1 - \varepsilon_i(\tau_k)}{1 - S_k(\mathbf{x}_i)}.$$

The optimal weights make S_k as close as possible to 1 if i remains failure-free by τ_k and to 0 otherwise, while outputs of 1 and 0 are definitely true and definitely false predictions, respectively. Our learning criterion is then to minimize $D_i(k)$ over time snapshots $\mathcal{K}_i^{\{1,0\}} = \mathcal{K}_i^0 \cup \mathcal{K}_i^1$ at which survival statuses are known, for all N_{tr} training entities.

3.3 TSNN Learning

It is worth mentioning the known fact that S_k descends from 1 to 0, as time goes by, from the beginning to the end of life. Hence, the minimization should be constrained by the monotonicity:

$$\Delta_i(k, k+1) = S_k(\mathbf{x}_i) - S_{k+1}(\mathbf{x}_i) > 0, \forall k = 1, 2 \ldots, K-1, \forall i = 1, 2, \ldots, N_{\mathrm{tr}}.$$

The proven penalty method converts the constrained optimization problem into a series of unconstrained optimization problems. Accordingly, we utilize the static penalty [14] that along with its parameter λ incurred for violating the inequality constraints and minimize the average error computed by

$$E = \frac{1}{N_{\mathrm{tr}}} \sum_{i=1}^{N_{\mathrm{tr}}} \left(\left| \mathcal{K}_i^{\{1,0\}} \right|^{-1} \sum_{k \in \mathcal{K}_i^{\{1,0\}}} D_i(k) + \frac{\lambda}{K-1} \sum_{k=1}^{K-1} \left(\min \left\{ 0, \Delta_i(k, k+1) \right\} \right)^2 \right).$$

We train the neural network using the forward-only Levenberg-Marquardt algorithm presented in [22], which inherits the speed advantage of the Gauss-Newton algorithm and the stability of the steepest descent method.

4 Experiments

4.1 Data and Pre-processing

Four lifetime datasets were drawn from the prognostics data repository provided by the PCoE of NASA Ames, the Surveillance, Epidemiology, and End Results

(SEER) statistics database, and Canadian Community Health Survey (CCHS) statistical surveys. In the Engine dataset, 388 engines' cycles were considered unobserved, for a 27.4% censoring rate. The objective was to predict the number of operational cycles remaining until pressure compressor and fan degradation. The randomized Battery usage dataset employed the first 20 cells, each with 42 galvanostatic voltage curves. Failure was censored for 45.8% of batteries and 10 prognostic variables were extracted from the time series of temperature and current (mA) every 30 s. For the breast Cancer dataset, RULs were computed by subtracting the date of diagnosis from the date of last contact (the study cutoff). The healthy Aging data were acquired directly between Dec 2008 and Nov 2009 from respondents in a survey, which focused on the health of Canadians aged 45 and over by examining the various factors that impact healthy aging. A total of 3,390 valid interviews covering the population living in the ten provinces were used. Table 1 summarizes the statistics, including data size N, dimensionality V, censoring rate C, missing-value percentage M and failure of interest. Categorical variables were transformed into numerical values by means of the probabilistic frequency estimator presented in [2]. Afterwards, missing values were filled in via a linear regression provided by [8]. In order to reduce data redundancy and improve data integrity, all values were normalized.

Table 1. Statistics of the four lifetime datasets

Dataset (source)	N	V	C	M	Failure of interest
Engine (NASA)	1,416	21	27.4%	11.3%	Compressor and fan degradation
Battery (NASA)	842	10	45.8%	5.9%	30% fade in rated battery capacity
Cancer (SEER)	3,390	18	19.3%	15.7%	Breast cancer caused death
Aging (CCHS)	7,611	35	34.5%	26.2%	Retirement and disability

4.2 Competitors

We compared TSNN against several state-of-the-art methods. CoxNN [4] replaces the linear exponent of the Cox hazard by a nonlinear artificial neural networks output; TD-Cox [5] extends the Cox model to time-varying variables; AFT [21] assumes a Weibull RUL distribution in our experiments; EN-BJ [1] extends the least squares estimator to the semi-parametric linear regression model in which the error distribution is completely unspecified; MTLR [13] models RUL distribution by combining multi-task logistic regression in a dependent manner, with the regularization parameter chosen via an additional 10-fold cross validation (10CV); RSF [6] estimates conditional cumulative failure hazard by aggregating tree-based Nelson-Aalen estimators.

We also studied TSNN with simplified configurations, yielding three models as follows. SNN does not estimate the latent risk. Rather, it predicts the output probabilities using Eq. 1 with the time-varying input $\mathbf{x}(t)$; KM-TSNN uses a Kaplan-Meier (KM) estimator to fill in the RULs for censored cases, according to the method introduced in [17]; KM-SNN uses a KM estimator to fill in the RULs for censored cases in SNN. The parameters for competitors were those used in the original papers. For TSNN, KM-TSNN, SNN and KM-SNN, we set the hidden layer to $H = 4$ neurons. An output layer with $K = 20$ was used in analyses of the Engine and Battery datasets, and $K = 12$ in the Cancer and Aging datasets. The penalty parameter λ was chosen through an independent 10CV on the training data. The decay coefficient $G = 1.5$ was used in TSNN and KM-TSNN.

4.3   Evaluation Metrics

Performance on the N_{te} test entities was evaluated in terms of three independent metrics: the failure AUC (FAUC), the concordance index (C-index) and the censoring Brier score (CBS), redefined as follows ($\mathbb{1}$ is the indicator function)

$$\text{FAUC} = \frac{\sum_{i:\epsilon_i(\tau_K)=0} \sum_{j:\epsilon_j(\tau_K)=1} \mathbb{1}\left\{S_K(\mathbf{x}_i) < S_K(\mathbf{x}_j)\right\}}{|\{i : \epsilon_i(\tau_K) = 0\}| \times |\{j : \epsilon_j(\tau_K) = 1\}|}$$

$$\text{C-index} = \frac{\sum_{i:\epsilon_i(\tau_K)=0} \sum_{j:T_i<T_j} \mathbb{1}\left\{S_{\min\{\mathcal{K}_i^0\}}(\mathbf{x}_i) < S_{\min\{\mathcal{K}_i^0\}}(\mathbf{x}_j)\right\}}{|\{i : \epsilon_i(\tau_K) = 0\}| \times |\{j : T_i < T_j\}|}$$

$$\text{CBS} = \frac{1}{N_{\text{te}}} \sum_{i=1}^{N_{\text{te}}} \left(1 - \varepsilon_i(\tau_K) - S_K(\mathbf{x}_i)\right)^2.$$

FAUC provides a probability measure of classification ability at a pre-specified time snapshot (e.g., at τ_K in our case). It qualifies the model's ability to address the issue *"Is i likely to remain failure-free by time t?"* C-index serves as a generalization of the FAUC, giving an estimate of how accurately to answer the question *"Which of i and j is more likely to remain failure-free?"* CBS measures an ensemble prediction error across the test data, i.e., the power of a model to address the issue *"How accurate is the prediction that i will remain failure-free?"*.

4.4   Results and Discussion

From the 10CV results on the test data, shown in Table 2, it is evident that TSNN outperforms all the other models but FAUC yielded by MTLR on the Cancer dataset. The alternatives SNN and KM-TSNN perform second-best, with the sole exception of FAUC on the Cancer dataset (second-best results yielded by TD-Cox) and FAUC on the Aging dataset (by EN-BJ). The superior performance of TSNN relative to KM-TSNN, and of SNN relative to KM-SNN, reveal that our survival learning approach to minimize the censoring KL divergence can effectively cope with censored data in comparison to the conventional survival estimator. Comparing TSNN with SNN and KM-TSNN with KM-SNN, we find that TSNN and KM-TSNN perform much better. This demonstrates the significance and effectiveness of estimating the latent failure risk. CoxNN yields even lower accuracies in comparison to TD-Cox, demonstrating that use of the risk nonlinearity property alone does not enhance the Cox model [7]. Note that TSNN and SNN take into account potential relationships between the classifications and therefore achieve a significant performance gain over the regression method MTLR which performs each prediction task independently [13]. Note also the extremely low CBS achieved by TSNN on the four datasets indicates high accuracy in predicting the absolute RUL-specific probability and high confidence in forecasting failure.

Table 2. Comparison of the 10CV FAUC, C-index and CBS results on the test data, in the form of mean (standard deviation). The best results are in bold and the second-best performances are underlined.

	FAUC	C-index	CBS	FAUC	C-index	CBS
	Engine			Battery		
TSNN	**.744**(.017)	**.753**(.028)	**.163**(.018)	**.810**(.022)	**.761**(.014)	**.212**(.029)
SNN	.719(.038)	_.724_(.026)	_.185_(.023)	_.769_(.015)	.710(.029)	_.229_(.038)
KM-TSNN	.731(.024)	.678(.040)	.248(.011)	.695(.032)	_.733_(.015)	.261(.034)
KM-SNN	.676(.022)	.639(.029)	.283(.016)	.674(.021)	.656(.019)	.255(.018)
CoxNN	.686(.036)	.613(.028)	.404(.025)	.664(.049)	.718(.013)	.332(.026)
TD-Cox	_.740_(.047)	.587(.029)	.276(.018)	.754(.028)	.686(.017)	.301(.048)
AFT	.682(.014)	.636(.053)	.241(.042)	.625(.030)	.674(.020)	.274(.022)
EN-BJ	.736(.029)	.688(.015)	.339(.012)	.718(.024)	.654(.034)	.237(.013)
MTLR	.708(.051)	.683(.023)	.215(.043)	.726(.020)	.670(.015)	.364(.019)
RSF	.695(.019)	.675(.031)	.268(.031)	.578(.029)	.520(.041)	.286(.031)
	Cancer			Aging		
TSNN	_.794_(.013)	**.782**(.029)	**.186**(.017)	**.787**(.028)	**.765**(.031)	**.151**(.019)
SNN	.785(.034)	_.756_(.017)	_.217_(.008)	.706(.020)	.722(.018)	.221(.015)
KM-TSNN	.694(.041)	.681(.024)	.226(.047)	.730(.016)	_.736_(.022)	_.166_(.027)
KM-SNN	.663(.032)	.639(.018)	.322(.014)	.707(.010)	.645(.029)	.224(.011)
CoxNN	.733(.038)	.674(.019)	.235(.034)	.721(.022)	.717(.016)	.301(.032)
TD-Cox	.753(.010)	.642(.025)	.297(.018)	.652(.045)	.628(.038)	.359(.007)
AFT	.689(.034)	.564(.028)	.263(.036)	.707(.037)	.660(.024)	.305(.026)
EN-BJ	.767(.023)	.745(.033)	.279(.014)	_.742_(.044)	.720(.022)	.235(.018)
MTLR	**.818**(.022)	.739(.025)	.243(.017)	.716(.017)	.734(.026)	.324(.030)
RSF	.732(.017)	.673(.037)	.272(.053)	.722(.035)	.684(.025)	.336(.027)

The censoring KL divergence based survival learning may enable TSNN (and SNN) to recommend the right moment for preventive intervention. For this investigation, we performed a case study on the Engine dataset. The engines that experienced failure were divided into 6 groups according to their times to failure. In each sub-figure of Fig. 2, we plotted an RUL curve according to the average RUL-specific probability predicted by each model on the corresponding group of engine failures. It can be seen from the respective gray areas that

TSNN (plotted by the salmon dashed curve) yields a significantly lower average probability over all data (i.e., all engine failures) in comparison to other models, mainly because latent risk estimation can help in amending the relationship between latent risk and RUL-specific probability. This means that, using our TSNN, the equipment crew could be issued a warning much earlier than in the other models, and offered advice on maintenance intervention in time to stave off potential failure.

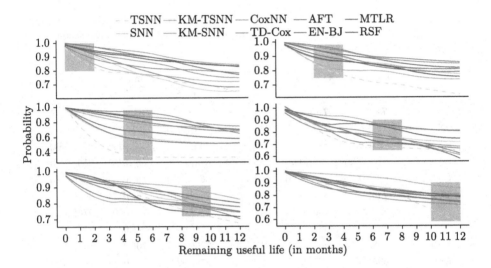

Fig. 2. Change in predicted RUL-specific probability curve for engines. The 6 sub-figures are plotted for the engines that failed at intervals of 2 month (see each gray rectangle), from the 1st month to the 12th month. Every curve in each sub-figure is the average predicted probability of the engines.

In order to provide a deeper insight into the functionality of TSNN, we set a varying K value of 1, 2, 3, 4, 6, 8, 12 and 24 when it runs on the Aging dataset (in 2-year study period), with the output time interval becoming 24, 12, 8, 6, 4, 3, 2 and 1 month(s), respectively. (Please keep in mind that K is a user-defined value and the time interval is not required to be equal.) The FAUC, C-index and CBS results shown in Fig. 3 change less than 8%, 11% and 9%, respectively; this demonstrates that users can count on TSNN as reliable, as it won't fluctuate enormously with change in the output layer of neural networks.

Figure 4 shows the average results of TSNN with a varying decay coefficient G, which might lead to an inaccurate risk estimate and therefore a poor predictive ability when it becomes extremely large or small. It can be seen clearly that TSNN achieves high FAUC and C-index results, and maintains a low CBS when it takes a value in the range [1,2].

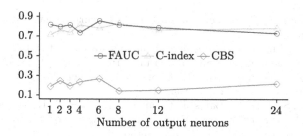

Fig. 3. Change in TSNN performance on the aging dataset with varying K

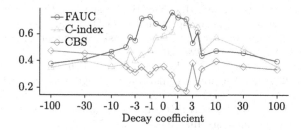

Fig. 4. Change in TSNN performance on the aging dataset with varying G

5   Conclusions

In this paper, we proposed a data-driven TSNN model for RUL prediction. TSNN performs an additive latent failure risk estimation and multiple binary classifications for predicting RUL-specific probabilities. The new survival learning approach optimizes a neural network by minimizing the censoring KL divergence between the resulting probabilities and the actual survival process. In addition, the learning criterion constrains the RUL-specific probability to decrease as time elapses. Experimental results on four lifetime datasets confirm that our model outperforms several state-of-the-art models and is therefore a good candidate for developing a decision-making assistance system to help with preventive intervention.

Acknowledgments. This work was supported by the National Natural Science Foundation of China (NSFC) under Grant No. 61672157, the Natural Sciences and Engineering Research Council of Canada (NSERC) under Grant No. 396097-2010, the program PAFI of Centre de Recherche du CHUS.

References

1. Buckley, J., James, I.: Linear regression with censored data. Biometrika **66**(3), 429–436 (1979)
2. Chen, L., Wang, S.: Central clustering of categorical data with automated feature weighting. In: IJCAI, pp. 1260–1266 (2013)
3. Cox, D.R.: Regression models and life tables. J. R. Stat. Soc. Ser. B. Stat. Methodol. **34**, 187–220 (1972)

4. Faraggi, D., Simon, R.: A neural network model for survival data. Stat. Med. **14**(1), 73–82 (1995)
5. Fisher, L.D., Lin, D.Y.: Time-dependent covariates in the Cox proportional-hazards regression model. Annu. Rev. Public Health **20**(1), 145–157 (1999)
6. Ishwaran, H., Kogalur, U.B., Blackstone, E.H., Lauer, M.S.: Random survival forests. Ann. Appl. Stat. **2**, 841–860 (2008)
7. Katzman, J., Shaham, U., Bates, J., Cloninger, A., Jiang, T., Kluger, Y.: DeepSurv: personalized treatment recommender system using a Cox proportional hazards deep neural network. BMC Med. Res. Methodol. **18**, 18–24 (2018)
8. Kim, H., Golub, G.H., Park, H.: Imputation of missing values in DNA microarray gene expression data. In: CSB, pp. 572–573 (2004)
9. Li, H., Ge, Y., Zhu, H., Xiong, H., Zhao, H.: Prospecting the career development of talents: a survival analysis perspective. In: KDD, pp. 917–925 (2017)
10. Li, Y., Wang, J., Ye, J., Reddy, C.K.: A multi-task learning formulation for survival analysis. In: KDD, pp. 1715–1724 (2016)
11. Li, Y., Wang, L., Wang, J., Ye, J., Reddy, C.K.: Transfer learning for survival analysis via efficient L2, 1-norm regularized Cox regression. In: ICDM, pp. 231–240 (2017)
12. Li, Y., Xu, K.S., Reddy, C.K.: Regularized parametric regression for high-dimensional survival analysis. In: SDM, pp. 765–773 (2016)
13. Lin, H.C., Baracos, V., Greiner, R., Chun-Nam, J.Y.: Learning patient-specific cancer survival distributions as a sequence of dependent regressors. In: NIPS, pp. 1845–1853 (2011)
14. Michalewicz, Z., Schoenauer, M.: Evolutionary algorithms for constrained parameter optimization problems. Evol. Comput. **4**(1), 1–32 (1996)
15. Moghaddass, R., Rudin, C.: The latent state hazard model, with application to wind turbine reliability. Ann. Appl. Stat. **9**(4), 1823–1863 (2014)
16. Sinha, D., Ibrahim, J.G., Chen, M.: A Bayesian justification of Cox's partial likelihood. Biometrika **90**(3), 629–641 (2003)
17. Street, W.N.: A neural network model for prognostic prediction. In: ICML, pp. 540–546 (1998)
18. Vinzamuri, B., Li, Y., Reddy, C.K.: Active learning based survival regression for censored data. In: CIKM, pp. 241–250 (2014)
19. Wang, L., Li, Y., Zhou, J., Zhu, D., Ye, J.: Multi-task survival analysis. In: ICDM, pp. 485–494 (2017)
20. Wang, P., Li, Y., Reddy, C.K.: Machine learning for survival analysis: a survey. ACM Comput. Surv. **51**(6), 1–36 (2019)
21. Wei, L.J.: The accelerated failure time model: a useful alternative to the Cox regression model in survival analysis. Stat. Med. **11**(14–15), 1871–1879 (1992)
22. Wilamowski, B.M., Yu, H.: Neural network learning without backpropagation. IEEE Trans. Neural Netw. **21**(11), 1793–1803 (2010)
23. Wu, Y., Yuan, M., Dong, S., Lin, L., Liu, Y.: Remaining useful life estimation of engineered systems using vanilla LSTM neural networks. Neural Comput. **275**, 167–179 (2018)
24. Yu, S., et al.: Privacy-preserving Cox regression for survival analysis. In: KDD, pp. 1034–1042 (2008)
25. Zhang, J., Chen, L., Vanasse, A., Courteau, J., Wang, S.: Survival prediction by an integrated learning criterion on intermittently varying healthcare data. In: AAAI, pp. 72–78 (2016)
26. Zhang, J., Wang, S., Courteau, J., Chen, L., Bach, A., Vanasse, A.: Predicting COPD failure by modeling hazard in longitudinal clinical data. In: ICDM, pp. 639–648 (2016)

ACNet: Aggregated Channels Network for Automated Mitosis Detection

Kaili Cheng, Jiarui Sun, Xuesong Chen, Yanbo Ma, Mengjie Bai,
and Yong Zhao$^{(\boxtimes)}$

School of Electronic and Computer Engineering,
Peking University Shenzhen Graduate School, Shenzhen, China
chengkaili@pku.edu.cn, yongzhao@pkusz.edu.cn

Abstract. Mitosis count is a critical predictor for invasive breast cancer grading using the Nottingham grading system. Nowadays mitotic count is mainly performed on high-power fields by pathologists manually under a microscope which is a highly tedious, time-consuming and subjective task. Therefore, it is necessary to develop automated mitosis detection methods that can save a large amount of time for pathologists and enhance the reliability of pathological examination. This paper proposes a powerful and effective novel framework named ACNet to count mitosis by aggregating auxiliary handcrafted features associated with tissue texture into CNN and jointly training neural network in an end-to-end way. Completed Local Binary Patterns (CLBP) features, Scale Invariant Feature Transform (SIFT) features and edge features are extracted and used in the classification task. In the process of network training, we expand the original training set by utilizing hard example mining, making our network focus on classification of the most difficult cases. We evaluate our ACNet by conducting experiments on the public MITOSIS dataset from MICCAI TUPAC 2016 competition and obtain state-of-the-art results.

Keywords: Mitosis detection · Breast histopathology · CLBP · SIFT · Edge

1 Introduction

Digital pathology is one of the important and challenging research areas in modern medicine. Pathological examination plays a crucial role in the diagnosis process. Histopathological grading of breast cancer is a quantitative and qualitative assessment which provides prior knowledge of the patient's prognosis and helps to develop further treatment plans. According to the Nottingham Grading System [1], there are three morphological features for the grading of breast cancer on Hematoxylin and Eosin (H&E) stained slides, including tubular differentiation, nuclear atypia and mitotic count. Among them, mitotic count is a critical predictor for breast cancer diagnosis [2]. Nowadays mitotic count is mainly performed on high-power fields (HPFs) by pathologists manually under a microscope. It is a very tedious, time-consuming and subjective task due to the large number of cells in the slides and great difference in the appearance of mitotic figures [3]. Therefore, it is important to develop a method for automatically

Q. Yang et al. (Eds.): PAKDD 2019, LNAI 11439, pp. 453–464, 2019.
https://doi.org/10.1007/978-3-030-16148-4_35

detecting mitosis, which can save a lot of time for pathologists, and also reduce the subjectivity of diagnosis and improve the reliability of pathological examination.

However, there are many challenges in automatically detecting mitotic figures from H&E stained slides for the following reasons. First, numerous structures of various shapes exist in mitotic figures. The development of mitotic cells can be divided into four main phases: prophase, metaphase, anaphase and telophase (see Fig. 1(a)). During the four phases, the shape of nucleus appears differently. Second, there are some other cell types (like apoptotic cells, lymphocytes) whose appearance is very similar to mitotic cells, resulting in a lot of false positives in the detection process (see Fig. 1(b) (c)). In addition, mitotic cells are considerably less than non-mitotic cells. The low probability of their occurrence makes the detection more challenging.

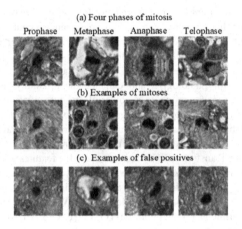

Fig. 1. Some examples of mitoses and false positives. It can be seen that they are very similar in appearance.

In recent years, some automatic methods have been proposed to detect mitosis in breast histological images. The existing approaches are generally divided into two types. Early studies usually employed handcrafted features which capture particular characteristics of mitotic cells for automatic detection [4, 6, 10, 18, 22]. The handcrafted features usually contain morphological (such as shape and nucleus contour), statistical and textural characteristics of mitosis. Some of methods combine two or more of these features to improve detection accuracy. Although these handcrafted features cannot describe the appearance of mitosis well enough as a large variation of mitotic cells, they still guide us to detect mitotic cells to some extent. The other is based on the abstract features automatically learned from deep convolutional neural network [9, 12, 13, 23, 24]. Handcrafted features correspond to what learned from the lower layers of CNN while features extracted from higher layers of CNN are abstract and comprehensive [17]. Both handcrafted and abstract features are important for mitosis detection. One drawback to CNN-based detectors is that although the convolution and pooling layers produce high-level semantic activation maps, they also obscure the boundaries between adjacent instances. An intuitive solution is to utilize additional low-level apparent features (such as edges) to address location defects by providing detailed appearance information to detectors. However, aggregating handcrafted features which can boost classification task into deep convolutional neural network and jointly training the network with extra channel features in an end-to-end way are seldom studied in mitosis detection.

Motivated by previous work which integrated extra features into deep CNN in generic object detection [14–16], in this paper, we propose an effective novel framework named ACNet—aggregated channels network for automated mitosis detection. Our method is to concatenate auxiliary feature associated with tissue texture to original stained normalized image patch in breast histopathology. We choose Completed Local Binary Patterns (CLBP) [19], Scale Invariant Feature Transform (SIFT) [21] and edge which are the most favorable textural features to better discriminate mitosis and other objects. The advantages of these features can be attributed to simple computation as well as specific statistical patterns.

What's more, class-imbalance is also a crucial problem affecting detection effect. The number of non-mitotic cells and the number of true mitosis on a Hematoxylin and Eosin (H&E) stained slide can reach more than 1000:1 in the mitosis detection task, which introduces unbalanced samples problem during neural network training. To tackle this problem, during the training stage, we expand our original training set with additional hard negative samples and hard positive samples predicted by a preliminary network in order to make our network to focus on classification of the most difficult cases. We evaluate our ACNet by conducting experiments on the public MITOSIS dataset from MICCAI TUPAC 2016 competition and obtain the highest F1-score exceeding ever records.

2   Related Work

Early studies usually employed handcrafted characteristics in specific fields to describe the morphological, statistical or textural features of mitosis [4, 6, 10, 18, 22]. They are designed on the basis of pathologists' knowledge of mitosis. Most proposed methods follow a two-step object detection method [11]. The first step is to extract the candidate objects from the original image and then usually classify it as either mitoses or non-mitoses by support vector machines (SVM), random forest and Adaboost in the second step. For instance, Sommer et al. [4] proposed a pixel-wise classifier including comprehensive analysis of texture and shape features for mitosis detection in breast histological images. Tashk et al. [18] used completed local binary pattern (CLBP) to extract texture features robust to rotation and color variation. Irshad et al. [22] extracted texture, SIFT features and modified biologically inspired model of HMAX respectively, then performed SVM and decision tree classifiers. However, since there are a large variation of morphologies and textures characteristics of mitoses, these handcrafted features are hard to describe all mitoses in a high detection accuracy.

The hierarchical feature that are automatically learned by the deep convolutional neural network can bring about better detection results for mitotic cells compared with the handcrafted features [9, 12, 13, 23, 24]. Ciresan et al. [23] adopted a sliding window way to directly apply the deep neural network to histological image which is very computationally intensive so that the method is not suitable for application in the clinic. Paeng et al. [12] presented a unified framework to predict tumor proliferation scores based on molecular data and mitotic counts. The framework consists of three modules. The first module is used to process the whole slide image, the second is a mitosis detection network based on deep learning, and the third is a proliferation scores prediction module. Their approach obtained 0.652 F1-score in mitosis detection which was the first place in the MICCAI TUPAC 2016 competition for mitosis detection task. Zerhouni et al. [13] applied Wide Residual Networks to detect mitosis in breast histology images and performed a post-processing operation to the network output to help filter out noise and select true mitosis.

It has been demonstrated that the aggregation of different types of channel features is useful in many decision-forest-based object detectors [14–16]. Park et al. [15] incorporate optical flow and temporal difference features into a boosted decision forest to improve both pedestrian detection and human pose estimation working on video clips. Yang et al. [16] proposed a method called CCF which uses the low-level features from pre-trained CNN models on ImageNet as channel features. CCF is proved to be a great approach to tailor pre-trained CNN models to various tasks by obtaining good performances in the field of face, pedestrian and edge detection. Mao et al. [14] conducted extensive experiments to explore how CNN-based pedestrian detectors can benefit from different types of extra channel features including ICF channel, edge channel, heatmap channel and depth channel. However, aggregating extra features that facilitate classification tasks into deep convolutional neural network and jointly learning network with the extra channel features are rarely studied in mitosis detection. We propose an effective aggregated channels network for automated mitosis detection and achieve a better detection performance.

3   Methods

The features learned from the lower layers of CNN correspond to general characters such as edges and textures, and the features from higher layer are more abstract and class-specific. An overview of the proposed ACNet framework is shown in Fig. 2. We respectively aggregate the features channel including CLBP, SIFT and edge into stained normalized image patch before the convolution layer of CNN. The parameters of the convolutional neural network are presented in Table 1.

3.1   CLBP

Ojala et al. [20] raised to employ Local Binary Pattern (LBP) which is a simple and effective operator in gray-scale and rotation invariant texture classification. LBP has been applied in the field of face recognition and shape location. The method we adopted in this paper to get additional feature map channels was an improved LBP named completed LBP (CLBP) proposed by Guo et al. [19]. CLBP consisted of three operators: CLBP_C, CLBP_S and CLBP_M, which respectively code the image local center gray level, the sign and magnitude features of local difference.

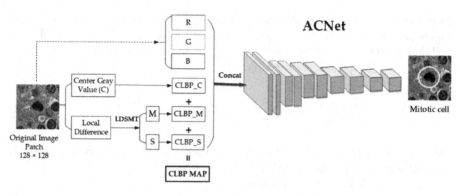

Fig. 2. An overview of the proposed ACNet framework aggregating CLBP features. When aggregating SIFT or edge features, simply replace their feature map with the CLBP map.

Table 1. Architecture of the ACNet

Layer name	Output size	Block (kernel size, output channel)
Input	128 × 128	–
Conv1	64 × 64	7 × 7,64, stride = 2
Residual block1	64 × 64	$\begin{bmatrix} 1 \times 1 \text{ conv}, 64 \\ 3 \times 3 \text{ conv}, 64 \\ 1 \times 1 \text{ conv}, 128 \end{bmatrix} \times 3$
Residual block2	32 × 32	$\begin{bmatrix} 1 \times 1 \text{ conv}, 64 \\ 3 \times 3 \text{ conv}, 64 \\ 1 \times 1 \text{ conv}, 128 \end{bmatrix} \times 3$
Residual block3	16 × 16	$\begin{bmatrix} 1 \times 1 \text{ conv}, 64 \\ 3 \times 3 \text{ conv}, 64 \\ 1 \times 1 \text{ conv}, 128 \end{bmatrix} \times 3$
Residual block4	8 × 8	$\begin{bmatrix} 1 \times 1 \text{ conv}, 64 \\ 3 \times 3 \text{ conv}, 64 \\ 1 \times 1 \text{ conv}, 128 \end{bmatrix} \times 2$
Avg_pool	2 × 2	4 × 4 average pool
Conv2	1 × 1	2 × 2, 128, stride = 1
Fc1	160D fully-connected	
Fc2	2D fully-connected, softmax	

* Note that each "conv" layer corresponds to the sequence BN-ReLU-Conv.

The CLBP_S operator is equivalent to the original LBP operator. Given a pixel in the image, calculate a CLBP_S code by comparing it with its neighbors:

$$CLBP\_S_{N,R} = \sum_{n=0}^{N-1} s(g_n - g_c)2^n, s(x) = \begin{cases} 1, x \geq 0 \\ 0, x < 0 \end{cases} \tag{1}$$

where g_c represents the gray value of the central pixel, g_n represents the value of its neighboring pixels, R is the radius of the neighborhood and N is all number of neighboring pixels.

The CLBP_M operator is defined in (2):

$$CLBP\_M_{N,R} = \sum_{n=0}^{N-1} t(m_n, c)2^n, t(x, c) = \begin{cases} 1, x \geq c \\ 0, x < c \end{cases} \tag{2}$$

where m_n is the magnitude of the difference between g_c and g_n as described above, c is the average of m_n from the whole image.

And the CLBP_C operator is defined in (3):

$$CLBP\_C_{N,R} = \sum_{n=0}^{N-1} t(g_c, a_I), t(x, c) = \begin{cases} 1, x \geq c \\ 0, x < c \end{cases} \tag{3}$$

where a_I represents the average gray value of the whole image.

The whole ACNet framework aggregating CLBP features is shown in Fig. 2. At first, the original image is represented by the local difference and the center pixel gray level (C). Then a local difference sign-magnitude transform (LDSMT) is used to decompose the local difference into two parts, including sign (S) and magnitude (M) components. After that, applying CLBP_S, CLBP_M and CLBP_C operators to code the S, M and C features, respectively. These three code maps have the same form so as to be easily fused into CLBP feature map. Finally, we concatenate the obtained CLBP map with the original input image and feed it into classification network.

3.2 SIFT

Scale Invariant Feature Transform (SIFT) feature extraction [21] is a widely known method which converts an image into a large set of local feature vectors. Each local feature vector is invariant to image scaling, translation and rotation. This is why we choose the SIFT features to apply in classification of mitosis patch. In SIFT methods, the difference-of-Gaussian (DoG) function is applied to calculate a series of features in scale space. After selecting a set of features, Euclidean distance and the full set of matching are used to compare the features of new images with these candidate regions. The scale-invariant features can be identified effectively by using a hierarchical filtering method. Then, a histogram of features is calculated as the image descriptors. The descriptors are concatenated with original image after performing reshape and normalization.

3.3 Edge

Since chromosomes condensation occurs at the onset of mitosis, the intensity patterns of mitoses and non-mitoses are usually different [5]. Therefore, in most cases, the mitotic nucleus appears denser and "darker" than the non-mitotic nucleus. We first filter the image patch, leaving the regions whose intensity is greater than the mean of the whole image, and then apply Sobel edge detector to generate edge map.

3.4   Hard Examples Mining

In the mitosis detection task, the number of non-mitotic cells and the number of samples of mitotic cells on a Hematoxylin and Eosin (H&E) stained slide have a very large ratio which can reach more than 1000:1. Therefore, it will introduce a new challenging problem of unbalanced samples during neural network training. We adopt hard examples mining method to solve this problem since it always contains a large number of simple samples and a handful of hard samples in detection datasets. Hard examples mining can focus the classification power of the network on the most difficult examples, making the training more effective and efficient.

During the training stage of the network, we first train a preliminary network on the whole training data until the loss function of neural network almost tends to be stable. Then we apply the preliminary trained network to make evaluations on all the training data. This yields a probability of each image patch being a mitosis. We apply a threshold p and the predicted probability above this threshold would be considered as a positive sample while others are deemed to negative samples. We choose these hardest image patches including negative samples predicted by the network as mitoses with the highest probability and positive samples missed by the network to expand our original training data set. Then we continue training the network.

4   Experimental Evaluation

We evaluate the proposed method in a public MITOSIS dataset from MICCAI TUPAC 2016 competition for mitosis detection task [8]. The evaluation measurements of algorithm is F1 score: $F_1 = \frac{2 \times R \times P}{R + P}$, where R is recall: $R = \frac{TP}{TP + FN}$ and P is precision: $P = \frac{TP}{TP + FP}$. TP, FP and FN represent the number of true positives, false positives and false negatives, respectively. The whole ACNet is implemented based on TensorFlow deep learning framework using Python. Experiments are carried out on a Linux server with NVIDIA Tesla P40 24 GB GPU.

4.1   Dataset

TUPAC2016 MITOSIS dataset contains 690 high-power fields (HPFs) at 40X magnification stained with H&E, of which 656 HPFs from 73 patients are used for training and 34 HPFs from 34 patients are used for testing. The first 23 training images are consistent with the AMIDA13 challenge dataset [11] collected from the Department of Pathology at the University Medical Center in Utrecht. The remaining 50 images come from two different pathology centers in The Netherlands. Images of TUPAC2016 are produced with Leica SCN400 leading to a spatial resolution of 0.25 μm per pixel. The size of each HPF is 2000 × 2000 pixels or 5657 × 5657 pixels. Note that the input to our framework is a cropped 128 × 128 patch. In this dataset, the annotation only contains the single centroid coordinate of each mitosis. According to evaluation criteria, a correct detection is that the Euclidean distance between the coordinates provided and the groundtruth is less than 7.5 μm (30 pixels) [11].

4.2 Implementation Details

Medical diagnosis usually stains slides by Hematoxylin and Eosin (H&E). Nucleic acids would be selectively dyed to a blue-purple hue with Hematoxylin while proteins would be dyed to a bright pink color with Eosin. Due to the differences existing in staining manufacturers, storage times and staining steps, slide preparation varies widely. The number of false positives may increase when slide is over-stained. So we employ staining normalization to all the images both in the training datasets and testing datasets at the first step. We apply the method described in [7] to normalize the staining of the whole images.

For the positive samples, we extract a 128×128-pixel patch from the whole images after staining normalization centered on the coordinates given in groundtruth. Since the number of positive samples is quite small and unbalanced relative to the number of negative samples, we need to augment the positive samples set by translation and rotation. In our experiments, we increased 30 times as the original number of positive samples through performing random translations at five times and rotations of $30, 60, 90, 120, 150, 180°$.

For the negative samples, we first extend a 64-pixel mirroring border in each boundary of the whole images. Then, we extract a 128×128-pixel patch by applying a sliding windows on the whole images with a stride of 32. Patches located at more than 30 pixels away from the center of the mitosis are considered as negative samples. We do not need to do augmentation for the negative samples.

All the networks are trained with stochastic gradient descent (SGD) with a batch size of 256. The initial learning rate is set to 0.1 and is reduced to 10 times at epoch 5, 10, 15 and 25. We train all models for total 30 epochs, of which 10 epochs are trained for the preliminary network and 20 epochs are continued after adding hard examples. We set weight decay to 10^{-5} and momentum of MomentumOptimizer to 0.9. We use dropout rate of 0.5 after convolutional layers to prevent overfitting.

4.3 Experimental Results and Comparison

To validate the entire proposed architecture, we split the dataset into five random and nearly equal subsets (S1-S5) and perform a 5-fold cross-validation same as the method used in [5]. We first quantitatively assess the effects of aggregating three different features (CLBP, SIFT and edge) on one subset (S1). Then a 5-fold cross-validation is carried out for the best performing features and the average is compared to other methods evaluated on TUPAC2016 MITOSIS dataset [8].

Usually models integrated with the extra channel features bring out improvement relative to baseline. So for a fair comparison, we need to add a controlled experiment where the original image patch is used as an extra channel. The F1 score for different features is reported in Table 2.

Table 2. F1-score on S1 subset by aggregating different features

Method	Precision	Recall	F1-score
Baseline	0.446	0.909	0.598
ACNet (+original image)	0.452	0.955	0.614
ACNet (+SIFT)	0.770	0.608	0.679
ACNet (+edge)	0.642	0.803	0.713
ACNet (+CLBP)	0.817	0.680	**0.742**

It can be seen that there is no obvious improvement by taking original image as extra channel input, confirming that the performance boost is indeed due to channel feature aggregation. Among three features, CLBP feature contributes the most to detection performance owing to it can convey so much discriminative information of local structure. A precision-recall curve (PR curve) of aggregating CLBP feature is plotted in Fig. 3. The classification threshold p is set to 0.90.

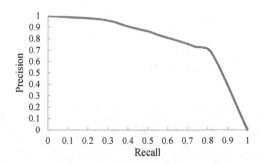

Fig. 3. A PR curve of aggregating CLBP feature for TUPAC2016 dataset.

Next, we perform a 5-fold cross validation for aggregating CLBP feature and compare its average with the top six results participating in MICCAI TUPAC 2016 competition for mitosis detection task [8] from the groups Lunit [12], IBM Research [13], Contextvison, CUHK, Microsoft Research Asia and Radboud UMC. The details are shown in Tables 3 and 4. Table 3 shows that F1 scores across different subsets of dataset remain almost consistent. Our method achieves the highest F1-score and 10% improvement over the best method available whereas the standard deviation is only 0.0298, demonstrating the effectiveness of the proposed algorithm.

Figure 4 shows some detection results examples by ACNet (+CLBP) on TUPAC2016 MITOSIS dataset. Despite a few false positives and false negatives still exist in the final results, most true mitoses could be successfully detected by our method.

Table 3. Average F1-score for 5-fold cross validation by ACNet (+CLBP)

Subset of dataset	Precision	Recall	F1-score
S1	0.817	0.680	0.742
S2	0.763	0.752	0.757
S3	0.681	0.667	0.674
S4	0.675	0.723	0.698
S5	0.718	0.712	0.715
Mean	0.731	0.707	0.717
Standard deviation	0.0533	0.0304	0.0298

Table 4. F1 scores of our method with other competing methods for TUPAC2016 MITOSIS dataset.

Method	F1-score
Radboud UMC	0.541
Microsoft Research Asia	0.596
CUHK	0.601
Contextvison	0.616
IBM Research [13]	0.648
Lunit [12]	0.652
ACNet (+CLBP)	**0.717**

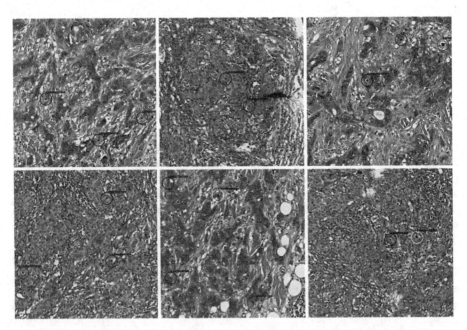

Fig. 4. Some detection results examples by ACNet (+CLBP) on TUPAC2016 MITOSIS dataset. Yellow, blue and green circles denote true positives, false negatives and false positives, respectively. The number near the circle represents the predicted probability. (Color figure online)

5 Conclusion

In this paper, we propose a novel automated mitosis detection method based on aggregating different features into deep CNN. Completed Local Binary Patterns (CLBP), Scale Invariant Feature Transform (SIFT) and edge features are evaluated through conducting experiments on TUPAC2016 MITOSIS dataset. Results show that CLBP feature contributes the most to detection performance and our method achieves significant performance improvement over all other competition methods. In future, we plan to research model based on more features for continuing improving the results of mitosis detection.

Acknowledgments. This work was supported by Science and Technology Planning Project of Shenzhen (No. NJYJ20170306091531561), Science and Technology Planning Project of Shenzhen (No. JCYJ20160506172651253), and National Science and Technology Support Plan, China (No. 2015BAK01B04).

References

1. Elston, C.W., Ellis, I.O.: Pathological prognostic factors in breast cancer. I. The value of histological grade in breast cancer: experience from a large study with long-term follow-up. Histopathology **19**(5), 403–410 (1991)
2. Genestie, C., et al.: Comparison of the prognostic value of Scarff–Bloom–Richardson and Nottingham histological grades in a series of 825 cases of breast cancer: major importance of the mitotic count as a component of both grading systems. Anticancer Res. **18**(1B), 571–576 (1998)
3. Veta, M., Van Diest, P.J., Jiwa, M., Al-Janabi, S., Pluim, J.P.: Mitosis counting in breast cancer: Object-level interobserver agreement and comparison to an automatic method. PLoS ONE **11**(8), e0161286 (2016)
4. Sommer, C., Fiaschi, L., Hamprecht, F.A., Gerlich, D.W.: Learning-based mitotic cell detection in histopathological images. In: Proceedings of the 21st International Conference on Pattern Recognition, ICPR 2012, pp. 2306–2309. IEEE (2012)
5. Paul, A., Mukherjee, D.P.: Mitosis detection for invasive breast cancer grading in histopathological images. IEEE Trans. Image Process. **24**(11), 4041–4054 (2015)
6. Irshad, H.: Automated mitosis detection in histopathology using morphological and multi-channel statistics features. J Pathol. Inform. **4**(1), 1–10 (2013)
7. Macenko, M., Niethammer, M., Marron, J.S., et al.: A method for normalizing histology slides for quantitative analysis. In: IEEE International Symposium on Biomedical Imaging: From Nano to Macro, ISBI 2009, pp. 1107–1110. IEEE (2009)
8. Tumor-proliferation-assessment-challenge (2016). http://tupac.tue-image.nl/
9. Wang, H., Cruz-Roa, A., Basavanhally, A., et al.: Cascaded ensemble of convolutional neural networks and handcrafted features for mitosis detection. In: Medical Imaging 2014: Digital Pathology, vol. 9041, pp. 90410B. International Society for Optics and Photonics (2014)
10. Tek, F.B.: Mitosis detection using generic features and an ensemble of cascade adaboosts. J Pathol. Inform. **4**(12), 1–12 (2013)
11. Veta, M., Van Diest, P.J., Willems, S.M., Wang, H., Madabhushi, A., Cruz-Roa, A., et al.: Assessment of algorithms for mitosis detection in breast cancer histopathology images. Med. Image Anal. **20**(1), 237–248 (2015)

12. Paeng, K., Hwang, S., Park, S., Kim, M.: A unified framework for tumor proliferation score prediction in breast histopathology. In: Cardoso, M.J., et al. (eds.) DLMIA/ML-CDS -2017. LNCS, vol. 10553, pp. 231–239. Springer, Cham (2017). https://doi.org/10.1007/978-3-319-67558-9_27

13. Zerhouni, E., Lányi, D., Viana, M., Gabrani, M.: Wide residual networks for mitosis detection. In: 2017 IEEE 14th International Symposium on Biomedical Imaging, ISBI 2017, pp. 924–928. IEEE (2017)

14. Mao, J., Xiao, T., Jiang, Y., Cao, Z.: What can help pedestrian detection? In: 2017 IEEE Conference on Computer Vision and Pattern Recognition, CVPR 2017, pp. 6034–6043. IEEE (2017)

15. Park, D., Zitnick, C.L., Ramanan, D., Dollár, P.: Exploring weak stabilization for motion feature extraction. In: Proceedings of the IEEE Conference on Computer Vision and Pattern Recognition, pp. 2882–2889 (2013)

16. Yang, B., Yan, J., Lei, Z., Li, S.Z.: Convolutional channel features. In: Proceedings of the IEEE International Conference on Computer Vision, pp. 82–90 (2015)

17. He, K., Zhang, X., Ren, S., Sun, J.: Deep residual learning for image recognition. In: Proceedings of the IEEE Conference on Computer Vision and Pattern Recognition, pp. 770–778 (2015)

18. Tashk, A., Helfroush, M.S., Danyali, H., Akbarzadeh, M.: An automatic mitosis detection method for breast cancer histopathology slide images based on objective and pixel-wise textural features classification. In: The 5th Conference on Information and Knowledge Technology, pp. 406–410. IEEE (2013)

19. Guo, Z., Zhang, L., Zhang, D.: A completed modeling of local binary pattern operator for texture classification. IEEE Trans. Image Process. 19(6), 1657–1663 (2010)

20. Ojala, T., Pietikainen, M., Maenpaa, T.: Multiresolution gray-scale and rotation invariant texture classification with local binary patterns. IEEE Trans. Pattern Anal. Mach. Intell. 24(7), 971–987 (2002)

21. Lowe, D.G.: Object recognition from local scale-invariant features. In: Proceedings of the Seventh IEEE International Conference on Computer Vision, vol. 2, pp. 1150–1157. IEEE (1999)

22. Irshad, H., et al.: Automated mitosis detection using texture, SIFT features and HMAX biologically inspired approach. J. Pathol. Inform. 4(Suppl) (2013)

23. Cireşan, D.C., Giusti, A., Gambardella, L.M., Schmidhuber, J.: Mitosis detection in breast cancer histology images with deep neural networks. In: Mori, K., Sakuma, I., Sato, Y., Barillot, C., Navab, N. (eds.) MICCAI 2013. LNCS, vol. 8150, pp. 411–418. Springer, Heidelberg (2013). https://doi.org/10.1007/978-3-642-40763-5_51

24. Chen, H., Dou, Q., Wang, X., Qin, J., Heng, P.A.: Mitosis detection in breast cancer histology images via deep cascaded networks. In: Thirtieth AAAI Conference on Artificial Intelligence, AAAI 2016, pp. 1160–1166 (2016)

Attention-Based Hierarchical Recurrent Neural Network for Phenotype Classification

Nan Xu[1,2], Yanyan Shen[2(✉)], and Yanmin Zhu[1,2(✉)]

[1] Shanghai Engineering Research Center of Digital Education Equipment,
Shanghai, China
[2] Department of Computer Science and Engineering, Shanghai Jiao Tong University,
Shanghai, China
{xunannancy,shenyy,yzhu}@sjtu.edu.cn

Abstract. This paper focuses on labeling phenotypes of patients in Intensive Care Unit given their records from admission to discharge. Recent works mainly rely on recurrent neural networks to process such temporal data. However, such prevalent practice, which leverages the last hidden state in the network for sequence representation, falls short when dealing with long sequences. Moreover, the memorizing strategy inside the recurrent units does not necessarily identify the key health records for each specific class. In this paper, we propose an attention-based hierarchical recurrent neural network (*AHRNN*) for phenotype classification. Our intuition is to remember all the past records by a hierarchical structure and make predictions based on crucial information in the label's perspective. To the best of our knowledge, it is the first work of applying attention-based hierarchical neural networks to clinical time series prediction. Experimental results show that our model outperforms the state-of-the-arts in accuracy, time efficiency and model interpretability.

Keywords: Temporal data · Classification · Attention mechanism

1 Introduction

Currently, the broad adoption of Electronic Health Record (EHR) makes it possible to access unprecedented amount of clinical data. Some relevant applications have improved the quality of health care and lowered the engendered cost [1, 20].

Phenotype classification, also known as "phenotyping", is a relatively new yet critical medical informatics problem. The main goal is to classify patients by analyzing a series of EHRs including heart rate, blood pressure, etc. The clinically meaningful categories, namely "phenotypes", are decided based on diagnoses assigned to patients at discharge. These categories can be acute conditions such as *cerebrovascular disease* or chronic conditions like *kidney disease*. According to a person's phenotypes, it is worth ranking hospitalizations and managing care plans for him. As precision medicine becomes an emerging approach to disease

© Springer Nature Switzerland AG 2019
Q. Yang et al. (Eds.): PAKDD 2019, LNAI 11439, pp. 465–476, 2019.
https://doi.org/10.1007/978-3-030-16148-4_36

prevention and treatment, patients' phenotypes also play an important role in triggering clinical decision support systems and predicting future health resource utilization [5].

To deal with variable-length clinical sequences, recent researches mainly leverage recurrent neural networks (RNNs) for phenotype classification. The recurrent units in the network, e.g., Long Short-Term Memory (LSTM) [8] and Gated Recurrent Unit (GRU) [4], determine how much of the past memory to preserve and what to collect from the current data. For example, Lipton et. al. established a simple RNN with LSTM units for classifying 128 phenotypes given 13 clinical variables [11]. By analyzing 17 physiologic features, Harutyunyan et. al. provided a phenotyping benchmark on 25 labels with a 2-layer LSTM network.

Although RNNs have achieved better classification results compared with traditional models, they still have limitations as follows. The first is **high time complexity**. The sequential computation in recurrent units makes the time complexity proportional to the length of sequence, which becomes problematic for long sequences. The second is incapacity to capture **long-range dependency**. To represent the whole sequence, RNNs utilize the hidden state at the last time step. But it has been proved insufficient in domains like machine translation [18] and speech recognition [3]. The third is **no interpretability**. As one of the neural network models, RNNs are black boxes that provide no explicit explanation of predictions by introspecting model parameters.

To overcome the above limitations, we introduce a novel hierarchical framework with two levels of RNNs for phenotype classification. After splitting the long sequence of records into small blocks, the first level is fed with individual block data and generates corresponding vector representations to reveal the short-term health status. The second level adopts block representations and develops an awareness of the overall health condition. Since serial computation is only performed intra-blocks, time complexity of our model is proportional to the number of records in each block, rather than the total sequence length. Moreover, the hierarchical model obtains a global awareness of the patient's health state after collecting local clinical condition information, which is quite a novel attempt for **recognizing long-range dependency**.

We have also observed that different phenotypes show different time sensitivities. For example, symptoms of *acute respiratory distress syndrome* may not appear until 24–48 h after lung injury, while symptoms of *asthma attack* may present shortly after admission but remit over time [11]. To take advantage of such divergence of temporal attention, the first level of AHRNN utilizes an extra attention mechanism for better block representation. Specifically, records with high attention weights to a certain phenotype are identified as key roles in predicting the chance of having this phenotype. Such easy-to-interpret attention mechanism makes the classification results more **explainable** and acceptable.

We evaluate *AHRNN* on a large-scale EHR dataset and it outperforms previous models with better prediction performance as well as higher time efficiency. A case study further shows model interpretability.

We summarize our main contributions as follows:

- We propose a novel hierarchical recurrent neural network for phenotype classification, where a global knowledge of the patient's health state is generated after analyzing his local clinical conditions. This is the first hierarchical model to analyze clinical time series.
- To exploit temporal sensitivities of distinct phenotypes, we incorporate the attention mechanism in the first level of *AHRNN* to recognize key records.
- We evaluate our model on a large-scale open dataset and it outperforms the state-of-the-arts with relative low time consumption. Model's interpretability owing to the attention module also show clearly.

2 Preliminary

2.1 Problem Statement

In this paper, we aim to predict the likely diseases the patients have suffered from admission to discharge. Note that phenotyping is a multi-label classification problem as patients in ICU are typically diagnosed with multiple conditions. Notations are defined in Table 1 and the problem is defined as follows:

Problem 1. Given records of one patient denoted by $\boldsymbol{X} \in \mathbb{R}^{N \times k}$, the goal is to estimate the binary vector $\boldsymbol{y} \in \mathbb{R}^L$, where $y_i = 1$ indicates the patient has the i-th disease and $y_i = 0$ means no related symptoms to such disease are observed.

2.2 A Basic RNN Solution

Vanilla RNN, LSTM, GRU are the most basic recurrent units in RNNs while the last two achieve better performance due to their complex internal structures. We here introduce a toy model equipped with LSTM units for demonstration while the other two follow the similar way.

LSTM. It is short for long short-term memory and consists of one cell to remember historical values and three gates to regulate the information flow into and out of the cell. The three gates are input gate for remembering current input, forget gate for removing past memory and output gate for selective update, respectively.

We denote the computation function for updating the LSTM internal state by $h_t = \text{LSTM}(\boldsymbol{x}_t, h_{t-1})$, where $\boldsymbol{x}_t$ and h_t are the input and hidden state at time t, respectively. Details of this procedure can be found in [8].

Phenotype Classification. Given records $\boldsymbol{X} \in \mathbb{R}^{N \times k}$, a RNN model with a single LSTM layer predicts the probability of having the i-th phenotype as: $\overline{\boldsymbol{y}} = \sigma_g(\boldsymbol{W}_p h_N + \boldsymbol{b}_p)$, where $\boldsymbol{W}_p \in \mathbb{R}^{L \times d}$, $\boldsymbol{b}_p \in \mathbb{R}^L$ are trainable variables.

As Fig. 1 shows, the LSTM unit updates its internal state recursively until the end of the sequence is encountered, which requires $O(N)$ serial computation. Thus time complexity is $O(N \cdot kd)$ for the first hidden layer and $O(Nd^2)$ for each following layer if there are more than one recurrent layer.

Table 1. Notations. Here '#' means 'the number of'.

Symbol	Meaning
k	Input dimension
L	#phenotypes to be classified
n	#records per block
N	#records in total
m	Dimension of label embedding
d	#units per RNN layer

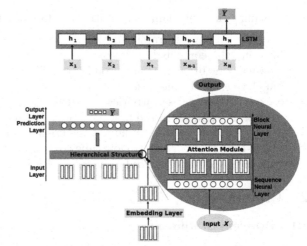

Fig. 1. Two models for phenotyping: LSTM (upper) and AHRNN (lower).

3   Methodology

In this section, we first present our neural network model for phenotype classification and then describe the learning process. We finally discuss the time complexity.

3.1   AHRNN Model

Given one patient's records $\boldsymbol{X} \in \mathbb{R}^{N \times k}$ and $\boldsymbol{l}_i$ as the one-hot representation of the i-th phenotype, the computations performed by *AHRNN* is shown in Fig. 1 and defined from bottom layer to top layer as:

$$Emb(\boldsymbol{l}_i) = Embedding(\boldsymbol{l}_i), \tag{1}$$

$$SNL_{i,b}(\boldsymbol{X}) = Attend(RNN_{seq}(\{\boldsymbol{x}_{b,t}\}_{t=1,\dots,n}), Emb(\boldsymbol{l}_i)), \tag{2}$$

$$BNL_i(\boldsymbol{X}) = RNN_{block}(\{SNL_{i,b}(\boldsymbol{X})\}_{b=1,\dots,\lceil \frac{N}{n} \rceil}), \tag{3}$$

$$\overline{y}_i(\boldsymbol{X}) = Prediction_i(BNL_i(\boldsymbol{X})), \tag{4}$$

where Eq. 1 lists the implementation of label representation in *embedding layer*, Eqs. 2 and 3 present the functionality of *sequence neural layer* and *block neural layer* in the hierarchical structure, Eq. 4 shows how *prediction layer* works.

In what follows, we elaborate the four equations with detailed network layer description.

Embedding Layer. Similar to word embedding, we map the i-th label's one hot representation $\boldsymbol{l}_i = \{l_{i,j} \mid 1 \leq j \leq L; l_{i,j} = 1 \; if \; j = i \; otherwise \; 0\}$ into an embedding vector $Emb(\boldsymbol{l}_i) \in R^m$ through one layer of multilayer perceptron

(MLP). This process is expressed by function *Embedding* in Eq. 1 and can be formulated as:

$$Embedding(l_i) = \boldsymbol{W}_i^e \cdot l_i, \tag{5}$$

where $\boldsymbol{W}_i^e \in \mathbb{R}^{m \times L}$ denotes the weight to learn for the i-th label.

Frequently co-occurred phenotypes could trace back to identical lesion locations and cause similar symptoms, resulting in coincident temporal sensitivities. Through this layer, representations for related phenotypes tend to be close in the embedding space during training and generate similar attention scores for records at each time step (see details in Sect. 3.1).

Bottom of Hierarchy: Sequence Neural Layer. Similar to the map procedure in MapReduce programs [6,13], we first split the sequential measurements into small groups containing equal amount of records. The last block, which normally has fewer records than earlier blocks, is padded with zeros for concurrent computation during training. The recurrent units update the hidden states until the last record in the block is encountered. Such serial computation only takes place inside each block. Thus the independent computation in distinct blocks can be carried out in parallel.

Attention Module. As shown in Fig. 1, conventional RNNs for sequence modeling normally preserve the last hidden state to represent the whole time series, which is insufficient for long sequences.

We employ the attention mechanism to alleviate the burden of remembering the whole input sequence, and let the critical constituent to pay different attention to inputs at different time steps. Specifically, the phenotype labels determine the attention score of each record by modeling the correlation between the label embedding and the hidden state of input. Here we consider two ways to learn the correlation:

- **Map-Attention**: the hidden state and the label embedding are multiplied with the help of a trainable matrix $\boldsymbol{W}_i^{map}$. That is, $\eta(\boldsymbol{h}_{b,t}, Emb(l_i)) = \boldsymbol{h}_{b,t}^T \cdot \boldsymbol{W}_i^{map} \cdot Emb(l_i)$,
- **Concat-Attention**: label embedding is concatenated to the hidden state and the concatenation is transformed to a scalar with a trainable matrix $\boldsymbol{W}_i^{con}$. That is, $\eta(\boldsymbol{h}_{b,t}, Emb(l_i)) = \boldsymbol{W}_i^{con} \cdot [\boldsymbol{h}_{b,t}; Emb(l_i)]$,

where $\boldsymbol{h}_{b,t} \in \mathbb{R}^d$ is the hidden state of the t-th record in block b and $\eta \in \mathbb{R}$ is a scalar denoting the attention weight of $x_{b,t}$ by the i-th label.

According to the attention scores, we leverage the weighted sum of all past hidden states in a block as a local health condition summary. *Attend* function in Eq. 2 can be expressed as:

$$\alpha_{i,b,t} = \frac{\exp^{\eta(h_{b,t}, Emb(l_i))}}{\sum_{s=1}^n \exp^{\eta(h_{b,s}, Emb(l_i))}}, \tag{6}$$

$$SNL_{i,b}(\boldsymbol{X}) = \sum_{t=1}^n \alpha_{i,b,t} \boldsymbol{h}_{b,t}. \tag{7}$$

Top of Hierarchy: Block Neural Layer. A basic recurrent neural layer is adopted here to acquire global health information by analyzing the input block vectors as Eq. 3 shows. Different from previous layers where distinct parameter sets are leveraged for different labels, trainable parameters like weight and bias in this layer are shared among all the classes. Such parameter sharing is expected to map the phenotype's awareness of overall condition with their attentive signals regardless of the label type and the signal content.

Prediction Layer. After establishing the hierarchical network with label-based attention, different phenotypes obtain different understandings of the patient's overall health condition. We then apply a dense layer denoted by *Prediction* in Eq. 4 to project the health condition's vector representation $BNL_i(\boldsymbol{X})$ to a final probability. This function is implemented by one layer of MLP:

$$Prediction_i(BNL_i(\boldsymbol{X})) = \sigma_g(\boldsymbol{w}_i^p \cdot BNL_i(\boldsymbol{X}) + b_i^p), \qquad (8)$$

where $\boldsymbol{w}_i^p \in \mathbb{R}^d$ and $b_i^p \in \mathbb{R}$ are the weight and bias, σ_g is a sigmoid function. The resulted scalar denotes the probability of this patient attacked by phenotype l_i.

3.2 Learning

For the multi-label classification task, we optimize the pointwise cross entropy to force the prediction score $\overline{y}_i(\boldsymbol{X})$ to be close to $y_i(\boldsymbol{X})$:

$$L = \sum_{1 \leq i \leq L, \boldsymbol{X} \in \mathcal{D}} -y_i(\boldsymbol{X}) \log(\overline{y}_i(\boldsymbol{X})) - (1 - y_i(\boldsymbol{X})) \log(1 - \overline{y}_i(\boldsymbol{X})), \qquad (9)$$

where $\mathcal{D}$ is the record dataset.

3.3 Time Complexity Analysis

AHRNN requires $O(n)$ sequential operations. Thus the time cost is $O(Lm)$ for *label embedding layer*, $O(nd^2)$ per *sequence neural layer*, $O(\lceil \frac{N}{n} \rceil d^2)$ per *block neural layer*. To compute attention scores, it takes $O(nL(2d + m))$ time for *concat-attention* or $O(nL(dm + m + d))$ for *map-attention*. If *AHRNN* is deep that L and m are much less than d, then time consumption for label embedding and attention is negligible and the overall time complexity is $O((n + \lceil \frac{N}{n} \rceil)d^2)$, which is much lower than $O(Nd^2)$ for plain RNNs as mentioned in Sect. 2.2.

In this paper, computation for *prediction layer* is not discussed as both models have this layer. We assume that the intra- and inter-unit computations in each layer are sequential. Superiority of *AHRNN* over plain RNN in time efficiency still exists if they are conducted in parallel.

4 Experimental Evaluation

In this section, we conduct experiments to answer the following questions:

RQ1 How does our model perform as compared to the state-of-the-art methods for the phenotyping task?

RQ2 Does the hierarchical framework really help speed up training?
RQ3 Can we interpret the predictions made by *AHRNN*?

4.1 Dataset

To analyze time series data in health care domain, most of the studies use their institutions' own private datasets, which are not beneficial to reproducing the reported results [15]. Instead, we use the publicly available MIMIC-III dataset [9]. It covers 53,423 distinct hospital admissions for adult patients admitted to critical care units between 2001 and 2012. For a fair comparison, we follow the same way of splitting and preprocessing data as [7]: 17 physiologic variables and 25 phenotype categories are selected, 70%, 15%, 15% of the total 41,902 stays are allocated for the training, validation and testing set, respectively. More data statistics could be found in [7].

4.2 Evaluation Metrics

Similar to [7,16], Area Under the ROC Curve (AUC) is utilized for performance evaluation. We also adopt F1 score in consideration of data imbalance. Both of the two metrics have three versions, i.e., micro-, macro- and weighted-average. F1 metrics require a thresholding strategy, and in experiments we select 0.5 as the threshold.

4.3 Compared Methods

We compare performance of the following models.

- **LR**: logistic regression [7].
- **LSTM**: this 2-layer LSTM was proposed for the single phenotyping task [7].
- **SAnD**: Simply Attend and Diagnose, this model makes predictions solely based on self-attention mechanism [16].
- **HRNN**: hierarchical recurrent neural network without the attention module.
- **AHRNN_map**: HRNN with *map-attention*.
- **AHRNN_concat**: HRNN with *concat-attention*.

4.4 Implementation

We use Tensorflow[1] for network construction, ADAM [10] optimizer with a 10^{-4} learning rate, and one TITAN Xp graphics card for all the experiments. There are two hidden layers prior to *prediction layer*: one is *sequence Neural Layer* with 1024 LSTM units and the other is *block Neural Layer* with 512 LSTM units. The batch size is 8 for model *LSTM* and *HRNN*, 4 for *AHRNN* due to memory limitation. The block size n for *HRNN*, *AHRNN_map* and *AHRNN_concat* are 5, 6 and 6, respectively. The label embedding dimension m for the latter two

---

[1] https://www.tensorflow.org/.

Table 2. Performance of different models. The number after ± means standard deviation. Results marked with * are from the original papers while the others are implemented by us. F1 scores for *SAnD* are '-' since the codes haven't been published. For all the measures, the higher the better, the best in boldface.

Model	AUC			F1 score		
	Micro	Macro	Weighted	Micro	Macro	Weighted
LR	0.801*	0.741*	0.732*	0.320 ± 0.0004	0.240 ± 0.0003	0.299 ± 0.0003
LSTM	0.821*	0.771*	0.759*	0.388 ± 0.026	0.300 ± 0.026	0.363 ± 0.027
SAnD	0.816*	0.766*	0.754*	-	-	-
HRNN	0.823 ± 0.001	0.775 ± 0.001	0.762 ± 0.001	0.391 ± 0.010	0.303 ± 0.008	0.365 ± 0.010
AHRNN_concat	$\mathbf{0.825 \pm 0.002}$	$\mathbf{0.777 \pm 0.003}$	$\mathbf{0.764 \pm 0.002}$	0.402 ± 0.010	0.312 ± 0.010	0.376 ± 0.010
AHRNN_map	$\mathbf{0.825 \pm 0.002}$	$\mathbf{0.777 \pm 0.002}$	$\mathbf{0.764 \pm 0.002}$	$\mathbf{0.408 \pm 0.012}$	$\mathbf{0.317 \pm 0.011}$	$\mathbf{0.382 \pm 0.012}$

AHRNN models are 25, 12, respectively. In our models, the learning rate, batch size, and network scale are the same as those in the LSTM model [7]. The other hyperparameters are selected by grid search according to their performance on the validation set. With different random seeds, all the models are trained and tested for 10 times. In each turn, the performance are evaluated on the testing set when the loss on the validation set reaches the lowest during training.

4.5   Main Results

Table 2 shows the average performance of different models evaluated with AUC and F1 score. To compare four recurrent models, Fig. 2 illustrates their performance on the validation set during training and Table 3 gives some time-consuming statistics prior to model convergence.

Performance Comparison (RQ1). In Table 2, either the *map-attention* or the *concat-attention* module assists the hierarchical recurrent model to outperform the existing state-of-the-arts when measured in AUC. From the view of F1 score, *AHRNN_map* is superior to the others with values 5.2%, 5.7% and 5.2% higher than the previously proposed recurrent model *LSTM* in micro-, macro- and weighted-average, respectively. Meanwhile, *HRNN* performs competitively or slightly better than the non-hierarchical *LSTM* in all the metrics. Figure 2 shows the superiority of model *HRNN*, *HARNN_concat* and *HARNN_map* than *LSTM* in jumpstart and asymptotic performance. With the observations above, we can safely come to the following main conclusions. Firstly, generating a global awareness of the patient's health state on the basis of the local clinical condition under the hierarchical framework won't hurt the overall performance. Secondly, allowing the diverse phenotypes to pay different attention to vital signs at different time steps will improve the accuracy of prediction.

Time Efficiency (RQ2). As listed in Table 3, the three hierarchical models takes far fewer seconds to accomplish the training for one epoch, e.g., *HRNN* spends nearly one fifth of time for *LSTM*. The two attention-based models consume more seconds than *HRNN* as extra computation for temporal concentration is executed. Until the end of the 12th epoch where the loss converges,

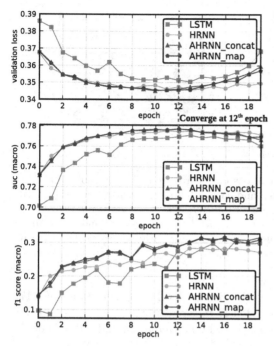

Fig. 2. Performance of recurrent models on the validation set during training. Macro-average results are shown as they reveal the overall performance. The loss of the models all converges at 12th epoch.

Table 3. Time-consuming statistics during training for recurrent models. #params means the number of trainable variables and is listed for model comparability. Epoch time in unit second (s) is the average training time for each epoch. Total time in unit hour (h) is the time the model spends until convergence (after the 12th epoch). For the two time measures, the lower the better, the best in boldface.

Model	Attribute	Value
LSTM	#params	7.67 M
	Epoch time	528 s
	Total time	1.9 h
HRNN	#params	7.67 M
	Epoch time	**117 s**
	Total time	**0.42 h**
AHRNN_concat	#params	7.72 M
	Epoch time	254 s
	Total time	0.91 h
AHRNN_map	#params	8.32 M
	Epoch time	253 s
	Total time	0.91 h

models with the hierarchical framework spends fewer than one hour while the non-hierarchical model consumes almost 2 h. The experimental results coincide with the theoretical time complexity analysis in Sect. 3.3 that the employment of the hierarchy structure speeds up training.

4.6   Case Study for Interpretability (RQ3)

We here focus on a sampled patient from the test set to demonstrate the explainability of *AHRNN*. This patient has 41 records in total and is labeled with two phenotypes which are frequently co-occur due to lesions in the cardiovascular system: *acute myocardia infraction* and *coronary atherosclerosis*. The probabilities of being diagnosed with these two phenotypes by our best model are 0.91 and 0.89, respectively.

In Fig. 3, the two heat maps reveal that the predictions of cardiovascular-related phenotypes pay the most attention to the last two records, where a sudden increase can be observed in blood-pressure-related vital signs, such as DBP, MBP and SBP. This is quite reasonable since extensive clinical researches have pointed out that having high blood pressures damages arteries, accelerates

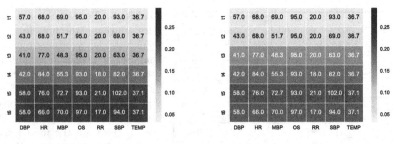

(a) Acute myocardia infraction   (b) Coronary atherosclerosis

Fig. 3. Attention scores of two phenotypes on records from the last block. Each row represents the record at one time step and each column represents a vital sign. DBP, HR, MBP, OS, RR, SBP and TEMP are *diastolic blood pressure, heart rate, mean blood pressure, oxygen saturation, respiratory rate, systolic blood pressure* and *temperature*, respectively. We only show these features as the others are either categorical or invariant along time.

the buildup of plaque and increases the risk of developing a heart attack [17]. Therefore, the prediction of the previously referred probabilities of two phenotypes are interpretable, since they are based on observation and analysis of the critical moments.

5   Related Work

Recent Phenotying Methods. Lipton et. al. were among the first to apply RNNs to classify ICU patients according to their time series vital signs [11]. Later on, a few works emerged to provide benchmarks with RNN models for different clinical tasks including phenotyping [7,14]. An attention-based sequence modeling architecture named SAnD was proposed and achieved similar performance as previous benchmarks [16]. Based on the conventional RNNs, our model utilize the hierarchical framework for time efficiency and the attention mechanism for accuracy improvement.

Hierarchical Recurrent Neural Networks. Hierarchical RNNs have been widely adopted in the natural language processing since a paragraph or a document is composed of closely linked sentences and a sentence is a combination of semantically related words [12,19,21]. Different from the well-organized text data, there are no apparent temporal boundaries for consecutive medical records. In our *AHRNN* model, different boundary settings are tried through the hyperparameter n for block size. Similarly, no clear shot boundaries exist among frames for hierarchical video representation. The model in [22] was trained only on dual-shot samples and the unique boundary for each video was manually labeled. Hence the proposed model is not practical for multi-shot videos.

(Soft) Attention-based Recurrent Neural Networks. The attention mechanism utilizes several variables (hidden states at different time steps in RNN) and a context to generate a summary of these variables according to their correlations with the context [2,23]. For instance, relevance between station-oriented measurements and local urban features [2], between local time-frequency features and global sequence trend [23] are considered to compute attention scores in air quality inference and time series forecasting, respectively.

6   Conclusion

This paper tackles the challenging task of predicting whether a patient in ICU suffers from several phenotypes, which is important for improving health care and predicting future resource utilization. We study the limitations of conventional RNNs, and then address the need of finding the distinguished temporal concentration of different phenotypes, which is overlooked by previous studies. We propose an attention-based hierarchical recurrent neural network model ($AHRNN$) to capture the long-range dependency of phenotype prediction on past measured physiologic features, and employ an attention module to generate label-oriented predictions. Experimental results show that our method outperforms the state-of-the-arts in accuracy, time efficiency and model interpretability.

Acknowledgments. This research is supported in part by NSFC (No. 61772341, 61472254) and STSCM (No. 18511103002). This work is also supported by the Program for Changjiang Young Scholars in University of China, the Program for China Top Young Talents, the Program for Shanghai Top Young Talents, and Shanghai Engineering Research Center of Digital Education Equipment.

References

1. Bayati, M.: Data-driven decision making in healthcare systems (2011)
2. Cheng, W., Shen, Y., Zhu, Y., Huang, L.: A neural attention model for urban air quality inference: learning the weights of monitoring stations. In: Thirty-Second AAAI Conference on Artificial Intelligence (2018)
3. Chiu, C.C., et al.: State-of-the-art speech recognition with sequence-to-sequence models. In: 2018 IEEE International Conference on Acoustics, Speech and Signal Processing (ICASSP), pp. 4774–4778. IEEE (2018)
4. Cho, K., et al.: Learning phrase representations using RNN encoder-decoder for statistical machine translation. arXiv preprint arXiv:1406.1078 (2014)
5. Healthcare Cost and Utilization Project: Clinical classifications software (CCS) for ICD-9-CM. https://www.hcup-us.ahrq.gov/toolssoftware/ccs/ccs.jsp. Accessed 11 May 2011
6. Dean, J., Ghemawat, S.: MapReduce: simplified data processing on large clusters. Commun. ACM **51**(1), 107–113 (2008)
7. Harutyunyan, H., Khachatrian, H., Kale, D.C., Galstyan, A.: Multitask learning and benchmarking with clinical time series data. arXiv preprint arXiv:1703.07771 (2017)
8. Hochreiter, S., Schmidhuber, J.: Long short-term memory. Neural Comput. **9**(8), 1735–1780 (1997)

9. Johnson, A.E., et al.: MIMIC-III, a freely accessible critical care database. Sci. Data **3**, 160035 (2016)

10. Kingma, D.P., Ba, J.: Adam: a method for stochastic optimization. arXiv preprint arXiv:1412.6980 (2014)

11. Lipton, Z.C., Kale, D.C., Elkan, C., Wetzell, R.: Learning to diagnose with LSTM recurrent neural networks. arXiv preprint arXiv:1511.03677 (2015)

12. Nallapati, R., Zhou, B., Gulcehre, C., Xiang, B., et al.: Abstractive text summarization using sequence-to-sequence RNNs and beyond. arXiv preprint arXiv:1602.06023 (2016)

13. Peng, Z., et al.: Mining frequent subgraphs from tremendous amount of small graphs using MapReduce. Knowl. Inf. Syst. **56**(3), 663–690 (2018)

14. Purushotham, S., Meng, C., Che, Z., Liu, Y.: Benchmark of deep learning models on large healthcare mimic datasets. arXiv preprint arXiv:1710.08531 (2017)

15. Shickel, B., Tighe, P.J., Bihorac, A., Rashidi, P.: Deep EHR: a survey of recent advances in deep learning techniques for electronic health record (EHR) analysis. IEEE J. Biomed. Health Inf. **22**(5), 1589–1604 (2018)

16. Song, H., Rajan, D., Thiagarajan, J.J., Spanias, A.: Attend and diagnose: clinical time series analysis using attention models. arXiv preprint arXiv:1711.03905 (2017)

17. Staessen, J.A., Wang, J., Bianchi, G., Birkenhäger, W.H.: Essential hypertension. Lancet **361**(9369), 1629–1641 (2003)

18. Sutskever, I., Vinyals, O., Le, Q.V.: Sequence to sequence learning with neural networks. In: Advances in Neural Information Processing Systems, pp. 3104–3112 (2014)

19. Tang, D., Qin, B., Liu, T.: Document modeling with gated recurrent neural network for sentiment classification. In: Proceedings of the 2015 Conference on Empirical Methods in Natural Language Processing, pp. 1422–1432 (2015)

20. Xierali, I.M., et al.: The rise of electronic health record adoption among family physicians. Ann. Fam. Med. **11**(1), 14–19 (2013)

21. Xiong, C., Merity, S., Socher, R.: Dynamic memory networks for visual and textual question answering. In: International Conference on Machine Learning, pp. 2397–2406 (2016)

22. Zhao, B., Li, X., Lu, X.: HSA-RNN: hierarchical structure-adaptive RNN for video summarization. In: Proceedings of the IEEE Conference on Computer Vision and Pattern Recognition, pp. 7405–7414 (2018)

23. Zhao, Y., Shen, Y., Zhu, Y., Yao, J.: Forecasting wavelet transformed time series with attentive neural networks. In: 2018 IEEE International Conference on Data Mining (ICDM), pp. 1452–1457. IEEE (2018)

Identifying Mobility of Drug Addicts with Multilevel Spatial-Temporal Convolutional Neural Network

Canghong Jin[1](✉), Haoqiang Liang[2], Dongkai Chen[1], Zhiwei Lin[2], and Minghui Wu[1]

[1] Zhejiang University City College, Huzhou Street 51, Hangzhou, China
{jinch,mhwu}@zucc.edu.cn, 31501324@stu.zucc.edu.cn
[2] Zhejiang University, Zheda Road 38, Hangzhou, China
{lhq,21721335}@zju.edu.cn

Abstract. Human identification according to their mobility patterns is of great importance for a wide spectrum of spatial-temporal based applications. For example, detecting drug addicts from normal residents in public security area. However, extracting and classifying user behaviors in massive amount of moving records is not trivial because of three challenges: (1) the complex transition records with noisy data; (2) the heterogeneity and sparsity of spatiotemporal trajectory features; and (3) extremely imbalanced data distribution of real world data. In this paper, we propose MST-CNN, a multi-level convolutional neural network with spatial and temporal features. We first embed the multiple factors on single trajectory level and then generate a behavior matrix to capture the user's mobility patterns. Finally, a CNN module is used to extract various features with different filters and classify user type. We perform experiments on real-life mobility datasets provided by public security office, and extensive evaluation results demonstrate that our method obtains significant improvement performance in identification accuracy and outperform all baseline methods.

Keywords: Convolutional neural network · Spatiotemporal embedding · Human trajectory pattern · Addict identification

1 Introduction

Addiction to drugs among men and women is an acute social problem faced by most of the countries worldwide. The National Narcotics Control Commission reported that the number of drug addicts is still slowly growing in China. There were more than 2.5 million drug users in the country by the end of 2016 with 6.8% growth [1], while in 2018 World Drug Report, about 275 million people worldwide, which is roughly 5.6 percent of global population aged 15–64 years, used drugs at least once during 2016 [2]. Drug addict not only deeply affects the individual health but also is thought to be guilty of crime or offense, who would lead to public safety concerns. The police department, as one of the most important roles in controlling drug problems, utilize common detection methods like call records analysis, trade records exploration or human face recognition. However, the performance of these methods is not promised

© Springer Nature Switzerland AG 2019
Q. Yang et al. (Eds.): PAKDD 2019, LNAI 11439, pp. 477–488, 2019.
https://doi.org/10.1007/978-3-030-16148-4_37

because they are all based on explicit, implicit or inferred relationships between arrested criminals and suspects, which needs restrict precondition and is limited.

On the other hand, as article [3] reports, drug addicts with low educational level have trouble in rejoining mainstream society and finding a gainful employment. A study on drug in Shandong province of China illustrates that around 75 percent of drug addicts are mobile population, in which most of them are unemployment or self-employment [4]. Since the addicts are special group of people, their daily life might be different from common residents, which could be used for identification task. In this paper, we collect people's movement trajectories to describe user daily life, because trajectory pattern mining is becoming increasingly popular with the development of ubiquitous computing technology and trajectory data contains abundant semantic and geographic information. For example, police turn to Google to find crime suspects by seeking data from mobile phones in target areas [5]. Pickpocket suspect identifying could be leveraged by could be achieved by mining public transit records in Beijing [6]. However, it is still challenging to detect addicts because moving records is trivial and massive and trajectory of each user is sparse. Moreover, unlike data of automated fare collection systems (AFC), people travel in the city do not follow the fixed routed but wander about freely, their movement behaviors are more complicated. Therefore, it is critical to propose a smart surveillance and tracking tool for identifying personal behavior and detecting drug addicts from normal residents.

In this paper, we propose MST-CNN, a multiple level spatial-temporal convolutional neural network model for detecting human identification from lengthy and sparse trajectories. We give the embedding module that converts sparse moving record features (e.g., geographical location, timestamp, number of activity, length of trajectory user) into dense representations, which are then fed into a convolutional neural network to classify user behaviors. Better still, the learned weights offer an easy-to-interpret way to describe the behavior characteristics of drug addicts.

Our contributions can be summarized as follows:

- We propose a spatial-temporal model, MST-CNN, to identify special users from massive trivial moving records. Our model could describe the factors of each trajectory and extract the implicit movement pattern shared in certain group.
- We design mixed mechanisms that are combine both location information and time information by two steps. The first is to directly embed location points into independent latent vectors and interpret trajectory with timestamp factors; while the second is to generate user movement characteristics from relevant trajectories by selection module.
- We evaluate the effectiveness and performance on real-life mobility dataset provided by police department during three months. To the best of our knowledge, there is no existing deep learning approach to solve the similar user identification problem. Compared with the results of traditional classification method and anomaly detection method, the evaluation results demonstrate that MST-CNN outperforms all baselines.

2 Related Work

Spatial-temporal pattern mining has emerged as an active research field, like urban traffic network analysis, automatic intersection recognition and movement behavior mining. In this section, we provide a brief review of the related works, including two categories: movement pattern mining and behavior understanding.

2.1 Movement Patterns Mining

The enormous amount of spatial-temporal data could be used to mine movement pattern. Gong [7] proposes a methodology to detect five travel models (walk, car, bus, subway and commuter rail) from amount of data generated by GPS in New York. In article [8], Pinelli proposes an extension of the sequential pattern mining paradigm to analyze the trajectories of moving objects. REMO (Relative Motion) [9] method is based on traditional cartographic approach of comparing snapshots and develops a comparison method based on motion parameters to reveal the movement patterns. Article [10] presents a complete and computationally tractable model for estimating and predicting trajectories based on sparsely sampled, anonymous GPS land-marks that called GPS snippets. For example, [11] identifies spatiotemporal patterns from GPS traces of taxis for night bus route planning. [12] tries to reflect the common routing preference of the past passengers by finding the most frequent path of a certain time period. [13] discovers and explains movement patterns of a set of moving objects (e.g. track management, bird migration, disease spreading).

These previous works give some inspiration for representation of a trajectory, and traditional machine learning approaches proposed in these works will build baselines for comparison in Sect. 5.

2.2 Behavior Understanding

A number of techniques for understanding user behaviors have also been proposed. For example, article extracts user features from subway transit records and explores abnormal traveling behaviors to discovery the pickpocket suspects [6]. Along the line of location-based anomaly detection, a framework that learns the context of different functional regions in a city is presented, which provides the basis of our feature extraction approach [14].

Traditional trajectory-based similarity calculations use the longest common substring to calculate the similarity of user history trajectories [15, 16]. Abul proposes a W4M (wait for me) method, which uses edit distance to measure the similarity of different paths [17]. Considering the mobility similarity between user group, Zhang et al. [18] proposes GMove modeling method to share significant movement regularity.

In recent research, some deep learning methods are applied to encode the trajectory. ST-ResNet [19] is designed to forecast the flow of crowd. DeepMove [20] model predicts human mobility with attentional recurrent network, while HST-LSTM [21] capture location prediction by Spatial-Temporal LSTM.

Our work refers some ideas to above mentioned embedding techniques and merges additional temporal factor to encode trajectory. Unlike traditional classification models, our model encodes the personal behaviors by its relevant trajectory vectors directly and identify user type by convolutional neural network without the need of effective features.

3 Preliminaries

In this section, we formulate addict classification problem, and briefly introduce the convolutional neural network in sequence classification.

3.1 Problem Formulation

Definition 1 (Moving point): A position point with timestamp and MAC (Media Access Control) which is named as a moving point $O = (p, m, t)$ where p is position information with latitude and longitude, m is a unique mac address and t is timestamp.

Definition 2 (Path): Given a set of moving points O and a special mac a, a path of certain people could be defined as $P = \{O_1, O_2, \ldots, O_n\}$ where $\forall O_i(m) = a$, and $O_i(t) < O_j(t)$ when $i < j$.

Definition 3 (Trajectory): A Trajectory T is a subset of a path with special certain scenario. We give two criterions to present trajectory. (1) $O_i(t) - O_{i-1}(t) < \tau$, where τ is less than 30 min. (2) $distince(O_i(p) - O_{i-1}(p)) > d$, where d is 0.3 km in our model.

In this paper, due to the equipment defects and path semantic context, we quantify the time interval τ is 30 min and distance d is 0.3 km.

Problem 1 (Addict identification): Given a user u and its relevant list of trajectories T_u^i during a period of time, detecting if u is a drug addict or not.

3.2 CNN in Sequence Classification

Unsupervised learning word embedding has achieved tremendous success in such NLP tasks since word2vec was introduced in [22]. Meanwhile, our problem is similar to sequence classification tasks such as sentence classification and sentiment analysis in some degree. To apply CNN in the task of text/sequence classification or sentiment analysis, words/sentences will first be passed to the embedding layer to generate low-dimensional representation vectors, then convolutions will be performed over the embedded word vectors, commonly with multiple filter sizes [23]. With the help of successful implementation of CNN in sequence classification, we are ready to present our model.

4   MST-CNN: Multiple Spatial-Temporal Convolutional Neural Network

4.1   Motivation and Overview

Here we draw two heat maps in Fig. 1. Figure 1(a) and (b) reflect to related points for different types of citizen, addicts and residents respectively. However, they cannot be separated because two maps are similar in macroscopic view, which means those factors may not be quite defining. Fortunately, compared to the residents' moving points, from a micro perspective, addicts visited some unique points, it is mainly because that those people live in different parts of the city. Thus, the residence address could not be treated as an effective feature, if we want the model performs robustly, since transferring a model to other individuals may encounter impasse due to their different latent characteristics.

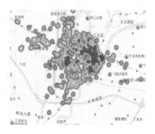

(a) Heat map of addicts   (b) Heat map of residents

Fig. 1.   Activity heat map of addicts and residents. In each figure, we select top-100 visited points for either addicts or residents by calculating total number of relevant records.

Secondly, the behaviors of people in the same group are not as similar as we expected. People living in a city have their own habits of commuting and lifestyle. In order to estimate people's living habits, we collect hundreds of people's visiting points in weekdays, then calculate similarity of different days by *Jaccard* functions. According to Table 1, for example, person m_1 traveled from p_1 to p_9 regions from Monday to Friday. The similarity value will be much higher if m_1 went to the same region every day.

dayLoc and *weekLoc* present the set of region places in weekdays and in a whole week. We define *weekLoc* as the union set of *dayLoc* as $weekLoc = Union(dayLoc(x))$, $x \in Mon, Tus, Wed, Thur, Fri$. Then the similarity could be calculated by

$$J(dayLoc, weekLoc) = \frac{|S(dayLoc) \cap S(weekLoc)|}{|S(dayLoc) \cup S(weekLoc)|} \tag{1}$$

The average similarity value similarity of m1 is $(0.6 + 0.4 + 0.4 + 0.5 + 0.5)/5 = 0.48$, which means its similarity is actually a little low. Although we only present one example here, we extract thousands of people's data to calculate their own similarities. And it turns out that only 20 percent of people have regular activities that could be predicted as Fig. 2 shows that only 20 percent of people are predictable and the others' behavior are irregular.

Table 1. The similarity of residents' transit pattern.

Mac	Day	*GeoLoc* sequence	Similarity
m_1	Monday	$p_1, p_2, p_3, p_4, p_6, p_7$	0.6
m_1	Tuesday	p_2, p_3, p_5, p_8	0.4
m_1	Wednesday	p_2, p_4, p_9, p_{10}	0.4
m_1	Thursday	p_5, p_6, p_7, p_8, p_9	0.5
m_1	Friday	p_1, p_3, p_4, p_7, p_9	0.5

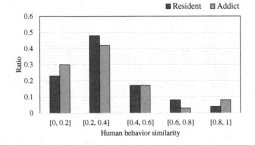

Fig. 2. Distribution of human behavior similarity. We calculate the self-similarity for every user and put them into corresponding groups. Then group distribution for different types of users can be generated.

Finding useful features in user's movement behaviors is an important but difficult job. Actually, it would be too narrow to classify the two different kinds of people by analyzing their moving points or residence, but still some excellent methods could have a breakthrough towards this problem. As the report states, more than half of addicts are less education and self-employment or even unemployment, therefore, which means the activity mode of this special group might be different from normal residents (i.e., trajectory scope, activity time, preferred transit tool). Thus, from those aspects, it may be rather useful and effective to overcome the classified puzzles through mobility patterns, which are based on the spatiotemporal relations and can provide sufficient sematic information, which means it plays a significant role when we want to train or transfer our model. Our extensive experiments in Sect. 5 verify our assumption.

4.2   The MST-CNN Framework

As illustrated in the left part of Fig. 3, we collect moving data from tagged individuals by Wi-Fi probes. We generate users' trajectories including values of locations and time through data cleaning and specific empirical tricks. After owning the trajectories, we embedding the value of positions and time in each trajectory by the methods we mention below. Briefly, on the first level, the trajectory data will be sliced according to a fixed length, and on the second level, we embed the trajectory segments by location embedding and time embedding and concatenate them to a trajectory tensor. And then we generate relevant trajectories, which contains the aforementioned trajectory features: location information, the length and shape of trajectory, activity time and preferred transit tools. Finally, data will be operated by a convolutional neural network to predict identification tag.

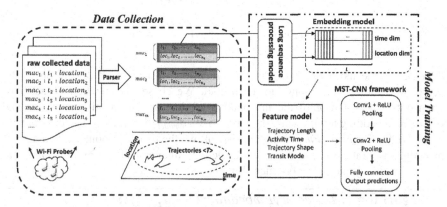

Fig. 3. Identification process with MST-CNN Framework. The process is comprised of four steps from left to right. First, collecting moving data with user tag, then generate trajectories for each user. After that, embedding user behavior into vector space and extract features through different filters. Finally classifying identification by a fully connected layer. MST-CNN proposes a novel embedding method to combine spatiotemporal information and describes user movement behaviors on trajectory and personal levels.

Structure of Trajectory Embedding. Given a trajectory $T_i^n = (O_1, O_2, \ldots O_n)$, where $O_i \in v$ (v is the WiFi Probe), we would like to maximize the $Pr(O_n | O_1, O_2, \ldots, O_{n-1})$ over all the training corpus. The similar problem also has been found in the sequence word2vec, where the input and output have the same resolution [23], thus we do not list its formulas again here but to illustrate our concept briefly. *PointEmb* function takes its input as a large corpus of trajectories and produces a vector space, with each unique location object O_i in the space corpus assigned a corresponding vector v_i in the space with dimension m. Besides the location information embedding, we need to map the activity time of each trajectory to a vector. Given trajectory $T_i^n = (O_1, O_2, \ldots, O_n)$, the activity time is $T_i^n(t) = (O_1(t), O_2(t), \ldots, O_n(t))$ which is a series of timestamps. A high-level view of the human mobility activities could be summarized by transition during time sections, where each time section might contain multiple transit timestamps. In our work, we use t as time section which has 24 dimensions and it presents people appears in O_x point at p th hour in a day, thus, each item of $T_i^n(t)$ can be represented by t_j through its definition.

$$t = [0 \ldots 1 \ldots 0], \; O_x(t) \in p \; \text{th hour} \; p \in [0, 23] \tag{2}$$

The j-th index position in t illustrates the activity behavior at that hour, which presents the human movements on that time dimension. Since the time sections are ordered in sequence, we add "positional encodings" by sine and cosine functions of different indexes [24]:

$$TimeEmb_{(pos,2i)} = sin\left(\frac{pos}{1000^{\frac{2i}{d_{model}}}}\right) \tag{3}$$

$$TimeEmb_{(pos,2i+1)} = cos\left(\frac{pos}{1000^{\frac{2i}{d_{model}}}}\right) \tag{4}$$

where pos is the index of t and i is the dimension. That is, each dimension of the positional encoding corresponds to a sinusoid. We chose this function because we assumed that it would learn the information of time and its relative position. The positional embedding has the same dimension with d_{model}, thus the two can be summed.

Algorithm *TrajectoryEmbedding* consists of two the main components, first a point embedding generated by *PointEmb*, and an activity time embedding by *TimeEmb*, and second, a padding and refine produce (Fig. 4).

Algorithm 1: *TrajectoryEmbedding* (T_i^n, d_p, d_t, l)

Input: trajectories T_i^n, d_p position embedding size,
   d_t time embedding size, l a length to split each trajectory
Output: matrix representations $\Phi \in \mathbb{R}^{|V| \times (l+t)}$
1.   Embed object location: $V_p = PointEmbedding(T_i^n, d_p)$
2.   Given a certain trajectory T_i
3.   **for** $i = 0$ to l **do**
4. $V = [V, V_p]$ or Padding
5.   Activity time embedding: $V_t = TimeEmbedding(T_i^n, d_t)$
6. $V = [V, V_t]$
7.   *return* V and cut off T_i, back to line 3

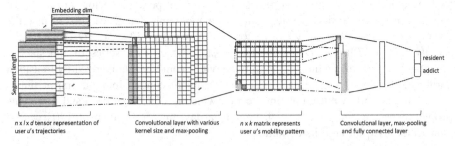

$n \times l \times d$ tensor representation of   Convolutional layer with various $n \times k$ matrix represents   Convolutional layer, max-pooling
user u's trajectories   kernel size and max-pooling   user u's mobility pattern   and fully connected layer

Fig. 4. Network architecture for MST-CNN. The network is comprised of two convolutional and max-pooling layers. Elements with various color are presented as different object location, which could generate factors by diversity filters.

As Algorithm 1 shows, we first embed the location point to vector with random path walk, and then, we split a long path into several short paths with equal length l (line3) in order to keep each trajectory can have the same dimension. Combining every vector V_p in sequence order and padding vector if the length of path less than l (line 4) and computing the activity time V_t and merge it into vector V (line 5–6). By repeating the embedding process, all the trajectories would be embedded to fixed length matrix vectors.

Structure of Personal Encode. A user u with n trajectories could be presented as $v_n^u = [v_1^u, v_2^u, \ldots, v_n^u]$, that we concatenate single trajectory embedding to generate the vector space of user. A convolution operation involves a filter $w \in \mathbb{R}^{hk}$, which is applied to a short segment of h moving objects to produce a new feature. A feature c_i is calculated from a segment $v_{i:i+h-1}^u$ by formula $c_i = f\left(\mathrm{w} \cdot v_{i:i+h-1}^u + b\right)$, where $b \in \mathbb{R}$ is a bias term and f is a non-linear function as ReLu. We then apply a max-pooling function over the feature map and extract the maximum value $\hat{c} = \max\{c\}$ as the particular selected feature. We select various filters (with varying segment length) to generate different features. Finally, a fully connected layer with dropout and softmax output is given to classify user behavior patterns.

5   Experiments

In this section, we evaluate the effectiveness of our model with traditional classifiers. All the experimental data is collected from a city in the east of China. There are more than thirty thousand WiFi Probes installed all over the city, which generate about two billion records every day.

5.1   Data Set and Experimental Platform

Data Sets. According to the top-10 best-selling and most popular phone in China, we first select android brands accounted for over 75% market in 2017. By verifying validity of MAC through MAC API website, we would discard records excluded above brands. And then, we need to distinguish a given MAC belongs to a resident or a tourist. In this work, we discard MAC if it contains records in few days during the month.

Experiment Platform. All the experiments are conducted on two environments. First one is a Cloudera platform with 24 physical machines, which is used to do pre-process and generate dataset. The other platform is a Dell server 64-bit system (16 core CPU, each with 2.6 GHz, GPU GTX 1080ti, 32G main memory). The algorithms and models in our paper were implemented by Python 3.

5.2   Baselines Setup

Our method is compared with a variety of competing methods grouped into the two categories: CM and AD. And since the positive instances (addicts) are extremely low in our experiment, we use under-sample method on negative instances (residents) to balance the data in training process. All the methods will repeat 10 times in train and validate processes with random selected data set and report the averaged results.

Classification Methods (CM). Any supervised machine learning algorithm requires a set of informative, discriminating and independent features. Since the moving data is too trivial, we do some preprocess operations to extract features. For each moving record, we first transfer location information to its related region by GeoHash function, and then split continuous transit time into 15-min slices. Through these steps, we get about 400 geo features and 96 time-window features for each trajectory.

After that, the classification methods, including Native Bayes (NB), Radom Forest (RF), Logistic Regression (LR), Gradient Boosting Decision Tree (GBDT) and k-Nearest Neighbor (KNN), are fitted with training set and evaluated with test data set.

Anomaly Detection (AD). Anomaly detection method is unsupervised and finding outliers by measure the deviation of a given data point with its neighbors. In this work, we use one-class SVM (OCSVM) to identify addicts, which only the negative instances in the training set.

Neural Network Methods (MST-CNN). As aforementioned, MST-CNN uses trajectory embedding and personal embedding in two levels to extract the information of moving data and leverage convolutional neural network to identify the positive instances. We split half data set for training and evaluate in the same test dataset as CM, AD methods did.

The default training settings for our models are presented as follows: learning rate 1e-3, dropout 0.5, kernel [2–4]. We use precision, recall, and F-measure computed with test data to evaluate the effectiveness of three models.

5.3   Result Summary

From Table 2, the precisions of traditional classifiers including RF, GBDT, KNN, LR and NB. The AD method performs somehow better in recall, but still worse in precision, which means that user movement patterns among residents are not as similar as we expect. In contrast, our deep learning approach MST-CNN significantly improve the precision and F score. This observation shows that MST-CNN could reduce the false-positive effectively and perform better in terms of CM and AD methods.

Table 2. Summarizes the performances of MST-CNN method and the baselines listed above.

Category	Classification	Precision	Recall	F1
CM	Random forest	0.0280	0.7851	0.0541
	GBDT	0.0301	0.8018	0.0580
	KNN	0.0369	0.7782	0.0704
	Logistic regression	0.0118	0.6843	0.0231
	Naïve Bayes	0.0165	0.7675	0.0323
AD	One Class SVM	0.0182	**0.9123**	0.0357
CNN	MST-CNN	**0.1906**	0.6500	**0.2948**

5.4   Parameter Sensitivity

In order to evaluate the effectiveness of various parameterization of MST-CNN, we conduct experiments on classification task with vary the number of latent dimensions (d), the length of segment (l). Except for the parameter being tested, all the other parameters assume the default values.

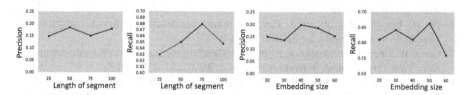

Fig. 5. The study of parameter sensitivity: calculate precision and recall on test dataset with various parameters.

We measure precision and recall as functions of parameters l and d. In Fig. 5(a), we observe that increase the length of segment cannot improve the performance at the same rate and set length to 50 is better than other values by considering both precision and recall, which means longer trajectory does not contain more effective information. We also examine how the embedding size affect the performance and choose 50 as our parameter in other experiments.

6   Conclusion

In this paper, we investigated the problem of user identification from massive and sparse trajectories in the city. We proposed a multiple-level convolutional neural network MST-CNN with both spatial and temporal factors. Both trajectory and user profile are embedded and various features extracted by different filters. Extensive experiments show that our model significantly outperformance all the baselines, including classification models and anomaly detection models in real dataset. Meanwhile, our model is able to effectively capture the identification task in addicts control scenarios.

There are several future directions for our work. First, we only use geographical information and time information to embed user behaviors. Other information like region function is not encoded this time. Second, our current work does not consider the influence of group activities, we detect addicts only based on their personal behavior. Actually, an addict prefers to travel with other addicts. We plan to add these features into our model to describe user movement patterns.

References

1. XINHUANET. http://www.xinhuanet.com/english/2017-03/27/c_136161743.htm. Accessed 10 Sept 2018
2. Supplementary document. http://www.unodc.org/doc/wdr2018/WDR_2018_Press_Release ENG.PDF

3. Zhonghua, S.U., et al.: A longitudinal survey of patterns and prevalence on addictive drug use in general population in five or six areas with high-prevalence in china from 1993 to 2000 part three: demographic characteristics of illicit drug users. Chinese J. Drug Depend. (2005)

4. Yan, W., Jiang, W.W., Zhang, D.S.: A study on drug-taking behavior based on big data: taking Guizhou province as an example. J. Shandong Police Coll. (2017)

5. WRAL.   https://www.wral.com/Raleigh-police-search-google-location-history/17377435. Accessed 10 Sept 2018

6. Du, B., Liu, C., Zhou, W., et al.: Catch me if you can: detecting pickpocket suspects from large-scale transit records. In: SIGKDD (2016)

7. Gong, H., Chen, C., Bialostozky, E., et al.: A GPS/GIS method for travel mode detection in New York City. Comput. Environ. Urban Syst. **36**(2), 131–139 (2012)

8. Pinelli, F., Pinelli, F., Pinelli, F., et al.: Trajectory pattern mining. In: SIGKDD (2007)

9. Laube, P., Imfeld, S.: Analyzing relative motion within groups oftrackable moving point objects. In: Egenhofer, M.J., Mark, D.M. (eds.) GIScience 2002. LNCS, vol. 2478, pp. 132–144. Springer, Heidelberg (2002). https://doi.org/10.1007/3-540-45799-2_10

10. Li, M., Ahmed, A., Smola A.J.: Inferring movement trajectories from GPS snippets. In: WSDM (2015)

11. Chen, C., Zhang, D., Zhou, Z-H., Li, N., Atmaca, T., Li, S.: B-planner: night bus route planning using large-scale taxi GPS traces. In: PerCom (2013)

12. Luo, W., Tan, H., Chen, L., Ni, L.M.: Finding time period-based most frequent path in big trajectory data. In: SIGMOD (2013)

13. Coelho da Silva, T.L., de Macêdo, J.A.F., Casanova, M.A.: Discovering frequent mobility pat-terns on moving object data. In: MobiGIS (2014)

14. Jing, Y., Zheng, Y., Xie, X.: Discovering regions of different functions in a city using human mobility and pois. In: KDD (2012)

15. Li, Q., Zheng, Y., Xie, X., et al.: Mining user similarity based on location history. In: SIGSPATIAL (2008)

16. Ying, J.C., Lu, H.C., Lee, W.C., et al.: Mining user similarity from semantic trajectories. In: SIGSPATIAL (2010)

17. Abul, O., Bonchi, F., Nanni, M.: Anonymization of moving objects databases by clustering and perturbation. Inf. Syst. **35**(8), 884–910 (2010)

18. Zhang, C., Zhang, K., Yuan, Q., et al.: GMove: group-level mobility modeling using geo-tagged social media. In: SIGKDD (2016)

19. Zhang, J., Zheng, Y., Qi, D.: Deep spatio-temporal residual networks for citywide crowd flows prediction. In: AAAI (2017)

20. Feng, J., et al.: DeepMove: predicting human mobility with attentional recurrent networks. In: WWW (2018)

21. Kong, D., Wu, F.: HST-LSTM: a hierarchical spatial-temporal long-short term memory network for location prediction. In: IJCAI (2018)

22. Mikolov, T., Chen, K., Corrado, G., Dean, J.: Efficient estimation of word representations in vector space. In: ICLR (2013)

23. Kim, Y.: Convolutional neural networks for sentence classification. arXiv preprint arXiv: 1408.5882, (2014)

24. Vaswani, A., et. al.: Attention is all you need. arXiv:1706.03762

MC-eLDA: Towards Pathogenesis Analysis in Traditional Chinese Medicine by Multi-Content Embedding LDA

Ying Zhang[1], Wendi Ji[2], Haofen Wang[3], Xiaoling Wang[1(✉)], and Jin Chen[4]

[1] Shanghai Key Laboratory of Trustworthy Computing,
East China Normal University, Shanghai, China
ying.zhang@ecnu.cn, xlwang@sei.ecnu.edu.cn
[2] Liaoning University, Shenyang, China
wdji@lnu.edu.cn
[3] Shanghai Leyan Technologies Co. Ltd., Shanghai, China
nobot@leyantech.com
[4] Department of Computer Science,
Institute for Biomedical Informatics Department of Internal Medicine,
University of Kentucky Lexington, Lexington, USA
chen.jin@uky.edu

Abstract. Traditional Chinese medicine (TCM) is well-known for its unique theory and effective treatment for complicated diseases. In TCM theory, "pathogenesis" is the cause of patient's disease symptoms and is the basis for prescribing herbs. However, the essence of pathogenesis analysis is not well depicted by current researches. In this paper, we propose a novel topic model called Multi-Content embedding LDA (MC-eLDA), aiming to collaboratively capture the relationships of symptom-pathogenesis-herb triples, relationship between symptom-symptom, and relationship between herb-herb, which can be used in auxiliary diagnosis and treatment. By projecting discrete symptom words and herb words into two continuous semantic spaces respectively, the semantic equivalence can be encoded by exploiting the contiguity of their corresponding embeddings. Compared with previous models, topic coherence in each pathogenesis cluster can be promoted. Pathogenesis structures that previous topic modeling can not capture can be discovered by MC-eLDA. Then a herb prescription recommendation method is conducted based on MC-eLDA. Experimental results on two real-world TCM medical cases datasets demonstrate the effectiveness of the proposed model for analyzing pathogenesis as well as helping make diagnosis and treatment in clinical practice.

Keywords: Topic modeling · Embedding ·
Traditional Chinese medicine

Q. Yang et al. (Eds.): PAKDD 2019, LNAI 11439, pp. 489–500, 2019.
https://doi.org/10.1007/978-3-030-16148-4_38

1   Introduction

Traditional Chinese medicine (TCM) is an important way for disease treatment in Chinese society and is increasingly adopted as a complementary therapy around the world. Actually, TCM has been successfully applied to the prevention and treatment of complicated diseases in western modern medical practice [14]. However, the essence of TCM theory is not well depicted by current researches. Induction of the key knowledge from clinical data is a vital task for TCM.

In TCM theory, "pathogenesis" is the latent cause of patient's disease symptoms and is the basis for prescribing herbs. Pathogenesis analysis is the key point for disease diagnosis and treatment. For example, symptoms such as "cough" are the appearance of inherent pathogenesis "Lung heat" and herbs are treatments that target at this pathogenesis. Learning how doctors analyze pathogenesis can be helpful in the diagnosis and treatment of TCM.

A number of models have been proposed for analyzing pathogenesis that connect symptoms and herbs in TCM [1,3,4,15]. A major shortcoming among these methods is that semantic similarity of symptoms and similarity of herbs are not considered in these models. For example, "dark tongue" and "purple tongue" are two semantic similar symptoms, if they are clustered into the same pathogenesis then they can be treated by the same herbs. As for herbs, jointly use similar herbs such as "mirabilite" and "rhubarb" can enhance the curative effect, which is called *mutual promotion* in TCM theory. Previous model which infers topic distributions only by word co-occurrence can not cluster similar symptoms or similar herbs into the same pathogenesis if they do not appear together in the same medical case. The relationships among symptoms, herbs and pathogenesises are not well explored.

Motivated by the aforementioned challenges, in this paper, based on Latent Dirichlet Allocation(LDA) [12], we propose a Multi-Content embedding LDA model (MC-eLDA) to analyze pathogenesis that associates symptoms with and herbs. Instead of discrete word terms, symptoms and herbs are two types of continuous space embeddings sharing the same topic. Since embeddings have been shown to be effective at capturing semantic regularities, and Gaussian distributions capture a notion of centrality in space, our MC-eLDA can encode the semantic similarity of symptoms and the semantic similarity of herbs. Based on this model, diagnosis and herb prescription recommendation can be given according to patient's symptoms.

Compared with previous models, our model can assign high probability to a symptom or herb which is similar to an existing topical symptom or herb by exploiting the similarity in the embedding space. Hence topic coherence in each pathogenesis cluster can be promoted. Further, completeness and distinctiveness of pathogenesis clusters are improved. Connections of symptoms and herbs that previous topic modeling can not captured can be exploited by MC-eLDA. The relationships among symptoms, herbs and pathogenesises are deeply depicted. Then diagnosis and herb recommendation based on pathogenesis can be more accurate. In addition, when there comes a new medical case containing symptoms or herbs haven't seen before, the continuous topic distribution makes it possible to assign topics to previously unseen symptoms or herbs.

The main contributions of this paper are as follows:

- The relationships of symptom-pathogenesis-herb triples, relationship between symptom-symptom, and relationship between herb-herb are collaboratively depicted. Topic coherence is promoted and the ability of topic pattern discovery is improved.
- To the best of our knowledge, we are the first to capture the similarity of symptoms and the similarity of herbs by projecting discrete symptoms and herbs into continuous semantic space. Multivariate Gaussian distribution have been defined to handle the continuous embeddings.
- Based on the MC-eLDA model, herb recommendation with higher accuracy can be realized to assist doctors make diagnosis and treatment in real clinical practice.
- We have conducted extensive experiments on two real-world medical case datasets to evaluate the performance of the proposed method. The results demonstrate that MC-eLDA as well as its corresponding herb prescription recommendation method outperforms all the compared methods.

2 Related Work

Recently, discovering and extracting medical knowledge have gained the popularity. Particularly, in TCM researches, a hierarchical clustering model named latent tree model [2] has been proposed for discovering latent structures of symptoms. The authors apply this model to establish objective standards for syndrome differentiation in TCM diagnosis of kidney deficiency syndromes. However, the pathogenesis is only based on symptoms without considering the corresponding herbs.

For the task of exploring relationships of symptoms and herbs in TCM medical cases, in [3], the authors propose a WSSH-MIML model. They formulate the prescription-symptoms-predicting as a multi-label learning problem. However, TCM diagnosis is a complicated problem. A herb along with different herbs to form formula may treat different diseases, and a symptom may be caused by different diseases. To address this, Wang et al. [4] propose a new asymmetric multinomial probabilistic model for the joint analysis of symptoms, diseases, and herbs in patient records to discover and extract latent TCM knowledge. Moreover, a topic model MC-LDA [1] is proposed for modeling relationship between the symptoms and herbs, which considers pathogenesis as the latent topic that connect symptoms and herbs. Both of these two models explore the relationships of symptoms and herbs with latent variables. However, these models find topic patterns only by word co-occurrence, without considering the semantic similarity of symptoms or herbs.

Using deep learning to facilitate and enhance medical analysis is a promising and important area. Several studies have been done to use neural network based models for feature representation. In [6], Autoencoder is used to extract deep features for cancer diagnosis. Followed by Denoising Autoencoder (DAE) [7] and Stacked Autoencoder (SAE) [8] being applied in feature learning for disease diagnosis.

3   The Proposed Framework

3.1   Problem Definition

We define our computational problem as follows: the input consists of a set of medical cases, where each medical case can be defined as a document that contains a sequence of symptoms and a sequence of corresponding herbs. Then, each symptom word and herb word is transferred as an embedding, denoted by $D = \{s_1, s_2, ..., s_n, h_1, h_2, ..., h_m\}$, where each embedding s_i or h_j is an $R-$dimensional vector. These embeddings are pre-computed from external TCM corpus. What is important is that similar symptoms or similar herbs in some appropriate sense (in terms of the semantic of symptoms and efficacy, property, or channel tropism of herbs) end up having similar embeddings. The output consists of the pathogenesis distribution of each medical case and the probability of each symptom and herb under each pathogenesis topic. The relationships among symptom words, symptom embeddings, herb words, herb embeddings and pathogenesises can be shown in Fig. 1.

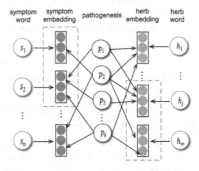

Fig. 1. Relationships of symptom words, symptom embeddings, herb words, herb embeddings and pathogenesises.

3.2   Multi-Content embedding LDA Model (MC-eLDA)

In this paper, we propose a Multi-Content embedding LDA (MC-eLDA) to take into account the similarity of symptoms and similarity of symptoms, aiming to improve the semantic coherence and explore the relationships among symptoms, herbs and pathogenesises, which can be used in auxiliary diagnosis and treatment. First, MC-eLDA contains one latent factor that is associated with

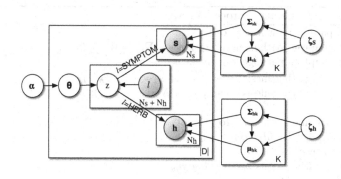

Fig. 2. Graphical model of MC-eLDA

both symptoms and herbs. Then, rather than assuming a medical case as an aggregation of sequences of independent item types, MC-eLDA assumes that medical case consists of a sequence of symptom embeddings and a sequence of herb embeddings. These embeddings are generated by external TCM data. In this way, similar symptoms and similar herbs can be captured by the distance of their corresponding embeddings.

The graphical model of MC-eLDA is shown in Fig. 2. As before, each document is generated by drawing a document-specific mixture factor θ over pathogenesis topic $1, ..., K$ and the mix θ is drawn from a Dirichlet prior α. To operate on continuous space of symptom embeddings and herb embeddings, based on [9], we define a multivariate Gaussian distribution for each pathogenesis topic k over embeddings, with mean μ_{sk}, covariance Σ_{sk} for symptom and mean μ_{hk}, covariance Σ_{hk} for herb. The mean and variance of each of these two Gaussian distributions are governed by a Gaussian-Inverse-Wishart (a.k.a Normal-Inverse-Wishart, NIW) prior, which can be denoted respectively as $(\mu_{sk}, \Sigma_{sk}) \sim NIW(\mu_s, \kappa_s, \Psi_s, \nu_s)$, $(\mu_{hk}, \Sigma_{hk}) \sim NIW(\mu_h, \kappa_h, \Psi_h, \nu_h)$, given by

$$p(\mu_{sk}, \Sigma_{sk}|\zeta_s) = N(\mu_{sk}|\mu_s, \frac{1}{\kappa_s}\Sigma_{sk})W^{-1}(\Sigma_{sk}|\Psi_s, \nu_s), \qquad (1)$$

$$p(\mu_{hk}, \Sigma_{hk}|\zeta_h) = N(\mu_{hk}|\mu_h, \frac{1}{\kappa_h}\Sigma_{hk})W^{-1}(\Sigma_{hk}|\Psi_h, \nu_h), \qquad (2)$$

where parameter set $\zeta_s = \{\mu_s, \kappa_s, \Psi_s, \nu_s\}$, and $\zeta_h = \{\mu_h, \kappa_h, \Psi_h, \nu_h\}$.

Note that there is an observable binary variable, indicator l, whose value is either $SYMPTOM$ or $HERB$. When $l = SYMPTOM$, a symptom embedding will be generated. When $l = HERB$, a herb embedding will be generated.

Thus, the generative process of MC-eLDA can be proceeded as it is shown in Algorithm 1.

3.3   Parameter Estimation

In MC-eLDA model, symptom embeddings and herb embeddings of pathogenesis topic k are generated by multivariate Gaussian distributions. Let

---

Algorithm 1. Generative process of MC-eLDA

1: **for** each pathogenesis $k \in K$ **do**
2:      Draw symptom pathogenesis covariance $\Sigma_{sk} \sim W^{-1}(\Psi_s, \nu_s)$;
3:      Draw symptom pathogenesis mean $\mu_{sk} \sim N(\mu_s, \frac{1}{\kappa_s}\Sigma_{sk})$;
4:      Draw herb pathogenesis covariance $\Sigma_{hk} \sim W^{-1}(\Psi_h, \nu_h)$;
5:      Draw herb pathogenesis mean $\mu_{hk} \sim N(\mu_h, \frac{1}{\kappa_h}\Sigma_{hk})$;
6: **end for**
7: **for** each document $d_m \in D$ **do**
8:      Draw $\theta_m \sim Dirichlet(\alpha)$;
9:      **for** each term in d_m **do**
10:          Draw pathogenesis $z_{m,i} \sim Disc(\theta_m)$;
11:          **if** l=SYMPTOM **then**
12:              Draw symptom vector $s_{m,i} \sim N(\mu_{sz_{m,i}}, \Sigma_{sz_{m,i}})$;
13:          **end if**
14:          **if** l=HERB **then**
15:              Draw herb vector $h_{m,i} \sim N(\mu_{hz_{m,i}}, \Sigma_{hz_{m,i}})$;
16:          **end if**
17:      **end for**
18: **end for**

$S^{(k)} = \{s_1^{(k)}, s_2^{(k)}, ... s_{n_{sk}}^{(k)}\}$ be the symptom embedding set, $H^{(k)} = \{h_1^{(k)}, h_2^{(k)}, ... h_{n_{hk}}^{(k)}\}$ be the herb embedding set where the embedding's topic is assigned to k in all documents. The likelihood can be written as:

$$p(S^{(k)}, H^{(k)}|\mu_{sk}, \mu_{hk}, \Sigma_{sk}, \Sigma_{hk}) = \prod_{i=1}^{s_k} p(s_i^{(k)}|\mu_{sk}, \Sigma_{sk}) \prod_{i=1}^{h_k} p(h_i^{(k)}|\mu_{hk}, \Sigma_{hk}). \quad (3)$$

The posterior distribution of symptom part also follows NIW distribution, whose parameters $\zeta_s' = \{\mu_s', \kappa_s', \Psi_s', \nu_s'\}$ is given by:

$$\kappa_s' = \kappa_s + N_{sk} \qquad \nu_s' = \nu_s + N_{sk} \qquad \mu_s' = \frac{\kappa_s \mu_s + N_{sk}\bar{s}^{(k)}}{\kappa_s'}$$

$$\Psi_s' = \Psi_s + C_{sk} + \frac{\kappa_s N_{sk}}{\kappa_s'}(\bar{s}^{(k)} - \mu_s)(\bar{s}^{(k)} - \mu_s)^{\mathrm{T}}, \quad (4)$$

in which

$$\bar{s}^{(k)} = \frac{1}{N_{sk}} \sum_{sd} \sum_{i:z_{sd,1}=k} (s_{d,i}) \qquad C_{sk} = \sum_{sd} \sum_{i:z_{sd,i}=k} (s_{d,i} - \bar{s}^{(k)})(s_{d,i} - \bar{s}^{(k)})^{\mathrm{T}}, \quad (5)$$

where N_{sk} is the count of symptom embeddings whose topics are assigned to k in all documents, $\bar{s}^{(k)}$ is the sample mean of the symptom vectors and C_{sk} is the scaled form of sample covariance. The physical meaning of κ_s' and ν_s' is the weight of the mean and covariance of the prior respectively.

Since generating symptom embeddings and herb embeddings are in the same way. We focus on the detail derivation process of topic-symptom part, the derivation of topic-herb part is omitted here.

We observe symptom embeddings and herb embeddings of different medical cases, aiming to infer the model's parameters $\Theta = \{\boldsymbol{\theta}_m, \boldsymbol{\mu}_{sk}, \boldsymbol{\Sigma}_{sk}, \boldsymbol{\mu}_{hk}, \boldsymbol{\Sigma}_{hk}\}$. Under the Bayesian framework, one of the reasonable ways to estimate parameters is using the expectation as parameters' value:

$$\hat{\boldsymbol{\theta}}_{dk} = \frac{N_{sd}^{(k)} + N_{hd}^{(k)} + \alpha_k}{\sum_{k=1}^{K}(N_{sd}^{(k)} + N_{hd}^{(k)} + \alpha_k)} \qquad \hat{\boldsymbol{\mu}}_{sk} = \frac{\kappa_s \boldsymbol{\mu}_s + N_{sk}\bar{\boldsymbol{s}}^{(k)}}{\kappa_s'} \qquad \hat{\boldsymbol{\Sigma}}_{sk}' = \frac{\boldsymbol{\Psi}_s'}{\nu_s' - R + 1}$$

$$\hat{\boldsymbol{\mu}}_{hk} = \frac{\kappa_h \boldsymbol{\mu}_h + N_{hk}\bar{\boldsymbol{h}}^{(k)}}{\kappa_h'} \qquad \hat{\boldsymbol{\Sigma}}_{hk}' = \frac{\boldsymbol{\Psi}_h'}{\nu_h' - R + 1}, \tag{6}$$

where $N_{sd}^{(k)}$ or $N_{hd}^{(k)}$ represents the number of symptom or herb embeddings that are assigned to topic k in symptom or herb part of medical case document d. R is the dimension of the embeddings. For the symptom part, model's mean is estimated from κ_s observations with sample mean $\boldsymbol{\mu}_s$, covariance matrix is estimated from ν_s observations with sample mean $\boldsymbol{\mu}_s$ and with sum of pairwise deviation products $\boldsymbol{\Psi}_s = \boldsymbol{\mu}_s \boldsymbol{\Sigma}_s$. The herb part is similar. With formula (4) and formula (5), the model's parameters can be computed.

Then, we adopt the Gibbs sampling algorithm [10] as the learning algorithm. When $l = SYMPTOM$, given conditional distribution for the ith symptom embedding in the dth document, the sampler draws the hidden topic $z_{sd,i}$ as follows:

$$p(z_{sd,i} = k | l = SYMPTOM, \boldsymbol{z}_{\neg(sd,i)}, \boldsymbol{S}, \boldsymbol{H}, \zeta_s, \zeta_h, \alpha) \propto$$

$$\frac{N_{sd,\neg i}^{(k)} + N_{hd}^{(k)} + \alpha_k}{\sum_{k=1}^{K}(N_{sd,\neg i}^{(k)} + N_{hd}^{(k)} + \alpha_k)} \cdot t_{\nu_s' - R + 1}\left(\boldsymbol{s}_{d,i} | \boldsymbol{\mu}_s', \frac{\kappa_s' + 1}{\kappa_s'(\nu_s' - R + 1)}\boldsymbol{\Psi}_s'\right). \tag{7}$$

When $l = HERB$, given conditional distribution for the ith herb embedding in the dth document, the sampler draws the hidden topic $z_{hd,i}$ as follows:

$$p(z_{hd,i} = k | l = HERB, \boldsymbol{z}_{\neg(hd,i)}, \boldsymbol{S}, \boldsymbol{H}, \zeta_s, \zeta_h, \alpha) \propto$$

$$\frac{N_{sd}^{(k)} + N_{hd,\neg i}^{(k)} + \alpha_k}{\sum_{k=1}^{K}(N_{sd}^{(k)} + N_{hd,\neg i}^{(k)} + \alpha_k)} \cdot t_{\nu_h' - R + 1}\left(\boldsymbol{h}_{d,i} | \boldsymbol{\mu}_h', \frac{\kappa_h' + 1}{\kappa_h'(\nu_h' - R + 1)}\boldsymbol{\Psi}_h'\right). \tag{8}$$

Here, $\boldsymbol{z}_{\neg(sd,i)}$ represents the topic assignments of all excluding the ith symptom embedding of document d. The posterior predictive is given by a multivariate t-distribution.

3.4   Embedding with Prior Knowledge

The feature embeddings used by the MC-eLDA model is pre-computed. In this work, we generate symptom and herb embeddings respectively based on deep learning technique.

For symptoms, we use a popular and effective tool called *word2vec*[1] as our *symptom2vec* method to generate symptom embeddings from unlabeled corpus.

---

[1]   https://code.google.com/archive/p/word2vec/.

The corpus is obtained from an online TCM encyclopaedia[2] in the form of natural language statements. Each word is used as an input to a log-linear classifier with continuous projection layer [13]. In this way, the semantic properties of symptoms can be captured hence semantically related symptoms can be identified.

As for herbs, on one hand, there are various types of semantic relationship among herbs. Therefore, it is not reasonable to simply learn the context of herb words to generate embeddings by *word2vec*. On the other hand, the semantic that we focus on for herbs is the structured feature information such as efficacy, property, channel tropism, which can be used to capture the *mutual promotion* relation. To facilitate this and obtain more discriminating features, we propose a *herb2vec* method. First, each herb is transformed to a V-dimensional binarized sparse vector, where each dimension represents a feature such as "moistening lung" efficacy. Then, denoising autoencoder (DAE) [11] is adopted, which is widely used for unsupervised learning. Compared with traditional feature extraction methods such as PCA [5], DAE can extract non-linear more complicated patterns more robustly. By computing the hidden layer representation of DAE, which contains R units, we can obtain a concise R-dimensional ($R \ll V$) embedding for each herb.

4   Experiments

4.1   Experimental Settings

In this section, we conduct experiments on two real-world TCM datasets. One is the medical cases from ancient Chinese medicine book for disease of amenorrhea, which contains 106 cases with 152 symptom terms and 248 herb terms. The other dataset is a collection of modern medicine clinical records which mainly focuses on lung cancer disease, provided by a famous TCM hospital in China. The dataset contains 952 medical cases, with 77 symptom terms and 356 herb terms.

Our proposed model is compared with the following models:

- LDA [12]. A topic model where symptoms and herbs are treated as the same type of word.
- MC-LDA [1]. A topic model where symptom words and herb words are treated as two types of content in a document.
- G-LDA [9]. A topic model where symptoms and herbs are not distinguished. Besides, each symptom and each herb in the medical case documents is presented as an embedding rather than a word term.
- User-based collaborative filtering (CF) [16]. A group recommendation method for recommending herbs for a list of symptoms, which uses the smallest rating score of group users as the recommendation score.

---

[2] http://www.a-hospital.com/w/.

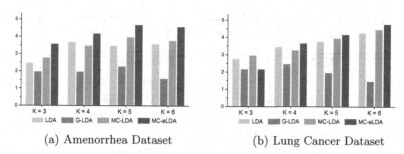

(a) Amenorrhea Dataset   (b) Lung Cancer Dataset

Fig. 3. Pathogenesis evaluation on two datasets.

4.2   Pathogenesis Evaluation

In this section, we evaluate the correspondence of symptoms and herbs under each pathogenesis cluster. We invite TCM specialists to score each cluster on a 6-point scale: 5 corresponding to completely reasonable clusters where symptoms and herbs can be completely linked by pathogenesises and 0 to completely unreasonable clusters.

Figure 3 shows the average score of MC-eLDA and comparison models when number of topics K range from 3 to 6. The result shows that symptoms and herbs under latent pathogenesises found by MC-eLDA are more reasonable based on the theory of TCM. For LDA and G-LDA, we make no distinction between symptoms and herbs. Therefore, in each cluster, the ratio of the number of symptoms to the number of herbs are not certain. Especially in G-LDA, some clusters contain only symptom words of herb words, which make no sense to learn the relationship of symptoms and herbs.

4.3   Topic Coherence Evaluation

Topic coherence measures the degree of semantic similarity between high scoring words in a topic. In this case, the coherence score of each topic is calculated by the following equation:

$$Coherence\ Score = \frac{\sum\limits_{s_1 \in S_t} \sum\limits_{s_2 \in S_t} D(s_1, s_2) + \sum\limits_{h_1 \in H_t} \sum\limits_{h_2 \in H_t} D(h_1, h_2)}{|S|^2 + |H|^2}, \tag{9}$$

where S_t and H_t are the sets of high scoring symptoms and high scoring herbs in topic t, D represents the euclidean distance between two vectors.

The mean and variance values of the topic coherence score are shown in Table 1. It can be observed that MC-eLDA and MC-LDA have better coherence score compared to the other two models. This result can verify the correctness and feasibility of multi-content models to analyze pathogenesis that is associated with both symptoms and herbs in TCM. Further more, MC-eLDA gives the best performance over all the other models, which demonstrates that considering the semantic similarity of both symptoms and herbs can improve topic coherence.

Table 1. Results of the mean and variance values of the coherence scores on two datasets. ↓ indicates the smaller the value is, the better the performance is.

Models	Amenorrhea		Lung cancer	
	Mean ↓	Variance	Mean ↓	Variance
LDA	3.1943	0.6052	3.4484	0.2822
G-LDA	3.0680	0.7126	4.0734	0.4388
MC-LDA	2.4064	0.1043	3.4583	0.2663
MC-eLDA	2.3221	0.2230	3.4213	0.2970

4.4   Qualitative Evaluation

We pick up some topics and list the top symptoms and herbs of the topics in Table 2, sorted by decreasing probability. We do not pick up topics generated by LDA and G-LDA since they can not distinguish symptoms and herbs as two types of words.

As we can see, the first topic of these two models both contains "cough", "phlegm", along with other symptoms like "red tongue", we can indicate that these two clusters represent lung-related diseases. According to the TCM doctors, the corresponding herbs, such as Rhizoma arisaematis, has the function of relieving cough and reducing sputum to relieve asthma, and it belongs to "lung channel". However, the herb "Fuling" in MC-LDA doesn't has the efficacy for lung-related diseases. Futhermore, in the result of MC-eLDA, we can see similar symptoms, such as "fatigue" and "feeble" are clustered into the same topic. In the herb part, "Angelica" and "Angelica root", "Rehmannia glutinosa" and "Radix rehmanniae", are in the same topic.

4.5   Herbs Recommendation Accuracy

We follow the hybrid recommendation method in [1] to recommend herbs given a list of symptoms. Due to the same reason mentioned in the above section that LDA and G-LDA can't distinguish symptoms and herbs, they fail in this task.

For each dataset, two-thirds of medical cases are used as training data and the others as test data. We employ Jaccard Coefficient to measure the performance, which is defined as:

$$\text{Jaccard} = \frac{1}{|U|} \sum_{u \in U} \frac{|R(u) \cap T(u)|}{|R(u) \cup T(u)|}, \tag{10}$$

where $|U|$ is the number of cases in test set, $R(u)$ is the list of recommended herbs given the symptoms of a patient and $T(u)$ is the prescription given by doctor.

Table 3 shows the accuracy of these three methods. Top frequent method performs worst, since it only recommends the most commonly prescribed herb for each symptom. MC-eLDA based method outperforms other methods on both

Table 2. Comparison between top symptoms and herbs of some topics from MC-eLDA and MC-LDA.

symptom	herb	symptom	herb
MC-eLDA			
phlegm (痰)	Rhizoma arisaematis (天南星)	amenorrhea (闭经)	Angelica (当归)
cough (咳嗽)	Hedyotis diffusa (蛇舌草)	white tongue fur (苔白)	Angelica root (当归尾)
sharp cough (咳剧)	Rhizoma acori (菖蒲)	thin pulse (脉细)	Cuscuta (菟丝子)
abundant sputum (痰多)	Rhizoma pinelliae (半夏)	deep thready pulse (脉沉细)	Rehmannia glutinosa (地黄)
red tongue (舌红)	Radix glehniae (北沙参)	fatigue (神疲)	Radix Rehmanniae (生地)
thin tongue fur (苔薄)	Ophiopogon japonicus (麦冬)	feeble (乏力)	Fructus lycii (枸杞)
no tongue fur (无舌苔)		poor sleep (寐差)	
MC-LDA			
cough (咳嗽)	Rhizoma pinelliae (半夏)	amenorrhea (闭经)	Angelica (当归)
phlegm (痰)	Liquorice (甘草)	feeble (乏力)	Atractylodes (白术)
breathe heavily (气喘)	Fuling (茯苓)	cold limbs (肢冷)	Liquorice (甘草)
headache (头痛)	Platycodon grandiflorum (桔梗)	slippery pulse (脉滑数)	Cyperus rotundus (香附)
fever (发热)		greasy tongue fur (苔腻)	Rhizoma pinelliae (半夏)

Table 3. Accuracy of herb prescription recommendation

	Top frequent	CF	MC-LDA	MC-eLDA
Amenorrhea	0.1465	0.2410	0.2573	0.3437
Lung cancer	0.0741	0.2090	0.4034	0.4811

dataset. This result shows the potential that our model learns from patient record to assist doctors make diagnosis and treatment in real clinical practice.

5   Conclusion

In this paper, we propose a novel model MC-eLDA to analyze pathogenesises that connect symptoms and herbs in TCM clinical data. In particular, symptoms and herbs are projected as two types of continuous space embeddings sharing the same topic. Therefore, the relationship among symptom-pathogenesis-herb triples, relationship between symptom-symptom, and relationship between herb-herb can be collaboratively captured. Experimental results on two real-world TCM medical cases datasets demonstrate that our model can deeply and rationally analyze pathogenesises and recommend herb prescription as references for patients and doctors. For the future work, we plan to learn the embeddings in an integrated way by solving an optimization problem with both representation learning and topic modeling.

Acknowledgments. This work was supported by National Key R&D Program of China (No. 2017YFC0803700), NSFC grants (No. 61532021 and 61472141), Shanghai Knowledge Service Platform Project (No. ZF1213), and SHEITC.

References

1. Ji, W., Zhang, Y., Wang, X., et al.: Latent semantic diagnosis in traditional Chinese medicine. World Wide Web **20**(5), 1071–1087 (2017)
2. Zhang, N.L., Yuan, S., Chen, T., et al.: Latent tree models and diagnosis in traditional Chinese medicine. Artif. Intell. Med. **42**(3), 229–245 (2008)
3. Li, Y., Li, H., Wang, Q., et al.: Traditional Chinese medicine formula evaluation using multi-instance multi-label framework. In: 2016 IEEE International Conference on Bioinformatics and Biomedicine (BIBM), pp. 484–488. IEEE (2016)
4. Wang, S., Huang, E.W., Zhang, R., et al.: A conditional probabilistic model for joint analysis of symptoms, diseases, and herbs in traditional Chinese medicine patient records. In: 2016 IEEE International Conference on Bioinformatics and Biomedicine (BIBM), pp. 411–418. IEEE (2016)
5. Everitt, B.S., Dunn, G.: Principal components analysis. Appl. Multivar. Data Anal. Second Ed. 48–73 (1993)
6. Fakoor, R., Ladhak, F., Nazi, A., et al.: Using deep learning to enhance cancer diagnosis and classification. In: Proceedings of the International Conference on Machine Learning (2013)
7. Li, J., Struzik, Z., Zhang, L., et al.: Feature learning from incomplete EEG with denoising autoencoder. Neurocomputing **165**, 23–31 (2015)
8. Suk, H.I., Lee, S.W., Shen, D., et al.: Latent feature representation with stacked auto-encoder for AD/MCI diagnosis. Brain Struct. Funct. **220**(2), 841–859 (2015)
9. Das, R., Zaheer, M., Dyer, C.: Gaussian LDA for topic models with word embeddings. In: ACL, vol. 1, pp. 795–804 (2015)
10. Porteous, I., Newman, D., Ihler, A., et al.: Fast collapsed Gibbs sampling for latent Dirichlet allocation. In: Proceedings of the 14th ACM SIGKDD International Conference on Knowledge Discovery and Data Mining, pp. 569–577. ACM (2008)
11. Vincent, P., Larochelle, H., Bengio, Y., et al.: Extracting and composing robust features with denoising autoencoders. In: Proceedings of the 25th International Conference on Machine Learning, pp. 1096–1103. ACM (2008)
12. Blei, D.M., Ng, A.Y., Jordan, M.I.: Latent Dirichlet allocation. J. Mach. Learn. Res. **3**(Jan), 993–1022 (2003)
13. Mikolov, T., Chen, K., Corrado, G., et al.: Efficient estimation of word representations in vector space. arXiv preprint arXiv:1301.3781 (2013)
14. Ling, C., Yue, X., Ling, C.: Three advantages of using traditional Chinese medicine to prevent and treat tumor. J. integr. Med. **12**(4), 331–335 (2014)
15. Jiang, Z., Zhou, X., Zhang, X., et al.: Using link topic model to analyze traditional Chinese medicine clinical symptom-herb regularities. In: 2012 IEEE 14th International Conference on e-Health Networking, Applications and Services (Healthcom), pp. 15–18. IEEE (2012)
16. Amer-Yahia, S., Roy, S.B., Chawlat, A., et al.: Group recommendation: semantics and efficiency. Proc. VLDB Endowment **2**(1), 754–765 (2009)

Enhancing the Healthcare Retrieval with a Self-adaptive Saturated Density Function

Yang Song[1]([✉]), Wenxin Hu[1], Liang He[1,2], and Liang Dou[1,2]

[1] Department of Computer Science and Technology, East China Normal University,
Shanghai 200241, China
ysong@ica.stc.sh.cn, wxhu@cc.ecnu.edu.cn, {lhe,ldou}@cs.ecnu.edu.cn
[2] NPPA Key Laboratory of Publishing Intergration Development, ECNUP,
Shanghai, China

Abstract. The proximity based information retrieval models usually use the same pre-define density function for all of terms in the collection to estimate their influence distribution. In healthcare domain, however, different terms in the same document have different influence distributions, the same term in different documents also has different influence distributions, and the pre-defined density function may not completely match the terms' actual influence distributions. In this paper, we define a saturated density function to measure the best suitable density function that fits the given term's influence distribution, and propose a self-adaptive approach on saturated density function building for each term in various circumstance. Particularly, our approach utilizing Gamma process is an unsupervised model with no requirements for external resources. Then, we construct a density based weighting method for the purpose of evaluating the effectiveness of our approach. Finally, we conduct our experiment on five standard CLEF and TREC datasets, and the experimental results show that our approach is promising and outperforms the pre-defined density functions in healthcare retrieval.

Keywords: Saturated density function · Information retrieval · Self-adaptive

1 Introduction

Embedding the context information of queries and documents into retrieval process is an effective technique for boosting the overall performance in healthcare Information Retrieval (IR), which has been drawn a lot of attention in recent years. Specifically, proximity based approaches considering the influence a query term acts on its surrounding text show significant improvements over basic probabilistic IR models [12–14, 22]. In these models, an occurrence of a term has been assumed to have an impact towards its neighboring texts, while the influence was decreasing with the increase of the distance to the place of occurrence. A density

© Springer Nature Switzerland AG 2019
Q. Yang et al. (Eds.): PAKDD 2019, LNAI 11439, pp. 501–513, 2019.
https://doi.org/10.1007/978-3-030-16148-4_39

function is utilized to estimate this influence distribution. Many different classical mathematic equations have been selected to serve as the density functions in the proximity based IR models.

However, many proximity based approaches assume that the terms in the whole collection share the same influence distribution over multiple documents. In other words, they use a pre-defined density function to fit every term's influence distribution without considering the contextual information within different documents. This may lead to a low performance in healthcare IR, since different word influence may mean different disease.

Many previous work [13,18,22] have shown the evidences that different density functions work in different situations and the pre-defined density functions are not always effective in retrieval tasks, especially in healthcare domain. Firstly, a sentence is composed by different sentence constituents, for example subject, predicate and object, such that different medical words in different sentence constituents have different influence distributions in a healthcare document. Then, the same medical word has different influence distribution within different document in terms of the scenario it has occurrence, such as celebrating plot and leaving plot. Finally, the existing kernel functions may not match the evolution of the medical term's actual influence distribution since authors have different writing styles. Those evidences imply that a single density function cannot completely match each medical term's actual influence distribution in multiple healthcare documents at each position in the term's impact scope. Thus, we are motivated to build a best suitable density function for each medical term according to its characteristics in a document, which ideally results in a self-adaptive density function building approach.

We propose the definition of Saturated Density Function (SDF) and provided a self-adaptive approach on constructing the SDF for probabilistic IR system. The major contribution is that the proposed approach builds completely new and suitable density function for each term in multiple healthcare documents, which is distinctly different with the pre-defined density function. The self-adaptive approach is implemented by maximizing the probability in the Gamma process, where the shape parameter α is configured as the distance away from the given term, and the scale parameter β is configured as the quotient of actual influence and α. Furthermore, our approach do not need any prior data, external resources or to train the terms in advance. Finally, a density based weighting method is introduced for the evaluation purpose.

We test our approach in the healthcare domain, where three CLEF eHealth datasets and two TREC CDS datasets are adopted. The experimental results show that our self-adaptive approach is promising and significantly outperforms the pre-defined density functions and the state-of-the-art models in the retrieval.

The remainder of this paper is organized as follows. We first discuss the related work in Sect. 2. In Sect. 3, we introduce the preliminary knowledge. Then, we propose the self-adaptive SDF building approach in Sect. 4 and apply it into probabilistic information retrieval. After that, we set up the experimental environment in Sect. 5, and analyze the experimental results in Sect. 6. Finally, the conclusions and future work are presented in Sect. 7.

2 Related Work

Term proximity in information retrieval has been investigated a lot since 1990s [3, 7,9,10]. Keen et al. [9,10] and Clarke et al. [3] introduced the "NEAR" operator to quantify the proximity of query terms. Then, the term span was first utilized to measure the proximity in Hawking's work [7].

In order to describe proximity specifically, many statistics concepts were adopted, such as the minimum distance, the average of the shortest distance, etc. In [20], Tao et al. listed and evaluated different proximity measures, then concluded that the minimum pair-wise distance was the most effective. Cummins et al. in [4] investigated several term-term proximity measures and developed a learning approach to combine those proximity measures.

Recently, the proximity-based density functions have been widely adopted to propagate term influence in the existing retrieval models [2,5,6,11,15,17,18]. In the early times, Kretser and Moffat [5] applied the $tf \cdot idf$ score of each query term to its positions, where the Triangle kernel, the Cosine kernel, the Circle kernel and the arc contribution functions were also discussed. The document score was the highest accumulated $tf \cdot idf$ in the influence scope. Similarly, the vector space model and Boolean model were utilized as density functions respectively in [2,11].

At present, most state-of-the-art proximity-based information retrieval models [12–14,16,19,21,22] usually utilized kernel functions to act as the density functions. Lv and Zhai proposed a positional language model in [13]. They defined a language model for each position of a document, and scored a document based on the scores of its PLMs. The PLM was estimated based on propagated counts of words within a document through a pre-defined density function. Zhao et al. defined a cross term as a virtual term in the document in [22]. A cross term occurred when two query terms appeared close to each other and their density function had an intersection. Embedding the cross term information into the BM25 model, they established a CRTER model to rank the retrieval documents. In this approach, kernel functions were adopted as the density functions to calculate the weight of cross term. They studied the Gaussian, Triangle, Cosine and Circle kernel density functions, and suggested Gaussian as the best candidate.

However, the existing proximity-based IR models usually use the pre-defined density function for all query terms without considering their characteristics. Our work is much different, where we design a self-adaptive approach to build new density functions for a term in different circumstance. Meanwhile, we apply this self-adaptive approach into a probabilistic weighting model to achieve a density based retrieval model.

3   Preliminaries

In this section, we provide the preliminary knowledge of information content, kernel density functions and unite influence, for the purpose of introducing the definition of saturated density function and the self-adaptive approach we proposed in the following sections. Suppose, $Q = \{q_1, q_2, ..., q_n\}$ is a query and D is a document, where q_i $(i = 1, 2, ..., n)$ is a query term occurring in document D.

3.1   Information Content

In semantic information theory, the information content of a term measures the expected value of the information contained in this term. Utilizing the Poisson process, the information content of term q_i in document D can be defined similarly as [1]:

$$Inf_c = (\lambda + \frac{1}{12 \cdot tf} - tf) \cdot log_2 e + tf \cdot log_2 \frac{tf}{\lambda} \\ + 0.5 \cdot log_2(2\pi \cdot tf), \tag{1}$$

where $\lambda = \frac{F}{N}$, F and N are the document number in the elite set (a set of documents in which the term t occurs) and in the collection respectively. tf is the word frequency of term t in document D.

Here $log_2\lambda$ is the inverse document frequency (IDF) of q_i. Therefore, Inf_c is actually determined by the TF and IDF of term q_i.

3.2   Kernel Density Functions and Influence Scope

There are four popular kernel density functions introduced as follows.

1. Gaussian Density: $Gaussian(u) = exp(\frac{-u^2}{2\sigma^2})$, $u \leq \sigma$.
2. Cosine Density: $Cosine(u) = \frac{1}{2}[1 + cos(\frac{\pi u}{\sigma})]$, $u \leq \sigma$.
3. Circle Density: $Circle(u) = \sqrt{1 - (\frac{u}{\sigma})^2}$, $u \leq \sigma$.
4. Triangle Density: $Triangle(u) = 1 - \frac{u}{\sigma}$, $u \leq \sigma$.

Where u is the distance away from the given query term. The parameter σ restricts the propagation scope of each term which is defined as the term's influence scope. Followed by the previous work [5], we define σ as:

$$\sigma = \lfloor \frac{n}{dl} \cdot \frac{dl}{tf} \rfloor, \tag{2}$$

where n is the number of unique words in the given document, dl is the length of the document, and tf is the word frequency.

3.3   Unit Influence

Density functions are to describe the occurrence impact of term q_i $(i = 1, 2, ..., n)$ on its neighboring text. Then, we define the unit influence of q_i at the distance u upon multiple density functions as:

$$Inf_{q_i}^u(f_{density}) = \int_{u-1}^{u} f_{density}(x)dx, \quad f_{density} \in \Omega, \tag{3}$$

where $u \leq \sigma$, and Ω is a density function space which is composed by pre-defined density functions.

4   Methodology

In this section, we first propose the definition of saturated density function as an approximating of the actual influence distribution of a given term. Then, we introduce the self-adaptive approach on automatically building the saturated density function for each term in multiple circumstance. Finally, we apply a density based weighting method to evaluate the saturated density function.

4.1   Saturated Density Function

The Saturated Density Function (SDF) is a function that best fits the actual influence distribution of a term. The SDF is various based on different terms in multiple documents, which is constructed by utilizing the pre-defined density functions in a density function space. Here, we present the mathematical definition of the SDF as follows.

Definition 1. *Suppose $I(x)$ is the actual influence distribution of the term q_i, Ω is the space of q_i's density functions. The piecewise function SDF_{q_i} which is constructed by density functions in Ω, satisfies*

$$SDF_{q_i} = arg \min_{f_{density} \in \Omega} | \int_{u-1}^{u} I(x)dx - Inf_{q_i}^u(f_{density})|, \tag{4}$$

*in each unit interval $[u-1, u]$ is called the **saturated density function** of q_i in Ω, where $u \leq \sigma$ is the distance away from q_i, and σ is the influence propagation scope of q_i.*

The definition implies that the large the density function space is the accurate the SDF is.

4.2   Self-adaptive SDF Building Approach

Under the above definition of saturated density function, we build the SDF in the density function space Ω with the Gamma process (Formula 5).

$$P(x; \alpha, \beta) = \frac{1}{\beta^\alpha \Gamma(\alpha)} x^{\alpha-1} e^{-\frac{x}{\beta}}, \quad x > 0. \tag{5}$$

The Gamma process expresses the waiting time distribution that an event utilized to occur α times if this event occurs with a known average rate [8]. Here, α is called the shape parameter and β is called the scale parameter. In fact, $\alpha\beta$ is the expectation of Gamma process.

Mathematically, if we take the term's unit influence calculated by the density function as a random variable, the actual influence of the term in the unit interval $[u-1, u]$ is exactly the best choice of the variables' expectation. In the following section, we will give the definition and the calculation method of the term's actual influence.

Here, we integrate the proximity information into Gamma process. Suppose the distance u away from the given term stands for the times the event occurs (i.e. α in Formula 5), and the term's unit influence $Inf_{q_i}^u(f_{density})$ in the interval $[u - 1, u]$ stands for the time consumed when event occurs u time (i.e. x in Formula 5). At the same time, the actual influence of q_i describes the real impact on its neighboring text. If we deem the unit influence $Inf_{q_i}^u(f_{density})$ as the random variable. Then, the term's actual influence I_{q_i} in the interval $[u - 1, u]$ is theoretically selected as the expected value of the associated random variable (i.e. $\alpha\beta$). Then, the larger the probability is, the closer the $Inf_{q_i}^u(f_{density})$ to the actual influence is. Therefore, the maximum probability of the SDF in the interval $[u - 1, u]$ can be:

$$SDF_{q_i} = \arg \max_{f_{density} \in \Omega} \left\{ \frac{1}{\beta^\alpha \Gamma(\alpha)} x^{\alpha-1} e^{-\frac{x}{\beta}} \right\}, \tag{6}$$

where $x = Inf_{q_i}^u(f_{density})$, $\alpha = u$, $\beta = \frac{I_{q_i}}{u}$.

4.3 Actual Influence

The total actual influence of q_i, noted as TAI_{q_i}, is the accumulation (or summation) of q_i's genuine impact towards its neighboring text, i.e. $TAI_{q_i} = \int_0^\sigma I(x)dx$. Thus, $I_{q_i} = \frac{TAI_{q_i}}{\sigma}$ can be seen as the actual influence at each unit interval $[u - 1, u]$, $u \leq \sigma$.

For sake of obtain the actual influence I_{q_i}, we should calculate the total actual influence TAI_{q_i} at first, since I_{q_i} is derived from TAI_{q_i}. The information content measures the expected value of the information contained in the term, which also reveals the term's total actual influence. Hence, we make the following assumption.

Assumption 1. *For a given term q_i, there exists a linear relationship between its total actual influence and the information content. That is:*

$$TAI_{q_i} = c_1 Inf_c + c_0, \tag{7}$$

where $c_1 = \frac{c_2}{tf}$. $c_0, c_2 \in \Re$ are constants, tf is the word frequency of q_i.

For a given term q_i, its information content Inf_c measures the total influence that q_i acts on the whole document D. Then, $\frac{Inf_c}{tf}$ is q_i's total influence in its impact scope at the position q_i. $\frac{Inf_c}{tf}$ can be treated as the approximate value of q_i's total actual influence, since it is derived from the information content. In order to better satisfy different datasets' characteristics, we make a linear transformation on $\frac{Inf_c}{tf}$ and obtain Formula 7.

4.4 Density Based Weighting Method

In this subsection, a density based weighting method is derived to evaluate the effectiveness of the proposed SDF in an IR system. In the consideration that the term's total influence on the document reflects the value of its information content, and the classical DFR-PL2 IR model

$$w(q_i, D) = (1 - Prob) \cdot Inf_c,$$

the density based weighting method is presented as:

$$w(q_i, D) = (1 - Prob) \cdot tf \cdot \sum_{u=1}^{\sigma} Inf_{q_i}^u(kernel), \tag{8}$$

where $Prob$ is the probability of occurrence of a term within a document with respect to its elite set (set of documents that contain the term), $1 - Prob$ measures the risk of accepting a term as a good descriptor of the document when the document is compared with the elite set of the term, $1 - Prob$ is measured by the Laplace's law of succession.

Here for $1 - Prob$, we refer the definition in [1] as:

$$1 - Prob = \frac{1}{tf + 1}.$$

Therefore, the final matching function of the relevant documents is

$$R(Q, D) = \sum_{q_i \in Q} qtf_i \cdot w(q_i, D), \tag{9}$$

where qtf_i stands for within-query term frequency.

5 Experiment

5.1 Datasets and Evaluation Metrics

We conduct the experiments on three standard CLEF collections and two TREC collections. The datasets concentrate on the medical domain. The statistics information is presented in Table 1, where "eHealth13", "eHealth14" and "eHealth15" denote the CLEF eHealth datasets utilized in the year of 2013, 2014 and 2015. These collections consist of web pages covering a broad range health topics, targeted at both the general public and healthcare professionals. Meanwhile, "CDS14" and "CDS15" denote the TREC Clinical Decision Support track datasets utilized in the year of 2014 and 2015. These collections are snapshots of the open access subset of PubMed Centeral. All of the topics utilized in the five collections are medical case narratives. For all the datasets, each term is stemmed by the Porter's English stemmer, and the standard English stop words are removed.

We adopt the TREC official evaluation measures in our experiments, namely the Mean Average Precision (MAP) and Normalized Discounted Cumulative Gain (NDCG@10). To emphasize the top retrieved documents, we also include P@10 in the evaluation measures. All statistical tests are based on the Wilcoxon Matched-pairs Signed-rank test.

Table 1. Overview of datasets

Collection	# of Docs	Topics	# of Topics
eHealth13	1,104,298	qtest1-qtest50	50
eHealth14	1,104,298	qtest2014.1-qtest2014.50	50
eHealth15	1,104,298	clef2015.test.1-clef2015.test.67	67
CDS14	733,138	1–30	30
CDS15	733,138	1–30	30

5.2 Experimental Results

Our experiment mainly investigate the effectiveness of the SDF over the five medical datasets. Since our density based IR model is derived from the classical DFR-PL2 model, we adopt the original DFR-PL2 model without incorporating proximity information as our baseline. At the same time, many start-of-the-art proximity based IR models suggested that kernel density functions, especially the Gaussian density function, Triangle density function, Cosine density function and Circle density function, are the better candidates for the density functions since they make significant improvements in IR task. Thus, we suppose our density function space is composed by Gaussian, Triangle, Cosine and Circle kernel density functions, and the SDF function is built based on this space. Finally, the relevant parameters in our approach are set as $c_1 = \frac{1}{tf}$ and $c_0 = 0$.

Table 2 presents the performance in terms of MAP, NDCG and P@10 on five datasets. Gaussian, Triangle, Cosine and Circle stand for the pre-defined density functions in the density space Ω for all the terms over the multiple documents. The experimental results show that our proposed SDF makes the significant achievements over the baseline and all of the pre-defined density functions.

Table 2. Overall performance: "$*$" indicates the significant improvement over DFR-PL2, and "$+$" means the significant improvement over the Gaussian density function.

	Eval Metric	eHealth13	eHealth14	eHealth15	CDS14	CDA15
DFR-PL2	MAP	0.2735	0.2989	0.2578	0.1684	0.1686
	P@10	0.4080	0.3414	0.4000	0.2967	0.2967
	NDCG@10	0.3762	0.3802	0.3888	0.2689	0.2613
SDF	MAP	0.3568^{*+} (+30.46%)	0.4140^{*+} (+38.51%)	0.2854^{*+} (+10.71%)	0.1883^{*+} (+11.82%)	0.1807^{*+} (+7.18%)
	P@10	0.548^{*+} (+34.31%)	0.6126^{*+} (+32.42%)	0.4316^{*+} (+7.90%)	0.3633^{*+} (+22.45%)	0.3963^{*+} (+33.57%)
	NDCG@10	0.4722^{*+} (+25.52%)	0.4968^{*+} (+30.67%)	0.4299^{*+} (+10.57%)	0.3329^{*+} (+23.80%)	0.3217^{*+} (+23.12%)
Gaussian	MAP	0.3222^{*} (+17.81%)	0.3858^{*} (+29.07%)	0.2660^{*} (+3.18%)	0.1804^{*} (+7.13%)	0.1722^{*} (+2.14%)
	P@10	0.4980^{*} (+22.06%)	0.4780^{*} (+15.46%)	0.4076^{*} (+1.90%)	0.3400^{*} (+14.59%)	0.3567^{*} (+20.22%)
	NDCG@10	0.4404^{*} (+17.07%)	0.4420^{*} (+16.25%)	0.4022^{*} (+3.45%)	0.3177^{*} (+18.15%)	0.3070^{*} (+17.49%)
Triangle	MAP	0.2971^{*} (+8.63%)	0.3052^{*} (+2.11%)	0.2496 (-3.18%)	0.1795^{*} (+6.59%)	0.1715 (+1.72%)
	P@10	0.4600^{*} (+12.75%)	0.4340^{*} (+4.83%)	0.3727 (-6.83%)	0.3400^{*} (+14.59%)	0.3600^{*} (+21.33%)
	NDCG@10	0.3981^{*} (+5.82%)	0.4188^{*} (+10.15%)	0.3707 (-4.66%)	0.3168^{*} (+17.81%)	0.3083^{*} (+17.99%)
Circle	MAP	0.3061^{*} (+11.92%)	0.3239^{*} (+8.36%)	0.2510 (-2.64%)	0.1801^{*} (+6.95%)	0.1716 (+1.78%)
	P@10	0.4740^{*} (+16.18%)	0.4640^{*} (+12.08%)	0.3833 (-4.18%)	0.3400^{*} (+14.59%)	0.3567^{*} (+20.22%)
	NDCG@10	0.4088^{*} (+8.67%)	0.4311^{*} (+13.39%)	0.3746 (-3.65%)	0.3171^{*} (+17.92%)	0.3047^{*} (+16.61%)
Cosine	MAP	0.2971^{*} (+8.63%)	0.3052^{*} (+2.11%)	0.2496 (-3.18%)	0.1795^{*} (+6.59%)	0.1715 (+1.72%)
	P@10	0.4600^{*} (+12.75%)	0.4340^{*} (+4.83%)	0.3727 (-6.83%)	0.3400^{*} (+14.59%)	0.3600^{*} (+21.33%)
	NDCG@10	0.3981^{*} (+5.82%)	0.4188^{*} (+10.15%)	0.3707 (-4.66%)	0.3168^{*} (+17.81%)	0.3083^{*} (+17.99%)

6 Analysis

6.1 Influence of Proximity with Density Functions

Table 2 shows that proximity with density function works very well in the healthcare IR system, since we can see that all the density functions, including the SDF, outperforms the baseline without proximity. At the same time, Gaussian density function achieves the best performance compared with other density function in the space Ω. These conclusions are consistent to those proposed by the previous work [13, 22].

According to our statistical numbers, the average length of documents in eHealth and CDS datasets are 1071 and 3785 respectively, which indicate that the proximity with density function is promising in improving long text retrieval.

Moreover, experimental results in eHealth15 show that only Gaussian density function have improvement in the IR system. This evidence indicates that the per-defined density functions are not always effective in IR task, and different density functions work in different circumstances.

6.2   Effectiveness of SDF

Our proposed SDF outperforms all of the pre-defined density functions in the density function space, especially in eHealth15 data, which implies that the SDF fuses the advantages of all candidate density functions for the terms. This finding further proves our motivation that the SDF optimizes the pre-defined density functions based on the situations of the terms in a document, instead of only applying a pre-defined kernel over a collection.

In particular, we find that the SDF is more sensitive in eHealth15 data since Triangle, Cosine and Circle density functions make no improvement. We go further into the data and find the main reason that topics of eHealth15 are described in colloquial language, without containing any standard medical terminology, while the topics of eHealth13, eHealth14, CDS14 and CDS15 are made of professional medical terms. This evidence indicates that the proposed SDF can better understand queries from patients without any medical backgrounds.

6.3   Triangle vs. Cosine

The experiment results show that the Triangle kernel and the Cosine kernel have the same performance in the five datasets. In fact, the documents are ranked by the score $R(Q, D)$ (Formula 9), where the only difference among the pre-defined kernel functions is the value of unit influence $Inf_{q_i}^u(f_{density})$. For Triangle and Cosine, we have

$$\sum_{u=1}^{\sigma} Inf_{q_i}^u(Triangle) = \int_0^{\sigma} Triangle(u)du = \frac{\sigma}{2},$$

and

$$\sum_{u=1}^{\sigma} Inf_{q_i}^u(Cosine) = \int_0^{\sigma} Cosine(u)du = \frac{\sigma}{2},$$

Here $\sum_{u=1}^{\sigma} Inf_{q_i}^u(Triangle) = \sum_{u=1}^{\sigma} Inf_{q_i}^u(Cosine)$, which means that a document has the same score with Triangle and Cosine. This explains why Triangle and Cosine get the same results.

Furthermore, the above conclusion suggests to replace Cosine with Triangle in the real applications, since Triangle has a lower computation complexity than Cosine.

6.4   Comparisons with the State-of-the-Art Approaches

Here we compare our proposed SDF with the state-of-the-art models PLM [13] and CRTER [22], and CLEF eHealth official best results in Table 3 in terms of MAP on three eHealth datasets. We can see that our SDF leads the best performance, where the significant tests have been done on CRTER results. However, the experimental results show that our SDF model makes higher improvements in eHalth13 and eHealth15 contrast with eHealth14. Since the eHealth13, eHealth14 and eHealth15 uses the same collection, we go further into the queries utilized in these three datasets and find that eHealth13 and eHealth15 tend to have a longer topics. In fact, queries utilized in eHealth13 and eHealth15 are sentences, whereas the topics utilized in eHealth14 are composed by seldom keywords. This evidence implies that our SDF based IR model has outstanding performance in long queries retrieval.

Table 3. MAP comparison with CRTER and official best results ("*" indicates significant improvement over CRTER, and the percentage indicates the promotion compared with the official best results.)

	eHealth13	eHealth14	eHealth15
PLM	0.2849	0.2918	0.2427
CRTER	0.3141	0.2802	0.2431
Official results	0.3040	0.4016	0.2549
SDF	0.3568* (+17.37%)	0.4140* (+3.09%)	0.2854* (+11.97%)

7   Conclusions and Future Work

In this paper, we proposed a saturated density function based method to promote the healthcare information retrieval. Firstly, we define a saturated density function, and propose a self-adaptive approach on saturated density function building, where the approach provides each term a freedom to build the best suitable density function at different circumstances, instead of universally utilizing a single pre-defined density function over the whole collection. Then, our experimental results on five standard CLEF and TREC datasets show that our proposed SDF makes significant improvement compared with all of the pre-defined density functions. Furthermore, SDF is effective in understanding queries with colloquial language. Finally, we suggest that Triangle is computationally a better candidate than Cosine when they have the same ranking scores.

In the future, we will consider more density functions and more application circumstances, for instance, taking the positions that query terms have occurrence in the sentence or the paragraph into consideration. At the same time, we will investigate whether the SDF performance is coincident with the increase of the density function space size.

Acknowledgements. We thank all viewers who provided the thoughtful and constructive comments on this paper. The second author is the corresponding author. This research is supported by the open funds of NPPA Key Laboratory of Publishing Integration Development, ECNUP, the Shanghai Municipal Commission of Economy and Informatization (No. 170513), and Xiaoi Research. The computation is performed in the Supercomputer Center of ECNU.

References

1. Amati, G., Van Rijsbergen, C.J.: Probabilistic models of information retrieval based on measuring the divergence from randomness. ACM Trans. Inf. Syst. **20**(4), 357–389 (2002)
2. Beigbeder, M., Mercier, A.: Fuzzy proximity ranking with Boolean queries. In: Fourteenth Text Retrieval Conference, Trec 2005, Gaithersburg, Maryland, November 2005
3. Clarke, C.L., Cormack, G.V., Burkowski, F.J.: Shortest substring ranking (multitext experiments for TREC-4). In: TREC, vol. 4, pp. 295–304. Citeseer (1995)
4. Cummins, R., O'Riordan, C., Lalmas, M.: An analysis of learned proximity functions. In: Adaptivity, Personalization and Fusion of Heterogeneous Information (2010)
5. De Kretser, O., Moffat, A.: Effective document presentation with a locality-based similarity heuristic. In: SIGIR 1999: Proceedings of the International ACM SIGIR Conference on Research and Development in Information Retrieval, Berkeley, 15–19 August 1999, pp. 113–120 (1999)
6. Gerani, S., Carman, M., Crestani, F.: Aggregation methods for proximity-based opinion retrieval. ACM Trans. Inf. Syst. **30**(4), 403–410 (2012)
7. Hawking, D., Thistlewaite, P.: Proximity operators-so near and yet so far. In: Proceedings of the 4th Text Retrieval Conference, pp. 131–143 (1995)
8. Hogg, R.V., Craig, A.T.: Introduction to Mathematical Statistics, 5th edn. Englewood Hills, New Jersey (1995)
9. Keen, E.M.: The use of term position devices in ranked output experiments. J. Doc. **47**(1), 1–22 (1991)
10. Keen, E.M.: Some aspects of proximity searching in text retrieval systems. J. Inf. Sci. **18**(2), 89–98 (1992)
11. Kise, K., Junker, M., Dengel, A., Matsumoto, K.: Passage retrieval based on density distributions of terms and its applications to document retrieval and question answering. In: Dengel, A., Junker, M., Weisbecker, A. (eds.) Reading and Learning. LNCS, vol. 2956, pp. 306–327. Springer, Heidelberg (2004). https://doi.org/10.1007/978-3-540-24642-8_17
12. Lu, X.: Improving search using proximity-based statistics. In: The International ACM SIGIR Conference, pp. 1065–1065 (2015)
13. Lv, Y., Zhai, C.X.: Positional language models for information retrieval. In: International ACM SIGIR Conference on Research and Development in Information Retrieval, pp. 299–306 (2009)
14. Lv, Y., Zhai, C.: Positional relevance model for pseudo-relevance feedback. In: Proceedings of the 33rd international ACM SIGIR Conference on Research and Development in Information Retrieval, pp. 579–586. ACM (2010)
15. Mahdabi, P., Gerani, S., Huang, J.X., Crestani, F.: Leveraging conceptual lexicon: query disambiguation using proximity information for patent retrieval. In: International ACM SIGIR Conference on Research and Development in Information Retrieval, pp. 113–122 (2013)
16. Metzler, D., Croft, W.B.: A Markov random field model for term dependencies. In: Proceedings of the 28th Annual International ACM SIGIR Conference on Research and Development In information Retrieval, pp. 472–479. ACM (2005)
17. Miao, J., Huang, J.X., Ye, Z.: Proximity-based Rocchio's model for pseudo relevance. In: Proceedings of the 35th International ACM SIGIR Conference on Research and Development in Information Retrieval, pp. 535–544 (2012)

18. Petkova, D., Croft, W.B.: Proximity-based document representation for named entity retrieval. In: Sixteenth ACM Conference on Information and Knowledge Management, CIKM 2007, Lisbon, pp. 731–740, November 2007

19. Song, Y., Hu, W., Chen, Q., Hu, Q., He, L.: Enhancing the recurrent neural networks with positional gates for sentence representation. In: Cheng, L., Leung, A.C.S., Ozawa, S. (eds.) ICONIP 2018. LNCS, vol. 11301, pp. 511–521. Springer, Cham (2018). https://doi.org/10.1007/978-3-030-04167-0_46

20. Tao, T., Zhai, C.X.: An exploration of proximity measures in information retrieval. In: Proceedings of the International ACM SIGIR Conference on Research and Development in Information Retrieval, SIGIR 2007, Amsterdam, pp. 295–302, July 2007

21. Zhao, J., Huang, J.X.: An enhanced context-sensitive proximity model for probabilistic information retrieval. In: International ACM SIGIR Conference on Research and Development in Information Retrieval, pp. 1131–1134 (2014)

22. Zhao, J., Huang, J.X., He, B.: CRTER: using cross terms to enhance probabilistic information retrieval. In: Proceeding of the International ACM SIGIR Conference on Research and Development in Information Retrieval, SIGIR 2011, Beijing, pp. 155–164, July 2011

CytoFA: Automated Gating of Mass Cytometry Data via Robust Skew Factor Analzyers

Sharon X. Lee[(✉)]

School of Mathematics and Physics, University of Queensland, Brisbane, Australia
s.lee11@uq.edu.au

Abstract. Cytometry plays an important role in clinical diagnosis and monitoring of lymphomas, leukaemia, and AIDS. However, analysis of modern-day cytometric data is challenging. Besides its high-throughput nature and high dimensionality, these data typically exhibit complex characteristics such as multimodality, asymmetry, heavy-tailness and other non-normal characteristics. This paper presents cytoFA, a novel data mining approach capable of clustering and performing dimensionality reduction of high-dimensional cytometry data. Our approach is also robust against non-normal features including heterogeneity, skewness, and outliers (dead cells) that are typical in flow and mass cytometry data. Based on a statistical approach with well-studied properties, cytoFA adopts a mixtures of factor analyzers (MFA) to learn latent nonlinear low-dimensional representations of the data and to provide an automatic segmentation of the data into its comprising cell populations. We also introduce a double trimming approach to help identify atypical observations and to reduce computation time. The effectiveness of our approach is demonstrated on two large mass cytometry data, outperforming existing benchmark algorithms. We note that while the approach is motivated by cytometric data analysis, it is applicable and useful for modelling data from other fields.

1 Introduction

Flow cytometry enables the study of physical and chemical properties of particles at the single-cell level, thus rendering it widely useful in many biomedical fields and is now routinely used in both clinical and research immunology. Modern cytometers has the ability to detect up to 30 markers simultaneously at a rate of 10,000 cells per second, thus generating datasets of massive size in a high-throughput manner. A more recent technology, mass cytometry or cytometry by time of flight (CyTOF), had the potential to detect up to 100 simultaneous measurements per cell [1], although at a slower speed than flow cytometer. This

Electronic supplementary material The online version of this chapter (https://doi.org/10.1007/978-3-030-16148-4_40) contains supplementary material, which is available to authorized users.

Q. Yang et al. (Eds.): PAKDD 2019, LNAI 11439, pp. 514–525, 2019.
https://doi.org/10.1007/978-3-030-16148-4_40

had surpassed the ability of manual analysis, the current standard procedure for analysing these data. This process is also subjective, error-prone, and time-consuming.

An important part of cytometric data analysis is the task of identifying different cell populations within the data. Traditionally, this is carried out by analysts in a manual process known as gating where regions of interests are identified by visually inspecting a series of bivariate projections of the data. Not only is this process time-consuming and subjective, but it can be very difficult to detect higher-dimensional inter-marker relationships in this way. In view of this, computational tools have been developed in recent times to assist or automate the gating process [2–4]. Many of these proposals adopt a mixture model-based approach. However, traditional mixture models are not well suited for modelling cytometric data as the symmetric component densities find it challenging to handle non-normal cluster shapes. This led to recent work on finite mixtures of skew distributions [5–11]. Although they demonstrated that skew mixture models are effective in segmenting cytometry data, these papers focused almost exclusively on small cytometric data sets due to the associated computational complexity.

In this paper, we propose a new computational tool for analyzing complex cytometry data sets. It is especially suited for large and high-dimensional data for which traditional skew mixture models find it challenging to analyse. We propose to adopt a factor analysis approach to simultaneously perform gating and dimension reduction of large cytometric data sets. This allows us to achieve local linear dimensionality reduction as well as global non-linear dimensionality reduction. Further to this, by adopting latent factors that follow skew distributions, our approach can provide a more realistic model for cytometric cell populations. Moreover, a double trimming approach is proposed to help discern atypical and dead cells from normal cells.

The rest of this paper is organised as follows. In Sect. 2, we present a small example to motivate the adoption of skew factor analysis model. In Sect. 3, we

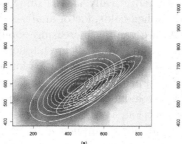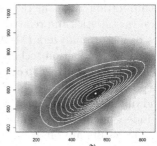

Fig. 1. Modelling an asymmetric cluster of cells. (a) Fitting a GMM would require two components to adequately model the cluster. (b) A single skew normal distribution would suffice for the asymmetric cluster.

build the cytoFA model by walking through its submodels. This includes a discussion of skew normal models, its mixture variants, their factor analytic variants, and a novel double trimming approach. In Sect. 4, we briefly outline how the cytoFA model can be fitted to a data. An ECM algorithm can be implemented in closed form for this task. Finally, the usefulness of the proposed approach is demonstrated in Sect. 5 on two large high-dimensional mass cytometry data containing 24 and 14 major immune cell populations and 13 and 32 markers, respectively.

2   Motivation

To begin, let us consider a small flow cytometric data derived from a peripheral blood sample of a Graft vs Host Disease (GvHD) patient. The sample, after staining with CD8β and CD8, reveals a unimodal and skewed cell population (Fig. 1). The inadequacy of the traditional Gaussian mixture model in this case is apparent in Fig. 1(a) where it splits the population into two clusters in order to accommodate for the asymmetry in the data. While many authors would choose to merge the two components in this situation, it will lead to unnecessary difficulties in downstream analysis (for example, in interpretation and derivation of population characteristics), let alone being computationally inefficient. Notably, a single skew normal distribution would suffice in this case and provides a far more accurate fit to the data (Fig. 1(b)). Moreover, the fitted skew normal distribution can correctly estimate the mode of the distribution of the cell population, whereas none of the mode of the components of the GMM coincides with the empirical mode of the data. This motivated the adoption of skew mixture models for clustering flow cytometric data in [5]. Their effectiveness have been demonstrated in a number of subsequent work including [12–14].

As mixture models are highly parametrized models, direct application to high-dimensional data may become computationally infeasible. This is even more pronounced for skew mixture models as they involve more parameters than their symmetric counterparts. It can be shown that the number of parameters, and hence the computation time, increases quadratically as the dimension of the data increases. In statistics and related applied fields, factor analysis (FA) model and mixtures of factor analyzers (MFA) are gaining increasing popularity as a powerful dimensional reduction technique. These models are implicit dimension reduction techniques, in contrast to commonly used variable selection approaches which are explicit. The former had the appealing advantage of simultaneous clustering and dimension reduction. Although one may carry out clustering subsequent and independently to variable selection, as is common practice in many applications, the results are often much inferior to implicit approaches; see [15] for an illustrative example.

As skew mixture models is a relatively new area of research, few authors have considered factor analytic extensions of these models. Furthermore, while explicit variable selection have recently been considered in the computational cytometry literature (such as the DensVM algorithm by [16]), implicit approaches for these data have not been explored. In this paper, we propose a new computational tool for simultaneous clustering and dimension reduction of cytometric data, based on a skew normal mixture of factor analyzers. By combining with skew normal component distributions, we extend the MFA model to provide a promising and more general tool that are robust to asymmetrically distributed clusters. The effectiveness of our approach will be demonstrated in Sect. 5.

3   The CytoFA Algorithm

The cytoFA model is designed to address five objectives at the same time. These are (i) to provide an automated method to segment cell populations in the data, (ii) to perform implicit dimension reduction, (iii) to identify atypical observations, (iv) be robust to non-normal cluster shapes, and (v) be computationally efficient. To achieve these, cytoFA adopts a four-in-one approach that combines mixture modelling with skew models, factor analysis, and double trimming.

3.1   Multivariate Skew Normal Distributions

We begin by defining the multivariate skew normal (MSN) density to be adopted in this paper. This distribution is employed to model a cell population within the sample. Let Y be a p-dimensional random vector that follows a skew normal distribution [17] with $p \times 1$ location vector μ, $p \times p$ scale matrix Σ, $p \times 1$ skewness vector δ. For cytometric data analysis, Y corresponds to the measurements of the p markers on a cell. Then its probability density function (pdf) can be expressed as a product of a multivariate normal density and a (univariate) normal distribution function, given by

$$f_p(y; \mu, \Sigma, \delta) = 2\phi_p(y; \mu, \Omega)\, \Phi(y^*; 0, \lambda), \tag{1}$$

where $\Omega = \Sigma + \delta\delta^T$, $y^* = \delta^T \Omega^{-1}(y - \mu)$, and $\lambda = 1 - \delta^T \Omega^{-1}\delta$. Here, we let $\phi_p(.; \mu, \Sigma)$ be the p-dimensional normal distribution with mean vector μ and covariance matrix Σ, and $\phi(.; \mu, \Sigma)$ is the corresponding (cumulative) distribution function. The notation $Y \sim \mathrm{MSN}_p(\mu, \Sigma, \delta)$ will be used. Note that when $\delta = 0$, (1) reduces to the symmetric normal density $\phi_p(y; \mu, \Sigma)$. Geometrically, the δ vector specifies the orientation or direction of the skewness of the distribution [18]. It is noted that various versions of the multivariate skew normal density exist in the literature. The version adopted here is equivalent to the well-known version formulated by [17]; see [19] for some discussions on this.

3.2   Finite Mixture Model

As a sample is comprised of multiple heterogeneous cell populations, it can be effectively modelled by a finite mixture of MSN distributions. Finite mixture model is a convex linear combination of (a finite number of) component densities.

In our case, the g-component finite mixture of MSN distributions has density given by

$$f(y; \Psi) = \sum_{i=1}^{g} \pi_i f_p(y; \mu_i, \Sigma_i, \delta_i), \qquad (2)$$

where $f_p(y; \mu_i, \Sigma_i, \delta_i)$ denotes the ith MSN component of the mixture model as defined by (1), with location parameter μ_i, scale matrix Σ_i, and skewness parameter δ_i. In the above, the π_i are the mixing proportions or weights and they satisfy $\pi_i \geq 0$ $(i = 1, \ldots, g)$ and $\sum_{i=1}^{g} \pi_i = 1$. The vector Ψ contains all the unknown parameters of the FMMSN model; that is, it consists of $\pi_1, \ldots, \pi_{g-1}$ and $\theta_1^T, \ldots, \theta_g^T$ where now θ_i is the vector of all the unknown parameters of the ith component density function (which includes μ_i, δ_i, and the distinct elements of Σ_i).

3.3   Multivariate Skew Normal Factor Analyzers

As the number of markers p increases, the model (2) can quickly become computationally infeasible to fit. We thus consider a commonly used alternative. A factor analysis (FA) model [20,21] assumes that the data space can be represented by a lower-dimensional subspace of the original feature space. More formally, it models the distribution of the data using

$$y = \mu + Bu + e, \qquad (3)$$

where u is a q-dimensional $(q < p)$ random vector of latent variables known as factors, B is a $p \times q$ matrix of factor loadings, and e are independently distributed error variables. The traditional FA model assumes that U follows a standard (multivariate) normal distribution, whereas e follows a centered normal distribution with a diagonal covariance matrix.

To extend the traditional FA model in (3), we let the latent factors u follows a MSN distribution. The error variables remain independent to u and follows $N_p(0, D)$, where D is a diagonal matrix. In addition, we choose a parameterization of the MSN distribution such that the factors have mean being the zero vector and covariance matrix being the identity matrix, thus preserving the property of u in the traditional FA model. To achieve this, we let $u \sim \mathrm{MSN}_q\left(-c\Lambda^{\frac{1}{2}}, \Lambda, \Lambda^{\frac{1}{2}}\delta\right)$, where $c = \frac{\pi}{2}$ and $\Lambda = \left(I_q + (1 - c^2)\delta\delta^T\right)^{-1}$. It follows that the marginal distribution of y in this case is also a MSN distribution and is given by $y \sim \mathrm{MSN}_p\left(\mu - cB\Lambda^{\frac{1}{2}}\delta, \Sigma, B\Lambda^{\frac{1}{2}}\right)$ which leads to y having mean μ and covariance matrix $BB^T + D$, conforming to the case of the traditional FA model. We shall adopt (3) to model a single population in a sample.

3.4   Finite Mixtures of Multivariate Skew Normal Factor Analyzers

Let y_j $(j = 1, \ldots, n)$ denote n random observations of Y, that is, measurements from n cells in a sample. To model all g cell populations in a sample, we consider the mixture version of (3). This leads to a g-component mixture of skew normal factor analyzers. In this case, the jth observation can be modelled by

$$\boldsymbol{y}_j = \boldsymbol{\mu}_i + \boldsymbol{B}_i \boldsymbol{u}_{ij} + \boldsymbol{e}_{ij}, \tag{4}$$

conditional on $\boldsymbol{y}_j$ belonging to the ith component of the mixture model ($i = 1, \ldots, g$). Thus, unconditionally, the density of $\boldsymbol{y}_j$ is given by

$$f(\boldsymbol{y}_j; \boldsymbol{\Psi}) = \sum_{i=1}^{g} \pi_i \, \mathrm{MSN}_p \left(\boldsymbol{\mu}_i - c\boldsymbol{B}_i \boldsymbol{\Lambda}_i^{\frac{1}{2}} \boldsymbol{\delta}_i, \ \boldsymbol{\Sigma}_i, \ \boldsymbol{B}_i \boldsymbol{\Lambda}_i^{\frac{1}{2}} \right). \tag{5}$$

We shall refer to the above model as the cytoFA model. Note that each component of the cytoFA model has its own specific matrix of factor loadings, hence allowing the latent subspace of each component to be different. The effect is thus a globally nonlinear and locally linear dimension reduction of the feature space.

3.5   Robust Double Trimming

Trimming refers to the technique of temporarily discarding certain observations during the model fitting process (to be described in the next Section). This trimmed likelihood approach [22] is based on the idea that a proportion of observations that are deemed as unlikely to occur can be temporarily removed so that they temporarily do not contribute to the model fitting procedure. This simple but powerful technique has been shown to be quite effective at handling outliers. It proceeds by first attaching a binary label t_j (known as a trimming label) to each observation $\boldsymbol{y}_j$. We then rank the contribution of each observation to the mixture model. The observations corresponding to the smallest $\lfloor nt \rfloor$ contributions are then trimmed and their trimming label are set to 0, where t is the trimming proportion. Note that these contributions are updated during each iteration of the fitting procedure and thus the trimming labels may change between iterations. When the fitting algorithm terminates, observations that are trimmed are labeled as 'outliers'. However, we can still compute a cluster label for these atypical observations by applying the maximum *a posteriori* rule to their estimated posterior probabilities of component membership.

The above technique can be interpreted as single trimming. In our case, n is typically quite large. To reduce the amount of calculations required, we introduce a second level of trimming. The idea is that we may temporarily not update observations that have a high probability τ of belonging to a component and have not been 'active' for the past α iterations. We proceed by monitoring the posterior probability of membership associated with each observation. If it satisfies the criterion of non-activity, we skip this observation during the E-step for the next τ iterations. Thus, it is similar to performing partial E-steps. When concurrently applying this strategy and the single trimming technique to an EM algorithm, we refer to it as a double trimming technique. It can be shown that with double trimming, the EM algorithm still convergences to a local stationary point of the observed log-likelihood function.

3.6   The Final Model

Combining the above approaches into one, we obtain a finite mixtures of skew normal factor analyzers with double trimming. The density of a sample is given by (5), and the density of a cell population with in the sample is a MSN distribution. When carrying out parameter estimation for the models of the cytoFA model, the double trimming approach is integrated into the fitting procedure as described above.

4   Parameter Estimation of the CytoFA Model via an EM Algorithm

The cytoFA model admits a convenient hierarchical characterization that facilitates the computation of the maximum likelihood estimator (MLE) of the unknown model parameters using the Expectation-Maximization (EM) algorithm [23]; see [24] and [25] for technical details. The implementation of the EM algorithm requires alternating repeatedly the E- and M-steps until convergence in the case where the changes in the log likelihood values are less than some specified small value. The E-step calculates the expectation of the complete-data log likelihood given the observed data y using the current estimate of the parameters, known as the Q-function. The M-step then maximizes the Q-function with respect to the parameters Ψ to obtain updated estimates of the model parameters. For technical details of the E- and M-steps expressions, see the derivations provided by [24]. However, we need to integrate the double trimming approach into the EM algorithm, similar to [25], when implementing the cytoFA algorithm.

5   Analysis of High-Dimensional CyTOF Data

To demonstrate the usefulness of the cytoFA model, we consider the gating of a 13-dimensional and a 32 dimensional mass cytometry data considered in [26]. In [27], these two data were used as a benchmark for the comparison of 13 unsupervised algorithms on this data. This includes some well-known and/or commonly used algorithms specialized for cytometric data, such as FlowSOM [28], immunoClust [29], and SWIFT [30]. We will use the results reported in [27] as a benchmark to compare the performance of cytoFA with these state-of-the-art algorithms.

Similar to [26], we applied the cytoFA model to the full data (that is, including ungated and dead cells) and its performance is assessed on the gated populations. We calculated the F-measure score as is typically used in evaluating the performance of cytometric data analysis algorithms. The F-measure is defined as the harmonic mean of precision and recall. It ranges between 0 and 1, with 1 indicating a perfect match with the true class labels and 0 is the worse match. In calculating the F-measure, we choose among the possible permutations of the cluster labels the one that gives the highest value.

5.1   CyTOF-13 Data

The first data contain measurements on 167,044 cells derived from a sample donated from a healthy human individual. There are 24 major immune cell populations identified by manual analysis in this data, and their relative population size varies considerably (ranges from 0.006% to 17.1%) This data is difficult to analyse [27] due to the large span in the abundance of the populations, ranging from the smallest Platelet population of 5 cells to the largest population of 13,964 mature CD4+ T-cells. Our task here is to identify and model all these cell populations and provide a predicted class label for each cell.

We fitted the cytoFA model to the mass cytometry data using the EM algorithm described in Sect. 4, with double trimming and without double trimming. The later is denoted by cytoFA*. It achieved a F-measure of 0.712 for this data when no trimming is applied and 0.831 when trimming is applied, both outperforming previous results reported on this data (see [27]). It was observed in this previous study that all considered methods performed poorly on this data, achieving very low F-measures. The very low abundance of some cell populations appears to make this data particularly challenging to model. In [27], the reported best results was 0.518, obtained by flowMeans [31]. This suggests the cytoFA models provides a notable improvement in gating performance compared to these methods for this data.

To pursue this further, we report the F-measure for each individual cell population in Fig. 2. The heatmap reveals that the cytoFA and cytoFA* models perform quite well across the populations and almost consistently better than competing methods, especially in the case of low abundance populations where other models seem to have difficulty with. To gain a better insight into the relative contribution of the skewing and factor analysis part of the model, we have experimented with the traditional MFA model and the skew normal mixture model. They obtained a F-measure of 0.53 and 0.61, respectively, suggesting the skewness plays a somewhat more important role in improving the performance. Inspection of the δ parameters of the fitted cytoFA* model reveals that skewness range from -2.70 to 3.08 across the populations (see Supplementary Fig. 1), with two-thirds of the populations exhibiting a (mean) magnitude above 1.0. The benefit of adopting skew distributions is also supported by the results from the other models in the comparison. Many of the competing models are based on symmetric mixture models (or k-means) combined with techniques such as merging component (for example, flowMeans, SWIFT, and immunoClust) and variable transformation (such as flowClust) to cater for asymmetric clusters. According to the results, their clustering performance is inferior to both the skew normal mixture model and the MFA model in this data.

In addition, we note that the computation time of our models (30 and 210 s for the cytoFA and cytoFA*, respectively) is comparable to many of the above mentioned methods, which range from 2 s (k-means) to 29469 s (DensVM). The median and mean computation time is 729 s (PhenoGraph) and 5997.32 s, respectively. Another remark is that when trimming is applied (10% in this case), not only does it runs seven times faster, but considerable improvements in F-measure can be observed across most of the cell populations.

cytoFA	cytoFA*	flowMeans	FlowSOM	PhenoGraph	DensVM	k-means	FLOCK	ACCENSE	immunoClust	SamSPECTRAL	SWIFT	
0.86	0.97	0.86	0.97	0.98	0.99	0.63	0.94	0.35	0.20	0.80	0.15	11
0.82	0.40	0.49	0.46	0.90	0.76	0.57	0.80	0.38	0.14	0.81	0.08	5
0.97	0.87	0.96	0.96	0.95	0.97	0.90	0.72	0.41	0.17	0.71	0.20	18
0.99	0.99	0.97	0.96	0.95	0.96	0.74	0.08	0.45	0.24	0.00	0.24	12
0.99	0.95	0.62	0.65	0.97	0.66	0.86	0.96	0.60	0.28	0.94	0.44	9
0.99	0.98	0.94	0.92	0.96	0.97	0.78	0.68	0.58	0.28	0.00	0.32	17
0.82	0.66	0.95	0.86	0.94	0.78	0.64	0.81	0.50	0.30	0.95	0.47	2
0.92	0.62	0.82	0.93	0.91	0.91	0.90	0.93	0.91	0.32	0.70	0.11	19
0.96	0.46	0.88	0.55	0.87	0.60	0.69	0.00	0.58	0.15	0.00	0.06	13
0.43	0.47	0.32	0.23	0.40	0.23	0.36	0.15	0.33	0.16	0.10	0.13	16
0.71	0.74	0.19	0.30	0.00	0.03	0.13	0.00	0.23	0.29	0.37	0.12	3
0.97	0.71	0.66	0.72	0.60	0.58	0.62	0.64	0.66	0.49	0.00	0.11	24
0.98	0.76	0.61	0.39	0.70	0.70	0.64	0.00	0.57	0.43	0.00	0.16	1
0.77	0.78	0.51	0.57	0.00	0.00	0.05	0.00	0.43	0.32	0.00	0.18	10
0.49	0.36	0.19	0.05	0.01	0.15	0.19	0.14	0.21	0.17	0.00	0.13	8
1.00	1.00	0.35	0.61	0.00	0.00	0.00	0.78	0.07	0.77	0.00	0.00	20
1.00	0.89	0.64	0.62	0.62	0.63	0.63	0.59	0.32	0.61	0.55	0.39	21
0.51	0.88	0.16	0.02	0.00	0.13	0.11	0.00	0.28	0.14	0.00	0.09	7
0.96	0.96	0.03	0.42	0.00	0.44	0.44	0.43	0.35	0.43	0.00	0.16	4
0.90	0.96	0.45	0.00	0.43	0.00	0.00	0.00	0.00	0.72	0.00	0.24	23
0.98	0.96	0.42	0.36	0.04	0.00	0.22	0.00	0.00	0.45	0.10	0.16	14
0.62	0.86	0.04	0.04	0.00	0.04	0.02	0.04	0.02	0.17	0.00	0.19	15
0.74	0.61	0.39	0.28	0.00	0.21	0.30	0.39	0.00	0.14	0.04	0.11	6
1.00	0.57	0.00	0.00	0.00	0.00	0.00	0.00	0.00	0.01	0.00	0.04	22

Fig. 2. F-measure (per cell population) of various methods applied to the CyTOF-13 data. The rows are ordered by decreasing abundance.

5.2 CyTOF-32 Data

Finally, we consider the larger CyTOF-32 data, containing measurements of 32 markers on 265,627 cells. There are a total of 14 immune cell populations gated by manual analysis. The abundance of the cell populations varies between 0.49% to 25.41%. However, the skewness in this data is relatively smaller compared to the CyTOF-13 data and hence we do not expect the other methods to be disadvantaged. On applying the cytoFA model (with trimming) to this data, we obtained an overall F-measure of 0.855. This is a notable improvement from the results reported in [27], in which the best F-measure was 0.78 obtained by flowSOM. An inspection of Fig. 3 reveals that cytoFA scores high F-measure across all the cell populations. For seven of the 14 cell populations, cytoFA achieved a perfect or near-perfect F-measure of 0.99 to 1.00. Of the remaining cell populations, its F-measure is above 0.80 with the exception of two populations. Looking at the low abundance populations (those with abundance under 1%), it is of interest to note that cytoFA performs significantly better than its competitors (with a F-measure at least 0.10 higher than the next best algorithm).

Concerning computation time, the fitting of cytoFA completed in 64 s in our experiment. This places it slightly slower than k-means (13 s) and flowSOM (41 s), but is significantly faster than the other methods (which ranges from 225 s to 30613 s). Thus, cytoFA compares favourably with these methods when considering both its clustering performance and computation time. Furthermore,

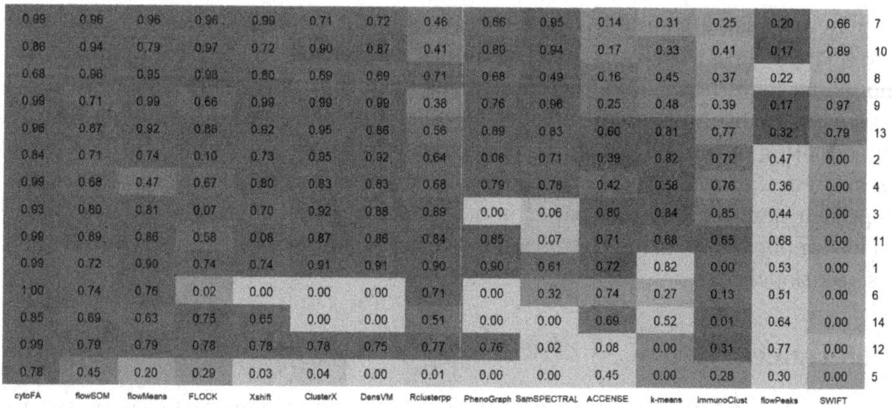

cytoFA	flowSOM	flowMeans	FLOCK	Xshift	ClusterX	DensVM	Rclusterpp	PhenoGraph	SamSPECTRAL	ACCENSE	k-means	immunoClust	flowPeaks	SWIFT	
0.99	0.96	0.96	0.96	0.99	0.71	0.72	0.46	0.66	0.95	0.14	0.31	0.25	0.20	0.66	7
0.86	0.94	0.79	0.97	0.72	0.90	0.87	0.41	0.80	0.94	0.17	0.33	0.41	0.17	0.89	10
0.68	0.98	0.95	0.98	0.80	0.59	0.69	0.71	0.68	0.49	0.16	0.45	0.37	0.22	0.00	8
0.99	0.71	0.99	0.66	0.99	0.99	0.99	0.38	0.76	0.96	0.25	0.48	0.39	0.17	0.97	9
0.96	0.87	0.92	0.88	0.92	0.95	0.86	0.56	0.89	0.83	0.60	0.81	0.77	0.32	0.79	13
0.84	0.71	0.74	0.10	0.73	0.95	0.92	0.64	0.08	0.71	0.39	0.82	0.72	0.47	0.00	2
0.99	0.68	0.47	0.67	0.80	0.83	0.83	0.68	0.79	0.78	0.42	0.58	0.76	0.36	0.00	4
0.93	0.80	0.81	0.07	0.70	0.92	0.88	0.89	0.00	0.06	0.80	0.84	0.85	0.44	0.00	3
0.99	0.89	0.86	0.58	0.08	0.87	0.86	0.84	0.85	0.07	0.71	0.68	0.65	0.68	0.00	11
0.99	0.72	0.90	0.74	0.74	0.91	0.91	0.90	0.90	0.61	0.72	0.82	0.00	0.53	0.00	1
1.00	0.74	0.76	0.02	0.00	0.00	0.00	0.71	0.00	0.32	0.74	0.27	0.13	0.51	0.00	6
0.85	0.69	0.63	0.75	0.65	0.00	0.00	0.51	0.00	0.00	0.69	0.52	0.01	0.64	0.00	14
0.99	0.79	0.79	0.78	0.78	0.78	0.75	0.77	0.76	0.02	0.08	0.00	0.31	0.77	0.00	12
0.78	0.45	0.20	0.29	0.03	0.04	0.00	0.01	0.00	0.00	0.45	0.00	0.28	0.30	0.00	5

Fig. 3. F-measure (per cell population) of various methods applied to the CyTOF-32 data. The rows are ordered by decreasing abundance.

cytoFA has the ability to identify and eliminate atypical observations which most of the competing methods considered here cannot. For this data, cytoFA identified a large proportion of cells (approximately 50%) as atypical. This may have contributed to its fast computation time. Surprisingly, the large proportion of trimmed observations corresponds well with manual analysis, which concluded that approximately 60% of the cells do not belong to any of the 14 major immune populations. Thus, cytoFA is effective at discriminating uninteresting and/or dead cells.

6 Conclusions

We have presented a new computational tool, cytoFA, for simultaneous clustering and dimension reduction of large cytometric data. Based on finite mixtures of skew normal factor analyzers, the approach can effectively accommodate clusters that are asymmetrically distributed. Adopting a factor analytic characterization of the component densities enables analysis of high-dimensional data to be performed within very reasonable time. The cytoFA model admits convenient stochastic representations which facilitates model fitting via the EM algorithm. An illustration on two large mass cytometry data shows that the cytoFA model compares favourably to other state-of-the-art specialized algorithms, achieving a higher F-measure than those reported in other analyses of these data. The computation time of our approach is also well within the reasonable range, being less than 4% of the mean computation time of the reported state-of-the-art algorithms. Accurate gating and quantification of cells is crucial for downstream analysis. The construction of class template and the supervised classification for new samples, for example, rely on quantitative features derived from these population statistics. By employing a more accurate model such as the cytoFA model, it can bring significant improvements to diagnosis and prognosis in downstream analysis.

The promising results in this paper shows that implicit dimension reduction can be combined with skew mixture models to provide an effective approach for modelling high-dimensional heterogeneous data that may exhibit distributional skewness. Future work may consider further improvements to the model such as the adopting of more flexible component distributions to cater for features such as heavy tail, outliers, and other non-elliptical shapes. Finally we note that the approach described in this paper is useful not only for cytometric data, but is widely applicable to other data from other fields that exhibit similar complicated distributional features.

References

1. Bendall, S.C., Simonds, E.F., Qiu, P., Amir, E.D., Krutzik, P.O., Finck, R.: Single-cell mass cytometry of differential immune and drug responses across a human hematopoietic continuum. Science **332**, 687–696 (2011)
2. Aghaeepour, N., et al.: Critical assessment of automated flow cytometry analysis techniques. Nat. Methods **10**, 228–238 (2013)
3. Saeys, Y., Van Gassen, S., Lambrecht, B.N.: Computational flow cytometry: helping to make sense of high-dimensional immunology data. Nat. Rev. Immunol. **16**, 449–462 (2016)
4. Weber, L.M., Robinson, M.D.: Comparison of clustering methods for high-dimensional single-cell flow and mass cytometry data. Cytom. A **89**, 1084–1096 (2016)
5. Pyne, S., et al.: Automated high-dimensional flow cytometric data analysis. Proc. Natl. Acad. Sci. USA **106**, 8519–8524 (2009)
6. Pyne, S., et al.: Joint modeling and registration of cell populations in cohorts of high-dimensional flow cytometric data. PloS One **9**, e100334 (2014)
7. Wang, K., Ng, S.K., McLachlan, G.J.: Multivariate skew t mixture models: applications to fluorescence-activated cell sorting data. In: Shi, H., Zhang, Y., Bottema, M.J., Lovell, B.C., Maeder, A.J. (eds.) Proceedings of Conference of Digital Image Computing: Techniques and Applications, Los Alamitos, California, pp. 526–531. IEEE (2009)
8. Frühwirth-Schnatter, S., Pyne, S.: Bayesian inference for finite mixtures of univariate and multivariate skew-normal and skew-t distributions. Biostatistics **11**, 317–336 (2010)
9. Lee, S.X., McLachlan, G.J.: Model-based clustering and classification with non-normal mixture distributions. Stat. Methods Appl. **22**, 427–454 (2013)
10. Lee, S.X., McLachlan, G.J.: Finite mixtures of canonical fundamental skew t-distributions: the unification of the restricted and unrestricted skew t-mixture models. Stat. Comput. **26**, 573–589 (2016)
11. Lee, S.X., McLachlan, G.J., Pyne, S.: Modelling of inter-sample variation in flow cytometric data with the joint clustering and matching (JCM) procedure. Cytom. A **89**, 30–43 (2016)
12. Pyne, S., Lee, S., McLachlan, G.: Nature and man: the goal of bio-security in the course of rapid and inevitable human development. J. Indian Soc. Agric. Stat. **69**, 117–125 (2015)
13. Rossin, E., Lin, T.I., Ho, H.J., Mentzer, S.J., Pyne, S.: A framework for analytical characterization of monoclonal antibodies based on reactivity profiles in different tissues. Bioinformatics **27**, 2746–2753 (2011)

14. Lee, S.X., McLachlan, G., Pyne, S.: Application of mixture models to large datasets. In: Pyne, S., Rao, B.L.S.P., Rao, S.B. (eds.) Big Data Analytics, pp. 57–74. Springer, New Delhi (2016). https://doi.org/10.1007/978-81-322-3628-3_4

15. Bouveyron, C., Brunet-Saumard, C.: Model-based clustering of high-dimensional data: a review. Comput. Stat. Data Anal. **71**, 52–78 (2014)

16. Becher, B., et al.: High-dimensional analysis of the murine myeloid cell system. Nat. Immunol. **15**, 1181–1189 (2014)

17. Azzalini, A., Dalla Valle, A.: The multivariate skew-normal distribution. Biometrika **83**, 715–726 (1996)

18. McLachlan, G.J., Lee, S.X.: Comment on "on nomenclature for, and the relative merits of, two formulations of skew distributions" by A. Azzalini, R. Browne, M. Genton, and P. McNicholas. Stat. Probab. Lett. **116**, 1–5 (2016)

19. Lee, S.X., McLachlan, G.J.: On mixtures of skew-normal and skew t-distributions. Adv. Data Anal. Classif. **7**, 241–266 (2013)

20. Ghahramani, Z., Beal, M.: Variational inference for Bayesian mixture of factor analysers. In: Solla, S., Leen, T., Muller, K.R. (eds.) Advances in Neural Information Processing Systems, pp. 449–455. MIT Press, Cambridge (2000)

21. McLachlan, G.J., Peel, D.: Mixtures of factor analyzers. In: Proceedings of the Seventeenth International Conference on Machine Learning, pp. 599–606. Morgan Kaufmann, San Francisco (2000)

22. Neykov, N., Filzmoser, P., Dimova, R., Neytchev, P.: Robust fitting of mixtures using the trimmed likelihood estimator. Comput. Stat. Data Anal. **52**, 299–308 (2007)

23. Dempster, A.P., Laird, N.M., Rubin, D.B.: Maximum likelihood from incomplete data via the EM algorithm. J. R. Stat. Soc. B **39**, 1–38 (1977)

24. Lin, T.I., McLachlan, G.J., Lee, S.X.: Extending mixtures of factor models using the restricted multivariate skew-normal distribution. J. Multivar. Anal. **143**, 398–413 (2016)

25. Lee, S.X.: Mining high-dimensional CyTOF data: concurrent gating, outlier removal, and dimension reduction. In: Huang, Z., Xiao, X., Cao, X. (eds.) ADC 2017. LNCS, vol. 10538, pp. 178–189. Springer, Cham (2017). https://doi.org/10. 1007/978-3-319-68155-9_14

26. Levine, J.H., et al.: Data driven phenotypic dissection of aml reveals progenitor-like cells that correlate with prognosis. Cell **162**, 184–197 (2015)

27. Weber, L.M., Robinson, M.D.: Comparison of clustering methods for high-dimensional single-cell flow and mass cytometry data. Cytom. A **89A**, 1084–1096 (2016)

28. Van Gassen, S., Callebaut, B., Van Helden, M.J., Lambrecht, B.N., Demeester, P., Dhaene, T.: FlowSOM: using self-organizing maps for visualization and interpretation of cytometry data. Cytom. A **87A**, 636–645 (2015)

29. Sorensen, T., Baumgart, S., Durek, P., Grutzkau, A., Haaupl, T.: immunoClust - an automated analysis pipeline for the identification of immunophenotypic signatures in high-dimensional cytometric datasets. Cytom. A **87A**, 603–615 (2015)

30. Mosmann, T.R., Naim, I., Rebhahn, J., Datta, S., Cavenaugh, J.S., Weaver, J.M.: SWIFT - scalable clustering for automated identification of rare cell populations in large, high-dimensional flow cytometry datasets. Cytom. A **85A**, 422–433 (2014)

31. Aghaeepour, N., Nikoloc, R., Hoos, H.H., Brinkman, R.R.: Rapid cell population identification in flow cytometry data. Cytom. A **79**, 6–13 (2011)

Clustering and Anomaly Detection

Consensus Graph Learning
for Incomplete Multi-view Clustering

Wei Zhou, Hao Wang, and Yan Yang$^{(\boxtimes)}$

School of Information Science and Technology,
Southwest Jiaotong University, Chengdu, China
18108045668@126.com, hwang@my.swjtu.edu.cn, yyang@swjtu.edu.cn

Abstract. Multi-view data clustering is a fundamental task in current machine learning, known as multi-view clustering. Existing multi-view clustering methods mostly assume that each data instance is sampled in all views. However, in real-world applications, it is common that certain views miss number of data instances, resulting in incomplete multi-view data. This paper concerns the task of clustering of incomplete multi-view data. We propose a novel Graph-based Incomplete Multi-view Clustering (GIMC) to perform this task. GIMC can effectively construct a complete graph for each view with the help of other view(s), and automatically weight each constructed graph to learn a consensus graph, which gives the final clusters. An alternating iterative optimization algorithm is proposed to optimize the objective function. Experimental results on real-world datasets show that the proposed method outperforms state-of-the-art baseline methods markedly.

Keywords: Multi-view clustering · Incomplete views ·
Graph-based clustering

1 Introduction

In many natural scenarios, data are collected from multiple perspectives or diverse domains. Each of these domains presents a particular view of the data, where each view may have its own individual properties. Such forms of data are referred to as multi-view data. This paper makes a focus contribution on clustering of multi-view data, known as multi-view clustering [1,27]. Most existing multi-view clustering methods assume that each data instance is sampled in all views. While, in real-world applications, it is frequently happened that certain data instances are missing in certain views because of sensor fault or machine down. That is, the collected multi-view data are incomplete, referred to as incomplete multi-view data. In this case, traditional multi-view clustering approaches, such as [12,14,22,24,32,35], perform poorly or even helplessly.

This paper studies clustering of incomplete multi-view data, called incomplete multi-view clustering. Recently, several incomplete multi-view clustering

---

W. Zhou and H. Wang—Authors contributed equally to this work.

© Springer Nature Switzerland AG 2019
Q. Yang et al. (Eds.): PAKDD 2019, LNAI 11439, pp. 529–540, 2019.
https://doi.org/10.1007/978-3-030-16148-4_41

approaches have been proposed. Existing works such as those in [6,8,21] performed matrix factorization as base technology to tackle incomplete views, and those in [11,26,29,30] projected incomplete multi-view data into a common subspace with the help of complete views. Semi-supervised clustering with the user-provided constrain information has also been studied in [3,28,31]. However, matrix factorization technologies and subspace methods depend on their initialization. Semi-supervised clustering asks the users to provide prior constrain information. In this work, we perform our task based on graph clustering. Graph-based clustering produced several state-of-the-art multi-view clustering methods, [5,14,16,23,25,34] to name a few. However, it is unclear about how to model graph method for partitioning incomplete multi-view data.

In this work, we make the first attempt to model a graph method for partitioning incomplete multi-view data. We call the proposed method G̲raph-based I̲ncomplete M̲ulti-view C̲lustering (GIMC). GIMC can not only handle incomplete multi-view setting, but also overcome the limitations of existing graph-based multi-view clustering. For example, existing graph-based multi-view clustering methods (a) do not give sufficient consideration to weights of different views, e.g., [18], (b) require an additional clustering algorithm to produce the final clusters, e.g., [5,15], and (c) need to tune parameters, e.g., [16,34]. Our proposed GIMC can tackle these problems. Specially, GIMC first constructs a graph for each view (a complete view or incomplete view). Then, GIMC fuses all constructed graph matrix to generate a unified/consensus graph matrix. The proposed fusion method can automatically weight each constructed graph matrix. A rank constraint is imposed on the Laplacian matrix of the unified graph matrix. Comparing with multi-view spectral clustering, thus our method can directly produce the final clustering results on the unified graph matrix without any additional clustering algorithms. To the best our knowledge, this is the first such formulation.

In summary, this paper makes the following contributions. (1) It studies a natural and general multi-view setting, i.e., incomplete multi-view clustering. (2) It proposes a novel incomplete multi-view clustering method, named Graph-based Incomplete Multi-view Clustering (GIMC), which can not only address incomplete multi-view setting, but also tackle the limitations in the existing graph-based multi-view clustering approaches. (3) Extensive experiments show that the proposed GIMC method makes considerable improvement over the state-of-the-art baseline methods.

2   Related Work

Our method is clearly related to incomplete multi-view clustering. To our knowledge, there is no existing graph-based incomplete multi-view clustering method. Existing works on incomplete multi-view clustering are mainly based on matrix factorization or subspace learning. For example, non-negative matrix factorization (NMF) was exploited in incomplete multi-view setting [8], but only for two-view data. [33] combined this NMF idea and manifold regularization, which still works for two-view data. Toward this end, [10] extended this NMF idea to data with more views. Later on, more advanced methods based on NMF were studied in [6,21]. Besides, [20] formulated a joint tensor factorization process by taking into account missing data. Apart from the above-mentioned approaches, [26,29,30] investigated subspace technologies to address incomplete multi-view setting. However, both matrix factorization and subspace learning methods rely on their initialization. In Sect. 5, we will compare with these methods experimentally and show that our method performs robustly. Given the user-provided constrain information (i.e., must-link and cannot-link), semi-supervised clustering on incomplete multi-view data has also been studied in [3,27,31]. However, asking the user(s) to provide these constrain information may be unrealistic in practice because achieving these constrain information is a laborious work. Our method is an unsupervised clustering method.

Our work is also related to graph-based clustering as the proposed method builds upon graph-based clustering. Given a data matrix $\mathbf{X} \in \mathbb{R}^{d \times n}$ in single-view setting, where d is the dimension of features and n is the number of data instances, graph-based clustering methods typically partition the n data instances into c clusters as follows:

Step 1. Construct the data graph matrix $\mathbf{G} \in \mathbb{R}^{n \times n}$, where each entry g_{ij} in $\mathbf{G}$ denotes the similarity between $\mathbf{x}_i$ and $\mathbf{x}_j$;
Step 2. Compute the graph Laplacian matrix $\mathbf{L}_G = \mathbf{D}_G - (\mathbf{G}^T + \mathbf{G})/2$, where $\mathbf{D}_G$ is a diagonal matrix whose i-th diagonal element is $\sum_j (g_{ij} + g_{ji})/2$;
Step 3. Solve $\min_{\tilde{\mathbf{G}} \in \mathbb{R}^{n \times n}} Tr(\mathbf{E}^T \mathbf{L}_G \mathbf{E})$ with the rank of $\mathbf{L}_G$ equal to $n - c$, where $\mathbf{E}$ is an embedding matrix and $\tilde{\mathbf{G}}$ is a new data graph matrix;
Step 4. Produce clusters on the resulting $\tilde{\mathbf{G}}$.

We note that spectral clustering methods usually perform similar steps. Spectral clustering methods require and perform an additional clustering algorithm (e.g., K-means) on the embedding matrix $\mathbf{E}$ to produce the final clusters. Our method produces the clustering results on the learned graph matrix (i.e., the unified graph matrix $\mathbf{U}$, which will be seen shortly) without any additional clustering algorithms.

3   Graph-Based Incomplete Multi-view Clustering

3.1   Graph Construction for Incomplete Multi-view Data

For an incomplete multi-view dataset with m views, let $\mathbf{X}^1, \cdots, \mathbf{X}^m$ be the data matrices of the m views and $\mathbf{X}^v = \{\mathbf{x}_1^v, ..., \mathbf{x}_{n_v}^v\} \in \mathbb{R}^{d_v \times n_v}$ be the data matrix of the v-th view, where d_v is the dimensionality of the v-th view and n_v is the number of data instances. In such a way, the number of complete data instances is $n = max\{n_1, ..., n_m\}$. In order to diminish the effect of missing instances and improve the clustering performance in incomplete multi-view setting, the proposed GIMC is modeled as a joint graph-based multi-view clustering and meanwhile integrates weighted mechanism. It assigns weights for each data sample/instance to reduce the impact of missing instances, and automatically weights each constructed graph $\mathbf{A}^v \in \mathbb{R}^{n \times n}$ which is a symmetric graph matrix to learn a unified graph $\mathbf{U} \in \mathbb{R}^{n \times n}$. To handle those missing instances, we define a binary indicator matrix $\mathbf{B} \in \mathbb{R}^{m \times n}$, whose v, j-th entry is defined as follows:

$$b_{v,j} = \begin{cases} 1, & \text{if } j\text{-th instance exists in the } v\text{-th view} \\ 0, & \text{otherwise} \end{cases} \tag{1}$$

where each row (e.g., v) in $\mathbf{B}$ denotes the instance sampled in view v. For a complete view (i.e., no data instance missing), we have $\sum_{j=1}^{n} \mathbf{B}_{v,j} = n$ as $\mathbf{B}$ is a matrix with all one. Similarly, for an incomplete view (i.e., some data instances missing), we have $\sum_{j=1}^{n} \mathbf{B}_{v,j} < n$.

In incomplete multi-view setting, missing instances may mislead information from each view. In our work, we first fill the missing data instances of each incomplete view by using the average features of sampled instances in that view as [6]. Then we introduce a weight mechanism to construct an effective graph similarity matrix $\mathbf{A}^v$, where any entry a_{ij}^v denotes the similarity between data instances x_i and x_j in v-th view. We compute the graph matrix $\mathbf{A}^v$ of each view (e.g., v) by solving the following problem:

$$\min_{\mathbf{A}^v} \sum_{i,j=1}^{n} ||h_{v,i}\mathbf{x}_i^v - h_{v,j}\mathbf{x}_j^v||_2^2 \, a_{ij}^v \qquad s.t. \ a_{ii}^v = 0, a_{ij}^v \geq 0, \mathbf{1}^T \mathbf{a}_i^v = 1 \tag{2}$$

where $h_{v,i}$ denotes the weight of i-th instances in the v-th view, denoted as

$$h_{v,i} = \begin{cases} 1, & \text{if } v\text{-th view contains } i\text{-th instance, i.e., } b_{v,i} = 1 \\ \tilde{h}_{v,i}, & \text{otherwise} \end{cases} \tag{3}$$

where $\tilde{h}_{v,i}$ is defined as the percentage of the presence instances for v-th view, shown as $\tilde{h}_{v,i} = \frac{\sum_{j=1}^{n} b_{v,j}}{n}$.

Note that Eq. (2) has a trivial solution. The solution is that only the data instance with the smallest distance to $\mathbf{x}_i^v$ has the value 1, while all the other data instances have the value 0. Following [17], we add a prior to Eq. (2), formulated as below

$$\min_{\mathbf{A}^v} \sum_{i,j=1}^{n} ||h_{v,i}\mathbf{x}_i^v - h_{v,j}\mathbf{x}_j^v||_2^2 \, a_{ij}^v + \gamma \sum_{i=1}^{n} ||\mathbf{a}_i^v||_2^2 \tag{4}$$
$$s.t. \ a_{ii}^v = 0, a_{ij}^v \geq 0, \mathbf{1}^T \mathbf{a}_i^v = 1.$$

where the prior can be seen as the similarity value of each data instance to $\mathbf{x}_i^v$, which is $\frac{1}{n}$, if we only use the second term of Eq. (4). Now we can construct a graph matrix for each view. Next, we present our graph fusion method.

3.2   Graph Fusion

Suppose we have obtained the graph matrix $\mathbf{A}^v$ ($v = 1, 2, ..., m$) for each view. GIMC integrates the information of all views into a unified graph matrix $\mathbf{U}$ by utilizing the complementary information among all views. We introduce an auto-weighted fusion technology. We denote the weights as $\mathbf{w} = [w_1, w_2, \cdots, w_v]^T$. Here we formulate the problem of graph fusion as follows:

$$\min_{\mathbf{U}} \sum_{v=1}^{m} w_v \|\mathbf{U} - \mathbf{A}^v\|_F^2 \qquad s.t. \ u_{ij} \geq 0, \mathbf{1}^T \mathbf{u}_i = 1, w_v \geq 0, \mathbf{1}^T \mathbf{w} = 1. \quad (5)$$

Then we utilize the Lagrange function of Eq. (5) and get its derivative with respect to $\mathbf{U}$ shown below:

$$\sum_{v=1}^{m} w_v \frac{\partial \|\mathbf{U} - \mathbf{A}^v\|_F^2}{\partial \mathbf{U}} + \frac{\partial \Phi(\Lambda, \mathbf{U})}{\partial \mathbf{U}} = 0 \quad (6)$$

where Λ is the Lagrange multiplier, $\Phi(\Lambda, \mathbf{U})$ is a proxy for the constraints $u_{ij} \geq 0$, $\mathbf{1}^T \mathbf{u}_i = 1$, and w_v is determined as

$$w_v = \frac{1}{2\sqrt{\|\mathbf{U} - \mathbf{A}^v\|_F^2}}. \quad (7)$$

Next we impose Laplacian rank constrain on the learned graph matrix $\mathbf{U}$, which helps directly produce the final clustering results on $\mathbf{U}$ without any additional clustering algorithm. The graph Laplacian matrix $\mathbf{L}_U$ of matrix $\mathbf{U}$ is denoted as $\mathbf{L}_U = \mathbf{D}_U - (\mathbf{U}^T + \mathbf{U})/2$, where $\mathbf{D}_U$ is a diagonal matrix whose i-th diagonal element is $\sum_j (u_{ij} + u_{ji})/2$. According to [2,13], we have Theorem 1.

Theorem 1. *The graph Laplacian matrix $\mathbf{L}_U$ has the following properties:*

1. *$\mathbf{L}_U$ is a symmetric positive semi-definite matrix. Thus all eigenvalues of $\mathbf{L}_U$ are real and non-negative, and $\mathbf{L}_U$ has a full set of n real and orthogonal eigenvectors.*
2. *$\mathbf{L}_U \mathbf{e} = 0$, where $\mathbf{e}$ is all ones column vector. Thus 0 is the eigenvalue of $\mathbf{L}_U$ and $\mathbf{e}$ is the corresponding eigenvector.*
3. *If $\mathbf{U}$ has r connected components, then $\mathbf{L}_U$ has r eigenvalues that equal 0.*

The proof of Theorem 1 is presented in [2]. As a conclusion, if $rank(\mathbf{L}_U) = n - c$ as $r = c$, we can cluster the corresponding graph matrix $\mathbf{U}$ into c groups directly. Thus, we add a rank constraint on Eq. (5). According to Ky Fan's theorem [4], normally, we formulate our objective function as

$$\min_{\mathbf{U}} \sum_{v=1}^{m} w_v \|\mathbf{U} - \mathbf{A}^v\|_F^2 + 2\beta Tr(\mathbf{F}^T \mathbf{L}_U \mathbf{F})$$
$$s.t. \ a_{ii}^v = 0, a_{ij}^v \geq 0, \mathbf{1}^T \mathbf{a}_i^v = 1, u_{ij} \geq 0, \mathbf{1}^T \mathbf{u}_i = 1, \quad (8)$$
$$w_v \geq 0, \mathbf{1}^T \mathbf{w} = 1, \mathbf{F}^T \mathbf{F} = \mathbf{I}.$$

4   Optimization Procedure

As introduced in the above section, we propose our GIMC with two parts. Below we introduce our algorithm for each part.

4.1   Optimization for Constructing Graph

As can be seen, Eq. (4) is independent for each data instance, so we can compute $\mathbf{a}_i^v$ separately for each data instance (e.g., i), formulated as

$$\min_{\mathbf{a}_i^v} ||\mathbf{a}_i^v + \frac{\mathbf{d}_i}{2\gamma}||_2^2, \qquad s.t.\ a_{ij}^v \geq 0, \mathbf{1}^T\mathbf{a}_i^v = 1 \qquad (9)$$

where $\mathbf{d}_i$ consists of the entry as d_{ij}, and $d_{ij} = ||h_{v,i}\mathbf{x}_i^v - h_{v,j}\mathbf{x}_j^v||_2$.

Then the Lagrangian function of Eq. (9) in terms of $\mathbf{a}_i^v$ is written as follows

$$l(\mathbf{a}_i^v, \eta, \boldsymbol{\xi}) = ||\mathbf{a}_i^v + \frac{\mathbf{d}_i}{2\gamma}||_2^2 - \eta(\mathbf{1}^T\mathbf{a}_i^v - 1) - \boldsymbol{\xi}^T\mathbf{a}_i^v \qquad (10)$$

where $\boldsymbol{\xi}$ is a Lagrangian coefficient vector and η is a coefficient scalar.

According to the Karush-Kuhn-Tucker conditions [7], the optimal solution is learned as follows:

$$a_{ij}^v = (-\frac{d_{ij}}{2\gamma} + \eta)_+ \qquad s.t.\ a_{ij}^v \geq 0, \mathbf{1}^T\mathbf{a}_i^v = 1. \qquad (11)$$

In practice, we aim to learn $\mathbf{a}_i^v$ with only k nonzero values, where k is the number of neighbors. Meanwhile, according to the constrain $\mathbf{1}^T\mathbf{a}_i^v = 1$, we have $\eta = \frac{1}{k} + \frac{1}{2k\gamma}\sum_{j=1}^k d_{ij}$. Thus, the similarity a_{ij}^v between x_i and x_j in v-th view is obtained by

$$a_{ij}^v = \begin{cases} \frac{d_{i,k+1}-d_{ij}}{kd_{i,k+1}-\sum_{q=1}^k d_{iq}} & j \leq k \\ 0 & j > k \end{cases} \qquad (12)$$

Here we give a summary of the algorithm for graph construction, which is shown in Algorithm 1.

Algorithm 1. Algorithm for constructing the v-th graph

Input: data matrix $\mathbf{X}^v$; indicator matrix $\mathbf{B}$; the number of neighbors k
Output: graph matrix $\mathbf{A}^v$
1: **for** i=1, 2, ..., n **do**
2:    Computing $\mathbf{h}_v$ by using Eq. (3);
3:    Calculating the Lagrangian function according to Eq. (10);
4:    Solving Eq. (11), and sorting $d_{i1}, \cdots, d_{in}$ in ascending order to obtain $\gamma = \frac{k}{2}d_{i,k+1} - \frac{1}{2}\sum j = 1^k d_{ij}$;
5:    Using Eq. (12) to get the similarity a_{ij}^v.
6: **end for**

4.2 Optimization for Graph Fusion

Suppose we have obtained the $\mathbf{A}^1, \mathbf{A}^2, ..., \mathbf{A}^m$ by using Algorithm 1, we now compute the unified graph matrix $\mathbf{U}$ by using Eq. (8). Next we propose an alternating iterative algorithm to solve Eq. (8) as follows:

Solving w with Fixed U and F. When $\mathbf{U}$ and $\mathbf{F}$ are fixed, solving $\mathbf{w}$ using Eq. (8) is equivalent to solving $\mathbf{w}$ using Eq. (5). Thus, w_v is updated by $w_v = \frac{1}{2\sqrt{||\mathbf{U}-\mathbf{A}^v||_F^2}}$, i.e., Eq. (7).

Solving U with Fixed w and F. Since $Tr(\mathbf{F}^T \mathbf{L}_U \mathbf{F}) = \frac{1}{2}\sum_{i,j}||\mathbf{f}_i - \mathbf{f}_j||_2^2 u_{ij}$, and we denote $q_{ij} = ||\mathbf{f}_i - \mathbf{f}_j||_2^2$ and further denote $\mathbf{q}_i$ as a vector with the j-th entry as q_{ij}. Therefore Eq. (8) can be rewritten as follows:

$$\min_{\mathbf{u}_i} \sum_{v=1}^{m} ||\mathbf{u}_i - \mathbf{p}^v||_2^2 \qquad s.t.\ u_{ij} \geq 0, \mathbf{1}^T \mathbf{u}_i = 1. \tag{13}$$

where $\mathbf{p}^v$ is a constant when $\mathbf{w}$ and $\mathbf{F}$ are fixed, and $\mathbf{p}^v = \mathbf{a}_i^v - \frac{\beta}{2mw_v}\mathbf{q}_i$.

We now take the Lagrange function of the Eq. (13) with respect to $\mathbf{u}_i$, shown as below:

$$l(\mathbf{u}_i, \rho, \boldsymbol{\mu}) = \frac{1}{2}\sum_{v=1}^{m} ||\mathbf{u}_i - \mathbf{p}^v||_2^2 - \rho(\mathbf{1}^T \mathbf{u}_i - 1) - \boldsymbol{\mu}^T \mathbf{u}_i \tag{14}$$

where ρ is a scalar and $\boldsymbol{\mu}$ is a Lagrangian coefficient vector.

In such a way, we can update $\mathbf{U}$ by solving the derivation of Eq. (14).

Solving F with Fixed w and U. Optimizing $\mathbf{F}$ is as follows:

$$\min Tr(\mathbf{F}^T \mathbf{L}_U \mathbf{F}) \quad s.t.\ \mathbf{F}^T \mathbf{F} = \mathbf{I}. \tag{15}$$

The optimal solution to $\mathbf{F}$ is formed by the c eigenvectors of $\mathbf{L}_U$ corresponding to the c smallest eigenvalues.

The proposed algorithm to solve Eq. (8) is summarized in Algorithm 2.

Algorithm 2. Algorithm for graph fusion

Input: Similarity matrix $\mathbf{A}^1, \mathbf{A}^2, ..., \mathbf{A}^m$ for each view; the number of clusters c; initial
    parameter β (tuned automatically)
Output: The unified matrix $\mathbf{U}$
  1: Initialize the weight of each view $w_v = \frac{1}{m}$;
  2: Initialize $\mathbf{U}$ and $\mathbf{F}$ by connecting $\mathbf{A}^1, \mathbf{A}^2, ..., \mathbf{A}^m$ with $\mathbf{w}$ and solving Eq. (15)
  3: **repeat**
  4:    Fixing $\mathbf{F}$ and $\mathbf{U}$, compute w_v by using Eq. (7);
  5:    Fixing $\mathbf{w}$ and $\mathbf{F}$, compute $\mathbf{U}$ by solving Eq. (14);
  6:    Fixing $\mathbf{w}$ and $\mathbf{U}$, compute $\mathbf{F}$ by using Eq. (15);
  7: **until** convergence

---

5   Experiments

In this section, we evaluate our GIMC method for incomplete multi-view clustering on real-world benchmark datasets. Accuracy (ACC) and Normalized Mutual Information (NMI) are used to measure the clustering performance.

Table 1. Constitutions of datasets in the experiments

Dataset	#instances	#views	#clusters	$\#d_1$	$\#d_2$	$\#d_3$	$\#d_4$	$\#d_5$	$\#d_6$
100leaves[a]	1600	3	100	64	64	64	-	-	-
Yale[b]	165	4	15	256	256	256	256	-	-
Mfeat[c]	2000	6	10	216	76	64	6	240	47
ORL[d]	400	4	40	256	256	256	256	-	-

[a] https://archive.ics.uci.edu/ml/datasets/One-hundred+plant+species+leaves+data+set
[b] www.cad.zju.edu.cn/home/dengcai/Data/FaceData.html
[c] http://archive.ics.uci.edu/ml/datasets/Multiple+Features
[d] http://www.cad.zju.edu.cn/home/dengcai/Data/FaceData.html

Datasets. We perfom evaluation using four real-world datasets. The datasets are summarized in Table 1. For each dataset, we randomly remove certain instances from each view with the missing rate increasing from 0.1 to 0.5 with 0.1 intervals to form incomplete multi-view data by following [21].

Baselines. As same to [6], we compared our GIMC method with the following algorithm: **Multi-NMF** [9], **PVC** [8], **IMG** [33], **MIC** [21] and **OMVC** [19]. We also compared with **DAIMC** [6].

Settings. Note that Multi-NMF only works on complete view data. Following [6], we first fill the missing instance(s) with average feature values of that view, then perform Multi-NMF on the filled data. PVC and IMG only work for two-view data. We also follow [6] to evaluate PVC and IMG on all the two-views combinations and report the best result. For all the baselines, we obtained the original systems from their authors and used its default parameter settings. For our GIMC method, we empirically set $k = 15$. The parameter β is set to 1 as an initial value. Then, we increase it with $\beta = \beta \times 2$ or decrease it with $\beta = \beta/2$ if the connected components of $\mathbf{U}$ is smaller or greater than the number of clusters c, respectively. In our experiments, the missing rates is changed from 0.1 to 0.5 with the interval of 0.1 to demonstrate how the clustering performance can be improved by our method.

5.1   Experiments on Real-World Data

We run each algorithm 10 times and calculate the means and standard deviations of the performance measures. The results are shown in Table 2 with missing rate 0.1, 0.2, 0.3, 0.4, and 0.5. The numbers in the parentheses are the standard deviations. From Table 2, we make the following observations:

- In most cases, our method GIMC outperforms all baseline algorithms. The results show that our GIMC method is a promising incomplete multi-view clustering method.
- All the baseline methods result in a high standard deviation as they are based on NMF and subspace learning, which depend on the intializations. Our method results in a very small standard deviation because we give an

Table 2. Clustering performance comparison in terms of ACC and NMI

Method	100leaves		Yale		Mfeat		ORL	
	ACC(%)	NMI(%)	ACC(%)	NMI(%)	ACC(%)	NMI(%)	ACC(%)	NMI(%)
Clustering performance comparison in terms of missing rate 0.1								
MultiNMF	41.75 (1.67)	68.16 (0.78)	44.85 (2.68)	49.35 (2.48)	68.83 (1.43)	67.99 (0.65)	33.75 (3.68)	58.87 (1.62)
PVC	22.39 (0.38)	53.27 (0.22)	40.64 (2.83)	44.08 (2.19)	62.53 (3.38)	62.46 (1.61)	60.68 (3.05)	75.76 (1.70)
IMG	68.71 (1.86)	83.99 (2.75)	43.88 (1.15)	48.53 (1.52)	74.63 (0.00)	79.10 (0.00)	60.83 (1.87)	75.32 (1.12)
MIC	56.90 (8.33)	76.65 (10.99)	42.59 (5.06)	51.40 (5.49)	79.45 (3.02)	73.12 (3.60)	55.99 (10.04)	73.95 (9.80)
OMVC	31.62 (0.94)	56.93 (0.81)	44.61 (2.77)	50.75 (2.50)	61.69 (2.25)	56.52 (0.61)	57.80 (2.40)	75.31 (1.39)
DAIMC	64.20 (2.38)	82.41 (0.70)	33.15 (2.57)	41.60 (1.72)	87.27 (4.10)	80.04 (2.57)	57.95 (3.10)	75.35 (1.38)
GIMC	**86.00** (0.00)	**92.01** (0.00)	**49.70** (0.00)	**54.98** (0.00)	**88.65** (0.00)	**91.21** (0.00)	**71.75** (0.00)	**83.70** (0.00)
Clustering performance comparison in terms of missing rate 0.2								
MultiNMF	35.81 (1.30)	63.09 (0.64)	35.15 (3.16)	40.65 (3.43)	67.95 (1.03)	64.12 (0.55)	30.25 (1.67)	54.39 (1.69)
PVC	18.11 (0.45)	51.25 (0.60)	46.36 (2.67)	49.39 (1.95)	60.67 (1.96)	59.55 (2.06)	60.74 (2.64)	75.70 (1.32)
IMG	64.21 (1.86)	80.85 (2.75)	41.76 (1.41)	45.96 (1.27)	66.15 (0.00)	73.23 (0.00)	57.10 (1.77)	71.72 (1.35)
MIC	50.47 (7.34)	71.51 (10.25)	41.93(5.61)	48.24 (5.76)	77.80 (4.16)	69.85 (4.50)	51.37 (9.23)	69.95 (9.36)
OMVC	31.73 (0.99)	58.33 (0.59)	41.79 (2.24)	47.61 (1.65)	60.45 (0.31)	57.44 (0.23)	54.44 (3.19)	73.01 (1.10)
DAIMC	53.78 (1.65)	75.00 (0.57)	32.67 (2.24)	41.27 (1.91)	**88.72** (1.08)	80.07 (1.44)	53.90 (4.19)	72.36 (1.82)
GIMC	**77.81** (0.00)	**86.04** (0.00)	**48.48** (0.00)	**54.28** (0.00)	88.40 (0.00)	**91.15** (0.00)	**67.25** (0.00)	**81.60** (0.00)
Clustering performance comparison in terms of missing rate 0.3								
MultiNMF	31.83 (0.79)	59.89 (0.61)	32.73 (2.46)	38.49 (3.37)	53.10 (0.71)	49.90 (0.58)	28.00 (1.22)	51.41(1.73)
PVC	33.44 (0.65)	64.12 (0.28)	36.21 (2.46)	40.97 (1.67)	53.77 (1.76)	52.92 (1.15)	59.79 (2.17)	75.13 (1.36)
IMG	60.02 (1.86)	78.73 (2.75)	37.76 (1.64)	41.93 (2.19)	65.20 (0.00)	62.37 (0.00)	54.63 (2.04)	70.95 (1.62)
MIC	42.64 (6.19)	66.37 (9.50)	40.94 (4.73)	46.87 (5.09)	71.21 (3.52)	61.12 (4.10)	47.90 (9.03)	67.02 (9.06)
OMVC	29.78 (0.98)	57.15 (0.59)	36.64 (2.44)	41.95 (2.06)	55.26 (3.26)	47.88 (1.55)	47.21 (1.61)	66.24 (1.34)
DAIMC	41.44 (1.35)	67.36 (0.63)	29.03 (2.22)	36.69 (2.64)	86.32 (2.49)	77.41 (1.31)	49.57 (3.15)	69.24 (1.72)
GIMC	**68.37** (0.00)	**79.66** (0.00)	**44.85** (0.00)	**49.64** (0.00)	**87.55** (0.00)	**90.63** (0.00)	**64.00** (0.00)	**77.69** (0.00)
Clustering performance comparison in terms of missing rate 0.4								
MultiNMF	30.06 (0.94)	57.23 (0.76)	33.33 (3.14)	39.10 (2.57)	45.78 (0.82)	46.41 (0.60)	28.50 (1.01)	49.43 (1.49)
PVC	40.99 (0.88)	67.71 (0.32)	42.58 (1.94)	47.31 (1.62)	65.96 (2.62)	62.30 (1.61)	58.26 (2.14)	74.32 (1.16)
IMG	54.96 (1.86)	**77.05** (2.75)	36.61 (1.88)	40.63 (1.68)	53.26 (0.28)	60.08 (0.15)	51.93 (1.88)	68.75 (1.50)
MIC	39.92 (5.85)	63.61 (9.13)	36.67 (4.81)	43.95 (5.04)	65.25 (3.72)	56.03 (4.20)	46.45 (8.48)	65.43 (8.74)
OMVC	21.96 (0.64)	48.91 (0.56)	34.67 (2.60)	43.50 (1.98)	45.98 (0.46)	43.52 (0.37)	36.90 (1.86)	58.07 (1.28)
DAIMC	32.14 (1.00)	61.21 (0.38)	28.67 (2.21)	34.70 (2.30)	82.81 (4.09)	72.11 (2.65)	43.35 (2.64)	63.25 (1.68)
GIMC	**61.31** (0.00)	74.20 (0.00)	**44.85** (0.00)	**49.31** (0.00)	**85.85** (0.00)	**89.55** (0.00)	**63.75** (0.00)	**75.69** (0.00)
Clustering performance comparison in terms of missing rate 0.5								
MultiNMF	24.25 (1.09)	52.60 (0.69)	30.91 (2.83)	38.33 (2.71)	37.22 (0.93)	31.11 (0.61)	25.75 (1.77)	46.90 (2.12)
PVC	43.98 (1.10)	68.01 (0.55)	39.73 (2.36)	46.97 (2.20)	50.13 (2.92)	48.17 (1.25)	56.56 (2.14)	74.18 (1.51)
IMG	12.19 (1.86)	37.38 (2.75)	22.73 (2.25)	27.95 (3.02)	18.45 (1.76)	10.34 (2.12)	18.40 (1.28)	39.04 (2.42)
MIC	35.30 (5.20)	60.31 (8.66)	34.73 (4.23)	41.56 (4.60)	45.01 (3.19)	45.98 (3.30)	41.54 (7.49)	61.15 (8.16)
OMVC	20.85 (0.74)	48.53 (0.63)	35.06 (2.06)	44.59 (1.56)	37.76 (0.36)	34.77 (0.81)	39.39 (1.82)	62.44 (1.10)
DAIMC	26.23 (1.02)	57.63 (0.57)	28.48 (3.00)	34.79 (2.79)	70.26 (5.66)	62.15 (3.41)	36.48 (2.33)	57.56 (1.13)
GIMC	**54.25** (0.00)	**68.19** (0.00)	**41.21** (0.00)	**48.54** (0.00)	**77.50** (0.00)	**86.60** (0.00)	**62.00** (0.00)	**74.67** (0.00)

optimized solution for the objective function. This indicates that our method performs robustly.

- The clustering performance of all algorithms decline with the missing rate changing from small to large. Our method performs smoothly.

5.2   Complexity and Convergence Study

In this section, we analyze the computational complexity and study how fast the proposed algorithm can converge. According to the procedure of optimization, the complexity of constructing graph is $O(mn^2)$, and that of graph fusion is $O(Knm)$ where K is the number of iterations until convergence. Specially, we conduct experiments with the missing rate of 0.1, 0.2, 0.3, 0.4, and 0.5 respectively on each dataset. The results are shown in Fig. 1. It is noted that the proposed algorithm converges within 10 iterations. The reason is that we provided an optimized solution for each subproblem.

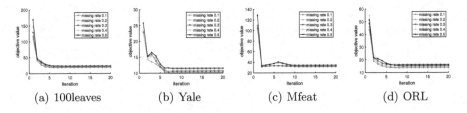

(a) 100leaves   (b) Yale   (c) Mfeat   (d) ORL

Fig. 1. Convergence curve of GIMC over each dataset

6   Conclusions

This paper proposed a novel incomplete multi-view clustering method, called Graph-based Incomplete Multi-view Clustering (GIMC). GIMC considers the influence of missing instances and introduces an auto-weighted mechanism to perform clustering on incomplete multi-view data. After this, GIMC uses graph-based method as the base approach to construct a graph matrix for each view first, extract the relationship among views, and then fuse the constructed graph matrix from each view data to learn a unified graph matrix, which produces the final clusters. The experimental results on four real-world benchmark datasets demonstrate the superiority of the proposed GIMC method.

In this work, we fill the missing data instances in the incomplete views using the average feature values of that view. If the missing rate is very large in practice, the filled values may decrease the clustering performance. So, our future work is to study efficient filling methods for the proposed framework.

Acknowledgment. This work was supported by the National Natural Science Foundation of China (No. 61572407). Working at the University of Illinois at Chicago supported by the China Scholarship Council (No. 20170700064) has also given Hao Wang a broader perspective on data mining and machine learning.

References

1. Chao, G., Sun, S., Bi, J.: A survey on multi-view clustering. CoRR abs/1712.06246 (2017)
2. Dhillon, I.S.: Co-clustering documents and words using bipartite spectral graph partitioning. In: Proceedings of the ACM SIGKDD International Conference on Knowledge Discovery and Data Mining, pp. 269–274 (2001)
3. Eaton, E., Desjardins, M., Jacob, S.: Multi-view constrained clustering with an incomplete mapping between views. Knowl. Inf. Syst. **38**(1), 231–257 (2014)
4. Fan, K.: On a theorem of Weyl concerning eigenvalues of linear transformations I. Proc. Natl. Acad. Sci. U. S. A. **35**(11), 652–655 (1949)
5. Hou, C., Nie, F., Tao, H., Yi, D.: Multi-view unsupervised feature selection with adaptive similarity and view weight. IEEE Trans. Knowl. Data Eng. **29**(9), 1998–2011 (2017)
6. Hu, M., Chen, S.: Doubly aligned incomplete multi-view clustering. In: Proceedings of the International Joint Conference on Artificial Intelligence, pp. 2262–2268 (2018)
7. Lemaréchal, C., Boyd, S., Vandenberghe, L.: Convex optimization. Eur. J. Oper. Res. **170**(1), 326–327 (2006)
8. Li, S.Y., Jiang, Y., Zhou, Z.H.: Partial multi-view clustering. In: Proceedings of the AAAI Conference on Artificial Intelligence, pp. 1968–1974 (2014)
9. Liu, J., Wang, C., Gao, J., Han, J.: Multi-view clustering via joint nonnegative matrix factorization. In: Proceedings of the SIAM International Conference on Data Mining, pp. 252–260 (2013)
10. Liu, J., Jiang, Y., Li, Z., Zhou, Z.H., Lu, H.: Partially shared latent factor learning with multiview data. IEEE Trans. Neural Netw. Learn. Syst. **26**(6), 1233–1246 (2015)
11. Liu, X., et al.: Late fusion incomplete multi-view clustering. IEEE Trans. Pattern Anal. Mach. Intell. 1 (2018). https://doi.org/10.1109/TPAMI.2018.2879108
12. Liu, X., et al.: Optimal neighborhood kernel clustering with multiple kernels. In: Proceedings of the AAAI International Conference on Artificial Intelligence, pp. 2266–2272 (2017)
13. Mohar, B., Alavi, Y., Chartrand, G., Oellermann, O.: The Laplacian spectrum of graphs. Graph Theory Comb. Appl. **2**(12), 871–898 (1991)
14. Nie, F., Cai, G., Li, J., Li, X.: Auto-weighted multi-view learning for image clustering and semi-supervised classification. IEEE Trans. Image Process. **27**(3), 1501–1511 (2018)
15. Nie, F., Li, J., Li, X.: Parameter-free auto-weighted multiple graph learning: a framework for multiview clustering and semi-supervised classification. In: Proceedings of the International Joint Conference on Artificial Intelligence, pp. 1881–1887 (2016)
16. Nie, F., Li, J., Li, X.: Self-weighted multiview clustering with multiple graphs. In: Proceedings of the International Joint Conference on Artificial Intelligence, pp. 2564–2570 (2017)
17. Nie, F., Wang, X., Jordan, M.I., Huang, H.: The constrained Laplacian rank algorithm for graph-based clustering. In: Proceedings of the AAAI Conference on Artificial Intelligence, pp. 1969–1976 (2016)
18. Saha, M.: A graph based approach to multiview clustering. In: Maji, P., Ghosh, A., Murty, M.N., Ghosh, K., Pal, S.K. (eds.) PReMI 2013. LNCS, vol. 8251, pp. 128–133. Springer, Heidelberg (2013). https://doi.org/10.1007/978-3-642-45062-4_17

19. Shao, W., He, L., Lu, C., Yu, P.S.: Online multi-view clustering with incomplete views. In: Proceedings of the 2016 IEEE International Conference on Big Data, pp. 1012–1017 (2016)
20. Shao, W., He, L., Yu, P.S.: Clustering on multi-source incomplete data via tensor modeling and factorization. In: Cao, T., Lim, E.-P., Zhou, Z.-H., Ho, T.-B., Cheung, D., Motoda, H. (eds.) PAKDD 2015. LNCS (LNAI), vol. 9078, pp. 485–497. Springer, Cham (2015). https://doi.org/10.1007/978-3-319-18032-8_38
21. Shao, W., He, L., Yu, P.S.: Multiple incomplete views clustering via weighted non-negative matrix factorization with $L_{2,1}$ regularization. In: Appice, A., Rodrigues, P.P., Santos Costa, V., Soares, C., Gama, J., Jorge, A. (eds.) ECML PKDD 2015. LNCS (LNAI), vol. 9284, pp. 318–334. Springer, Cham (2015). https://doi.org/10.1007/978-3-319-23528-8_20
22. Tao, H., Hou, C., Liu, X., Liu, T., Yi, D., Zhu, J.: Reliable multi-view clustering. In: Proceedings of the AAAI Conference on Artificial Intelligence, pp. 4123–4130 (2018)
23. Tao, H., Hou, C., Zhu, J., Yi, D.: Multi-view clustering with adaptively learned graph. In: Proceedings of the Asian Conference on Machine Learning, pp. 113–128 (2017)
24. Wang, H., Yang, Y., Li, T.: Multi-view clustering via concept factorization with local manifold regularization. In: Proceedings of the IEEE International Conference on Data Mining, pp. 1245–1250 (2016)
25. Wang, H., Yang, Y., Liu, B., Fujita, H.: A study of graph-based system for multi-view clustering. Knowl.-Based Syst. **163**, 1009–1019 (2019)
26. Xu, C., Tao, D., Xu, C.: Multi-view learning with incomplete views. IEEE Trans. Image Process. **24**(12), 5812–5825 (2015)
27. Yang, Y., Wang, H.: Multi-view clustering: a survey. Big Data Min. Anal. **1**(2), 83–107 (2018)
28. Yang, Y., Zhan, D.C., Sheng, X.R., Jiang, Y.: Semi-supervised multi-modal learning with incomplete modalities. In: Proceedings of the International Joint Conference on Artificial Intelligence, pp. 2998–3004 (2018)
29. Yin, Q., Wu, S., Wang, L.: Incomplete multi-view clustering via subspace learning. In: Proceedings of the ACM International on Conference on Information and Knowledge Management, pp. 383–392 (2015)
30. Yin, Q., Wu, S., Wang, L.: Unified subspace learning for incomplete and unlabeled multi-view data. Pattern Recognit. **67**, 313–327 (2017)
31. Zhang, X., Zong, L., Liu, X., Yu, H.: Constrained NMF-based multi-view clustering on unmapped data. In: Proceedings of the AAAI Conference on Artificial Intelligence, pp. 3174–3180 (2015)
32. Zhang, Y., Yang, Y., Li, T., Fujita, H.: A multitask multiview clustering algorithm in heterogeneous situations based on LLE and LE. Knowl.-Based Syst. **163**, 776–786 (2019)
33. Zhao, H., Liu, H., Fu, Y.: Incomplete multi-modal visual data grouping. In: Proceedings of the International Joint Conference on Artificial Intelligence, pp. 2392–2398 (2016)
34. Zhuge, W., Nie, F., Hou, C., Yi, D.: Unsupervised single and multiple views feature extraction with structured graph. IEEE Trans. Knowl. Data Eng. **29**(10), 2347–2359 (2017)
35. Zong, L., Zhang, X., Liu, X., Yu, H.: Weighted multi-view spectral clustering based on spectral perturbation. In: Proceedings of the AAAI Conference on Artificial Intelligence, pp. 4621–4628 (2018)

Beyond Outliers and on to Micro-clusters: Vision-Guided Anomaly Detection

Wenjie Feng[1,2](✉), Shenghua Liu[1,2](✉), Christos Faloutsos[3](✉),
Bryan Hooi[3](✉), Huawei Shen[1,2](✉), and Xueqi Cheng[1,2](✉)

[1] CAS Key Laboratory of Network Data Science and Technology,
Institute of Computing Technology, Chinese Academy of Sciences, Beijing, China
{fengwenjie,shenhuawei,cxq}@ict.ac.cn, liu.shengh@gmail.com
[2] University of Chinese Academy of Sciences, Beijing 100190, China
[3] School of Computer Science, Carnegie Mellon University, Pittsburgh, PA, USA
christos@cs.cmu.edu, bhooi@andrew.cmu.edu

Abstract. Given a heatmap for millions of points, what patterns exist
in the distributions of point characteristics, and how can we detect them
and separate anomalies in a way similar to human vision? In this paper,
we propose a vision-guided algorithm, EagleMine, to recognize and summarize point groups in the feature spaces. EagleMine utilizes a water-
level tree to capture group structures according to vision-based intuition at multiple resolutions, and adopts statistical hypothesis tests to
determine the optimal groups along the tree. Moreover, EagleMine can
identify anomalous micro-clusters (i.e., micro-size groups), which exhibit
very similar behavior but deviate away from the majority. Extensive
experiments are conducted for large graph scenario, and show that our
method can recognize intuitive node groups as human vision does, and
achieves the best performance in summarization compared to baselines.
In terms of anomaly detection, EagleMine also outperforms state-of-the-
art graph-based methods by significantly improving accuracy in synthetic
and microblog datasets.

1 Introduction

Given real-world graphs with millions of nodes and connections, the most intuitive way to explore the graphs is to construct a correlation plot [25] based on
the features of graph nodes. Usually a heatmap of those scatter points is used
to depict their density, which is a two-dimensional histogram [20]. In the histogram, people can visually recognize nodes gathering into disjointed dense areas
separately as groups (see Fig. 1), which help to explore patterns (like communities, co-author association behaviors) and detect anomalies (e.g., fraudsters,
attackers, fake-reviews, outlier etc.) in an interpretable way [22].

In particular, a graph can represent friendships in Facebook, ratings from
users to items in Amazon, or retweets from users to messages in Twitter, even
they are time-evolving. Numerous correlated features can be extracted from

© Springer Nature Switzerland AG 2019
Q. Yang et al. (Eds.): PAKDD 2019, LNAI 11439, pp. 541–554, 2019.
https://doi.org/10.1007/978-3-030-16148-4_42

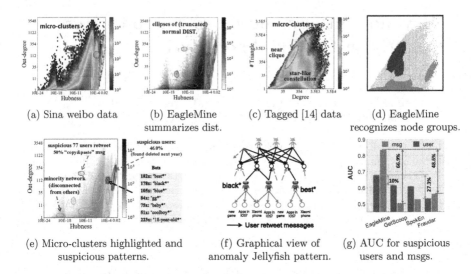

(a) Sina weibo data   (b) EagleMine   (c) Tagged [14] data   (d) EagleMine
                         summarizes dist.   recognizes node groups.

(e) Micro-clusters highlighted and   (f) Graphical view of   (g) AUC for suspicious
    suspicious patterns.   anomaly Jellyfish pattern.   users and msgs.

Fig. 1. Heatmaps of correlation plots for some feature spaces of two real datasets and the performance of **EagleMine** algorithm. The bottom figures focus on the Sina weibo data. (a) Out-degree vs. Hubness feature space for weibo. (b) EagleMine summarizes the distribution of graph nodes for (a) with truncated Gaussian distributions. The ellipses denote the 1.5 and 3 times covariance of corresponding Gaussian. (c) # Triangle vs. Degree feature space for Tagged. (d) Depicts the recognized node groups for (c). (e) Highlights some micro-clusters in (b), including a disconnected small network, and very suspicious ones. A username list on the right side shows the name patterns of bots in a micro-cluster, where 182x: "best*" means 182 bots share prefix "best". (f) The structure of identified anomalous Jellyfish patterns. (g) Shows the AUC performance for detecting suspicious users and msgs compared with state-of-the-art competitors.

graph, like degree, triangles, spectral vectors, and PageRank etc. and combination of these generate correlation plots. It becomes, even, labor-intensive to manually monitor and recognize patterns from heatmaps of the snapshots of temporal graphs. So, this raises the following questions: ***Given*** *a heatmap (i.e., histogram) of the scatter points in some feature space, how can we design an algorithm to automatically* ***recognize*** *and* ***monitor*** *the point groups as human vision does,* ***summarize*** *the points distribution in the feature space and* ***identify*** *suspicious micro-clusters?*

'Micro-cluster' refers to relatively small group of points (like users, items) with similar behavior in the feature space. Here we demonstrate some possible feature spaces, namely

i. out-degree vs hubness - Fig. 1a - this can spot nodes with high out-degree, but low hubness score (i.e., fraudsters, which have many outgoing edges to non-important nodes, probably, customers, that paid them) [24].

ii. #triangle vs degree - spotting a near-clique group (too many triangles, for their degree), as well as star-like constellations (too few triangles for such high degree) [23].

Table 1. Comparison between algorithms.

	Density-based clustering	SPOKEN	GetScoop	Fraudar	EagleMine
Micro-cluster detection	✓	✓	✓		✓
Micro-cluster suspiciousness			✓		✓
Linear scalability	?			✓	✓

In this paper, we propose EagleMine, a novel tree-based mining approach to recognize and summarize the point groups in the heatmap of scatter plots, and can also identify anomalous micro-clusters. Experiments show that EagleMine outperforms baselines and achieves better performance both in quantitative (i.e., the code length for compact model description) and qualitative (i.e., consistent with vision-based judgment) comparisons, detects a micro-cluster of hundreds of bots in microblog data, Sina weibo[1], which presents strong signs of sharing unusual login-name prefixes, e.g., 'best*', 'black*' and '18-year-old*', and exhibiting very similar behavior in the feature space (see Fig. 1e).

In summary, the proposed EagleMine has the following advantages:

- **Anomaly detection:** can spot and explain anomalies on real data by identifying suspicious micro-clusters. Compared with the graph-based anomaly detection methods, EagleMine achieves higher accuracy for finding suspiciousness in Sina weibo.
- **Automated summarization:** automatically summarizes a histogram plot derived from correlated graph features (see Fig. 1b), and recognizes node groups forming disjonted dense areas as human vision does (see Fig. 3e).
- **Effectiveness:** detects interpretable groups, and outperforms the baselines and even those with *manually tuned parameters* in qualitative experiments (see Fig. 3).
- **Scalability:** is scalable with nearly linear time complexity in the number of graph nodes, and can deal with more correlated features in multi-dimensional space.

Our code is open-sourced at https://github.com/wenchieh/eaglemine, and most of the datasets we use are publicly available online. The supplementary material [1] provides proof, detailed information and additional experiments.

2   Related Work

Supported by human vision theory, including visual saliency, color sensitive, depth perception and attention of vision system [17], visualization techniques [5,38] and HCI tools help to get insight into data [2,35]. SCAGNOSTIC [6,35] diagnoses the anomalies from the plots of scattered points. [39] improves the detection by statistical features derived from graph-theoretic measures. Net-Ray [22] visualizes and mines adjacency matrices and scatter plots of a large graph, and discovers some interesting patterns.

---

[1] One of the largest microblog websites in China.

For graph anomaly detection, [21,30] find communities and suspicious clusters with spectral-subspace plots. SPOKEN [30] considers the "eigenspokes" on EE-plot produced by pairs of eigenvectors, and is later generalized for fraud detection. As more recent works, dense block detection has been proposed to identify anomalous patterns and suspicious behaviors [18,26]. Fraudar [18] proposed a densest subgraph-detection method that incorporates the suspiciousness of nodes and edges during optimization.

Density based methods, like DBSCAN [13] can detect clusters of arbitrary shape and data distribution, while the clustering performance relies on density threshold. STING [37] hierarchically merges grids in lower layers to find clusters with a given density threshold; Clustering algorithms [31] derived from the watershed transformation [36], treat pixel region between watersheds as one cluster, and only focus on the final results and ignores the hierarchical structure of clusters. [7] compared different clustering algorithms and proposed a hierarchical clustering method, "HDBSCAN", while its complexity is prohibitive for very large dataset (like graphs) and the "outlierness" score is not line with our expectations. Community detection algorithms [27], modularity-driven clustering, and cut-based methods [32] usually can't handle large graphs with million nodes or fail to provide intuitive and interpretable result when applying to graph clustering.

A comparison between EagleMine and the majority of the above methods is summarized in Table 1. Our method EagleMine is the only one that matches all specifications.

3   Proposed Model

Consider a graph $\mathcal{G}$ with node set V and edge set E. $\mathcal{G}$ can be either homogeneous, such as friendship/following relations, or bipartite as users rating restaurants. In some feature space of graph nodes, our goal is to optimize the consistent node-group assignment with human visual recognition, and the goodness-of-fit (GoF) of node distribution in groups. So we map the node into a (multi-dimensional) histogram constructed based on a feature space, which can include multiple node features. Considering the histogram $\mathcal{H}$ with dimension $dim(\mathcal{H})$, we use h to denote the number of nodes in a bin, and b to denote a bin, when without ambiguity.

Model: To summarize the histogram $\mathcal{H}$ in a feature space of graph nodes, we utilize some statistical distributions as vocabulary to describe the node groups in $\mathcal{H}$. Therefore, our vocabulary-based summarization model consists of **Configurable vocabulary:** statistical distributions $\mathcal{Y}$ for describing node groups of $\mathcal{H}$ in a feature space; **Assignment variables:** $\mathcal{S} = \{s_1, \cdots, s_C\}$ for the distribution assignment of C node groups; **Model parameters:** $\Theta = \{\theta_1, \cdots, \theta_C\}$ for distributions in each node group, e.g. the mean and variance for normal distribution. **Outliers:** unassigned bins $\mathcal{O}$ in $\mathcal{H}$ for outlier nodes.

In terms of the configurable vocabulary $\mathcal{Y}$, it may include any suitable distribution, such as Uniform, Gaussian, Laplace, and exponential distributions or others, which can be tailored to the data and characteristics to be described.

4   Our Proposed Method

In human vision and cognitive system, connected components can be rapidly captured [11,28] with a top-to-bottom recognition and hierarchical segmentation manner [3]. Therefore, this motivates us to identify each node group as an inter-connected and intra-disjointed dense area in heat map, which guides the refinement for smoothing, and to organize and explore connected node groups by a hierarchical structure, as we will do.

Our proposed **EagleMine** algorithm consists of two steps:

- Build a hierarchical tree $\mathcal{T}$ of node groups for $\mathcal{H}$ in some feature space with WATERLEVELTREE algorithm.
- Search the tree $\mathcal{T}$ and get summarization of $\mathcal{H}$ with TREEEXPLORE algorithm.

EagleMine hierarchically detects micro-clusters in the $\mathcal{H}$, then computes the optimal summarization including the model parameters Θ, and the assignment S for each node group, and outliers indices $\mathcal{O}$ in final. We elaborate each step in the following subsections.

4.1   Water-Level Tree Algorithm

In the histogram $\mathcal{H}$, we imagine an area consisting of jointed positive bins ($h > 0$) as an *island*, and the other bins as **water area**. Then we can flood the island areas, making those bins with $h < r$ to be underwater, i.e., setting those $h = 0$, where r is a water level. Afterwards, the remaining positive bins form new islands in condition of water level r.

To organize all the islands in different water levels, we propose a water-level tree structure, where each node represents an island and each edge represents the relationship: where a child island at a higher water level comes from a parent island at a lower water level. Note that increasing r from 0 corresponds to raising the water level and moving from root to leaves.

The WATERLEVELTREE algorithm is shown in Algorithm 1. We start from the root, and raise water level r in logarithmic scale from 0 to $\log h_{max}$ with step ρ, to account for the power-law-like distribution of h, where $h_{max} = \max \mathcal{H}$. We use the binary opening[2] operator ($\circ$) [15] for smoothing each internally jointed island, which is able to remove small isolated bins (treated as noise), and separate weakly-connected islands with a specific structure element. Afterwards, we link each island at current level r_{curr} to its parent at lower water level r_{prev} of the tree. The flooding process stops until r reaches the maximum level—$\log h_{max}$. Subsequently, we propose following steps to refine the raw tree $\mathcal{T}$ (the pictorial explanation for each step are given in the supplementary [1]):

---

[2] Binary opening is a basic workhorse of morphological noise removal in computer vision and image processing. Here we use $\underbrace{2 \times \cdots \times 2}_{dim(\mathcal{H})}$ square-shape "probe".

Algorithm 1. WATERLEVELTREE Algorithm

Input: Histogram $\mathcal{H}$.
Output: Water-level tree $\mathcal{T}$.
  1: $\mathcal{T} = \{$positive bins in $\mathcal{H}$ as root$\}$.
  2: **for** $r = 0$ to $\log h_{max}$ by step ρ **do**
  3:     $\mathcal{H}^r$: assign $h \in \mathcal{H}$ to zero if $\log h < r$.
  4:     $\mathcal{H}^r = \mathcal{H}^r \circ \mathbf{E}$.   ▷ *binary opening* to smooth.
  5:     islands $\mathcal{A}^r = \{$jointed bin areas in $\mathcal{H}^r\}$.
  6:     link each island in $\mathcal{A}^r$ to its parent in $\mathcal{T}$.
  7: **end for**
  8: **Contract** $\mathcal{T}$: iteratively remove each single-child island and link its children to its parent.
  9: **Prune** $\mathcal{T}$: heuristically remove noise nodes.
 10: **Expand** islands in $\mathcal{T}$ with extra neighbors.
 11: **return** $\mathcal{T}$

Contract: The current tree $\mathcal{T}$ may contain many ties, meaning no new islands separated, which are redundant. Hence we *search the tree using depth-first search; once a single-child node is found, we remove it and link its children to its parent.*

Prune: The purpose of pruning is to smooth away noisy peaks on top of each island, arising from fluctuations of h between neighbor bins. Hence we *prune such child branches (including children's descendants) based on their total area size: the ratio of the sum of h in child bins to the sum of h in parent bins, is no less than 95%.*

Expand: We include additional surrounding bins into each island to avoid over-fitting for learning distribution parameters and to eliminate the possible effect of uniform step ρ for logarithmic scale. Hence we *iteratively augment towards positive bins around each island by a step of one-bin size until islands touch each other, or doubling the number of bins as contained in original island.*

Comparably in the Watershed formalization [36], the foreground of $\mathcal{H}$ are defined as catchment basins for clustering purpose, and can capture the boundaries between clusters as segmentation. We will see in experiments

---

Algorithm 2. TREEEXPLORE Algorithm

Input: WATERLEVELTREE $\mathcal{T}$
Output: summarization $\{\mathcal{S}, \Theta, \mathcal{O}\}$.
  1: $\Theta = \emptyset$.
  2: $\mathcal{S} = $ decide the distribution type s_α from vocabulary for each island in $\mathcal{T}$.
  3: Search $\mathcal{T}$ with BFS to iteratively conduct following to each node: use *DistributionFit* to determine the parameter; apply Hypothesis test to select optimal one; and insert result into Θ and update $\mathcal{S}$.
  4: **Stitch** and replace promising distributions in $\mathcal{S}$, then update Θ.
  5: Decide outliers $\mathcal{O}$ deviating from the recognized groups.
  6: **return** summarization $\{\mathcal{S}, \Theta, \mathcal{O}\}$.

(Sect. 5.3 and Fig. 3), the segmentation in Watershed approximates the islands in one level of tree $\mathcal{T}$, with a threshold parameter for background. STING also selects clusters in the same level, and needs a density threshold; HDBSCAN extracts hierarchies with MST that can not capture trees with any branches. However, EagleMine has no tuning parameters, and then searches the water-level tree to find the best combination of islands, which may come from different levels (see Sect. 4.2).

4.2   Tree Explore Algorithm

With the water-level tree and describing vocabulary, we can then determine the optimal node groups and their summarization. The main procedure is described in Algorithm 2, where we decide the distribution vocabulary s_α for each tree node (island) α, search the tree with BFS, select the optimal islands with some criteria, and refine the final results using stitching. In addition, we believe the pictorial illustration in supplement [1] will offer intuitive explanation for the algorithm.

We now describe our vocabulary Θ. Truncated Gaussian distribution [34] is a flexible model for capturing clusters of different shapes, like line, circle, and ellipse, or their truncation in 2D case. Due to the discrete unit bins in $\mathcal{H}$, the *discretized, truncated, multivariate* Gaussian distribution (DTM Gaussian for short) with the mean $\boldsymbol{\mu}$ and co-variance $\boldsymbol{\Sigma}$ as parameter is used as one of the vocabulary. Observing the multi-mode distribution of islands (skewed triangle-like island in Fig. 1a) which exist in many different histogram plots and contains the majority of graph nodes, we add Mixture of DTM Gaussians as another vocabulary term to capture these complex structures.

In general, to decide the assignment $\mathcal{S}$ of vocabulary to each island, we can use distribution-free hypothesis test, like Pearson's χ^2 test, or other distribution specified approaches. Here, we heuristically assign Mixture of DTM Gaussians to the island containing the largest number of graph nodes at each tree level, and DTM Gaussian to other islands for simplicity. After vocabulary assignment, we use the maximum likelihood estimation to learn the parameters $\theta_\alpha \in \Theta$ for a island α, which $\theta_\alpha = \{\boldsymbol{\mu}_\alpha, \boldsymbol{\Sigma}_\alpha, \tilde{N}_\alpha\}$ and $\tilde{N}_\alpha = \sum_{(i_1,\cdots,i_F)\in\alpha} \log h_{i_1,\cdots,i_F}$. Let $DistributionFit(\alpha, s_\alpha)$ denote the step of learning the parameter θ_α.

Afterwards, we search along the tree $\mathcal{T}$ with BFS to select the optimal combination of clusters. In principle, metrics like AIC and BIC in machine learning and Pearson's χ^2 test and K-S test in statistics, can be adopted to determine whether to explore the children of $\mathcal{T}$. Here we utilize statistical hypothesis test to select models for its better adaptation and performance in experiments, which measure the statistical significance of the null hypothesis [10,16]. The null hypothesis for searching the children of island α in $\mathcal{T}$ is:

$\boldsymbol{H}_0$: the bins of island α come from distribution s_α.

If $\boldsymbol{H}_0$ is not rejected, we stop searching the island's children. Otherwise, we further explore the children of α.

Specifically, We apply this hypothesis test to an island based on its binary image, which focuses on whether the island's shape looks like a truncated Gaussian or mixture. Simply, we project the bin data to some dimensions and apply the test according to projection pursuit [19] and G-means [16]. We implement the Quadratic class 'upper tail' Anderson-Darling Statistic test[3] [9,33] (with 1% significance level) due to the truncation. And we accept the null hypothesis H_0 only when the test is true for all dimension projections. If one of them is rejected, H_0 will be rejected. Finally, we get the node groups to summarize the histogram until the BFS stops.

Stitch: some islands from different parents are physically close to each other. In such case, those islands can probably be summarized by the same distribution. So we use *stitch* process in step 4 to merge them by hypothesis test as well. The stitch process stops until no changes occur. When there are multiple pairs of islands to be merged at the same time, we choose the pair with the least average log-likelihood reduction after stitching:

$$(\alpha_{i^*}, \alpha_{j^*}) = \arg \min_{i,j} \frac{\mathcal{L}_i + \mathcal{L}_j - \mathcal{L}_{ij}}{\#\text{points of } \alpha_i \text{ and } \alpha_j}$$

where α_i and α_j are the pairs of islands to be merged, $\mathcal{L}_{(.)}$ is log-likelihood of a island, and $\mathcal{L}_{ij}$ is the log-likelihood of the merged island.

Outliers and Suspiciousness Score: The outliers comprise of the bins far away from any distribution of the identified node groups (i.e. with probability $< 10^{-4}$). Intuitively, the majority island containing the most nodes is normal, so we define the weighted KL-divergence of an island from the majority island as its suspiciousness score.

Definition 1 (Suspiciousness). *Given the parameter θ_m for the majority island, the suspiciousness of the island α_i described by distribution with parameter θ_i is:*

$$\kappa(\theta_i) = \log \bar{d}_i \cdot \sum_{b \in \alpha_i} N_i \cdot KL\left(P(b \mid \theta_i) \| P(b \mid \theta_m)\right)$$

where $P(b \mid \theta)$ is the probability in the bin b for the distribution with θ as parameter, N_i is the number of nodes in the island i, and we use the logarithm of $\bar{d}_i$, average degree of all graph nodes in the island i, as the weight based on the domain knowledge that if other features are the same, higher-degree nodes are more suspicious.

Time Complexity: Given features associated with nodes V, generating the histogram takes $O(|V|)$ time. Let $nnz(\mathcal{H})$ be the number of non-empty bins in $\mathcal{H}$ and C be the number of clusters. Assume the number of iterations for learning parameters in $DistributionFit(\cdot)$ is T, then we have (proofs are in our supplementary material [1]):

---

[3] This measures the goodness-of-fit of the left-truncated Gaussian distribution.

Theorem 1. *The time complexity of EagleMine is* $O(\dfrac{\log h_{max}}{\rho} \cdot nnz(\mathcal{H}) +$ $\mathcal{C} \cdot T \cdot nnz(\mathcal{H})$.

Table 2. Dataset statistics summary and synthetic settings.

	# of nodes	# of edges	Content	Injected block
BeerAdvocate [29]	(33.37K, 65.91K)	1.57M	rate	1k × 500, 2k × 1k
Flickr	(1.4M, 466K)	1.89M	user to group	2k × 2k, 4k × 2k
Amazon	(2.14M, 1.23M)	5.84M	rate	-
Yelp	(686K, 85.54K)	2.68M	rate	-
Tagged	(2.73M, 4.65M)	150.8M	anonymized Links	-
Youtube	(3.22M, 3.22M)	9.37M	who-follow-who	-
Sina weibo	(2.75M, 8.08M)	50.1M	user-retweet-msg	-

5   Experiments

We design the experiments to answer the following questions: **[Q1] Anomaly detection**: How does EagleMine's performance on anomaly detection compare with the state-of-art methods? How much improvement does the visual-inspired information bring? **[Q2] Summarization**: Does EagleMine give significant improvement in concisely summarizing the graph? Does it accurately identify micro-clusters that agree with human vision? **[Q3] Scalability**: Is EagleMine scalable with regard to the data size?

The dataset[4] information used in our experiments is illustrated in Table 2. The Tagged [14] dataset was collected from Tagged.com social network website. It contains 7 anonymized types of links between users, and here we only choose the links of type-6, which is a homogeneous graph. The microblog Sina Weibo dataset was crawled in November 2013 from weibo.com, consisting of user-retweeting-message (bipartite) graph.

5.1   Q1. Anomaly Detection

To demonstrate EagleMine can effectively detect anomalous, we conduct experiments on both synthetic and real data, and compare the performance with state-of-the-art fraud detection algorithms GetScoop [21], SPOKEN [30], and Fraudar [18].

---

[4] The public datasets are available at: Amazon: http://konect.uni-koblenz.de/networ ks/amazon-ratings, Yelp: https://www.yelp.com/dataset_challenge, Flickr: https:// www.aminer.cn/data-sna#Flickr-large, Youtube: http://networkrepository.com/ soc-youtube.php, Tagged: https://linqs-data.soe.ucsc.edu/public/social_spammer/.

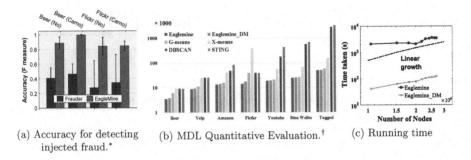

(a) Accuracy for detecting injected fraud.*

(b) MDL Quantitative Evaluation.[†]

(c) Running time

Fig. 2. EagleMine performance for anomaly detection, summarization, and scalability. (a) EagleMine achieves best accuracy for detecting injected fraud for Beer-Advocate ('Beer' as the abbr.) and Flickr data. *Note that GetScoop and spokEn are omitted for failing to catch any injected object. (b) MDL is compared on different datasets. EagleMine achieves the shortest description code length, which means concise summarization, and outperforms all other baselines ([†]Watershed clustering method is omitted due to its MDL results is even much larger than the worst case). (c) *blue* curve shows the running time of EagleMine v.s. # of node in graph in log-log scale. (Color figure online)

In the synthetic case, we inject different size fraud (as a block) with and without random camouflage into real datasets as Table 2 shows, where the ratio of camouflage is set to 50%, i.e. randomly selecting different objects as the same size as the targets. For BeerAdovate, the density of injected fraud is 0.05. For Flickr, the density of injected fraud are 0.05, 0.1, 0.2. We use F score for nodes on both sides of injected block to test the detection accuracy, and report the averaged result over above trials for each dataset in Fig. 2a. GetScoop and SPOKEN are omitted since they fail to catch any injected object. It is obvious that EagleMine consistently outperforms Fraudar and achieves less variance for the injection cases with and without camouflages.

To verify that EagleMine accurately detects anomalies in Sina weibo data, we labeled these nodes, both user and message, from the results of baselines, and sampled nodes of our suspicious clusters from EagleMine, since that it is impossible to label all the nodes. Our labels were based on the following rules (like [18]): **(1)** deleted user-accounts/messages by the online system[5] **(2)** a lot of users that share unusual login-names prefixes, and other suspicious signals: approximately the same sign-up time, friends and followers count. **(3)** messages about advertisement or retweeting promotion, and having lots of *copy-and-paste* text content. In total, we labeled 5,474 suspicious users and 4,890 suspicious messages.

The anomaly detection results are reported in Fig. 1g. Using AUC to quantify the quality of the ordered result from the algorithm, the sampled nodes from micro-clusters are ranked in descendant order of hubness or authority. The results

---

[5] The status is checked three years later (May 2017) with API provided by Sina weibo service.

show that EagleMine achieves more than 10% and about 50% improvement for anomalous user and msg detection resp., outperforming the baselines. The anomalous users detected by Fraudar and SpokEn only fall in the micro-cluster ① in Fig. 3e, since their algorithms can only focus on densest core in a graph. But EagleMine detects suspicious users by recognizing noteworthy micro-clusters in the whole feature space. Simply put, EagleMine detects more anomalies than the baselines, identifying more extra micro-clusters ②, ③, and ④.

5.2 Case Study and Found Patterns

As discussed above, the micro-clusters ③ and ④ in out-degree vs hubness Fig. 3e contains those users frequently rewtweet non-important messages. Here we study the behavior patterns of micro-clusters ① and ② on the right side of the majority group. Note that almost half of the users are deleted by system operators, and many existing users share unusual name prefixes as Fig. 1e shown.

What patterns have we found? The Fig. 1f shows the 'Jellyfish' structure of the subgraph consisting of users from micro-clusters ① and ②. The head of 'Jellyfish' is the densest core (①), where the users created unusual dense connections to a group of messages, showing high hubness. The users (spammers or bots) aggressively 'copy-and-paste' many advertising messages a few times, which includes 'new game', 'apps in IOS7', and 'Xiaomi Phone', Their structure looks like 'Jellyfish' tail. Thus the bots in ② shows lower hubness than those in ①, due to the different spamming strategies, which are overlooked by density-based detection methods.

5.3 Q2. Summarization Evaluation on Real Data

We select X-means, G-means, DBSCAN and STING as the comparisons, the setting details are described in supplements. We also include EagleMine (DM) by using multivariate Gaussian description. We chose the feature spaces as degree vs pagerank and degree vs triangle for Tagged dataset, and choose in-degree vs authority and out-degree vs hubness for the rest.

We use Minimum Description Length (MDL) to measure the summarization as [4] do, by envisioning the problem of clustering as a compression problem. In short, it follows the assumption that the more we can compress the data, the more we can learn about its underlying patterns. The best model has the smallest MDL length. The MDL lengths for the baselines are calculated as [4, 8, 12]. With the same principle, the MDL of EagleMine is: $L = \log^*(C) + L_\mathcal{S} + L_\Theta + L_\mathcal{O} + L_\epsilon$; details are listed in the supplementary [1].

The comparison results of MDL are reported in Fig. 2b. We can see that EagleMine achieves the shortest description length, indicating a concise and good summarization. Compared with the competitors, EagleMine reduces the MDL code length more 26.2% at least and even 81.1% than G-means and STING resp. on average, it also outperforms EagleMine (DM) over 6.4%, benefiting from a proper vocabulary selection. Therefore, EagleMine summarizes histogram with recognized groups in the best description length.

Besides the quantitative evaluation for summarization, we illustrate the results on 2D histogram for vision-based qualitative comparison. Due the space limit, here we only exhibit the results for Sina Weibo dataset. As Fig. 3 shows, the plot features are user's out-degree and hubness indicating how many important messages retweeted. Without removing some low-density bins as background, Watershed algorithm easily identified all the groups into one or two huge ones. Hence we manually tuned the threshold of background to attain a better result, which is similar to the groups in a level of our water-level tree. The background for Watershed is shown with gray color in Fig. 3b. As we can see, Watershed only recognized a few very dense groups while failing to separate the two groups on the right and leaving other micro-clusters unidentified. Our EagleMine recognized groups in a more intuitive way, and identify those micro-clusters missed by DBSCAN and STING. Note that the user deletion ratio in the missed micro-clusters ① and ③ is unusually high, and they were suspended by the system operators for anti-spam. Besides, those micro-clusters ③ and ④ include the users have high out-degree but low-hubness, i.e., users retweeting many non-important messages (e.g., advertisements). Hence, EagleMine identify very useful micro-clusters automatically as human vision does.

5.4 Q3. Scalability

Figure 2c shows the near-linear scaling of EagleMine's running time in the numbers graph nodes. Here we used Sina weibo dataset, we selected the snapshot of the graph, i.e., the reduced subgraph, according to the timestamp of edge creation in first $3, 6, \ldots, 30$ days. Slope of *black dot* line indicates linear growth.

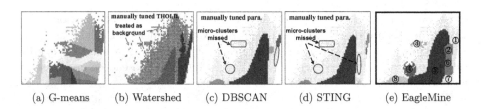

(a) G-means   (b) Watershed   (c) DBSCAN   (d) STING   (e) EagleMine

Fig. 3. EagleMine visually recognizes better node groups than clustering algorithms for the feature space in Fig. 1a. Watershed (with a threshold for image background), DBSCAN, and STING are mannaully tuned to have a relatively better results. The blue scattering points in (c)–(e) denote individual outliers. Even though DBSCAN and STING are extensively manually tunned, some micro-clusters of low density are missed. (Color figure online)

6 Conclusions

We propose a tree-based approach EagleMine to mine and summarize all point groups in a heatmap of scatter plots. EagleMine finds optimal clusters based on a water-level tree and statistical hypothesis tests, and describes them with a configurable model vocabulary. EagleMine can automatically and effectively summarize the histogram and node groups, detects explainable anomalies on synthetic and real data, and can scale up linearly. In general, the algorithm is applicable to any two/multi-dimensional heatmap.

Acknowledgments. This material is based upon work supported by the Strategic Priority Research Program of CAS (XDA19020400), NSF of China (61772498, 61425016, 91746301, 61872206), and the Beijing NSF (4172059).

References

1. Supplementary document (proof and additional experiments). https://goo.gl/ ZjMwYe
2. Akoglu, L., Chau, D.H., Kang, U., Koutra, D., Faloutsos, C.: OPAvion: mining and visualization in large graphs. In: SIGMOD, pp. 717–720 (2012)
3. Arbelaez, P., Maire, M., Fowlkes, C., Malik, J.: Contour detection and hierarchical image segmentation. PAMI **33**, 898–916 (2011)
4. Böhm, C., Faloutsos, C., Pan, J.Y., Plant, C.: Robust information-theoretic clustering. In: KDD, pp. 65–75. ACM (2006)
5. Borkin, M., et al.: Evaluation of artery visualizations for heart disease diagnosis. IEEE Trans. Vis. Comput. Graph. **17**, 2479–2488 (2011)
6. Buja, A., Tukey, P.A.: Computing and Graphics in Statistics. Springer, New York (1991)
7. Campello, R.J.G.B., Moulavi, D., Zimek, A., Sander, J.: Hierarchical density estimates for data clustering, visualization, and outlier detection. ACM TKDD **10**(1), 5:1–5:51 (2015). https://doi.org/10.1145/2733381
8. Chakrabarti, D., Papadimitriou, S., Modha, D.S., Faloutsos, C.: Fully automatic cross-associations. In: SIGKDD, pp. 79–88 (2004)
9. Chernobai, A., Rachev, S.T., Fabozzi, F.J.: Composite goodness-of-fit tests for left-truncated loss samples. In: Lee, C.-F., Lee, J.C. (eds.) Handbook of Financial Econometrics and Statistics, pp. 575–596. Springer, New York (2015). https://doi. org/10.1007/978-1-4614-7750-1_20
10. Cubedo, M., Oller, J.M.: Hypothesis testing: a model selection approach (2002)
11. DiCarlo, J.J., Zoccolan, D., Rust, N.C.: How does the brain solve visual object recognition? Neuron **73**, 415–434 (2012)
12. Elias, P.: Universal codeword sets and representations of the integers. IEEE Trans. Inf. Theory **21**, 194–203 (1975)
13. Ester, M., Kriegel, H.P., Sander, J., Xu, X.: A density-based algorithm for discovering clusters in large spatial databases with noise. In: KDD (1996)
14. Fakhraei, S., Foulds, J., Shashanka, M., Getoor, L.: Collective spammer detection in evolving multi-relational social networks. In: SIGKDD, KDD 2015. ACM (2015)
15. Gonzalez, R.C., Woods, R.E.: Digital image processing (2007)
16. Hamerly, G., Elkan, C.: Learning the k in k-means. In: NIPS (2004)
17. Heynckes, M.: The predictive vs. the simulating brain: a literature review on the mechanisms behind mimicry. Maastricht Stud. J. Psychol. Neurosci. **4**(15) (2016)
18. Hooi, B., Song, H.A., Beutel, A., Shah, N., Shin, K., Faloutsos, C.: FRAUDAR: bounding graph fraud in the face of camouflage. In: SIGKDD, pp. 895–904 (2016)
19. Huber, P.J.: Projection pursuit. Ann. Stat. **13**(2), 435–475 (1985)

20. Jiang, M., Cui, P., Beutel, A., Faloutsos, C., Yang, S.: CatchSync: catching synchronized behavior in large directed graphs. In: SIGKDD (2014)
21. Jiang, M., Cui, P., Beutel, A., Faloutsos, C., Yang, S.: Inferring strange behavior from connectivity pattern in social networks. In: Tseng, V.S., Ho, T.B., Zhou, Z.-H., Chen, A.L.P., Kao, H.-Y. (eds.) PAKDD 2014. LNCS (LNAI), vol. 8443, pp. 126–138. Springer, Cham (2014). https://doi.org/10.1007/978-3-319-06608-0_11
22. Kang, U., Lee, J.-Y., Koutra, D., Faloutsos, C.: Net-ray: visualizing and mining billion-scale graphs. In: Tseng, V.S., Ho, T.B., Zhou, Z.-H., Chen, A.L.P., Kao, H.-Y. (eds.) PAKDD 2014. LNCS (LNAI), vol. 8443, pp. 348–361. Springer, Cham (2014). https://doi.org/10.1007/978-3-319-06608-0_29
23. Kang, U., Meeder, B., Faloutsos, C.: Spectral analysis for billion-scale graphs: discoveries and implementation. In: Huang, J.Z., Cao, L., Srivastava, J. (eds.) PAKDD 2011. LNCS (LNAI), vol. 6635, pp. 13–25. Springer, Heidelberg (2011). https://doi.org/10.1007/978-3-642-20847-8_2
24. Kleinberg, J.M.: Authoritative sources in a hyperlinked environment. JACM **46**(5), 604–632 (1999). https://doi.org/10.1145/324133.324140
25. Koutra, D., Jin, D., Ning, Y., Faloutsos, C.: Perseus: an interactive large-scale graph mining and visualization tool. VLDB **8**(12), 1924–1927 (2015)
26. Kumar, R., Novak, J., Tomkins, A.: Structure and evolution of online social networks. In: Yu, P., Han, J., Faloutsos, C. (eds.) Link Mining: Models, Algorithms, and Applications, pp. 337–357. Springer, New York (2010). https://doi.org/10.1007/978-1-4419-6515-8_13
27. Lancichinetti, A., Fortunato, S.: Community detection algorithms: a comparative analysis. Phys. Rev. E **80**, 056117 (2009)
28. Liu, X.M., Ji, R., Wang, C., Liu, W., Zhong, B., Huang, T.S.: Understanding image structure via hierarchical shape parsing. In: CVPR (2015)
29. McAuley, J.J., Leskovec, J.: From amateurs to connoisseurs: modeling the evolution of user expertise through online reviews. In: WWW (2013)
30. Prakash, B.A., Sridharan, A., Seshadri, M., Machiraju, S., Faloutsos, C.: EigenSpokes: surprising patterns and scalable community chipping in large graphs. In: Zaki, M.J., Yu, J.X., Ravindran, B., Pudi, V. (eds.) PAKDD 2010. LNCS (LNAI), vol. 6119, pp. 435–448. Springer, Heidelberg (2010). https://doi.org/10.1007/978-3-642-13672-6_42
31. Roerdink, J.B., Meijster, A.: The watershed transform: definitions, algorithms and parallelization strategies. Fundam. Informaticae **41**, 187–228 (2000)
32. Schaeffer, S.E.: Graph clustering. Comput. Sci. Rev. **1**, 27–64 (2007)
33. Stephens, M.A.: EDF statistics for goodness of fit and some comparisons. J. Am. Stat. Assoc. **63**, 730–737 (1974)
34. Thompson, H.R.: Truncated normal distributions. Nature **165**, 444–445 (1950)
35. Tukey, J.W., Tukey, P.A.: Computer graphics and exploratory data analysis: an introduction. National Computer Graphics Association (1985)
36. Vincent, L., Soille, P.: Watersheds in digital spaces: an efficient algorithm based on immersion simulations. PAMI **13**, 583–598 (1991)
37. Wang, W., Yang, J., Muntz, R., et al.: STING: a statistical information grid approach to spatial data mining. In: VLDB, pp. 186–195 (1997)
38. Ware, C.: Color sequences for univariate maps: theory, experiments and principles. IEEE Comput. Graph. Appl. **8**, 41–49 (1988)
39. Wilkinson, L., Anand, A., Grossman, R.: Graph-theoretic scagnostics. In: Proceedings - IEEE Symposium on Information Visualization, INFO VIS, pp. 157–164 (2005)

Clustering of Mixed-Type Data Considering Concept Hierarchies

Sahar Behzadi[1(✉)], Nikola S. Müller[2], Claudia Plant[1,3], and Christian Böhm[4]

[1] Faculty of Computer Science, Data Mining, University of Vienna, Vienna, Austria
sahar.behzadi@univie.ac.at
[2] Institute of Computational Biology, Helmholtz Zentrum München,
Munich, Germany
[3] ds:UniVie, University of Vienna, Vienna, Austria
[4] Ludwig-Maximilians-Universität München, Munich, Germany

Abstract. Most clustering algorithms have been designed only for pure numerical or pure categorical data sets while nowadays many applications generate mixed data. It arises the question how to integrate various types of attributes so that one could efficiently group objects without loss of information. It is already well understood that a simple conversion of categorical attributes into a numerical domain is not sufficient since relationships between values such as a certain order are artificially introduced. Leveraging the natural conceptual hierarchy among categorical information, concept trees summarize the categorical attributes. In this paper we propose the algorithm *ClicoT* (**CL**ustering mixed-type data **I**ncluding **CO**ncept **T**rees) which is based on the Minimum Description Length (MDL) principle. Profiting of the conceptual hierarchies, ClicoT integrates categorical and numerical attributes by means of a MDL based objective function. The result of ClicoT is well interpretable since concept trees provide insights of categorical data. Extensive experiments on synthetic and real data set illustrate that ClicoT is noise-robust and yields well interpretable results in a short runtime.

1 Introduction

Clustering mixed-data is a non-trivial task and typically is not achieved by well-known clustering algorithms designed for a specific type. It is already well-understood that converting one type to another one is not sufficient since it might lead to information loss. Moreover, relations among values (e.g. a certain order) are artificially introduced. Let Fig. 1 show a mixed-type data where three different clusters are illustrated by different shapes. The data set comprises of two numerical attributes concerning the position of objects and a categorical attribute representing the color. We simply converted the color to a numerical attribute by mapping numbers to various colors. Considering the *Normalized Mutual Information* (NMI) [12] as an evaluation measure, Fig. 1 depicts the inefficiency of applying K-means and DBSCAN, two popular clustering algorithms, on the converted data. Therefore, integrating categorical and numerical

© Springer Nature Switzerland AG 2019
Q. Yang et al. (Eds.): PAKDD 2019, LNAI 11439, pp. 555–573, 2019.
https://doi.org/10.1007/978-3-030-16148-4_43

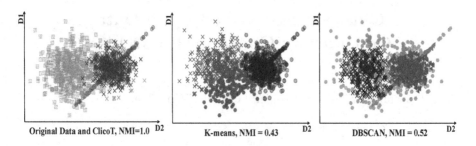

Fig. 1. Clustering results after converting categorical attribute *Color* to numerical. (Color figure online)

attributes without any conversion is required since it preserves the original format of any attribute.

Utilizing the MDL principle we regard the clustering task as a data compression problem so that the best clustering is linked to the strongest data set compression. MDL allows integrative clustering by relating the concepts of likelihood and data compression while for any attribute a representative model is required. Although for solely numerical data sets a *Probability Distribution Function* (PDF) represents an approximation of data, finding an appropriate approximation for categorical attributes is not straight-forward. Considering the natural hierarchy among categorical values we introduce *concept hierarchy* to summarize the categorical information. Back to the running example, considering pink as a higher-level hierarchy for the objects in the cluster consisting of rose and purple points with the shape ×, more accurately represents the characteristics of the cluster.

Beyond the clustering approaches, detecting the most relevant attributes during this process improves the quality of clustering. However, considering a data set with an unknown distribution where only few subgroups in the data space are actually relevant to characterize a cluster, it is not trivial to recognize the cluster-specific attributes. Thus, we employ an information-theoretic greedy approach to specify the most relevant attributes. As a result, our novel parameter-free **CL**ustering algorithm for mixed-type data **I**ncluding **CO**ncept **T**ress, shortly *ClicoT*, provides a natural interpretation avoiding any conversion which leads to an effective clustering (c.f. Fig. 1). Our approach consists of several contributions:

- **Integration:** ClicoT integrates two types of information considering data compression as an optimization goal. ClicoT flexibly learns the relative importance of the two different sources of information for clustering without requiring the user to specify input parameters which are usually difficult to estimate.
- **Interpretation:** In contrast to most clustering algorithms, ClicoT not only provides information about *which* objects are assigned to which clusters, but also gives an answer to the central question *why* objects are clustered together. As a result of ClicoT, each cluster is characterized by a signature of cluster-specific relevant attributes providing appropriate interpretations.

- **Robustness:** The compression-based objective function ensures that only the truly relevant attributes are marked as cluster-specific attributes. Thereby, we avoid over-fitting, enhance the interpretability and guarantee the validity of the result.
- **Usability:** ClicoT is convenient to be used in practice since our algorithm scales well to large data sets. Moreover, our compression-based approach avoids difficult estimation of input parameters e.g. the number or the size of clusters.

2   Clustering Mixed Data Types

To design a mixed-type clustering algorithm we need to address three fundamental questions: How to model numerical attributes to properly characterize a cluster? How to model categorical attributes? And finally how to efficiently integrate heterogeneous attributes when the most relevant attributes are specified? In principle, a PDF summarizes values by approximating meaningful parameters. However, the idea of using a background PDF for categorical attributes is not intuitive at first, therefore we employ concept hierarchies.

2.1   Concept Hierarchy

As mentioned, concept hierarchies allow us to express conceptual interchangeable values by selecting an inner node of a concept hierarchy to describe a cluster. Concept hierarchies not only capture more relevant categories for each cluster but also help to interpret the clustering result appropriately. Let $\mathcal{DB}$ denote a database consisting of n objects. An object o comprises m categorical attributes $\mathcal{A} = \{A_1, A_2, ..., A_m\}$ and d numerical attributes $\mathcal{X} = \{x_1, x_2, ..., x_d\}$. For a categorical attribute A_i, we denote different categorical values by $A_i{}^{(j)}$. An *Element* represents a categorical value or a numerical attribute and we denote the number of all *Elements* by E. Considering the natural hierarchy between different categories, for each categorical attribute A_i a concept hierarchy is already available as follows:

Definition 1 *Concept Hierarchy. Let $T_{A_i} = (N, \mathcal{E})$ be a tree with root A_i denoting the concept hierarchy corresponding to the categorical attribute A_i with the following properties:*

1. *T_{A_i} consists of a set of nodes $N = \{n_1, ..., n_s\}$ where any node is corresponding to a categorical concept. $\mathcal{E}$ is a set of directed edges $\mathcal{E} = \{e_1, ..., e_{s-1}\}$, where n_j is a parent of n_z if there is an edge $e_l \in \mathcal{E}$ so that $e_l = (n_j, n_z)$.*
2. *The level $l(n_j)$ of a node n_j is the height of the descendant sub-tree. If n_j is a leaf, then $l(n_j) = 0$. In a concept tree leaf nodes are categorical values existing in the dataset. The root node is the attribute A_i which has the highest level, also called the height of the concept hierarchy.*
3. *Each node $n_j \in N$ is associated with a probability $p(n_j)$ which is the frequency of the corresponding category in a dataset.*
4. *Each node n_j represents a sub-category of its parent therefore all probabilities of the children sum up to the probability of the parent node.*

2.2 Cluster-Specific Elements

Beside an efficient clustering approach, finding relevant attributes to capture the best fitting model is important. Usually the clustering result is disturbed by irrelevant attributes. To make the model for each cluster more precise we distinguish between relevant and irrelevant attributes. Each cluster c is associated with a subset of the numerical and categorical relevant elements denoted by *cluster-specific elements*. Categorical cluster-specific elements are represented by a specific concept hierarchy which diverges from the background hierarchy (i.e. the concept hierarchy of the entire database).

Definition 2 *Cluster.* *A cluster c is described by:*

1. *A set of objects $\mathcal{O}_c \subset \mathcal{DB}$.*
2. *A cluster-specific subspace $I = \mathcal{X}_c \cup \mathcal{A}_c$, where $\mathcal{X}_c \subseteq \mathcal{X}$ and $\mathcal{A}_c \subseteq \mathcal{A}$.*
3. *For any categorical attribute $A_i \in \mathcal{A}_c$, the corresponding cluster-specific concept hierarchy is a tree $T^c{}_{A_i} = (N_c, \mathcal{E}_c)$ with nodes and edges as specified in Definition 1. $N_c \subset N$ indicates the cluster-specific nodes. For computing the probabilities associated with the cluster-specific nodes instead of all n objects, only the objects $\mathcal{O}_c$ in cluster c are applied, i.e. $p(n_j) = \frac{|n_j|}{|\mathcal{O}_c|}$.*

2.3 Integrative Objective Function

Given the appropriate model corresponding to any attribute, MDL allows a unified view on mixed data. The better the model matches major characteristics of the data, the better the result is. Following the MDL principle [11], we encode not only the data but also the model itself and minimize the overall description length. Simultaneously we avoid over-fitting since the MDL principle tends to a natural trade-off between model complexity and goodness-of-fit.

Definition 3 *Objective Function.* *Considering the cluster c the description length (DL) corresponding to this cluster defined as:*

$$DL(c) = DL_c(\mathcal{X}) + DL_c(\mathcal{A}) + DL(model(c))$$

The first two terms represent coding costs concerning numerical and categorical attributes, respectively while the last term is the model encoding cost. Our proposed objective function minimizes the overall description length of the database which is defined as:

$$DL(\mathcal{DB}) = \sum_{c \in \mathcal{C}} DL(c)$$

Coding Numerical Attributes: Considering Huffman coding scheme, the description length of a numerical value o_i is defined by $-\log_2 \mathrm{PDF}(o_i)$. We assume the same PDF to encode the objects in various clusters and clusters compete for an object while the description length is computed by means of the same PDF for evrey cluster. Therefore any PDF would be applicable and using a specific model is not a restriction [3]. For simplicity we select Gaussian PDF, $\mathcal{N}(\mu, \sigma)$. Moreover, we distinguish between the cluster-specific attributes in any cluster c, denoted by $\mathcal{X}_c$, and the remaining attributes $\mathcal{X} \setminus \mathcal{X}_c$ (Definition 2). Let μ_i and σ_i denote the mean and variance corresponding to the numerical attribute x_i in cluster c. If x_i is a cluster-specific element ($x_i \in \mathcal{X}_c$), we consider only cluster points to compute the parameters otherwise ($x_j \in \mathcal{X} \setminus \mathcal{X}_c$) the overall data points will be considered. Thus, the coding cost for numerical attributes in cluster c is provided by:

$$DL_c(\mathcal{X}) = \sum_{x_i \in \mathcal{X}} \sum_{o_i \in \mathcal{O}_c} -\log_2 \left(\mathcal{N}(\mu_i, \sigma_i) \right)$$

Coding Categorical Attributes: Analogously, we employ Huffman coding scheme for categorical attributes. The associated probability to a category is its frequency w.r.t. either the specific or the background hierarchy (Definition 1). Similar to numerical attributes, we assume $\mathcal{A}_c$ as the set of cluster-specific categorical attributes and $\mathcal{A} \setminus \mathcal{A}_c$ for the rest. Let o_j denote a categorical object value corresponding to the attribute A_j. We define $f(A_j, o_j)$ as a function which maps o_j to a node in either a specific or a background hierarchy depending on A_j. Thus, the categorical coding cost for a cluster c is given by:

$$DL_c(\mathcal{A}) = \sum_{A_j \in \mathcal{A}} \sum_{o_j \in \mathcal{O}_c} -\log_2 \left(p(f(A_j, o_j)) \right)$$

Model Complexity: Without taking the model complexity into account, the best result will be a clustering consisting of singleton clusters. This result is completely useless in terms of the interpretation. Focusing on cluster c, the model complexity is defined as:

$$DL(model(c)) = idCosts(c) + SpecificIdCosts(c) + paramCosts(c)$$

The idCosts are required to specify which cluster is assigned to a object while balancing the size of clusters. Employing the Huffman coding scheme, idCosts are defined by $|\mathcal{O}_c| \cdot log_2 \frac{n}{|\mathcal{O}_c|}$ where $|\mathcal{O}_c|$ denotes the number of objects assigned to cluster c. Moreover, in order to avoid information loss we need to specify whether an attribute is a cluster-specific attribute or not. That is, given the number of specific elements s in cluster c, the coding costs corresponding to these elements, *SpecificIdCosts*, is defined as:

$$SpecificIdCosts(c) = s \cdot \log_2 \frac{E}{s} + (E - s) \cdot \log_2 \frac{E}{(E - s)}$$

Following fundamental results from information theory [11], the costs for encoding the model parameters is reliably estimated by:

$$paramCosts(c) = \frac{numParams(c)}{2} \cdot \log_2 |\mathcal{O}_c|$$

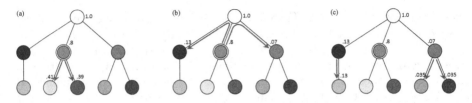

Fig. 2. Update concept hierarchies considering *pink* as a cluster-specific node. (Color figure online)

For any numerical cluster-specific attribute we need to encode its mean and variance while for a categorical one the probability deviations to the default concept hierarchy need to be encoded, i.e. $numParams(c) = |\mathcal{X}| \cdot 2 + \sum_{A_i \in \mathcal{A}} |N_c|$. Moreover, we need to encode the probabilities associated with the default concept hierarchy, as well as the default (global) means and variances for all numerical attributes. However, these costs are summarized to a constant term which does not influence our subspace selection and clustering technique.

3   Algorithm

Together with the main building blocks of ClicoT, two other steps are required to achieve an appropriate parameter free clustering: (1) recognizing the cluster-specific elements and (2) probability adjustments.

Cluster-Specific Elements: Let the *specific coding cost* denote the cost where an element is marked as specific and the *non-specific coding cost* indicate the cost otherwise. Consulting the idea that cluster-specific elements have the most deviation of specific and non-specific cost and therefore saves more coding costs, we introduce a greedy method to recognize them. We iteratively sort the elements according to their deviations and specify the first element as a cluster-specific element. We continue marking elements until marking more elements does not pay off in terms of the coding cost. Note that different nodes of a concept hierarchy have the same opportunity to be specific.

Probability Adjustment: To adjust the probabilities for a numerical cluster-specific attribute we can safely use mean and variance corresponding to the cluster. In contrast, learning the cluster-specific concept hierarchy is more challenging since we need to maintain the integrity of a hierarchy. According to Definition 1 we assure that node probabilities of siblings in each level sum up to the probability of the parent node. Moreover node probabilities should sum up to one for each level. we provide a pseudocode concerning this procedure in appendix. To clarify, let Fig. 2 show the procedure on the concept hierarchy corresponding to the running example (Fig. 1) where labels denote the frequencies. Moreover, let pink be a cluster-specific node for the cluster with the shape ×. The adjustment starts with the root node and processes its children. Then it continues computing the relative probabilities for the specific concept hierarchy rather

Algorithm 1. ClicoT

---

input $\mathcal{DB}$
learn background distributions of each attribute
$C' = \{C_0\}$ with $C'_0 = O_i \in \mathcal{DB}$
repeat
   // try to split until convergence
   $C = C'$
   cost $= DL(\mathcal{DB}|C)$ // current cost
   $C' = \{C'_1 \ldots C'_{k-1}\}$ split worst $C_i \in C$ to $\{C'_i, C'_k\}$
   while clustering C' changes **do**
      $C'_i = \{O_j : \min_i DL(O_j|C'_i)\}$ // assign objects
      Select cluster-specific elements by a greedy method for each cluster and compute
      costs
      Update each attribute of C'_i
   end while
   cost' $= DL(\mathcal{DB}|C')$ // split cost
until cost $>$ cost'
$k = |C|$
return C, k

---

by background probability fraction (Fig. 2a). 80% relative probability should be distributed between two children, rose and purple, based on the computed propagation factor. During the next step the remaining 20% probability is assigned level-wise to blue and green to assure that probabilities in each level sum up to 1 (Fig. 2b). Again each parent propagates down its probability (Fig. 2c). The result is a concept hierarchy best fitting to the objects when the background distributions are preserved.

ClicoT Algorithm: ClicoT is a top-down parameter-free clustering algorithm. That is, we start from a cluster consisting of all objects and iteratively split down the most expensive cluster c in terms of the coding cost to two new clusters $\{c'_a, c'_b\}$. Then, we apply a k-Means-like strategy and assign every point to closest cluster which is nothing else than the cluster with the lowest increase in the coding cost. Employing the greedy algorithm, we determine the cluster-specific elements and finally we compute the compression cost for clustering results in two cases, before and after splitting (Definition 1). If the compression cost after splitting, i.e. $\mathcal{C}'$ with $|\mathcal{C}'| = k + 1$, is cheaper than the cost of already accepted clustering $\mathcal{C}$ with $|\mathcal{C}| = k$ then we continue splitting the clusters. Otherwise the termination condition is reached and the algorithm will be stopped.

4   Related Work

Driven by the need of real applications, the topic of clustering mixed-type data represented by numerical and categorical attributes has attracted attentions, e.g. CFIKP [13], CAVE [7], CEBMDC [5]. In between, most of the algorithms are designed based on the algorithmic paradigm of k-Means, K-means-mixed

(KMM) [1], k-Prototypes [8]. Often in this category not only the number of clusters k but also the weighting between numerical and categorical attributes in clustering has to be specified by the user. Among them, KMM avoids weighting parameters by an optimization scheme learning the relative importance of the single attributes during runtime, although it needs the number of clusters k as input parameter. Following a mixture of Gaussian distributions, model based clustering algorithms have been also proposed for mixed-type data. In between, clustMD [9] is developed using a latent variable model and employing an expectation maximisation (EM) algorithm to estimate the mixture model. However this algorithm has a certain Gaussian assumption which does not have to be necessarily fulfilled. Some of the approaches utiliz the unique characteristics of any data type to avoid the drawbacks of converting a data type to another one. Profiting of the concept hierarchy, these algorithms introduce an integrative distance measure applicable for both numerical and categorical attributes. The algorithm DH [6] proposes a hierarchical clustering algorithm using a distance hierarchy which facilitates expressing the similarity between categorical and numerical values. As another method, MDBSCAN [2] employs a hierarchical distance measure to introduce a general integrative framework applicable for the algorithms which require a distance measure .e.g. DBSCAN. On the other hand, information-theoretic approaches have been proposed to avoid the difficulty of estimating input parameters. These algorithms regard the clustering as a data compression problem by hiering the Minimum Description Length (MDL). The cluster model of these algorithms comprises joint coding schemes supporting numerical and categorical data. The MDL principle allows balancing model complexity and goodness-of-fit. INCONCO [10] and Integrate [4] are two representative for mixed-type clustering algorithms in this family. While Integrate has been designed for general integrative clustering, INCONCO also supports detecting mixed-type attribute dependency patterns.

5   Evaluation

In this section we assess the performance of ClicoT comparing to other clustering algorithms in terms of NMI which is a common evaluation measure for clustering results. NMI numerically evaluates pairwise mutual information between ground truth and resulted clusters scaling between zero and one. We conducted several experiments evaluating ClicoT in comparison to KMM [1], INCONCO [10], DH [6], ClustMD [9], Integrate [4] and MDBSCAN [2]. In order to be fair in any experiment, we input the corresponding concept hierarchy to the algorithms which are not designed for dealing with it. That is, we encode the concept hierarchy as an extra attribute so that categorical values belonging to the same category have the same value in this extra attribute. Our algorithm is implemented in Java and the source code as well as the data sets are publicly available[1].

---

[1] https://bit.ly/2FkUB3Q.

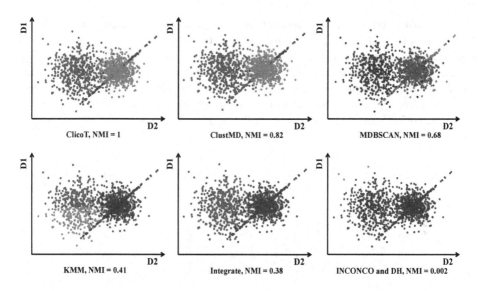

Fig. 3. Clustering results on the running example. (Color figure online)

5.1 Mixed-Type Clustering of Synthetic Data

In order to cover all aspects of ClicoT we first consider a synthetic data set. Then we continue experiments by comparing all algorithms in terms of the noise-robustness. Finally we will discuss the runtime efficiency.

Clustering Results: In this experiment we evaluate the performance of all the algorithms on the running example (Fig. 1) while all parametric algorithms are set up with the right number of clusters. The data has two numerical attributes concerning the position of any data point and a categorical attribute showing the color of the points. Figure 3 shows the result of applying the algorithms where different clusters are illustrated by different colors. As it is explicitly shown in this figure ClicoT, with NMI 1, appropriately finds the initially sampled three clusters where green, pink and blue are cluster-specific elements. Setting the correct number of cluster and trying various Gaussian mixture models, ClustMD results the next accurate clustering. Although MDBSCAN utilizes the distance hierarchy, but it is not able to capture the pink and green clusters. KMM can not distinguish among various colors. Since two clusters pink and green are heavily overlapped, Integrate can not distinguish among them. DH and INCONCO poorly result on this data set and they found almost only one cluster.

Noise-Robustness: In this section we benchmark noise-robustness of ClicoT w.r.t the other algorithms in terms of NMI by increasing the noise factor. To address this issue we generate a data set with the same structure as the running example and we add another category, brown, to the categorical attribute color as noise. Regarding numerical attributes we increase the variance of any cluster. We start from 5% noise (noise factor = 1) and iteratively increase the noise factor ranging to 5. Figure 4 clearly illustrates noise-robustness of ClicoT comparing to others.

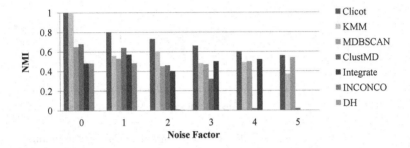

Fig. 4. Comparing noise-robustness of ClicoT to other algorithms.

Scalability: To evaluate the efficiency of ClicoT w.r.t the other algorithms, we generated a 10 dimensional data set (5 numerical and 5 categorical attributes) with three Gaussian clusters. Then respectively we increased the number of objects ranging from 2,000 to 10,000. In the other case we generated different data sets of various dimensionality ranging from 10 to 50 where the number of objects is fixed. Figure 5 depicts the efficiency of all algorithms in terms of the runtime complexity. Regarding the first experiment on the number of objects, ClicoT is slightly faster than others while increasing the dimensionality Integrate performs faster. However, the runtime of this algorithm highly depends on the number of clusters k initialized in the beginning (we set $k = 20$). That is, this algorithm tries a rang of k and outputs the best results. Therefore, by increasing k the runtime is also increasing.

5.2 Real Experiments

Finally, we evaluate clustering quality and interpretability of ClicoT on real world data sets. We used *MPG*, *Automobile* and *Adult* data sets from the UCI Repository as well as *Airport* data set from the public project *Open Flights*[2].

---

[2] http://openflights.org/data.html.

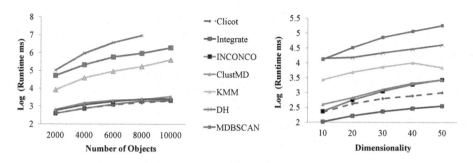

Fig. 5. Runtime experiment.

MPG: MPG is a slightly modified version of the data set provided in the StatLib library. The data concerns city-cycle fuel consumption in miles per gallon (MPG) in terms of 3 categorical and 5 numerical attributes consisting of different characteristics of 397 cars. We consider MPG ranging from 10 to 46.6 as the ground truth and divide the range to 7 intervals of the same length. The information about the concept hierarchy is provided in the appendix. Comparing ClicoT (NMI = 0.4) to the other algorithms INCONCO(0.17), KMM(0.37), DH(0.14), MDBSCAN(0.02), ClustMD(0.33) and Integrate(0). ClicoT correctly finds 7 clusters each of which compatible with one of the MPG groups. Cluster 2, for instance, is compatible with the first group of MPGs since the frequency of the first group in this cluster is 0.9. In this cluster American cars with the frequency of 1.0, cars with 8 cylinders with the frequency of 1 and model year in first group (70–74) with the frequency of 0.88 are selected as cluster-specific elements.

Automobile: This data set provides 205 instances with 26 categorical and numerical attributes. The first attribute defining the risk factor of an automobile has been used as class label. Altogether there are 6 different classes. Due to many missing values we used only 17 attributes. Comparing the best NMI captured by every algorithm, ClicoT (NMI = 0.38) outperforms kMM(0.23), INCONCO(0.20), Integrate(0.17), DH(0.04), ClusterMD(0.16) and MDBSCAN(0.02). Furthermore, ClicoT gives an insight in the interpretability of the clusters where Cluster 12, for instance, is characterized mostly by the fuel system of *2bbl*, but also by *1bbl* and *4bbl*. Also we see that Cluster 26 is consisting of both *mpfi* and slightly of *mfi*, too. Concerning the risk analysis this clustering serves, ClicoT allows to recognize which fuel systems share the same insurance risk.

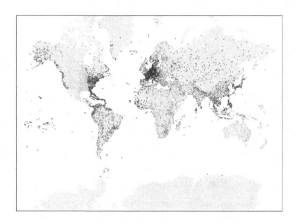

Fig. 6. Result of ClicoT on Open Flight data set. (Color figure online)

Adult Data Set: Adult data set without missing values, extracted from the census bureau database, consists of 48,842 instances of 11 attributes. The class attribute Salary indicates whether the salary is over 50K or lower. Categorical attributes consist of different information e.g. work-class, education, occupation. A detailed concept hierarchy is provided in appendix. Although comparing to INCONCO(0.05), ClustMD(0.0003), MDBSCAN(0.004), DH(0) and Integrate(0), our algorithm ClicoT(0.15) outperforms all other algorithms except KMM(0.16) which is slightly better. But it seems that NMI does not sound a reasonable evaluation measure for this data set since there are only two classes in ground truth. ClicoT found 4 clusters in which Cluster 2, the biggest cluster consisting of almost 56% of objects, specifies *Husband* as the cluster-specific element, since it has the most deviation, but negative. The probability of instances having *Husband* as categorical value and the salary $\leq 50K$ is zero in this cluster. Therefore along with the negative deviation this means that in Cluster 2 persons with the role as husband in a family earn more than $50K$.

Open Flights Data Set: The public project Open Flights provides world wide information about airports, flights and airlines. Here we consider instances of airports in order to carry out a cluster analysis. The data set consists of 8107 instances each of which represents an airport. The numeric attributes show the longitude and latitude, the sea height in meters and the time zone. Categorical attributes consist of the country, where the airport is located and the day light saving time. We constructed the concept hierarchy of the country attribute so that each country belongs to a continent. Since there is no ground truth provided for this data set we interpret the result of ClicoT (Fig. 6) and we refer the reader to the appendix for more results regarding other algorithms.

Clustering results illustrated in Fig. 6 consists of 15 clusters and shows that ClicoT appropriately grouped almost geographically similar regions in the clusters. Starting from west to east, North American continent divided into five clusters. Obviously here the attribute of the time zone was chosen as specific because the clusters are uniquely made according to this attribute. Moving to the south, ClicoT pulled a plausible separation between South and North America. Considering South America as cluster-specific element and due to the rather low remaining airport density of South America ClicoT combined almost all of the airports to a cluster (red). In Western Europe there are some clusters, which can be distinguished by their geographic location. Additionally many airports around and in Germany are be grouped together.

6   Conclusion

To conclude, we have developed and demonstrated that ClicoT is not only able to cluster mixed-typed data in a noise-robust manner, but also yields most interpretable cluster descriptions. By using data compression as the general principle ClicoT automatically detects the number of clusters within any data set without any prior knowledge. Moreover, the experiments impressively demonstrated that clustering can greatly benefit from a concept hierarchy. Therefore, ClicoT excellently complements the approaches for mining mixed-type data.

Appendix

A   Probability Adjustment

To adjust the probabilities for a numerical cluster-specific attribute we can safely use mean and variance corresponding to the cluster. In contrast, learning the cluster-specific concept hierarchy is more challenging since we need to maintain the integrity of a hierarchy. We need to assure that node probabilities of siblings in each level sum up to the probability of the parent node. Moreover node probabilities should sum up to one for each level. *ProcessHierarchy()* in Algorithm 2 is a recursive function to update the concept tree assuming marked cluster-specific elements. Simultaneously in this function, Propagatedown() tries to preserve the concept tree properties by propagating down the parents probabilities to their children.

Algorithm 2. Concept tree updates

ProcessHierarchy $(Vertex\ V)$
ssp := sum of specific probabilities
sup := sum of unspecific probabilities
if V is a leaf **then**
  if V is specific **then**
    return $(V.probability, 0)$
  end if
  return $(0, V.backgroundProbability)$
end if
// now V is not a leaf
$(ssp, sup) := (0, 0)$
for all C in $children(V)$ **do**
  $(s, u) := processHierarchy(C)$
  $(ssp, sup) := (ssp + s, sup + u)$
end for
if V is specific or root **then**
  $factor := (V.probability - ssp)/sup$
  for all C in $children(V)$ **do**
    $propagateDownFactor(C, factor)$
  end for
  return $(V.probability, 0)$
end if
return (ssp, sup)

Algorithm 3. Down-propagation of the adjustment factor

PropagateDownFactor (Vertex V, **double** $factor$)
if V is unspecific **then**
  $V.probability := V.probability \cdot factor$
  if V is not leaf **then**
    for all $C \in V.children$ **do**
      PropagateDownFactor$(C, factor)$
    end for
  end if
end if

B   MPG

MPG is a slightly modified version of the data set provided in the StatLib library. The data concerns city-cycle fuel consumption in miles per gallon (MPG) in terms of 3 categorical and 5 numerical attributes consisting of different characteristics of 397 cars. We consider MPG ranging from 10 to 46.6 as the ground truth and divide the range to 7 intervals of the same length. Considering a concept hierarchy for the name of cars we group all the cars so that we have three branches: European, American, Japanese cars. Moreover we divide the range of model year attribute to three intervals: 70–74, 75–80, after 80. We leave the third attribute as a flat concept hierarchy since there is no meaningful hierarchy between variation of cylinders.

C   Adult Dataset

Adult data set, extracted from the census bureau database, consists of 48,842 instances of 11 attributes excluding the attributes with missing values (six numerical and 5 categorical). The class attribute Salary indicates whether the salary is over 50K or lower. Categorical attributes consist of different information e.g. work-class, education, occupation and so on. Figure 7 indicates concept hierarchies for three selected categorical attributes, including work-class, relationship and education.

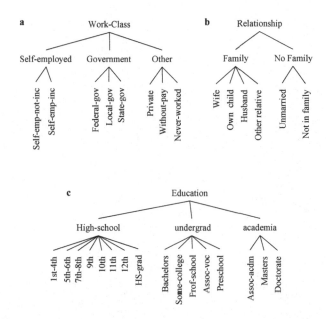

Fig. 7. Concept tree for 3 categorical attributes of adult dataset.

D   Open Flights Dataset

Clustering results applying various algorithms with a better resolution illustrating is provided here (Figs. 8, 9, 10, 11 and 12).

Fig. 8. ClicoT.

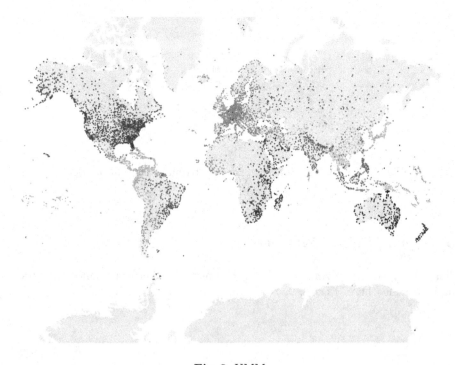

Fig. 9. KMM.

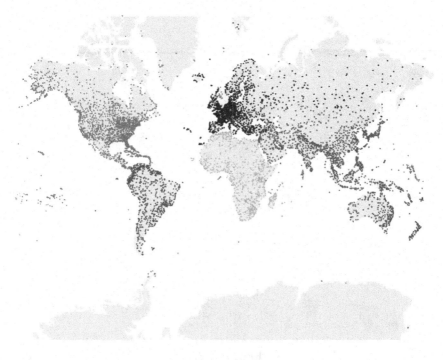

Fig. 10. MDBSCAN.

Fig. 11. INCONCO and integrate.

Fig. 12. DH.

References

1. Ahmad, A., Dey, L.: A k-mean clustering algorithm for mixed numeric and categorical data. Data Knowl. Eng. **63**, 503–527 (2007)
2. Behzadi, S., Ibrahim, M.A., Plant, C.: Parameter free mixed-type density-based clustering. In: Hartmann, S., Ma, H., Hameurlain, A., Pernul, G., Wagner, R.R. (eds.) DEXA 2018. LNCS, vol. 11030, pp. 19–34. Springer, Cham (2018). https://doi.org/10.1007/978-3-319-98812-2_2
3. Böhm, C., Faloutsos, C., Pan, J., Plant, C.: Robust information-theoretic clustering. In: KDD (2006)
4. Böhm, C., Goebl, S., Oswald, A., Plant, C., Plavinski, M., Wackersreuther, B.: Integrative parameter-free clustering of data with mixed type attributes. In: Zaki, M.J., Yu, J.X., Ravindran, B., Pudi, V. (eds.) PAKDD 2010. LNCS (LNAI), vol. 6118, pp. 38–47. Springer, Heidelberg (2010). https://doi.org/10.1007/978-3-642-13657-3_7
5. He, Z., Xu, X., Deng, S.: Clustering mixed numeric and categorical data: a cluster ensemble approach. CoRR abs/cs/0509011 (2005)
6. Hsu, C.C., Chen, C.L., Su, Y.W.: Hierarchical clustering of mixed data based on distance hierarchy. Inf. Sci. **177**(20), 4474–4492 (2007)
7. Hsu, C.C., Chen, Y.C.: Mining of mixed data with application to catalog marketing. Expert Syst. Appl. **32**(1), 12–23 (2007)
8. Huang, Z.: Extensions to the k-means algorithm for clustering large data sets with categorical values. Data Min. Knowl. Discov. **2**, 283–304 (1998)

9. McParland, D., Gormley, I.C.: Model based clustering for mixed data: ClustMD. Adv. Data Anal. Classif. **10**(2), 155–169 (2016)
10. Plant, C., Böhm, C.: INCONCO: interpretable clustering of numerical and categorical objects. In: KDD, pp. 1127–1135 (2011)
11. Rissanen, J.: A universal prior for integers and estimation by minimum description length. Ann. Stat. **11**(2), 416–31 (1983)
12. Vinh, N.X., Epps, J., Bailey, J.: Information theoretic measures for clusterings comparison: is a correction for chance necessary? In: ICML (2009)
13. Yin, J., Tan, Z.: Clustering mixed type attributes in large dataset. In: Pan, Y., Chen, D., Guo, M., Cao, J., Dongarra, J. (eds.) ISPA 2005. LNCS, vol. 3758, pp. 655–661. Springer, Heidelberg (2005). https://doi.org/10.1007/11576235_66

DMNAED: A Novel Framework Based on Dynamic Memory Network for Abnormal Event Detection in Enterprise Networks

Xueshuang Ren[1,2] and Liming Wang[1(✉)]

[1] Institute of Information Engineering, Chinese Academy of Sciences, Beijing, China
{renxueshuang,wangliming}@iie.ac.cn
[2] School of Cyber Security, University of Chinese Academy of Sciences, Beijing, China

Abstract. Abnormal event detection is a crucial step towards discovering insider threat in enterprise networks. However, most existing anomaly detection approaches fail to capture latent correlations between disparate events in different domains due to the lack of a panoramic view or the disability of iterative attention. In light of this, this paper presents DMNAED, a novel framework based on dynamic memory network for abnormal event detection in enterprise networks. Inspired by question answering systems in natural language processing, DMNAED considers the event to be inspected as a question, and a sequence of multi-domain historical events serve as a context. Through an iterative attention process, DMNAED captures the context-question interrelation and aggregates relevant historical events to make more accurate anomaly detection. The experimental results on the CERT insider threat dataset r4.2 demonstrate that DMNAED exhibits more stable and superior performance compared with three baseline methods in identifying aberrant events in multi-user and multi-domain environments.

Keywords: Abnormal event detection · Dynamic memory network · Iterative attention · Relevant historical events

1 Introduction

With the evolution of insider threat, abnormal event detection plays a more and more significant role in protecting information assets of enterprise networks. Due to increasing complexities in business intelligence sharing, contractor relationships, and geographical distribution [16], the footprints left by malicious insiders tend to disperse across different domains. As a result, it is a mounting challenge to identify stealthy malicious activities from large numbers of multi-domain heterogenous event records.

In order to accurately detect abnormal events, analysts need to piece together fragments of contextual information for an event, and perform iterative reasoning based on the synthetic context. However, existing approaches [3,4,6,11] mainly

© Springer Nature Switzerland AG 2019
Q. Yang et al. (Eds.): PAKDD 2019, LNAI 11439, pp. 574–586, 2019.
https://doi.org/10.1007/978-3-030-16148-4_44

focus on learning the patterns of user behaviors based on multi-dimensional features to identify suspicious events by comparing behaviors of different individuals or comparing behaviors of the same individual in different time periods. These approaches have the following two limitations: (1) Lacking a panoramic view required to capture event correlation patterns across multiple domains because they typically target a particular domain or specific log data. (2) Lacking the ability of iterative reasoning over the synthetic contextual information which is a patchwork of multi-domain heterogeneous event records. Although many researches have attempted to solve the limitations with various deep learning techniques, they are still insufficient to provide stable and reliable detection performance in multi-user and multi-domain environments.

Dynamic memory network (DMN) is a unified neural network that uses some ideas from neuroscience such as semantic and episodic memories [9]. It is characterized by a recurrent attention mechanism which enables iterative reasoning. Given an input sequence and a question, DMN can focus on specific input relative to the question, and form episodic memories for generating the corresponding answer. Motivated by the excellent performance of DMN in many varied tasks, this paper proposes DMNAED, a novel DMN-based framework for abnormal event detection in enterprise networks, which models an event record as a structured language sentence that consists of a set of fields. Specifically, DMNAED takes the event to be inspected (referred to as current event) as a question, and takes a synthetic sequence of multi-domain historical events as a context. Each event is mapped into a multi-dimensional vector by field-level embedding and event-level encoding. Through an iterative attention process, DMNAED iteratively retrieves the historical events conditioned on current event to aggregate meaningful contextual information, thus paving the way for further prediction and anomaly detection. The contributions of this paper are summarized as follows:

1. We present a novel framework called DMNAED that is able to process multi-domain heterogeneous event records and detect abnormal events taking into consideration the interaction between different domains. DMNAED is an innovative extension of DMN to anomaly detection.
2. We propose a temporary storage mechanism to preserve the indices of related historical events such that once an anomaly is detected, DMNAED can provide valuable clues for provenance tracking and forensic analysis.
3. We conduct a series of comparative experiments on the CERT dataset r4.2, and investigate the parameter sensitivity of DMNAED. The results demonstrate that DMNAED achieves more stable and superior performance compared with three baseline anomaly detection methods.

2   Related Work

The relevant efforts on abnormal event detection in enterprise networks mainly encompass three aspects as follows:

Data pre-processing: Raw logs and event records are always manifested in various formats. To deal with inconsistent log data from different sources, Yen et al. [16] employed a security information and event management (SIEM) system, and developed efficient techniques to remove noise in the logs. Lee et al. [10] presented a two-stage framework that includes data normalization and graph-based data fusion for unifying different datasets. Pei et al. [14] leveraged raw log parsers to extract specific fields from each input entry for log correlation. To facilitate event correlation analysis, DMNAED follows a similar way to [14] that utilizes a set of pre-defined fields to capture pivotal information of each event record, but unlike previous efforts, DMNAED aims at normalizing heterogenous event records into a common concise representation for multi-domain event correlation.

Anomaly detection: A large number of anomaly detection techniques have been illustrated in [3,4], including classification based models, clustering based models, and statistical models etc. With the boosting of deep learning in recent years, deep neural networks, such as *long short-term memory* (LSTM) networks, have been widely used in anomaly detection. For instance, Buda et al. [2] employed various LSTMs on streaming data, and merged the predictions of multiple models to detect anomalies. Du et al. [5] proposed a LSTM based deep neural network that models system logs as natural language sequences for detecting execution path anomalies and parameter value anomalies. Meng et al. [11] leveraged LSTM to learn the pattern of user behaviors and extract temporal features. These approaches are meritorious in handling particular time series which contain strong temporal dependencies. Whereas, the synthetic sequence of multi-domain event records is equivalent to a hybrid of multiple time series, in which the dependencies between adjacent events are notably diminished. The previous approaches become ineffective in this context. DMNAED addresses the issue by leveraging an iterative attention process which concentrates on relevant historical events and skips over interfering incidents.

Provenance tracking: Too many uncorrelated alerts may be either deemed false positives or overlooked by security officers [14]. Thus, it is imperative to provide enough actionable intelligence or pertinent evidence for provenance tracking. King et al. [8] built dependency graphs by correlating events to trace the root causes of intrusions. Hossain et al. [7] developed tag-based techniques for attack reconstruction, relying on dependency graphs. Besides, workflow construction has also been studied largely for anomaly diagnosis, such as in [1,5]. However, whether dependency graphs or workflows, they are both prone to high cost of overhead and severe delay. DMNAED avoids the pitfall by leveraging a temporary storage mechanism, which preserves the indices of relevant historical events during iterative attention process. This allows DMNAED to provide historical clues about an anomaly to security officers in real time for further investigation.

To the best of our knowledge, applying DMN to anomaly detection has not been studied yet. The proposed framework DMNAED employs a modified DMN to detect anomalous events in enterprise networks, which involves the three collaborative aspects mentioned above.

3 DMNAED Framework

Figure 1 shows an overview of the DMNAED framework. Different from the architecture of original DMN which has four modules (input module, question module, episodic memory module and answer module, respectively) [9], DMNAED is structured hierarchically, which consists of five layers, detailed in the following subsections:

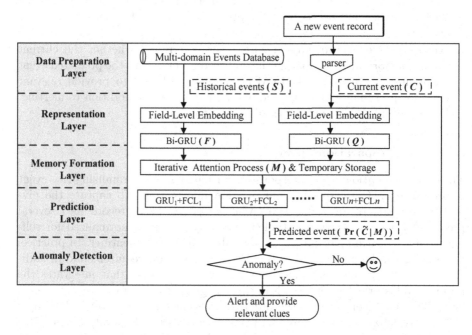

Fig. 1. Overview of DMNAED. The flow of execution is indicated by arrows. The output of each layer except the last layer is shown in the parentheses (i.e. S, C, F, Q, M, $\Pr(\tilde{C}|M)$).

- **Data Preparation Layer:** This layer preprocesses raw event records and maintains a multi-domain events database which arranges all the normalized historical events in chronological order. When a new event record (i.e. current event) arrives, it is normalized into a set of pre-defined fields, and a window of the k most recent events are fetched from the database as the context. For notational convenience, current event and historical events are denoted by C and S respectively.
- **Representation Layer:** This layer transforms C and S into multi-dimensional numeric vectors via field-level embedding and event-level encoding. The event-level encoding is implemented by bi-directional GRUs (Bi-GRUs). The outputs are represented by Q and F for current event and historical events respectively.

– **Memory Formation Layer:** This layer retrieves historical events conditioned on current event to aggregate related facts via an iterative attention process. Each iteration produces an episode, finally forming a compositive memory M. During the process, the indices of relevant historical events are preserved in buffer for later queries once needed.
– **Prediction Layer:** This layer employs a series of GRUs combined with full connected layers (FCL) to decode the memory M, and predict the event $\tilde{C}$ that comes right after the given historical sequence. Particularly, a separate $GRU_j + FCL_j$ network is built for predicting the jth field of the expected event $\tilde{C}$.
– **Anomaly Detection Layer:** This layer determines whether the current event is abnormal by comparing the observed event C with the predicted event candidates. If abnormal, DMNAED queries the buffer to provide relevant clues for provenance tracking and anomaly diagnosis in addition to alerting. If normal, DMNAED automatically clears the buffer.

3.1 Data Preparation Layer

The data preparation layer is responsible for parsing and normalizing raw event records using network-specific configuration information. To capture the pivotal information of an event, DMNAED parses an event record into several pre-defined fields as shown in Table 1, where each field is normalized for self-identifying. For example, subject is normalized as a user identifier. In practice, these fields can be augmented or rewritten to meet new requirements. In addition, this layer maintains a multi-domain events database that integrates the normalized historical event records in chronological order. For concurrent events, which lead to uncertainty in the ordering of event entities, DMNAED adopts several customized policies based on principal priority and operation logic [7].

Inspired by question answering in NLP tasks [9], DMNAED considers current event to be inspected as a question, denoted by C. The corresponding historical event sequence, which is a window of the T most recent events across multiple domains, serves as the context, denoted by S.

Table 1. Pre-defined fields for an event

Field	Explanation
Subject	The initiator of actions (i.e. a user)
Action	Operations performed by the subject on the object
Object	The receptors of actions (e.g. files, devices, messages, programs)
Device	Device where the action took place
Timestamp	Time when the action took place

3.2   Representation Layer

The representation layer maps each event into a multi-dimensional vector by two phases for exploring the hidden interconnectivity of disparate events.

Phase 1: Field-Level Embedding. Different from traditional NLP tasks, the input of this layer are structured event records, whose basic elements are fields, rather than words. Therefore, field-level embedding is imperative. To this end, we pretrain a recurrent Continuous Bag of Words (CBOW) model using a fraction of normal event records to compute the embedding vector of each field. The recurrent CBOW model is able to circularly utilize the hidden states, fusing rich contextual information in the field embedding [15]. For fields with continuous value such as timestamp, we partition the value range into several segments such that a large amount of continuous value is reduced to a smaller set of discrete intervals.

Phase 2: Event-Level Encoding. We apply event-level encoding to capture more comprehensive information of event sequence, implemented by Bi-GRU. The field-level embedding vectors are fed into a Bi-GRU, whose hidden states are obtained by concatenating the hidden states of the forward and backward GRUs. We subsample the output of the Bi-GRU, and leverage the hidden unit representations that correspond to end-of-event markers as the event encoding vectors for the sequence of historical events, while the event-level encoding for current event is defined as the final hidden unit representation of the Bi-GRU. We represent the encoding vectors as $F = [f_1, f_2, \cdots, f_T]$ and Q for historical events and current event respectively, where f_t represents the encoding vector of the tth historical event.

3.3   Memory Formation Layer

Analogous to the episodic memory module in original DMN [9], the memory formation layer is characterized by an iterative attention process, concentrating on relevant historical events in regard to current event, and yielding a compositive memory: M.

We initialize the memory to the encoding vector of current event. i.e. $M^0 = Q$. For each iteration i, an attention vector $A^i = [a_1^i, a_2^i, \cdots, a_T^i]$ is calculated with respect to the given historical events. a_t^i is an attention weight which measures the correlation between a historical event f_t and the current event Q in iteration i. The attention weight is computed using a two-layer feed forward neural network, which takes as input a pre-generated vector: z_t. The equations are expressed as:

$$z_t = [f_t, \ M^{i-1}, \ f_t \circ Q, \ f_t \circ M^{i-1}, \\ |f_t - Q|, \ |f_t - M^{i-1}|, \ f_t^{\mathrm{T}} W Q, \ f_t^{\mathrm{T}} W M^{i-1}], \tag{1}$$

$$a_t^i = \sigma(W_2 tanh(W_1 z_t + b_1) + b_2), \tag{2}$$

where ∘ is the element-wise product. W, W_1, W_2, b_1, b_2 are parameters to be learned. z_t captures the interior connection between a historical event f_t, the previous memory M^{i-1}, and the current event Q.

Given the attention vector, we compute an episode E_{att} by running a modified GRU, denoted by attGRU. The equation to update the hidden states of the attGRU at time t for iteration i is given as:

$$h_t^i = a_t^i GRU(f_t, h_{t-1}^i) + (1 - a_t^i)h_{t-1}^i, \tag{3}$$

where the input parameters include last hidden state h_{t-1}^i, a historical event f_t, and the corresponding attention weight a_t^i. The episode is defined as the final hidden state of the attGRU, i.e. $E_{att}^i = h_T^i$. At the end of iteration i, the memory M is updated as:

$$M^i = GRU(E_{att}^i, M^{i-1}). \tag{4}$$

Due to the lack of explicit supervision, we set the maximum number of iterations as r. When the iterations come to end, the final memory is formed, which summarizes the episodes produced by every iteration.

Temporary storage: The iterative attention process enables the model to focus on different events during each iteration, which realizes incessant reasoning. As a matter of fact, this process implicitly encodes the workflow path of the underlying system, which can provide important clues about the likelihood of an event being a potential part of a malicious activity. To facilitate provenance tracking, when the final memory is obtained, DMNAED puts the indices of relevant facts into a buffer for temporary storage. The relevant facts refer to the historical events whose attention weights surpass a threshold: λ. In this way, once an anomaly is detected, DMNAED can query the buffer to obtain the relevant anterior events, providing valuable contextual information about anomalies in real time to security officers for forensic analysis. If current event is judged as normal, DMNAED automatically clears the buffer. Note that the threshold λ has no impact on the result of anomaly detection. It only affects the number of anomaly-associated clues provided to security officers.

3.4   Prediction Layer

Considering that for an enterprise network, each field (e.g. subject, object, and device) has a finite number of optional values, the prediction of a field can be cast into a multi-class classification problem. We postulate that different fields are mutually independent, thus the probability of an event can be calculated by taking the product of the probabilities of its fields.

The prediction layer employs a series of GRUs and fully connected layers (FCLs) to decode the compositive memory M and predict the subsequent event $\tilde{C}$ that is expected to occur after the given historical events. Suppose the pre-defined fields for an event are $u_1, u_2, \cdots, u_n$, a separate GRU+FCL network is built for predicting the field u_j of the expected event, represented by

$GRU_j + FCL_j$ as shown in Fig. 1. The probability distribution for the field u_j of the expected event given the memory M is calculated as:

$$g_t^j = GRU_j(M, g_{t-1}^j), \tag{5}$$

$$y_t^j = relu(W_j^{(1)} g_t^j + b_j^{(1)}), \tag{6}$$

$$\Pr(u_j|M) = softmax(W_j^{(2)} y_t^j + b_j^{(2)}), \tag{7}$$

where g_t^j is the hidden state of GRU_j at time t. $W_j^{(1)}$, $W_j^{(2)}$, $b_j^{(1)}$, and $b_j^{(2)}$ are the weights and biases of FCL_j. y_t^j is the output of the first fully connected layer, while the conditional probability of u_j is obtained at the end of the second fully connected layer using a $softmax$ function.

Overall, the prediction layer assembles n GRU+FCL models to predict the expected event $\tilde{C}$, where n is the number of pre-defined fields for an event. The conditional probability for $\tilde{C}$ is expressed as:

$$\Pr(\tilde{C}|M) = \prod_{j=1}^{n} \Pr(u_j|M). \tag{8}$$

To train these models, we minimize the cross-entropy loss between the predicted event and the observed event over the training event sequences. Moreover, we adopt a variety of techniques, such as L_2 regularization, dropout, adding gradient noise, to avoid over-fitting.

3.5 Anomaly Detection Layer

As of now, we have obtained the probability distribution of the expected event. The last step is to determine whether current event is abnormal by comparing the observed event C and the predicted event candidates. First, we need to set a threshold as the cutoff in the prediction output. In this work, we consider the event as normal if it is among the top k predicted events with high probabilities, i.e. $C \in \{\tilde{C}_1, \tilde{C}_2, \cdots, \tilde{C}_k\}$. Under the circumstances, DMNAED clears the buffer, and adds the normalized representation of current event into multi-domain events database. Otherwise, the current event is flagged as abnormal, in which case DMNAED alerts an alarm immediately, and reads the buffer to provide relevant historical events as clues to security officers for provenance tracking, followed by clearing the buffer.

In addition, DMNAED allows the security officers to return feedback for updating the models as in [5]. If a false positive is reported, DMNAED re-trains the weights of its model in an online manner to adapt to new patterns.

4   Experimental Evaluation

In this section, we are dedicated to evaluating the performance of the proposed framework. All of our experiments have been performed on a machine with an Intel Xeon E5-1603 2.80 GHz CPU, 12 GB RAM and a Windows 7 Ultimate 64-bit operating system. DMNAED is implemented by Python language with Tensorflow as the backend.

4.1   Dataset

The CERT dataset[1] was published by Carnegie Mellon University for insider threat detection. We utilized the r4.2 dataset for our evaluation. The dataset contains various event records (HTTP, logon, device, file, and email) reflecting the behaviors of 1000 employees over a 17-month period. According to the dataset, the employees, i.e. users, come from 22 departments. A department can be viewed as a workgroup role, where the users exhibit a specific set of expected behaviors and potentially interact with a specific pool of resources [13]. Among the different departments, we select four to conduct experiments. The event records generated by a department constitute a dataset, denoted by D1, D2, D3, D4 respectively. Table 2 summarizes the four datasets. Note that the size and complexity of the datasets are increasing gradually.

Table 2. Dataset statistics

Dataset	Department	Number of users	Total number of events
D1	Payroll	4	150569
D2	Manufacturing engineering	12	425120
D3	Medical	60	2023865
D4	Software management	101	3542668

4.2   Baselines

For a comprehensive evaluation, we experimented with DMNAED as well as three existing anomaly detection methods as follows:

LSTM-based method: LSTM networks are renowned for their ability to remember history information in temporal sequence data using memory gates [2]. For the LSTM-based abnormal event detection method, the input is only a history window w of the h most recent events: $w = \{f_{t-h}, \cdots, f_{t-2}, f_{t-1}\}$. The output is a conditional probability distribution of the subsequent expected event: $\Pr(f_t|w)$, which is used for anomaly detection as DMNAED does.

SOM-based method: Self-organizing map (SOM) is a fully connected, single-layer neural network that maps a multi-dimensional data set onto a one or two-dimensional space [12]. The SOM-based model clusters events by soft

---

[1] https://resources.sei.cmu.edu/library/asset-view.cfm?assetid=508099.

competition. If the minimum distance between a new event and the neurons of the trained SOM exceeds a pre-set threshold, the event is labeled as anomalous.

Graph-based method: We built a dependency graph based on interrelationships between subjects and objects as in [7]. Events are represented by edges. To prioritize our analysis, we leveraged tags to identify subjects and objects. Finally, the graph properties are computed to feed into isolation forest algorithm to detect anomalous events.

4.3   Results and Discussion

For all the approaches, we explored their parameter space and reported their best results. We use the standard metrics of precision, recall and F_1-score to evaluate the performance of different anomaly detection methods. Figure 2 shows the performance of these methods on D1, D2, D3 and D4 respectively.

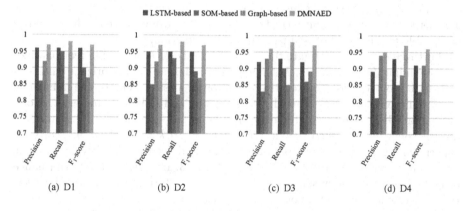

(a) D1   (b) D2   (c) D3   (d) D4

Fig. 2. The performance of DMNAED and three baseline methods on D1–D4 datasets

Evidently, DMNAED overwhelms the three baseline methods in various degree across all the datasets. For example, for the dataset D1 (Fig. 2a), the precision of DMNAED is 97%, improved by 1%, 11%, and 5%, compared with the LSTM-based, SOM-based and graph-based methods respectively. The recall of DMNAED is 98%, improved by 2%, 3%, and 16% in comparison. Besides, we observe that, for the datasets (D1 and D2) which contain event records of fewer users, the LSTM-based method exhibits comparable performance to DMNAED. The SOM-based method manifests a reasonable recall but a pathetic precision. The graph-based approach shows an acceptable precision but a pitiful recall. With the expansion of the scale of users (from D1 to D4), the performance of the LSTM-based method degrades obviously. The reason is easy to understand: more users bring more various event records and more complex interaction patterns. For a new event to be inspected, the corresponding history window would contain more irrelevant information, resembling background noise, which causes interference for capturing truly aberrant traces. Without a selective attention mechanism, the LSTM-based method treats the events within the whole history window as related facts. Hence, its performance tends to fluctuate with the sophistication of input data. Similarly, the SOM-based method and the graph-based method are also biased towards particular datasets, but not effective as a generic anomaly detection method under multi-user and multi-domain environments, due to the disability of selective attention and intelligent reasoning over a hybrid sequence of various event records. In comparison, DMNAED performs more stable and superior on the datasets of different sizes. The F_1-score of DMNAED only declines by 2% from D1 to D4. The slippage reveals that the performance of DMNAED is also affected by the characteristics of input data, but the impact is fairly slight. Moreover, it is important to note that, except the graph-based method, the other methods are all designed for online anomaly detection in a streaming fashion. The running of SOM is the quickest because of its simple structure, but its precision is worst. Even though the architecture of DMNAED is a bit complicated, its actual running is quite smoothly, since the adoption of GRU greatly reduces the computation cost compared with LSTM. The detection cost per event record of DMNAED is around 1.2 ms, which is quite acceptable for the need of real-time anomaly detection.

Furthermore, we investigate the sensitivity of DMNAED to different parameters, including: T, r, k and H. T denotes the window size of historical events. r is the number of iterations in iterative attention process. k denotes the number of predicted event candidates considered as normal, and H is the number of hidden units in attGRU. We conducted experiments on the dataset D2. By default, we use the following parameter values: $T = 60, r = 10, k = 20, H = 192$. For each experiment, we only alter one parameter value and use the default values for the other parameters. The results are shown in Fig. 3. Intuitively, the performance of DMNAED is not very sensitive to T, r and k in terms of F_1-score. However, the adjustment of H exerts noticeable impact on the effectiveness of DMNAED. When H is less than 128, the F_1-score of DMNAED is lower than 90% and drops dramatically. This is because H implicitly determines the memory capacity of DMNAED. A smaller H value leads to a more narrow space for storing useful historical information. Therefore, it is crucial to assign a proper value to H in practice.

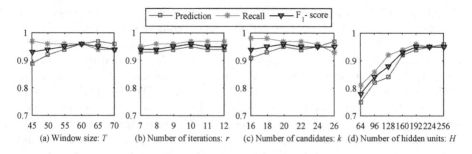

Fig. 3. The impact of different parameters for DMNAED

5   Conclusion

This paper presents DMNAED, a novel framework for abnormal event detection in enterprise networks. DMNAED is an innovative extension of dynamic memory network, which views the event to be inspected as a question, and a sequence of multi-domain historical events as a context. DMNAED retrieves the historical events conditioned on current event to aggregate useful contextual information for forward prediction and anomaly detection. Once an anomaly is identified, DMNAED is able to provide relevant clues in real time for forensic analysis. Extensive experiments have demonstrated the superior performance of DMNAED compared with three baseline approaches. One of our future work is to combine DMNAED with distributed architecture, dividing big data into multiple blocks for parallel processing to improve the executive efficiency of the framework.

Acknowledgments. This research was supported by National Research and Development Program of China (No. 2017YFB1010000).

References

1. Beschastnikh, I., Brun, Y., Ernst, M.D., Krishnamurthy, A.: Inferring models of concurrent systems from logs of their behavior with CSight. In: 36th ICSE, pp. 468–479 (2014)
2. Buda, T.S., Caglayan, B., Assem, H.: DeepAD: a generic framework based on deep learning for time series anomaly detection. In: Phung, D., Tseng, V.S., Webb, G.I., Ho, B., Ganji, M., Rashidi, L. (eds.) PAKDD 2018. LNCS (LNAI), vol. 10937, pp. 577–588. Springer, Cham (2018). https://doi.org/10.1007/978-3-319-93034-3_46
3. Chandola, V., Banerjee, A., Kumar, V.: Anomaly detection: a survey. ACM Comput. Surv. **41**(3), 15:1–15:58 (2009)
4. Denning, D.E.: An intrusion-detection model. IEEE Trans. Softw. Eng. **13**(2), 222–232 (1987)
5. Du, M., Li, F., Zheng, G., Srikumar, V.: Deeplog: anomaly detection and diagnosis from system logs through deep learning. In: Proceedings of the 2017 ACM SIGSAC, CCS, pp. 1285–1298 (2017)
6. Gamachchi, A., Sun, L., Boztas, S.: Graph based framework for malicious insider threat detection. In: HICSS (2017)
7. Hossain, M.N., et al.: SLEUTH: real-time attack scenario reconstruction from COTS audit data. CoRR abs/1801.02062 (2018)
8. King, S.T., Chen, P.M.: Backtracking intrusions. ACM Trans. Comput. Syst. **23**(1), 51–76 (2005)
9. Kumar, A., et al.: Ask me anything: dynamic memory networks for natural language processing. In: ICML, pp. 1378–1387 (2016)
10. Lee, W., Hsu, W.H., Satoh, S.: Learning from cross-domain media streams for event-of-interest discovery. IEEE Trans. Multimed. **20**(1), 142–154 (2018)
11. Meng, F., Lou, F., Fu, Y., Tian, Z.: Deep learning based attribute classification insider threat detection for data security. In: DSC, pp. 576–581 (2018)
12. Nam, T.M., et al.: Self-organizing map-based approaches in DDoS flooding detection using SDN. In: ICOIN, pp. 249–254 (2018)

13. Nance, K., Marty, R.: Identifying and visualizing the malicious insider threat using bipartite graphs. In: HICSS, pp. 1–9 (2011)

14. Pei, K., et al.: HERCULE: attack story reconstruction via community discovery on correlated log graph. In: ACSAC, pp. 583–595 (2016)

15. Wang, Q., Xu, J., Chen, H., He, B.: Two improved continuous bag-of-word models. In: IJCNN (2017)

16. Yen, T.F., et al.: Beehive: large-scale log analysis for detecting suspicious activity in enterprise networks. In: Proceedings of the 29th ACSAC, pp. 199–208 (2013)

NeoLOD: A Novel Generalized Coupled Local Outlier Detection Model Embedded Non-IID Similarity Metric

Fan Meng, Yang Gao$^{(\boxtimes)}$, Jing Huo, Xiaolong Qi, and Shichao Yi

State Key Lab for Novel Software Technology, Nanjing University, Nanjing, China
mengf.nju@gmail.com, {gaoy,huojing}@nju.edu.cn,
178668985@qq.com, shichaoyi@just.edu.cn

Abstract. Traditional generalized local outlier detection model (TraLOD) unifies the abstract methods and steps for classic local outlier detection approaches that are able to capture local behavior to improve detection performance compared to global outlier detection techniques. However, TraLOD still suffers from an inherent limitation for rational data: it uses traditional (Euclidean) similarity metric to pick out the context/reference set ignoring the effect of attribute structure. i.e., it is with the fundamental assumption that attributes and attribute values are independent and identically distributed (*IID*). To address the issue above, this paper introduces a novel Non-*IID* generalized coupled local outlier detection model (NeoLOD) and its instance (NeoLOF) for identifying local outliers with strong couplings. Concretely, this paper mainly includes three aspects: (i) captures the underlying attribute relations automatically by using the Bayesian network. (ii) proposes a novel Non-*IID* similarity metric to capture the intra-coupling and inter-coupling between attributes and attribute values. (iii) unifies the generalized local outlier detection model by incorporating the Non-*IID* similarity metric and instantiates a novel NeoLOF algorithm. Results obtained from 13 data sets show the proposed similarity metric can utilize the attribute structure effectively and NeoLOF can improve the performance in local outlier detection tasks.

Keywords: Local outlier detection · Non-*IID* · Attribute structure · Coupled similarity metric

1 Introduction

Traditional generalized local outlier detection model (TraLOD) provides a formalized method to improve understanding of the shared properties and the differences of local outlier detection algorithms for rational data in RDBMS[1] [1–5]. However, to grasp the meaning of "local" is probably the most difficult part

---

[1] RDBMS refers to the database management system based on the relational model.

© Springer Nature Switzerland AG 2019
Q. Yang et al. (Eds.): PAKDD 2019, LNAI 11439, pp. 587–599, 2019.
https://doi.org/10.1007/978-3-030-16148-4_45

that heavily depends on an effective similarity metric, which plays a pivot role to pick out the context/reference set of an object. Here, we will scrutinize the challenges of traditional generalized local outlier detection model in handling data by combining the classic LOF algorithm [1].

Problem Analysis. In general, TraLOD model [5] is coming with a fundamental assumption. It is with a *IID* assumption for all attributes and attribute values. i.e., the traditional (Euclidean) similarity metric that plays a pivotal role to indicate "locality" in local outlier detection tasks does not incorporate the inner attribute structure.

Therefore, the generalized model of Non-*IID* local outlier detection addressed in this study is based on the following three objectives:

- **Objective 1.** To detect the inner structure of attributes automatically rather than need strong prior domain knowledge to pick it out.
- **Objective 2.** To design a proper (Non-*IID*) similarity metric that utilizes the attribute structure to measure the similarity between objects rather than treat attributes and attribute values equally.
- **Objective 3.** To unify the (Non-*IID*) similarity metric into traditional generalized local outlier detection model rather than deal with it separately.

Our Design and Contributions. Based on the above analysis, solely relying on the similarity methods with the *IID* assumption may not enough to discriminate between outliers and normals in a local outlier detection task. Instead, exploring an attribute structure and embedding it into a proper similarity measure may yield better justifying.

This paper first captures the underlying attribute relations automatically by using the Bayesian network learning method. Secondly, based on the attribute structure, proposes the Non-*IID* similarity metric incorporating intra-coupling and inter-coupling between attributes and attribute values. Finally, unifies a novel Non-*IID* gEneralized cOupled based Local Outlier Detection model (NeoLOD) and its instance NeoLOF algorithm shown in Figs. 1(b) and 2. In summary, this paper includes the following three major contributions:

- We propose a novel NeoLOD model to abstract the methods in local outlier detection tasks. In contrast to TraLOD model, our framework is not "(*IID*) attributes + local outlier detection algorithm" as previous methods, but three stages as "Non-*IID* attribute structure learning + Non-*IID* coupled similarity metric + Non-*IID* local outlier detection algorithm".
- The performance of our model is verified by well-known methods and demonstrated with 13 data sets by three criteria: the effectiveness of the similarity metric, the accuracy in local outlier detection tasks and the stability test.
- The NeoLOD model can flexibly instantiate different classic local outlier detection algorithms and is easy to implement.

The paper is organized as follows. In Sect. 2, we briefly review the related work in local outlier detection tasks. Section 3 shows the preliminaries of the proposed model. Section 4 describes the details of the proposed method. Next, we summarize the results of experiments in Sect. 5. Finally, we conclude this paper in Sect. 6.

2   Related Work

Local outlier detection has been widely studied in literature from a number of different views [1–3,5]. Many objects can be considered as outliers with respect to their local neighbors rather than the entire data set. In a rather general sense, the very nature of local outlier detection requires the comparison of an object with a set of its neighbors w.r.t. some properties. Breuning et al. proposed the classic notion of local outlier detection (LOF) [1]. This algorithm assigns a local outlier factor for each data object to indicate its abnormal degree. LoOP [3] presented a density-based outlier scoring with a probabilistic, statistically-oriented approach. Further, Schubert et al. abstracted the notion of locality and proposed a generalized local outlier detection model [5].

Usually, most of the existing local outlier detection studies focus on finding outliers that are significantly different from the nearest neighbors of an object. However, the attribute structure of an object, which provides crucial information for outlier analysis, is often missed due to the complexity of capturing the attribute structure from a data set.

As far as we know, limited studies consider the attribute structure in local outlier detection, but conditional outlier detection and spatial outlier detection exhibit the similarity to some extent [6–8]. Wang and Davidson used random walks to find context and outliers [7]. It identifies a context as a 2-coloring of a random walk graph. Song et al. proposed the notion of conditional outliers to model the outliers manifested by a set of behavioral attributes (e.g. temperature) conditionally depending on a subset of attributes (e.g. longitude and latitude) [6].

While these studies of local outlier detection present two key challenges according to the scenarios of many real applications. Firstly, these methods tackle different local outlier detection tasks under the *IID* assumption and ignore the relations between attributes and attribute values. Secondly, most of those methods require prior knowledge to pre-defined the attribute structure. However, the strong prior knowledge restricts the practicability in many real-world applications. Therefore, it is a promising work to capture an attribute structure automatically and make full use of the Non-*IID* relations to detect the outliers.

More broadly, learning from Non-*IID* data is a recent topic to address the intrinsic data complexities, with preliminary work reported such as for metric similarity [9,10], and outlier detection [11,12]. However, it is seldom exploited in local outlier detection tasks. In this paper, we intend to explore Non-*IID* similarity metric method and embed it into the generalized local outlier detection model.

3   Preliminary

In this section, we develop a formal definition of Non-*IID* local outlier by incorporating a novel Non-*IID* similarity metric into local outlier detection tasks, which avoids the shortcomings presented in the previous section. The main notations in this paper are listed in Table 1.

Table 1. Summary of definitions and frequently used notions.

Notation	Description
$T = \{A_1, A_2, ..., A_n\}$	The base table in RDBMS. A_n is the nth attribute in table T
$X = \{x_1, x_2, ..., x_n\}$	A collection of n objects in table T
$v_n^{x_k}, v_n^{x_q}$	Specific values of attribute A_n for objects x_k, x_q
$c(x_k)$	Context function: a context set of an object x_k is used for model building
$f(x_k)$	Model function: assigns a model to each object $x_k \in X$ based on the set $c(x_k) \subseteq X$
deg	The threshold of local outlier degree which is defined to pick out the outliers

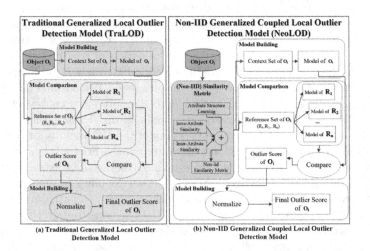

Fig. 1. The illustration shows the comparison between traditional generalized local outlier detection model (TraLOD) and Non-*IID* generalized coupled local outlier detection model (NeoLOD).

Definition 1. Non-*IID* Local Outlier. *Let T be a base table in RDBMS. Given a local outlier degree threshold deg > 0, and a series of model functions, context functions and similarity functions* $[(f_1, c_1, s_1), (f_2, c_2, s_2), ..., (f_i, c_i, s_i)]$, *the object is an outlier if the outlier degree $f_i(x_k) \geq deg$, where the $f_i(x_k)$ values are sorted in descend. i.e., these object x_q whose $f_i(x_q) \geq f_i(x_k)$ are called the Non-IID local outliers w.r.t its attribute structure.*

Note that context function c_i maps x_k to its context set $c_i(x_k, s(x_k)) \subseteq X$ by using the similarity function s_i and model function is $f_i(x_k, c_i(x_k, s_i(x_k)))$.

4   NeoLOD Model and Instantiation

In this section, we discuss the details of attribute structure learning, Non-IID coupled similarity metric (NeoDis) and NeoLOD model. Firstly, the attribute structure learning utilizes Bayesian network to capture the dependency of attributes and constructs an attribute dependency graph (AG) and attribute-value dependency triple (AVT) to represent the inner structure. Then, proposes a Non-IID similarity metric (NeoDis) incorporating the intra- and inter-coupling method to measure the similarity between objects by aggregating the attribute structure. Finally, introduces a novel NeoLOD model by embedding the similarity metric into TraLOD model and instantiates an algorithm, called NeoLOF.

4.1   Attribute Structure Learning

In this subsection, we capture the attribute structure by using the structure learning method of Bayesian network, whose nodes represent variables (attributes) and edges represent the direct dependent relationship between variables (attributes). According to the local Markov property, the probability distribution P over the variable set A can be factorized as follows:

$$P(A_1, A_2, ..., A_n) = \prod_{i=1}^{n} P(A_i | Pa(A_i)) \tag{1}$$

Note that $Pa(A_i)$ represents the set of parent nodes of a node A_i. Incorporating the classic MMHC algorithm for discovering the structure information [13], which first reconstructs the skeleton of a Bayesian network and then performs a Bayesian-scoring greedy hill-climbing search to orient the edges, we define of Attributes dependency Graph (AG) and Attribute-Value dependency Triple (AVT) to represent the attribute structure.

Definition 2. *Attributes Dependency Graph (AG)*. *Given a base table and its Bayesian network structure of attributes, the attributes dependency graph (AG) is defined as follows:*

$$AG(i, j) = \begin{bmatrix} W_{11} & W_{12} & \cdots & W_{1n} \\ W_{21} & W_{22} & \cdots & W_{2n} \\ \vdots & \vdots & \ddots & \vdots \\ W_{n1} & W_{n2} & \cdots & W_{nn} \end{bmatrix} \tag{2}$$

where W_{ij} is defined to represent the relationship of two attributes as follows:

$$W_{ij} = \begin{cases} 1, & \text{if } Pa(A_i) = A_j \\ 0, & \text{otherwise} \end{cases} \tag{3}$$

Definition 3. Attribute-Value Dependecy Triple (AVT). *Given the AG and an attribute value of an object x_k on attribute A_i (denoted $v_i^{x_k}$), the attribute-value dependency triple (AVT) is defined as follows:*

$$AVT = \{Pa(v_i^{x_k}), \delta(v_i^{x_k}), \varphi(v_i^{x_k}, Pa(v_i^{x_k}))\} \tag{4}$$

where $Pa(v_i^{x_k}) = \{v_j^{x_k}|W_{ij} = 1, j \in (1, 2, ..., n)\}$ is the parent set of attribute value $v_i^{x_k}$ and $\varphi(v_i^{x_k}, Pa(v_i^{x_k})) = p(v_i^{x_k}|Pa(v_i^{x_k}))$ measures the dependent degree of attribute values from parent attribute set, $\delta(v_i^{x_k}) = \frac{1}{2}(\frac{p_i(m) - p(v_i^{x_k})}{p_i(m)} + \frac{1}{p_i(m)})$ measures the dependent degree of attribute values within single attribute and $p_i(m)$ is the mode of the attribute A_i.

4.2   Neo-Based Coupled Similarity

Attribute Dependency Graph (AG) and Attribute Value Triple (AVT) have been successfully derived in 4.1. Inspired by [9], we defined a novel Non-*IID* similarity metric consisting of intra- and inter-attribute similarity to capture the relationship between attributes and attribute values based on the attribute structure.

Definition 4. Neo-Based Intra-attribute Similarity. *Given the AG and AVT, the Neo-based intra-attribute similarity between two attribute values $v_i^{x_k}$, $v_i^{x_q}$ of objects x_k and x_q on attribute A_i is defined as follows:*

$$S_{Ia}^i(v_i^{x_k}, v_i^{x_q}) = \begin{cases} 1, & \text{if } v_i^{x_k} = v_i^{x_q} \\ \frac{max(G_{x_k}^i, G_{x_q}^i)}{2 \cdot max(G_{x_k}^i, G_{x_q}^i) - min(G_{x_k}^i, G_{x_q}^i)}, & \text{otherwise} \end{cases} \tag{5}$$

Definition 5. Neo-Based Inter-attribute Similarity. *Given the AG and AVT, the inter-attribute similarity between two attribute values $v_i^{x_k}$ and $v_i^{x_q}$ of attribute A_i with its parent attribute values is defined as follows:*

$$S_{Ie}^i(v_i^{x_k}, v_i^{x_q}) = \begin{cases} 1, & \text{if } v_i^{x_k} = v_i^{x_q} \\ \frac{max(Q_{x_k}^i, Q_{x_q}^i)}{2 \cdot max(Q_{x_k}^i, Q_{x_q}^i) - min(Q_{x_k}^i, Q_{x_q}^i)}, & \text{otherwise} \end{cases} \tag{6}$$

Note that $G_{x_k}^i = \delta(v_i^{x_k})$ and $Q_{x_k}^i = \varphi(v_i^{x_k}, Pa(v_i^{x_k}))$. If the attribute values are identical, the similarity between them should be 1. The otherwise, their similarity depends on the dependency of attribute values. Further, we define the Neo-based attribute-value coupled similarity for attribute A_i.

Definition 6. Neo-Based Attribute-Value Coupled Similarity (Neo-AVS). *The Neo-based Attribute-value Coupled Similarity (NeoAVS) between attribute values $v_i^{x_k}$ and $v_i^{x_q}$ of attribute A_i is defined as follows:*

$$S^i(v_i^{x_k}, v_i^{x_q}) = \frac{1}{\theta \cdot \frac{1}{s_{Ia}^i} + (1-\theta) \cdot \frac{1}{s_{Ie}^i}} \tag{7}$$

Note that different $\theta \in [0,1]$ values reflect the different proportions of s_{Ia}^i and s_{Ie}^i in forming the overall object similarity. Finally, we calculate the similarity between two objects x_k and x_q defined in Eq. 8.

Definition 7. Neo-Based Coupled Metric Similarity (NeoDis). *The Neo-based coupled metric similarity (NeoDis) between two objects x_k and x_q is defined as follows:*

$$NeoDis(x_k, x_q) = \sum_{i=1}^{n} \alpha_i \cdot S^i(v_i^{x_k}, v_i^{x_q}) \tag{8}$$

where $\alpha_i \in [0,1]$ represents the weight of the coupled metric attribute value similarity of an attribute A_i and $\sum_{i=1}^{n} \alpha_i = 1$.

4.3 NeoLOD Model and Instantiation

Inspired by TraLOD [5], we propose a novel Non-*IID* generalized coupled local outlier detection model (NeoLOD) consisting of the above Non-iid similarity metric to capture the inner structure of attributes and pick out neighbors accurately. Figure 1 shows the overview of TraLOD Model and NeoLOD Model. Though not every component is actually used or presented in every instance of local outlier detection methods, many existing methods can be unified using this algorithmic framework. Further, we show its instance NeoLOF in Fig. 2.

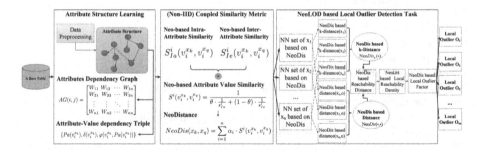

Fig. 2. The illustration of the instance NeoLOF. The proposed method first detects the attribute structure automatically using the method of MMHC; then defines a novel Non-*IID* coupled similarity metric by capturing the intra-coupling and inter-coupling of attributes and attribute values; final utilizes the NeoDis to derives the kNN sets and reachability distance and feeds it into a local outlier detection task.

Definition 8. *Non-IID Generalized Coupled Local Outlier Detection Model (NeoLOD).* NeoLOD model is a series of model functions, context functions, and similarity functions as follows:

$$[(f_0, c_0, s_0), (f_1, c_1, s_1), ..., (f_i, c_i, s_i)] \tag{9}$$

Note that the context function c_i is a function $c_i(x_k, s_i)$ (such as kNN) and maps objects x_k to their context set based on the similarity metric function $NeoDis(\cdot, \cdot)$. the model function $f_i(x_k, c_i)$ computes key properties. Further, based on the definition of NeoLOD model, its instance NeoLOF shows as follows:

$$NeoLOF(NeoDis, k) = [(maxdist_{0,NeoDis}, kNN_{NeoDis_k}, NeoDis),$$
$$(lrd_{0,1,NeoDis}, kNN_{NeoDis_k}, NeoDis), (mean_2, kNN_{NeoDis_k}, NeoDis), (frac_{3,2}, \varnothing, \varnothing)] \tag{10}$$

Algorithm 1. NeoLOF algorithm

Require: X - Data set, k - The number of neighbors, deg - Outlier threshold
Ensure: AG - Attributes Dependency Graph, AVT - Attribute-Value dependency
   Triple, O - Outlier set
1: **procedure** Neo-Similarity(X)
2:    **Compute** MMHC algorithm and derives the $AG \leftarrow W(i,j)$
3:    **Compute** $AVT \leftarrow Pa(v_i^{x_k}), \delta(v_i^{x_k})$ and $\varphi(v_i^{x_k}, Pa(v_i^{x_k}))$
4:    **for all** $x_k \in X$ set **do**
5:       **for all** $x_q \in X \setminus x_k$ set **do**
6:          **for all** $A_i \in A$ set **do**
7:             **Compute** $S_{Ia}^i(v_i^{x_k}, v_i^{x_q})$ for attribute A_i
8:             **Compute** $S_{Ie}^i(v_i^{x_k}, v_i^{x_q})$ for attribute A_i on $Pa(A_i)$
9:             **Compute** $S^i(v_i^{x_k}, v_i^{x_q})$ combining S_{Ia}^i and S_{Ie}^i
10:          **end for**
11:          **Compute** $NeoDis(x_k, x_q) \leftarrow \sum_{i=1}^{n} \alpha_i \cdot S^i(v_i^{x_k}, v_i^{x_q})$
12:       **end for**
13:    **end for**
14:    **Return** the set of all objects $NeoDis$.
15: **end procedure**
16: **procedure** Neo-LocalOutlierDetection($NeoDis, k, deg$)
17:    **for all** $x_k \in X$ set **do**
18:       **Compute** $NeoMaxdist(x_k), Neolrd(x_k)$, and $NeoLOF(x_k)$
19:       **if** $NeoLOF(x_k) > deg$ **then**
20:          $O \leftarrow x_k$
21:       **end if**
22:    **end for**
23:    **Return** the set of outlier O.
24: **end procedure**

Note that $NeoDis_k$ is used to derive the Neo-based k-distance of object x_k. The advantage of using the $NeoDis(\cdot, \cdot)$ instead of traditional (Euclidean) distance is that, if existing underlying attribute structure, the former can capture the effect of the relationship of attributes. The modified local outlier factor captures the outlier degree based on the attribute structure of a data set.

Algorithm 1 presents the main process of NeoLOF and includes two procedures: the first one is to derive the Non-*IID* similarity metric (NeoDis) between objects according to the definitions in Subsects. 4.1 and 4.2; the second one is to embed the NeoDis into LOF detection task and returns the set of outliers.

5   Experiments and Evaluation

In this section, we empirically evaluate the proposed NeoLOD model and its instance NeoLOF w.r.t. the following three criteria:

- **The similarity metric performance:** whether (Non-*IID*) similarity metric (NeoDis) enables a model to obtain better results.
- **The outlier detection performance:** whether the instance NeoLOF presents effectiveness in local outlier detection tasks compared to classic local outlier detection methods.
- **The stability performance:** whether the NeoLOF performance is stable under different parameter settings.

5.1   Experiment Environment

NeoLOF and its competitors were implemented in Python. The implementations of all the competitors are obtained from their authors or the open-source platform. All the experiments are executed on an i7-5600U CPU with 8 GB memory.

5.2   Experiment Design and Evaluation Method

According to the above analysis, we design the three experiments to verify the above three criteria as follows:

- **Similarity Performance.** The Non-*IID* similarity metric (NeoDis) is embedded into a kNN classification task, which is sensitive to distance measure and compared it with three distance measures, i.e., NeoDis-enabled Classification Performance with Euclidean distance, COS [14] and DILCA [15]. To demonstrate the NeoDis-enabled classification performance, we evaluate the performance by F-score, which is a combination of recall and precision.
- **Outlier Detection Accuracy.** The instance NeoLOF is evaluated against two well-known local outlier detection methods LOF [1], and LoOP [3]. We use the most popular evaluation methods to measure the local outlier detection performance: the area under ROC curve (AUC), precision at n, i.e., P@n (where we set n as the number of outliers in a data set), and recall at n, i.e., R@n (the fraction of total true anomalies the detected).
- **Stability Test.** We empirically evaluate the scalability of parameter k, which plays a pivot role to show the "locality" in local outlier detection tasks.

5.3 Data Sets

A 3-dimension synthetical data set and 12 public data sets from the UCI repository and the CMU statlib[2] are used to evaluate the detection performance. For each data set, we discretized the continuous attributes using liner bin if necessary when capturing the intra-coupling and inter-coupling of attributes and attribute values.

Before the experiments, several data sets are directly transformed from highly imbalanced classification data, where the smallest class is treated as outliers and the largest class is normal. The rest are randomly injected 1–5% objects as outliers. The cross-validation is taken to partition a data set to training and test sets in following experiments.

5.4 Results and Analysis

According to the above experiment design and evaluation methods, we report the empirical evaluation in three experiments as follows:

- **Similarity Performance.** The effectiveness of the similarity metric (NeoDis) is shown in Table 2. NeoDis-enabled Classification Performance obtains the best performance on 3 data sets; On average, it obtains about 8%, 5%, and 2% improvement over EucDis, COS, and DILCA. Note that without any domain knowledge, it is a promising work to identify and capture the relations using Non-*IID* methods.
- **Outlier Detection Accuracy.** We compare NeoLOF with 2 classic local outlier detection methods in terms of AUC, P@n, and R@n in Table 3. In terms of AUC, NeoLOF obtains the best performance on 9 data sets; and on average, it obtains about 1%, and 5% improvement over LOF, and LoOP, respectively. NeoLOF outperforms LOF and LoOP in AUC on 5 data sets. In terms of P@n, NeoLOF performs better than LOF and LoOP in 4 data sets and obtains more than 2% and 4% improvements on average, respectively. In terms of R@n, NeoLOF outperforms LOF and LoOP on 5 data sets; and on average, it obtains about 3% and 5% improvement over LOF, and LoOP, respectively. Further, we explore the reason why NeoLOF presents error-detection in the rest data sets. It seemly related to the incorrect attribute structure which amplifies the impact of structure information.
- **Stability Test.** We conduct the sensitivity test of the parameter k which is a parameter that controls the number of nearest-neighbors of a given object on 5 data sets as shown in Fig. 3. Here we illustrate representative trends in its AUC and R@n performance w.r.t. a wide range of k on these data sets due to space limits. NeoLOF shows stable performance in most of these data sets.

---

[2] They are downloaded from: http://archive.ics.uci.edu/ml/datasets.html; http://lib.stat.cmu.edu/index.php.

Table 2. KNN classification F-score with different distance measures on 4 data sets. $|N|$ and $|A|$ indicates the number of objects and attributes respectively.

Data sets	$|N|$	$|A|$	NeoDis	EucDis	COS	DILCA
Lymphography	148	18	0.77	0.80	0.78	**0.84**
Prim	339	17	**0.33**	0.15	0.23	0.27
Monk	432	6	**0.39**	0.32	0.35	0.35
BalanceScale	625	4	**0.28**	0.19	0.21	0.21
Avg.	386	11	**0.44**	0.36	0.39	0.42

Table 3. The results of AUC, P@n and R@n on 9 data sets.

	Stats			AUC			P@n			R@n		
Data sets	$|N|$	$|A|$	$|O|$	LOF	LoOP	NeoLOF	LOF	LoOP	NeoLOF	LOF	LoOP	NeoLOF
SynExample	505	3	5	0.65	0.69	**0.76**	0.56	0.60	**0.66**	0.55	0.60	**0.63**
Seeds	214	8	5	0.92	0.93	**0.94**	1.00	1.00	1.00	1.00	1.00	1.00
Bodyfat	262	15	10	0.93	0.91	0.92	1.00	1.00	1.00	0.90	0.90	0.86
Boston	516	14	10	0.89	0.80	0.87	1.00	1.00	1.00	0.45	0.48	**0.52**
HTRU	563	9	10	0.91	0.89	**0.93**	0.80	0.78	**0.85**	0.57	0.60	**0.66**
Houses	653	9	20	0.72	0.69	**0.78**	1.00	1.00	1.00	0.50	0.46	**0.53**
Credit	698	7	8	0.89	0.78	0.87	0.90	0.75	0.88	0.45	0.44	**0.47**
Energy	778	10	10	0.83	0.76	0.81	0.95	0.95	**0.97**	1.00	0.89	**0.96**
ElNino	1081	8	20	0.87	0.86	**0.89**	0.85	0.85	**0.88**	1.00	0.85	1.00
Avg.	586	9	11	0.85	0.81	**0.86**	0.90	0.88	**0.92**	0.71	0.69	**0.74**

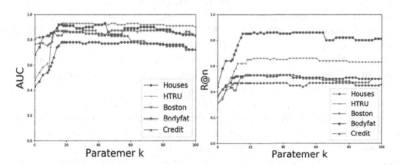

Fig. 3. Sensitivity Test of parameter k on 5 data sets.

6   Conclusions and Future Work

A local outlier detection task usually handles data with the *IID* assumption, which is inconsistent with the fact that attributes often exist an inner structure. This paper proposes a novel Non-*IID* similarity metric to capture the attribute structure and embed it into TraLOD model. The proposed approach includes three aspects: Firstly, it incorporates a Non-*IID* method to derive the attribute structure automatically by using Bayesian Learning method; Then, a novel

Non-*IID* similarity measure is proposed by capturing the intra-coupling and inter-coupling of attributes and attribute values; Finally, unifies a novel generalized coupled local detection model (NeoLOD) and its stance NeoLOF. Experiments show the effectiveness of the proposed model in three criteria: the first one is the performance of the similarity metric; the second one is the performance of local outlier detection; the final one is the stability under different parameter settings.

In the future, we are working on how to identify attribute structure on the larger scale and more complex data sets. One promising choice is to explore the problem of sequential outlier detection in RDBMS. Besides, we can exploit a deep structure model to generate the outliers according to a small amount of labeled data.

Acknowledgements. This work was supported by the National Key R&D Program of China (2017YFB0702600, 2017YFB0702601), the National Natural Science Foundation of China (61432008, U1435214, 61503178, 61806092) and Jiangsu Natural Science Foundation (BK20180326).

References

1. Breunig, M.M., Kriegel, H.-P., Ng, R.T., Sander, J.: LOF: identifying density-based local outliers. In: Proceedings of the 2000 ACM SIGMOD International Conference on Management of Data, pp. 1–12 (2000)
2. Ernst, M., Haesbroeck, G.: Comparison of local outlier detection techniques in spatial multivariate data. Data Min. Knowl. Discov. **31**(2), 371–399 (2017)
3. Kriegel, H.-P., Kröger, P., Schubert, E., Zimek, A.: LoOP: local outlier probabilities. In: Proceedings of the 18th ACM Conference on Information and Knowledge Management, pp. 1649–1652. ACM (2009)
4. Zhang, K., Hutter, M., Jin, H.: A new local distance-based outlier detection approach for scattered real-world data. In: Theeramunkong, T., Kijsirikul, B., Cercone, N., Ho, T.-B. (eds.) PAKDD 2009. LNCS (LNAI), vol. 5476, pp. 813–822. Springer, Heidelberg (2009). https://doi.org/10.1007/978-3-642-01307-2_84
5. Schubert, E., Zimek, A., Kriegel, H.-P.: Local outlier detection reconsidered: a generalized view on locality with applications to spatial, video, and network outlier detection. Data Min. Knowl. Discov. **28**(1), 190–237 (2014)
6. Song, X., Wu, M., Jermaine, C., Ranka, S.: Conditional anomaly detection. IEEE Trans. Knowl. Data Eng. **19**(5), 631–645 (2007)
7. Wang, X., Davidson, I.: Discovering contexts and contextual outliers using random walks in graphs. In: 2009 Ninth IEEE International Conference on Data Mining, ICDM 2009, pp. 1034–1039. IEEE (2009)
8. Zheng, G., Brantley, S.L., Lauvaux, T., Li, Z.: Contextual spatial outlier detection with metric learning. In: Proceedings of the 23rd ACM SIGKDD International Conference on Knowledge Discovery and Data Mining, pp. 2161–2170. ACM (2017)
9. Jian, S., Cao, L., Lu, K., Gao, H.: Unsupervised coupled metric similarity for non-IID categorical data. IEEE Trans. Knowl. Data Eng. **30**, 1810–1823 (2018)
10. Zhu, C., Cao, L., Liu, Q., Yin, J., Kumar, V.: Heterogeneous metric learning of categorical data with hierarchical couplings. IEEE Trans. Knowl. Data Eng. **30**, 1254–1267 (2018)

11. Chen, L., Liu, H., Pang, G., Cao, L.: Learning homophily couplings from non-IID data for joint feature selection and noise-resilient outlier detection. In: Proceedings of the Twenty-Sixth International Joint Conference on Artificial Intelligence, IJCAI-2017, pp. 2585–2591 (2017)
12. Pang, G., Cao, L., Chen, L., Liu, H.: Learning homophily couplings from non-IID data for joint feature selection and noise-resilient outlier detection. In: Proceedings of the 26th International Joint Conference on Artificial Intelligence, pp. 2585–2591. AAAI Press (2017)
13. Tsamardinos, I., Brown, L.E., Aliferis, C.F.: The max-min hill-climbing Bayesian network structure learning algorithm. Mach. Learn. $65(1)$, 31–78 (2006)
14. Wang, C., Dong, X., Zhou, F., Cao, L., Chi, C.H.: Coupled attribute similarity learning on categorical data. IEEE Trans. Neural Netw. Learn. Syst. $26(4)$, 781–797 (2015)
15. Ienco, D., Pensa, R.G., Meo, R.: From context to distance: learning dissimilarity for categorical data clustering. ACM Trans. Knowl. Discov. Data $6(1)$, 1–25 (2012)

Dynamic Anomaly Detection Using Vector Autoregressive Model

Yuemeng Li[1], Aidong Lu[1], Xintao Wu[2(✉)], and Shuhan Yuan[2]

[1] University of North Carolina at Charlotte, Charlotte, USA
{yli60,alu1}@uncc.edu
[2] University of Arkansas, Fayetteville, USA
{xintaowu,sy005}@uark.edu

Abstract. Identifying vandal users or attackers hidden in dynamic online social network data has been shown a challenging problem. In this work, we develop a dynamic attack/anomaly detection approach using a novel combination of the graph spectral features and the restricted Vector Autoregressive (rVAR) model. Our approach utilizes the time series modeling method on the non-randomness metric derived from the graph spectral features to capture the abnormal activities and interactions of individuals. Furthermore, we demonstrate how to utilize Granger causality test on the fitted rVAR model to identify causal relationships of user activities, which could be further translated to endogenous and/or exogenous influences for each individual's anomaly measures. We conduct empirical evaluations on the Wikipedia vandal detection dataset to demonstrate efficacy of our proposed approach.

Keywords: Anomaly detection · Vector autoregression · Granger causality · Dynamic graph · Matrix perturbation · Spectral graph analysis

1 Introduction

Anomalies and outliers refer to data points that behave differently from predefined normal behaviours. Detecting anomalies in a network under the dynamic setting belongs to sequential anomaly detection, where detection methods try to find abnormal observations from sequential data. There have been plenty of works studying the spectral properties of dynamic network data such as incremental spectral clustering [9], Nystrom low rank approximation [17], and matrix sketching [8], but there are few works on applying spectral analysis for anomaly detection on dynamic graphs. In [10], the authors derived a threshold based on the anomaly metric from the spectral features of the robust Principal Component Analysis for classification. Similarly, the authors in the work [4] proposed a threshold based on the anomaly metric derived from the principal eigenpairs of the associated adjacency matrix. In another work [11], the authors proposed using compact matrix decomposition (CMD) to compute the sparse low rank

© Springer Nature Switzerland AG 2019
Q. Yang et al. (Eds.): PAKDD 2019, LNAI 11439, pp. 600–611, 2019.
https://doi.org/10.1007/978-3-030-16148-4_46

approximations of the adjacency matrix. The approximation error of CMD and the observed matrix was used to quantify the anomaly. However, these works have two major shortcomings. First, instead of using statistical modeling approach to analyze the underlying structural correlations of the time series data systematically, each of the works only derived a threshold to evaluate the data points at each time frame individually. The other drawback is the lack of the ability to analyze the endogenous and/or exogenous causes for the observed anomalies. Since most relationship graphs are generated from the interaction information of the streaming OSN data, such interactions could cause the observed anomaly metrics to be correlated. Therefore, both endogenous and exogenous influences in the observed time series data need to be analyzed simultaneously so that the underlying casual relationships could be identified.

There exist extensive studies on the use of time series analysis methods such as Autoregressive (AR), Autoregressive Moving Average (ARMA), Vector Autoregressive (VAR), and Vector Autoregressive Integrated Moving Average (VARIMA) models in outlier detection [1,12]. However, their applications in anomaly detection in streaming online social network data have been limited. In this work, we propose to use the restricted Vector Autoregressive (rVAR) model to study the interactions and correlations of the observed anomaly measures of nodes. The fitted model then serves as the input for the subsequent casuality analysis. We adopt Granger causality [3] to analyze the fitted rVAR model and identify both endogenous and exogenous influences in the observed anomaly measures for each node.

To summarize, we incorporate the dynamic spectral features from the steaming network data with the rVAR model to develop an automatic fraud/attack analysis method. We develop a modified anomaly metric based on the node non-randomness measure derived from the adjacency spectral coordinates [15] to quantify how randomly nodes link to each other in signed and weighted graphs. We then propose to use the Granger causality analysis to identify the causal relationships amongst individuals. Several case studies on a partial WikiSigned dataset are conducted to demonstrate how the Granger causality analysis could be used to interpret the fitted rVAR model.

2 Preliminary

2.1 Graph Spectral Projections

$$
\alpha_u \rightarrow
\begin{array}{cccc}
\boldsymbol{v}_1 & \boldsymbol{v}_i & \boldsymbol{v}_K & \boldsymbol{v}_n \\
& \downarrow & & \\
\end{array}
\left(
\begin{array}{cc|cc|cc}
\boldsymbol{v}_{11} & \cdots & \boldsymbol{v}_{i1} & \cdots & \boldsymbol{v}_{K1} & \cdots & \boldsymbol{v}_{n1} \\
\vdots & & \vdots & & \vdots & & \vdots \\
\hline
\boldsymbol{v}_{1u} & \cdots & \boldsymbol{v}_{iu} & \cdots & \boldsymbol{v}_{Ku} & \cdots & \boldsymbol{v}_{nu} \\
\vdots & & \vdots & & \vdots & & \vdots \\
\boldsymbol{v}_{1n} & \cdots & \boldsymbol{v}_{in} & \cdots & \boldsymbol{v}_{Kn} & \cdots & \boldsymbol{v}_{nn}
\end{array}
\right)
\tag{1}
$$

For a given network, the spectral coordinates of nodes derived from the adjacency eigenspace are illustrated in Eq. (1). The eigenvalues $(\lambda_1, \cdots, \lambda_n)$ of a given adjacency matrix A are assumed to be in descending order when real. The corresponding eigenvectors $(v_1, \cdots, v_n)$ are sorted accordingly. The spectral decomposition of A takes the form $A = \sum_i \lambda_i v_i v_i'$. The row vector $\alpha_u = (v_{1u}, v_{2u}, \cdots, v_{Ku})$ is the spectral coordinate used for the projection of node u. The algebraic properties of the adjacency matrix are closely related to the underlying graph connectivity. Therefore, when the nodes are projected into the associated spectral space spanned by the chosen eigenvectors, such properties could be used to analyze the graph structure related problems. For example, [7,13] examined the line orthogonality of spectral coordinates formed by nodes from different clusters in the adjacency eigenspace and developed spectral clustering algorithms for community detection. Spectral coordinates were also used to detect random link attacks [16] and subtle anomalies [14] in social networks.

2.2 Non-randomness Measure

The node non-randomness is derived from the spectral coordinates to quantify how random a node is in terms of its connections in an unsigned network [15]. The edge and node non-randomness measures are defined as:

1. The edge non-randomness $R(w, u)$:

$$R(w, u) = \alpha_w \alpha_u' = \|\alpha_w\|_2 \|\alpha_u\|_2 \cos(\alpha_w, \alpha_u). \tag{2}$$

2. The node non-randomness $R(w)$:

$$R_w = \sum_{u \in \Gamma(w)} R(w, u), \tag{3}$$

where $\Gamma(w)$ denotes the set of neighbor nodes of w.

The measure was shown to effectively identify collaborative random link attacks in the spectral space of static unsigned graphs [16]. In this paper, we adapt the measure to the signed networks, apply it in the dynamic spectral space, and detect anomalies in streaming network data.

2.3 Vector Autoregression

Vector Autoregressive model is a time series analysis approach for analyzing multivariate data. It tires to capture the changes and interferences of multiple variables over time, where each variable is explained by the lagged values of itself and those of other variables. Equation (4) shows the general form of a n-variable VAR model with lag p:

$$\begin{pmatrix} x_{1,t} \\ \vdots \\ x_{n,t} \end{pmatrix} = \begin{pmatrix} c_1 \\ \vdots \\ c_n \end{pmatrix} + \sum_{i=1}^{p} \begin{pmatrix} \beta_{11,i} & \cdots & \beta_{1n,i} \\ \vdots & \ddots & \vdots \\ \beta_{n1,i} & \cdots & \beta_{nn,i} \end{pmatrix} \begin{pmatrix} x_{1,t-i} \\ \vdots \\ x_{n,t-i} \end{pmatrix} + \begin{pmatrix} \varepsilon_{1,t} \\ \vdots \\ \varepsilon_{n,t} \end{pmatrix}. \tag{4}$$

It can be written in a vector form as:

$$X_t = c + \sum_{i=1}^{p} \beta_i X_{t-i} + \varepsilon_t \equiv \Pi' Z_t + \varepsilon_t, \tag{5}$$

where $X_t = (x_{1,t}, \cdots, x_{n,t})'$, $\varepsilon_t = (\varepsilon_{1,t}, \cdots, \varepsilon_{n,t})'$, $c = (c_1, \cdots, c_n)'$ is the vector of constants, β_is are the matrices of parameters as shown in Eq. (4), $Z_t' = (1_{n \times 1}, X_{t-1}', \cdots, X_{t-p}')$, and $\Pi' = (c, \beta_1, \cdots, \beta_p)$. The existence of the estimators of the VAR model parameters requires that $np < T$, where n is the number of variables, p is the lag chosen, and T is the observation length.

3   Methodology

3.1   Overview

Formally, we model a dynamic network dataset as a sequence of graphs along the time dimension as G_t, where $t = 1, \cdots, T$. Each graph could be viewed as a snapshot of the network at time t. Hence, if we treat each snapshot at time t as a perturbation from the previous time $t-1$, the associated adjacency matrix can be written as $A_t = A_{t-1} + E_t$, where E_t contains the changes between two adjacent snapshots of G_{t-1} and G_t. There are three challenges involved in identifying dynamic attacks. The first challenge is to identify the correct snapshot time windows when the suspicious activities occur. The second challenge is to distinguish anomalies due to attacks and significant changes due to normal activities. The third challenge is to identify the endogenous and/or endogenous sources of the causes for the anomalies. Therefore, the task for detecting anomalies could be achieved by addressing the above challenges.

Dynamic networks focus on cognitive and social processes of users and can model the addition and removal of relations and interactions in networks. The dynamic changes of user activities are assumed to follow some particular probabilistic model such as the random walk or preferential attachment. When the perturbation E_t contains changes that deviate from the expected statistics under the assumed probabilistic model of normal behaviors, such events could be captured and treated as suspicious. We use the rVAR model to analyze the underlying correlations amongst individual's anomaly measures and make subsequent casuality inferences.

Algorithm 1 shows our algorithm for applying the rVAR method on streaming network data. The algorithm takes threes steps to complete the task. Firstly, the node nonrandomness measures are calculated from the spectral coordinates at each snapshot of the network. Secondly, for each target node, the chosen neighbors are incorporated to fit the rVAR(p) model. Johansen cointegration test from the work [5] is also performed at this step to prevent spurious regression in case where the associated time series data are integrated. Lastly, stepwise backward elimination Granger causality analysis is used to perform Granger casuality analysis of the node nonrandomness time series. For Algorithm 1, in lines 1–6, we calculate the node nonrandomness for each node at each network

---

Algorithm 1. *OSN_rVAR_Granger*: Anomaly analysis of the streaming OSN data using stepwise backward elimination Ganger casuality on the rVAR model of node nonrandomness measures

---

Input: $(A_1, \cdots, A_T), n, T, K, p, m$

Output: B, B_Ind

I/O: The inputs are the adjacency matrices $(A_1, \cdots, A_T)$, size of the users n, observation length T, number of the eigenpairs K, lag p, number of the steps of neighbors m, causality analysis method Arg. The outputs are parameters for fitted rVAR models B, and causality indicators B_Ind

---

1: **for** t from 1 to T **do**
2:     Compute eigenvectors $(v_1, \cdots, v_K)$ of A_t corresponding to the largest K eigenvalues $(\lambda_1, \cdots, \lambda_K)$;
3:     **for** w from 1 to n **do**
4:         Calculate the node nonrandomness score normalized by its number of connections $\check{R}_{w,t} = \frac{R_{w,t}}{\sum_{u \neq w} 1_{[A_{wu,t} \neq 0]}}$;
5:     **end for**
6: **end for**
7: **for** w from 1 to n **do**
8:     $S \leftarrow w \cup \Gamma(w)_m$;
9:     **for** u from the $m-$step neighbor set $\Gamma(w)_m$ **do**
10:         Perform Johansen cointegration test on the time series $\check{R}_{w,\cdot}$ and $\check{R}_{u,\cdot}$;
11:         **if** not cointegrated **then**
12:             $S \leftarrow S \setminus u$;
13:         **end if**
14:     **end for**
15:     Fit rVAR(p) model on the restricted set of nodes S with their corresponding time series $\check{R}_{s,\cdot}$, where $s \in S$;
16:     Extract $B_w \leftarrow (\beta'_{w,1}, \cdots, \beta'_{w,p})$;
17:     $B \leftarrow B_w$
18:     **for** Each $\beta \in B_w$ **do**
19:         Get P_β, the p-value for the F-statistic from the Granger causality test;
20:         **if** β is significant **then**
21:             $B\_Ind_{w,\beta} = 1$;
22:         **end if**
23:     **end for**
24:     $B\_Ind \leftarrow B\_Ind_w$
25: **end for**
26: Return B, B_Ind;

---

snapshot. In lines 8–14, we remove from the target node's neighbor set the nodes that are not cointegrated with the target node. We fit the rVAR model for each node in line 15 and evaluate its Granger causality in lines 18–24. The significance of the calculated F-statistic is determined by looking up the F-statistic table, where it is common to choose 0.05 alpha level.

The final outputs of the algorithm are two cell arrays, B, which contains the parameters for the rVAR model of all nodes, and B_Ind, which contains the corresponding indicators based on the causality analysis. Further inferences could be conducted based on the causality analysis results.

3.2 Adjusted Node Nonrandomness Measure

For signed and weighted graphs, the node nonrandomness measure in Sect. 2.2 may no longer be accurate, since the degree of a node can be negative. In order for the measure to work, we propose the following adjusted node nonrandomness.

Let w be a node and $\Gamma(w)$ be its neighbors from a signed graph. The adjusted node nonrandomness measure is

$$\check{R}_w = \frac{\sum_{u \in \Gamma(w)} R(w, u)}{\sum_{v \neq w} 1_{[A_{w,v} \neq 0]}}, \tag{6}$$

where $R(w, u)$ is the edge nonrandomness and A is the adjacency matrix.

This modification normalizes the node nonrandomness metric by its number of connections, since the edge nonrandomness metric holds true for directed signed and weighted graphs with an error term. If the node nonrandomness metric is normalized as shown above, the error term will shrink as well. As a result, the new metric could better approximate the true metric than the old metric would do.

Under the dynamic OSN setting, the past behaviours of nodes and their correlated ones could be incorporated in the rVAR model, so the influences of suspicious activities such as random link attacks could be studied through multiple snapshots of the network to provide an analysis over the time dimension. For a given node w, it has a sequence of observed node nonrandomness measures $(\check{R}_{w,1}, \cdots, \check{R}_{w,T})$ based on network snapshots. The observed values could change according to how the node and its neighbors act. By fitting the time series of any selected set of nodes into the rVAR model, we can identify the causal and dependency relationships amongst individuals' suspiciousness measures.

3.3 Variable and Model Selection

Due to the large sizes and long time spans of streaming network data, the observed time series data tend to cause the VAR model to have a large number of explanatory variables. Such datasets could cause the model to overfit. Hence, it is necessary to utilize some variable and model selection methods to obtain reliable and efficient estimation results. In this subsection, we explore the rVAR model which uses prior knowledge to regulate the parameters.

The rVAR method could take a binary restriction matrix $\dot{r}$ and remove variables corresponding to the zero locations. The vector form of the rVAR(P) model is:

$$X_t = \Pi'(\dot{r} \odot Z_t) + \varepsilon_t, \tag{7}$$

where $\dot{r}' = (1_{n \times 1}, r_1', \cdots, r_P')$.

When analyzing the network data, the restriction matrix could be the concatenation of any matrix representing the desired node connectivity such as

1-step or 2-step neighbor connectivity matrix at a specific lag. In applications, A_t, which is the most recent observed adjacency matrix for a given rVAR model, could be used as r_ps for $p \in (1, \cdots, P)$. The only drawback is that some previously unconnected nodes could be included in the model for some certain lags. However, as long as the number of added variables are small, those extra variables would not influence the model too much. Therefore, during the rVAR model estimation process, only the variables representing connected nodes could have nonzero parameters. As a result, the restricted model based on the network connectivity could greatly reduce the ambiguities caused by correlated variables representing disconnected nodes, reduce the number of variables entering the model, and reduce the risk of having rank deficient data.

3.4   Causal Analysis with Granger Causality

After fitting the rVAR model on each individual and its neighbors, the dependencies and casual relationships of their anomaly measures could be analyzed. The classical Granger causality test [3] is an F test to validate if by adding an extra explanatory variable could better explain the current response variable. That is, for models:

$$\text{Model 1:} \quad y_t = \alpha y_{t-1} + \varepsilon_t \tag{8}$$

$$\text{Model 2:} \quad y_t = \alpha y_{t-1} + \beta x_{t-1} + \varepsilon_t, \tag{9}$$

the hypothesis $H_0 : \beta = 0$ and $H_1 : \beta \neq 0$, are tested against each other. Then, the F-statistics

$$F = \frac{(RSS_1 - RSS_2)/(p_2 - p_1)}{RSS_2/(T - 1 - p_2)} \sim \mathcal{F}(p_2 - p_1, T - 1 - p_2),$$

where RSS_i and p_i are the residual sum of squares and the number of parameters of model i respectively, has a F-distribution with $(p_2 - p_1, T - 1 - p_2)$ degrees of freedom if the null hypothesis holds.

When H_1 holds, it simply suggests that X_{t-1} "Granger causes" Y_t, which means that it helps forecast Y_t, but it does not conclude that X_{t-1} causes Y_t. Two adaptations of Granger causality test for multivariate regressions are stepwise forward selection and stepwise backward elimination of the explanatory variables. In both cases, each variable's lagged terms are tested one by one using the models in Eqs. (8) and (9). Both methods use np tests in total, which are time consuming but can provide more specific casuality analysis for each individual at each lag. Based on the causality analysis results, the sources of endogenous and exogenous causes for each node's anomaly measures could be identified. Therefore, we can distinguish whether the node itself is anomalous or it is caused by adjacent neighbors's behaviours.

4   Empirical Evaluation

We conduct evaluations using a partial UMDWikipedia dataset from [6]. The dataset contains 770,040 edits of Wikipedia pages made by both vandal and benign users between January 01, 2013 and July 31, 2014. Since we focus on analyzing the dynamic interactions of the user behaviors, only pages edited by more than 3 unique users and users editing more than 3 unique pages are kept. After preprocessing, we have 17,733 edits and 805 users spanning over 10,451 unique event times, where there are 456 benign users and 349 vandal users.

We use this partial dataset to evaluate our dynamic anomaly detection algorithm. We use the rVAR(5), which is the rVAR model of lag 5, on 1-step neighbors for all the case study examples unless further specified. We use 5 time events as the interval to build the time series data, so we have 2,091 time frames. Due to space limits, we skip the results of showing the effectiveness of the derived node nonrandomness metric for signed networks and the efficiency of the use of the rVAR model. In this section, we focus on several case studies to demonstrate the effectiveness of using the Granger causality to identify the causes for the observed anomaly measures.

4.1   Case Study I

The target node is 7 and its 1-step neighbors are 48, 232, 281 and 378. The node anomaly measure variables are relabeled as X_1 to X_5 respectively. The adjusted model using stepwise backward elimination multivariate Granger causality analysis method is

$$
\begin{aligned}
X_{1,t} = {} & 0.288X_{1,t-1} + 0.443X_{1,t-3} - 0.648X_{1,t-4} \\
& + 0.56X_{3,t-4} + 0.467X_{1,t-5} - 0.186X_{3,t-5},
\end{aligned}
\tag{10}
$$

where $c = 0.00007$ is not significant with t-statistic value of 1.1679. The parameter vector and the associated significance indicators from Granger casuality results are shown in Fig. 1(a) and (b). For the parameter vector figures, each row represents the parameters for all the variables of a certain lag and each column represents the parameters of all the lags for a certain variable. For the causality indicators figures, each shaded location suggests that the corresponding column variable Granger causes the row variable. For example, Fig. 1(b) shows the exogenous sources of influence towards node 7 (X_1) include its early lags (except lag 2) and the lag 4 and 5 terms of the variable representing node 232 (X_3).

We also show the parameter vector for node 232 (X_3) in Fig. 1(c). The associated significance indicator grid is shown in Fig. 1(d). The causality result suggests that node 7 is also an exogenous source of cause for the anomaly measures of node 232. Therefore, the anomaly measures of node 7 and node 232 are closely correlated. By checking the original data, we find that user $Jodosma_7$ co-edited with user $Bnseagreen_{232}$ three times on the page titled "Dhani Matang Dev". Furthermore, none of the edits made by user 7 and user 232 were reverted. Therefore, both users are considered normal users who have edited the same page. The labels for both users are benign.

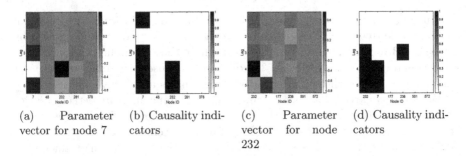

(a)   Parameter vector for node 7

(b) Causality indicators

(c)   Parameter vector for node 232

(d) Causality indicators

Fig. 1. (a) The parameters of all 5 lags for the rVAR model of node 7. (b) The anomaly measures of node 232 Granger cause those of node 7. (c) The parameters for the rVAR model of node 232. (d) The anomaly measures of node 7 Granger cause those of node 232.

4.2 Case Study II

The target node is 466 and its 1-step neighbors are 67, 312, 330, 421, 563, 605 and 683. The node anomaly measure variables are relabeled as X_1 to X_8 respectively. The adjusted model using stepwise backward elimination multivariate Granger causality analysis method is

$$X_{1,t} = 0.562X_{1,t-1} - 0.396X_{7,t-1} + 0.603X_{7,t-2} - 0.572X_{7,t-3} + 0.517X_{1,t-4}$$
$$- 0.231X_{6,t-4} - 0.387X_{7,t-4} - 0.229X_{8,t-4} + 0.119X_{1,t-5} + 0.014X_{6,t-5}$$
$$- 0.114X_{7,t-5} + 0.083X_{8,t-5},$$

$$(11)$$

where $c = 0.033$ is significant with t-statistic value of 2.3034. The parameter vector and the associated from Granger casuality indicators are shown in Fig. 2(a) and (b).

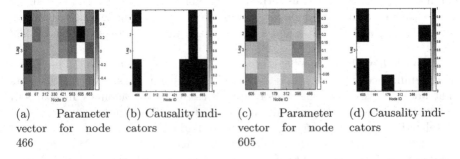

(a)   Parameter vector for node 466

(b) Causality indicators

(c)   Parameter vector for node 605

(d) Causality indicators

Fig. 2. (a) The parameters of all 5 lags for the rVAR model of node 466. (b) The anomaly measures of node 605 Granger cause those of node 466. (c) The parameters for the rVAR model of node 605. (d) The anomaly measures of node 466 Granger cause those of node 605.

In this example, both users $Grobelaar0811_{466}$ (X_1) and $Bobcalderon_{605}$ (X_7) edited the page titled "Sofia Vergara" together. In fact, there were co-edits at four different times between these two users. The casuality analysis suggests that the activities of user 563 Granger cause the anomaly measures of user 466.

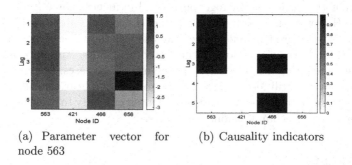

(a) Parameter vector for node 563

(b) Causality indicators

Fig. 3. (a) The parameters of all 5 lags for the rVAR model of node 563. (b) Node 466 is an exogenous source of influence to the anomaly measures of node 563.

By looking at the model and Granger Causality results for node 605 in Fig. 2(c) and (d), we notice that the causal relationship between user 466 and user 605 is bidirectional and conclude that they have a relatively close relationship. By further checking the edit history of both users, we find that all 7 edits of user 466 and all 5 edits of user 605 were reverted, which indicate that both users could be vandals with very high probability. Combining all the observations with the causality analysis results, we suspect that user 466 and user 605 may have attacked the page collaboratively. The labels for both users are vandal.

Another neighbor, user $Themaxandpeter_{563}$ also has similar causality results with node 466 as node 605 does, ss shown in Fig. 3(a) and (b). Nodes 466 and 563 edited two pages titled "Fulham F.C." and "Dynamo (magician)" together. According to the edit history, all of the edits made by user 563 were reverted. According to all the above results, we suspect that all 3 users 466, 563 and 605 collaborated their attacks on Wikipedia pages.

4.3   Case Study III

In this case study, we examine the rVAR(2) model of node 466 with its 2-step neighbors. Node 466 has 7 1-step neighbors and 115 2-step neighbors. The meaning for 2-step neighbors in this particular dataset is that those nodes edited some pages together with the 1-step neighbors of the target node. For any relatively well connected graph, the number of 2-step neighbors tends to grow very large. As mentioned before, fitting a rVAR model on a large number of variables requires high computational power and may have rank deficiency problems. After the rank test, only 73 neighbors and the target have time series data that are not linearly dependent. The Johansen cointegration test suggests that all 73 neighbors' time series are cointegrated with the target node's time series.

As more variables entering the model, the causality analysis results become more complicated. Since the model contains more variables, only a few lags could be incorporated to make the computational time manageable. The results indicate that the majority of exogenous causal influences come from the 2-step neighbors. However, the weights of the influences of any 2-step neighbors should be significantly lowered to reflect their weaker conductivities (in terms of distances) with node 466 than those of the 1-step neighbors. An important issue is that, if the number of variables entering the model increases, the chance for the model to be overfit will also increase. Therefore, in applications, the number of variables and the size of the lag need to be chosen carefully to prevent overfitting a model and to save computational resources.

5 Summary

We have presented a novel approach for dynamic fraud detection by using the rVAR model. We have introduced a new measure, node nonrandomness for signed and weighted graphs, to quantify node anomaly. We have also proposed to use the stepwise backward elimination Granger casuality method to analyze the casual relationships of node activities from the fitted rVAR model. To our knowledge, this is the first work to systematically analyze the graph spectrum based anomaly metric time series data simultaneously using a multivariate statistical modeling tool. This is also the first work to use a strict statistical inference method for identifying the endogenous and/or exogenous sources of casual influence of node interactions in dynamic graphs. As demonstrated in the case studies, by quantifying the randomness of node activities into the node nonrandomness measures and analyzing the resulting time series data, the proposed method could help us identify different activity patterns such as collaborative attacks, benign users sharing a common interest, and benign users being attacked by vandals.

For future works, we will explore different anomaly metrics under different scenarios for a more extensive coverage of anomaly detection. This is feasible as the method proposed in this paper is modularized where different types of anomaly metrics could be plug in to analyze different anomaly behaviors. We will also investigate the conditional Granger causality [2] to analyze the multivariable dependencies for dynamic attack detection. We plan to study how to improve the efficiency of our approach especially for large graphs.

Acknowledgments. This work was supported in part by NSF 1564250 and 1564039.

References

1. Bianco, A.M., Garcia Ben, M., Martinez, E., Yohai, V.J.: Outlier detection in regression models with arima errors using robust estimates. J. Forecast. **20**(8), 565–579 (2001)
2. Geweke, J.F.: Measures of conditional linear dependence and feedback between time series. J. Am. Stat. Assoc. **79**(388), 907–915 (1984)
3. Granger, C.W.: Investigating causal relations by econometric models and cross-spectral methods. Econometrica: J. Econometric Soc. **37**(3), 424–438 (1969)
4. Idé, T., Kashima, H.: Eigenspace-based anomaly detection in computer systems. In: Proceedings of the Tenth ACM SIGKDD International Conference on Knowledge Discovery and Data Mining, pp. 440–449. ACM (2004)
5. Johansen, S.: Estimation and hypothesis testing of cointegration vectors in Gaussian vector autoregressive models. Econometrica: J. Econometric Soc. **59**(6), 1551–1580 (1991)
6. Kumar, S., Spezzano, F., Subrahmanian, V.: VEWS: a Wikipedia vandal early warning system. In: Proceedings of the 21th ACM SIGKDD International Conference on Knowledge Discovery and Data Mining, pp. 607–616. ACM (2015)
7. Li, Y., Wu, X., Lu, A.: Analysis of spectral space properties of directed graphs using matrix perturbation theory with application in graph partition. In: Proceedings of IEEE International Conference on Data Mining, pp. 847–852. IEEE (2015)
8. Liberty, E.: Simple and deterministic matrix sketching. In: Proceedings of the 19th ACM SIGKDD International Conference on Knowledge Discovery and Data Mining, pp. 581–588. ACM (2013)
9. Ning, H., Xu, W., Chi, Y., Gong, Y., Huang, T.: Incremental spectral clustering with application to monitoring of evolving blog communities. In: Proceedings of the 2007 SIAM International Conference on Data Mining, pp. 261–272. SIAM (2007)
10. Shyu, M.L., Chen, S.C., Sarinnapakorn, K., Chang, L.: A novel anomaly detection scheme based on principal component classifier. In: Proceedings of the IEEE Foundations and New Directions of Data Mining Workshop, pp. 171–179. IEEE (2003)
11. Sun, J., Xie, Y., Zhang, H., Faloutsos, C.: Less is more: compact matrix representation of large sparse graphs. In: Proceedings of 7th SIAM International Conference on Data Mining (2007)
12. Tsay, R.S., Peña, D., Pankratz, A.E.: Outliers in multivariate time series. Biometrika **87**(4), 789–804 (2000)
13. Wu, L., Ying, X., Wu, X., Zhou, Z.H.: Line orthogonality in adjacency eigenspace with application to community partition. In: Proceedings of the 22nd International Joint Conference on Artificial Intelligence, pp. 2349–2354 (2011)
14. Wu, L., Wu, X., Lu, A., Zhou, Z.H.: A spectral approach to detecting subtle anomalies in graphs. J. Intell. Inf. Syst. **41**(2), 313–337 (2013)
15. Ying, X., Wu, X.: On randomness measures for social networks. In: Proceedings of 9th SIAM International Conference on Data Mining (2009)
16. Ying, X., Wu, X., Barbará, D.: Spectrum based fraud detection in social networks. In: 2011 IEEE 27th International Conference on Data Engineering, pp. 912–923. IEEE (2011)
17. Zhang, K., Tsang, I.W., Kwok, J.T.: Improved nyström low-rank approximation and error analysis. In: Proceedings of the 25th International Conference on Machine Learning, pp. 1232–1239. ACM (2008)

A Convergent Differentially Private k-Means Clustering Algorithm

Zhigang Lu[1] and Hong Shen[1,2(✉)]

[1] School of Computer Science, The University of Adelaide, Adelaide, Australia
{zhigang.lu,hong.shen}@adelaide.edu.au
[2] School of Data and Computer Science, Sun Yat-sen University, Guangzhou, China

Abstract. Preserving differential privacy (DP) for the iterative clustering algorithms has been extensively studied in the interactive and the non-interactive settings. However, existing interactive differentially private clustering algorithms suffer from a non-convergence problem, i.e., these algorithms may not terminate without a predefined number of iterations. This problem severely impacts the clustering quality and the efficiency of the algorithm. To resolve this problem, we propose a novel iterative approach in the interactive settings which controls the orientation of the centroids movement over the iterations to ensure the convergence by injecting DP noise in a selected area. We prove that, in the expected case, our approach converges to the same centroids as Lloyd's algorithm in at most twice the iterations of Lloyd's algorithm. We perform experimental evaluations on real-world datasets to show that our algorithm outperforms the state-of-the-art of the interactive differentially private clustering algorithms with a guaranteed convergence and better clustering quality to meet the same DP requirement.

Keywords: Differential privacy · Adversarial machine learning · k-means clustering

1 Introduction

Clustering algorithms help us to learn the insights behind the data. Nevertheless, the privacy releasing risk thwarts people's willingness to contribute personal data to the clustering algorithms. Assume there is an adversary who knows $n-1$ out of n items of a dataset, and a set of centroids at any arbitrary iteration, it is easy to infer the missing item with such information. To prevent such an adversary, differential privacy (DP) [4] has been applied in iterative clustering algorithm (i.e., Lloyd's algorithm [11]) in both the interactive and the non-interactive settings. In this paper, we consider the differentially private k-means clustering in the interactive settings [17] whereby random differential privacy (DP) noises were injected into each iteration when running Lloyd's algorithm.

In a nutshell, there are a long line of works [2,5,14,16–18] guarantee DP while achieve acceptable clustering quality in the interactive settings via three DP

© Springer Nature Switzerland AG 2019
Q. Yang et al. (Eds.): PAKDD 2019, LNAI 11439, pp. 612–624, 2019.
https://doi.org/10.1007/978-3-030-16148-4_47

mechanisms: the sample and aggregation framework of DP [15], the exponential mechanism of DP (ExpDP) [13], and the Laplace mechanism of DP (LapDP) [6]. We observed two weaknesses from existing work. Particularly, the work [14], with the sample and aggregation framework, showed unsatisfactory clustering quality due to its large amount of noises injection in both sampling and aggregation stage. The works [2,5,16–18] applied the ExpDP or the LapDP suffered from a non-convergence problem. That is, a predefined iteration number was required to terminate the clustering. This problem severely impacts the clustering quality (as the clustering result would have a large distance to the true centroids) and the efficiency of the algorithm (as it takes a large computational cost to determine this predefined parameter).

Therefore, to fill the gaps, the research challenge is to guarantee convergence and better clustering quality with the same DP requirement as existing work in the interactive settings of the k-means problem (take Lloyd's algorithm as the base clustering algorithm). In summary, our main contributions are:

- To the best of our knowledge, this is the first work to explore the convergence of a differentially private k-means clustering algorithm in the interactive settings.
- We propose a novel approach to inject DP noise into the iterations by applying the ExpDP. Our framework addresses the non-convergence problem in existing work by controlling the orientation of the convergence while keeping DP for each iteration.
- We mathematically evaluate the key properties of our differentially private k-means algorithm, i.e., the convergence, the convergence rate, the clustering quality, and the bound of DP.
- We experimentally evaluate the performance of clustering quality across various experimental settings on six real-world datasets. With the same DP guarantee, our approach achieves better clustering quality than the state-of-the-art differentially private k-means algorithms.

2   Related Work

In this section, we briefly summarise the long line of works on the differentially private k-means clustering [2,5,14,16–18] in the interactive settings. To sum up, these works guaranteed DP with three major mechanisms of DP: the Laplace mechanism (LapDP) [6], the sample and aggregation framework [15], and the exponential mechanism (ExpDP) [13].

There is a group of works [2,5,17] injected Laplace noise to the iterations of Lloyd's algorithm directly to ensure DP. The difference among these works is the way to allocate privacy budget to each iteration. Blum et al. [2] split the overall privacy budget uniformly to each iteration, prior to that, a total number of iterations was determined empirically. In spite of its simplicity, this scheme requires significant computational resources as it has to repeatedly run the algorithm on the target dataset to have a suitable number of iterations. Su et al. [17] improved the weaknesses of [2] by allocating the privacy budget with a theoretically guaranteed optimal allocation method. However, this optimal allocation scheme may not fit all real-world datasets, as it assumes that all the clusters always have the same size. Dwork [5] allocated the privacy budget with a decreasing exponential distribution, that is, assigned $1/2^i$ of the overall privacy budget at iteration i until using up the overall privacy budget. Unfortunately, this scheme results in unsatisfactory clustering quality since the injected noises keep increasing when the allocated privacy budget is decreasing.

The sample and aggregation framework and the ExpDP were also used to ensure DP for an interactive k-means algorithm. Mohan et al. [14] proposed GUPT applied the sample and aggregation framework of DP with Lloyd's algorithm. Briefly, GUPT uniformly samples items from an input dataset to different buckets, where local clustering result of each bucket is generated by Lloyd's algorithm. The final clustering result is the mean of those local ones with Laplace noise. Although GUPT is convergent, the clustering quality is unsatisfactory due to its large amount of noises involved in both sampling and aggregation steps. Zhang et al. [18] proposed a genetic algorithm (GA) based differentially private k-means algorithm, PrivGene. Unlike the traditional GA, PrivGene randomly sampled the candidates for the next iteration with the ExpDP rather than selecting the top-quality ones. PrivGene achieves fair clustering quality if the input dataset is relatively small because in this case, it produces global optimal clustering result with high probability. However, similar to [2], PrivGene also requires a predefined iteration number to terminate the algorithm. So efficiency would be a major problem to it. Differing from the above algorithms, Park et al. [16] achieved (ϵ, δ)-DP, rather than ϵ-DP, with given assumption on the distribution of the input dataset which narrows its applicability in the real-world scenarios.

3   Preliminaries

3.1   Lloyd's k-Means Algorithm

The k-means clustering aims to split a dataset with N items into k clusters where each item is allocated into a cluster with the nearest cluster centroid to itself. The formal cost function of the k-means problem is:

$$\underset{\mathbf{C}}{\arg\min}\, J = \sum_{i=1}^{k} \sum_{x \in C_i} ||x - S_i||^2, \tag{1}$$

where $\mathbf{C} = \{C_1, C_2, \ldots, C_k\}$ is the set of k clusters, x is an item in the dataset $X = \{x_1, x_2, \ldots, x_N\}$, S_i is the centroid of C_i. Equation 1 calculates the total cost of a set of centroids.

The most well known k-means algorithm is an iterative refinement algorithm called Lloyd's k-means algorithm [11]. In brief, Lloyd's algorithm improves the quality of centroids by iteratively running a *re-assignment* step and a *re-centroid* step. In the *re-assignment* step, it assigns each item to its nearest centroid to build the k clusters. In the *re-centroid* step, it re-calculates the centroid (mean) for each cluster. This new/updated k centroids are used for the next *re-assignment* step. Lloyd's algorithm terminates itself when the k centroids keep the same in two neighbouring iterations. Namely, this algorithm is guaranteed to converge within finite iterations.

3.2 Differential Privacy

Differential privacy (DP) is a famous notion of privacy in the current privacy-preserving research field [9,12], which was first introduced and defined by Dwork et al. [6]. It guarantees the presence or absence of any item in a dataset is concealed to the adversary.

Definition 1 (ϵ-DP [6]). *A randomised mechanism $\mathcal{T}$ is ϵ-differentially private if for all neighbouring datasets X and X', and for an arbitrary answer $s \in Range(\mathcal{T})$, $\mathcal{T}$ satisfies:*

$$\Pr[\mathcal{T}(X) = s] \leq \exp(\epsilon) \cdot \Pr[\mathcal{T}(X') = s]$$

where ϵ is the privacy budget.

Two parameters are essential to DP: the privacy budget ϵ and the local function sensitivity Δf, i.e. $\Delta f(X)$, where f is the query function to the dataset X. The reason why we use local sensitivity is that it offers better utility to respond query f when guaranteeing DP. Δf is calculated by the following equation,

$$\Delta f(X) = \max_{\forall X'} |f(X) - f(X')|.$$

In this paper, we mainly use two main mechanisms of DP: the LapDP and the ExpDP. In general, the LapDP adds random noise with Laplace distribution for the numeric computation to satisfy Definition 1. While for the non-numeric computation, the ExpDP introduces a scoring function $q(X,x)$ which reflects how appealing the pair (X,x) is, where X denotes a dataset and x is the random response to a query function on the dataset X. Technically, the ExpDP is a process of weighted sampling, where the scoring function assigns weights to the sample space.

Definition 2 (ExpDP [13]). *Given a scoring function of a dataset X, $q(X,x)$, which reflects the quality of query respond x. The exponential mechanism $\mathcal{T}$ provides ϵ-DP, if $\mathcal{T}(X) = \{\Pr[x] \propto \exp(\frac{\epsilon \cdot q(X,x)}{2\Delta q})\}$, where Δq is the sensitivity of $q(X,x)$, ϵ is the privacy budget.*

4 Algorithm and Analysis

4.1 Approach Overview

Figure 1 illustrates the overview of our approach. In $C_i^{(t)}$ (cluster i at iteration t), we use an $\hat{S}_i^{(t)}$ (the differentially private centroid of $C_i^{(t)}$) to replace the $S_i^{(t)}$

(the real centroid (mean) of $C_i^{(t)}$) as the result of *re-centroid*. Differing from the existing work where the $\hat{S}_i^{(t)}$ was arbitrarily produced by a DP mechanism, the key idea of our approach is that we bound a *sampling zone* to sample the $\hat{S}_i^{(t)}$ for the sake of convergence. To have a good convergence rate, we further control the orientation when sampling the $\hat{S}_i^{(t)}$ from the *sampling zone*. Specifically, the sampled $\hat{S}_i^{(t)}$ orients to an $S_i^{(t+r_t)}$ as close as possible, where r_t is the *orientation controller* at iteration t, the $S_i^{(t+r_t)}$ is produced by running Lloyd's algorithm for r_t times from the $S_i^{(t-1)}$ (which is actually the $\hat{S}_i^{(t-1)}$). The challenge in our approach to fill the research gap is designing a suitable *sampling zone* to guarantee the convergence and better clustering quality while meeting the same DP requirement in the interactive settings as existing work.

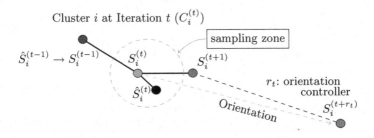

Fig. 1. Overview of orientation controlling.

4.2 Preliminary Analysis

In this section, we provide the preliminary analysis which help us to build up the our algorithm in the next section. We first study the convergence for a randomised iterative clustering algorithm in Lemma 1.

Lemma 1. *A randomised iterative clustering algorithm is convergent if, in $C_i^{(t)}$, the sampled $\hat{S}_i^{(t)}$ satisfies $||\hat{S}_i^{(t)} - S_i^{(t)}|| < ||S_i^{(t)} - S_i^{(t-1)}||$ in Euclidean distance, $\forall\, t,\, i$.*

Proof. In Lloyd's algorithm, after *re-assignment*, prior to *re-centroid*, we build $C_i^{(t)}$ and have $J^{(S_i^{(t-1)})} = \sum_{x \in C_i^{(t)}} ||x - S_i^{(t-1)}||^2$, where $S_i^{(t-1)}$ is the mean of $C_i^{(t-1)}$ which is used for the *re-assignment* to generate $C_i^{(t)}$. Similarly, after *re-centroid* (members in $C_i^{(t)}$ did not change), we have $J^{(S_i^{(t)})} = \sum_{x \in C_i^{(t)}} ||x - S_i^{(t)}||^2$.

Assume the Euclidean distance between $S_i^{(t-1)}$ and $S_i^{(t)}$ is $a_i^{(t)} = ||S_i^{(t-1)} - S_i^{(t)}||$, then we have $J^{(S_i^{(t-1)})} - J^{(S_i^{(t)})} = ||C_i^{(t)}|| \times (a_i^{(t)})^2$, where $||C_i^{(t)}||$ is the number of items in $C_i^{(t)}$. Note that, in Lloyd's algorithm, $J^{(S_i^{(t)})}$ is the minimum cost in $C_i^{(t)}$. If we pick a random node $\hat{S}_i^{(t)}$ from $C_i^{(t)}$ as the centroid for $C_i^{(t)}$

which satisfies $||\hat{S}_i^{(t)} - S_i^{(t)}|| = \delta_i^{(t)} a_i^{(t)} < ||S_i^{(t-1)} - S_i^{(t)}|| = a_i^{(t)}$ $(0 < \delta_i^{(t)} < 1)$, then we have $J^{(S_i^{(t-1)})} - J^{(\hat{S}_i^{(t)})} = ||C_i^{(t)}|| \times (1 - (\delta_i^{(t)})^2) \times (a_i^{(t)})^2 > 0$.

So by updating the centroids to this set $\hat{S}^{(t)} = \{\hat{S}_1^{(t)}, \hat{S}_2^{(t)}, \ldots, \hat{S}_k^{(t)}\}$ (rather than the mean of clusters, $S^{(t)}$), the value of every item $\sum_{x \in C_i} ||x - S_i||^2$ can be further decreased, then the value of the cost function (Eq. 1) can be further decreased.

In addition, since we have a finite set of all possible clustering solutions (at most k^N), and we decrease the cost in each iteration of a randomised iterative algorithm, the algorithm satisfies the properties from the above proof must converge (not approach) to a fixed value of the cost function.   □

According to Lemma 1, in this paper, we have a *convergent zone* $= \{Node\ S : ||S - S_i^{(t)}|| < ||S_i^{(t-1)} - S_i^{(t)}||\}$, where $S_i^{(t)}$ is the mean of $C_i^{(t)}$, and a *sampling zone* is a subset of the *convergent zone*.

Next, we shall study the convergence and the convergence rate for a special case of $\hat{S}_i^{(t)}$ in Lemmas 2 and 3, respectively. This special $\hat{S}_i^{(t)}$ is in the line segment of $\overline{S^{(t-1)}S^{(t)}}$, where $||\hat{S}_i^{(t)} - S_i^{(t)}|| = \delta_i^{(t)} \times ||S_i^{(t-1)} - S_i^{(t)}||$, $\delta_i^{(t)} < 1$. These two lemmas assist us to prove Theorems 1 and 2 in the next section.

Lemma 2. *An algorithm ALG, randomly selects an $\hat{S}_i^{(t)}$ in the line segment of $\overline{S^{(t-1)}S^{(t)}}$ in $C_i^{(t)}$, and Lloyd's algorithm are convergent to the same set of centroids, if they both use the same initial set of centroids.*

Proof. We know that the k-means problem has a set of local optimal solutions, $\mathbb{S} = \{\mathbf{S}_1, \mathbf{S}_2, \cdots, \mathbf{S}_n\}$, where $\mathbf{S}_i$ is one local optimum (the one that Lloyd's algorithm converges to) contains k centroids of the clusters, $\mathbf{S}_i = \{S_{i,1}, S_{i,2}, \cdots, S_{i,k}\}$. According to Lemma 1, assume ALG is convergent to $\hat{\mathbf{S}} = \{\hat{S}_1, \hat{S}_2, \cdots, \hat{S}_k\} \notin \mathbb{S}$. Then we must have room to further reduce the cost by either the *re-assignment* or the *re-centroid*. Therefore, $\hat{\mathbf{S}}$ is not the set of centroids which makes ALG convergent, unless $\hat{\mathbf{S}} \in \mathbb{S}$. So ALG is convergent to, at least, one local optimum of the k-means problem.

We say a set of k nodes (each cluster contributes one node) belongs to a local optimum, $\mathbf{S}_i$, if Lloyd's algorithm converges to $\mathbf{S}_i$ by taking such a set of nodes as the initial set of centroids. Because the two ends of the line segment $\overline{S^{(t-1)}S^{(t)}}$ belong to the same local optimum, then it is guaranteed that $\mathbf{S}^{(t-1)}$, $\mathbf{S}^{(t)}$, and $\hat{\mathbf{S}}^{(t)}$ always belong to the same local optimum, for all iterations. Therefore, this lemma holds.   □

Lemma 3. *The algorithm ALG in Lemma 2 has at most $\frac{1}{1-\delta^2}$ times iterations of Lloyd's algorithm in the expected case, where δ is the expectation of $\delta_i^{(t)}$, $\delta \in (0, 1)$.*

Proof. Based on Lemma 2, the overall value difference of Eq. 1 from the first iteration to the last iteration, $J = \sum_{i=1}^k J^{S_i^{(0)}} - \sum_{i=1}^k J^{S_i^{(I)}}$, is the same in both ALG and Lloyd's algorithm, where I is the total iterations. In each iteration, the cost is decreased by two steps: *re-assignment* and *re-centroid*. Then, without loss of generality, we have $J = \sum_{t=1}^{I}(\Delta_t^{(ra)} + \Delta_t^{(rc)})$ for Lloyd's algorithm, and $J = \sum_{t=1}^{\hat{I}}(\hat{\Delta}_t^{(ra)} + \hat{\Delta}_t^{(rc)})$ for ALG. Because of the properties of Lloyd's algorithm, we know that $\Delta_t = \sum_{i=1}^{k}\Delta_i^{(t)}$ for all clusters at iteration t. According to Lemma 1, when *re-assignment*, we have $\hat{\Delta}_i^{(t)} = (1-\delta^2) \times \Delta_i^{(t)}$, where $\delta = \mathbb{E}(\delta_i^{(t)})$. So $\hat{\Delta}_t^{(ra)} = \sum_{i=1}^{k}[(1-\delta^2)\times\Delta_i^{(t)}] \in [\min_{i=1}^{k}\{1-(\delta_i^{(t)})^2\}, \max_{i=1}^{k}\{1-(\delta_i^{(t)})^2\}] \times \Delta_t^{(ra)}$. In the expected case, $\hat{\Delta}_t^{(ra)} = (1-\delta^2) \times \Delta_t^{(ra)}$, $\delta = \mathbb{E}(\delta_i^{(t)})$. In the worst case, $\hat{I} < \frac{1}{\min_{i,t}\{1-(\delta_i^{(t)})^2\}} \times I$. As $\hat{\Delta}_t^{(rc)} > \Delta_i^{(rc)}$, we have

$$J = (\overline{\Delta^{(ra)}} + \overline{\Delta^{(rc)}}) \times I = (\overline{\hat{\Delta}^{(ra)}} + \overline{\hat{\Delta}^{(rc)}}) \times \hat{I} > [(1-\delta^2)\overline{\Delta^{(ra)}} + \overline{\Delta^{(rc)}}] \times \hat{I}$$
$$> (1-\delta^2) \times (\overline{\Delta^{(ra)}} + \overline{\Delta^{(rc)}}) \times \hat{I}.$$

Therefore, $\hat{I} < \frac{1}{1-\delta^2} \times I$ in the expected case.   □

4.3 Our Approach and Its Analysis

In this section, we firstly show how we build our *sampling zone*, then describe how we sample the $\hat{S}_i^{(t)}$ from such a *sampling zone*, finally provide our approach in Algorithm 1 and its analysis.

Ideally, in our *convergent zone*, when applying LapDP, the probability of a node S as the $\hat{S}_i^{(t)}$ need follow a monotonous decreasing function of the distance between S and $S_i^{(t)}$. However, a truncated LapDP [1] will result in an unsatisfactory utility problem as the nodes in the border of the *convergent zone* may have a higher probability (sum of the probabilities of the nodes outside the *convergent zone*) than the ones more close to the $S_i^{(t)}$. Therefore, in this paper, we apply the ExpDP in the *sampling zone* for the $\hat{S}_i^{(t)}$.

When designing a *sampling zone*, we should have the following conditions. Firstly, there should be a single *sampling zone* in $C_i^{(t)}$ for all parties: the trusted data curator and the adversaries. Otherwise the differences among the *sampling zones* in different parties will result in significant gap among their clustering results, which could be used for privacy inference. Secondly, the single *sampling zone* should not have an explicit relationship to the $S_i^{(t)}$, the real mean of $C_i^{(t)}$. Thirdly, to control the convergence orientation, $S_i^{(t+r_t)}$ should be involved when building the *sampling zone*. In this paper, considering the computational cost and privacy disclosure risk, we choose the *orientation controller* $r_t = 1$.

Lines 7 to 9 in Algorithm 1 show how we build the *sampling zone*. In our implementation, the centre of the *sampling zone* $P_i^{(t)}$ is determined by a random number $\lambda_i^{(t)} \in (1/2, 1)$ which is the off-set in $\overline{S_i^{(t)}S_i^{(t+1)}}$. Because a larger *sampling zone* provides more choices for the $\hat{S}_i^{(t)}$, we use the following probability function for sampling $\lambda_i^{(t)}$ as $\Pr[P_i^{(t)}] = \Pr[\lambda_i^{(t)} = r] \propto \exp(2 - 2r) = p, r \in (1/2, 1)$.

Once having the *sampling zone*, each party samples their own $\hat{S}_i^{(t)}$ from this *sampling zone* with the ExpDP. In the implementation, we sample the $\hat{S}_i^{(t)}$ by sampling a pair $(\delta_i^{(t)} = \frac{||S_i^{(t)} - \hat{S}_i^{(t)}||}{||S_i^{(t)} - S_i^{(t+1)}||}, \alpha_i^{(t)} = \angle \hat{S}_i^{(t)} S_i^{(t)} S_i^{(t+1)})$, where $\delta_i^{(t)} \in (0, 1)$, $\alpha_i^{(t)} \in (-\pi/2, \pi/2)$. Because an $\hat{S}_i^{(t)}$, that is close to the $S_i^{(t)}$, has better clustering quality for iterations in the interactive settings, that is, the scoring function should be monotonous decreasing to both $\delta_i^{(t)}$ and $\alpha_i^{(t)}$. In this paper, we use the following scoring function for the pair $(\delta_i^{(t)}, \alpha_i^{(t)})$ because of its simplicity: $q(\delta_i^{(t)}, \alpha_i^{(t)}) = (1 - \delta_i^{(t)}) + (1 - 2|\alpha_i^{(t)}|/\pi)$. It is easy to have the local sensitivity of the scoring function is 2, i.e. $\Delta q = 2$.

Finally, when the clusters converge (to a real local optimum as Lloyd's algorithm), we apply the LapDP to inject noise to the final clustering result. Specifically, to have good clustering quality, we inject the Laplace noise to the counts when calculating the mean of each cluster (Line 15 in Algorithm 1). The local sensitivity of this counting function is 1. Algorithm 1 shows how our approach works. Figure 2 depicts the key idea of Algorithm 1.

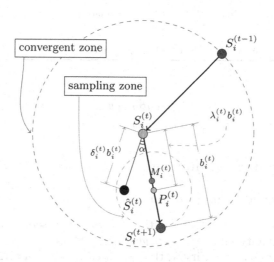

Fig. 2. The key idea of Algorithm 1: centroids updating.

According to Lemmas 1, 2, and 3, we have Theorems 1 and 2 to study the convergence and the convergence rate of Algorithm 1, respectively. Theorem 3 studies the privacy bound of Algorithm 1.

Theorem 1. *Algorithm 1 and Lloyd's algorithm are convergent to the same set of centroids with high probability if they both use the same initial set of centroids.*

Theorem 2. *Algorithm 1 is convergent with at most* $\frac{2}{-\delta^2 + 2\delta\cos\alpha + 1} \in (1, 2)$ *times iterations of Lloyd's algorithm in the expected case, where δ and α are the expectations of $\delta_i^{(t)}$ and $\alpha_i^{(t)}$.*

Algorithm 1. A Convergent Differentially Private k-Means Clustering Algorithm.

Input : $X = \{x_1, x_2, \ldots, x_N\}$: dataset in size N.
$\qquad\quad$ k: number of clusters $(< N)$.
$\qquad\quad$ $\epsilon_i^{(t)}$: privacy budget for Cluster i at Iteration t, $C_i^{(t)}$.
$\qquad\quad$ ϵ_0: privacy budget for the final output.
$\qquad\quad$ $\Pr[P_i^{(t)}]$: probability to generate $SamplingZone_i^{(t)}$ for $C_i^{(t)}$.
$\qquad\quad$ q: scoring function for the ExpDP when sampling the $\hat{S}_i^{(t)}$.
Output: S: set of the final k centroids.

1   Initialisation: Uniformly sample k initial centroids $\mathbf{S}^{(0)} = (S_1^{(0)}, S_2^{(0)}, \ldots, S_k^{(0)})$ from X;
2   **while** *clusters do not converge* **do**
3   **for** *each Cluster i at Iteration t* **do**
4         $C_i^{(t)} \leftarrow$ assign each x_j to its closest centroid $S_i^{(t-1)}$;
5         $S_i^{(t)} \leftarrow$ mean of $C_i^{(t)}$;
6         $S_i^{(t+1)} \leftarrow$ mean of $C_i^{(t+1)}$ based on $C_i^{(t)}$ (by running Lloyd's algorithm);
7         $M_i^{(t)} \leftarrow$ midpoint of the line segment $\overline{S_i^{(t)} S_i^{(t+1)}}$;
8         $P_i^{(t)} \leftarrow$ sample from the line segment $\overline{M_i^{(t)} S_i^{(t+1)}}$ by $\Pr[P_i^{(t)}]$;
9         $SamplingZone_i^{(t)} \leftarrow$ centre: $P_i^{(t)}$, radius: $r_i^{(t)} = ||S_i^{(t+1)} - P_i^{(t)}||$;
10        $\hat{S}_i^{(t)} \leftarrow$ sample from $SamplingZone_i^{(t)}$ by the ExpDP with q and $\epsilon_i^{(t)}$;
11        $S_i^{(t)} \leftarrow \hat{S}_i^{(t)}$;
12   Publish: $SamplingZone_i^{(t)}$, q, $\epsilon_i^{(t)}$, $S_i^{(t)}$(optional);
13   **end**
14 **end**
15 $\mathbf{S} \leftarrow$ add noise to $\mathbf{S}^{(t)}$ by the LapDP with ϵ_0, publish ϵ_0.

Theorem 3. *Algorithm 1 is ϵ-differentially private, where $\epsilon = \epsilon_0 + \sum_{t=1}^{\hat{I}} \max_{i=1}^{k} \{\epsilon_i^{(t)}\}$, $\hat{I}$ is its total iterations to converge.*

5 Experimental Evaluation

5.1 Datasets and Configuration

Table 1 illustrates the key features of the real-world datasets we used to evaluate the clustering quality and the convergence rate of Algorithm 1. We use these datasets with two reasons. Firstly, they are used for the clustering experiments in several research papers, e.g., [18] and [17]. Secondly, their sizes are in different orders of magnitude, which help us to show the performance stability and the scalability of an algorithm over different datasets.

We compare the clustering quality of Algorithm 1 with that of the state-of-the-art ϵ-differentially private k-means algorithms and the non-private Lloyd's algorithm. The clustering quality is measured by the difference/gap of the cost (Eq. 1) at the final iteration between a differentially private k-means algorithm and Lloyd's algorithm. A smaller gap indicates better clustering quality. In the experiments, we implement and name them as ThisWork (Algorithm 1), SU [17], PrivGene [18], GUPT [14], DWORK [5], BLUM [2], and LLOYD [11].

Because the six algorithms achieve ϵ-DP are randomised, we report their expected clustering quality. According to the law of large numbers, we run all

Table 1. Descriptions of datasets.

Dataset	#Records	#Dims	#Clusters
Iris [3]	150	4	3
House [8]	1837	3	3
S1 [7]	5000	2	15
Birch2 [19]	12000	2	5
Image [8]	34112	3	3
Lifesci [10]	26733	10	3

the seven algorithms 300 times and take the average results as the expectations. The initial set of centroids is randomly selected for all methods in each run. For those relying on a predefined iteration number, we take the corresponding value (or function) from the original papers. In addition, we normalise the data in all the datasets to $[0, 1]$. Furthermore, we normalise the final cost for all involved algorithms, i.e., the final cost of Lloyd's algorithm is always one.

Note that calculating the overall privacy budget depends on whether a method converges. ThisWork and GUPT calculate the overall privacy budget bottom-up. That is, once it terminates, we sum all the privacy budgets used in each iteration to have the overall privacy budget. SU, PrivGene, DWORK, and BLUM calculate it top-down. Namely, the given overall privacy budget is split to each iteration (the overall number of iterations were predefined) at the initialisation step. Therefore, in the experiments, we first allocate the same privacy budget to each atom step for ThisWork and GUPT, then calculate their overall privacy budgets. Next we take the overall privacy budget of ThisWork as the overall privacy budget for the methods cannot converge. In the experiments, local sensitivity is applied for all DP algorithms.

5.2   Experimental Results

Figure 3 reports the expected clustering quality of each algorithm, where the cost gap is in log scale, the privacy budget is varied in $[0.1, 1.0]$. Generally, Algorithm 1 shows a better clustering quality with the same DP requirement in the six datasets. Additionally, the performance gap between Algorithm 1 and the existing algorithms increases when increasing ϵ, which indicates the better trade-off between privacy and utility with our algorithm. Furthermore, Algorithm 1 performs much better than other algorithms in the larger datasets (e.g., Image and Lifesci), which reflects the potentially good scalability of our algorithm.

Figure 4 shows the iteration ratio of Algorithm 1 and Lloyd's algorithm to converge, which confirms the theoretical analysis in Theorem 2. Note that, in the experiments, the privacy budget does not impacts the number of iterations significantly because the experimental performance of the ExpDP is not as good as its theoretical guarantee with a relatively small *sampling zone*.

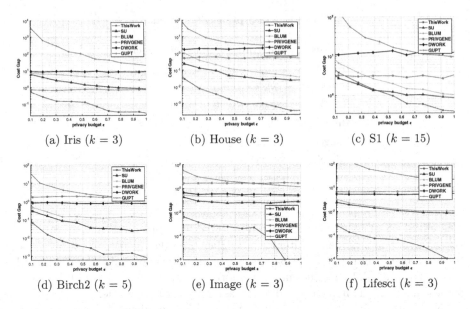

(a) Iris ($k = 3$)   (b) House ($k = 3$)   (c) S1 ($k = 15$)

(d) Birch2 ($k = 5$)   (e) Image ($k = 3$)   (f) Lifesci ($k = 3$)

Fig. 3. Clustering quality comparisons.

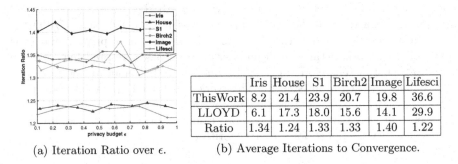

(a) Iteration Ratio over ϵ.

	Iris	House	S1	Birch2	Image	Lifesci
ThisWork	8.2	21.4	23.9	20.7	19.8	36.6
LLOYD	6.1	17.3	18.0	15.6	14.1	29.9
Ratio	1.34	1.24	1.33	1.33	1.40	1.22

(b) Average Iterations to Convergence.

Fig. 4. Iterations to convergence.

6 Conclusion

To address the non-convergence problem in the existing differentially private k-means algorithms in the interactive settings, in this paper, we proposed a novel centroids updating approach by applying the exponential mechanism of differential privacy in a selected area. The novelty of our approach is the orientation controlling for the sake of convergence. Both mathematical and experimental evaluations support that with the same DP guarantee, our algorithm achieves better clustering quality than the state-of-the-art differentially private algorithms in the interactive settings.

Acknowledgements. The authors would like to thank the anonymous reviewers for their valuable comments. This work is supported by Australian Government Research Training Program Scholarship, Australian Research Council Discovery Project DP150104871, National Key R & D Program of China Project #2017YFB0203201, and supported with supercomputing resources provided by the Phoenix HPC service at the University of Adelaide. The corresponding author is Hong Shen.

References

1. Andrés, M.E., Bordenabe, N.E., Chatzikokolakis, K., Palamidessi, C.: Geo-indistinguishability: differential privacy for location-based systems. In: Proceedings of the 2013 ACM SIGSAC Conference on Computer & Communications Security, pp. 901–914. ACM (2013)
2. Blum, A., Dwork, C., McSherry, F., Nissim, K.: Practical privacy: the SuLQ framework. In: Proceedings of the Twenty-Fourth ACM SIGMOD-SIGACT-SIGART Symposium on Principles of Database Systems, pp. 128–138. ACM (2005)
3. Dheeru, D., Karra Taniskidou, E.: UCI machine learning repository (2017). http://archive.ics.uci.edu/ml
4. Dwork, C.: Differential privacy. In: Bugliesi, M., Preneel, B., Sassone, V., Wegener, I. (eds.) ICALP 2006. LNCS, vol. 4052, pp. 1–12. Springer, Heidelberg (2006). https://doi.org/10.1007/11787006_1
5. Dwork, C.: A firm foundation for private data analysis. Commun. ACM **54**(1), 86–95 (2011)
6. Dwork, C., McSherry, F., Nissim, K., Smith, A.: Calibrating noise to sensitivity in private data analysis. In: Halevi, S., Rabin, T. (eds.) TCC 2006. LNCS, vol. 3876, pp. 265–284. Springer, Heidelberg (2006). https://doi.org/10.1007/11681878_14
7. Fränti, P., Virmajoki, O.: Iterative shrinking method for clustering problems. Pattern Recogn. **39**(5), 761–765 (2006). http://cs.uef.fi/sipu/datasets/
8. Fränti, P., Virmajoki, O.: Clustering datasets (2018). http://cs.uef.fi/sipu/datasets/
9. Gupta, A., Ligett, K., McSherry, F., Roth, A., Talwar, K.: Differentially private combinatorial optimization. In: Proceedings of the 21st Annual ACM-SIAM Symposium on Discrete Algorithms, pp. 1106–1125 (2010)
10. Komarek, P.: Komarix datasets (2018). http://komarix.org/ac/ds/
11. Lloyd, S.: Least squares quantization in PCM. IEEE Trans. Inf. Theor. **28**(2), 129–137 (1982)
12. McSherry, F.: Privacy integrated queries. In: Proceedings of the 2009 ACM SIGMOD International Conference on Management of Data. ACM (2009)
13. McSherry, F., Talwar, K.: Mechanism design via differential privacy. In: 2007 48th Annual IEEE Symposium on Foundations of Computer Science, pp. 94–103. IEEE (2007)
14. Mohan, P., Thakurta, A., Shi, E., Song, D., Culler, D.: GUPT: privacy preserving data analysis made easy. In: Proceedings of the 2012 ACM SIGMOD International Conference on Management of Data, pp. 349–360. ACM (2012)
15. Nissim, K., Raskhodnikova, S., Smith, A.: Smooth sensitivity and sampling in private data analysis. In: Proceedings of the Thirty-Ninth Annual ACM Symposium on Theory of Computing, pp. 75–84. ACM (2007)
16. Park, M., Foulds, J., Choudhary, K., Welling, M.: DP-EM: differentially private expectation maximization. In: Artificial Intelligence and Statistics, pp. 896–904 (2017)

17. Su, D., Cao, J., Li, N., Bertino, E., Lyu, M., Jin, H.: Differentially private k-means clustering and a hybrid approach to private optimization. ACM Trans. Priv. Secur. **20**(4), 16 (2017)
18. Zhang, J., Xiao, X., Yang, Y., Zhang, Z., Winslett, M.: PrivGene: differentially private model fitting using genetic algorithms. In: Proceedings of the 2013 ACM SIGMOD International Conference on Management of Data, pp. 665–676. ACM (2013)
19. Zhang, T., Ramakrishnan, R., Livny, M.: Birch: a new data clustering algorithm and its applications. Data Min. Knowl. Discov. **1**(2), 141–182 (1997). http://cs.uef.fi/sipu/datasets/

Author Index

Printed in the United States
By Bookmasters